AF906079

Signals and Communication Technology

More information about this series at http://www.springer.com/series/4748

Jose Maria Giron-Sierra

Digital Signal Processing with Matlab Examples, Volume 2

Decomposition, Recovery, Data-Based Actions

Springer

Jose Maria Giron-Sierra
Systems Engineering and Automatic Control
Universidad Complutense de Madrid
Madrid
Spain

ISSN 1860-4862 ISSN 1860-4870 (electronic)
Signals and Communication Technology
ISBN 978-981-10-2536-5 ISBN 978-981-10-2537-2 (eBook)
DOI 10.1007/978-981-10-2537-2

Library of Congress Control Number: 2016951678

MATLAB® is a registered trademark of The MathWorks, Inc., and is used with permission. The MathWorks does not warrant the accuracy of the text or exercises in this book. This book's use or discussion of MATLAB software or related products does not constitute endorsement or sponsorship by the MathWorks of a particular pedagogical approach or particular use of the MATLAB software.

© Springer Science+Business Media Singapore 2017
This work is subject to copyright. All rights are reserved by the Publisher, whether the whole or part of the material is concerned, specifically the rights of translation, reprinting, reuse of illustrations, recitation, broadcasting, reproduction on microfilms or in any other physical way, and transmission or information storage and retrieval, electronic adaptation, computer software, or by similar or dissimilar methodology now known or hereafter developed.
The use of general descriptive names, registered names, trademarks, service marks, etc. in this publication does not imply, even in the absence of a specific statement, that such names are exempt from the relevant protective laws and regulations and therefore free for general use.
The publisher, the authors and the editors are safe to assume that the advice and information in this book are believed to be true and accurate at the date of publication. Neither the publisher nor the authors or the editors give a warranty, express or implied, with respect to the material contained herein or for any errors or omissions that may have been made.

Printed on acid-free paper

This Springer imprint is published by Springer Nature
The registered company is Springer Nature Singapore Pte Ltd.
The registered company address is: 152 Beach Road, #22-06/08 Gateway East, Singapore 189721, Singapore

Preface

Probably the most important technological invention of the previous century was the transistor. And another very important invention was the digital computer, which got a definitive boost with the advent of integrated transistor-based circuits. In the days when these creatures were born, it was foreseen a bright future for communication, control and computing. The three C's. What could be certainly said today is that there is an impressive plethora of digital processing devices being used by millions of people: smartphones, GPS, cameras, computers in all formats. Part of the future will be also service robots, internet of things, autonomous vehicles, etc. Most of these devices depend critically on digital signal processing, which is the theme of this book.

This is the second book of a trilogy. In the first book we took in consideration common signal processing functions, like filtering, or modulation, and a first view of non-stationary signals. Now, in this book, we pay attention to recent topics that respond to the needs of new digital technologies. As in the other book, a series of MATLAB programs are embedded in the chapters for several purposes: to illustrate the techniques, to provide implementation examples, to encourage for personal exploration departing from a successful start. The programs can be downloaded from the web, but it would be rewarding for the reader to re-type the programs (at least the shorter listings) in order to get more familiar with MATLAB coding.

The book has two parts. The first part includes four chapters and is devoted to decomposition and recovery of signals, with emphasis on images. The second part includes three chapters and deals with important data-based actions, such are adaptive filtering, experimental modeling, and classification.

The main contents of the first part are filter banks, wavelets, and images. While the Fourier transform is adequate for periodic signals, wavelets are more suitable for other cases, like for instance short-duration signals: bursts, spikes, tweets, lung sounds, etc. Both Fourier and wavelets transforms decompose signals into components. Both are also invertible, so original signals can be recovered from their components.

In general, the wavelets can be introduced emphasizing more or less the point of view of filtering, which is frequency-domain, or the point of view of mathematics, which is mostly time-domain. We preferred to start with a filtering approach: signals are decomposed into sub-bands using filter banks. This is Chap. 1 of the book.

In Chap. 2, wavelets are introduced using an important example: the Haar wavelet. This wavelet can be implemented with a filter bank for real-time application. The same happens with other types of wavelets: they can be implemented with filter banks. Thus, Chaps. 1 and 2 are closely linked together.

One of the revolutions we are witnessing is digital images and videos. Wavelets are used in this context for compression, denoising, and certain analysis tasks (for instance in medical applications). Chapters 3 and 4 introduce fundamentals of digital image processing, and then apply filter banks and wavelets to images according with peculiar aspects, like for instance directional filtering, or JPEG compression.

Inside the matters covered in the first part of the book there is a basic problem that is analog to a problem of tiling. Suppose you want to cover with tiles the floor of a room. This could be done with triangular, rectangular, hexagonal tiles, but no with pentagonal tiles. Actually, there are not so many tile geometries that allow for exact covering. With signals something similar happens, they can be decomposed into components by certain types of filters, and then they can be recovered by re-combining these components. Not any filter can be used for that. This is a main aspect that motivates the title of the Part I.

Part II of the book embodies several topics having in common the protagonism conceded to data. The characteristics of signals or data are taken into account for adaptive filtering. The processing of input/output data can be used for establishing a dynamic model. And the classification of data can help to extract relevant information. Main topics of the second part are the Wiener filter, the recursive adaptive filters, the estimation of transfer functions or ARMA models, the estimation of principal or independent components of data, the Support Vector Machine for classification, the K-means algorithm, the Expectation–Maximization algorithm, the use of neurons, etc. Experiments about face recognition were included. Illustrative applications of adaptive filters are modem equalization, echo cancellation for the use of mobile phones inside cars, or the correction of motion-blurred pictures.

For easier reading of the book, the longer programs have been put in an appendix.

The reader is invited to discover the profound interconnections and commonalities that exist behind the variety of topics in this book. This common ground would become surely the humus for the next signal processing future.

As said in the preface of the first book, our particular expertise on signal processing has two main roots: research and teaching. I belong to the Faculty of Physics, University Complutense of Madrid, Spain. During our experimental research on autonomous vehicles, maritime drones, satellite control, etc., we practiced main methods of digital signal processing, for the use of a variety of sensors and for prediction of vehicle motions. From years ago, I teach Signal Processing in a Master of Biomedical Physics, and a Master on New technologies.

The style of the programs included in the book is purposively simple enough. The reader is invited to typeset the programs included in the book, for it would help for catching coding details. Anyway, all programs are available from the book web page: http://www.dacya.ucm.es/giron/SPBook2/Programs.

A lot of different materials have been used to erect this book: articles, blogs, codes, experimentation. I tried to cite with adequate references all the pieces that have been useful. If someone has been forgotten, please contact me. Most of the references cited in the text are available from Internet. We have to express our gratitude to the public information available in this way.

Please, send feedback and suggestions for further improvement and support.

Acknowledgments

Thanks to my University, my colleagues and my students. Since this and the other book required a lot of time taken from nights, weekends and holidays, I have to sincerely express my gratitude to my family.

Madrid, Spain Jose Maria Giron-Sierra

Contents

List of Figures

Listings

Part I
Decomposition and Recovery. Images

Chapter 1
Filter Banks

1.1 Introduction

Filter banks allow signals to be decomposed into subbands. In this way, parallel powerful processing can be easily applied. Also, the decomposition paves the way for signal compression procedures. Due to these reasons, the interest on filter banks has significantly grown along years, so today there is large body of theory on this matter.

The chapter starts with an introduction to new concepts and to architectural elements associated to filter banks. Most of the chapter focuses on FIR filters, so after the introductory section, the next one treats in detail the aspects of FIR filters that are relevant for their use in filter banks.

The fourth section attacks the main issue, which is *perfect reconstruction*. The question is that the original signal should be recovered after decomposition into subbands. A series of mathematical conditions for this to happen are discovered, and then a classification of filter banks is derived. This is done in the most simple context: 2-channel filter banks.

The chapter continues with the introduction of filter bank structures and design approaches, and then with extensions to M-channel filter banks and to multidimensional signals (for instance, images).

An important point is that wavelets, which will be the theme of the next chapter, are strongly related to filter banks, at least for real-time implementation.

Concerning notation, the filter coefficients would be $h(0) \ldots h(N-1)$, so it has N terms (filter length), or they would be $h(-L) \ldots h(L)$ with $L = (N-1)/2$. In some cases we use $2k+1$ instead of $(N-1)$ to highlight that N is even. The frequency responses of the filters are between 0 and π.

The basic reference literature for this chapter is [24, 27, 44, 45, 49]. Other more specific references will be given in appropriate places.

© Springer Science+Business Media Singapore 2017

J.M. Giron-Sierra, *Digital Signal Processing with Matlab Examples, Volume 2*,
Signals and Communication Technology, DOI 10.1007/978-981-10-2537-2_1

1.2 Filter Banks and Multirate Systems

Nowadays a lot of people use compressed information. This is the case of MP3, MP4, etc., for audio and video. It can be recognized as a problem of efficient use of a limited resource: the storage capacity, or perhaps the bandwidth of Internet.

In real life signal processing applications, like in mobile phones, generic or specific microprocessors are employed. These units do have processing rate limits, and this should be taken into account.

Thus there is a problem of efficiency. And there are several alternatives to deal with it. Processing work could be distributed. Resources could be shared. Tasks could be tailored.

Distribution could be done by decomposition in the frequency domain, by using sets of filters. It also can be done with decomposition in the time domain, for instance in multiplexing style.

Tailoring could be done by adapting the sampling rate to the signal frequencies, thus leading to multirate systems.

Figure 1.1 presents a basic example of single input-multiple output (SIMO) filter bank made with three band-pass filters. The figure includes a diagram of the filter frequency bands.

Given a sampled signal $y(n)$, many signal processing ideas are based on handling a set $y(k), y(k-1), y(k-2)\ldots y(k-N)$ of signal samples. Chains of phase shifters (each one corresponding to z^{-1}) could be used to record this set, thus obtaining polyphase systems.

Figure 1.2 depicts a chain of three phase shifters to be used for polyphase systems. Each block means a unit delay.

The section starts with a fundamental piece of signal processing: the Fourier transform. It would serve as the entrance to filter banks and multirate systems.

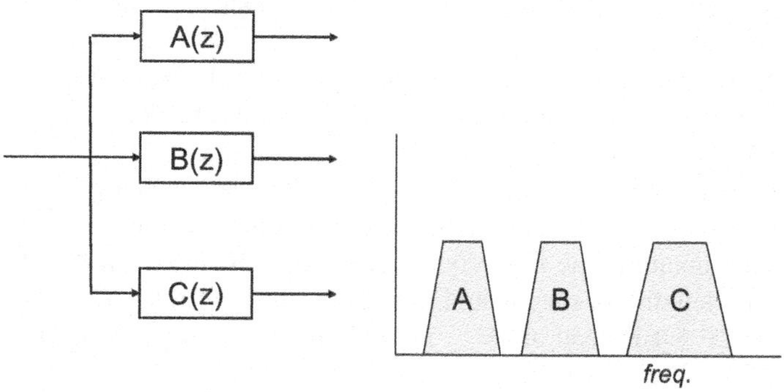

Fig. 1.1 A basic filter bank

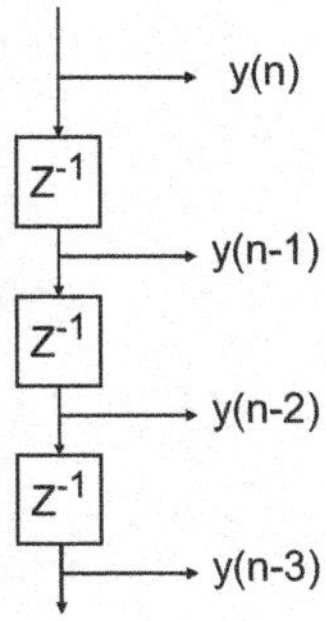

Fig. 1.2 A basic chain of phase shifters

1.2.1 Discrete Fourier Transforms

1.2.1.1 Discrete Fourier Transforms

Let us fix terms. The Discrete Time Fourier Transform (DTFT) of a sequence $y(n)$ is:

$$Y(\omega) = \sum_{n=\infty}^{\infty} y(n)\, e^{-j\,\omega\, n} \tag{1.1}$$

$Y(\omega)$ can be also described as the evaluation of $Y(z)$ on the unit circle. Then $Y(\omega)$ is periodic, and it is sufficient to study it in the Nyquist interval $-\pi \leq \omega \leq \pi$.

The inverse DTFT is:

$$y(n) = \frac{1}{2\pi} \int_{-\pi}^{\pi} Y(\omega)\, e^{j\,\omega\, n} d\omega \tag{1.2}$$

Given a finite duration sampled signal $y(n)$, its Discrete Fourier Transform (DFT) is:

$$Y(k) = \sum_{n=0}^{N-1} y(n)\, e^{-j\,\omega_k\, n} \qquad (k = 0,\, 1,\, 2\ldots,\, N-1) \tag{1.3}$$

with ('DFT frequencies'):

$$\omega_k = \frac{2\,\pi\,k}{N} \quad (k = 0,\, 1,\, 2\ldots,\, N-1) \tag{1.4}$$

In the same line as the DTFT, the DFT $Y(k)$ can be also described as the evaluation of $Y(z)$ on the unit circle at points z_k. $Y(k)$ is periodic with period N. The points z_k are the *Nth roots of unity*:

$$z_k = e^{\,j\,\omega_k} = e^{\frac{j\,2\pi\,k}{N}} \qquad (k = 0,\, 1,\, 2\ldots,\, N-1) \tag{1.5}$$

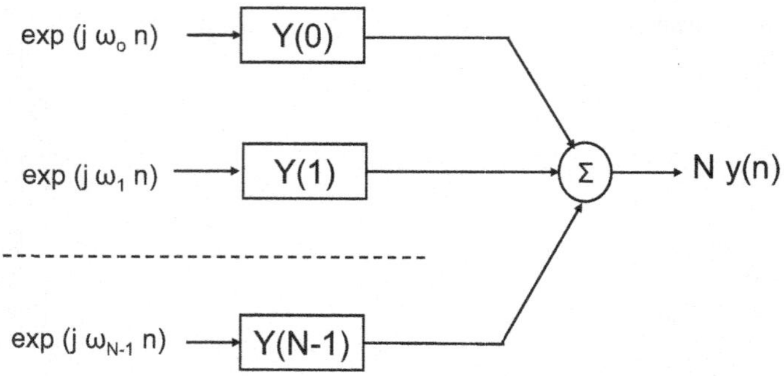

Fig. 1.3 Inverse DFT as a synthesizer Filter Bank

The DFT for a real $y(n)$ is symmetrical around $k = N/2$, as follows:

$$\begin{aligned} \text{Real } Y(k) &= \text{Real } Y(N-k) \\ \text{Imag } Y(k) &= -\text{Imag } Y(N-k) \end{aligned} \tag{1.6}$$

The inverse DFT is:

$$y(n) = \frac{1}{N} \sum_{k=0}^{N-1} Y(k)\, e^{j\,\omega_k\, n} \tag{1.7}$$

This last equation can be considered as the sum of the outputs of N filters; the k-filter would have *exp(j* $\omega_k\, n)$ as complex sinusoidal input and $Y(k)$ *exp(j* $\omega_k\, n)$ as output. Figure 1.3 visualizes this. Such type of configurations is denoted as *synthesizer filter bank.* It is a multiple input-single output (MISO) filter bank.

1.2.1.2 The DFT Matrix

Let us introduce a slight change of notation. Instead of $Y(k)$ we shall write Y_k. In consequence the DFT is computed with:

$$Y_k = \sum_{n=0}^{N-1} y(n)\, e^{-j\,\omega_k\, n} = \sum_{n=0}^{N-1} y(n)\, e^{\frac{-j\,2\pi\, k\, n}{N}} \qquad (k = 0,\, 1,\, 2 \ldots,\, N-1) \tag{1.8}$$

Denote:

$$W_N = e^{\frac{-j\,2\pi}{N}} \tag{1.9}$$

Then the DFT could we written as:

$$Y_k = \sum_{n=0}^{N-1} y(n)\, W_N^{k\,n} \qquad (k = 0,\, 1,\, 2 \ldots,\, N-1) \tag{1.10}$$

Fig. 1.4 Example of the values that W_N^{kn} can take (for $k = 1, N = 5, n = 0 \ldots N-1$)

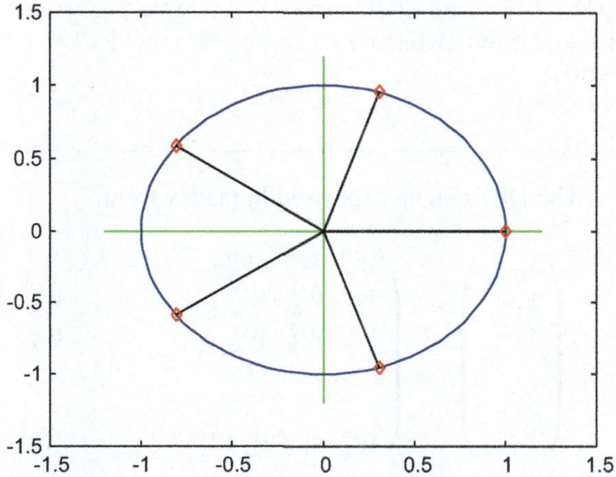

And the inverse DFT (IDFT):

$$y(n) = \frac{1}{N} \sum_{n=0}^{N-1} Y_k \, W_N^{-kn} \qquad (n = 0, 1, 2 \ldots, N-1) \tag{1.11}$$

Figure 1.4 shows on a complex plane with the unit circle an example of the values W_N^{kn} can take (for $k = 1, N = 5, n = 0 \ldots N-1$). The figure has been generated with the Program 1.1, which also gives the numerical complex values of W_N^{kn}. The reader is invited to change the values of k and N to see other results.

Program 1.1 values of WN(kn)

```
% Values of WN(kn)
k=1; %the index of Yk (please edit as you wish)
N=5; %number of samples
WN=zeros(1,N); %space for the samples
%taking the samples
for nn=1:N,
WN(nn)=exp((-i*((2*pi)*(nn-1)*k))/N);
end;
%display on a circle
figure(1)
k=1:(1/100):100; t=2*pi*k;
x=zeros(1,length(t)); y=zeros(1,length(t));
x=cos(t); y=sin(t);
plot([-1.2 1.2],[0 0],'g'); hold on; %x axis
plot([0 0],[-1.2 1.2],'g'); %y axis
plot(x,y,'b'); %the circle
for nn=1:N,
```

```
plot([0 real(WN(nn))],[0 imag(WN(nn))],'k');
plot(real(WN(nn)),imag(WN(nn)),'rd');
end;
title('Values of WN(kn) with k=constant, n=0..N-1');
WN
```

The DFT can be expressed in matrix form:

$$\begin{pmatrix} Y_0 \\ Y_1 \\ Y_2 \\ . \\ . \\ Y_{N-1} \end{pmatrix} = \begin{pmatrix} W_N^0 & W_N^0 & W_N^0 \ldots & W_N^0 \\ W_N^0 & W_N^1 & W_N^2 \ldots & W_N^{N-1} \\ W_N^0 & W_N^2 & W_N^4 \ldots & W_N^{2(N-1)} \\ \ldots\ldots & & & \\ \ldots. & & & \\ W_N^0 & W_N^{N-1} & W_N^{(N-1)2} \ldots & W_N^{(N-1)(N-1)} \end{pmatrix} \begin{pmatrix} y(0) \\ y(1) \\ y(2) \\ . \\ . \\ y(N-1) \end{pmatrix} \tag{1.12}$$

Briefly,

$$\bar{Y} = \mathrm{W}\,\bar{y} \tag{1.13}$$

The matrix **W** has the first row and the first column all ones. The matrix **W** is symmetric, that is, $\mathbf{W}^T = \mathbf{W}$.

Of course the matrix **W** can be pre-computed and stored or put into firmware.

The inverse DFT can be written as:

$$\bar{y} = \frac{1}{N}\,\mathrm{W}^*\,\bar{Y} \tag{1.14}$$

where **W*** is the complex conjugate of **W**, obtained by conjugating every element of **W**. No matrix inversion is required.

A direct way to obtain with MATLAB the matrix W is by using *fft(eye(N))*. The Program 1.2 computes an example, which is shown in Fig. 1.5.

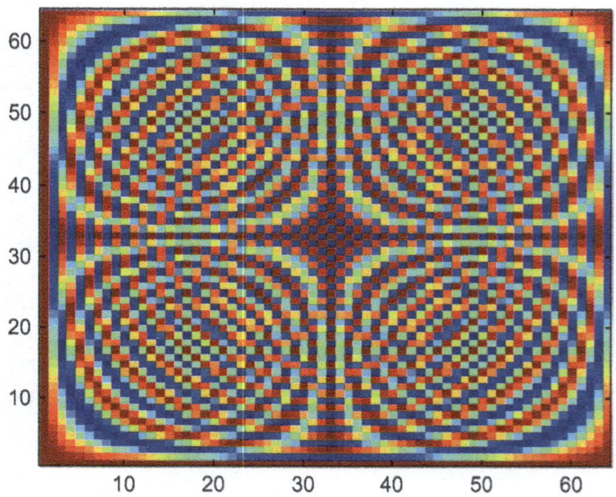

Fig. 1.5 64 × 64 DFT matrix

Program 1.2 DFT matrix example

```
% DFT matrix example
N=64;
I=eye(N); %NxN identity matrix
D=fft(I); %the DFT matrix
R=real(D); %real part, for display purposes
imagesc(R); axis xy; %matrix display as image
title('the 64x64 DFT matrix');
% test with a square signal
t=0:0.1:6.3;
y=square(t)';
Y1=fft(y); %the MATLAB fft
Y2=D*y; %the fft via matrix
dif=Y1-Y2;
max_diference=max(dif)
```

1.2.2 Modulated Filter Banks

A typical processing structure consists in a first filter bank that decomposes the signal into bands, which is called the *analyzer filter bank*, and then a second synthesizer filter bank that recovers the (processed) signal. A central block may be included for processing purposes. Figure 1.6 shows a sketch of this structure.

Consider a low-pass filter with $h_0(n)$ impulse response. Suppose we have a signal $y(n)$ with 10 KHz bandwidth and that the low-pass filter has a 1 KHz bandwidth. The signal can be decomposed into 10 components using copies of the same filter $H_0(z)$ as shown in Fig. 1.7. The filter bank has 1 input and 10 outputs that may be connected to communication channels.

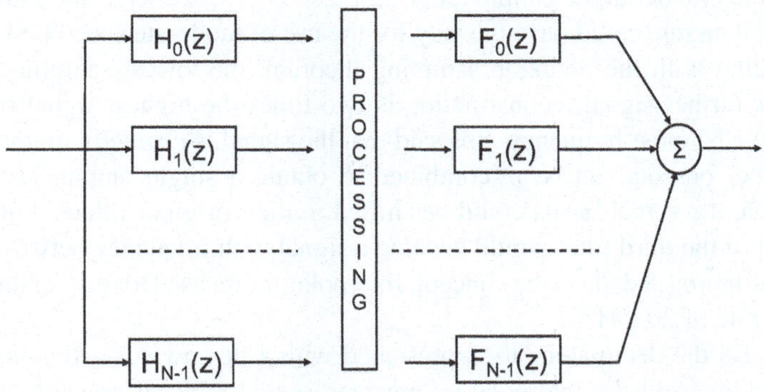

Fig. 1.6 Analysis and synthesis through filter banks

Fig. 1.7 Modulated analyzer filter bank

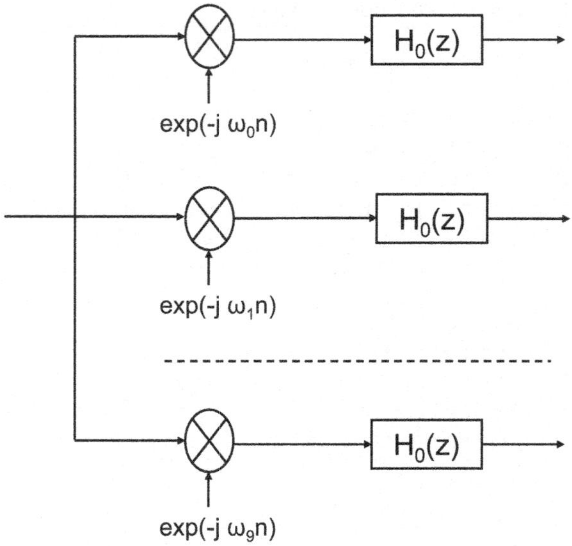

As described in Fig. 1.7, the signal $y(n)$ is multiplied by the complex sinusoids *exp(-j* $\omega_k n$), $k = 0, 1, 2, \ldots 9$, resulting in 10 modulated signals. Recall that this modulation causes a frequency shift. The first filter will select the part of $y(n)$ with frequencies between 0 and 1 KHz, the second filter will select the part between 1 and 2 KHz (which is shifted to between 0 and 1 KHz), the third filter will select between 2 and 3 KHz (which is shifted to between 0 and 1 KHz), and so on.

Notice that all filters would have the same delay.

Suppose that the signal $y(n)$ has been sampled at 20 KHz rate (strictly Nyquist). It is not necessary to keep this sampling rate for signals with frequency between 0 and 1 KHz. To make a slower data rate we can use decimators, taking in this case 1 of every 10 samples, so the sampling rate becomes 2 KHz. The input sample rate is 20 KHz, the global output sample rate is $2 + 2 + 2 \ldots = 20$ KHz, the same.

Slower data rate could pave the way for the use of moderate speed DSP cores.

According with the Shannon sampling theorem, the lowest sampling rate that allows for further signal reconstruction is two times the highest signal frequency, which is the Nyquist frequency. It is said that the signal is *critically sampled*.

The filter outputs can be re-combined to obtain a single output. Using again modulation, the signal bands could be shifted to their original values. For instance the output of the third filter would become a signal with frequency between 2 and 3 KHz, etc. Figure 1.8 depicts the concept. Interpolators are used to recover the original sampling rate of 20 KHz.

In Fig. 1.8 the decimators are represented with a downwards arrow, to indicate downsampling, and the interpolators are represented with an upwards arrow, to indicate upsampling. In this example, the downsampling ratio is N = 10, and the upsampling ratio is also N = 10.

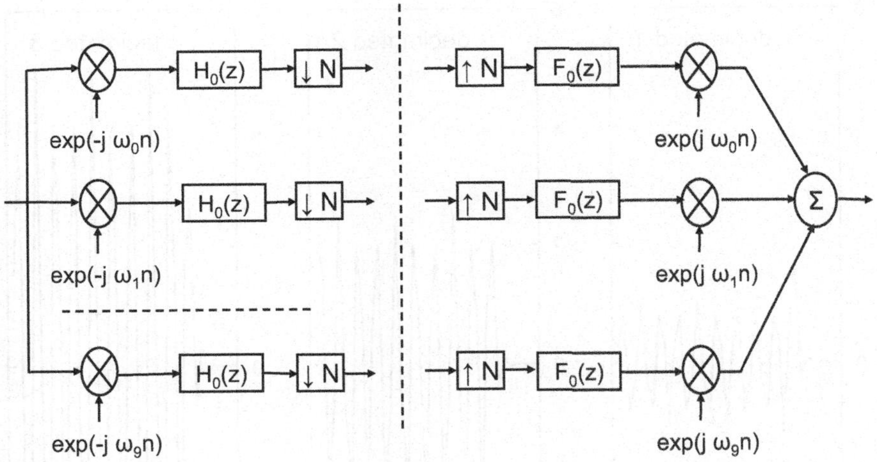

Fig. 1.8 Modulated analysis and synthesis

Fig. 1.9 The input signal used as example

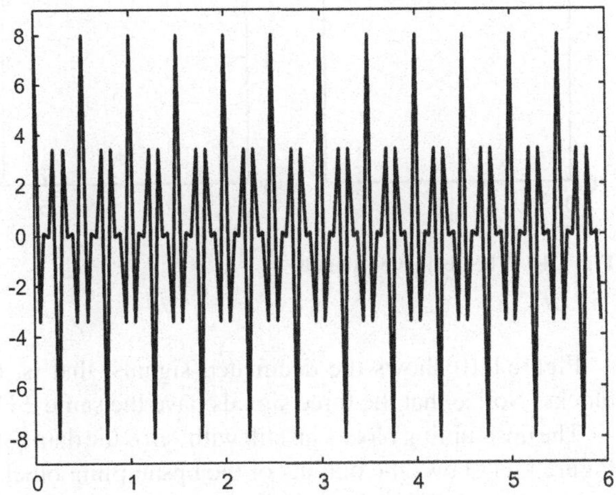

Notice that the decimators and interpolators just mentioned do not include any filtering.

A simple and direct implementation of the ideas just introduced has been done in the Program 1.3. The case of just three channels has been chosen. Therefore, there are three filters in the analysis part of the filter bank, and another three filters in the synthesis part. The filter bank uses FIR filters.

An example of input signal has been built, in order to be processed by the Program 1.3. The signal contains three sinusoidal harmonics, with frequencies of 2, 6 and 10 Hz. Figure 1.9 plots this input signal.

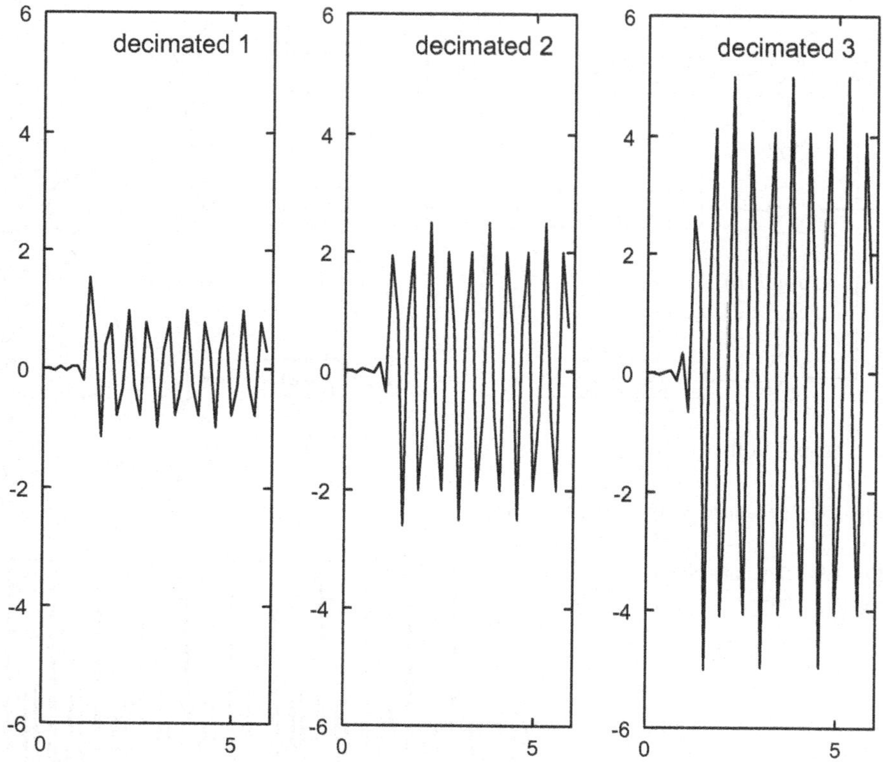

Fig. 1.10 The decimated signals

Figure 1.10 shows the decimated signals; that is, the outputs of the decimator blocks. Notice that the three signals have the same 2 Hz. frequency.

The upsampling blocks just fill with zeros the data between the decimated samples. Figure 1.11 shows the outputs of the upsampling blocks.

Figure 1.12 shows the outputs of the filters belonging to the synthesis part. Good sinusoids are obtained.

Finally, Fig. 1.13 shows the results after demodulation and summation. Compare with the input signal. Clearly, there is a delay of 50 samples caused by the chain of FIR filters.

The Program 1.3 listing only includes the signal processing code. A complete version of the program, including a long part devoted to display of figures, has been included in Appendix A. The prototype filter that has been chosen for the example is a FIR filter with Kaiser window.

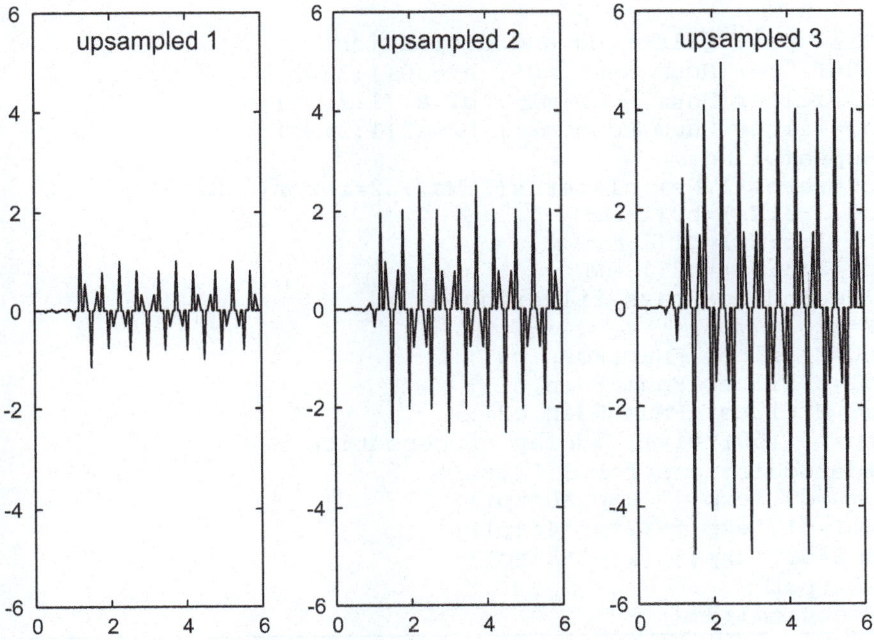

Fig. 1.11 The upsampled signals

Program 1.3 Modulated analysis and synthesis

```
% Modulated analysis and synthesis
fs=20; %sampling frequency
tiv=1/fs; %sampling period
Ns=120; %number of samples
t=0:tiv:((Ns-1)*tiv); %time vector,
%signal components
y0=cos(2*2*pi*t);
y1=cos(6*2*pi*t);
y2=cos(10*2*pi*t);
%added signal
y=(1*y0)+(4*y1)+(3*y2);
%prototype filters (Kaiser, FIR)
fc=3/(fs/2); %cut-off frequency at 3 Hz
L=50;beta=5;
hw=kaiser(L+1,beta); %Kaiser window
Hnum=fir1(L,fc,hw); %FIR coeffs
Hden=[1]; %denominator
Fnum=Hnum;
Fden=Hden;
%modulations for 3 filters
m0=y.*exp(-j*(2*pi*0*t));
m1=y.*exp(-j*(2*pi*4*t));
m2=y.*exp(-j*(2*pi*8*t));
```

```
%low-pass filtering and decimation
a0=filter(Hnum,Hden,m0); b0=a0(1:3:Ns);
a1=filter(Hnum,Hden,m1); b1=a1(1:3:Ns);
a2=filter(Hnum,Hden,m2); b2=a2(1:3:Ns);
%upsampling
c0=zeros(1,Ns);c1=zeros(1,Ns);c2=zeros(1,Ns);
c0(1:3:Ns)=b0(1:end);
c1(1:3:Ns)=b1(1:end);
c2(1:3:Ns)=b2(1:end);
%second low-pass filtering
M=3;
d0=M*filter(Fnum,Fden,c0);
d1=M*filter(Fnum,Fden,c1);
d2=M*filter(Fnum,Fden,c2);
tp=t-((50)*tiv); %delay compensation
%demodulations for 3 filters
dm0=d0.*exp(j*(2*pi*0*tp));
dm1=d1.*exp(j*(2*pi*4*tp));
dm2=d2.*exp(j*(2*pi*8*tp));
%output
x=dm0+dm1+dm2;
```

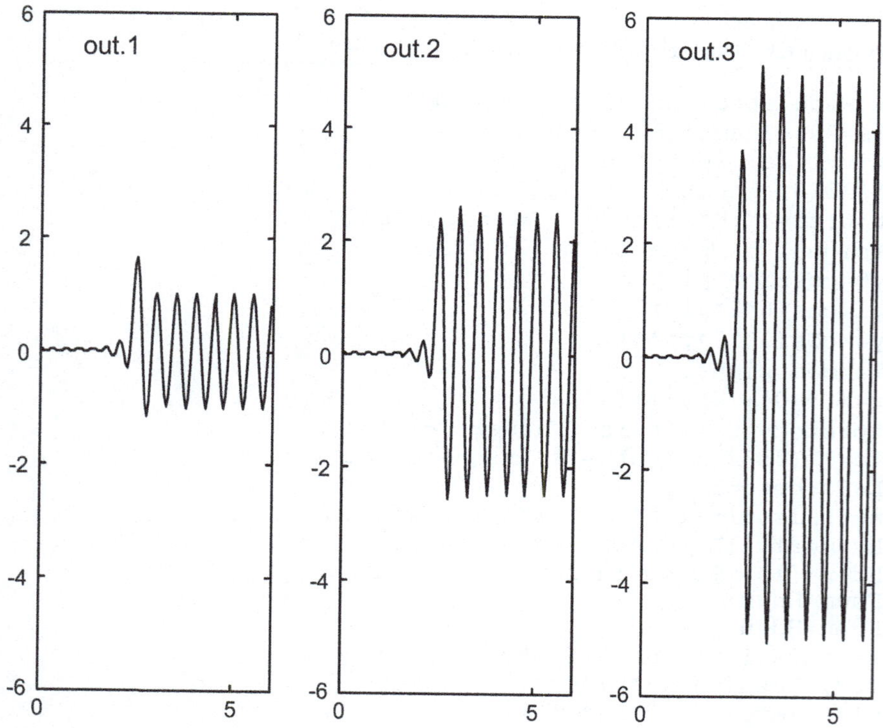

Fig. 1.12 The outputs of the synthesis filters

Fig. 1.13 The output signal

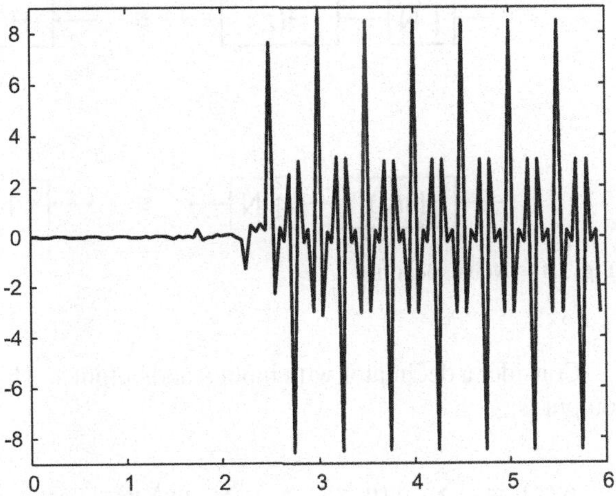

Notice that instead of using modulation to shift the signal frequency bands, it is possible to devise '*modulated*' filters, shifting their pass band properly. Again, the starting point is a low-pass filter $H_0(z)$, which is called *the prototype filter*. Let us obtain from $H_0(z)$ several contiguous band-pass filters. The *kth* filter is obtained as follows (impulse response):

$$h_k(n) = h_0(n)\, e^{\,j\,\omega_k\, n} \quad (k = 0,\, 1,\, 2\ldots,\, N-1) \tag{1.15}$$

Consequently the frequency response of the *kth* filter is:

$$H_k(\omega) = H_0(\omega - \omega_k) \quad (k = 0,\, 1,\, 2\ldots,\, N-1) \tag{1.16}$$

Since (ideally) the frequency response of $H_0(\omega)$ is only $\neq 0$ for $0 \leq \omega \leq \omega_h$ the frequency response of $H_k(\omega)$ is only $\neq 0$ for $\omega_k \leq \omega \leq \omega_h + \omega_k$. Then $H_k(\omega)$ is a band-pass filter. The modulation shifts the pass-band to the right by an amount of ω_k.

1.2.3 Decimators and Interpolators

1.2.3.1 Noble Identities

During the design of filter banks it is useful to consider the identities depicted in Fig. 1.14:

1.2.3.2 Decimators and Interpolators; Images, Alias

An N decimator retains only the input samples with sample numbers equal to a multiple of N. It causes a sample rate reduction by a factor of N.

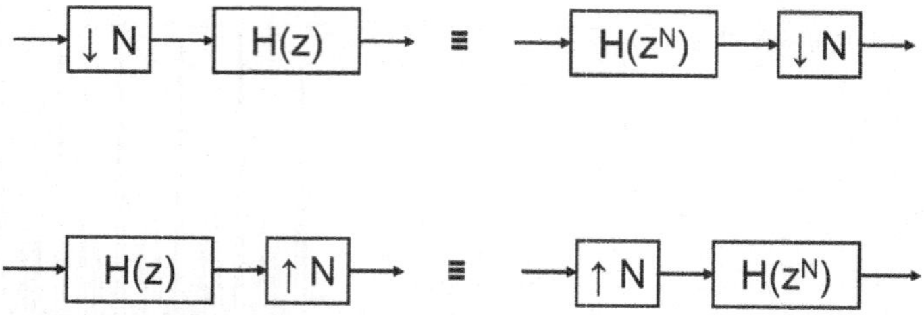

Fig. 1.14 Noble identities

Consider a decimator with input v and output x. The z-transform of the decimator output is:

$$\begin{aligned} X(z) &= \sum_{n=-\infty}^{\infty} x(n)\, z^{-n} = \sum_{n=-\infty}^{\infty} v(Nn)\, z^{-n} = \sum_{n=-\infty}^{\infty} v_1(Nn)\, z^{-n} = \\ &= \sum_{m=-\infty}^{\infty} v_1(m)\, z^{-m/N} = V_1(z^{1/N}) \end{aligned} \tag{1.17}$$

where:

$$v_1(j) = v(j) \sum_{n=-\infty}^{\infty} \delta(j - nN) \qquad (j = 0,\, 1,\, 2 \ldots) \tag{1.18}$$

Then:

$$\begin{aligned} V_1(e^{j\omega}) &= \tfrac{1}{2\pi}\, V(e^{j\omega})\, F\left\{ \sum_{n=-\infty}^{\infty} \delta(j - nN) \right\} \\ &= \tfrac{1}{2\pi}\, V(e^{j\omega})\, \tfrac{2\pi}{N} \sum_{k=0}^{N-1} \delta(\omega - \tfrac{2\pi k}{N}) = \tfrac{1}{N} \sum_{k=0}^{N-1} V(e^{j(\omega - \frac{2\pi k}{N})}) \end{aligned} \tag{1.19}$$

Therefore:

$$V_1(z) = \frac{1}{N} \sum_{k=0}^{N-1} V(W_N^k\, z) \tag{1.20}$$

Finally:

$$X(z) = \frac{1}{N} \sum_{k=0}^{N-1} V(W_N^k\, z^{1/N}) \tag{1.21}$$

The spectrum of the output x is a modified version of the input spectrum. Figure 1.15 depicts a sketch of this, with reference to normalized frequency. The figure represents the spectra of the original signal, of a N = 2 downsampled signal, and of a N = 3 downsampled signal. The central band is the main band, the other are lateral copies. If there was overlapping of contiguous bands; aliasing can cause loss of information. The figure corresponds to a case where N = 2 do not cause problems, but N = 3 cause overlapping and therefore problems. Take into account that the decimator output

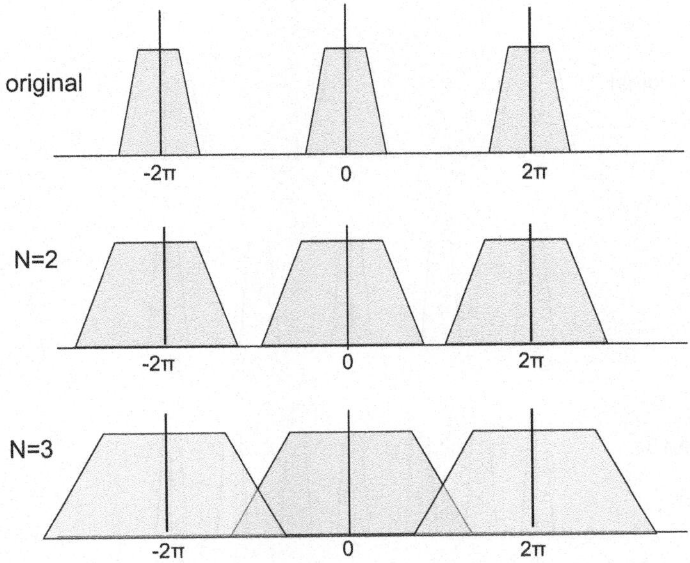

Fig. 1.15 Spectra of downsampled signals

data rate is 1/N times the input data rate. Clearly, a band-pass signal with bandwidth $\Delta\omega \leq 2\,\pi/N$ can be N-decimated without alias overlapping.

The interpolator inserts N − 1 samples with zero value between adjacent samples of the input signal.

Consider an interpolator with input x and output y. The z-transform of the interpolator output is:

$$y(n) = \begin{cases} x(\frac{n}{N}), n = 0, \pm N, \pm 2N \ldots \\ 0 \end{cases} \tag{1.22}$$

The z-transform is:

$$\begin{aligned} Y(z) &= \sum_n y(n)\, z^{-n} = \sum_{n=0,\,\pm N,\,\pm 2N \ldots} x(\tfrac{n}{N})\, z^{-n} \\ &= \sum_k x(k)\,(z^N)^{-k} = X(z^N) \end{aligned} \tag{1.23}$$

The interpolator causes multiple copies of the same input spectrum. This is called imaging effect; the extra copies are 'images' created by the interpolator. Figure 1.16 shows a representation of this effect. Note that the output data rate of the interpolator is N times the input data rate.

Suppose a decimator and an interpolator in cascade, as in Fig. 1.17.

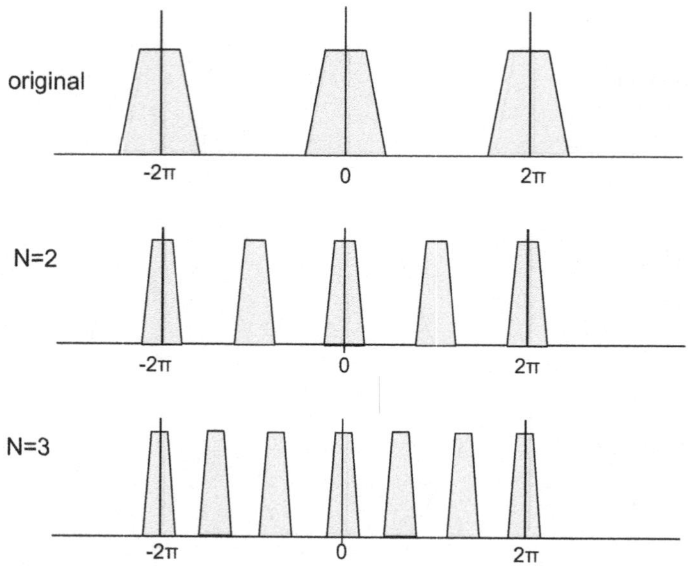

Fig. 1.16 Spectra of upsampled signals

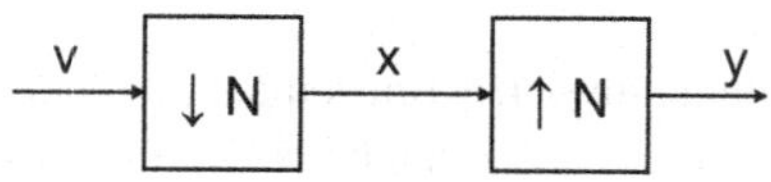

Fig. 1.17 Decimator and interpolator in cascade

Then:

$$Y(z) = \frac{1}{N} \sum_{k=0}^{N-1} V(W_N^{kN}\, z) \tag{1.24}$$

For the particular case with $N = 2$,

$$\begin{aligned} W_N^{kN} &= e^{-j\,k\,\pi} \quad (k = 0,\ 1) \\ Y(z) &= \tfrac{1}{2}\,(\,V(z) + V(-z)) \end{aligned} \tag{1.25}$$

1.2.4 *The Polyphase Representation*

1.2.4.1 The Polyphase Representation

Consider a filter:

$$H(z) = \sum_{n=-\infty}^{\infty} h(n)\, z^{-n} \tag{1.26}$$

Let us decompose the expression into even numbered samples and odd numbered samples:

$$\begin{aligned}H(z) &= [\cdots + h(-4)z^4 + h(-2)\,z^2 + h(0) + h(2)\,z^{-2} + h(4)\,z^{-4} + \cdots]\\ &+[\cdots + h(-3)z^3 + h(-1)\,z^1 + h(1)\,z^{-1} + h(3)\,z^{-3} + \cdots]\end{aligned} \tag{1.27}$$

That is:

$$H(z) = \sum_{n=-\infty}^{\infty} h(2n)\,z^{-2n} + z^{-1} \sum_{n=-\infty}^{\infty} h(2n+1)\,z^{-2n} \tag{1.28}$$

Denote:

$$E_0(z) = \sum_{n=-\infty}^{\infty} h(2n)\,z^{-n}; \qquad E_1(z) = \sum_{n=-\infty}^{\infty} h(2n+1)\,z^{-n} \tag{1.29}$$

Using $E_0(z)$ and $E_1(z)$, the filter $H(z)$ can be expressed as,

$$H(z) = E_0(z^2) + z^{-1}\,E_1\,(z^2) \tag{1.30}$$

Previous equation is a polyphase decomposition of $H(z)$ into two components. It corresponds to a decomposition of the impulse response $h(n)$ into two groups $h(2n)$, even terms, and $h(2n+1)$, odd terms.

For example, suppose that the filter is:

$$H(z) = 3\,z^4 + 2\,z^3 + 7\,z^2 - z + 33 + 11\,z^{-1} - 4\,z^{-2} + 9\,z^{-3} + 5\,z^{-4}$$

Then:

$$\begin{aligned}E_0(z^2) &= 3\,z^4 + 7\,z^2 + 33 - 4\,z^{-2} + 5\,z^{-4}\\ E_1(z^2) &= 2\,z^4 - z^2 + 11 + 9\,z^{-2}\end{aligned}$$

Applying more decompositions, it is possible to obtain:

$$H(z) = \sum_{k=0}^{N-1} z^{-k}\,E_k(z^N) \tag{1.31}$$

where,

$$E_k(z) = \sum_{n=-\infty}^{\infty} h(nN+k)\,z^{-n} \qquad (k = 0,\ 1,\ 2 \ldots,\ N-1) \tag{1.32}$$

A second alternative is:

$$H(z) = \sum_{k=0}^{N-1} z^{-(N-1-k)} R_k(z^N) \tag{1.33}$$

with:

$$R_k(z) = E_{N-1-k}(z) \qquad (k = 0,\, 1,\, 2 \ldots,\, N-1) \tag{1.34}$$

This last representation is more convenient for the computation of synthesis filter banks.

Continuing with the example, here are some more decompositions:

$$\begin{aligned} E_0(z^3) &= 2\,z^3 + 33 + 9\,z^{-3} \\ E_1(z^3) &= 7\,z^3 + 11 + 5\,z^{-3} \\ E_2(z^3) &= 3\,z^6 - z^3 - 4 \end{aligned}$$

$$\begin{aligned} E_0(z^4) &= 3\,z^4 + 33 + 5\,z^{-4} \\ E_1(z^4) &= 2\,z^4 + 11 \\ E_2(z^4) &= 7\,z^4 - 4 \\ E_3(z^4) &= -\,z^4 + 9 \end{aligned}$$

The last one is:

$$E_0(z^9) = 33;\; E_1(z^9) = 11;\; E_2(z^9) = -4;\; E_3(z^9) = 9;\; E_4(z^9) = 5;$$
$$E_5(z^9) = 3\,z^9;\; E_6(z^9) = 2\,z^9;\; E_7(z^9) = 7\,z^9;\; E_8(z^9) = -1\,z^9;$$

And:

$$\begin{aligned} R_0(z^2) &= E_1(z^2) \\ R_1(z^2) &= E_0(z^2) \end{aligned}$$

$$\begin{aligned} R_0(z^3) &= E_2(z^3) \\ R_1(z^3) &= E_1(z^3) \\ R_2(z^3) &= E_0(z^3) \end{aligned}$$

1.2.4.2 An Implementation of the Modulated Filter Bank

Let us consider again the filter system in Fig. 1.8 and the possible use of 'modulated' filters. Recall that the impulse response of the modulated *kth* filter is:

$$h_k(n) = h_0(n)\, e^{j\,\omega_k n} = h_0(n)\, e^{j\,\frac{2\pi k n}{N}} = h_0(n)\, W_N^{-k\,n} \tag{1.35}$$

Coming now to the z-domain we have:

$$H_k(z) = \sum_{n=-\infty}^{\infty} h_k(n)\, z^{-n} = \sum_{n=-\infty}^{\infty} h_0(n)\, W_N^{-kn} z^{-n} = H_0(z\, W_N^k) \tag{1.36}$$

and,

$$H_0(z) = \sum_{l=0}^{N-1} z^{-l}\, E_l(z^N) \tag{1.37}$$

Then:

$$\begin{aligned} H_k(z) &= H_0(z\, W_N^k) = \sum_{kl=0}^{N-1} z^{-l}\, W^{-kl}\, E_l(z^N\, W_N^{kN}) = \\ &= \sum_{l=0}^{N-1} z^{-l}\, W_N^{-kl}\, E_l(z^N) \end{aligned} \tag{1.38}$$

since $W_N^{KN} = 1$.

On the basis of this last equation, the complete set of filters for the analysis filter bank is given by:

$$\begin{pmatrix} H_0(z) \\ H_1(z) \\ H_2(z) \\ . \\ . \\ H_{N-1}(z) \end{pmatrix} = \begin{pmatrix} 1 & 1 & 1 \ldots & 1 \\ 1 & W_N^{-1} & W_N^{-2} \ldots & W_N^{-(N-1)} \\ 1 & W_N^{-2} & W_N^{-4} \ldots & W_N^{-2(N-1)} \\ \ldots\ldots & & & \\ \ldots & & & \\ 1 & W_N^{-(N-1)} & W_N^{-(N-1)2} \ldots & W_N^{(N-1)(N-1)} \end{pmatrix} \cdot$$

$$\cdot \begin{pmatrix} E_0(z^N) \\ z^{-1} E_1(z^N) \\ z^{-2} E_2(z^N) \\ . \\ . \\ z^{-(N-1)} E_{N-1}(z^N) \end{pmatrix} = \mathbf{W}^* \cdot \begin{pmatrix} E_0(z^N) \\ z^{-1} E_1(z^N) \\ z^{-2} E_2(z^N) \\ . \\ . \\ z^{-(N-1)} E_{N-1}(z^N) \end{pmatrix} = \mathbf{W}^* \cdot \widehat{E} \tag{1.39}$$

were **W*** is the matrix for the IDFT (Eq. 1.14), and we introduced $\hat{E}$ to represent the vector of polyphase components.

Using Noble identities the decimators can be moved to the left, so the modulated analyzer filter bank in Fig. 1.8 can be implemented as shown in Fig. 1.18.

Let us take as filter prototype for the synthesis filter bank the following:

$$F_0(z) = \sum_{l=0}^{N-1} z^{-(N-1-l)}\, R_l(z^N) \tag{1.40}$$

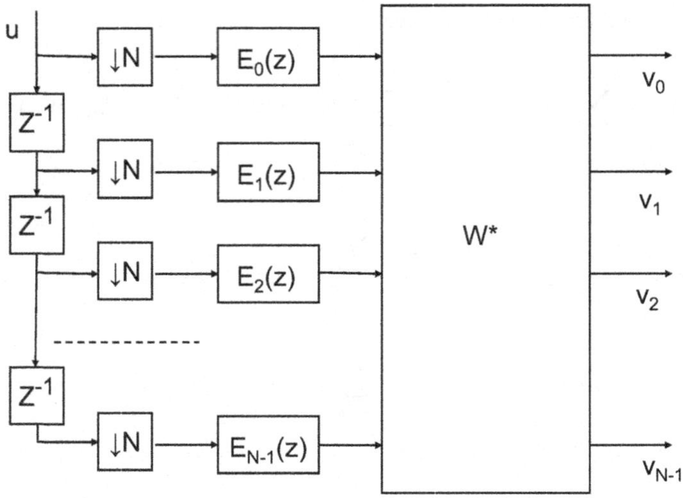

Fig. 1.18 Modulated analyzer filter bank

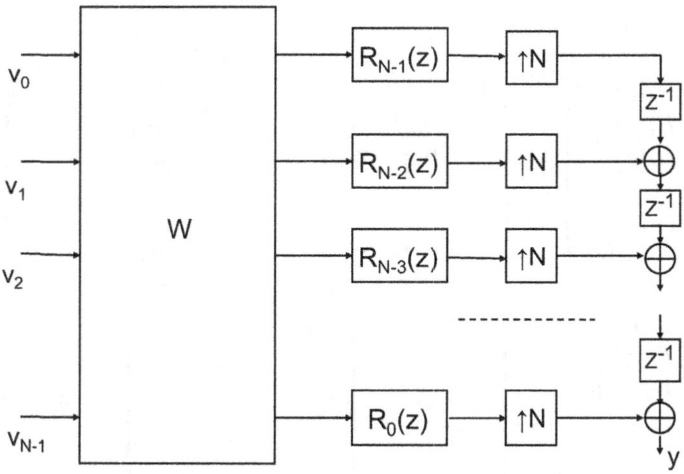

Fig. 1.19 Modulated synthesizer filter bank

and let us apply demodulation:

$$F_k(z) = F_0(z\, W_N^{-k}) = \sum_{kl=0}^{N-1} z^{-(N-1-l)}\, W^{k(N-1-l)}\, R_l(z^N\, W_N^{-kN}) = \\ = \sum_{l=0}^{N-1} z^{-(N-1-l)}\, W_N^{k(N-1-l)}\, R_l(z^N) \tag{1.41}$$

The complete set of filters for the synthesis bank is:

$$\begin{pmatrix} F_0(z) \\ F_1(z) \\ F_2(z) \\ . \\ . \\ F_{N-1}(z) \end{pmatrix} = \begin{pmatrix} 1 & 1 & 1 \ldots & 1 \\ 1 & W_N^1 & W_N^2 \ldots & W_N^{(N-1)} \\ 1 & W_N^2 & W_N^4 \ldots & W_N^{2\,(N-1)} \\ \ldots\ldots & & & \\ \ldots & & & \\ 1 & W_N^{(N-1)} & W_N^{(N-1)\,2} \ldots & W_N^{(N-1)(N-1)} \end{pmatrix} \cdot$$

$$\cdot \begin{pmatrix} R_{N-1}(z^N) \\ z^{-1} R_{N-2}(z^N) \\ z^{-2} R_{N-3}(z^N) \\ . \\ . \\ z^{-(N-1)} R_0(z^N) \end{pmatrix} = \mathbf{W} \cdot \begin{pmatrix} R_{N-1}(z^N) \\ z^{-1} R_{N-2}(z^N) \\ z^{-2} R_{N-3}(z^N) \\ . \\ . \\ z^{-(N-1)} R_0(z^N) \end{pmatrix} = \mathbf{W} \cdot \widehat{R} \tag{1.42}$$

where $\mathbf{W}$ is the DFT matrix, and we introduced $\widehat{R}$ to represent the vector of polyphase components.

Interpolators can be moved to the right, with Noble identities. Consequently the synthesizer filter bank of Fig. 1.8 can be implemented as shown in Fig. 1.19.

In order to draw Fig. 1.19, it has been taken into account that:

$$Y(z) = [F_0 \; F_1 \ldots F_{N-1}] \begin{bmatrix} v_0 \\ v_1 . \\ . \\ . \\ v_{N-1} \end{bmatrix} = (\mathrm{W}\,\widehat{R})^T\,\bar{V} = \widehat{R}^T\,\mathrm{W}^T\,\bar{V} = \widehat{R}^T\,\mathrm{W}\,\bar{V} \tag{1.43}$$

See [23, 53] and references therein for more details on modulated filter banks.

1.3 Symmetries and Filter Types

The theory of filter banks, and the theory of wavelets, frequently deal with symmetries of filter coefficients and of frequency responses. Therefore, in view of the matters to be considered soon, it is convenient to examine basic aspects of symmetries and filter characteristics.

A sequence:

$$x(n), \; -\infty \le n \le \infty \tag{1.44}$$

- Is symmetric if $x(n) = x(-n)$
- Is antisymmetric if $x(n) = -x(-n)$

A complex sequence:

- Is conjugate symmetric if $x(n) = x^*(-n)$
- Is conjugate antisymmetric if $x(n) = -x^*(-n)$.

Let $X(k)$ be the DFT of the sequence $x(n)$. Then:

- If $x(n)$ real and symmetric, $X(k)$ real and symmetric
- If $x(n)$ real and antisymmetric, $X(k)$ imaginary and antisymmetric
- If $x(n)$ conjugate symmetric, $X(k)$ real
- If $x(n)$ conjugate antisymmetric, $X(k)$ imaginary
- If $x(n)$ imaginary and symmetric, $X(k)$ imaginary and symmetric
- If $x(n)$ imaginary and antisymmetric, $X(k)$ real and antisymmetric.

In the context of filters, the sequences of interest are commonly filter impulse responses (note that in FIR filters the impulse response and the filter coefficients are the same). It can be observed in the scientific literature that there are two mentalities.

Some scientists usually do an off-line work, with recorded data or signal vectors. In this case it is possible to consider past and future in order to have no filtering delay. The *non-causal filters* they use have impulse responses from $h(-L)$ to $h(L)$.

Other scientists do practice on-line filtering. In this case there is unavoidable delay in the filtering process. The *causal filters* they use have impulse responses from $h(0)$ to $h(N-1)$.

In both causal and non-causal filters, the symmetries are related to the middle of the $h(i)$ series.

Example of FIR non-causal filter with symmetrical coefficients

$$z^{-3} + 4z^{-2} + 3z^{-1} + 7 + 3z + 4z^2 + z^3$$

Example of FIR causal filter with symmetrical coefficients (just a delayed version of the previous filter):

$$1 + 4z^{-1} + 3z^{-2} + 7z^{-3} + 3z^{-4} + 4z^{-5} + z^{-6}$$

1.3.1 Linear Phase

Most filter designers wish to have linear phase in their applications, so the filters had constant group delay (this feature is usually an advantage).

A digital filter has linear phase if its frequency response is:

$$H(e^{j\omega}) = B(\omega)\, e^{-j\omega\tau + j\varphi} \tag{1.45}$$

With $B(\omega)$ real, and ϕ and τ (the group delay) are constants.

Equation (1.41) implies that:

$$h(n) = e^{j\varphi}\, b(n - \tau) \tag{1.46}$$

($b(n)$ the inverse Fourier transform of $B(\omega)$).

Then:

$$b(n) = b^*(n) \;\; (B(\omega) \; is \; real)$$

$$b(n) = e^{-j\varphi}\, h(n + \tau)$$

$$e^{-j\varphi}\, h(n + \tau) = e^{j\varphi}\, h^*(-n + \tau)$$

Finally:

$$h(n + \tau) = e^{2j\varphi}\, h^*(-n + \tau) \tag{1.47}$$

Notice that, in particular:

$$h(0) = e^{2j\varphi}\, h^*(2\tau) \tag{1.48}$$

In the case of causal FIR filters of finite duration ($0 \leq n \leq N - 1$) we have:

$$\tau = \frac{N-1}{2}, \; and \; h(n) = e^{2j\varphi}\, h^*(N - 1 - n) \tag{1.49}$$

Filters with $h(n) = h^*(-n)$ (conjugate symmetry) are *zero-phase filters* (their frequency responses are real).

Linear phase filters must be FIR filters, because a causal stable IIR filter cannot have a symmetric impulse response.

1.3.2 *FIR Filters with Linear Phase*

FIR filters with linear phase are a subclass of FIR filters.

As seen just before, a causal FIR filter ($0 \leq n \leq N - 1$) is linear phase if its coefficients are:

$$h(n) = e^{2j\varphi}\, h^*(N - 1 - n) \tag{1.50}$$

1.3.2.1 Linear Phase FIR Filters with Real Coefficients:

Most FIR applications use real filter coefficients. In this case:

$$h(n) = h^*(n) \tag{1.51}$$

So $e^{2j\varphi}$must be real and then:

$$\varphi = \frac{q\,\pi}{2}, \quad (q = 0,\, 1,\, 2\, \ldots)$$

In consequence, for linear phase:

$$h(n) = (-1)^q\, h(N-1-n) \tag{1.52}$$

Equation (1.48) says that for linear phase the filter impulse response must be either symmetric or antisymmetric.

The literature distinguishes four types of linear phase FIR filters with real coefficients:

- Type I: $q = 0$ ($h(n)$ symmetric), N odd
- Type II: $q = 0$ ($h(n)$ symmetric), N even
- Type III: $q = 1$ ($h(n)$ antisymmetric), N odd
- Type IV: $q = 1$ ($h(n)$ antisymmetric), N even

The frequency response of type II filters has $H(e^{j\pi}) = 0$, therefore high-pass filters cannot be of this type.

Type III filters have $h((N-1)/2) = 0$, and $H(e^{j0}) = H(e^{j\pi}) = 0$, so the filter cannot be low-pass, nor high-pass. This type of filters is suitable for band-pass filters, differentiators and Hilbert transform.

The frequency response of type IV filters has $H(e^{j0}) = 0$, therefore low-pass filters cannot be of this type.

In all causal four types the group delay is $(N-1)/2$, and the phase is:

- Types I and II: $-\omega\, (N\_1)/2$
- Types III and IV: $-\omega\, (N-1)/2 + (\pi/2)$

The preferred type for most applications is type I.

The next Table 1.1 can be useful for a quick look of main filter characteristics:

Here are some simple examples of causal linear phase FIR filters with 7 real coefficients:

- $N = 7$, type I:

$$1 + 3\,z^{-1} + 2\,z^{-2} + 5\,z^{-3} + 2\,z^{-4} + 3\,z^{-5} + z^{-6} \tag{1.53}$$

Table 1.1 Linear phase FIR filter characteristics

Type	Sym./Antisym.	Filter length	Gain at DC	High pass	Low pass
1	Sym	Odd	Any	Yes	Yes
2	Sym	Even	Any	No	Yes
3	Anti	Odd	Zero	No	No
4	Anti	Even	Zero	Yes	No

- $N = 6$, type II:

$$1 + 2z^{-1} + 5z^{-2} + 5z^{-3} + 2z^{-4} + z^{-5} \tag{1.54}$$

- $N = 7$, type III:

$$-1 - 3z^{-1} - 2z^{-2} + 0z^{-3} + 2z^{-4} + 3z^{-5} + z^{-6} \tag{1.55}$$

- $N = 6$, type IV:

$$-1 - 2z^{-1} - 5z^{-2} + 5z^{-3} + 2z^{-4} + z^{-5} \tag{1.56}$$

In the case of non-causal filters it is interesting to note that non-causal type I and type II filters are zero-phase.

1.3.3 Complementary Filters

A set of transfer functions $H_0(z), H_1(z) \ldots H_{N-1}(z)$ is *strictly complementary* if:

$$\sum_{k=0}^{N-1} H_k(z) = c\, z^{-m} \tag{1.57}$$

where c is a constant.

An example is the set of *Nth* band filters, with $H_k(z) = H(z\, W_N^k)$.

An *Nth* band filter is a zero-phase filter with:

$$h\,(Nn) = \begin{cases} c & n = 0 \\ 0 & n \neq 0 \end{cases} \tag{1.58}$$

With *Nth* band filters we have:

$$\sum_{k=0}^{N-1} H(zW_N^k) = N \cdot c \tag{1.59}$$

A way to obtain *Nth* band filters is:

$$h(n) = \frac{\sin\left(\frac{\pi n}{N}\right)}{\pi n}\, w(n) \tag{1.60}$$

with $w(n)$ a window function (for instance the Kaiser window).

A set of transfer functions $H_0(z), H_1(z) \ldots H_{N-1}(z)$ is *power complementary* if:

$$\sum_{k=0}^{N-1} \left| H_k(e^{j\omega}) \right|^2 = c \tag{1.61}$$

Equation (1.57) is equivalent to:

$$\sum_{k=0}^{N-1} H_k^*(-z^{-1})\, H_k(z) = c \tag{1.62}$$

Two FIR filters $H_0(z)$ and $H_1(z)$ are *Euclidean complementary* if the polynomials $H_0(z)$ and $H_1(z)$ are relatively prime (they do not share any common factor $(1 - \beta z^{-1}),\ \beta > 0)$). If it is so, then there exist polynomials $F_0(z)$ and $F_1(z)$ such that:

$$H_0(z)\, F_0(z) + H_1(z)\, F_1(z) = c \tag{1.63}$$

(the Euclidean algorithm can be used to find $F_0(z)$ and $F_1(z)$)

1.3.4 Symmetries in the Frequency Response

A basic symmetry of the frequency magnitude response is given by the symmetry of the DFT of a filter $H(z)$ with real coefficients. Thus we can write:

$$|H(e^{j\omega})| = |H(e^{-j\omega})| \tag{1.64}$$

Likewise:

$$|H(e^{j(\omega-\pi)})| = |H(e^{j(\pi-\omega)})| \tag{1.65}$$

A ***half-band*** filter is a particular case of the *Nth* band filter for N = 2. Consequently, the half-band filter is a zero-phase filter with:

$$h(2n) = \begin{cases} c & n = 0 \\ 0 & n \neq 0 \end{cases} \tag{1.66}$$

Example of half-band filter with real coefficients:

$$z^{-5} - 3z^{-3} + 2z^{-1} + 15 + 2z - 3z^3 + z^5$$

The example could be useful to check that making $z = e^{j\omega}$ or $z = e^{-j\omega}$, both give the same result.

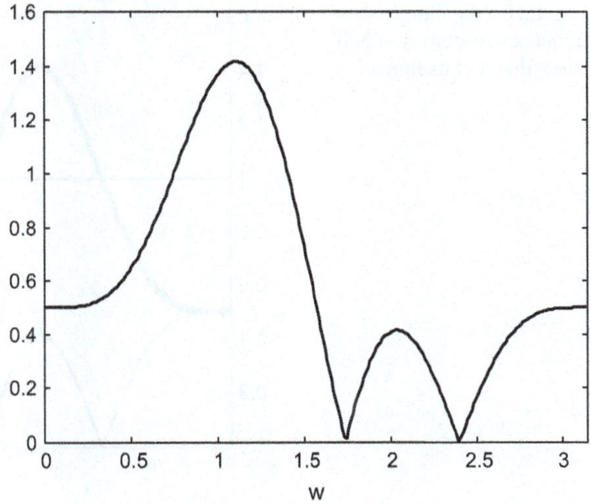

Fig. 1.20 Magnitude of the frequency response of a FIR half band filter

Figure 1.20 shows the magnitude of the frequency response of a half band example (the coefficients are smaller than the just given example). The figure has been generated with the Program 1.4.

Program 1.4 Simple half-band FIR example

```
% Simple half-band FIR example
%
%original causal half-band filter:
h0=[0.1 0 -0.3 0 0.2 0.5 0.2 0 -0.3 0 0.1];
w=0:(2*pi/511):pi;
H0=(fft(h0,512)); %discrete Fourier transform
MH0=abs(H0);
plot(w,MH0(1:256),'k');
axis([0 pi 0 1.6]);
title('frequency response (magnitude)');
xlabel('w');
```

Using polyphase representation, one can write:

$$H(z) = c + z^{-1}\, E_1(z^2) \tag{1.67}$$

From which one obtains:

$$H(z) + H(-z) = 2\,c \tag{1.68}$$

Taking $z = e^{j\omega}$ and $c = 0.5$:

$$H(e^{j\omega}) + H(e^{j(\omega-\pi)}) = H(e^{j\omega}) + H(e^{j(\pi-\omega)}) = 1 \tag{1.69}$$

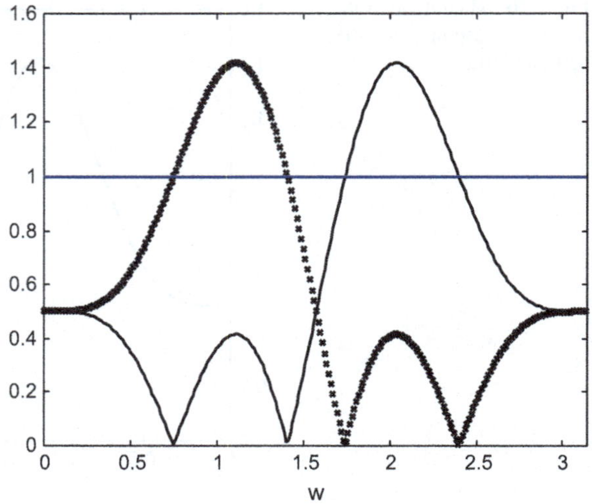

Fig. 1.21 Magnitude of the frequency responses of half band filter and its mirror

Another, simpler way to write this expression is:

$$H(\omega) + H(\pi - \omega) = 1 \tag{1.70}$$

Equation (1.64) shows that there is symmetry of both frequency responses with respect to $\pi/2$. This is why the name half-band filter.

A causal version of the half-band filter is:

$$H_c(z) = z^{-k} H(z) \tag{1.71}$$

with N an odd number and $k = (N\text{-}1)/2$.

Then:

$$H_c(z) - H_c(-z) = 2c \cdot z^{-k} \tag{1.72}$$

Fig. 1.21 shows the magnitude of the frequency responses of the two filters in (1.66). The result of Eq. (1.64) has been also drawn. The figure has been generated with the Program 1.6.

Program 1.5 Simple half-band FIR pair example

```
% Simple half-band FIR pair example
%
%original causal half-band filter:
h0=[0.1 0 -0.3 0 0.2 0.5 0.2 0 -0.3 0 0.1];
%a mirror filter is obtained by sign alternation:
h1=[-0.1 0 0.3 0 -0.2 0.5 -0.2 0 0.3 0 -0.1];
w=0:(2*pi/511):pi;
H0=(fft(h0,512)); %discrete Fourier transform
```

```
H1=(fft(h1,512)); %discrete Fourier transform
HT=H0+H1;
MH0=abs(H0);
MH1=abs(H1);
MHT=abs(HT);
plot(w,MH0(1:256),'kx'); hold on;
plot(w,MH1(1:256),'k');
plot(w,MHT(1:256),'b');
axis([0 pi 0 1.6]);
title('frequency response (magnitude)');
xlabel('w');
```

The filter $H_0(z)$ is *power symmetric* if:

$$|H_0(\omega)|^2 + |H_0(\pi - \omega)|^2 = 1 \tag{1.73}$$

Fig. 1.22 shows the frequency response of the power of two power symmetrical filters. The result of Eq. (1.67) has been computed and also represented in the same figure (it results in a horizontal line). The figure has been generated with the Program 1.6.

Program 1.6 Frequency response of the power of power symmetric filters

```
% Frequency response of the power of power symmetric filters
c=1/sqrt(2);
h0=[c c]; %low-pass filter
h1=[c -c]; %high-pass filter
w=0:(2*pi/511):pi;
H0=abs(fft(h0,512)); %discrete Fourier transform
H1=abs(fft(h1,512)); % """
PH0=H0.*H0;
PH1=H1.*H1;
PHT=PH0+PH1;
plot(w,PH0(1:256),'kx'); hold on;
plot(w,PH1(1:256),'k');
plot(w,PHT(1:256),'b');
axis([0 pi 0 2.5]);
title('Power symmetric filters: frequency response of power');
xlabel('w');
```

1.3.5 Orthogonal FIR Filters

Two filters $A(z)$ and $B(z)$ are orthogonal if:

$$\langle A, B \rangle = \sum_n a(n)\, b(n) = 0 \tag{1.74}$$

where $a(n)$ are the coefficients of $A(z)$, and $b(n)$ are the coefficients of $B(z)$.

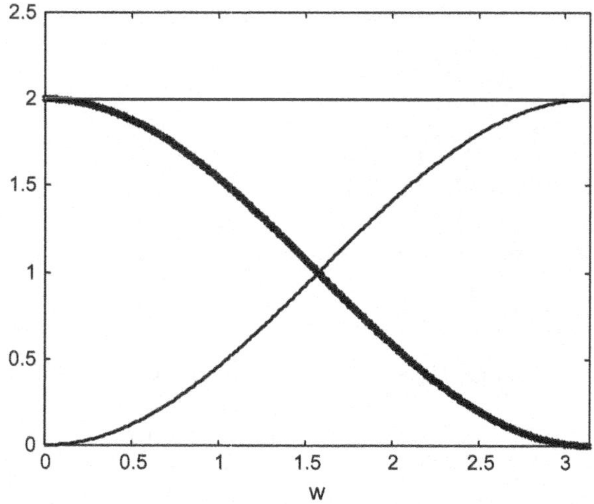

Fig. 1.22 Frequency response of the power of two power symmetric filters

Given a FIR filter $A(z)$ there is a simple way to obtain another FIR filter $B(z)$ that is orthogonal to $A(z)$. The orthogonal filter $B(z)$ is obtained as follows:

$$B(z) = (-z)^{-2k+1} A(-z^{-1}) \tag{1.75}$$

Let us consider in detail an example:

$$A(z) = 1 + 2z^{-1} + 3z^{-2} + 4z^{-3} + 5z^{-4} + 6z^{-5}$$

Then:

$$A(z^{-1}) = 1 + 2z + 3z^2 + 4z^3 + 5z^4 + 6z^5$$

And:

$$A(-z^{-1}) = 1 - 2z + 3z^2 - 4z^3 + 5z^4 - 6z^5$$

Now multiply by $(-z^{-5})$ to obtain:

$$(-z^{-5})\, A(-z^{-1}) = -z^{-5} + 2z^{-4} - 3z^{-3} + 4z^{-2} - 5z^{-1} + 6$$

Finally, reordering terms:

$$B(z) = 6 - 5z^{-1} + 4z^{-2} - 3z^{-3} + 2z^{-4} - z^{-5}$$

In words the sequence of terms is time-reversed and then the sign of odd terms is changed. It is easy to check that the inner product (1.74) is zero.

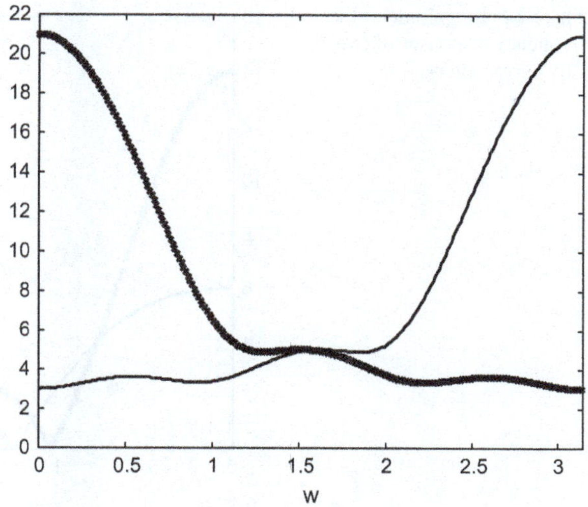

Fig. 1.23 Magnitude of the frequency responses of two FIR orthogonal filters

Figure 1.23 shows the magnitude of the frequency response of the filters given in the example. The curve with crosses corresponds to $A(z)$. The figure has been generated with the Program 1.7.

Program 1.7 Simple orthogonal FIR pair example

```
% Simple orthogonal FIR pair example
% a second FIR filter is obtained by flipping
% and sign alternation
h0=[1 2 3 4 5 6]; %original non-symmetrical filter.
h1=[6 -5 4 -3 2 -1]; %a second filter is obtained.
w=0:(2*pi/511):pi;
H0=abs(fft(h0,512)); %discrete Fourier transform
H1=abs(fft(h1,512)); %discrete Fourier transform
plot(w,H0(1:256),'kx'); hold on;
plot(w,H1(1:256),'k');
axis([0 pi 0 22]);
title('frequency responses (magnitude)');
xlabel('w');
```

1.3.6 *Mirror FIR Filters*

Given a symmetric FIR filter $H_0(z)$, a mirror filter $H_1(z)$ can be easily obtained by sign alternation:

$$h_1(n) = (-1)^n \, h_0(n) \tag{1.76}$$

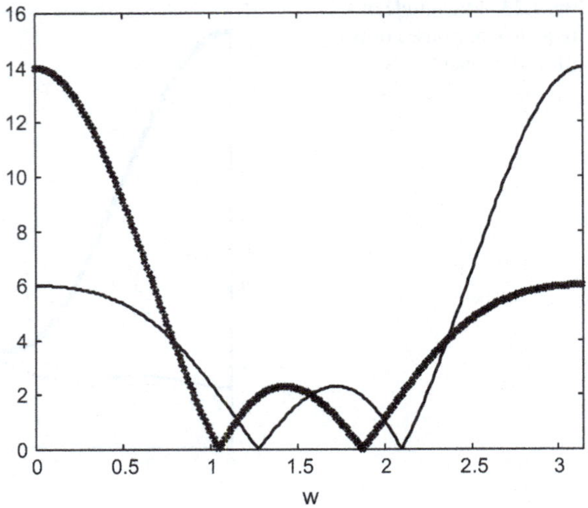

Fig. 1.24 Magnitude of the frequency responses of two FIR mirror filters

This is equivalent to:

$$H_1(z) = H_0(-z) \tag{1.77}$$

Therefore the frequency response of H_1 is a reflection of H_0 about $\pi/2$ (Fig. 1.24):

$$H_1(e^{j\omega}) = H_0(e^{j(\omega+\pi)}) \tag{1.78}$$

Program 1.8 Simple mirror FIR example

```
% Simple mirror FIR example
% a second FIR filter is obtained by sign alternation
h0=[1 3 1 4 1 3 1]; %original filter with symmetrical coeffs.
h1=[1 -3 1 -4 1 -3 1]; %a second filter is obtained.
w=0:(2*pi/511):pi;
H0=abs(fft(h0,512)); %discrete Fourier transform
H1=abs(fft(h1,512)); %discrete Fourier transform
plot(w,H0(1:256),'kx'); hold on;
plot(w,H1(1:256),'k');
axis([0 pi 0 16]);
title('frequency response (magnitude)');
xlabel('w');
```

1.3.7 *Zeros of FIR Filters. Spectral Factorization*

Consider a FIR filter with real coefficients. If this filter has a complex zero z_i, then there must be a conjugate zero z_i^*.

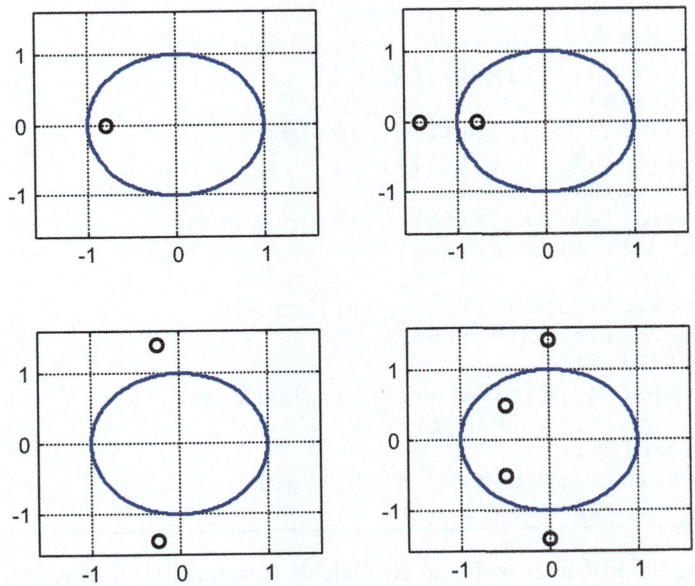

Fig. 1.25 Zeros of non-symmetric FIR filters: four cases

The Program 1.9 draws the unit circle and the zeros for 4 examples of causal non symmetric FIR filters with real coefficients. The result is shown in Fig. 1.25. The examples are the following:

$$1 + 0.8\,z^{-1}$$

$$1 + 2.2\,z^{-1} + 1.1\,z^{-2}$$

$$1 + 0.5\,z^{-1} + 2\,z^{-2}$$

$$1 + z^{-1} + 2.5\,z^{-2} + 2\,z^{-3} + z^{-4}$$

Program 1.9 Zeros of non symmetric FIR filters

```
% Zeros of non symmetric FIR filters
% 4 cases
alfa=0:0.01:(2*pi);
subplot(2,2,1)
plot(cos(alfa),sin(alfa),'b'); hold on;
F=[1 0.8]; %FIR filter
[zi]=roots(F);
plot(real(zi),imag(zi),'ko'); grid;
axis([-1.6 1.6 -1.6 1.6]);
title('zeros of non-symmetric FIR filters');
subplot(2,2,2)
```

```
plot(cos(alfa),sin(alfa),'b'); hold on;
F=[1 2.2 1.1]; %FIR filter
[zi]=roots(F)
plot(real(zi),imag(zi),'ko');; grid;
axis([-1.6 1.6 -1.6 1.6]);
subplot(2,2,3)
plot(cos(alfa),sin(alfa),'b'); hold on;
F=[1 0.5 2]; %FIR filter
[zi]=roots(F);
plot(real(zi),imag(zi),'ko');; grid;
axis([-1.6 1.6 -1.6 1.6]);
subplot(2,2,4)
plot(cos(alfa),sin(alfa),'b'); hold on;
F=[1 1 2.5 2 1]; %FIR filter
[zi]=roots(F);
plot(real(zi),imag(zi),'ko');; grid;
axis([-1.6 1.6 -1.6 1.6]);
```

Suppose the FIR filter with real coefficients is symmetric, if there is a zero z_i, then $1/z_i$ is also a zero. Notice that if z_i is inside the unit circle, then $1/z_i$ is outside.

In consequence, symmetric FIR filters with real coefficients can have zeros in quadruplets (some authors use the term '*constellation*').

The Program 1.10 is very similar to the previous one, only that now it deals with symmetric FIR filters. It represents the zeros of 4 examples. Figure 1.26 shows the results. The examples are the following:

$$0.5 + 0.5\,z^{-1}$$

$$1 + 2.1\,z^{-1} + z^{-2}$$

$$1 + 0.5\,z^{-1} + z^{-2}$$

$$1 + z^{-1} + 2.5\,z^{-2} + z^{-3} + z^{-4}$$

Program 1.10 Zeros of symmetric FIR filters

```
% Zeros of symmetric FIR filters
% 4 cases
alfa=0:0.01:(2*pi);
subplot(2,2,1)
plot(cos(alfa),sin(alfa),'b'); hold on;
F=[0.5 0.5]; %FIR filter
[zi]=roots(F);
plot(real(zi),imag(zi),'ko'); grid;
axis([-1.6 1.6 -1.6 1.6]);
title('zeros of symmetric FIR filters');
```

```
subplot(2,2,2)
plot(cos(alfa),sin(alfa),'b'); hold on;
F=[1 2.1 1]; %FIR filter
[zi]=roots(F);
plot(real(zi),imag(zi),'ko');; grid;
axis([-1.6 1.6 -1.6 1.6]);
subplot(2,2,3)
plot(cos(alfa),sin(alfa),'b'); hold on;
F=[1 0.5 1]; %FIR filter
[zi]=roots(F);
plot(real(zi),imag(zi),'ko');; grid;
axis([-1.6 1.6 -1.6 1.6]);
subplot(2,2,4)
plot(cos(alfa),sin(alfa),'b'); hold on;
F=[1 1 2.5 1 1]; %symmetrical FIR filter
[zi]=roots(F);
plot(real(zi),imag(zi),'ko');; grid;
axis([-1.6 1.6 -1.6 1.6]);
```

Given a certain symmetrical FIR filter $G(z)$, it may be convenient for certain design purposes to factorize $G(z)$into two FIR filters $A(z)$ and $B(z)$, taking the zeros inside the unit circle to form $A(z)$, and the zeros outside the unit circle to form $B(z)$. This factorization $G(z) = A(z)\, B(z)$ is called '*spectral factorization*'.

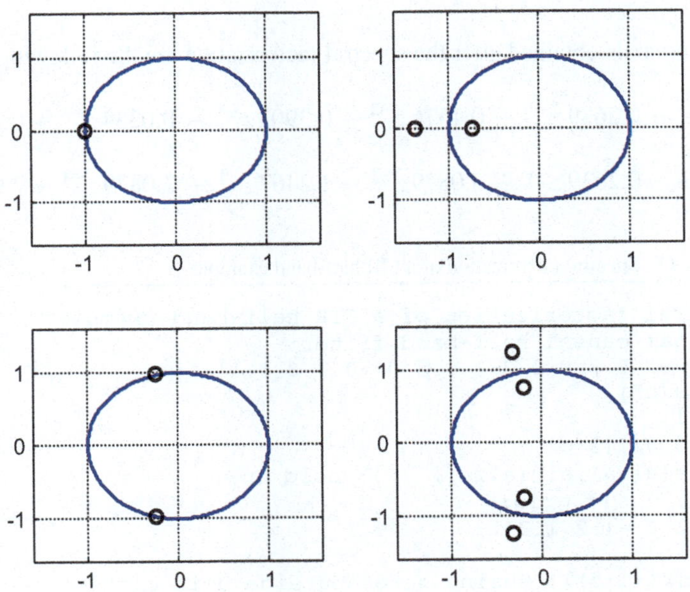

Fig. 1.26 Zeros of symmetric FIR filters: four cases

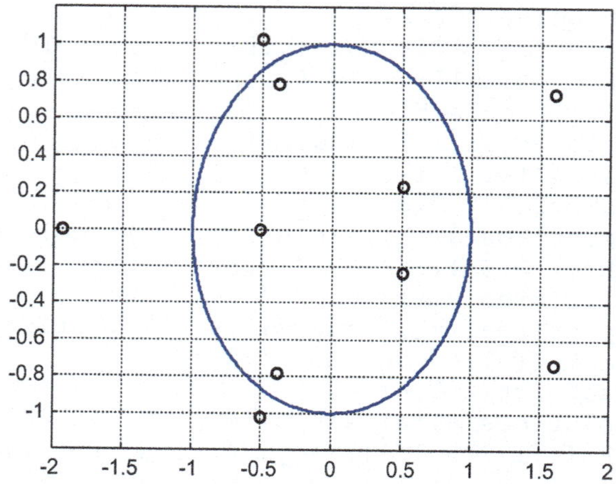

Fig. 1.27 Zeros of a FIR half band filter

The Program 1.11 takes an example of half band FIR filter with real coefficients and plots two figures. Figure 1.27 shows the zeros of the half band filter. Figure 1.28 shows the magnitude of the frequency response of the two filters obtained by the spectral factorization.

The example of causal FIR half band filter is the following:

$$0.1 - 0.3\,z^{-2} + 0.2\,z^{-4} + z^{-5} + 0.2\,z^{-6} - 0.3\,z^{-8} + 0.1\,z^{-10}$$

The polynomials obtained after the spectral factorization are the following:

$$A(z) = 1 + 0.2639\,z^{-1} + 0.1576\,z^{-2} - 0.3903\,z^{-3} - 0.0334\,z^{-4} + 0.1264\,z^{-5}$$

$$B(z) = 1 - 0.2639\,z^{-1} - 3.0879\,z^{-2} + 1.2467\,z^{-3} + 2.0879\,z^{-4} + 7.9123z^{-5}$$

Program 1.11 Spectral factorization of a FIR half-band example

```
% Spectral factorization of a FIR half-band example
% original causal half-band filter:
hf=[0.1 0 -0.3 0 0.2 1 0.2 0 -0.3 0 0.1];
r1=roots(hf);
figure(1)
alfa=0:0.01:(2*pi);
plot(cos(alfa),sin(alfa),'b'); hold on;
plot(r1,'ko'); grid;
axis([-2 2 -1.2 1.2]);
title('zeros of FIR half band filter');
h1=poly(r1(1:5)); %using zeros outside unit circle
h0=poly(r1(6:10)); %using zeros inside unit circle
figure(2)
w=0:(2*pi/511):pi;
H0=abs(fft(h0,512)); %discrete Fourier transform
```

```
H1=abs(fft(h1,512)); %"""
plot(w,H0(1:256),'kx'); hold on;
plot(w,H1(1:256),'k');
axis([0 pi 0 15]);
title('frequency response (magnitude) of factor filters');
xlabel('w');
```

A FIR filter is said to be '*minimum phase*' if it has no zeros outside the unit circle. In most cases, linear phase FIR filters have zeros outside the unit circle, and so they are not minimum phase.

1.4 Two-Channel Filters and Perfect Reconstruction

Many applications of filter banks want to obtain at the output a signal $y(n)$ equal to the input $u(n-d)$ for some integer d (in off-line work it may be possible to have $d = 0$). This is called perfect reconstruction (PR).

The two-channel filter banks are the simplest case of N-channel filter banks, and serve as a good introduction to key aspects of the PR topic.

1.4.1 *Automatic Aliasing Cancellation, Perfect Reconstruction*

A main problem for PR is aliasing caused from downsampling and upsampling. This aliasing can be cancelled by the filter bank, if proper design is done.

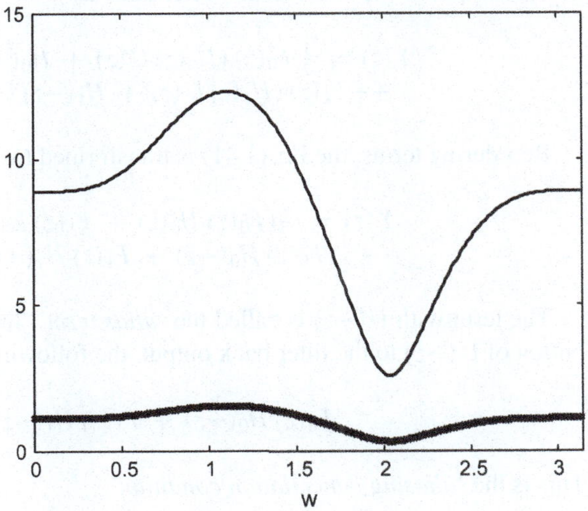

Fig. 1.28 Magnitude of the frequency responses of spectral factors of the FIR half band filter

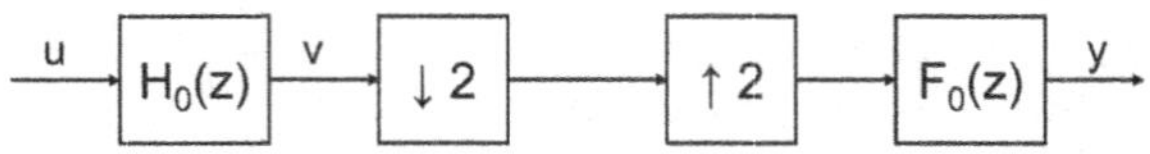

Fig. 1.29 A filter branch

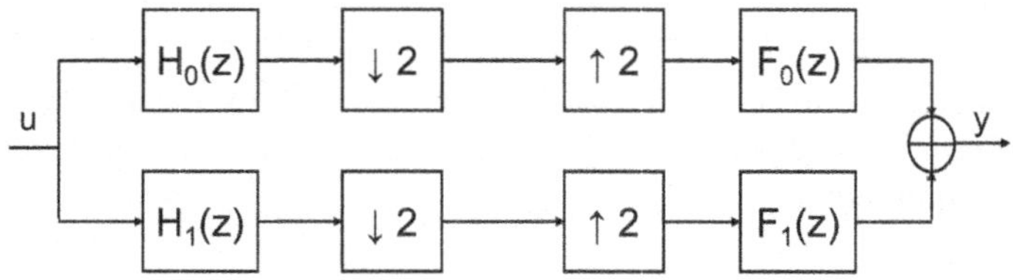

Fig. 1.30 A two-channel filter bank

Consider a filter branch as given in Fig. 1.29.

The output of the filter branch is given by:

$$Y(z) = F_0(z)\,\frac{1}{2}(V(z) + V(-z)) = \frac{1}{2}\,F_0(z)\,(H_0(z)\,U(z) + H_0(-z)\,U(-z)) \tag{1.79}$$

It can be noticed that the output has alias contents, since it is influenced by $U(-z)$.

1.4.1.1 Basic Equations for the Two-Channel Filter Bank

Let us study the two-channel filter bank represented in Fig. 1.30.

$H_0(z)$ and $H_1(z)$ form an analysis filter bank, and $F_0(z)$ and $F_1(z)$ form a synthesis filter bank. $H_0(z)$ and $F_0(z)$ are low-pass filters, while $H_1(z)$ and $F_1(z)$ are high-pass filters.

The output of the two-channel filter bank is:

$$\begin{aligned} Y(z) &= \tfrac{1}{2}\,F_0(z)\,(H_0(z)\,U(z) + H_0(-z)\,U(-z)) + \\ &+ \tfrac{1}{2}\,F_1(z)\,(H_1(z)\,U(z) + H_1(-z)\,U(-z)) \end{aligned} \tag{1.80}$$

Reordering terms, the Eq. (1.41) is transformed to:

$$\begin{aligned} Y(z) &= \tfrac{1}{2}(F_0(z)\,H_0(z) + F_1(z)\,H_1(z))\,U(z) + \\ &+ \tfrac{1}{2}(F_0(z)\,H_0(-z) + F_1(z)\,H_1(-z))\,U(-z) \end{aligned} \tag{1.81}$$

The term with $U(-z)$ is called the '*alias term*'. In order not to have any contribution of $U(-z)$ to the filter bank output, the following condition must be met:

$$F_0(z)\,H_0(-z) + F_1(z)\,H_1(-z) = 0 \tag{1.82}$$

This is the '*aliasing cancellation condition*'.

This condition is fulfilled with:

$$\frac{F_0(z)}{F_1(z)} = -\frac{H_1(-z)}{H_0(-z)} \tag{1.83}$$

Or,

$$\begin{aligned} F_0(z) &= K(z)\, H_1(-z) \\ F_1(z) &= -\, K(z)\, H_0(-z) \end{aligned} \tag{1.84}$$

for any rational $K(z)$.

The term with $U(z)$ in Eq. (1.70) is called the '*distortion term*' and for perfect reconstruction must be a constant or a pure delay (depending on we were working with non-causal filters or with causal filters). Therefore, the distortion term must be one of the following:

- Off-line work, non-causal filters:

$$F_0(z)\, H_0(z) + F_1(z)\, H_1(z) = 2 \tag{1.85}$$

- On-line work, causal filters:

$$F_0(z)\, H_0(z) + F_1(z)\, H_1(z) = 2\, z^{-d} \tag{1.86}$$

There are many possible designs for the PR filter bank, achieving both the aliasing cancellation condition, and the no-distortion condition.

Coming back to Eq. (1.72), if the filters satisfy this equation, then:

$$\begin{aligned} Y(z) &= \tfrac{1}{2}\,[H_0(z)\, H_1(-z) - H_1(z)\, H_0(-z)]\; K(z)\, U(z) = \\ &= T(z)\, U(z) \end{aligned} \tag{1.87}$$

where $T(z)$ is the **filter bank transfer function**. For perfect reconstruction, the transfer function must be:

$$T(z) = z^{-d} \tag{1.88}$$

Or, in the frequency domain:

$$T(\omega) = e^{-j\, d\, \omega} \tag{1.89}$$

which means flat magnitude response and linear phase response. If there is no perfect reconstruction, then the transfer function would be:

$$T'(\omega) = A(\omega)\; e^{-j\, \Phi(\omega)} \tag{1.90}$$

so there will be amplitude and phase distortion.

1.4.1.2 A Factorization Approach

The aliasing cancellation condition (1.82) can be easily satisfied if:

$$F_0(z) = H_1(-z) \tag{1.91}$$

$$F_1(z) = -H_0(-z) \tag{1.92}$$

This is equivalent to:

$$f_0(n) = (-1)^n h_1(n) \tag{1.93}$$

$$n = 0, 1, 2, \ldots, N_1 - 1$$

$$f_1(n) = (-1)^{n+1} h_0(n) \tag{1.94}$$

$$n = 0, 1, 2, \ldots, N_0 - 1$$

Equation (1.91) can be rewritten as:

$$H_1(z) = F_0(-z) \tag{1.95}$$

The no distortion condition is:

$$F_0(z)\, H_0(z) + F_1(z)\, H_1(z) = 2\, z^{-d} \tag{1.96}$$

Using (1.92) and (1.95), this condition can be written as:

$$\begin{aligned} F_0(z)\, H_0(z) - F_0(-z)\, H_0(-z) = 2\, z^{-d} = \\ = P_0(z) - P_0(-z) \end{aligned} \tag{1.97}$$

where:

$$P_0(z) = F_0(z)\, H_0(z) \tag{1.98}$$

According with Eq. (1.97):

$$p_0(n) + (-1)^{n+1}\, p_0(n) = 0, \quad \forall n \neq d \;\; (d = (N-1)/2) \tag{1.99}$$

This implies that odd terms must be zero. In consequence:

$$p_0(n) = \begin{cases} 0, & n\ odd\ \neq d \\ 1, & n = d \\ any, & n\ even \end{cases} \tag{1.100}$$

Notice that $p_0(n)$ is a causal half band filter of length $N_0 + N_1 - 1$.

The design of the PR filter can proceed in two steps: find a $P_0(z)$ according to (1.100), and then factorize $P_0(z)$ into $H_0(z)$ and $F_0(z)$ (and obtain $H_1(z)$ and $G_1(z)$ with (1.91) and (1.92).

A popular design of $P_0(z)$ is the following:

$$P_0(z) = (1 + z^{-1})^{2k}\, Q(z) \tag{1.101}$$

where $Q(z)$ is a polynomial of degree $(2k - 2)$ and $P_0(z)$ is type I FIR.

The factor $(1 + z^{-1})^{2k}$ is called '*binomial filter*' and it is a spline filter (splines will be covered in the next chapter).

$P_o(z)$, as given by (1.101), is named the '*maxflat filter*'. This filter has $2k$ zeros at -1. It has a maximally flat frequency response at $\omega = \pi$, and its first $2k - 1$ derivatives with respect to ω are zero.

A frequently cited example is the following:

$$P_0(z) = (1 + z^{-1})^4\, Q(z) \tag{1.102}$$

$$Q(z) = (-1/16)\,(1 - 4\,z^{-1} + z^{-2}) = (-1/16)\,R(z) \tag{1.103}$$

$$P_0(z) = \frac{1}{16}(-1 + 9\,z^{-2} + 16\,z^{-3} + 9\,z^{-4} - z^{-6}) \tag{1.104}$$

Since the roots of $R(z)$ are $r = 2 - \sqrt{3},\ \frac{1}{r} = 2 + \sqrt{3}$:

$$R(z) = (r - z^{-1})\,(\frac{1}{r} - z^{-1}) \tag{1.105}$$

And there are several simple alternative factorizations of $P_0(z)$:

$$H_0(z) = (1 + z^{-1})^2,\ F_0(z) = (1 + z^{-1})^2 \cdot Q(z) \tag{1.106}$$

$$H_0(z) = (1 + z^{-1})^3,\ F_0(z) = (1 + z^{-1}) \cdot Q(z) \tag{1.107}$$

$$H_0(z) = (1/c_1)(1 + z^{-1})^2\,(r - z^{-1}),\ F_0(z) = (1/c_2)(1 + z^{-1})^2 \cdot (\frac{1}{r} - z^{-1}) \tag{1.108}$$

(where c_1 and c_2 are scale constants)

1.4.1.3 Alternative Expressions for the Two-Channel Filter Bank

Several design techniques have been tried for the PR problem. Many of them are based on matrix formulations. It is opportune to assemble most representative alternatives.

Consider again the equation for the filter bank output:

$$Y(z) = \frac{1}{2} F_0(z)\,(H_0(z)\,U(z) + H_0(-z)\,U(-z)) \\ + \frac{1}{2} F_1(z)\,(H_1(z)\,U(z) + H_1(-z)\,U(-z))$$

It can be put in the following form:

$$Y(z) = \frac{1}{2}\,[U(z)\;\; U(-z)] \begin{pmatrix} H_0(z) & H_1(z) \\ H_0(-z) & H_1(-z) \end{pmatrix} \begin{bmatrix} F_0(z) \\ F_1(z) \end{bmatrix} \tag{1.109}$$

Perfect reconstruction means:

$$\begin{bmatrix} 2\,z^{-d} \\ 0 \end{bmatrix} = \begin{pmatrix} H_0(z) & H_1(z) \\ H_0(-z) & H_1(-z) \end{pmatrix} \begin{bmatrix} F_0(z) \\ F_1(z) \end{bmatrix} = \mathbf{H}_a \cdot \begin{bmatrix} F_0(z) \\ F_1(z) \end{bmatrix} \tag{1.110}$$

The matrix ***Ha*** is called the '*aliasing component matrix*'.

Then:

$$\begin{bmatrix} F_0(z) \\ F_1(z) \end{bmatrix} = \mathrm{H}_a^{-1} \cdot \begin{bmatrix} 2\,z^{-d} \\ 0 \end{bmatrix} = \frac{1}{\det(\mathrm{H}_a)} \begin{pmatrix} H_1(-z) & -H_1(z) \\ -H_0(-z) & H_0(z) \end{pmatrix} \begin{bmatrix} 2\,z^{-d} \\ 0 \end{bmatrix} \\ = \frac{2}{\det(\mathrm{H}_a)} \begin{bmatrix} z^{-d} H_1(-z) \\ -z^{-d} H_0(-z) \end{bmatrix} \tag{1.111}$$

where:

$$\det(\mathrm{H}_a) = [H_0(z)\,H_1(-z) - H_1(z)\,H_0(-z)] \tag{1.112}$$

Notice that according with Eq. (1.82):

$$\det(\mathbf{H}_a) = 2\,T(z)/K(z) \tag{1.113}$$

Other authors prefer to put the equation of the filter bank output in the following way:

$$Y(z) = \frac{1}{2}\,[F_0(z)\;\; F_1(z)] \begin{pmatrix} H_0(z) & H_0(-z) \\ H_1(z) & H_1(-z) \end{pmatrix} \begin{bmatrix} U(z) \\ U(-z) \end{bmatrix} \tag{1.114}$$

where:

$$\mathbf{H}_m = \mathbf{H}_a^T = \begin{pmatrix} H_0(z) & H_0(-z) \\ H_1(z) & H_1(-z) \end{pmatrix} \tag{1.115}$$

Is called the '*modulation matrix*'.

Clearly,

$$\det(\mathbf{H}_m) = \det(\mathbf{H}_a) \tag{1.116}$$

This modulation matrix corresponds to the analysis part of the filter bank. For the synthesis part, the modulation matrix is:

$$\mathbf{F_m} = \begin{pmatrix} F_0(z) & F_0(-z) \\ F_1(z) & F_1(-z) \end{pmatrix} \quad (1.117)$$

Based on (1.114) one can write:

$$Y(-z) = \frac{1}{2}\,[F_0(-z)\;\; F_1(-z)] \begin{pmatrix} H_0(z) & H_0(-z) \\ H_1(z) & H_1(-z) \end{pmatrix} \begin{bmatrix} U(z) \\ U(-z) \end{bmatrix} \quad (1.118)$$

Combining (1.114) and (1.118):

$$\begin{bmatrix} Y(z) \\ Y(-z) \end{bmatrix} = \frac{1}{2} \begin{bmatrix} F_0(z) & F_1(z) \\ F_0(-z) & F_1(-z) \end{bmatrix} \begin{pmatrix} H_0(z) & H_0(-z) \\ H_1(z) & H_1(-z) \end{pmatrix} \begin{bmatrix} U(z) \\ U(-z) \end{bmatrix} \quad (1.119)$$

Therefore:

$$\begin{bmatrix} Y(z) \\ Y(-z) \end{bmatrix} = \frac{1}{2}\,\mathbf{F_m^T}\,\mathbf{H_m} \begin{bmatrix} U(z) \\ U(-z) \end{bmatrix} \quad (1.120)$$

In case of PR, one has that: $Y(z) = z^{-d}\,X(z);\;\; Y(-z) = (-z)^{-d}\,X(-z)$
In consequence, the PR condition is satisfied if:

$$\frac{1}{2}\,\mathbf{F_m^T}\,\mathbf{H_m} = \begin{bmatrix} z^{-d} & 0 \\ 0 & (-z)^{-d} \end{bmatrix} \quad (1.121)$$

An important part of the research on filter banks makes use of *the polyphase representation*. According with (1.30):

$$H_0(z) = E_0(z^2) + z^{-1}\,E_1\,(z^2) \quad (1.122)$$

Let us introduce a notation change, to be continued later in the chapter of wavelets. Eq. (1.122) is rewritten as:

$$H_0(z) = H_{00}(z^2) + z^{-1}\,H_{01}\,(z^2) \quad (1.123)$$

For the analysis part of the two-channel filter bank, the following '*polyphase matrix*' is considered:

$$\mathbf{H}_p = \begin{pmatrix} H_{00}(z^2) & H_{01}(z^2) \\ H_{10}(z^2) & H_{11}(z^2) \end{pmatrix} \quad (1.124)$$

Clearly, the filters of the analysis part are related to the polyphase matrix as follows:

$$\begin{bmatrix} H_0(z) \\ H_1(z) \end{bmatrix} = \begin{pmatrix} H_{00}(z^2) & H_{01}(z^2) \\ H_{10}(z^2) & H_{11}(z^2) \end{pmatrix} \begin{bmatrix} 1 \\ z^1 \end{bmatrix} = \mathbf{H_p} \begin{bmatrix} 1 \\ z^1 \end{bmatrix} \tag{1.125}$$

For the synthesis part, another polyphase matrix is considered:

$$\mathbf{F}_p = \begin{pmatrix} F_{00}(z^2) & F_{01}(z^2) \\ F_{10}(z^2) & F_{11}(z^2) \end{pmatrix} \tag{1.126}$$

And so, concerning the filters of the synthesis part:

$$[F_0(z) \;\; F_1(z)] = [z^{-1} \;\; 1] \begin{pmatrix} F_{00}(z^2) & F_{01}(z^2) \\ F_{10}(z^2) & F_{11}(z^2) \end{pmatrix} = [z^{-1} \;\; 1]\, \mathbf{F_p} \tag{1.127}$$

The two-channel filter bank can be equivalently expressed as in Fig. 1.31.

Using Noble identities, it is possible to derive another equivalent diagram of the two-channel filter bank, as depicted in Fig. 1.32.

In this last diagram, ***B(z)*** is:

$$\mathbf{B(z)} = \mathbf{F_p(z)} \cdot \mathbf{H_p(z)} \tag{1.128}$$

If the analysis and synthesis filters are such:

$$\mathbf{F_p(z)} = \mathbf{H_p(z)}^{-1} \tag{1.129}$$

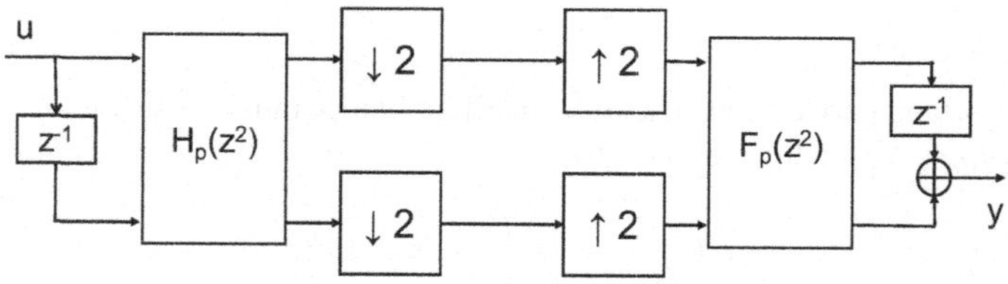

Fig. 1.31 Alternative representation of the two-channel filter bank

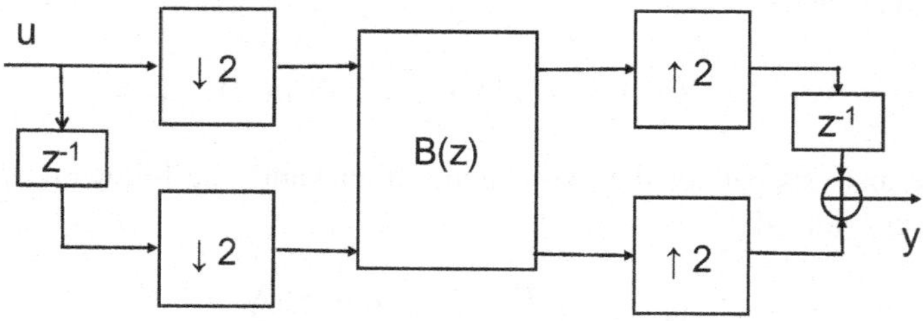

Fig. 1.32 Another equivalent representation of the two-channel filter bank

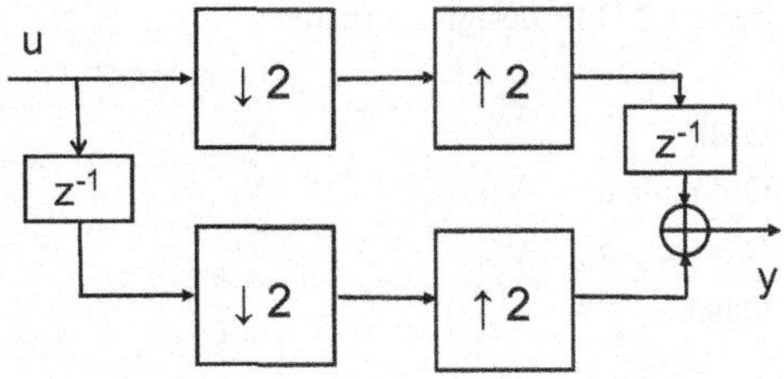

Fig. 1.33 The two-channel filter corresponding to ***B(z)* = *I***

then the filter represented in Fig. 1.32 becomes the simple system represented in Fig. 1.33, which is a PR system. Actually, this system is called the *'lazy wavelet'* and will appear again in the next chapter.

It can be shown that a necessary and sufficient condition for PR is that ***B(z)*** had one of the following expressions:

$$\mathbf{B(z)} = k \begin{bmatrix} z^{-d} & 0 \\ 0 & z^{-d} \end{bmatrix}; \quad \mathbf{B(z)} = k \begin{bmatrix} 0 & z^{-d} \\ z^{-d-1} & 0 \end{bmatrix} \tag{1.130}$$

The modulation matrix and the polyphase matrix are mutually related as follows:

$$\begin{aligned} \mathbf{H}_p &= \begin{pmatrix} H_{00}(z^2) & H_{01}(z^2) \\ H_{10}(z^2) & H_{11}(z^2) \end{pmatrix} \\ &= \frac{1}{2} \begin{pmatrix} H_0(z) & H_0(-z) \\ H_1(z) & H_1(-z) \end{pmatrix} \begin{pmatrix} 1 & 1 \\ 1 & -1 \end{pmatrix} \begin{pmatrix} 1 & 0 \\ 0 & z \end{pmatrix} \end{aligned} \tag{1.131}$$

That is:

$$\mathbf{H}_p = \frac{1}{2} \mathbf{H}_m \begin{pmatrix} 1 & 1 \\ 1 & -1 \end{pmatrix} \begin{pmatrix} 1 & 0 \\ 0 & z \end{pmatrix} \tag{1.132}$$

Hence:

$$\det(\mathbf{H}_m) = -2\,z^{-1} \det(\mathbf{H}_p) \tag{1.133}$$

1.4.2 Design Approaches for Two-Channel Filter Banks with PR

Let us advance that except for the QMF 2-tap filter solution, which will be described next, there are no two-channel PR filter banks that have all three of the following characteristics: FIR, orthogonal, linear phase.

By relaxing orthogonality it is possible to obtain FIR linear phase two-channel PR filter banks.

Table 1.2 FIR design alternatives

	Orthogonal	Linear phase
QMF	X	X
Orthogonal	X	
Biorthogonal		X
Other		

Table 1.2 summarizes the FIR design alternatives:
Let us consider in some detail the three first alternatives.

1.4.2.1 Quadrature Mirror Filter Bank (QMF)

One of the first approaches for the design of two-channel filters with PR, was proposed in the seventies. The suggested design choice was two mirror filters:

$$H_1(z) = H_0(-z) \tag{1.134}$$

(filters with real coefficients)

For the synthesis bank, an option that keeps $F_0(z)$ as low-pass and $F_1(z)$ as high-pass, and that cancels aliasing (1.72), is the following:

$$\begin{aligned} F_0(z) &= H_1(-z) \\ F_1(z) &= -\,H_0(-z) \end{aligned} \tag{1.135}$$

With these choices the transfer function (1.82) of the filter is:

$$T(z) = \frac{1}{2}(H_0^2(z) - H_1^2(z)) = \frac{1}{2}(H_0^2(z) - H_0^2(-z)) \tag{1.136}$$

The '*prototype filter*' $H_0(z)$ is designed to make $T(z) = z^{-d}$ for PR.

The magnitude of the frequency response of H_0 and H_1 form a mirror image pair, since:

$$\left|H_1\left(e^{j\,\omega}\right)\right| = \left|H_0\left(-e^{j\,\omega}\right)\right| = \left|H_0\left(e^{j\,(\omega-\pi)}\right)\right| = \left|H_0\left(e^{j\,(\pi-\omega)}\right)\right| \tag{1.137}$$

Figure 1.34 depicts the magnitude frequency response of H_0 and H_1, forming a mirror image pair (this is the reason for the name QMF). This figure has been generated with the Program 1.12.

Program 1.12 Frequency response of a QMF

```
% Frequency response of a QMF
c=1/sqrt(2);
h0=[c c]; %low-pass filter
h1=[c -c]; %high-pass filter
w=0:(2*pi/511):pi;
H0=real(fft(h0,512)); %discrete Fourier transform
H1=real(fft(h1,512)); % """
plot(w,H0(1:256),'k',w(1:8:256),H0(1:8:256),'kx'); hold on;
plot(w,H1(1:256),'k');
axis([0 pi 0 1.5]);
title('frequency response of QMF H0 and H1 filters');
xlabel('w');
```

It is interesting to see the implementation of a QMF, using polyphase representation

$$H_0(z) = H_{00}(z^2) + z^{-1}\, H_{01}\,(z^2) \tag{1.138}$$

From here the rest of the filters can be obtained as follows:

$$\begin{aligned} H_1(z) &= H_0(-z) = H_{00}(z^2) - z^{-1}\, H_{01}\,(z^2) \\ F_0(z) &= 2H_1(-z) = 2H_{00}(z^2) + 2\, z^{-1}\, H_{01}\,(z^2) \\ F_1(z) &= -2H_0(-z) = -2H_{00}(z^2) + 2\, z^{-1}\, H_{01}\,(z^2) \end{aligned} \tag{1.139}$$

Using Noble identities the QMF can be represented as shown in Fig. 1.35 Notice the lattice structure. This type of structure has a long tradition in the area of filters.

Fig. 1.34 Magnitude of the frequency response of QMF H_0 and H_1 filters

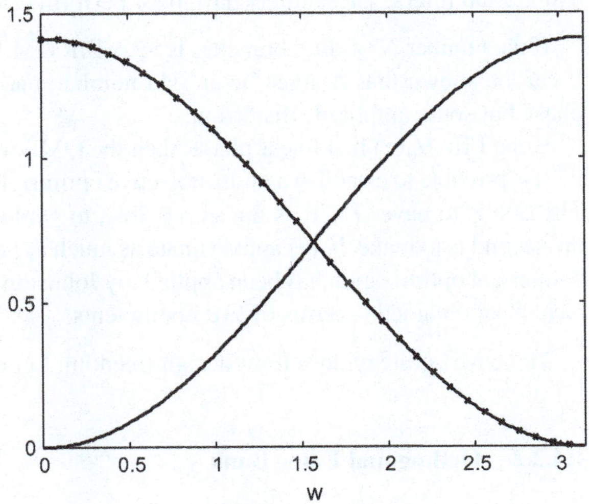

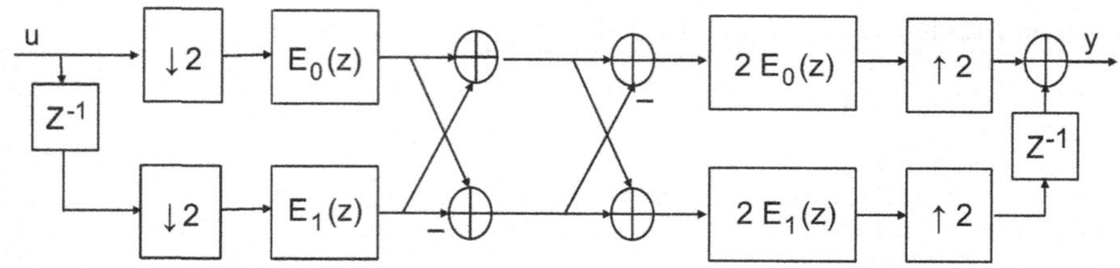

Fig. 1.35 QMF polyphase representation

Coming back to the QMF transfer function, it can be written as:

$$T(z) = \frac{1}{2}(H_0^2(z) - H_0^2(-z)) = \frac{1}{2}[H_0(z) + H_0(-z)]\,[H_0(z) - H_0(-z)] \tag{1.140}$$

For perfect reconstruction $T(z) = z^{-d}$. If we want a solution with FIR filters, then:

$$\begin{aligned} \tfrac{1}{\sqrt{2}}(H_0(z) + H_0(-z)) &= z^{-d_1} \\ \tfrac{1}{\sqrt{2}}(H_0(z) - H_0(-z)) &= z^{-d_2} \end{aligned} \tag{1.141}$$

Solving the pair of equations:

$$\begin{aligned} H_0(z) &= \tfrac{1}{\sqrt{2}}(z^{-d_1} + z^{-d_2}) \\ H_0(-z) &= \tfrac{1}{\sqrt{2}}(z^{-d_1} - z^{-d_2}) \end{aligned} \tag{1.142}$$

So the only way to get perfect reconstruction in FIR QMF with two branches is by using 2-tap filters. These filters have poor performance.

If the number N of filter branches is >2, a FIR QMF can have better performance. It can be shown that N must be an odd number, and that the QMF possess linear phase but some amplitude distortion.

If the FIR $H_0(z)$ had linear phase, then the QMF will have linear phase.

It is possible to establish a multi-objective optimization problem, with two targets. The first is to have $|T(\omega)|$ as flat as possible, to get little amplitude distortion. And the second is to make $H_0(z)$ approximate as much as possible an ideal low pass filter. Numerical optimization has been applied, by Johnston and other authors, and results have been obtained in terms of FIR coefficients.

The QMF strategy does limit design freedom. Let us look for other strategies.

1.4.2.2 Orthogonal Filter Bank

Let us design $H_0(z)$ as a power symmetric filter, so it satisfies the following equation:

$$1 = H_0(z)\,H_0(z^{-1}) + H_0(-z)\,H_0(-z^{-1}) \tag{1.143}$$

Consider frequency response, to confirm the power symmetry:

$$\begin{aligned} 1 &= H_0(e^{j\omega})\, H_0(e^{-j\omega}) + H_0(e^{j(\omega-\pi)})\, H_0(e^{-j(\omega-\pi)}) = \\ &= |H_0(e^{j\omega})|^2 + |H_0(e^{j(\omega-\pi)})|^2 = |H_0(e^{j\omega})|^2 + |H_0(e^{j(\pi-\omega)})|^2 \end{aligned} \tag{1.144}$$

Two filters satisfying the following equation are called *conjugate quadrature* filters (CQF).

$$H_1(z) = (-z)^{-2k+1}\, H_0(-z^{-1}) \tag{1.145}$$

(therefore both filters are orthogonal)

Let us choose $H_1(z)$ as given by Eq. (1.135).

Now, according with (1.135) the frequency magnitude responses of $H_0(z)$ and $H_1(z)$ would be:

$$|H_1(e^{j\omega})| = |H_0(e^{j(\pi-\omega)})| \tag{1.146}$$

Then, by substitution in (1.134) it can be shown that $H_0(z)$ and $H_1(z)$ are power complementary:

$$1 = |H_0(e^{j\omega})|^2 + |H_1(e^{j\omega})|^2 \tag{1.147}$$

To obtain PR, the other two filters can be:

$$\begin{aligned} F_0(z) &= 2\, H_1(-z) \\ F_1(z) &= -2\, H_0(-z) \end{aligned} \tag{1.148}$$

with this design, the aliasing is cancelled, and the no-distortion condition is satisfied as follows:

$$\begin{aligned} & F_0(z)\, H_0(z) + F_1(z)\, H_1(z) = 2H_1(-z)\, H_0(z) - 2H_0(-z)\, H_1(z) = \\ & = 2\,(z)^{-2k+1}\, H_0(z^{-1})\, H_0(z) - 2\,(-z)^{-2k+1}\, H_0(-z)\, H_0(-z^{-1}) \\ & = 2\,(z)^{-2k+1} \end{aligned} \tag{1.149}$$

where (1.133) has been applied, and it has been taken into account that from (1.135):

$$H_1(-z) = (z)^{-2k+1}\, H_0(z^{-1}) \tag{1.150}$$

Let us give an example, based on the factorization example (1.108). Choose as prototype filter the following:

$$H_0(z) = (1+z^{-1})^2\,(1/c_1)\,(r - z^{-1}) \tag{1.151}$$

with:

$$r = 2 - \sqrt{3};\;\; c_1 = 4r - 4 = 4(1 - \sqrt{3}) \tag{1.152}$$

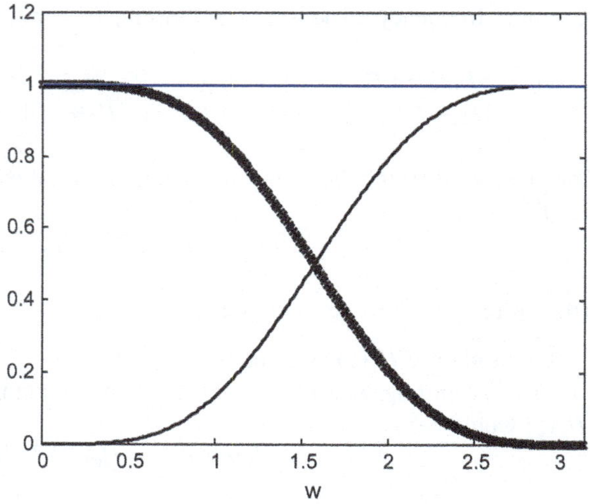

Fig. 1.36 H_0 and H_1 are power complementary

The scaling constant has been computed using the coefficients of:

$$(1+z^{-1})^2\,(r\ -z^{-1}) = r + (2r-1)\,z^{-1} + (r-2)\,z^{-2} - z^{-3}$$

The sum of the coefficients is:

$$r + (2r-1) + (r-2) - 1 = 4r - 4$$

And this value is given to c_1. Therefore:

$$H_0(z) = -0.0915 + 0.1585\,z^{-1} + 0.5915\,z^{-2} + 0.3415\,z^{-3}$$

The Program 1.13 computes the filter $H_0(z)$ and the corresponding orthogonal CQF filter $H_1(z)$, and then it displays the frequency response of the power of these two filters, to show that they are power complementary. The sum of both responses, which results in a horizontal line, is also displayed. Figure 1.36 depicts the results.

Program 1.13 Example of orthogonal filter bank

```
% Example of orthogonal filter bank
% H0 and H1 are power complementary
r=2-sqrt(3);
h0=conv([1 2 1],[r -1]);
sc=4*(1-sqrt(3)); %scaling with sum(h0(i))
h0=(1/sc)*h0; %prototype filter H0
for n=1:4, h1(n)=((-1)^n)*h0(5-n); end; % the CQF H1
w=0:(2*pi/511):pi;
H0=abs(fft(h0,512)); %discrete Fourier transform
H1=abs(fft(h1,512)); %"""
```

```
PH0=H0.*H0;
PH1=H1.*H1;
PHT=PH0+PH1;
plot(w,PH0(1:256),'kx'); hold on;
plot(w,PH1(1:256),'k');
plot(w,PHT(1:256),'b');
axis([0 pi 0 1.2]);
title('Prototype filter and the CQF:
frequency response of power');
xlabel('w');
```

Figure 1.37 shows the magnitude of the frequency response of the four filters forming the analysis and the synthesis filter banks. The figure has been generated with the Program 1.14, which also includes a check of the non-distortion condition: it is confirmed that the complete filter bank gives an exact delay of 4 time tics.

As a recommendation to run the programs of this section, do use '*clear*' in the MATLAB window in order to avoid inheritance of previous variable values, in particular the coefficient vectors corresponding to filters.

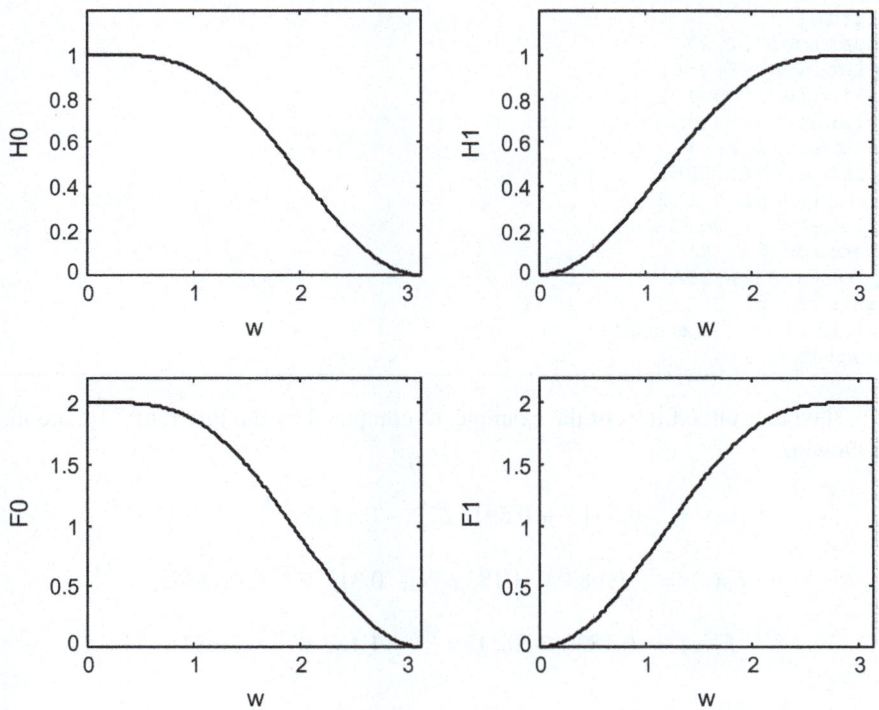

Fig. 1.37 Magnitude of the frequency responses of the filters forming the orthogonal filter bank

Program 1.14 Example of orthogonal filter bank

```
% Example of orthogonal filter bank
% the four filters, and a check of PR
r=2-sqrt(3);
h0=conv([1 2 1],[r -1]);
sc=4*(1-sqrt(3)); %scaling with sum(h0(i))
h0=(1/sc)*h0; %prototype filter H0
for n=1:4, h1(n)=((-1)^n)*h0(5-n); end; % the CQF H1
for n=1:4, f0(n)=2*((-1)^(n-1))*h1(n); end; % the filter F0
for n=1:4, f1(n)=-2*((-1)^(n-1))*h0(n); end; % the filter F1
% Check of PR
prod1=conv(f0,h0);
prod2=conv(f1,h1);
nodist=prod1+prod2;
w=0:(2*pi/511):pi;
H0=abs(fft(h0,512)); %discrete Fourier transform
H1=abs(fft(h1,512)); % """
F0=abs(fft(f0,512)); %"""
F1=abs(fft(f1,512)); % """
subplot(2,2,1)
plot(w,H0(1:256),'k');
axis([0 pi 0 1.2]);
ylabel('H0');xlabel('w');
title('frequency response (magnitude) of the four filters');
subplot(2,2,2)
plot(w,H1(1:256),'k');
axis([0 pi 0 1.2]);
ylabel('H1');xlabel('w');
subplot(2,2,3)
plot(w,F0(1:256),'k');
axis([0 pi 0 2.2]);
ylabel('F0');xlabel('w');
subplot(2,2,4)
plot(w,F1(1:256),'k');
axis([0 pi 0 2.2]);
ylabel('F1');xlabel('w');
nodist
```

The other three filters of the example, as computed by the Program 1.14, are the following:

$$H_1(z) = -0.3415 + 0.5915\,z^{-1} - 0.1585\,z^{-2} - 0.0915\,z^{-3}$$

$$F_0(z) = -0.683 - 1.183\,z^{-1} - 0.317\,z^{-2} + 0.1830\,z^{-3}$$

$$F_1(z) = 0.1830 + 0.317\,z^{-1} - 1.183\,z^{-2} + 0.683\,z^{-3}$$

1.4.2.3 Biorthogonal Filter Bank

While the previous alternative, with orthogonal filters, yields non linear phase characteristics, the bi-orthogonal filters approach obtains linear phase.

The idea is to start from a zero-phase odd-length real-coefficient half band filter $G(z)$, and then design $H_0(z)$ and $H_1(z)$ such that:

$$H_0(z)\, H_1(-z) = z^{-2k+1}\, G(z) \tag{1.153}$$

The causal filters $H_0(z)$ and $H_1(z)$ can be of different length.

Again, for PR the other two filters can be:

$$\begin{aligned} F_0(z) &= 2\, H_1(-z) \\ F_1(z) &= -2\, H_0(-z) \end{aligned} \tag{1.154}$$

Recall that zero-phase half band filters are such:

$$G(z) + G(-z) = 2c \quad (= 1 \;\; for \;\; c = 0.5) \tag{1.155}$$

The no-distortion condition is satisfied as follows (taking $c = 0.5$):

$$\begin{aligned} & F_0(z)\, H_0(z) + F_1(z)\, H_1(z) = 2H_1(-z)\, H_0(z) - 2H_0(-z)\, H_1(z) \\ & = 2\,(z)^{-2k+1}\, G(z) - 2\,(-z)^{-2k+1}\, G(-z) = 2\,(z)^{-2k+1} \end{aligned} \tag{1.156}$$

As before let us bring an example, now based on the factorization example (1.106). Choose as prototype filter the following:

$$H_0(z) = (1 + z^{-1})^2\, (1/c_1) \tag{1.157}$$

And choose the filter $F_0(z)$ (this is another way of choosing $H_1(-z)$) as:

$$F_0(z) = (1 + z^{-1})^2\, (1/c_2)\, (-1 + 4\, z^{-1} - z^{-2}) \tag{1.158}$$

After giving values for c_1 and c_2 to balance filter gains (the reader may try other values), the filters are:

$$\begin{aligned} H_0(z) &= 0.25 + 0.5\, z^{-1} + 0.25\, z^{-2}; \\ F_0(z) &= -0.25 + 0.5\, z^{-1} + 1.5\, z^{-2} + 0.5\, z^{-3} - 0.25\, z^{-4} \end{aligned}$$

Notice that the two filters are of different length.

The Program 1.15 computes the filter $H_0(z)$ and the corresponding biorthogonal filter $H_1(z)$. Then it displays the frequency response of the power of both filters, to show that they are not power complementary. The sum of both responses is also displayed, being not a horizontal line. Figure 1.38. depicts the results.

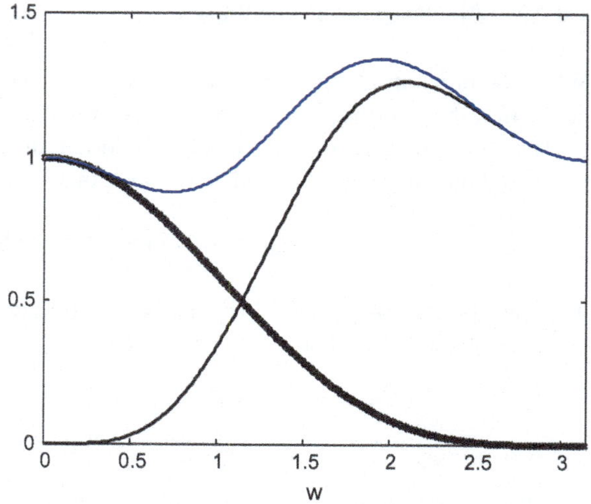

Fig. 1.38 H_0 and H_1 are not power complementary

Program 1.15 Example of biorthogonal filter bank

```
% Example of biorthogonal filter bank
% H0 and H1 are not power complementary
h0=[1 2 1]; %the filter H0
h0=(1/4)*h0; %scaling
fx=conv([1 2 1],[-1 4 -1]);
fx=(1/8)*fx; %scaling
f0=2*fx; %the filter F0
for n=1:5, h1(n)=((-1)^(n-1))*fx(n); end; % the filter H1
w=0:(2*pi/511):pi;
H0=abs(fft(h0,512)); %discrete Fourier transform
H1=abs(fft(h1,512)); %"""
PH0=H0.*H0;
PH1=H1.*H1;
PHT=PH0+PH1;
plot(w,PH0(1:256),'kx'); hold on;
plot(w,PH1(1:256),'k');
plot(w,PHT(1:256),'b');
axis([0 pi 0 1.5]);
title('Prototype filter and the biorthogonal:
frequency response of power');
xlabel('w');
```

Figure 1.39 shows the magnitude of the frequency response of the four filters forming the biorthogonal filter bank. The figure has been generated with the Program 1.16. The program includes a check of the non-distortion condition: as in the case of orthogonal filter bank it is confirmed that the complete filter bank gives an exact delay of 4 time tics.

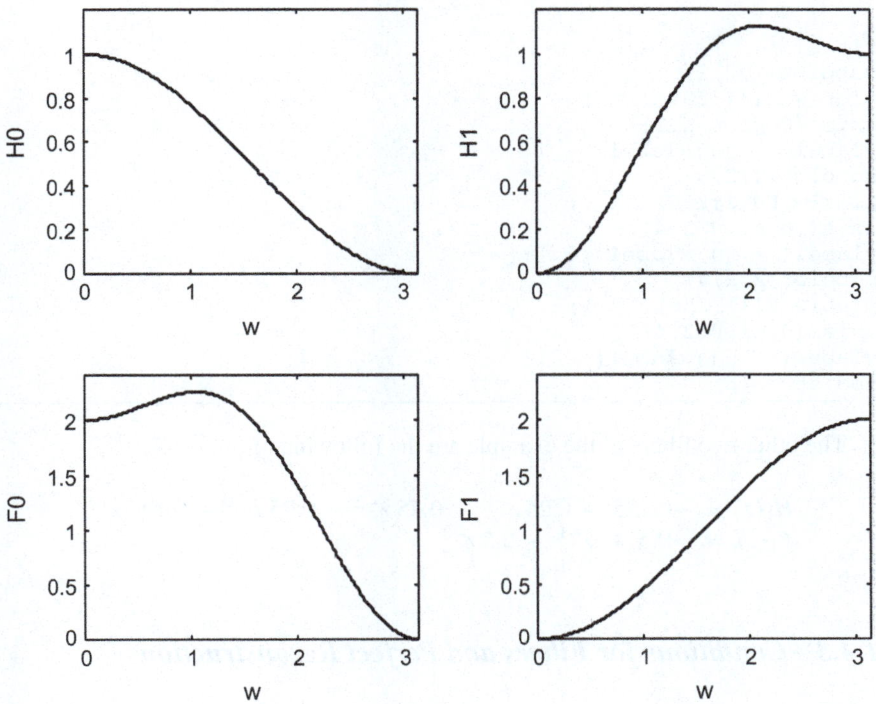

Fig. 1.39 Magnitude of the frequency responses of the filters forming the biorthogonal filter bank

Program 1.16 Example of biorthogonal filter bank

```
% Example of biorthogonal filter bank
% the four filters, and a check of PR
h0=[1 2 1]; %the filter H0
h0=(1/4)*h0; %scaling
fx=conv([1 2 1],[-1 4 -1]);
fx=(1/8)*fx; %scaling
f0=2*fx; %the filter F0
for n=1:5, h1(n)=((-1)^(n-1))*fx(n); end; % the filter H1
for n=1:3, f1(n)=-2*((-1)^(n-1))*h0(n); end; % the filter F1
% Check of PR
prod1=conv(f0,h0);
prod2=conv(f1,h1);
nodist=prod1+prod2
w=0:(2*pi/511):pi;
H0=abs(fft(h0,512)); %discrete Fourier transform
H1=abs(fft(h1,512)); % """
F0=abs(fft(f0,512)); % """
F1=abs(fft(f1,512)); % """
subplot(2,2,1)
plot(w,H0(1:256),'k');
axis([0 pi 0 1.2]);
ylabel('H0');xlabel('w');
```

```
title('frequency response (magnitude) of the four filters');
subplot(2,2,2)
plot(w,H1(1:256),'k');
axis([0 pi 0 1.2]);
ylabel('H1');xlabel('w');
subplot(2,2,3)
plot(w,F0(1:256),'k');
axis([0 pi 0 2.4]);
ylabel('F0');xlabel('w');
subplot(2,2,4)
plot(w,F1(1:256),'k');
axis([0 pi 0 2.4]);
ylabel('F1');xlabel('w');
nodist
```

The other two filters of the example are the following:

$$H_1(z) = -0.125 - 0.25\,z^{-1} + 0.75\,z^{-2} - 0.25\,z^{-3} - 0.125\,z^{-4};$$
$$F_1(z) = -0.5 + z^{-1} - 0.5\,z^{-2}$$

1.4.3 Conditions for Filters and Perfect Reconstruction

Further mathematical analysis shed more light about the conditions that filters must satisfy for perfect reconstruction. We will say that $H_0(z)$ and $H_1(z)$ are complementary if they form a PR pair.

1.4.3.1 Linear Phase with Longer Real FIR Filters

Using non-orthogonal filters, it is possible to have linear phase with longer real FIR filters.

There are two main classes of linear phase solutions:

- Both filters, H_0 and H_1, are type I FIR, with lengths:
  $N_0 = 2K + 1$, $N_1 = 2K + 3 + 4L$
  $(K = 0, 1, 2, \ldots; L = 0, 1, 2, \ldots)$
- One filter is type II and the other type IV, with lengths:
  $N_0 = 2K$, $N_1 = 2K + 4L$

There is a third alternative, with both filters having all zeros on the unit circle, with no practical interest.

1.4.3.2 Bezout's Theorem

Given two polynomials $p_1(x)$ and $p_2(x)$, the greatest common divisor (gcd) of the polynomials can be computed with the Euclidean algorithm (this topic will be seen in more detail in the next chapter). Let us write:

$$g(x) = \alpha \cdot \gcd(p_1(x),\ p_2(x)) \tag{1.159}$$

The Bezout's theorem states that there exist two polynomials $q_1(x)$ and $q_2(x)$, such that (Bezout's identity):

$$g(x) = q_1(x)p_1(x) + q_2(x)p_2(x) \tag{1.160}$$

If $p_1(x)$ and $p_2(x)$ are coprime then $g(x)$ has zeros at 0 or ∞ only.

$$g(x) = q_1(x)p_1(x) + q_2(x)p_2(x) \tag{1.161}$$

If one considers now the following equation:

$$H_0(z)H_1(-z) - H_0(-z)H_1(z) = 2\,z^{-d} \tag{1.162}$$

which is a combination of the aliasing cancellation and the non-distortion conditions. It is clear that this is a version of the Bezout's identity.

In consequence, $H_0(z)$ and $H_1(z)$ must be coprime for PR.

Other consequences are that:

- A filter $H_0(z)$ has a complementary filter $H_1(z)$ if and only if it has no zero pairs $(\alpha, -\alpha)$
- There is always a complementary filter $H_1(z)$ to the binomial filter:

$$H_0(z) = (1 + z^{-1})^K \tag{1.163}$$

1.4.3.3 Diophantine Equation

One of the interesting questions about filter banks is if one could use longer filters, looking for better filtering performance.

Suppose a Diophantine equation:

$$a\,x + b\,y = c \tag{1.164}$$

If one knows a solution (x_0, y_0), then many other solutions of the form $(x_0 + x', y_0 + y')$ can be obtained with:

$$a\,x' + b\,y' = 0 \tag{1.165}$$

This procedure can be applied to polynomials. By taking $a = -H_0(z)$, $b = H_0(-z)$, [48] shows that once a complementary filter $H_1(z)$ is found, others can be obtained by adding $C(z)H_0(z)$, where $C(z) = C(-z)$. In particular, given a complementary pair $H_0(z)$ and $H_1(z)$, all longer complementary $H_1^{'}(z)$ filters can be obtained with:

$$H_1^{'}(z) = z^{-2d} H_1(z) + C(z)\, H_0(z) \tag{1.166}$$

where $H_0(z)$, $H_1(z)$ and $H_1^{'}(z)$ are linear phase, odd length filters. $H_0(z)$ of length N, and $H_1(z)$ of length $N-2$. And:

$$C(z) = \sum_{i=1}^{d} \alpha_i (z^{-2(i-1)} + z^{-(4d-2i)}) \tag{1.167}$$

Moreover, all filters that are complementary to a length N filter $H_0(z)$ can be obtained with:

$$H_1^{'}(z) = z^{-2k} H_1(z) + C(z)\, H_0(z) \tag{1.168}$$

where $H_0(z)$ of length N, $H_1(z)$ of length $N-2$, $H_1^{'}(z)$ of length $N+2d-2$; $k \in \{0, 1, \ldots, d\}$. And:

$$C(z) = \sum_{i=0}^{d-1} \beta_i z^{-2i} \tag{1.169}$$

1.5 Aspects of Unity Gain Systems

If we want filter banks with PR, we want unit gain. Therefore, it is opportune to take a look of aspects of unit gain systems that have been considered by the literature, along time, from different points of view.

For instance, one could represent a unit gain filter using a phasor with length one, as represented in Fig. 1.40.

Taking as reference the phasor of Fig. 1.40, other unit gain filters could be obtained by rotations or reflections.

Fig. 1.40 A phasor with magnitude =1, and phase =β

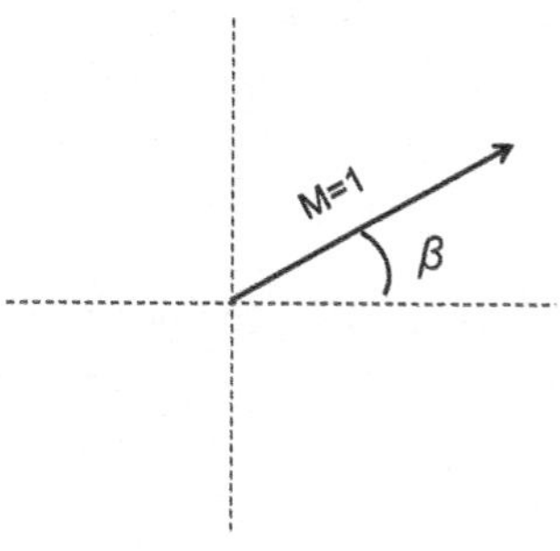

In general, the phase β would change in function of the frequency. A linear dependency is usually desired, since it means a constant delay. This is why FIR linear phase filters is the case commonly considered when dealing with filter banks.

If one deals with analog filters, the first attention is driven to passive circuits, made with resistors, capacitors and inductances. These circuits can be dissipative if there are resistors (which dissipate energy), or conservative (LC circuits). Of course, conservative passive circuits are lossless and have unit gain. If you put transistors in a circuit, it becomes an active circuit, in which energy could increase so unstability may appear.

Digital filters are based on computation, numbers, so one cannot speak of real energy but still it is possible to argue making analogies.

If the filter gain is related to the product of two matrices, unit gain would involve an issue of matrix inversion. Numerical problems may arise when computing the matrix inverse. Several mathematical approaches have been proposed to alleviate difficulties; like, for instance, different types of factorizations.

In the case of filters banks, typically one uses polynomial matrices for analysis and design. This section includes a definition of polynomial matrix, and then introduces concepts pertaining unit gain in this context.

There is a class of filters called allpass filters. This name is very clear, and is equivalent to unit gain for all frequencies. They are used for application where phase is the important thing.

When one faces the design of a filter bank with PR, it is good to know some structures that provide a kind of standard way for the design. These structures are linked to factorizations.

This section deals with the aspects just enounced in this introduction. It is mainly based on [9, 49].

1.5.1 Matrices of Interest

Some types of matrices are of special interest in the ambient of this chapter. It would be helpful to remind here some definitions and concepts.

The conjugate of a matrix A is the matrix A^* obtained by conjugating each element. The *conjugate transpose* of A will be denoted as follows:

$$A^+ = (A^*)^T \tag{1.170}$$

1.5.1.1 Unitary Matrix

A complex matrix U is *unitary* if:

$$U \cdot U^+ = I \tag{1.171}$$

If U is a real matrix, then U is called an orthogonal matrix.

If U is a unitary matrix, then: $\det(U) = 1$.

- The columns of a unitary matrix form an orthonormal set
- Unitary matrices have eigenvalues of unit modulus
- Matrices of reflection and of rotations are unitary matrices
- The product of unitary matrices is a unitary matrix

1.5.1.2 Hermitian Matrix

A complex matrix H is *hermitian* if:

$$H^+ = H \tag{1.172}$$

If H is a real matrix, then H is a symmetric matrix.

- The diagonal elements of H are real numbers, and elements on opposite sides of the main diagonal are conjugates
- The eigenvalues of H are real numbers. The eigenvectors corresponding to different eigenvalues are orthogonal.

Next Table 1.3 summarizes real versus complex correspondences:

1.5.1.3 Other Matrices

A (right) *circulant* matrix is a matrix where each row is a circular shift of the previous row:

$$\begin{pmatrix} c_1 & c_2 & c_3 & \ldots & c_n \\ c_n & c_1 & c_2 & \ldots & c_{n-1} \\ \vdots & & & & \\ c_2 & c_3 & c_4 & \ldots & c_1 \end{pmatrix} \tag{1.173}$$

A *Toeplitz* matrix is a matrix where the value of each element (i, j) depends only on $(i - j)$, so diagonals have equal values:

Table 1.3 Real versus complex correspondences

Real	Complex
Transpose $()^T$	Conjugate transpose $()^+$
Orthogonal matrix $A\,A^T = I$	Unitary matrix $U\,U^+ = I$
Symmetric matrix $A = A^T$	Hermitian matrix $H = H^+$

$$\begin{pmatrix} t_0 & t_1 & t_2 & \dots & t_{n-1} \\ t_{-1} & t_0 & t_1 & \dots & t_{n-2} \\ t_{-2} & t_{-1} & t_0 & \dots & t_{n-3} \\ \vdots & & & & \\ t_{-n+1} & t_{-n+2} & t_{-n+3} & \dots & t_0 \end{pmatrix} \tag{1.174}$$

MATLAB includes the *toeplitz( )* function for building Toeplitz matrices. See [37] for interesting applications.

1.5.1.4 Factorizations

For a non-singular square matrix A there exists a unique pair of unitary matrix Q and upper triangular matrix R with positive diagonal elements, such that:

$$A = Q\,R \tag{1.175}$$

An effective method for obtaining this QR decomposition is to multiply A on the left by some orthogonal matrices Q_i that eliminate elements of A below the main diagonal. After a series of products, a triangular matrix R is obtained:

$$Q_k \dots Q_2\, Q_1\, A = R \tag{1.176}$$

the matrix Q is $Q = Q_k \dots Q_2\, Q_1$.

Let us consider matrices Q_i corresponding to rotations or reflections.

The so-called *Givens rotations* have the form:

$$G_\alpha = \begin{pmatrix} \cos\alpha & -\sin\alpha \\ \sin\alpha & \cos\alpha \end{pmatrix} \tag{1.177}$$

Based on a Givens rotation, one can modify the I matrix inserting the four trigonometric entries in appropriate places, so the following matrix is built:

$$\Omega_{i,j} = \begin{pmatrix} \ddots & & i & & j & \\ & 1 & \vdots & 0 & \vdots & 0 \\ i & \dots & \cos\alpha & \dots & -\sin\alpha & \dots \\ & 0 & \vdots & \ddots & \vdots & 0 \\ j & \dots & \sin\alpha & \dots & \cos\alpha & \dots \\ & 0 & \vdots & 0 & \vdots & 1 \\ & & & & & \ddots \end{pmatrix} \tag{1.178}$$

The result of $\Omega_{i,j}\, A$ is a matrix with the same rows of A except for the rows *ith* and *jth*, which in $\Omega_{i,j}\, A$ are linear combinations of the *ith* and *jth* rows of A. By choosing a suitable rotation angle, a zero can be inserted on one of these rows. For example, suppose that A is:

$$A = \begin{bmatrix} 6 & 6 & 1 \\ 3 & 6 & 1 \\ 2 & 1 & 1 \end{bmatrix} \tag{1.179}$$

It is possible to obtain with an appropriate $\Omega_{1,2}$, a product with a zero under the main diagonal:

$$\Omega_{1,2} A = \begin{bmatrix} 6 & 6 & 1 \\ 0 & X & X \\ X & X & X \end{bmatrix} \tag{1.180}$$

With another suitable $\Omega_{1,3}$ one obtains:

$$\Omega_{1,3}\Omega_{1,2} A = \begin{bmatrix} X & X & X \\ 0 & X & X \\ 0 & X & X \end{bmatrix} \tag{1.181}$$

And finally, with another $\Omega_{2,3}$

$$\Omega_{2,3}\Omega_{1,3}\Omega_{1,2} A = \begin{bmatrix} X & X & X \\ 0 & X & X \\ 0 & 0 & X \end{bmatrix} \tag{1.182}$$

In this way a QR factorization has been achieved.

Alternatively one can use *Householder reflections*. A reflection in the direction of the vector V can be expressed with the following hermitian unitary matrix:

$$H_V = I - \frac{2}{\| V \|^2}\, V\, V^* \tag{1.183}$$

Using the previous example, let us select the first column of A: $c_1 = [6\ 3\ 2]^T$. It can be transformed to $[7\ 0\ 0]^T$ by using a reflection in the direction $V_1 = c_1 - [7\ 0\ 0]^T = [-1\ 3\ 2]^T$. Then, one obtains:

$$H_{V1} A = \begin{bmatrix} 7 & X & X \\ 0 & X & X \\ 0 & X & X \end{bmatrix} \tag{1.184}$$

And the process can be repeated with the second column in order to insert a zero.

1.5.1.5 Paraunitary Matrix

Let us introduce '*paraconjugation*'. The paraconjugate $\tilde{H}$ of a matrix H is obtained by conjugation of the coefficients, taking the transpose, and replacing z with z^{-1}.

A *polynomial matrix* is a matrix whose entries are polynomials. For instance, a matrix in which each entry is a FIR filter $H_{i,j}(z)$.

An interesting class of polynomial matrices are *unimodular* matrices, whose determinant is a constant (not a function of a variable). It can be shown that a matrix is unimodular iif its inverse is a polynomial matrix.

According with [49], the extension of the concept of unitary matrices to polynomial matrices leads to '*paraunitary*' matrices, which provide an important path for the design of multi-channel filter banks. A polynomial matrix $H(z)$ is paraunitary if:

$$\tilde{H}(z)\, H(z) \;=\; I \tag{1.185}$$

These matrices are unitary on the unit circle (for $z = e^{j\omega}$) and, in general, for any z.

Notice that the inverse of the paraunitary matrix $H(z)$ is just $\tilde{H}(z)$ (the paraconjugate).

The product of paraunitary systems is paraunitary.

There are also *rational matrices*, whose entries are ratios of two polynomials. For instance, a matrix in which each entry is an IIR filter.

1.5.2 Allpass Filters

An '*allpass*' filter is a filter with unit gain, causing only phase changes.

A simple example in the continuous time domain is a filter with the following transfer function:

$$G(s) \;=\; \frac{1 - as}{1 + as} \tag{1.186}$$

Another simple example, now in the discrete time domain is:

$$G(z) \;=\; \frac{z^{-1} - b^*}{1 - b\, z^{-1}} \tag{1.187}$$

where b is a complex pole. This filter has a zero at $1/b^*$. Both pole and zero are at mirror locations with respect to the unit circle.

The following cascade of two allpass filters give an allpasss filter with real coefficients:

$$G(z) \;=\; \frac{z^{-1} - b^*}{1 - b\, z^{-1}} \cdot \frac{z^{-1} - b}{1 - b^*\, z^{-1}} \tag{1.188}$$

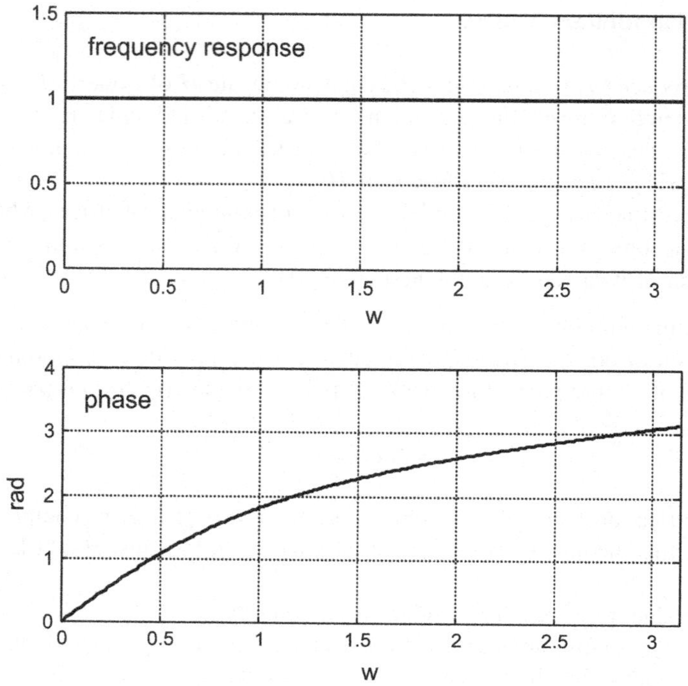

Fig. 1.41 Frequency response of a first-order allpass filter

A general allpass filter can be built as follows:

$$G(z) = \prod_{i=1}^{n} \frac{z^{-1} - b_i^*}{1 - b_i\, z^{-1}} = \frac{\tilde{B}(z)}{B(z)} \tag{1.189}$$

where $\tilde{B}(z) = z^{-N} B_*(z^{-1})$ is obtained by conjugation of the coefficients of $B(z)$ and replacing z by z^{-1}.

- The determinant of a square paraunitary matrix is an allpass filter
- A linear, time invariant filter $G(z)$ is **lossless** if it preserves signal energy for every input signal (only stable filters can be lossless)
- A paraunitary matrix $H(z)$ is lossless if all entries are stable.

Figure 1.41 shows the frequency response of a simple first-order allpass filter. The figure has been generated with the Program 1.17.

Program 1.17 Simple allpass filter example

```
% Simple Allpass filter example
a=0.4;
w=0:(2*pi/511):pi;
Anum=[1 -a];
Aden=[-a 1];
G=freqz(Anum,Aden,w);
%display
figure(1)
subplot(2,1,1)
plot(w,abs(G),'k'); grid;
axis([0 pi 0 1.5]);
title('frequency response of simple Allpass example');
xlabel('w');
subplot(2,1,2)
plot(w,angle(G),'k'); grid;
axis([0 pi 0 4]);
title('phase');
xlabel('w'); ylabel('rad')
```

1.5.3 *Lattice Structure*

Causal , FIR, 2 × 2 lossless systems of real coefficients can be written as follows:

$$H(z) = \begin{pmatrix} H_0(z) & H_2(z) \\ H_1(z) & H_3(z) \end{pmatrix} = \begin{pmatrix} H_0(z) & c\, z^{-N} \tilde{H}_1(z) \\ H_1(z) & c\, z^{-N} \tilde{H}_0(z) \end{pmatrix} \tag{1.190}$$

where $H_0(z)$ and $H_1(z)$ are power-complementary, and N is large enough for the right column to be causal.

It can be shown that, given two power-complementary polynomials $P_{K-1}(z)$ and $Q_{K-1}(z)$ with real coefficients and degree $(K-1)$, and with:

$$p_{K-1}(0)\; p_{K-1}(K-1) \neq 0 \tag{1.191}$$

then, there exists another two polynomials $P_{K-2}(z)$ and $Q_{K-2}(z)$ such that:

$$\begin{pmatrix} P_{K-1}(z) \\ Q_{K-1}(z) \end{pmatrix} = \begin{pmatrix} \cos\alpha & -\sin\alpha \\ \sin\alpha & \cos\alpha \end{pmatrix} \begin{pmatrix} P_{K-2}(z) \\ z^{-1} Q_{K-2}(z) \end{pmatrix} \tag{1.192}$$

By repeated use of this result, a *lattice* factorization of $H(z)$ can be obtained:

$$\begin{pmatrix} H_0(z) & H_2(z) \\ H_1(z) & H_3(z) \end{pmatrix} = \begin{pmatrix} \cos\alpha & -\sin\alpha \\ \sin\alpha & \cos\alpha \end{pmatrix} \cdot \prod_{i=1}^{N-1} \begin{pmatrix} 1 & 0 \\ 0 & z^{-1} \end{pmatrix} \begin{pmatrix} \cos\alpha_i & -\sin\alpha_i \\ \sin\alpha_i & \cos\alpha_i \end{pmatrix} \tag{1.193}$$

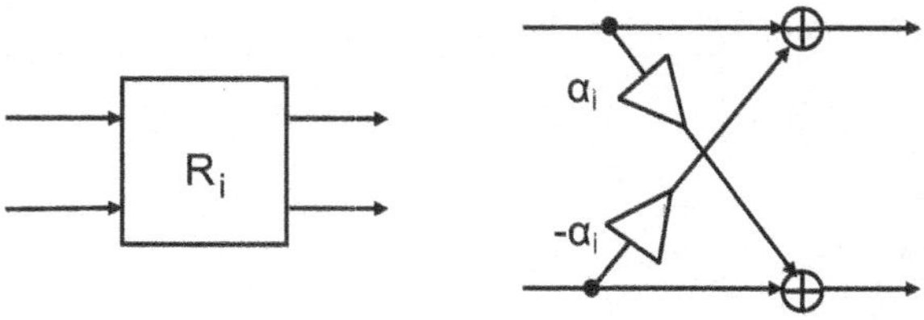

Fig. 1.42 Equivalent diagrams of the factor R_i

All orthogonal systems with filters of length $2K$ can be generated with this factorization.

Obviously, this lattice factorization is related to Givens rotations. Another lattice factorization can be obtained based on the Householder factorization:

$$H(z) = c\, R_N\, \Lambda(z)\, R_{N-1}\, \Lambda(z) \ldots R_1\, \Lambda(z) R_0 \tag{1.194}$$

with:

$$R_i = \begin{pmatrix} 1 & -\alpha_i \\ \alpha_i & 1 \end{pmatrix}; \quad \Lambda(z) = \begin{pmatrix} 1 & 0 \\ 0 & z^{-1} \end{pmatrix}; \quad c = \prod_{i=0}^{N} \frac{1}{\sqrt{1+\alpha_i^2}} \tag{1.195}$$

Figure 1.42 shows two equivalent representations of the factor R_i. These representations are frequently used in block diagrams corresponding to lattice decompositions.

1.5.4 The Case of 2-Channel Filter Banks

Let us focus on two types of 2-channel filter banks: orthogonal and biorthogonal.

1.5.4.1 Orthogonal Filter Banks

Suppose you choose the analysis filter so its polyphase matrix $\mathbf{H_p(z)}$ is FIR lossless. Then, PR is guaranteed by simply taking for the synthesis filter:

$$\mathbf{F_p(z)} = k\, z^{-d}\, \bar{\mathbf{H}}_\mathbf{p}(\mathbf{z}) \tag{1.196}$$

Thus, no matrix inversion is needed, which is a great advantage for the design. The result would be a *paraunitary filter bank.*

It can be shown that paraunitary 2-channel filters are orthogonal, the filter lengths are equal with N even, and the synthesis filters are time-reversed versions of the analysis filters [44].

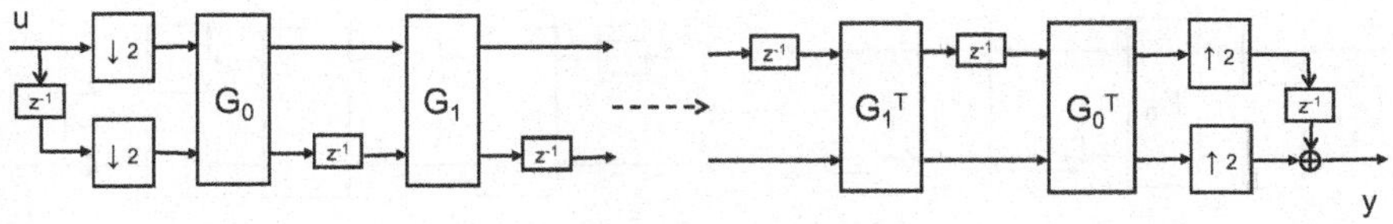

Fig. 1.43 A paraunitary filter in lattice structure

Now, let us study in more detail this type of filter bank. Let the polyphase matrix $\mathbf{H_p(z)}$ be FIR lossless, then:

$$\mathbf{H_p^{-1}(z)} = \tilde{\mathbf{H}}_\mathbf{p}(\mathbf{z}) \tag{1.197}$$

Recalling that:

$$\mathbf{H}_p = \frac{1}{2}\mathbf{H}_m \begin{pmatrix} 1 & 1 \\ 1 & -1 \end{pmatrix} \begin{pmatrix} 1 & 0 \\ 0 & z \end{pmatrix} \tag{1.198}$$

Then:

$$\bar{\mathbf{H}}_m(\mathbf{z}) \cdot \mathbf{H}_m(\mathbf{z}) = 2I \tag{1.199}$$

From this equation, one can deduce that for PR, the analysis and synthesis filters are mutually related as follows:

$$h_k(n) = f_k^*(-n); \;\; k = 0, 1 \tag{1.200}$$

It can also be deduced, from Eq. (1.199), that the filters $H_0(z)$ and $H_1(z)$ are power-complementary (and the filters $F_0(z)$ and $F_1(z)$).

The polyphase matrix $\mathbf{H_p(z)}$ has the form:

$$\mathbf{H_p(z)} = \begin{pmatrix} H_{00}(z) & c\,z^{-N}\tilde{H}_{10}(z) \\ H_{10}(z) & c\,z^{-N}\tilde{H}_{00}(z) \end{pmatrix} \tag{1.201}$$

Once a power symmetric filter is proposed for $H_0(z)$, the rest of the filters can be easily obtained.

Figure 1.43 shows a paraunitary filter in lattice structure, using Givens rotations.

For example, the Daubechies D6 wavelet (see next chapter) has lattice angles $\alpha_0 = \frac{\pi}{3}$ and $\alpha_1 = -\frac{\pi}{12}$.

Figure 1.44 shows an alternative lattice structure for the paraunitary filter, based on Householder reflections.

Lattice structures are much appreciated since they provide an "automatic" way for obtaining PR.

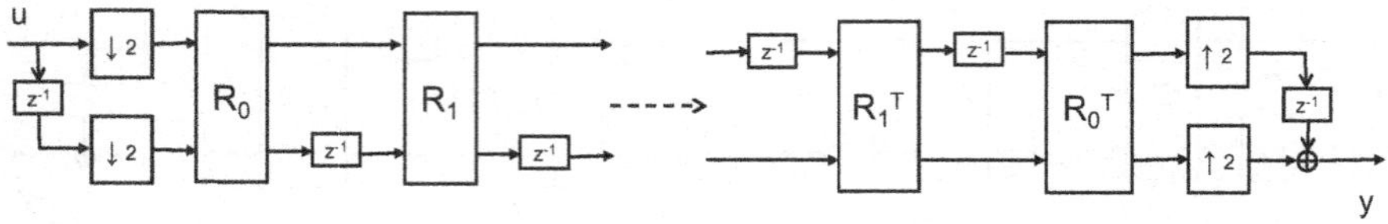

Fig. 1.44 An alternative lattice structure for the paraunitary filter

Table 1.4 Types of biorthogonal filter banks

	N_0	N_1	$H_0(z)$ & $F_0(z)$	$H_1(z)$ & $F_1(z)$
Type A	$2L$	$2L+4K$	Symmetric	Antisymmetric
Type B	$2L+1$	$2L+4K+3$	Symmetric	Symmetric

1.5.4.2 Biorthogonal Filter Banks

There exist lattice structures that produce linear phase filter banks with PR [27, 49]. There are two main cases, as described in next Table 1.4.

In case of type A filter bank, the polyphase matrix can be factorized as:

$$\mathbf{H_p(z)} = c\begin{pmatrix} 1 & 1 \\ -1 & 1 \end{pmatrix} \cdot \prod_{i=0}^{L-1}\left[\begin{pmatrix} 1 & 0 \\ 0 & z^{-1} \end{pmatrix}\begin{pmatrix} 1 & \alpha_i \\ \alpha_i & 1 \end{pmatrix}\right] \tag{1.202}$$

with $c = -\frac{1}{2}\prod_{i=1}^{K-1}\frac{1}{1-\alpha_i^2}$.

In case of type B filter bank, the polyphase matrix can be factorized as:

$$\mathbf{H_p(z)} = \begin{pmatrix} \alpha_0 & 0 \\ 0 & \alpha_1 \end{pmatrix} \cdot \prod_{i=2}^{L}\left[\begin{pmatrix} 1+z^{-1} & 1 \\ 1+\beta_i z^{-1}+z^{-2} & 1+z^{-1} \end{pmatrix}\begin{pmatrix} \alpha_i & 0 \\ 0 & 1 \end{pmatrix}\right] \tag{1.203}$$

Usually all β_i are set to 0, except for β_2. See [4, 28] for more details.

1.6 Tree-Structured Filter Banks

Certain applications use tree-structured filter banks based on cascades of 2-channel filter banks. Figure 1.45 shows an example of analysis filter bank.

PR is easily obtained with a synthesis filter bank being symmetric to the analysis filter bank.

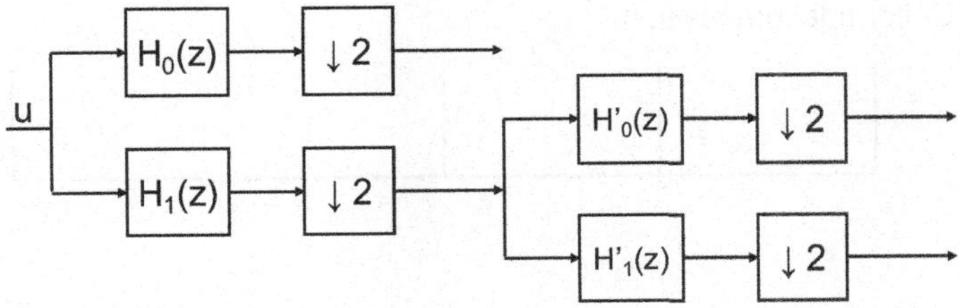

Fig. 1.45 An example of tree-structured filter bank

Fig. 1.46 An schematic view of the previous filter bank tree

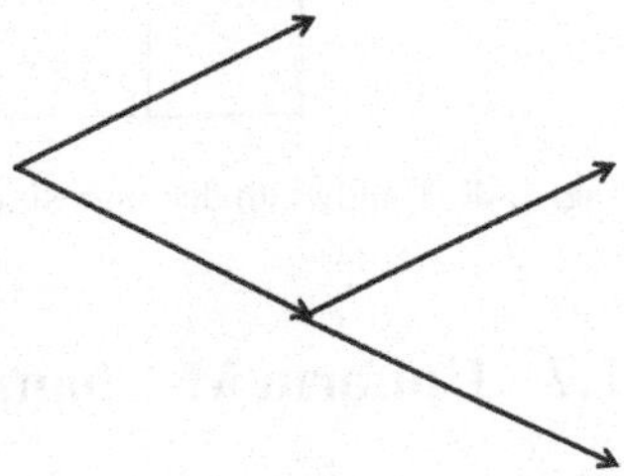

Fig. 1.47 Some other filter bank trees

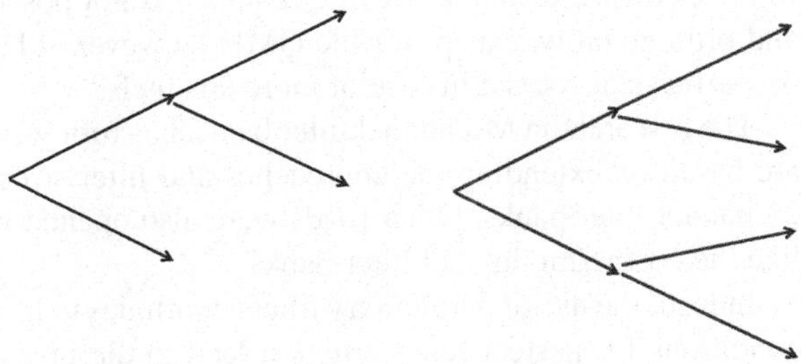

The structure of this filter tree example can be schematically expressed as in Fig. 1.46.

There are many other tree structures based on cascades of 2-channel filter banks. Figure 1.47 shows some examples.

The filter banks decompose the specified bandwidth into sub-bands. This can be done in several ways; Fig. 1.48 shows two typical examples:

As it will be explained in the next chapter, wavelets are intrinsically related to tree-structured filter banks.

The reader interested in audio applications and psychoacoustic models of human perception would find interesting details in [16]

From the perspective of signal processing, the tree structure, which is easy to design, has a drawback: it can cause excessive delays.

For a more complete view of tree-structure filter banks, see [3, 41].

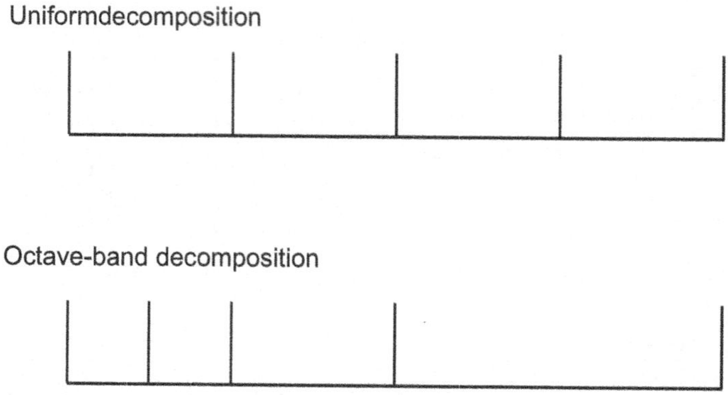

Fig. 1.48 Bandwidth decompositions

1.7 Uniform M-Channel Filter Banks

In the case of 2-channel FIR filter banks it is not possible to have both linear phase and orthogonality, except for the QMF. However, it has been shown that these two properties can coexist in case of more channels.

The research on M-channel filter banks has followed several paths. Some of them are based on extending the approaches and filter structures already established for 2-channel filter banks. Other roads were also opened with new or not so new ideas, like the cosine-modulated filter banks.

Indeed, the use of paraunitary filters continues to be attractive. This is because the conditions for perfect reconstruction lead to the inversion of polynomial matrices, but the inversion of a polynomial is not a polynomial, and then IIR filters would be needed. In the case of paraunitary filters, matrix inversion is not required and all filters can be FIR.

A number of proposed design approaches start with a certain filter structure (that may provide linear phase), and then impose conditions to obtain a paraunitary version.

In contrast with the tree-structured filter banks, uniform M-channel filter banks have less delay.

This section is guided by [27] and uses materials from [24]

1.7.1 Basic Equations

Figure 1.49 shows a diagram of a M-channel filter bank.

The intermediate variables are related with the input as follows:

$$V_j(z) = \frac{1}{N}\sum_{k=0}^{N-1} H_j(W_N^k\, z^{1/N})\, U(W_N^k\, z^{1/N})\,; \quad j = 0, \ldots, M-1 \tag{1.204}$$

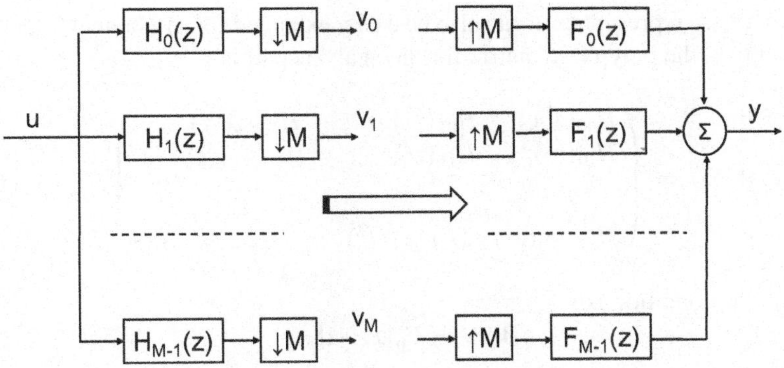

Fig. 1.49 An M-channel filter bank

The output is:

$$Y(z) = \frac{1}{N} \sum_{j=0}^{M-1} \sum_{k=0}^{N-1} F_j(z) H_j(W_N^k\, z)\, U(W_N^k\, z) \tag{1.205}$$

Or, equivalently:

$$Y(z) = \frac{1}{N} \sum_{k=0}^{N-1} U(W_N^k\, z) \sum_{j=0}^{M-1} F_j(z) H_j(W_N^k\, z) \tag{1.206}$$

For perfect reconstruction, $Y(z) = z^{-d}\, U(z)$, then:

$$\sum_{j=0}^{M-1} F_j(z)\, H_j(W_N^k\, z) = N\, z^{-d}\, \delta_{k0}; \quad 0 \leq k \leq N-1 \tag{1.207}$$

In this case, the aliasing component matrix would be:

$$\mathbf{H_a(z)} = \begin{pmatrix} H_0(z) & H_1(z) & \ldots & H_{M-1}(z) \\ H_0(zW_N) & H_1(zW_N) & \ldots & H_{M-1}(zW_N) \\ \vdots & \vdots & & \vdots \\ H_0(zW_N^{N-1}) & H_1(zW_N^{N-1}) & \ldots & H_{M-1}(zW_N^{N-1}) \end{pmatrix} \tag{1.208}$$

The PR condition can be written as follows:

$$\begin{bmatrix} N\, z^{-d} \\ 0 \\ \vdots \\ 0 \end{bmatrix} = \mathbf{H_a(z)} \begin{bmatrix} F_0(z) \\ F_1(z) \\ \vdots \\ F_{M-1}(z) \end{bmatrix} \tag{1.209}$$

The polyphase representation can also be easily extended for M-channel filter banks. For example, the polyphase matrix for the analysis part is:

$$\mathbf{H}_p(\mathbf{z}) = \begin{pmatrix} H_{00}(z^2) & H_{01}(z^2) & \ldots & H_{0,N-1}(z) \\ H_{10}(z^2) & H_{11}(z^2) & \ldots & H_{1,N-1}(z) \\ \vdots & \vdots & \vdots & \\ H_{M-1,0}(z^2) & H_{M-1,1}(z^2) & \ldots & H_{M-1,N-1}(z) \end{pmatrix} \tag{1.210}$$

And the PR condition is:

$$\mathbf{B(z)} = \mathbf{F_p(z)} \cdot \mathbf{H_p(z)} \tag{1.211}$$

where ***B(z)*** has one of the following expressions:

$$\mathbf{B(z)} = z^{-d}\, I; \quad \mathbf{B(z)} = z^{-d} \begin{bmatrix} 0 & I_{M-K} \\ z^{-1}\, I_K & 0 \end{bmatrix}; \; 0 \leq K \leq M-1 \tag{1.212}$$

One of the possible factorizations of $\mathbf{H_p(z)}$ could be the following:

$$\mathbf{H_p(z)} = A_K\, D(z)\, A_{K-1} D(z) \cdots D(z)\, A_0 \tag{1.213}$$

with:

$$D(z) = \begin{pmatrix} I_{M-1} & 0 \\ 0 & z^{-1} \end{pmatrix} \tag{1.214}$$

and the matrices $A_K, A_{K-1}, \ldots, A_0$ are arbitrary non-singular matrices.

The corresponding $\mathbf{F_p(z)}$ would be:

$$\mathbf{F_p(z)} = A_0^{-1} L(z)\, A_1^{-1} L(z) \cdots L(z)\, A_K^{-1} \tag{1.215}$$

with:

$$L(z) = \begin{pmatrix} z^{-1} I_{M-1} & 0 \\ 0 & 1 \end{pmatrix} \tag{1.216}$$

Some specific versions of this factorization will be introduced below.

1.7.2 Paraunitary M-Channel Filter Banks

As in 2-channel filter banks, if one chooses the analysis filters so $\mathbf{H_p(z)}$ is FIR lossless, it is easy to obtain the corresponding synthesis filters for PR: the analysis and synthesis filters should be related as follows:

$$h_k(n) = f_k^*(-n);\ k = 0, \ldots, M-1 \qquad (1.217)$$

Paraunitary filter banks are easily derived from the factorization just described, by restricting the matrices $A_K, A_{K-1}, \ldots, A_0$ to be unitary. This can be done with Givens rotations.

Also, the desired $\mathbf{H_p(z)}$ can be obtained using a factorization based on Householder reflections.

See [11] and references therein for a general view of paraunitary M-channel filter bank factorizations.

1.7.3 Cosine-Modulated Filter Banks

Cosine-modulated filter banks are frequently used in audio compression and telecommunications.

Given prototype filters $H(z)$ and $F(z)$ for the analysis and synthesis banks respectively, the rest of the filters are cosine-modulated versions:

$$h_k(n) = 2\,h(n)\,\cos\left(\frac{\pi}{M}(k+\frac{1}{2})(n+\frac{1}{2}-\frac{N}{2}) + (-1)^k\frac{\pi}{4}\right) \qquad (1.218)$$

$$f_k(n) = 2\,f(n)\,\cos\left(\frac{\pi}{M}(k+\frac{1}{2})(n+\frac{1}{2}-\frac{N}{2}) - (-1)^k\frac{\pi}{4}\right) \qquad (1.219)$$

The number of parameters to be designed are $h(n)$ and $f(n)$, with $0 \leq n \leq N-1$.

The target of the modulation is to obtain a set of filters with bandwidths as depicted in Fig. 1.50.

An example of cosine-modulated filter bank is the *Pseudo-QMF* bank, in which $H(z)$ is constrained to have a cutoff frequency such that there is no overlap between $H_K(z)$ and $H_{K\pm2}(z)$.

Another example is the *Near-Perfect-Reconstruction (NPR) Pseudo-QMF* bank, in which $H(z)$ is a linear-phase spectral factor of a 2M band filter.

For PR studies, the polyphase representation is used as follows:

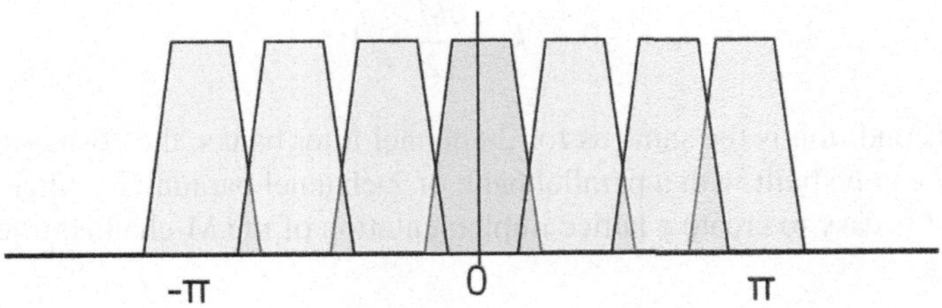

Fig. 1.50 Bandwidths of a cosine-modulated filter bank

- The linear phase prototype filter $H(z)$ is written in function of polyphase components:

$$H(z) = \sum_{k=0}^{2M-1} z^{-k}\, G_k(z^{2M}) \tag{1.220}$$

- The modulated cosines are represented with the matrix C. The entries of this matrix are:

$$c_{ij} = 2\cos\left(\frac{\pi}{M}(i+\frac{1}{2})(j+\frac{1}{2}-\frac{N}{2}) + (-1)^i\frac{\pi}{4}\right) \tag{1.221}$$

- Two auxiliary matrices are built:

$$\mathbf{g}_0(-\mathbf{z}^2) = diag\,[G_0(-z^2),\ \ldots,\ G_{M-1}(-z^2)] \tag{1.222}$$

$$\mathbf{g}_1(-\mathbf{z}^2) = diag\,[G_M(-z^2),\ \ldots,\ G_{2M-1}(-z^2)] \tag{1.223}$$

- The polyphase matrix of the analysis side can now be written as follows:

$$\mathbf{H_p(z)} = C\begin{pmatrix}\mathbf{g}_0(-\mathbf{z}^2)\\ z^{-1}\,\mathbf{g}_1(-\mathbf{z}^2)\end{pmatrix} \tag{1.224}$$

The cosine-modulated filter bank can be designed to be *paraunitary*. In this case, it is pertinent for PR to consider the product:

$$\mathbf{P(z)} = \bar{\mathbf{H}}_\mathbf{p}\mathbf{(z)}\,\mathbf{H_p(z)} \tag{1.225}$$

From this product, it can be established that the condition for the filter bank to be paraunitary is:

$$\bar{G}_k(z)G_k(z) + \bar{G}_{M+k}(z)G_{M+k}(z) = \frac{1}{2M}(-z)^{-d} \tag{1.226}$$

$$0 \le k \le \frac{M}{2} - 1 \tag{1.227}$$

Since this condition is the same as for 2-channel filter banks, the cosine-modulated filter bank can be built with a parallel bank of 2-channel paraunitary filter banks. In this way, it is easy to create a lattice implementation of the M-channel filter bank.

In order to illustrate with an example how to design a cosine modulated filter bank, it is interesting to choose the procedure proposed by [19]. The main issue is to design the prototype filters. In the case of [19] the idea is to use a FIR filter with Kaiser window. According with a formula developed by Kaiser, the order N of the filter can be estimated with:

$$N \approx \frac{A_s - 7.95}{14.36\ \Delta\omega / 2\pi} \tag{1.228}$$

where A_s is the stopband attenuation, and $\Delta\omega$ is the chosen transition width. The parameter β of the Kaiser window can be obtained from A_s. In particular, for $A_s > 50$:

$$\beta = 0.1102 * (A_s - 8.7) \tag{1.229}$$

To complete the specification of the prototype filter an adequate value of the cut-off frequency ω_c is sought. Denote the prototype filter as $p(n)$. According with [19], an objective function is proposed in the following terms:

$$\phi = \max_n |g(2Mn)| \tag{1.230}$$

where $g()$ is a filter such that $G(e^{j\omega}) = |P(e^{j\omega})|^2$.

A simple search can be applied to obtain the ω_c that minimizes ϕ.

Figure 1.51 shows the evolution of ϕ in function of ω_c. Clearly, there is a minimum. This figure has been generated with the Program 1.18.

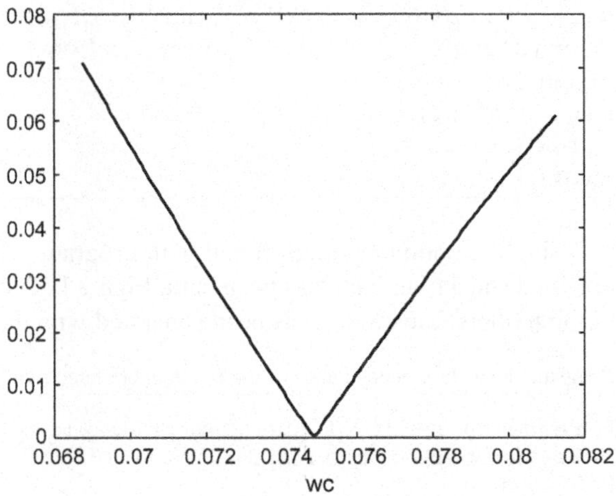

Fig. 1.51 Evolution of the objective function to be minimized

Program 1.18 Design of cosine modulated filter bank

```
%Design of cosine modulated filter bank
% Searching for minimum
M=8; %number of channels
As=80; %stopband attenuation
tband=.5; %transition band
%Kaiser formula for filter order:
N=floor((As-7.95)/14.36*M/tband);
disp(strcat('Prototype order: ',num2str(N)))
beta=0.1102*(As-8.7);
%set exploration range
w1=0.55/M;w2=0.65/M;
dw=(w2-w1)/200;
phimin=1000; nj=1;
%exploration
for wx=w1:dw:w2,
rwc(nj)=wx;
p=fir1(N,wx,kaiser(N+1,beta));
g=conv(p,p);
aux=g(N+1:(-2*M):1); axl=length(aux);
auxp= max(abs(aux(2:axl)))/aux(1);
rphi(nj)=auxp;
if auxp<phimin,
phimin=auxp;
wmin=wx;
end;
nj=nj+1;
end;
disp(strcat('phi_min: ',num2str(phimin)))
disp(strcat('wc_min: ',num2str(wmin)))
%display
plot(rwc,rphi,'k');
title('evolution of the objective function')
xlabel('wc');
```

Using the optimum values found with Program 1.18, a prototype filter has been obtained and a filter bank has been built. Figure 1.52 shows the frequency responses of the 8 filters. This figure has been generated with the Program 1.19.

Program 1.19 Frequency bands of the cosine modulated filter bank

```
%Frequency bands of the cosine modulated filter bank
M=8; %number of channels
N=80; %prototype order
wc0=0.074813; %cut-off freq. (prototype)
beta=7.8573; %beta (prototype)
h=fir1(N,wc0,kaiser(N+1,beta)); %the prototype filter
mxH=zeros(M,N+1); %matrix of filter coeffs
aux1=pi/M; aux2=pi/4; axN=(N+1)/2;
```

```
for nk=0:M-1,
for nj=0:N,
mxH(nk+1,nj+1)=2*h(nj+1)*cos(aux1*(nk+0.5)*(nj+0.5-axN)+...
aux2*(-1^nk));
end;
end;
figure(1)
wiv=pi/512;
w=0:wiv:pi-wiv;
G=freqz(mxH(1,:),1);
plot(w,abs(G),'k'); grid; hold on;
axis([0 pi 0 1.2]);
title('Frequency response of the filters');
xlabel('w');
for nk=2:M,
G=freqz(mxH(nk,:),1);
plot(w,abs(G),'k');
end;
```

1.7.4 Linear-Phase Filter Banks

We are going to consider a popular M-channel filter bank, the DCT. Then, we continue with an improvement over the DCT, the LOT. Finally, we focus on factorizations and lattice structures corresponding to paraunitary linear-phase filter banks (recall that this was not possible with 2-channel filter banks) [42].

See [43] and references therein for factorizations and lattice structures corresponding to linear-phase filter banks.

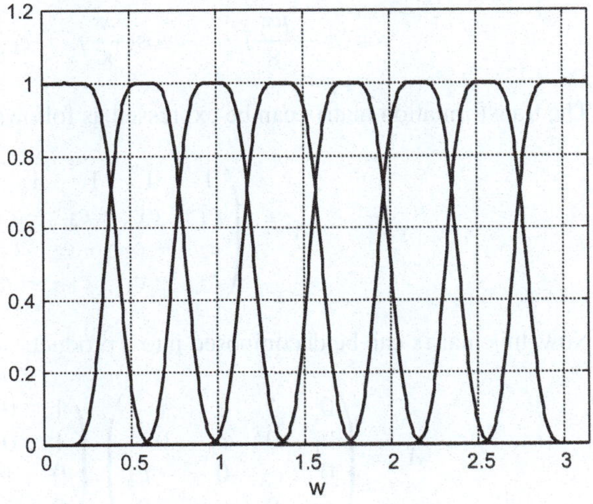

Fig. 1.52 Frequency responses of the 8 filters

1.7.4.1 Discrete Cosine Transform (DCT)

The one-dimensional (1D) discrete cosine transform (DCT) of a signal $x(n)$, is:

$$C(m) = \alpha(m) \sum_{j=0}^{N-1} x(n) \cos\left[\frac{\pi}{2N}(2j+1)m\right]; \quad m = 0, 1, \ldots, N-1 \quad (1.231)$$

with:

$$\alpha(m) = \left[\begin{array}{l} \sqrt{1/N} \;\; for\;\; m = 0 \\ \sqrt{2/N} \;\; for\;\; m \neq 0 \end{array}\right. \quad (1.232)$$

Notice the $C(0)$is the average value of the signal. The cosine terms are called *cosine basis functions.*

The DCT is an unitary transform, with real values. It can be regarded as a "real" version of the DFT.

The 1D DCT can be expressed as the product of a transformation matrix A and the signal:

$$C = A\,x \quad (1.233)$$

It has been observed that this transformation has some redundancy, and that the number of DCT operations can be reduced in several ways. For instance, suppose that $N = 4$, and so:

$$A = \begin{pmatrix} 1 & 1 & 1 & 1 \\ \cos(\pi/8) & \cos(3\pi/8) & \cos(5\pi/8) & \cos(7\pi/8) \\ \cos(2\pi/8) & \cos(6\pi/8) & \cos(10\pi/8) & \cos(14\pi/8) \\ \cos(3\pi/8) & \cos(9\pi/8) & \cos(15\pi/8) & \cos(21\pi/8) \end{pmatrix} \quad (1.234)$$

Using:

$$\cos\left(\pi - \frac{n\pi}{8}\right) = -\cos\left(\frac{n\pi}{8}\right); \quad c_k = \cos\left(\frac{k\pi}{8}\right) \quad (1.235)$$

The transformation matrix can be expressed as follows:

$$A = \begin{pmatrix} 1 & 1 & 1 & 1 \\ c_1 & c_3 & -c_3 & -c_1 \\ c_2 & -c_2 & -c_2 & c_2 \\ c_3 & -c_1 & c_1 & -c_3 \end{pmatrix} \quad (1.236)$$

Now, this matrix can be decomposed into a product:

$$A = \begin{pmatrix} 0 & 1 & 0 & 1 \\ c_1 & 0 & c_3 & 0 \\ 0 & c_2 & 0 & -c_2 \\ c_3 & 0 & -c_1 & 0 \end{pmatrix} \begin{pmatrix} 1 & 0 & 0 & -1 \\ 1 & 0 & 0 & 1 \\ 0 & 1 & -1 & 0 \\ 0 & 1 & 1 & 0 \end{pmatrix} \quad (1.237)$$

With this factorization, the number of operations for the DCT reduces from 16 products and 12 additions, to 5 products and 8 additions. A similar method can be successfully applied for $N = 8$.

It is possible to compute the DCT using DFT, after generating a symmetric signal mirroring $x(n)$ about the origin.

The 2D DCT can be expressed as:

$$C_{p\,q} = \alpha_p\,\alpha_q \sum_{i=0}^{M-1}\sum_{j=0}^{N-1} X_{i\,j} \cos\left[\frac{\pi}{2M}(2i+1)p\right] \cos\left[\frac{\pi}{2N}(2j+1)q\right]; \quad (1.238)$$

with:

$$0 \le p \le M-1;\;\; 0 \le q \le N-1; \quad (1.239)$$

$$\alpha_p = \begin{bmatrix} \sqrt{1/M} & for\;\; p = 0 \\ \sqrt{2/M} & for\;\; p \neq 0 \end{bmatrix}; \qquad \alpha_q = \begin{bmatrix} \sqrt{1/N} & for\;\; q = 0 \\ \sqrt{2/N} & for\;\; q \neq 0 \end{bmatrix} \quad (1.240)$$

The 2D DCT is a symmetric and orthogonal transformation.

The inverse of the 2D DCT transform is given by:

$$X_{i\,j} = \sum_{p=0}^{M-1}\sum_{q=0}^{N-1} \alpha_p\,\alpha_q\,C_{p\,q} \cos\left[\frac{\pi}{2M}(2i+1)p\right] \cos\left[\frac{\pi}{2N}(2j+1)q\right]; \quad (1.241)$$

with:

$$0 \le i \le M-1;\;\; 0 \le j \le N-1; \quad (1.242)$$

$$\alpha_p = \begin{bmatrix} \sqrt{1/M} & for\;\; p = 0 \\ \sqrt{2/M} & for\;\; p \neq 0 \end{bmatrix}; \qquad \alpha_q = \begin{bmatrix} \sqrt{1/N} & for\;\; q = 0 \\ \sqrt{2/N} & for\;\; q \neq 0 \end{bmatrix} \quad (1.243)$$

The inverse 2D DCT can be considered as a matrix that can be obtained as a weighted sum of the functions:

$$\alpha_p\,\alpha_q \cos\left[\frac{\pi}{2M}(2i+1)p\right] \cos\left[\frac{\pi}{2N}(2j+1)q\right]; \quad (1.244)$$

which are called the basis functions of the DCT.

In order to visualize the 2D DCT basis functions, a program has been written using basic MATLAB instructions, except for the last one, *imagesc()*, which displays matrices as images (more on that in a next chapter on Images). Figure 1.53 shows the result for $N = M = 8$. It is a set of 64 matrices, each one with 8×8 entries.

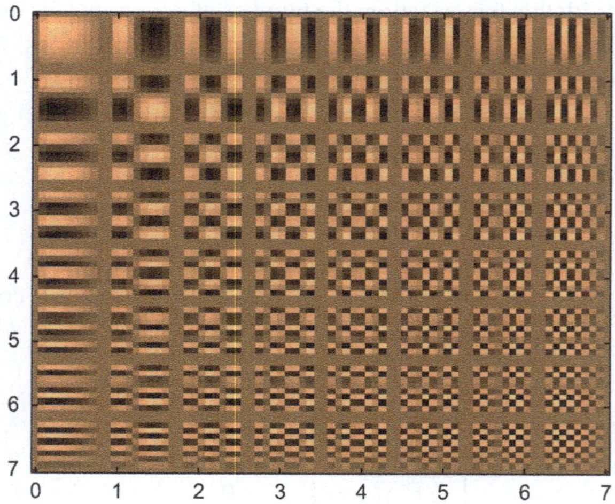

Fig. 1.53 2D DCT basis functions

Program 1.20 Display of 2D DCT bases

```
% Display of 2D DCT bases
% example of 8x8
%
N=8;
% computation of bases
i=0:N-1;
j=0:N-1;
[ni,nj]=meshgrid(i,j);
A=sqrt(2/N)*cos(((2.*ni+1).*nj*pi)/(N*2));
A(1,:)=A(1,:)./sqrt(2);
A=A';
B=zeros(N,N,N,N);
for i=1:N,
for j=1:N,
B(:,:,i,j)=A(:,i)*A(:,j)';
end;
end;
%image composition
L=2+N; %side of a square
Bimg=zeros(N*L,N*L); %space for the image
aux=zeros(8,8);
for i=0:N-1,
for j=0:N-1,
ix=((i*L)+2):(((i+1)*L)-1);
jx=((j*L)+2):(((j+1)*L)-1);
aux=B(i+1,j+1,:,:);
Bimg(ix,jx)=squeeze(aux);
end;
end;
```

```
%display
imagesc(0:7,0:7,Bimg);
colormap('copper');
title('DCT2 bases');
```

Actually, in the compression standards MPEG and JPEG, the image is decomposed into 8 × 8 blocks (or 16 × 16 blocks), and then the 2D DCT transform of each block is obtained by 1D DCT of each row followed by 1D DCT of each column.

Due to the popularity of those standards, the DCT is the *de-facto* image transformation in most visual systems.

The block by block processing of signals is often referred as block filtering. It can be represented with block diagonal transforms. Since 2D processing can always be done with 1D transforms, let us keep the 1D notation. Suppose a signal decomposed into blocks $x_0,\ x_1,\ \ldots,\ x_L$, the transform of the signal can be written as follows:

$$C = \begin{bmatrix} C_0 \\ C_1 \\ \vdots \\ C_{L-1} \end{bmatrix} = \begin{pmatrix} \ddots & & & 0 \\ & A & & \\ & & A & \\ & & & A \\ 0 & & & \ddots \end{pmatrix} \begin{bmatrix} x_0 \\ x_1 \\ \vdots \\ x_{L-1} \end{bmatrix} = T\,x \tag{1.245}$$

The inverse transform would be:

$$x = T^T C = diag(\,A^T,\ \ldots,\ A^T) \tag{1.246}$$

In many images, a lot of coefficients C_{pq} have insignificant values (those corresponding to high frequencies), and therefore they can be discarded with almost unnoticeable image quality degradation. In this way, considerable compression of images is available.

The MATLAB Image Processing Toolbox includes the function *dct()* for 1D DCT and *idct()* for the inverse 1D DCT. Likewise, it includes *dct2()* for 2D DCT, and *idct2()* for the inverse. In certain cases, it might be more efficient to use the function *dctmtx()*, which returns the transformation matrix A. Also, it could be useful to employ the *blkproc()* function for block filtering (there is a new version of this function, named *blockproc()*) .

The DCT was introduced in [2]. Faster DCT algorithms have been proposed by [5, 17].

Some simple examples may help to dig into an important aspect [12]. Consider a simple ordered sequence of numbers:

$$x = [1,\ 2,\ 3,\ \ldots,9] \tag{1.247}$$

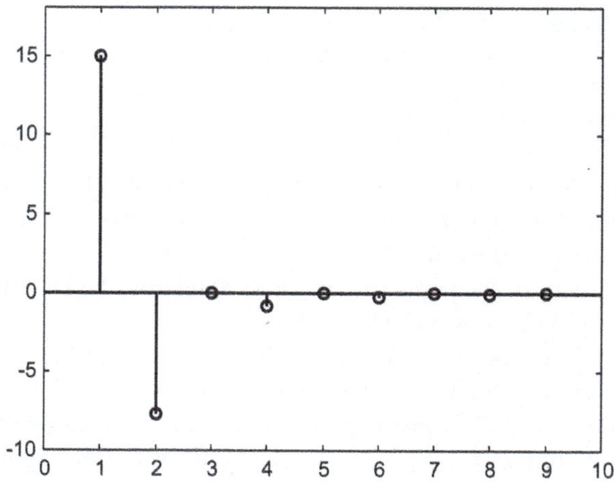

Fig. 1.54 1D DCT conversion of an ordered sequence

Since it is an ordered sequence, the numbers are correlated. Figure 1.54 shows the result of the 1D DCT of this sequence:

Program 1.21 1D DCT simple example, correlated data

```
% 1D DCT simple example, correlated data
%
x=[1 2 3 4 5 6 7 8 9]; %original data
WX=dct(x); %the DCT transform
E=sum(abs(WX)); %energy of transformation result
%display
figure(1)
stem(WX,'k'); hold on;
plot([0 10],[0 0],'k-');
axis([0 10 -10 18]);
title('Transform of correlated data');
disp('correlated data transform:');
WX
disp('correlated data transform energy:');
E
```

Notice that Program 1.21 computes, by simply adding absolute values, a kind of 'energy' of the 1D DCT transform result. This energy has a value of 23.8639 for the data considered.

Now, the same numbers as before, but with no sequential order, is used as the target of the 1D DCT, see Program 1.22. Figure 1.55 shows the result of the transform. Again, the energy is computed, and the result is higher: 38.1659.

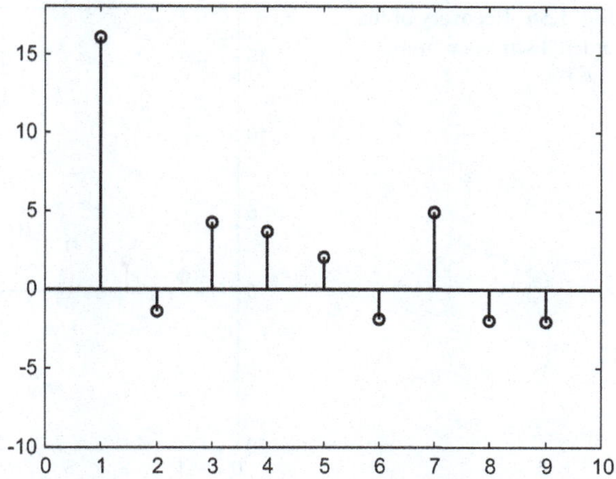

Fig. 1.55 1D DCT conversion of an unordered sequence

Program 1.22 1D DCT simple example, uncorrelated data

```
% 1D DCT simple example, uncorrelated data
%
x=[9 5 2 4 1 8 7 3 9]; %original data
WX=dct(x); %the DCT transform
E=sum(abs(WX)); %energy of transformation result
%display
figure(1)
stem(WX,'k'); hold on;
plot([0 10],[0 0],'k-');
axis([0 10 -10 18]);
title('Transform of uncorrelated data');
disp('uncorrelated data transform:');
WX
disp('uncorrelated data transform energy:');
E
```

When data are correlated, some mutual prediction is possible and better compression (in terms of energy) can be achieved. On the contrary, uncorrelated data means more entropy and less compressibility.

Now, let us do an interesting experiment related to the ordered sequence [1, 2, 3, …, 9]. Just after the application of the 1D DCT, which obtains a series of coefficients, let us try to recover the ordered sequence by taking only the first 3 coefficients and using the inverse 1D DCT transform. This is done with the Program 1.23. The result is shown in Fig. 1.56, which compares the original sequence (circles) and the recovered data (crosses). A good agreement is observed.

This example indicates that a lossy information compression is possible, by just retaining 3 of the coefficients obtained with the 1D DCT.

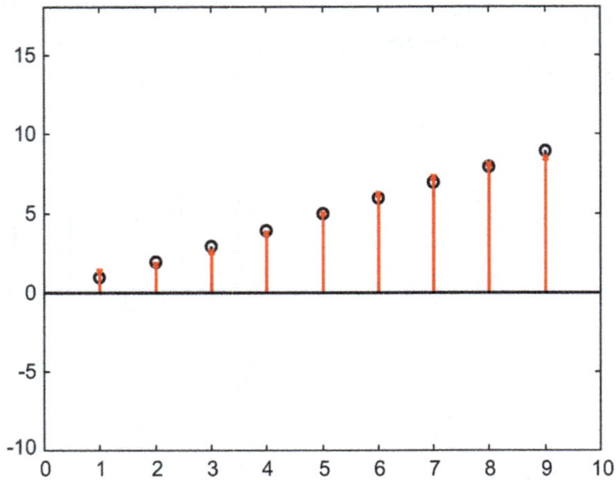

Fig. 1.56 Recovery of the ordered sequence from 3 coeffs

Program 1.23 1D DCT simple example, correlated data

```
% 1D DCT simple example, correlated data
% Recovering from first 3 coeffs
%
RWX=zeros(1,9); %space for recovery
x=[1 2 3 4 5 6 7 8 9]; %original data
WX=dct(x); %the DCT transform
RWX(1:3)=WX(1:3); %select 3 first coeffs
rx=idct(RWX); %recovery of data
%display
figure(1)
stem(x,'k'); hold on;
stem(rx,'rx');
plot([0 10],[0 0],'k-');
axis([0 10 -10 18]);
title('Comparison of original and recovered data');
disp('recovered data');
rx
```

Indeed, a main field of application of the DCT is 2D signals and images. Let us include an example: the photograph shown in Fig. 1.57. The Program 1.24 applies the 2D DCT to this image, obtaining the result shown in Fig. 1.58. Notice the use of the *imread()* and *imshow()* MATLAB functions, for reading a file with the image and showing it on screen. There is a chapter, later on, that focuses on image processing.

Fig. 1.57 A photography

Program 1.24 2D DCT of a photograph

```
% 2D DCT of a photograph
P=imread('elef.jpg');
D=dct2(P);
figure(1)
imshow(P);
title('original photo');
figure(2)
imshow(log(abs(D)));
title('2D DCT of the photo');
```

Notice in Fig. 1.58 that most energy of the 2D DCT comes to the top left corner, while darker pixels—corresponding to less energy—are on the bottom right corner. Again, this suggests a way for lossy information compression. A good treatment of these aspects is found in the documentation of MATLAB.

As said before, there are standard compression methods that are based on block by block processing. For instance, the image to be processed can be decomposed into blocks of 8×8 pixels, and then 2D DCT would be applied to each block. By using a simple mask, only the coefficients near the top left corner are selected, for lossy information compression.

Let us take again the image shown in Fig. 1.57. The Program 1.25 applies the block by block processing just described, obtaining the result shown in Fig. 1.59. In a second step, the program recovers the image from these data, and shows the result, Fig. 1.60. The result of this compression procedure seems fairly acceptable.

Fig. 1.58 2D DCT of the photography

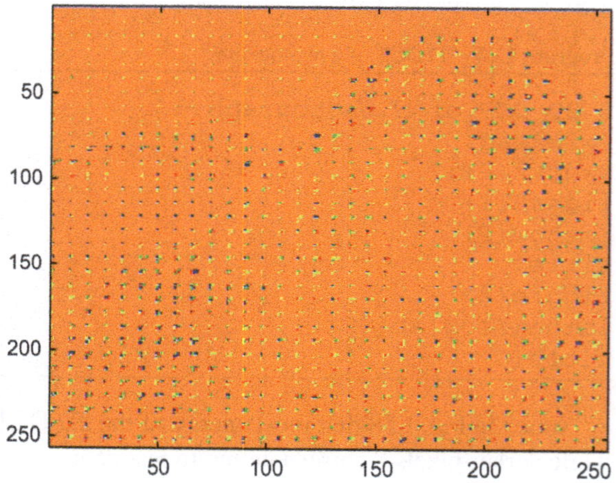

Fig. 1.59 The block by block DCT transform

Program 1.25 Compression of the photograph using 2D DCT

```
% Compression of the photograph using 2D DCT
P=imread('elef.jpg');
F=im2double(P);
M=dctmtx(8);
B=blkproc(F,[8 8],'P1*x*P2',M,M');
mask= [1 1 1 1 0 0 0 0
1 1 1 0 0 0 0 0
1 1 0 0 0 0 0 0
1 0 0 0 0 0 0 0
0 0 0 0 0 0 0 0
0 0 0 0 0 0 0 0
0 0 0 0 0 0 0 0
0 0 0 0 0 0 0 0];
```

```
C=blkproc(B,[8 8],'P1.*x',mask);
RP=blkproc(C,[8 8], 'P1*x*P2',M',M);
figure(1)
imagesc(C); colormap('prism');
title('the block by block transform');
figure(2)
imshow(RP);
title('recovered photo');
```

More details on the DCT theory and applications can be found in [14, 35, 50].

1.7.4.2 Lapped Ortogonal Transform (LOT)

Independent block processing may cause artifacts at the block boundaries. In particular, this can be observed when applying the DCT to images: the borders between 8 × 8 blocks may be noticeable. These artifacts are due to discontinuous transitions to zero at the tails of the transform basis functions. It can be avoided by allowing the basis functions of contiguous blocks to overlap.

In order to achieve the overlapping, the block filtering would be done according with the following expression:

$$C = \begin{bmatrix} C_0 \\ C_1 \\ \vdots \\ C_{L-1} \end{bmatrix} = \begin{pmatrix} \ddots & & & & 0 \\ & [A\ B]\ 0 & & & \\ & 0\ [A\ B]\ 0 & & & \\ & 0 \quad 0 & [A\ B] & & 0 \\ 0 & & & \ddots & B \\ B & & \cdots & & A \end{pmatrix} \begin{bmatrix} x_0 \\ x_1 \\ \vdots \\ x_{L-1} \end{bmatrix} = \Gamma x \tag{1.248}$$

Fig. 1.60 Recovered compressed image

where the N × N square matrix A has been replaced by a non-square matrix P built with the concatenation of two matrices, $P = [A\ B]$. Usually, B is chosen to be another N × N square matrix. Note that the last row of Γ corresponds to a periodic repetition of the signal.

By design, the transform is orthogonal: $\Gamma \cdot \Gamma^T = I$.

The original signal can be reconstructed with $x = \Gamma^T\ C$. It involves two steps: in the first each Cj is multiplied by P^T, yielding a 2L dimensional signal vector; in the second, contiguous (overlapping) signal blocks are added by pairs.

From $\Gamma \cdot \Gamma^T = I$, the following conditions are derived:

$$P \cdot P^T = A \cdot A^T + B \cdot B^T = I \ \ and \ \ A \cdot B^T = B \cdot A^T = 0 \tag{1.249}$$

If P_0 is a LOT matrix, other LOT matrices can be obtained with $P = Z \cdot P_0$, where Z is an orthogonal matrix. It has been proposed to build P_0 on the basis of the DCT. Half of the DCT basis functions are even symmetric, and the other half is odd. Two matrices are built, D_e and D_o.The rows of the $N/2 \times N$ matrix D_e are the even symmetric basis functions; and similarly D_o is made with the odd ones. The proposed P_0 is:

$$P_0 = \frac{1}{2}\begin{pmatrix} D_e - D_o & (D_e - D_o)\,J \\ D_e - D_o & -\,(D_e - D_o)\,J \end{pmatrix} \tag{1.250}$$

where J is the antidiagonal unit matrix.

Further developments concerning the LOT can be found in [1, 21, 22, 33].

The following fragment of MATLAB code shows an example of computation of the matrix P_0 (of course, you can change the value of N).

Fragment 1.26 the LOT matrix Po

```
% the LOT matrix Po
N=8; %number of rows of DCT matrix (edit this)
D=idct(eye(N));
De=D(1:2:N,:);
Do=D(2:2:N,:);
R=(De-Do);
J=flipud(eye(N));
Po=0.5*[R,R*J;R, -R*J];
```

1.7.4.3 Paraunitary

The paraunitary condition imposes constraints on the number of symmetric and antisymmetric filters in the M-channel filter bank:

- If M is even, there are M/2 symmetric, and M/2 antisymmetric filters
- If M is odd, there are (M + 1)/2 symmetric, and (M − 1)/2 antisymmetric filters.

There is a factorization with the name *GenLOT* [32]. The *DCT* and the *LOT* are special cases of *GenLOT*, with lengths M and 2M respectively. The *GenLOT* factorization is as follows:

$$\mathbf{H_p(z)} = K_{N-1}(z) K_{N-2}(z) \ldots K_1(z)\, E_0 \tag{1.251}$$

where:

$$K_j(z) = \Phi_j\, W\, \Lambda(z)\, W \tag{1.252}$$

$$W = \begin{pmatrix} I & I \\ I & -I \end{pmatrix};\; \Phi_j = \begin{pmatrix} U_j & 0 \\ 0 & V_j \end{pmatrix};\; \Lambda(z) = \begin{pmatrix} I & 0 \\ 0 & z^{-1} I \end{pmatrix} \tag{1.253}$$

$$E_0 = \frac{1}{\sqrt{2}} \begin{pmatrix} U_0 & 0 \\ 0 & V_0 \end{pmatrix} \begin{pmatrix} I & J \\ I & -J \end{pmatrix} \tag{1.254}$$

where U_j and V_j are arbitrary orthogonal matrices. The matrix J is the antidiagonal unit matrix.

Figure 1.61 shows a lattice structure corresponding to the GenLOT factorization, see [29] for more details.

The following factorization is for systems with an *even number of channels*.

Let $\mathbf{H_p(z)}$ be a FIR linear phase paraunitary matrix. In this case, the following factorization is always possible:

$$\mathbf{H_p(z)} = S\, Q\, T_N\, \Lambda(z)\, T_{N-1}\, \Lambda(z) \cdots \Lambda(z)\, T_0\, Q \tag{1.255}$$

where T_j are $M \times M$ orthogonal matrices, and:

$$\Lambda(z) = \begin{pmatrix} I_{M/2} & 0 \\ 0 & z^{-1} I_{M/2} \end{pmatrix} \tag{1.256}$$

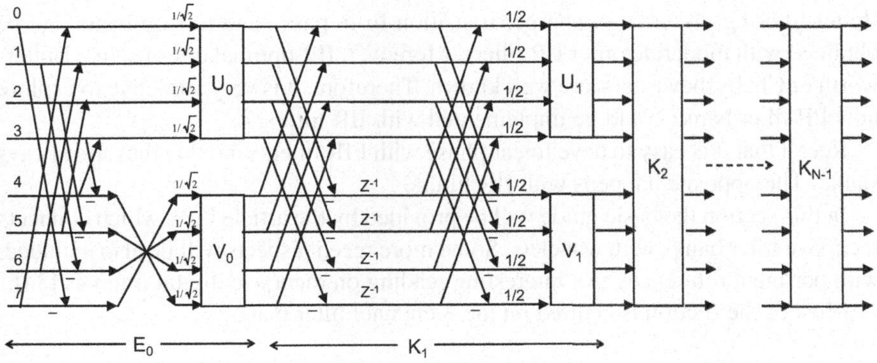

Fig. 1.61 Lattice structure corresponding to GenLOT factorization

and the orthogonal matrices S, Q, T_j are:

$$Q = \begin{pmatrix} I_{M/2} & 0 \\ 0 & J_{M/2} \end{pmatrix}; \quad T_j = \begin{pmatrix} U_j & V_j \\ V_j & U_j \end{pmatrix} \tag{1.257}$$

$$S = \frac{1}{\sqrt{2}} \begin{pmatrix} U_0 & 0 \\ 0 & V_0 \end{pmatrix} \begin{pmatrix} I_{M/2} & J_{M/2} \\ I_{M/2} & -J_{M/2} \end{pmatrix} \tag{1.258}$$

It can be shown that T_j and S are characterized by $2 \binom{M/2}{2}$ rotations.

The following factorization is for systems with an *odd number of channels*. $\mathbf{H_p(z)}$ is a FIR linear phase paraunitary matrix:

$$\mathbf{H_p(z)} = Q\, T_N\, \Lambda(z)\, T_{N-1}\, \Lambda(z) \cdots \Lambda(z)\, T_0\, Q \tag{1.259}$$

with:

$$\Lambda(z) = \begin{pmatrix} I_{(M+1)/2} & 0 \\ 0 & z^{-1} I_{(M-1)/2} \end{pmatrix} \tag{1.260}$$

and:

$$Q = \begin{pmatrix} I_{(M+1)/2} & 0 \\ 0 & J_{(M-1)/2} \end{pmatrix}; \quad T_j = \begin{pmatrix} U_j & 0 & V_j \\ 0^T & 1 & 0^T \\ V_j & 0 & U_j \end{pmatrix} \tag{1.261}$$

See [42] for an extended treatment of the last two factorizations. Lattice structures are subject to intense research, as it is reviewed and treated in [51].

1.8 IIR Filter Banks

Better filtering, in terms of sharper transition from pass-band to stop-band, can be obtained with moderate order IIR filters. Moreover, IIR optimal designs, like Butterworth or Chebyshev filters, are well known. Therefore, it is very attractive to explore how PR filter banks could be implemented with IIR filters.

Recall that it is easy to have linear phase with FIR filters, and that they are always stable. The opposite happens with IIR filters.

In this section the basic guide will be provided by the article [13], which connects recursive filter banks with wavelets. Some more recent aspects will be also included, with pertinent references. An interesting reading on recursive digital filters is [34].

Most of the section is centred on the 2-channel filter bank.

1.8.1 Orthogonal IIR Filter Banks

In order to have orthogonal filters, the following assignment is chosen:

$$H_1(z) = z^{2k-1}\, H_0(-z^{-1}) \tag{1.262}$$

$$F_0(z) = H_0(z^{-1});\;\; F_1(z) = H_1(z^{-1}) \tag{1.263}$$

Recall that the output of the analysis and synthesis cascade is:

$$Y(z) = \frac{1}{2}\,[F_0(z)\;\; F_1(z)] \begin{pmatrix} H_0(z) & H_0(-z) \\ H_1(z) & H_1(-z) \end{pmatrix} \begin{bmatrix} U(z) \\ U(-z) \end{bmatrix} \tag{1.264}$$

Hence:

$$Y(z) = \frac{1}{2}\,\left[H_0(z)\, H_0(z^{-1}) \;+\; H_0(-z)\, H_0(-z^{-1})\right]\; U(z) \tag{1.265}$$

The function $P(z) = H_0(z)\, H_0(z^{-1})$ is the '*autocorrelation*' function. From the previous equation, is clear that for PR:

$$P(z) + P(-z) = 2 \tag{1.266}$$

(for off-line operation).

Any function satisfying Eq. (1.249) is said to be *valid*. In our case, it is also pertinent to have rational valid functions, so they can be expressed as the product of two rational functions $H_0(z)$ and $H_0(z^{-1})$.

The task of building an IIR filter bank reduces to find a suitable valid function. It can be shown [13] that if $P(z)$ has no common factors between the numerator and the denominator, then the form of $P(z)$ in terms of polyphase components, is:

$$P(z) = \frac{z^{-k} D(z^2) + z^{-(k+1)} N(z^2)}{z^{-k} D(z^2)} \tag{1.267}$$

Now, it has been established [13] that all orthogonal rational 2-channel filter banks with PR can be built as follows:

- Choose an arbitrary polynomial $R(z)$, and then:

$$P(z) = \frac{2\, R(z)\, R(z^{-1})}{R(z)\, R(z^{-1}) + R(-z) R(z^{-1})} \tag{1.268}$$

- Factorize: $P(z) = H(z)\, H(z^{-1})$
- Build the filter $H_0(z) = A_0(z)\, H(z)$, where $A_0(z)$ is an arbitrary allpass filter

- Build the filter $H_1(z) = z^{2k-1} A_1(z)\, H_0(-z^{-1})$, where $A_1(z)$ is an arbitrary allpass filter
- Complete the filter bank with:

$$F_0(z) = H_0(z^{-1}); \;\; F_1(z) = -H_1(z^{-1})$$

It is also shown in [13] that in the case that $R(z)$ is symmetric and of even length, then $P(z)$ can be easily factorized since in this case:

$$P(z) = \frac{R(z)\, R(z^{-1})}{2\, R_0(z^2) R_0(z^{-2})} \tag{1.269}$$

using: $R(z) = R_0(z^2) + z^{-1} R_1(z^2)$

For an evident factorization, one could choose:

$$H(z) = \frac{R(z)}{\sqrt{2}\, R_0(z^2)} \tag{1.270}$$

Observe that the digital Butterworth, Chebyshev and elliptic filters have symmetric numerators.

As it will be seen in the next chapter, the typical wavelet design tries to use filters with a maximum number of zeros at $z = -1$, for maximum flatness. So, one could choose $R(z) = (1 + z^{-1})^N$ and then:

$$P(z) = \frac{2(1 + z^{-1})^N (1 + z)^N}{(z^{-1} + 2 + z)^N + (-z^{-1} + 2 - z)^N} \tag{1.271}$$

which is a halfband digital Butterworth filter.

For example, taking $N = 5$:

$$P(z) = \frac{(1 + z^{-1})^5 (1 + z)^5}{10\, z^4 + 120\, z^3 + 252 + 120\, z^{-2} + 10\, z^{-4}} \tag{1.272}$$

This $P(z)$ can be factorized as follows:

$$H_0(z) = \frac{1 + 5z^{-1} + 10z^{-2} + 10z^{-3} + 5z^{-4} + z^{-5}}{\sqrt{2}\,(1 + 10z^{-2} + 5z^{-4})} \tag{1.273}$$

$$H_1(z) = z^{-1}\, \frac{1 - 5z + 10z^2 - 10z^3 + 5z^4 - z^5}{\sqrt{2}\,(1 + 10z^2 + 5z^4)} \tag{1.274}$$

Based on this approach, *Butterworth wavelets* were proposed in [49]. For N odd, the filter $H_0(z)$ can be obtained with:

$$H_0(z) = \frac{\sum_{k=0}^{N} \binom{N}{k} z^{-k}}{\sqrt{2} \cdot \sum_{j=0}^{M} \binom{N}{2j} z^{-2j}} \tag{1.275}$$

where: $M = (N-1)/2$.

Other wavelets can be obtained based on a more general expression of $P(z)$:

$$P(z) = \frac{(1+z^{-1})^N (1+z)^N \, F_N(z)}{F_D(z)} \tag{1.276}$$

(note that $P(z)$ must be valid).

In [38], some aspects of orthogonal IIR filter banks were completed, and design formulae were introduced.

1.8.2 Linear Phase Orthogonal IIR Filter Banks

It can be shown that the solutions based on (1.276) cannot lead to a linear phase filter bank [13].

However, there are ways for obtaining linear phase orthogonal filters. In preparation for that, it is convenient to make some distinctions, and to introduce some nomenclature according with [13]. Asymmetric or symmetric filters that have a central term are called whole sample asymmetric (WSA) or symmetric (WSS) filters. In case of not having a central term, they are called half sample asymmetric (HSA) or symmetric (HSS) filters.

Suppose a linear phase IIR filter with numerator of length $N$and denominator of length D. Then, if $N - D$ is odd the filter is WSS or WSA; or, if $N - D$ is even, the filter is HSS or HAS.

An orthogonal filter bank with HSS filters has linear phase iif:

$$\mathbf{H_p(z)} = \begin{pmatrix} A(z) & z^{-k}\, A(z^{-1}) \\ -z^{k-n}\, A(z^{-1}) & z^{-n}\, A(z^{-1}) \end{pmatrix} \tag{1.277}$$

where: $A(z)\, A(z^{-1}) = I$ (that is, the filter $A(z)$ is all pass).

A very simple example is, with $k = n = 0$, the following:

$$H_0(z) = A\,(z^2) + z^{-1} A\,(z^{-2});\;\; H_1(z) = -A\,(z^2) + z^{-1} A\,(z^{-2}) \tag{1.278}$$

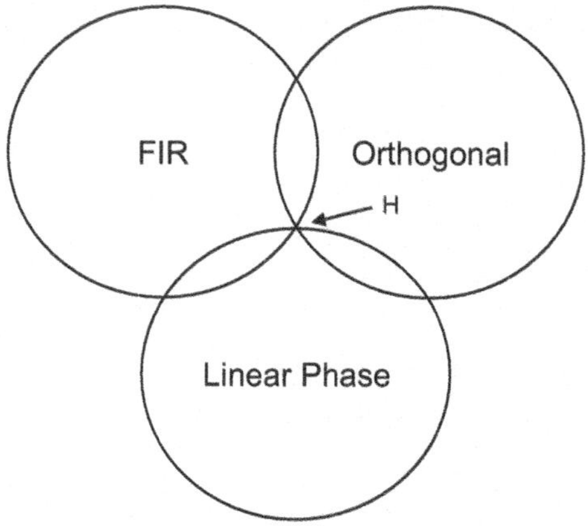

Fig. 1.62 Venn diagram with the 2-channel filter bank design alternatives

which is an orthogonal pair with linear phase. A particular solution used by [49] for the construction of wavelets, is:

$$A(z) = \frac{1 + 6z^{-1} + (15/7)z^{-2}}{(15/7) + 6z^{-1} + z^{-2}} \tag{1.279}$$

The case of WSS filters is more involved; refer to [13] for a detailed discussion.

Figure 1.62 shows a Venn diagram with the design alternatives concerning 2-channel filter banks. The point labeled H corresponds to the Haar wavelet, which will be the first to be studied in the next chapter.

1.8.3 Implementation Aspects

While IIR filters offer evident advantages of being more selective with small order polynomials, there are some implementation issues. In general it is desired a moderate processing effort, in terms of multiplications and other operations. Also, it is desired that numerical round-off had little impact on result quality.

Part of the research activity has focused on obtaining linear phase. In addition to proposed IIR filters with (approximately) linear phase, a way of solution has been offered by using phase correction after the IIR filter (this solution increases the delay). Of course, another alternative is to approximate and substitute the IIR filter by a FIR linear phase filter, but this filter could be quite long.

One of the difficulties when using IIR in filter banks is precisely that they have infinite impulse response, and that filters in the synthesis part usually are non-causal.

The aspects just mentioned will be succinctly examined next.

1.8.3.1 Using Allpass Filters

Suppose two filters such that:

$$|H(e^{j\omega}) + F(e^{j\omega})| = 1, \ \forall\omega \tag{1.280}$$

and:

$$|H(e^{j\omega})|^2 + |F(e^{j\omega})|^2 = 1, \ \forall\omega \tag{1.281}$$

It is said that these two filters are *doubly complementary*.

From the two equations, one can write:

$$|H(e^{j\omega}) + F(e^{j\omega})|^2 = |H(e^{j\omega})|^2 + |F(e^{j\omega})|^2 = 1, \ \forall\omega \tag{1.282}$$

Thinking in terms of phasors, it can be deduced that $H(e^{j\omega})$ and $F(e^{j\omega})$ are in quadrature, and that the phasors $|H(e^{j\omega}) + (F(e^{j\omega})|$ and $|H(e^{j\omega}) - (F(e^{j\omega})|$ both have magnitude one. Therefore:

$$H(z) + F(z) = A_1(z) \tag{1.283}$$

$$H(z) - F(z) = A_2(z) \tag{1.284}$$

where $A_1(z)$ and $A_2(z)$ are stable allpass functions.

Then, it is possible to write:

$$H(z) = \frac{1}{2}\,(A_1(z) + A_2(z)) \tag{1.285}$$

$$F(z) = \frac{1}{2}\,(A_1(z) - A_2(z)) \tag{1.286}$$

Figure 1.63 shows a diagram of the 2-filter structure.

This structure was first considered in relation with the wave digital lattice filters [8]. In the case that the two allpass filters are real functions, of order M_1 and M_2 respectively, it is found that for a low-pass high-pass pair:

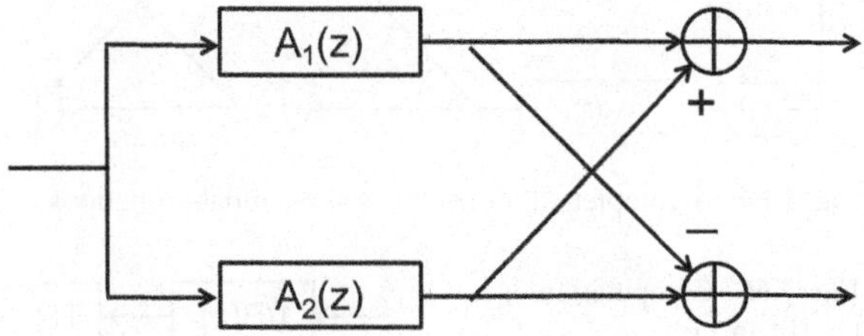

Fig. 1.63 Structure for the implementation of the doubly complementary filters using allpass filters

$$M_1 - M_2 = \pm 1 \tag{1.287}$$

It also can be observed that $H(z)$ will have all the poles of both allpass filters, so its order is $M_1 + M_2$. Then, because of (1.287), the order of $H(z)$ is odd.

In case of allowing the allpass functions to be complex, even-ordered filters can be obtained.

According with the method introduced in [47], each of the two allpass filters can be implemented as a cascade of first and second order blocks. The first order block requires only one multiplier, and the second order block two multipliers. The structure depicted in Fig. 1.63 is an optimum in terms of computational efficiency [36]. The corresponding implementation has extremely low coefficient sensitivity in the passband. The structure has favourable finite word-length properties.

An interesting and comprehensive overview of the use of allpass filters is given in [36], with many pertinent references. See also the Thesis [10].

1.8.3.2 Filter Inversion

Suppose you use the structure of Fig. 1.63 as the basis for the analysis part of a filter bank. It is possible to use a simple synthesis part to complete the system so you obtain PR. Figure 1.64 shows the complete filter bank.

Notice that you need the reciprocals of the allpass functions. Likewise, there are other filter systems that require the inversion of IIR filters.

There is a problem, when the inversion of the IIR filter is done, poles become zeros and zeros become poles. If the IIR filter has zeros outside the unit circle, then the inverse will have poles outside the unit circle, so it is unstable.

There is a way to solve this problem. Suppose your IIR filter is $H(z)$. First, factorize as $H(z) = B(z)\,G(z)$, so all zeros outside the unit circle go to $G(z)$ and the rest of zeros to $B(z)$. Then, you use the cascade represented in Fig. 1.65.

In Fig. 1.65 the unit TR does a time reversal of signal. Typically the computation is made signal block by signal block, using two buffers (one for each TR unit). See [40, 46], and references therein for more details.

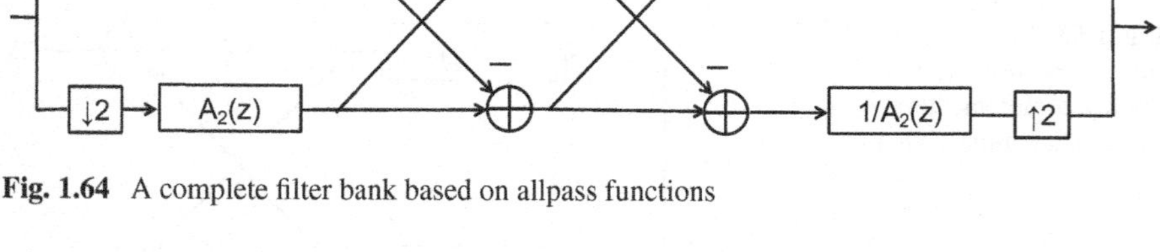

Fig. 1.64 A complete filter bank based on allpass functions

Fig. 1.65 Computation of the IIR inverse

Some aspects of the IIR inversion can be better seen by using a state-space representation. This kind of system representation is typically used in the context of automatic control theory, and has two main advantages: it takes explicitly into account the internal variables of the system, and it is suitable for computers since it is based on the use of matrices and vectors (the same can be said about MATLAB). Assume that the filter $G(z)$ has N taps, the input is $u(n)$ and the output is $y(n)$. The state-space representation is:

$$\begin{bmatrix} \mathbf{x}(n+1) \\ y(n) \end{bmatrix} = \begin{pmatrix} A & B \\ C & D \end{pmatrix} \cdot \begin{bmatrix} \mathbf{x}(n) \\ u(n) \end{bmatrix} \tag{1.288}$$

where A, B, C and D are matrices, and the vector

$$\mathbf{x}(n) = [x_1(n),\, x_2(n),\, \ldots,\, x_N(n)]$$

contains the states of the filter taps.

The state-space representation of the anticausal filter $\hat{G}(z) = 1/G(z^{-1})$ would be the following:

$$\begin{bmatrix} \hat{x}(n+1) \\ \hat{y}(n) \end{bmatrix} = \begin{pmatrix} \hat{A} & \hat{B} \\ \hat{C} & \hat{D} \end{pmatrix} \cdot \begin{bmatrix} \hat{x}(n) \\ \hat{u}(n) \end{bmatrix} \tag{1.289}$$

where:

$$\begin{pmatrix} \hat{A} & \hat{B} \\ \hat{C} & \hat{D} \end{pmatrix} = \begin{pmatrix} A & B \\ C & D \end{pmatrix}^{-1} \tag{1.290}$$

The work of the anticausal filter, for any signal block, starts with $\hat{x}(L) = \mathbf{x}(L)$, and then:

$$\hat{u}(L+k) = y(L-1-k), \quad for\, k = 0,\, 1,\, \ldots,\, L-1 \tag{1.291}$$

during this process, a series of outputs are obtained that reconstruct the $G(z)$ inputs:

$$\hat{y}(L+k) = u(L-1-k), \quad for\, k = 0,\, 1,\, \ldots,\, L-1 \tag{1.292}$$

and the final state will be $\hat{x}(2L) = \mathbf{x}(0)$.

It is important to correctly set the initial states of the anticausal filter. One of the proposed solutions for it is the embedded filter states method [39].

The research has provided several approaches to improve the IIR inversion. An interesting example of recent work is [25], which obtains a real time version with reduced delay.

1.8.3.3 FIR and IIR Combinations

In order to circumvent the difficulties related to the inversion of the allpass filters in the synthesis part of the bank, it has been proposed to use FIR filters instead of inversion. A main target is to compensate the phase distortion caused by the allpass filters. It is usually called *phase equalization.*

To capture the essence of the problem, consider the following equation:

$$A(z)\,F(z) = z^{-d} + \varepsilon \tag{1.293}$$

where the filter $F(z)$ in the synthesis part should minimize the error ε, to achieve near-perfect reconstruction.

Supposing that first order allpass filters are used in the analysis part, and a FIR filter is used in the synthesis part a possible approach would be:

$$\underbrace{\frac{1-\alpha z}{z-\alpha}}_{A(z)} \cdot \underbrace{[(z-\alpha)\sum_{k=0}^{d-1} z^{k-d}(\alpha)^k]}_{F(z)} = z^{-d} + (\alpha)^d \tag{1.294}$$

This expression is the basis for a set of analytical design alternatives derived by [15].

An interesting solution has been proposed in [20], by using a type of allpass filters for the phase equalization. The result is an IIR/IIR filter bank.

As reviewed in [15, 20], previous research has proposed some optimization-based methods for the design of the FIR filter, in order to minimize the error.

1.8.3.4 Lifting

Recall the PR condition:

$$\mathbf{B(z)} = \mathbf{F_p(z)} \cdot \mathbf{H_p(z)} \tag{1.295}$$

where one of the two possible values of ***B(z)*** is:

$$\mathbf{B(z)} = z^{-d}\, I; \tag{1.296}$$

Now, observe that for any $A(z)$ and $B(z)$ arbitrary transfer functions, one can write:

$$\begin{pmatrix} z^{-N} & 0 \\ -A(z) & 1 \end{pmatrix} \cdot \begin{pmatrix} 1 & 0 \\ A(z) & z^{-N} \end{pmatrix} = z^{-N} I \tag{1.297}$$

$$\begin{pmatrix} 1 & B(z) \\ 0 & z^{-M} \end{pmatrix} \cdot \begin{pmatrix} z^{-M} & -B(z) \\ 0 & 1 \end{pmatrix} = z^{-M} I \tag{1.298}$$

Therefore, if one chooses the following polyphase matrices:

$$\mathbf{H_p(z)} = \begin{pmatrix} z^{-M} & -B(z) \\ 0 & 1 \end{pmatrix} \begin{pmatrix} 1 & 0 \\ A(z) & z^{-N} \end{pmatrix} \tag{1.299}$$

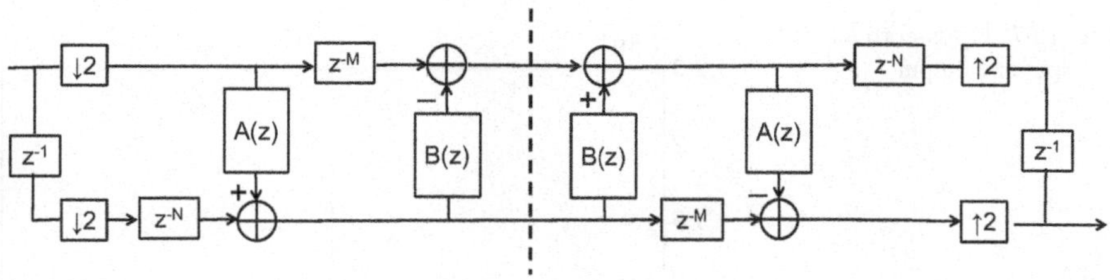

Fig. 1.66 Structure of the filter bank based on lifting

$$\mathbf{F_p(z)} = \begin{pmatrix} z^{-N} & 0 \\ -A(z) & 1 \end{pmatrix} \cdot \begin{pmatrix} 1 & B(z) \\ 0 & z^{-M} \end{pmatrix} \quad (1.300)$$

Then the PR condition is satisfied, regardless of the values of $A(z)$ and $B(z)$.

Concerning the actual filter bank, the filters will be:

$$H_0(z) = z^{-2M} - B(z^2)\, H_1(z) \quad (1.301)$$

$$H_1(z) = z^{-2N-1} + A(z^2) \quad (1.302)$$

and:

$$F_0(z) = -H_1(-z); \quad F_1(z) = H_0(-z) \quad (1.303)$$

Figure 1.66 shows the structure of the filter bank corresponding to this design method.

Some of the initial articles on this approach refers to it as *structural PR* (for instance, [52]). Later on it was named *the lifting method* (for instance in [10]), which is directly linked to a wavelet design methodology that will be described in the next chapter.

With this method, it is relatively easy to design IIR linear phase filter banks [52]. See [10] and references therein for interesting choices of $A(z)$ and $B(z)$.

1.9 Experiments

This section includes examples of perfect reconstruction, signal compression via DCT, and watermarking techniques. The accompanying programs may serve as a basis for further exploration by the reader.

1.9.1 Perfect Reconstruction, Music

Obviously, part of the experiments concerning this chapter should be devoted to perfect reconstruction. We preferred to select a musical example, since it would provide an intuitive opportunity for personal assessment.

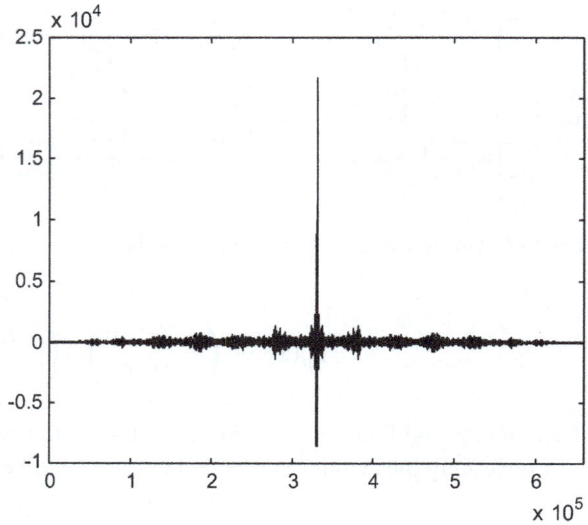

Fig. 1.67 Cross-correlation of input and output

A rather simple program—Program 1.27—has been developed for implementation of a two-channel filter bank. A biorthogonal filter with PR has been chosen, and a brief piece of well-known music is used. The program let you hear the reconstructed signal. In addition, the program generates a figure—Fig. 1.67—that shows the cross-correlation between the original and the reconstructed signal. The neat peak in the center of the figure indicates a strong similarity.

The reader is invited to change the melody, or to make audible the signals that cross one or another of the filter channels.

Program 1.27 Auditive check of PR

```
% Auditive check of PR
% 2-channel filter bank, and music
[u,fs]=wavread('B_allyou.wav'); %read wav file
L=length(u);
%biorthogonal filter bank
h0=[1 2 1]; %the filter H0
h0=(1/4)*h0; %scaling
fx=conv([1 2 1],[-1 4 -1]);
fx=(1/8)*fx; %scaling
f0=2*fx; %the filter F0
for n=1:5, h1(n)=((-1)^(n-1))*fx(n); end; % the filter H1
for n=1:3, f1(n)=-2*((-1)^(n-1))*h0(n); end; % the filter F1
%filtering and downsampling
v0=filter(h0,1,u); dv0=decimate(v0,2);
v1=filter(h1,1,u); dv1=decimate(v1,2);
%upsamplng and filtering
uv0=zeros(L,1); uv0(1:2:L)=dv0;
uv1=zeros(L,1); uv1(1:2:L)=dv1;
y0=filter(f0,1,uv0);
y1=filter(f1,1,uv1);
```

```
%final adding
yr=y0+y1;
sound(v0,fs);
xcr=xcorr(u,yr); %cross-correlation of output and input
%display
figure(1)
plot(xcr,'k');
axis([0 2*L -1e4 2.5e4]);
title('cross-correlation of input and output');
```

1.9.2 JPEG

The image compression methods fall into two general categories: lossless and lossy methods. The JPEG ( ) process is a form of lossy image compression based on 2D DCT. The process divides the image into 8×8 blocks, applying the 2D DCT to each block. Once on the DCT domain, a step called quantization comes. During this step, the less important frequencies are discarded (and so it is a lossy method).

The 2D DCT of the image block F can be expressed as:

$$P = A\,F\,A' \tag{1.304}$$

where A is the DCT matrix.

The quantization step uses a standard quantization matrix Q that has been obtained from subjective experiments involving the human visual system. There is a parameter that you can use to adjust the quality/compression ratio. The quantization is achieved by dividing each element of P by the corresponding element of Q, and then rounding. The result will be a quantized matrix C.

Now, the matrix C is encoded in a zig-zag manner, as sketched in Fig. 1.68 for the initial 4×4 sub-block (the zig-zag continues to cover the complete 8×8 block). With the zig-zag, the matrix is converted to a vector, which usually has many negligible components. More details on JPEG can be found in [7, 24, 30].

A simulation of the JPEG process has been developed. It is a program that has been included in Appendix A. Its coding in MATLAB has interesting aspects, although it does not include the zig-zag encoding. It is based on [18].

Figure 1.69 shows a colourful image that has been chosen for this exercise. It is compressed through the JPEG steps of DCT followed by quantization, and then it is recovered by the inverse process. Figure 1.70 shows the result.

The program with the JPEG simulation visualizes also the "energies" (sum of absolute values of matrix elements) of the original image, its compressed representation in DCT domain, and the recovered image (Fig. 1.71).

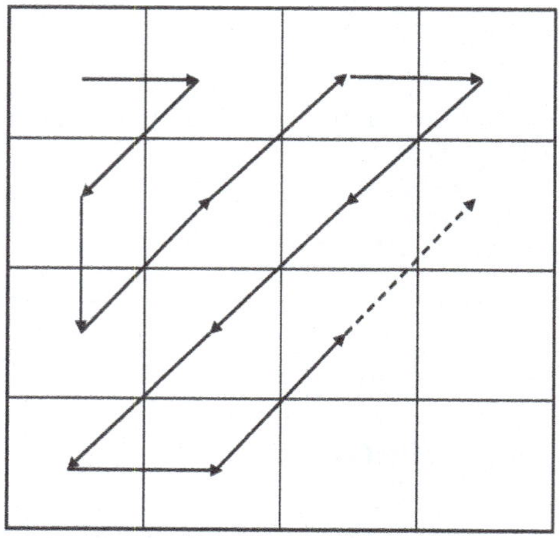

Fig. 1.68 Zig-zag encoding

Fig. 1.69 Original photograph

Fig. 1.70 Recovered photograph

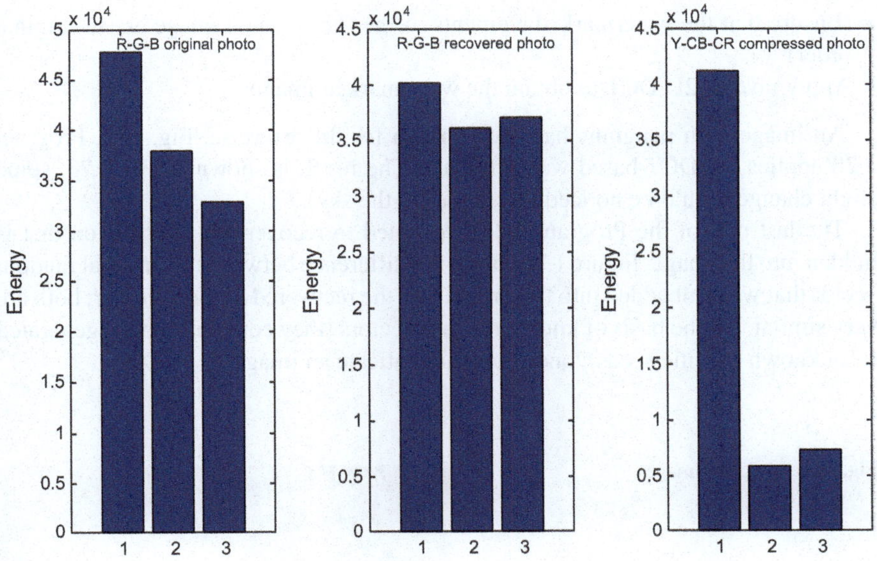

Fig. 1.71 "Energies" of representations

1.9.3 Watermarking

There are methods for embedding information into another signal. This can be used for authentication of images or for copyright protection; and it is called watermarking (which can be visible or not). On the other hand, the purpose could be to transmit some information under disguise; this is called steganography.

To illustrate such methods, two basic techniques have been selected. The first one is based on DCT, and the second on least significant bits (LSB).

1.9.4 Watermarking in Spectral Domain

Let us summarize the basic steps of DCT-based watermarking of an image:

- Read the image F
- Generate a watermark vector W of a certain length L. In our example, this will be a vector of random numbers
- Apply the 2D DCT to the image. You obtain DF
- Locate the L largest elements of DF (using the MATLAB function *sort(.)*)
- Watermarking of these L largest elements:

$$\mathbf{x}' = \mathbf{x} \cdot (1 + \alpha \cdot W) \tag{1.305}$$

- (Ensure that the watermarked elements of the transformed image are in original order)
- Apply inverse 2D DCT to obtain the watermarked image.

An image with penguins has been chosen for this exercise, Fig. 1.72. Program 1.28 applies the DCT-based watermarking. The result is shown in Fig. 1.73, some slight changes could be noticed (for instance, the sky).

The last part of the Program 1.28 is devoted to recover the information that is hidden into the image. Figure 1.74 shows the difference between the original random vector that was embedded into the image, and the recovered random vector; both are very similar. On the basis of known random vectors (they could have been generated from known specific seeds), one could authenticate an image.

Fig. 1.72 Original image

Fig. 1.73 Image with watermark

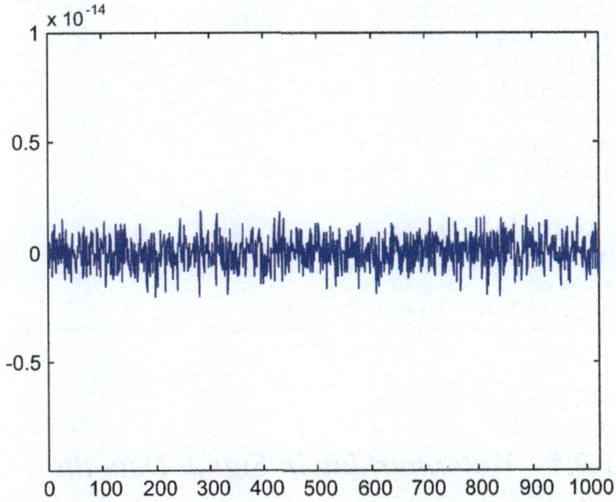

Fig. 1.74 Comparison between original and extracted watermark

Program 1.28 Watermarking via 2D DCT

```
% Watermarking via 2D DCT
%
P=imread('pengs.jpg'); %the cover
F=im2double(P);
[r,c]=size(F);
L=1024;
W=randn(1,L); %the watermark
DF=dct2(F); %2D DCT of the cover
vDF=reshape(DF,1,r*c); %convert to vector format
lv=length(vDF);
[v,ix]=sort(abs(vDF)); %vector reordering
cx=ix(lv-L:lv-1); %select L vector components (except DC entry)
%find which DF entries correspond to selected components
iDF=zeros(L,2);
for ne=1:L,
ic=floor(cx(ne)/r)+1; %column
ir=mod(cx(ne),r); %row
iDF(ne,1)=ir; iDF(ne,2)=ic;
end;
%Insert the watermark
alpha=0.1; %parameter
DFW=DF;
for ne=1:L,
ir=iDF(ne,1); ic=iDF(ne,2);
DFW(ir,ic)=DFW(ir,ic)+ (alpha*DFW(ir,ic)*W(ne));
end;
%inverse 2D DCT
FW=idct2(DFW);
%extract the watermark
Wex=zeros(1,L);
for ne=1:L,
ir=iDF(ne,1); ic=iDF(ne,2);
Wex(ne)=((DFW(ir,ic)/DF(ir,ic))-1)/alpha;
end;
```

```
difW=W-Wex; %for checking
%display
figure(1)
imshow(F);
title('original image');
figure(2)
imshow(FW)
title('image with watermark');
figure(3)
plot(difW);
title('W-Wex');
axis([0 L -1e-14 1e-14]);
```

1.9.5 Watermarking in Signal Domain

The LSB watermarking technique is simple: the least significant bits of each pixel of the cover image are replaced by the information or image to be embedded. These changes on the least significant bits would be almost unnoticeable.

For instance, Fig. 1.75 shows an image of the penguins that contains hidden information. The LSB watermarking has been done with the Program 1.29.

The last part of the Program 1.29 is used to extract the hidden information from the watermarked penguins image. The result is shown in Fig. 1.76.

Fig. 1.75 Image with watermark

Program 1.29 Watermarking via image bits

```
% Watermarking via image bits
%
P=imread('pengs.jpg'); %the cover (256x256 image)
F=im2uint8(P);
S=imread('sign1.jpg'); %the message (256x256 binary image)
M=im2uint8(S>30); %threshold
%embed watermark
for nr=1:256,
for nc=1:256,
aux=bitget(M(nr,nc),1); %read bit of message
%apply that bit to LSB of cover:
Q(nr,nc)=bitset(F(nr,nc),1,aux);
end;
end;
%extract watermark
W=zeros(256,256);
for nr=1:256,
for nc=1:256,
aux=bitget(Q(nr,nc),1); %read LSB of watermarked image
if aux==1, W(nr,nc)=256; end; %recover message
end;
end;
%display
figure(1)
imshow(Q);
title('image with watermark');
figure(2)
imshow(W*256);
title('the hidden message');
```

There are other watermarking techniques, of which some are variations or hybridizations of the two basic techniques just introduced. The topic has attracted much attention, and so there is a lot of literature on watermarking [6, 26, 31].

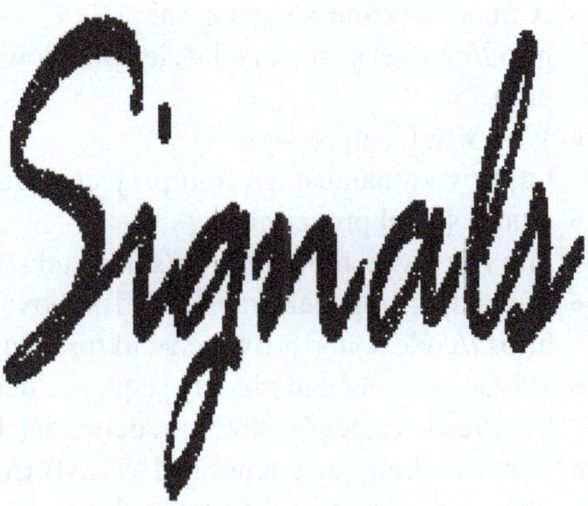

Fig. 1.76 The hidden message (the watermark)

1.10 Resources

1.10.1 MATLAB

Usually the toolboxes include tables of coefficients for the different wavelet families.

1.10.1.1 Toolboxes

- Filter-bank and wavelet toolbox at MATLAB Central:
  http://www.mathworks.se/matlabcentral/linkexchange/links/11-filter-bank-and-wavelet-toolbox
- Auditory Toolbox:
  http://cnmat.berkeley.edu/link/3630
- Auditory Modelling Toolbox:
  http://amtoolbox.sourceforge.net/
- Chroma Toolbox:
  http://www.mpi-inf.mpg.de/resources/MIR/chromatoolbox/
- LT-Toolbox:
  http://www.students.tut.fi/~jalhava/lt/intro.html
- The Large Time-Frequency Analysis Toolbox (LTFAT):
  http://ltfat.sourceforge.net/doc/index.php
- DirLOT Toolbox:
  http://telecom0.eng.niigata-u.ac.jp/index.php?DirLOT
- Image Compression Tools for MATLAB (ICTools):
  http://www.ux.uis.no/~karlsk/ICTools/ictools.html

1.10.1.2 Matlab Code

- A filter bank implementation:
  http://citeseerx.ist.psu.edu/viewdoc/download?doi=10.1.1.57.5728&rep=rep1&type=pdf
- W.A.V.S. Compression:
  http://www.aamusings.com/project-documentation/wavs/index.html
- Audio signal processing:
  http://www.tech.plym.ac.uk/spmc/links/matlab/matlab_examples.html
- Lapped Orthogonal Transform Toolbox:
  https://code.soundsoftware.ac.uk/projects/lots/embedded/lappedwindow.html
- JPEG encoding and zig-zag sequence demo:
  http://cs.uccs.edu/~cs525/dct/dctmatlab.html
- Watermarking programs at MATLAB Central:
  http://www.mathworks.com/matlabcentral/fileexchange/index?term=tag3Awatermarking

1.10.2 Internet

The web site of V. Kepuska, Florida Inst. Tech., contains extensive material related to signal and speech processing (https://www.my.fit.edu/~vkepuska/ece5525/).

See the Paolo D'Incau's blog for watermarking code and details.

It is also interesting to see the web site of the PHYDYAS Project on cognitive radio.

- Laurent Duval:
  http://www.laurent-duval.eu/misc-research-codes.html
- Marek Parfieniuk:
  http://aragorn.pb.bialystok.pl/~marekpk/
- Julius Orion Smith III:
  https://ccrma.stanford.edu/~jos/

References

1. T. Aach, Fourier, block and lapped transforms, in *Advances in Imaging and Electron Physics*, ed.by P.W. Hawkes (Academic Press, 2003)
2. N. Ahmed, T. Natarajan, K.R. Rao, Discrete cosine transform. IEEE T. Comput. 90–93, 1974
3. I. Balasingham, T. Ramstad, J. Lervik, Survey of odd and even length filters in tree-structured filter banks for subband image compression, in *Proceedings of the IEEE International Conference Acoustics, Speech and Signal Processing*, vol. 14, pp. 3073–3076 (1997)
4. C. Brislawn, A simple lattice architecture for even-order linear-phase perfect reconstruction filter banks, in *Proceedings of the IEEE International Symposium Time-Frequency and Time-Scale Analysis*, pp 124–127 (1994)
5. W.H. Chen, C.H. Smith, S.C. Fralick, A fast computational algorithm for the discrete cosine transform. IEEE T. Commun. **125**(9):1004–1009 (1977)
6. J. Cox, M.L. Miller, J.A. Boom, J. Fridrich, T. Kalker, *Digital Watermarking and Stegonography* (Morgan Kaufmann, 2007)
7. H.V. Dwivedi, *Design of JPEG Compressor* (National Institute of Technology, Rourkela, India, 2009) Bachl. thesis
8. A. Fettweis, H. Levin, A. Sedlmeyer, Wave digital lattice filters. Intl. J. Circuit Theory Appl. **2**, 203–211 (1974)
9. S. Foucart, *Linear Algebra and Matrix Analysis* (Math 504 Lectures Notes, Drexel University, 2010). http://www.math.drexel.edu/foucart/teaching.htm
10. F. Galijasevic, *Allpass-Based Near-Perfect-Reconstruction Filter Banks*. PhD thesis (Cristian-Albrechts University, Kiel, Germany, 2002)
11. X. Gao, T.Q. Nguyen, G. Strang, On factorization of M-channel paraunitary filterbanks. IEEE T. Sign. Process. **49**(7), 1433–1446 (2001)
12. H. Haberdar, *Discrete Cosine Transform Tutorial* (2013). www.haberdar.org/Discrete-Cosine_Transform-Tutorial.htm
13. C. Herley, M. Vetterli, Wavelets and recursive filter banks. IEEE T. Sign. Process. **41**(8), 2536–2556 (1993)
14. S.A. Khayam, *The Discrete Cosine Transform (DCT): Theory and Applications*. (Michigan State University, 2003). Notes of the ECE 802-602 course
15. J. Kliewer, E. Brka, Near-perfect reconstruction low-complexity two-band IIR/FIR QMF banks with FIR phase-compensation filters. Sig. Process. **86**, 171–181 (2005)

16. F. Kurth, M. Clausen, Filter bank tree and M-band wavelet packet algorithms in audio signal processing. IEEE T. Sign. Process. **47**(2), 549–554 (1999)
17. B.G. Lee, A new algorithm to compute the discrete cosine transform. IEEE T. Acoust., Speech, Sign. Process. **32**(6), 1243–1245 (1984)
18. A.B. Lewis, *JPEG Tutorial* (The Computer Laboratory: Topics in Security: Forensic Signal Analysis, University of Cambridge, UK., 2010). https://www.cl.cam.ac.uk/teaching/1011/R08/jpeg/acs10-jpeg.pdf
19. Y.P. Lin, P.P. Vaidyanathan, A Kaiser window approach for the design of prototype filters of cosine modulated filter banks. IEEE Sign. Process. Lett. **5**(6), 132–134 (1998)
20. H.W. Löllmann, P. Vary, Design of IIR QMF filter banks with near-perfect reconstruction and low complexity, in *Proceedings of the IEEE International Conference Acoustics, Speech and Signal Processing*, pp. 3521–3524 (2008)
21. H.S. Malvar, Lapped transforms for efficient transform/subband coding. IEEE T. Acoust., Speech, Sign. Process. **38**(6), 969–978 (1990)
22. H.S. Malvar, D.H. Staelin, The LOT: Transform coding without blocking effects. IEEE T. Acoust., Speech, Sign. Process. **37**(4), 553–559 (1989)
23. J. Mau, Perfect reconstruction modulated filter banks: Fast algorithms and attractive new properties, in *Proceedings of the IEEE International Conference Acoustics, Speech and Signal Processing*, pp. 225–228 (1993)
24. A. Mertins, *Signal Analysis, Filter Banks, Time-Frequency Transforms and Applications* (John Wiley, 1999)
25. A. Mouffak, M.F. Belbachir, Noncausal forward/backward two-pass IIR digital filters in real time. Turk J. Elec. Eng. Comput. Sci. **20**(5), 769–786 (2012)
26. P.M. Naini, Digital watermarking using MATLAB. in *Engineering Education and Research Using MATLAB*, ed. by A. Assi (InTech, 2011) Chap. 20
27. T.Q. Nguyen, *A Tutorial on Filter Banks and Wavelets* (University of Wisconsin, ECE Department, 1995). http://www.cs.tau.ac.il/~amir1/WAVELET/PAPERS/nguyen95tutorial.pdf
28. T.Q. Nguyen, P.P. Vaidyanathan, Two channel perfect reconstruction IR QMF structures which yield linear phase analysis and synthesis filters. IEEE T. Acoust., Speech, Sign. Process. 476–492 (1989)
29. S. Oraintara, D. Trans, P.N. Heller, T.Q. Nguyen, Lattice structure for regular paraunitary linear-phase filterbanks and M-band orthogonal symmetric wavelets. IEEE T. Sign. Process. **49**(11), 2659–2672 (2001)
30. W. Pennebaker, J. Mitchell, *JPEG: Still Image Data Compression Standard* (Springer, 1992)
31. C.I. Podilchuk, E.J. Delp, Digital watermarking algorithms and applications. IEEE Sign. Process. Mgz. **18**(4), 33–46 (2001)
32. R.L. Queiroz, T.Q. Nguyen, K.R. Rao, The GenLOT: Generalized linear-phase lapped orthogonal transform. IEEE T. Sign. Process. **44**(3), 497–507 (1996)
33. R.L. Queiroz, T.D. Tran, Lapped transforms for image compression, in *Handbook on Transforms and Data Compression* (CRC, 2000), pp. 1–64
34. C.M. Rader, The rise and fall of recursive digital filters. IEEE Sign. Process. Mgz. 46–49 (2006)
35. K.R. Rao, P. Yip, *Discrete Cosine Transform. Algorithms, Advantages, Applications* (Academic Press, 1990)
36. P.A. Regalia, S.K. Mitra, P.P. Vaidyanathan, The digital all-pass filter: A versatile signal processing building block. Proc. IEEE **76**(1), 19–37 (1988)
37. K. Sankar, *Using Toeplitz Matrices in MATLAB* (2007). http://www.dsplog.com/2007/04/21/using-toeplitz-matrices-in-matlab/
38. I.W. Selesnick, Formulas for orthogonal IIR wavelet filters. IEEE T. Sign. Process. **46**(4), 1138–1141 (1998)
39. U. Sezen, Perfect reconstruction IIR digital filter banks supporting nonexpansive linear signal extensions. IEEE T. Sign. Process., **57**(6), 2140–2150 (2009)
40. U. Sezen, S.S. Lawson, Anticausal inverses for digital filter banks, in Proc. Eur. Conf. Circuit Theory Des. 1–229 to 1–232 (2001)

41. M.J.T. Smith, T.P.III Barnwell, Exact reconstruction techniques for tree-structured subband coders. IEEE T. Acoust., Speech Sign. Process. **34**(3), 434–441 (1986)
42. A.K. Soman, P.P. Vaidyanathan, T.Q. Nguyen, Linear phase paraunitary filter banks: Theory, factorizations and design. IEEE T. Sign. Process. **41**(12), 3480–3496 (1993)
43. T.D. Tran, R.L. Queiroz, T.Q. Nguyen, Linear-phase perfect reconstruction filter bank: Lattice structure, design, and application in image coding. IEEE T. Sign. Process. **48**(1), 133–147 (2000)
44. P.P. Vaidyanathan, Multirate digital filters, filter banks, polyphase networks, and applications: A tutorial. Proc IEEE **78**(1), 56–93 (1990)
45. P.P. Vaidyanathan, *Multirate Systems and Filter Banks* (Prentice-Hall, 1992)
46. P.P. Vaidyanathan, T. Chen, Structures for anticausal inverses and application in multirate filter banks. IEEE T. Sign. Process. **46**(2), 507–514 (1998)
47. P.P. Vaidyanathan, S.K. Mitra, Y. Neuvo, A new approach to the realization of low-sensitivity IIR digital filters. IEEE T. Acoust., Speech, Sign. Process. **34**(2), 350–361 (1986)
48. M. Vetterli, C. Herley, Wavelets and filter banks; theory and design. IEEE T. Sign. Process. **40**(9), 2207–2232 (1992)
49. M. Vetterli, J. Kovacevic, *Wavelets and Subband Coding* (Prentice Hall, 1995)
50. A.B. Watson, Image compression using the discrete cosine transform. Math. J. **4**(1), 81–88 (1994)
51. Z. Xu, A. Makur, On the closeness of the space spanned by the lattice structures for a class of linear phase perfect reconstruction filter banks. Sign. Process. **90**, 292–302 (2010)
52. X. Zhang, T. Yoshikawa, Design of two-channel IIR Linear phase PR filter banks. Sign. Process. **72**, 167–175 (1999)
53. D. Zhou, A review of polyphase filter banks and their application. Technical report, Air Force Technical USA, 2006. AFRL-IF-RS-TR-277

Chapter 2
Wavelets

2.1 Introduction

Wavelets have attracted a lot of attention from people involved in time-frequency analysis of signals. The literature on wavelets, books, papers, is quite extensive. Many practical applications of wavelets have been found.

The purpose of this chapter is to introduce the fundamentals of wavelets. As in previous chapters, it is preferred to start with basic examples, and then to proceed with generalization and formalization of concepts.

Continuous periodic signals can be well represented as Fourier expansions. Other signals may be better represented with alternative expansions. Wavelets are functions that are localized both in time and in frequency, providing pieces for a new kind of expansions.

The main idea can be illustrated with the tiling of the time-frequency plane. When using the STFT, the tiling is as depicted in Fig. 2.1.

The tiles have an area $\Delta t \cdot \Delta f = A = constant$. Due to uncertainty, if you want to narrow Δt then you have to enlarge Δf, and vice-versa.

If you are analyzing a signal with significant low frequency components, then Δt must be relatively large (to accommodate at least one period of the low frequency component), and so you cannot study possible variations of high frequency components inside a Δt.

With wavelets the time-frequency tiling is as depicted in Fig. 2.2.

Tiles adapt to the signal components. Large Δt and narrow Δf for low-frequency components. Narrow Δt and wide Δf for high-frequency components. Therefore, a better study of the variations of high frequency components is allowed.

Now, let us introduce some formal concepts, which are obviously related with the tiling just described.

This chapter does not restrict to periodic functions but, instead, to the space L^2 of square-integrable functions. This refers to finite-energy signals.

A function $y(x)$ is said to have *support* in the interval $[x_1, x_2]$ if $y(x) = 0$ when x is outside the interval and $[x_1, x_2]$ is the smallest closed interval for which this

© Springer Science+Business Media Singapore 2017

J.M. Giron-Sierra, *Digital Signal Processing with Matlab Examples, Volume 2*, Signals and Communication Technology, DOI 10.1007/978-981-10-2537-2_2

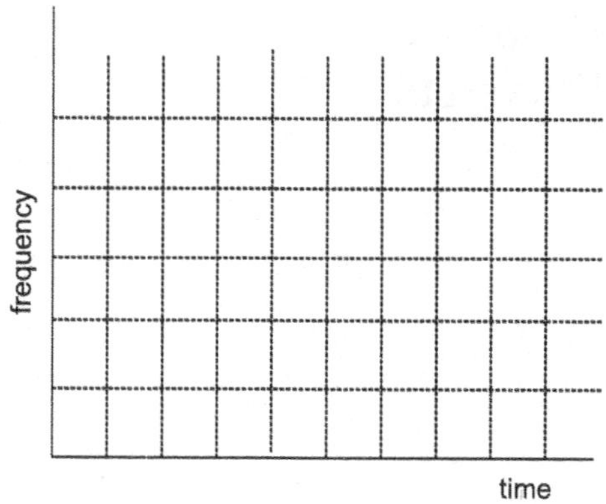

Fig. 2.1 Tiling of the TF plane corresponding to the STFT

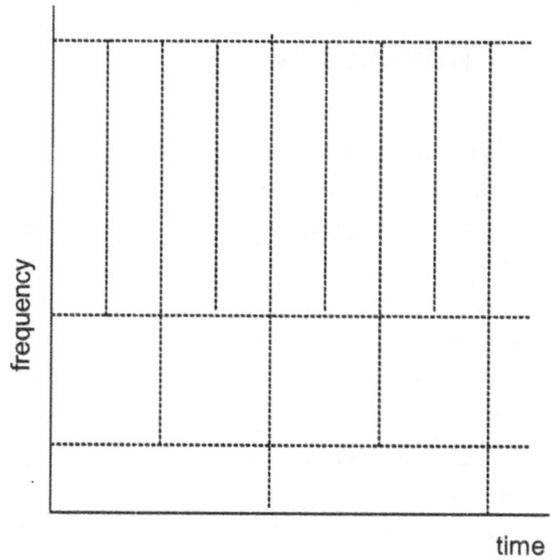

Fig. 2.2 Tiling of the TF plane corresponding to wavelets

holds. Compact support or finite support corresponds to finite intervals. This refers to finite-time signals, $y(t)$ in $[t_1, t_2]$, and to band-limited signals, $Y(\omega)$ in $[\omega_1, \omega_2]$.

Wavelets $\psi(t)$ are zero-mean real valued functions of time, which correspond in the frequency domain to band-pass behaviour:

$$\int_{-\infty}^{\infty} \psi(t)\, dt = 0 \tag{2.1}$$

It is convenient for many applications that ψ(t) had several vanishing moments:

$$m_1(k) = \int_{-\infty}^{\infty} t^k\, \psi(t) dt = 0;\, k = 0,\ 1,\ \ldots m \tag{2.2}$$

As it will be described in this chapter, a signal can be decomposed into wavelets (analysis) and then this signal can be recovered from these wavelets (synthesis).

The wavelet analysis of signals can be obtained in real time with filter banks.

Filter banks provide a way of understanding wavelets. In particular, the analysis-synthesis of signals via wavelets is related with the perfect reconstruction (PR) using filter banks.

In the next section, the Haar wavelet will be introduced. It is direcly related with the QMF filter bank, which is the only PR filter bank being FIR, orthogonal, and linear phase. The Haar wavelet is also unique in terms of compact support, orthogonality, and symmetry.

In addition to the Haar wavelet, the chapter will explore other orthogonal and biorthogonal wavelets, like it was done in the previous chapter with orthogonal and biorthogonal filter banks.

During the 1980s and 1990s, a lot of research work focused on the STFT. It was in this context that the wavelet analysis was introduced, in 1984, by Grossman and Morlet [49]. It soon awakened large interest, and then important contributions to a theory of wavelets came from Meyer [47], Daubechies [24], Mallat [45], and others. Initial efforts were oriented to orthogonal bases. During the 1990s the research departed from the limitations imposed by orthogonality, and many special versions and structures of wavelets appeared.

One of the main fields of applications is image processing. There are special versions of wavelets for this field. Next chapters will consider image processing, and will include pertinent wavelet variants.

There is a MATLAB toolbox for wavelets. This toolbox will be considered by the end of the chapter. For didactical reasons, it was preferred not to use this toolbox along the chapter, but instead to include our own MATLAB programs.

A lot of papers and books about wavelets have been published. This chapter is mostly based on [14, 25, 44, 48], with the complementary support of publications to be mentioned on proper places.

2.2 An Important Example: The Haar Wavelets

The Haar wavelets were introduced by Alfred Haar in 1909 (the term '*wavelet*' was not coined till later on).

2.2.1 Definitions

The Haar *wavelet function* is:

$$\psi(t) = \begin{cases} 1, & 0 \leq t < 0.5 \\ -1, & 0.5 \leq t < 1 \\ 0, & elsewhere \end{cases} \tag{2.3}$$

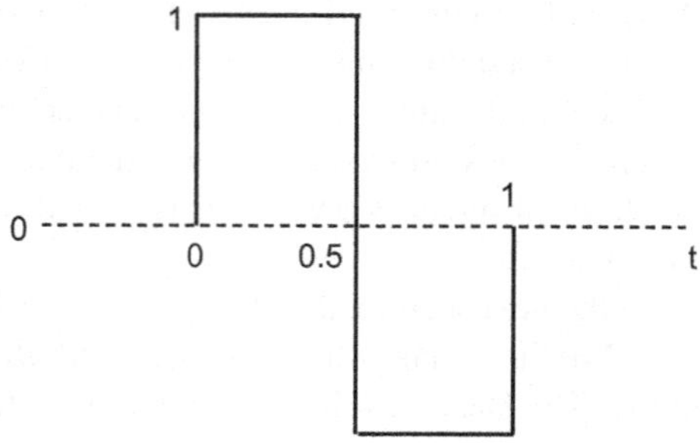

Fig. 2.3 The Haar wavelet

The function $\psi(t)$ is a 'mother wavelet'.

A plot of the Haar mother wavelet is shown in Fig. 2.3. Notice that the integral of $\psi(t)$ over its domain is 0.

'Baby wavelets' are obtained with scaling and shifting as follows:

$$\psi_{j,k}(t) = \sqrt{2^j}\,\psi(2^j t - k), \quad j = 0, 1, 2, \ldots\ (scale);\ k = 0, 1, 2, \ldots, 2^j - 1\ (shift) \tag{2.4}$$

Figure 2.4 shows examples of scaling and shifting.

Every baby wavelet has:

$$\int_0^1 (\psi_{j,k}(t))^2 dt = 1 \tag{2.5}$$

It can be shown that the set $\psi_{j,k}(t)$ is an orthonormal basis for L^2. A function $y(t)$ can be represented in terms of the Haar basis as follows:

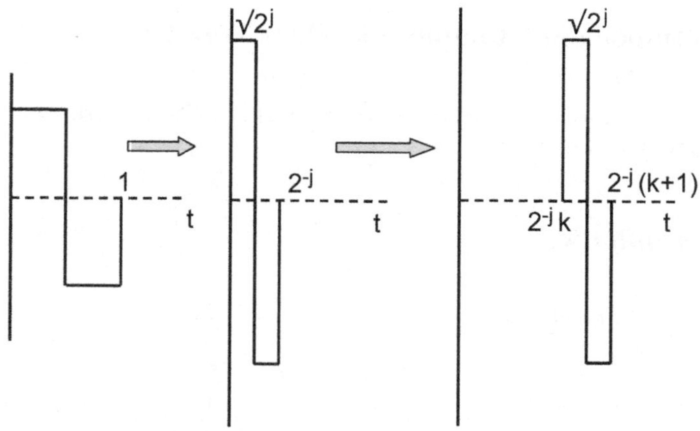

Fig. 2.4 Baby wavelets: scaling and shifting

Fig. 2.5 The Haar scaling function

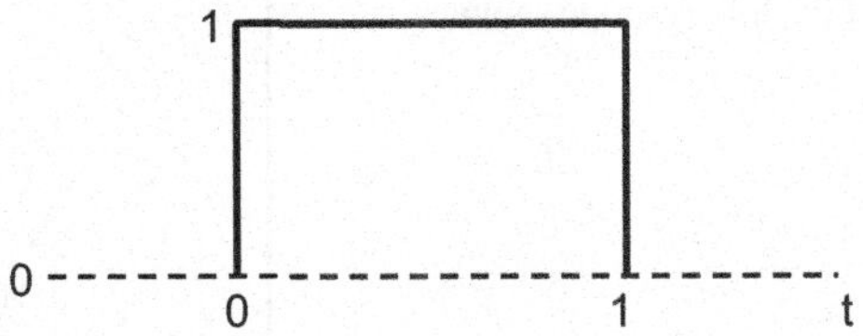

$$y(t) = \sum_{j,k} < y,\, \psi_{j,k} > \cdot \psi_{j,k}(t) = \sum_{j,k} d_{j,k} \cdot \psi_{j,k}(t) \tag{2.6}$$

where $d_{j,k}$ are the Haar wavelet coefficients.

The Haar wavelet has a companion function, named *scaling function* $\varphi(t)$, which is defined by (Fig. 2.5).

$$\varphi(t) = \begin{cases} 1, & 0 \le t < 1 \\ 0, & elsewhere \end{cases} \tag{2.7}$$

Also:

$$\varphi_{j,k}(t) = \sqrt{2^j}\,\varphi(2^j t - k), \quad j = 0, 1, 2, \ldots; \quad k = 0, 1, 2, \ldots, 2^j - 1 \tag{2.8}$$

and:

$$\int_0^1 (\varphi_{j,k}(t))^2 dt = 1 \tag{2.9}$$

The Haar wavelet and scaling function satisfy the identities:

$$\begin{aligned} \psi(t) &= \varphi(2t) - \varphi(2t - 1) \\ \varphi(t) &= \varphi(2t) + \varphi(2t - 1) \end{aligned} \tag{2.10}$$

Suppose a signal $y(t)$, then:

$$a_{j,k} = < y,\, \varphi_{j,k} > = \int y(t)\, \varphi_{j,k}(t)\, dt \tag{2.11}$$

is a set of 'average samples' of the signal (each $a_{j,k}$ is an average value of $y(t)$ over adjacent intervals). Figure 2.6 shows an example of average samples.

2.2.2 *Multiresolution Analysis*

There is a simple procedure to obtain function representations in terms of wavelets. The idea is to consider a set of subspaces corresponding to several resolution levels-that is multiresolution. Higher scale values, which corresponds to higher frequencies, allow for more time resolution (more values of k).

Fig. 2.6 Average samples

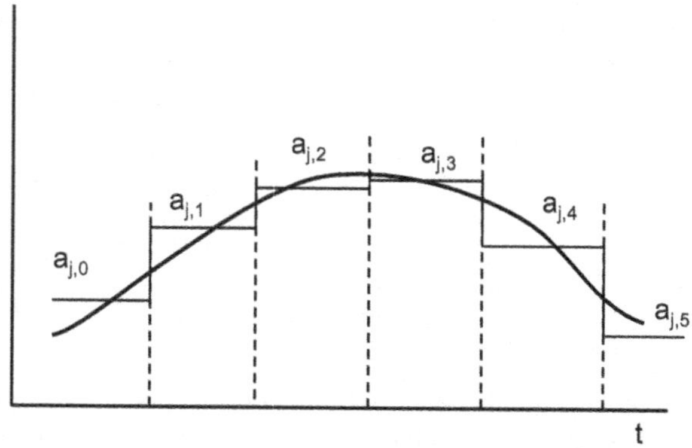

Let us advance step by step, with theory and examples.

Define the function space V_j as follows:

$$V_j = span\ \{\varphi_{j,k}\}_{k=0,\,1,\,2,\ldots,\,2^j-1} \tag{2.12}$$

where '*span*' is the linear span.

It can be shown that (Fig. 2.7): $V_j \subseteq V_{j+1}$

The set:

$$\{\varphi_{j,k}(t)\}_{k=0,\,1,\,2,\ldots,\,2^j-1} \tag{2.13}$$

is an orthonormal basis for V_j.

But the set:

$$\{\varphi_{j,k}(t)\}_{j=0,\,1,\,2\ldots;\,k=0,\,1,\,2,\ldots,\,2^j-1} \tag{2.14}$$

is not an orthonormal basis for L^2 because the spaces V_j are not mutually orthogonal.

Define the wavelet subspace W_j as the orthogonal complement of V_j in V_{j+1}. That is (Fig. 2.8):

$$V_{j+1} = V_j \oplus W_j \quad (\oplus\ orthogonal\, sum) \tag{2.15}$$

It can be shown that the set:

Fig. 2.7 Function spaces

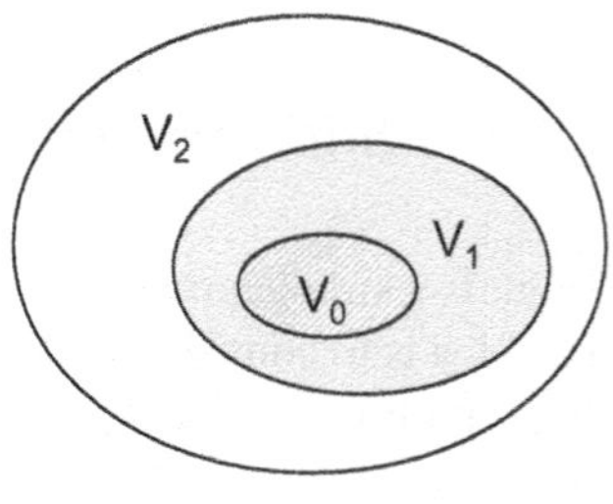

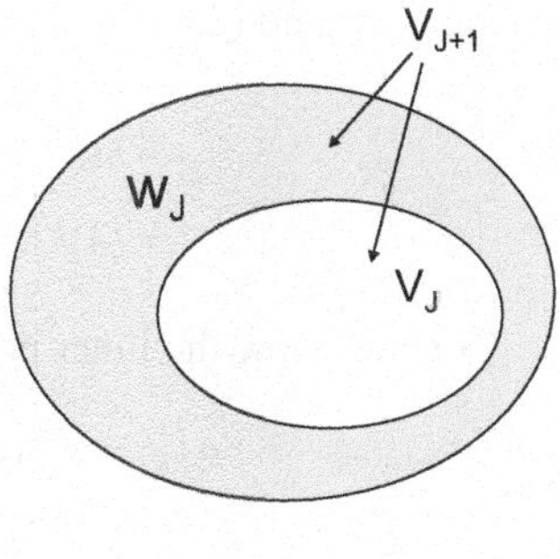

Fig. 2.8 Complementary spaces V_j and W_j

$$\{\psi_{j,k}(t)\}_{k=0,1,2,\ldots,2^j-1} \tag{2.16}$$

is an orthonormal basis for W_j.

In consequence it is possible to combine wavelets and scaling functions to generate V_{j+1} from V_j and W_j.

Notice that the decomposition could be iterated, so as:

$$V_{j+1} = V_0 \oplus W_0 \oplus W_1 \oplus W_2 \ldots \oplus W_j \tag{2.17}$$

For example, suppose we have a signal y with 8 samples (2^3 samples). The number of samples gives the maximum resolution; thus $j+1=3$. Let us expand this signal y into $V_3 = V_0 \oplus W_0 \oplus W_1 \oplus W_2$:

$$\begin{aligned} y =&< y, \varphi_{0,0} > \varphi_{0,0} + < y, \psi_{0,0} > \psi_{0,0} + < y, \psi_{1,0} > \psi_{1,0} + \\ &+ < y, \psi_{1,1} > \psi_{1,1} + < y, \psi_{2,0} > \psi_{2,0} + < y, \psi_{2,1} > \psi_{2,1} + \\ &+ < y, \psi_{2,2} > \psi_{2,2} + < y, \psi_{2,3} > \psi_{2,3} \end{aligned} \tag{2.18}$$

Fortunately it is not necessary to evaluate each inner product in the previous equation using integrals. There is a useful alternative that will be described now with some theory continuation and a practical example.

To simplify the notation, the Eq. (2.18) is written as follows:

$$\begin{aligned} y =& a_{0,0}\,\varphi_{0,0} + d_{0,0}\,\psi_{0,0} + d_{1,0}\,\psi_{1,0} + d_{1,1}\,\psi_{1,1} + \\ &+ d_{2,0}\,\psi_{2,0} + d_{2,1}\,\psi_{2,1} + d_{2,2}\,\psi_{2,2} + d_{2,3}\,\psi_{2,3} \end{aligned} \tag{2.19}$$

where $a_{j,k}$ correspond to average samples, and $d_{j,k}$ correspond to details or differences (this will be explained later on).

Substitute $2^j t - k$ in lieu of t in the identities (2.10). Also, multiply both sides of equations by $\sqrt{2^j}$:

$$\begin{aligned} \sqrt{2^j}\,\psi(2^j t - k) &= \tfrac{1}{\sqrt{2}}[\sqrt{2^{j+1}}\,\varphi(2^{j+1} t - 2k) - \sqrt{2^{j+1}}\,\varphi(2^{j+1} t - 2k - 1)] \\ \sqrt{2^j}\,\varphi(2^j t - k) &= \tfrac{1}{\sqrt{2}}[\sqrt{2^{j+1}}\,\varphi(2^{j+1} t - 2k) + \sqrt{2^{j+1}}\,\varphi(2^{j+1} t - 2k - 1)] \end{aligned} \tag{2.20}$$

Using (2.4) and (2.8) the above equation can be expressed in a simpler form:

$$\begin{aligned}\psi_{j,k}(t) &= \tfrac{1}{\sqrt{2}}[\varphi_{j+1,2k}(t) - \varphi_{j+1,2k+1}(t)] \\ \varphi_{j,k}(t) &= \tfrac{1}{\sqrt{2}}[\varphi_{j+1,2k}(t) + \varphi_{j+1,2k+1}(t)]\end{aligned} \tag{2.21}$$

Consider now the inner products of (2.18):

$$\begin{aligned}a_{j,k} &=< y,\, \varphi_{j,k} >= \int y(t)\, \varphi_{j,k}(t)\, dt = \\ &= \int y(t) \tfrac{1}{\sqrt{2}}[\varphi_{j+1,2k}(t) + \varphi_{j+1,2k+1}(t)]\, dt = \\ &= \tfrac{1}{\sqrt{2}}\, a_{j+1,2k} + \tfrac{1}{\sqrt{2}}\, a_{j+1,2k+1}\end{aligned} \tag{2.22}$$

$$\begin{aligned}d_{j,k} &=< y,\, \psi_{j,k} >= \int y(t)\, \psi_{j,k}(t)\, dt = \\ &= \int y(t) \tfrac{1}{\sqrt{2}}[\varphi_{j+1,2k}(t) - \varphi_{j+1,2k+1}(t)]\, dt = \\ &= \tfrac{1}{\sqrt{2}}\, a_{j+1,2k} - \tfrac{1}{\sqrt{2}}\, a_{j+1,2k+1}\end{aligned} \tag{2.23}$$

Equations (2.22) and (2.23) tell us that it is possible to obtain from the $a_{j,k}$ coefficients corresponding to scale j, the $a_{j-1,k}$ and $d_{j-1,k}$ coefficients corresponding to scale $j - 1$. Again, from the coefficients corresponding to scale $j - 1$ it is possible to obtain the coefficients for scale $j - 2$. And so on. This results in an easy iterative procedure to compute the wavelet expansion (2.18).

Let us illustrate the iterative method putting numbers for the 8 samples of the signal being considered so far:

$$y = \{3,\ 7,\ 1,\ 15,\ 2,\ 3,\ 6,\ 9\} \tag{2.24}$$

where integers have been used for simplicity.

The set of numbers (2.24) is also the set of $a_{j,k}$ coefficients corresponding to the scale $j = 3$. Let us compute the coefficients for the scale $j - 1$:

$$\begin{aligned}a_{2,0} &= \tfrac{1}{\sqrt{2}}\,(3 + 7) = \tfrac{10}{\sqrt{2}}; \quad a_{2,1} = \tfrac{1}{\sqrt{2}}\,(1 + 15) = \tfrac{16}{\sqrt{2}} \\ a_{2,2} &= \tfrac{1}{\sqrt{2}}\,(2 + 3) = \tfrac{5}{\sqrt{2}}; \quad a_{2,3} = \tfrac{1}{\sqrt{2}}\,(6 + 9) = \tfrac{15}{\sqrt{2}}\end{aligned} \tag{2.25}$$

$$\begin{aligned}d_{2,0} &= \tfrac{1}{\sqrt{2}}\,(3 - 7) = \tfrac{-4}{\sqrt{2}}; \quad d_{2,1} = \tfrac{1}{\sqrt{2}}\,(1 - 15) = \tfrac{-14}{\sqrt{2}} \\ d_{2,2} &= \tfrac{1}{\sqrt{2}}\,(2 - 3) = \tfrac{-1}{\sqrt{2}}; \quad d_{2,3} = \tfrac{1}{\sqrt{2}}\,(6 - 9) = \tfrac{-3}{\sqrt{2}}\end{aligned} \tag{2.26}$$

Expressions (2.25) and (2.26) show why we refer to the coefficients $a_{j,k}$ as averages and $d_{j,k}$ as differences.

Let us continue with the next lower scale:

$$a_{1,0} = \frac{1}{2}\,(10 + 16) = \frac{26}{2}; \quad a_{1,1} = \frac{1}{2}\,(5 + 15) = \frac{20}{2} \tag{2.27}$$

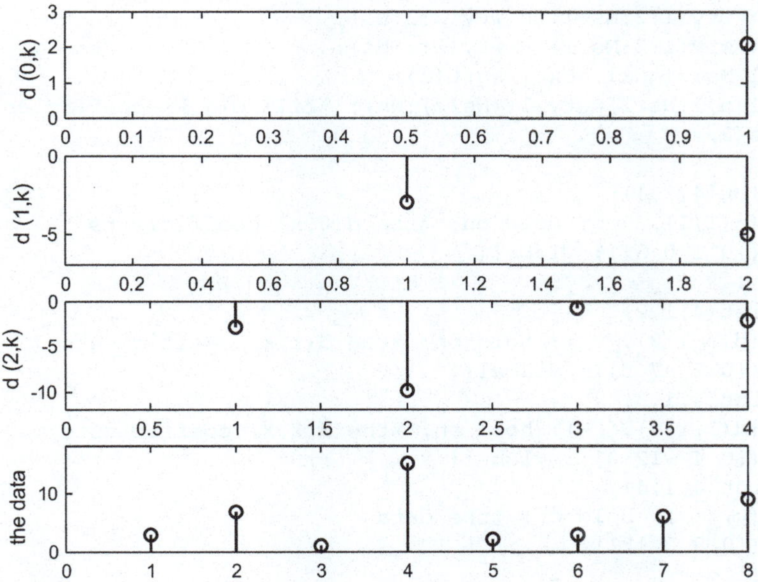

Fig. 2.9 Example of Haar transform

$$d_{1,0} = \frac{1}{2}(10 - 16) = \frac{-6}{2}; \quad d_{1,1} = \frac{1}{2}(5 - 15) = \frac{-10}{2} \tag{2.28}$$

and finally:

$$a_{0,0} = \frac{1}{2\sqrt{2}}(26 + 20) = \frac{46}{2\sqrt{2}} \tag{2.29}$$

$$d_{0,0} = \frac{1}{2\sqrt{2}}(26 - 20) = \frac{6}{2\sqrt{2}} \tag{2.30}$$

In this way, all coefficients in (2.19) have been determined.

The iterative coefficient computation method can be expressed in matrix format or with operators.

The Program 2.1 implements the procedure just described, with the same example. Figure 2.9 depicts the input data and the corresponding d(j, k) coefficients.

Program 2.1 Haar wavelet transform of a data set

```
% Haar wavelet transform of a data set
y=[3,7,1,15,2,3,6,9]; %data set
Ns=8; %number of samples
K=3; %exponent, 8=2^3
wty=y; %space for the wavelet transform
d=zeros(K,Ns/2); %space for d(j,k) coefficients
for n=1:K,
```

```
aux1= wty(1:2:Ns-1) + wty(2:2:Ns);
aux2= wty(1:2:Ns-1) - wty(2:2:Ns);
wty(1:Ns)=[aux1,aux2]/sqrt(2);
d(K+1-n,1:Ns/2)=wty(1+(Ns/2):Ns); %fill d(j,k) coefficients
Ns=Ns/2;
end;
subplot(4,1,1)
stem(d(1,1),'k'); hold on; %the d(0,k) coefficients
axis([0 1 0 3]); ylabel('d(0,k)');
title('Example of Haar wavelet transform');
subplot(4,1,2)
stem(d(2,1:2),'k'); hold on; %the d(1,k) coefficients
axis([0 2 -7 0]); ylabel('d(1,k)');
subplot(4,1,3)
stem(d(3,1:4),'k'); hold on; %the d(2,k) coefficients
axis([0 4 -12 0]); ylabel('d(2,k)');
subplot(4,1,4)
stem(y,'k'); hold on; %the data
axis([0 8 0 18]);
ylabel('the data');
d
wty
```

It is possible to recover the original signal samples from the computed coefficients. Notice that adding of subtracting Eqs. (2.22) and (2.23) the following is obtained:

$$\begin{aligned} a_{j,k} + d_{j,k} &= \frac{2}{\sqrt{2}}\, a_{j+1,2k} \\ a_{j,k} - d_{j,k} &= \frac{2}{\sqrt{2}}\, a_{j+1,2k+1} \end{aligned} \tag{2.31}$$

Consequently the iterative method can be reversed, computing the coefficients at scale 1 from coefficients at scale 0, coefficients at scale 2 from those at scale 1, etc. The last iteration gives the $a_{j,k}$ coefficients at scale j, which are the original signal samples.

The Program 2.2 recovers the original data from the Haar transform for the above example. Figure 2.10 depicts the a(j, k) coefficients.

Program 2.2 Recover data set from Haar wavelet transform

```
% recover data set from Haar wavelet transform
%the wavelet transform data:
wty=[16.2635,2.1213,-3.0000,-5.0000,-2.8284,...
-9.8995,-0.7071,-2.1213];
K=3; %exponent, 8=2^3, for 8 samples
J=K+1; %to adapt to MATLAB indexing
y=wty; %space for the recovered data
a=zeros(J,(2^K)); %space for coefficients
m=1;
a(1,1)=y(1);
for n=1:K,
```

```
a(n+1,1:2:(2*m-1))=(a(n,1:m)+y((1+m):(2*m)))/sqrt(2);
a(n+1,2:2:(2*m))=(a(n,1:m)-y((1+m):(2*m)))/sqrt(2);
m=m*2;
end;
y=a(4,1:8); %the recovered data
subplot(4,1,1)
stem(a(1,1),'k'); hold on; %the a(0,k) coefficients
axis([0 1 0 20]); ylabel('a(0,k)');
title('Example of data recovery');
subplot(4,1,2)
stem(a(2,1:2),'k'); hold on; %the a(1,k) coefficients
axis([0 2 0 15]); ylabel('a(1,k)');
subplot(4,1,3)
stem(a(3,1:4),'k'); hold on; %the a(2,k) coefficients
axis([0 4 0 15]); ylabel('a(2,k)');
subplot(4,1,4)
stem(y,'k'); hold on; %the data
axis([0 8 0 18]);
ylabel('the data');
```

As an exercise let us select a sawtooth signal and obtain its Haar wavelet transform. The Program 2.3 repeats the transform procedure for this case. Figure 2.11 shows the input signal and the corresponding d(j, k) coefficients. Most of the coefficients

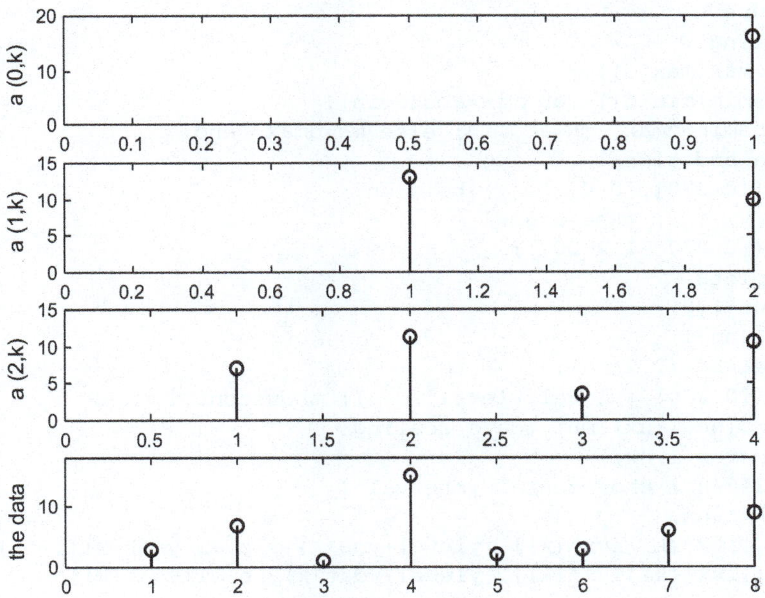

Fig. 2.10 Example of data recovery from Haar transform

are small and difficult to see, but the coefficients corresponding to the brisk changes of the sawtooth are clearly seen.

Program 2.3 Haar wavelet transform of a sawtooth signal

```
% Haar wavelet transform of a signal
% Sawtooth signal
fy=300; %signal frequency in Hz
wy=2*pi*fy; %signal frequency in rad/s
fs=20000; %sampling frequency in Hz
tiv=1/fs; %time interval between samples;
Ns=256; %let us take 256 signal samples
duy=Ns*tiv; %time for 256 samples
t=0:tiv:(duy-tiv); %time intervals set
y=sawtooth(wy*t); %signal data set (256 samples)
K=8; %exponent, 256=2^8
wty=y; %space for the wavelet transform
d=zeros(K,Ns/2); %space for d(j,k) coefficients
%the Haar wavelet transform
for n=1:K,
aux1= wty(1:2:Ns-1) + wty(2:2:Ns);
aux2= wty(1:2:Ns-1) - wty(2:2:Ns);
wty(1:Ns)=[aux1,aux2]/sqrt(2);
d(K+1-n,1:Ns/2)=wty(1+(Ns/2):Ns); %fill d(j,k) coefficients
Ns=Ns/2;
end;
%figure
%scaling
dmax=max(max(d));
dmin=min(min(d)); abdmin=abs(dmin);
if abdmin>dmax, mh=abdmin; else mh=dmax; end;
%area and signal
plot([0 270],[0 0],'b'); hold on;
plot(y,'k'); %the signal
axis([0 270 -1.2 20]);
%subplots
for nn=1:8,
nx=2^(nn-1);
ylevel=20-(2.2*nn);
plot([0 270],[ylevel ylevel],'b'); %horizontal axes
ydat=d(nn,1:nx)/mh; %data scaling
yno=zeros(1,nx);
ivx=256/nx; %horizontal interval
for k=1:nx,
plot([ivx*(k) ivx*(k)],[ylevel+yno(k) ylevel+ydat(k)],'k');
plot([ivx*(k) ivx*(k)],[ylevel+ydat(k)ylevel+ydat(k)],'rx');
end;
end;
title('Haar wavelet transform of sawtooth signal');
```

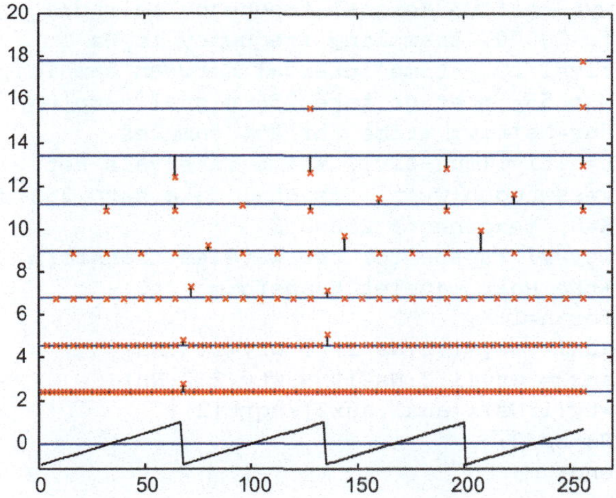

Fig. 2.11 Haar transform of sawtooth signal

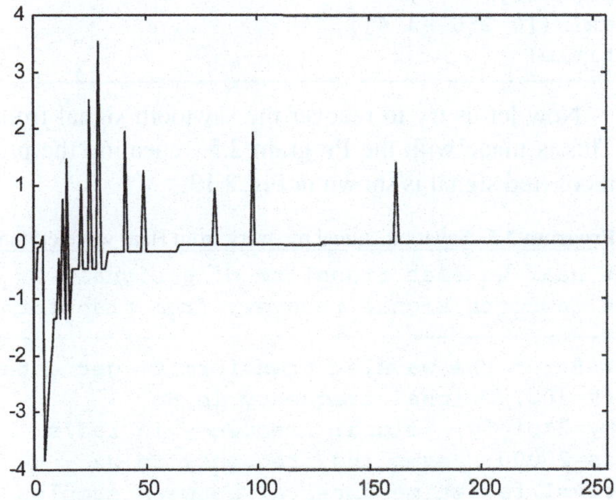

Fig. 2.12 The vector *wty(n)* corresponding to the sawtooth signal

Notice that in the wavelet transform examples just considered, a vector *wty(n)* has been used to contain $a(1, 1), d(1, 1), d(2, 1), d(2, 2), d(3, 1)\ldots$ (indexes according to MATLAB). It is convenient to plot this vector, since it gives quick information of the values, variance, etc. Likewise it may be useful to compare wavelet transform alternatives.

Figure 2.12 shows the vector *wty(n)* obtained by the Haar wavelet transform of the sawtooth signal just considered. The figure has been generated with the Program 2.4.

Program 2.4 Haar wavelet transform of a sawtooth signal: wty

```
% Haar wavelet transform of a signal
% Sawtooth signal
% Plot of wty
fy=300; %signal frequency in Hz
```

```
wy=2*pi*fy; %signal frequency in rad/s
fs=20000; %sampling frequency in Hz
tiv=1/fs; %time interval between samples;
Ns=256; %let us take 256 signal samples
duy=Ns*tiv; %time for 256 samples
t=0:tiv:(duy-tiv); %time intervals set
y=sawtooth(wy*t); %signal data set (256 samples)
K=8; %exponent, 256=2^8
wty=y; %space for the wavelet transform
%the Haar wavelet transform
for n=1:K,
aux1= wty(1:2:Ns-1) + wty(2:2:Ns);
aux2= wty(1:2:Ns-1) - wty(2:2:Ns);
wty(1:Ns)=[aux1,aux2]/sqrt(2);
Ns=Ns/2;
end;
%figure
plot(wty,'k');
axis([0 256 -4 4]);
title('Vector wty obtained by sawtooth Haar wavelet transform')
```

Now let us try to recover the sawtooth signal from its Haar wavelet transform. This is made with the Program 2.5, repeating the procedure of Program 2.2. The recovered signal is shown in Fig. 2.13.

Program 2.5 Sawtooth signal recovery from Haar wavelet transform

```
% Haar wavelet transform of a signal
% Sawtooth signal recovery from transform
%---------------------------------------------------
% First the wavelet transform to get data
fy=300; %signal frequency in Hz
wy=2*pi*fy; %signal frequency in rad/s
fs=20000; %sampling frequency in Hz
tiv=1/fs; %time interval between samples;
Ns=256; %let us take 256 signal samples
duy=Ns*tiv; %time for 256 samples
t=0:tiv:(duy-tiv); %time intervals set
y=sawtooth(wy*t); %signal data set (256 samples)
K=8; %exponent, 256=2^8
wty=y; %space for the wavelet transform
%the Haar wavelet transform
for n=1:K,
aux1= wty(1:2:Ns-1) + wty(2:2:Ns);
aux2= wty(1:2:Ns-1) - wty(2:2:Ns);
wty(1:Ns)=[aux1,aux2]/sqrt(2);
Ns=Ns/2;
end;
%------------------------------------------------------------
% Second the signal recovery from the wavelet transform data
```

```
J=K+1;
z=wty; %space for recovered data
a=zeros(J,(2^K)); %space for a(j,k) coefficients
m=1;
a(1,1)=z(1);
for n=1:K,
a(n+1,1:2:(2*m-1))=(a(n,1:m)+z((1+m):(2*m)))/sqrt(2);
a(n+1,2:2:(2*m))=(a(n,1:m)-z((1+m):(2*m)))/sqrt(2);
m=m*2;
end;
y=a(J,1:256); %the recovered data
%figure
plot(y,'k');
axis([0 256 -1.2 1.2]);
title('Recovered sawtooth signal,from Haar wavelet transform');
```

There is an intuitive visualization of wavelet coeffcients called '*scalogram*'. The x-axis of this representation is discrete time (samples), and the y-axis is the scale. Figure 2.14 shows the scalogram of a sawtooth signal, corresponding to the decomposition of this signal into Haar wavelets. In strict terms, the scalogram should show squared (positive) values of the coefficients, but in our case we preferred to show the original values, negative or positive, letting to MATLAB normalization tasks. The values are painted in pseudocolors or in gray scale, for instance. The figure has been generated with the Program 2.6, which contains interesting code. The figure clearly shows the correspondence of signal peaks and the detection of high frequencies by the wavelets (the higher scales are associated to high frequencies).

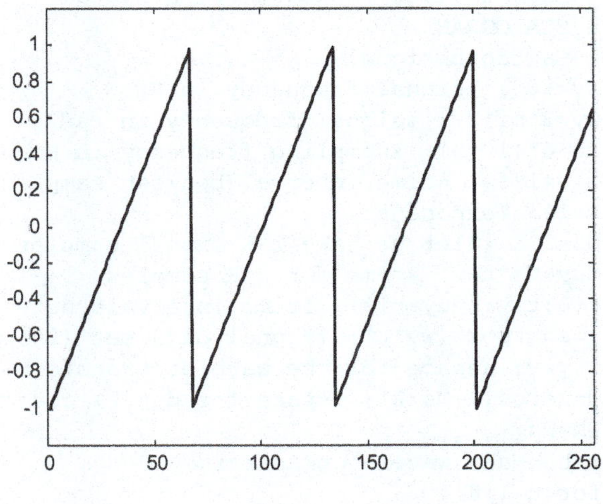

Fig. 2.13 Recovering the sawtooth signal from its Haar wavelet transform

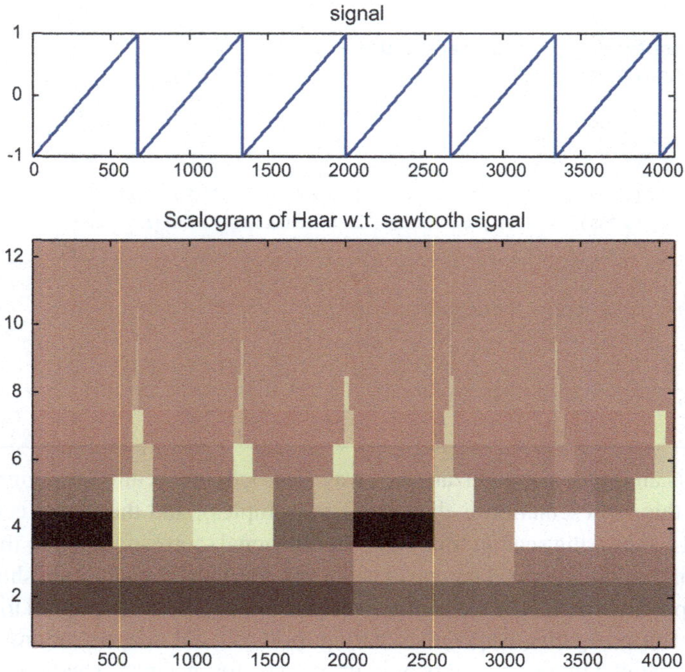

Fig. 2.14 Scalogram of sawtooth signal using Haar wavelets

Program 2.6 Scalogram of sawtooth signal, Haar wavelet

```
% Haar wavelet transform of a signal
% SCALOGRAM
% Sawtooth signal
fy=600; %signal frequency in Hz
wy=2*pi*fy; %signal frequency in rad/s
fs=4*(10^5); %sampling frequency in Hz
tiv=1/fs; %time interval between samples;
K=12; %exponent
Ns=2^K; %let us take 2^K signal samples
duy=Ns*tiv; %time for 2^K samples
t=0:tiv:(duy-tiv); %time intervals set
y=sawtooth(wy*t); %signal data set (256 samples)
wty=y; %space for the wavelet transform
d=zeros(K,Ns/2); %space for d(j,k) coefficients
NN=Ns;
%the Haar wavelet transform
for n=1:K,
aux1= wty(1:2:NN-1) + wty(2:2:NN);
aux2= wty(1:2:NN-1) - wty(2:2:NN);
wty(1:NN)=[aux1,aux2]/sqrt(2);
d(K+1-n,1:NN/2)=wty(1+(NN/2):NN); %fill d(j,k) coefficients
```

```
NN=NN/2;
end;
%preparing for scalogram
S=zeros(K,Ns); %space for S(j,k) scalogram coefficients
for n=1:K,
q=2^(n-1); L=Ns/q;
for m=1:q,
R=(1+(L*(m-1))):(L*m); %index range
S(n,R)=d(n,m);
end;
end;
%figure
subplot('position',[0.04 0.77 0.92 0.18])
plot(y);
axis([0 4096 -1 1]);
title('signal');
subplot('position',[0.04 0.05 0.92 0.6])
imagesc(S); colormap('pink');
title('Scalogram of Haar w.t. sawtooth signal');
h=gca; set(h,'YDir','normal');
```

2.2.3 Wavelets and Filter Banks

The iterative method just explained can be expressed as filtering the data. The computation of $a_{j,k}$ coefficients implies averaging, which is low pass filtering. Let us denote this low pass filter as *LP(z)*. Likewise, the computation of $d_{j,k}$ coefficients implies differencing, which is high pass filtering with the filter *HP(z)*. Both filters are moved along the input data. The expressions of these Haar filters are the following:

$$LP(z) = \frac{1}{\sqrt{2}}(z^{-1} + 1) \tag{2.32}$$

$$HP(z) = \frac{1}{\sqrt{2}}(z^{-1} - 1) \tag{2.33}$$

The frequency responses of the Haar filters are:

$$|LP(\omega)| = \sqrt{2}\,\cos(\frac{\omega}{2}) \tag{2.34}$$

$$|HP(\omega)| = \sqrt{2}\,\sin(\frac{\omega}{2}) \tag{2.35}$$

Figure 2.15 shows the frequency magnitude responses of the Haar LP and HP filters. According with the nomenclature of the filter banks, LP is $H_0(z)$ and HP is $H_1(z)$.

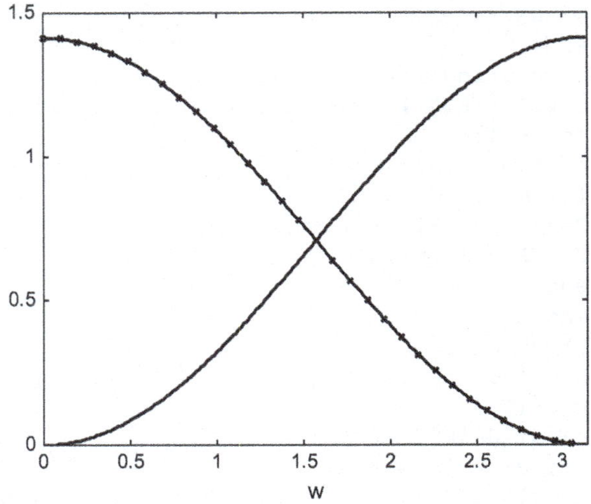

Fig. 2.15 Frequency magnitude responses of the LP and HP Haar filters

The Fig. 2.15 has been generated with the Program 2.7, which is almost the same as the program used in the previous chapter for the QMF filter bank (Sect. 1.4.2).

Program 2.7 Frequency magnitude response of Haar filters H0 and H1

```
% Frequency magnitude response of Haar filters H0 and H1
c=1/sqrt(2);
h0=[c c]; %low-pass filter
h1=[-c c]; %high-pass filter
w=0:(2*pi/511):pi;
H0=real(fft(h0,512)); %discrete Fourier transform
H1=real(fft(h1,512)); % """
plot(w,H0(1:256),'k',w(1:8:256),H0(1:8:256),'kx'); hold on;
plot(w,abs(H1(1:256)),'k');
axis([0 pi 0 1.5]);
title('frequency response of QMF H0 and H1 filters');
xlabel('w');
```

Consider again the two-channel filter bank, as in Fig. 2.16.

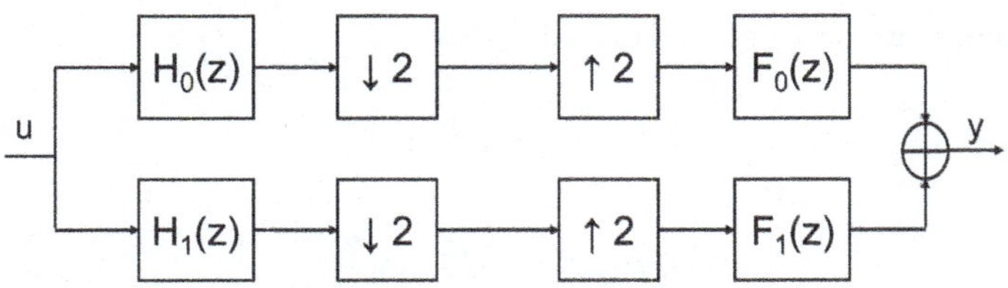

Fig. 2.16 A two-channel filter bank

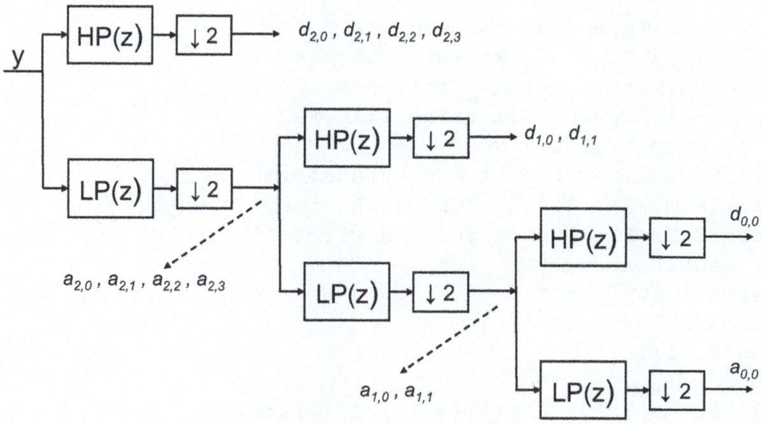

Fig. 2.17 Wavelet expansion using filter bank

If the Haar filters are chosen for $H_0(z)$ and $H_1(z)$, and the other two filters are designed as:

$$\begin{aligned} F_0(z) &= H_1(-z) \\ F_1(z) &= -H_0(-z) \end{aligned} \tag{2.36}$$

Then a QMF filter bank is obtained, with perfect reconstruction (see Sect. 1.4.2).

Actually, the iterative method for the wavelet decomposition can be expressed with two-channel filter banks. Figure 2.17 shows a filter bank corresponding to the example considered in Eq. (2.24) and thereafter. This is in clear connection with the topics of the previous chapter.

Let us visit again the example of the sawtooth signal for comparison purposes. The Program 2.8 applies Haar filters to obtain the wavelet transform of the sawtooth signal. Figure 2.18 shows the vector *wty(n)* that is obtained: it is exactly the same obtained by Program 2.4 (Fig. 2.12).

Program 2.8 Haar wavelet transform of sawtooth signal using filters

```
% Haar wavelet transform of a signal using filters
% Sawtooth signal
% Plot of wty
% The Haar filters
c=1/sqrt(2);
h0=[c c]; %low-pass filter
h1=[-c c]; %high-pass filter
%The sawtooth signal
fy=300; %signal frequency in Hz
wy=2*pi*fy; %signal frequency in rad/s
duy=0.03; %signal duration in seconds
fs=20000; %sampling frequency in Hz
tiv=1/fs; %time interval between samples;
Ns=256; %let us take 256 signal samples
```

```
duy=Ns*tiv; %time for 256 samples
t=0:tiv:(duy-tiv); %time intervals set
y=sawtooth(wy*t); %signal data set (256 samples)
%Haar wavelet transform using filters
K=8; %exponent, 256=2^8
wty=y; %space for the wavelet transform
d=zeros(K,Ns/2); %space for d(j,k) coefficients
a=zeros(K,Ns/2); %space for a(j,k) coefficients
My=y; %auxiliar vector
Ls=zeros(1,128); %"""
Hs=zeros(1,128); %"""
for nn=K:-1:1,
m=2^nn;
lx=filter(h0,1,My); Ls(1:(m/2))=lx(2:2:m);
a(nn,1:(m/2))=Ls(1:(m/2)); %LP and subsampling
hx=filter(h1,1,My); Hs(1:(m/2))=hx(2:2:m);
d(nn,1:(m/2))=Hs(1:(m/2)); %HP and subsampling
My=Ls(1:(m/2));
wty((1+(m/2)):m)=d(nn,1:(m/2)); %append to wty
end;
wty(1)=a(1,1);
%figure
plot(wty,'k');
axis([0 256 -4 4]);
title('Vector wty with Haar wavelet transform using filters');
```

The original data can be recovered with another filter bank. Figure 2.19 shows the filter bank that reverses the action of the filter bank represented in Fig. 2.17.

The Program 2.9 applies Haar filters to recover the sawtooth signal from its wavelet transform. Figure 2.20 shows the result.

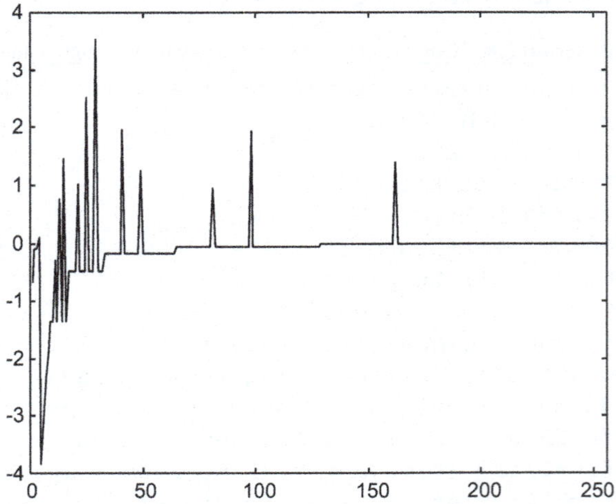

Fig. 2.18 The vector *wty(n)* corresponding to the sawtooth signal

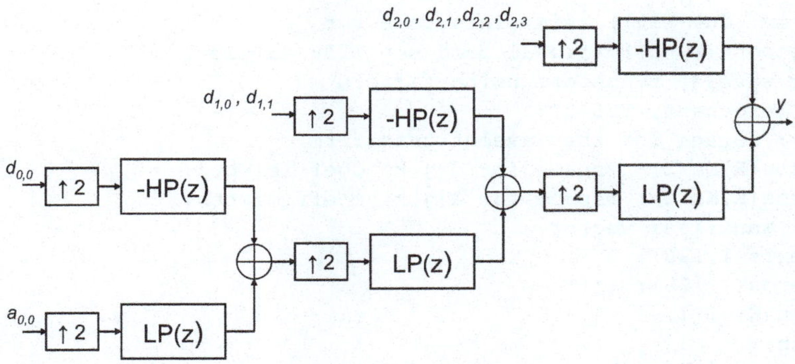

Fig. 2.19 Recovering original data from wavelet expansion

Fig. 2.20 Recovering the sawtooth signal from its Haar wavelet transform

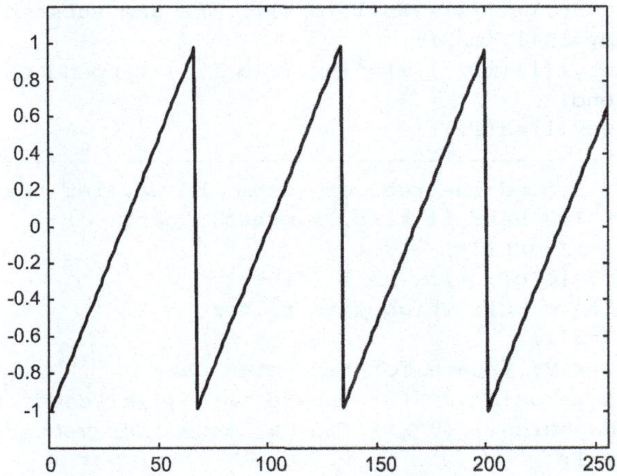

Program 2.9 Recovering of sawtooth signal from its Haar wavelet transform, using filters

```
% Haar wavelet transform of a signal using filters
% Recovering of sawtooth signal from its transform
% The Haar filters, analysis part
c=1/sqrt(2);
h0=[c c]; %low-pass filter
h1=[-c c]; %high-pass filter
%-------------------------------------------------------
% First the wavelet transform to get the data
%The sawtooth signal
fy=300; %signal frequency in Hz
wy=2*pi*fy; %signal frequency in rad/s
duy=0.03; %signal duration in seconds
fs=20000; %sampling frequency in Hz
tiv=1/fs; %time interval between samples;
Ns=256; %let us take 256 signal samples
duy=Ns*tiv; %time for 256 samples
```

```
t=0:tiv:(duy-tiv); %time intervals set
y=sawtooth(wy*t); %signal data set (256 samples)
%Haar wavelet transform using filters
K=8; %exponent, 256=2^8
wty=y; %space for the wavelet transform
d=zeros(K,Ns/2); %space for d(j,k) coefficients
a=zeros(K,Ns/2); %space for a(j,k) coefficients
My=y; %auxiliar vector
Ls=zeros(1,128); %"""
Hs=zeros(1,128); %"""
for nn=K:-1:1,
m=2^nn;
lx=filter(h0,1,My); Ls(1:(m/2))=lx(2:2:m);
a(nn,1:(m/2))=Ls(1:(m/2)); %LP and subsampling
hx=filter(h1,1,My); Hs(1:(m/2))=hx(2:2:m);
d(nn,1:(m/2))=Hs(1:(m/2)); %HP and subsampling
My=Ls(1:(m/2));
wty((1+(m/2)):m)=d(nn,1:(m/2)); %append to wty
end;
wty(1)=a(1,1);
%---------------------------------------------------------
% Second the recovery from the wavelet transform
% The Haar filters, synthesis part
c=1/sqrt(2);
f0=[c c]; %low-pass filter
f1=[c -c]; %high-pass filter
J=K+1;
z=wty; %space for recovered data
a=zeros(J,(2^K)); %space for a(j,k) coefficients
Ma=zeros(1,(2^K)); Md=Ma; %auxiliar vectors
m=1;
a(1,1)=z(1); d(1,1)=z(2);
Ma(1)=a(1,1); Md(1)=d(1,1);
for nn=1:(K-1),
m=2^nn;
lx=filter(f0,1,Ma); hx=filter(f1,1,Md);
a(nn+1,1:m)=lx(1:m)+hx(1:m);
Ma(1:2:(m*2))=a(nn+1,1:m); %upsampling
Md(1:2:(m*2))=z((1+m):(m*2)); %"""
end;
nn=8; m=2^nn;
lx=filter(f0,1,Ma); hx=filter(f1,1,Md);
a(nn+1,1:m)=lx(1:m)+hx(1:m);
y=a(J,1:256); %the recovered data
%figure
plot(y,'k');
axis([0 256 -1.2 1.2]);
title('Recovered sawtooth from Haar wavelet transform
using filters');
```

2.3 The Multiresolution Analysis Equation

After the introductory example of Haar wavelets, it is convenient to generalize the main ideas.

Signal processing with wavelets is based on the use of scaling functions and wavelet functions. A scaling function $\varphi(t)$ is given by the recursive equation:

$$\varphi(t) = \sum_n h_0(n)\,\sqrt{2}\,\varphi(2t - n); \quad \varphi(t) \in L^2; \quad n = 0, 1, 2, .N - 1 \tag{2.37}$$

This is an important equation, which receives several names like refinement equation, multiresolution analysis equation (MAE), dilation equation, etc.

In general, given a certain type of signals to be processed, certain properties of the wavelets and the scaling functions are sought. This usually implies bounds for the values of the coefficients $h_0(n)$. Once a set of $h_0(n)$ values is specified, the scaling function $\varphi(t)$ can be determined by recursion.

A family of functions is generated from the scaling function as follows:

$$\varphi_{j,k}(t) = \sqrt{2^j}\,\varphi(2^j\,t - k) \tag{2.38}$$

Denote the Fourier transform of $\varphi(t)$ as $\Phi(\omega)$, and:

$$H_0(\omega) = \sum_{n=-\infty}^{\infty} h_0(n)\,e^{j\omega n} \tag{2.39}$$

(the frequency response of the LP(z) filter)

Then the frequency domain version of the MAE is:

$$\Phi(\omega) = \frac{1}{\sqrt{2}} H_0(\frac{\omega}{2})\,\Phi(\frac{\omega}{2}) \tag{2.40}$$

which after iteration becomes:

$$\Phi(\omega) = \prod_{k=1}^{\infty} [\frac{1}{\sqrt{2}} H_0(\frac{\omega}{2^k})]\,\Phi(0) \tag{2.41}$$

2.3.1 Solving the MAE

2.3.1.1 Necessary Conditions

There is a set of necessary conditions for $\varphi(t)$ to be a solution of the MAE. These conditions can be used to deduce the $h_0(n)$ coefficients. Let us list some of them:

• If $\int \varphi(t)\, dt \neq 0$:

$$\sum_n h_0(n) = H_0(0) = \sqrt{2} \tag{2.42}$$

Notice that $H_0(0)$ is the frequency response of the filter LP(z) at DC.

• If $\int \varphi(t)\, dt = 1$ and

$$\sum_k \varphi(t - k) = \sum_k \varphi(k) = 1 \tag{2.43}$$

then

$$\sum_n h_0(2n) = \sum_n h_0(2n + 1) \tag{2.44}$$

and

$$H_0(\pi) = 0 \tag{2.45}$$

Equation (2.43) is called a partitioning of unity . Also, in the opposite direction, if (2.44) is true then (2.43) is true.

• If the integer translates of φ(t) are **orthogonal**:

$$\sum_n h_0(n)\, h_0(n - 2k) = \begin{cases} 1, \ if\ k = 0 \\ 0, \ otherwise \end{cases} \tag{2.46}$$

$$\sum_n |h_0(n)|^2 = 1 \tag{2.47}$$

$$\sum_n |\Phi(\omega + 2\pi\, k|^2 = 1 \tag{2.48}$$

$$\sum_n h_0(2n) = \sum_n h_0(2n + 1) = \frac{1}{\sqrt{2}} \tag{2.49}$$

$$|H_0(\omega)|^2 + |H_0(\omega + \pi)|^2 = 2 \tag{2.50}$$

The coefficients that satisfy (2.46) are coefficients of a QMF filter.

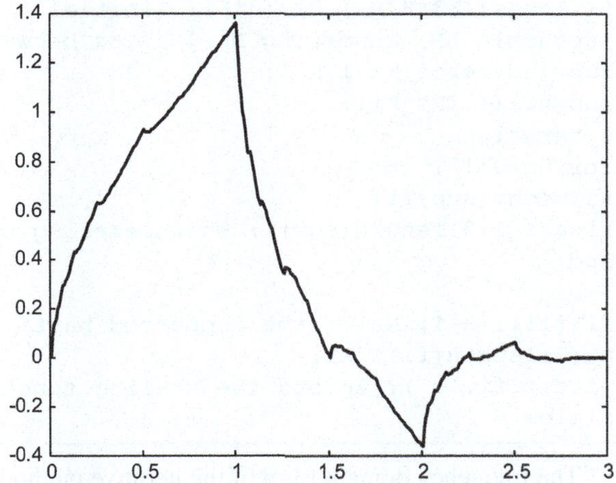

Fig. 2.21 Daubechies 4 scaling function

2.3.1.2 Iterative Computation of the Scaling Function

Usually there is no need for using scaling functions or wavelets explicitly. Instead wavelet transforms are applied by filtering, so the important issue is to know the values of $h_0(n)$ and $h_1(n)$.

Anyway, it is interesting to introduce two iterative approaches to compute the scaling function.

The first method is called the cascade algorithm, and consists of using the MAE for a series of iterations:

$$\varphi^{(l+1)}(t) = \sum_n h_0(n)\,\sqrt{2}\,\varphi^{(l)}(2t - n), \;\; l = 0,\, 1,\, 2,\, \ldots \tag{2.51}$$

An initial $\varphi^{(0)}(t)$ must be given.

Program 2.10 implements the cascade algorithm for the Daubechies 4 case. It generates the Fig. 2.21, which shows the famous Daubechies 4 scaling function. Notice its fractal nature.

Program 2.10 Compute a scaling function from MAE

```
% Compute a scaling function
% from MAE
% MAE coefficients
hden=4*sqrt(2); %coeff. denominator
hsq=sqrt(3); %factor
%Daubechies 4:
h=[(1+hsq)/hden, (3+hsq)/hden, (3-hsq)/hden, (1-hsq)/hden];
hN=(h*2)/sum(h); %normalization
K=length(hN);
Ns=128; %number of fi samples
```

```
fi=[ones(1,3*K*Ns),0]/(3*K); %initial iteration
%upsample hN, inserting Ns-1 zeros between samples
hup=[hN;zeros(Ns-1,K)];
hup=hup(1:(Ns*K));
%iteration
for nn=0:12,
aux=conv(hup,fi);
fi=aux(1:2:length(aux)); %downsampling by 2
end
%result
fi=fi(1:(K-1)*Ns); %the supported part
x=(1:length(fi))/Ns;
plot(x,fi,'k'); %plots the scaling function
title('Daubechies 4 scaling function');
```

The frequency domain form of the iterative method is:

$$\Phi^{(l+1)}(\omega) = \frac{1}{\sqrt{2}} H(\frac{\omega}{2})\, \Phi^{(l)}(\frac{\omega}{2}), \quad l = 0, 1, 2, \ldots \tag{2.52}$$

If there is a solution for the equation, the result of the iterations is the Fourier transform of the scaling function, and this can be written as follows:

$$\Phi(\omega) = \left\{ \prod_{k=1}^{\infty} \left(\frac{1}{\sqrt{2}} H(\frac{\omega}{2^k}) \right) \right\} \Phi(0) \tag{2.53}$$

Based on this last equation, a very effective procedure can be implemented, which proceeds from the largest value of k to the lowest. Figure 2.22 shows the first six successive iterations, converging to the target scaling function. The figure has been generated with the Program 2.11. Notice that *fft()* is required only once.

Program 2.11 Computing a scaling function using FFT and MAE

```
% Computing a scaling function using FFT and MAE
% Display of first iterations
%Daubechies 4, coeffs.
hden=4*sqrt(2); %coeff. denominator
hsq=sqrt(3); %factor
%Daubechies 4:
h=[(1+hsq)/hden, (3+hsq)/hden, (3-hsq)/hden, (1-hsq)/hden];
hN=(h*2)/sum(h); %normalization
Ns=2^12; %number of samples
Ffi=fft(hN,Ns); %Initial Fourier transform
aux=Ffi;
for nn=1:6,
%display
subplot(2,3,nn);
plot(abs(aux(1:(Ns/2))),'k'); axis([0 Ns/2 0 2*nn]);
```

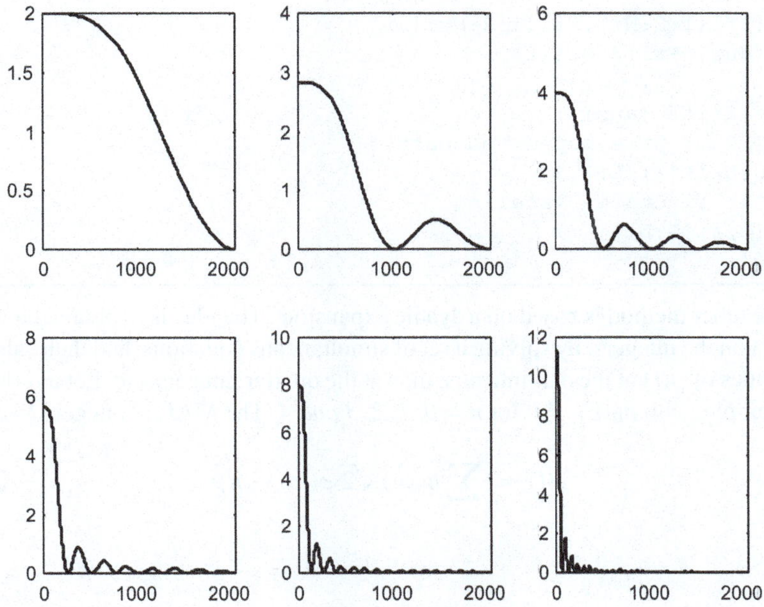

Fig. 2.22 Successive approximations to the scaling function (Daubechies case)

```
%iteration
Ffi=[Ffi(1:2:Ns),Ffi(1:2:Ns)];
aux=(aux.*Ffi)/sqrt(2);
end;
```

Once a good approximation of the frequency domain scaling function is obtained, it is a matter of inverse Fourier transform to get the time domain scaling function. Program 2.12 implements this method, and the result is shown in Fig. 2.23. The result is similar to Fig. 2.21, with a definition improvement since we get 1537 curve points instead of 384 points.

Program 2.12 Computing a scaling function using FFT and MAE: final result

```
% Computing a scaling function using FFT and MAE
% Display of final result
%Daubechies 4, coeffs.
hden=4*sqrt(2); %coeff. denominator
hsq=sqrt(3); %factor
%Daubechies 4:
h=[(1+hsq)/hden, (3+hsq)/hden, (3-hsq)/hden, (1-hsq)/hden];
hN=(h*2)/sum(h); %normalization
Ns=2^12; %number of samples
Ffi=fft(hN,Ns); %Initial Fourier transform
aux=Ffi;
for nn=1:8,
```

```
Ffi=[Ffi(1:2:Ns),Ffi(1:2:Ns)];
aux=(aux.*Ffi)/sqrt(2);
end;
fi=real(ifft(aux));
L=6*(2^8); %the supported part
t=((0:L-1)*3)/L;
ct=2^4; fi=ct*fi; %scaling
plot(t,fi(1:L),'k')
title('Daubechies 4 scaling function');
```

The other method is based on a dyadic expansion . The idea is to obtain the values of $\varphi(t)$ on the integers, by solving a set of simultaneous equations, and then calculate the values of $\varphi(t)$ at the half integers, then at the quarter integers, etc. Let us consider an example, with $h_0(n) \neq 0$ for n = *0, 1, 2, 3 and 4*. The MAE at integers l is:

$$\varphi(l) = \sum_n h_0(n)\,\sqrt{2}\,\varphi(2l - n) \tag{2.54}$$

Thus:

$$\begin{pmatrix} h_0(0) & 0 & 0 & 0 & 0 \\ h_0(2) & h_0(1) & h_0(0) & 0 & 0 \\ h_0(4) & h_0(3) & h_0(2) & h_0(1) & h_0(0) \\ 0 & 0 & h_0(4) & h_0(3) & h_0(2) \\ 0 & 0 & 0 & 0 & h_0(4) \end{pmatrix} \begin{pmatrix} \varphi(0) \\ \varphi(1) \\ \varphi(2) \\ \varphi(3) \\ \varphi(4) \end{pmatrix} = \begin{pmatrix} \varphi(0) \\ \varphi(1) \\ \varphi(2) \\ \varphi(3) \\ \varphi(4) \end{pmatrix} \tag{2.55}$$

This equation can be written in matrix form:

$$M_0\,\bar{\varphi} = \bar{\varphi} \tag{2.56}$$

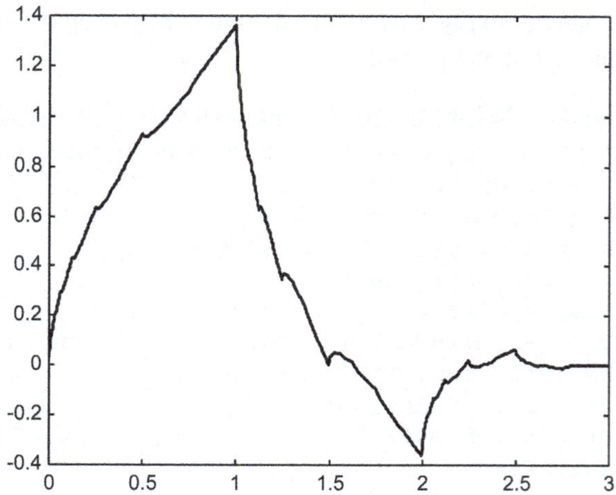

Fig. 2.23 Daubechies 4 scaling function

where the matrix M_0 has an eigenvalue of unity; this is guaranteed by the condition:

$$\sum_n h_0(2n) = \sum_n h_0(2n+1) \tag{2.57}$$

The system of equations can be solved to obtain $\varphi(0), \ldots, \varphi(4)$. Now, the MAE can also be written as:

$$\varphi(l/2) = \sum_n h_0(n)\,\sqrt{2}\,\varphi(l-n) \tag{2.58}$$

Consequently:

$$\begin{pmatrix} h_0(1) & h_0(0) & 0 & 0 & 0 \\ h_0(3) & h_0(2) & h_0(1) & h_0(0) & 0 \\ 0 & 0 & h_0(4) & h_0(3) & h_0(2) \\ 0 & 0 & 0 & 0 & h_0(4) \\ 0 & 0 & 0 & 0 & 0 \end{pmatrix} \begin{pmatrix} \varphi(0) \\ \varphi(1) \\ \varphi(2) \\ \varphi(3) \\ \varphi(4) \end{pmatrix} = \begin{pmatrix} \varphi(1/2) \\ \varphi(3/2) \\ \varphi(5/2) \\ \varphi(7/2) \\ \varphi(9/2) \end{pmatrix} \tag{2.59}$$

(the last row is not needed)

In matrix form:

$$M_1\,\bar{\varphi} = \bar{\varphi}_2 \tag{2.60}$$

The equation can be iterated to obtain the values of the scaling functions at quarter integers, etc.

Figure 2.24 shows the first iterations of the procedure, for the Daubechies 4 example again. The figure has been generated with the Program 2.13.

Program 2.13 Compute a scaling function with dyadic approach

```
% Compute a scaling function with dyadic approach
% display of first iterations
% Daubechies 4,coeffs.
hden=4*sqrt(2); %coeff. denominator
hsq=sqrt(3); %factor
%Daubechies 4:
h=[(1+hsq)/hden, (3+hsq)/hden, (3-hsq)/hden, (1-hsq)/hden];
hN=(h*2)/sum(h); %normalization
K=length(hN);
hrev=hN(K:-1:1); %reverse hN
%M0 matrix
MA=[hrev,zeros(1,(2*K)-2)];
MB=MA;
for nn=1:K-1,
MA=[0,0,MA(1:(3*K)-4)];
MB=[MB; MA];
end
M0=MB(:,K:(2*K)-1);
```

```
%Solving the first system of equations, for fi(0)..fi(3)
MF=M0-eye(K);
MG=[MF(1:K-1,:);ones(1,K)];
nfi=MG\[zeros(K-1,1);1];
%display
subplot(2,3,1);
x=(1:length(nfi))*(K-1)/length(nfi);
plot(x,nfi,'k',x,nfi,'dk');
axis([0 3 -0.5 1.5]);
%getting middle {&} quarter values
fi=nfi(2:length(nfi)-1);
fi=conv(hN,fi);
aux=fi(1:2:length(fi)); %downsampling by 2
%quarter values
y=conv(hN,aux);
%merge y and fi
aux=[y;fi,0]; aux=aux(:)';
fi=aux(1:length(aux)-1);
%display
subplot(2,3,2);
x=(1:length(fi))*(K-1)/length(fi); plot(x,fi,'k',x,fi,'dk');
axis([0 3 -0.5 1.5]);
%iteration
hup=hN; Nx=4;
for nn=1:Nx,
%upsample by 2
L=length(hup);
aux=[hup;zeros(1,L)]; aux=aux(:)';
hup=aux(1:2*L-1);
%intermediate terms
y=conv(hup,y);
%merge y and fi
aux=[y;fi,0]; aux=aux(:)';
fi=aux(1:length(aux)-1);
%display
subplot(2,3,2+nn);
x=(1:length(fi))*(K-1)/length(fi); plot(x,fi,'k',x,fi,'.k');
axis([0 3 -0.5 1.5]);
end
```

To complete the example, a program has been written (Program A.3) that implements the procedure to obtain 6143 curve points for the Daubechies 4 scaling function. Figure 2.25 shows the result. The program has been included in Appendix A.

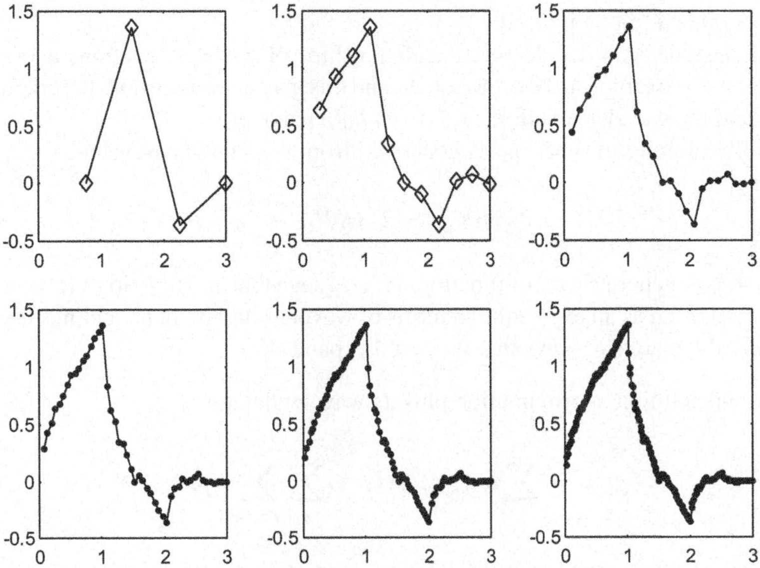

Fig. 2.24 First dyadic iterations to compute the Daubechies 4 scaling function

Fig. 2.25 Daubechies 4 scaling function, again

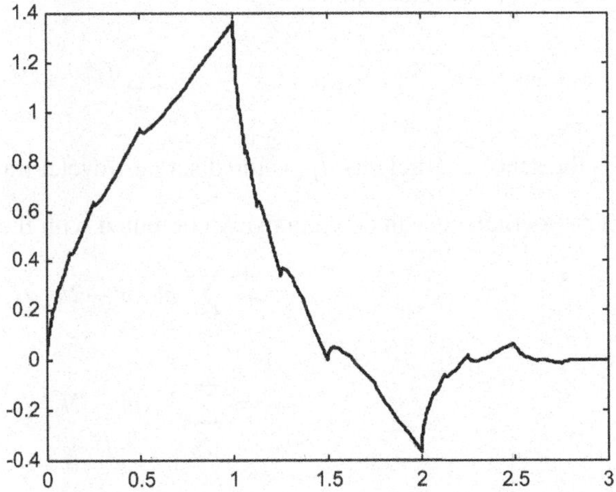

2.3.2 *Scaling Functions, Wavelets, and Function Expansions*

A wavelet function $\psi(t)$ can be obtained from scaling functions with the following equation:

$$\psi(t) = \sum_n h_1(n)\,\sqrt{2}\,\varphi(2t - n); \quad n = 0, 1, 2, .N - 1 \tag{2.61}$$

where $\psi(t)$ is a 'mother wavelet'.

The designer can decide certain relationships of wavelet functions and scaling functions –for example to be orthogonal-, and this may determine both the coefficients $h_1(n)$ and the wavelet function $\psi(t)$, from $h_0(n)$ and $\varphi(t)$.

Another family of functions is generated from the 'mother wavelet':

$$\psi_{j,k}(t) = \sqrt{2^j}\,\psi(2^j\,t - k) \tag{2.62}$$

The goal is to generate a set of functions $\psi_{j,k}(t)$ such that any function $y(t)$ of interest could be written as an expansion in terms of wavelets and perhaps scaling functions. Actually there are two ways that we could expand $y(t)$:

- A low-resolution approximation plus its wavelet details:

$$y(t) = \sum_k a_{jo,k}\varphi_{jo,k}(t) + \sum_k \sum_{j=jo}^{\infty} d_{j,k}\,\psi_{j,k}(t) \tag{2.63}$$

  this is the most useful in practice; also in practice the upper value of j is finite $(= M - 1)$.

- Only the wavelet details:

$$y(t) = \sum_{j,k} d_{j,k}\,\psi_{j,k}(t) \tag{2.64}$$

  the set of coefficients $d_{j,k}$ is the discrete wavelet transform of $y(t)$.

The coefficients can be recursively computed with the following equations:

$$a_{j,k} = \sum_m h_0(m - 2k)\,a_{j+1,m} \tag{2.65}$$

$$d_{j,k} = \sum_m h_1(m - 2k)\,a_{j+1,m} \tag{2.66}$$

with:

$$m = 2k,\ 2k+1,\ 2k+2,\ 2k+N-1 \tag{2.67}$$

The filter banks of Fig. 2.17 can be used, with LP(z) corresponding to (2.65) and HP(z) corresponding to (2.66). The series of values $h_0(m - 2k)$ and $h_1(m - 2k)$ are filter impulse responses.

2.3.2.1 Wavelet Properties

If the integer translates of φ(t) are orthogonal the corresponding wavelet has the following properties:

- The wavelet is orthogonal to the scaling function at the same scale if and only if:

$$h_1(n) = \pm(-1)^n \, h_0\,(L - n) \tag{2.68}$$

where L is an arbitrary odd integer.

Orthogonal at the same scale is:

$$\int \varphi(t-n)\,\psi(t-m)dt = 0 \tag{2.69}$$

- If the wavelet is orthogonal to the scaling function at the same scale then:

$$\sum_n h_0(n)\,h_1(n-2k) = 0 \tag{2.70}$$

- Also,

$$\sum_n h_1(n) = H_1(0) = 0 \tag{2.71}$$

$$|H_1(\omega)| = |H_0(\omega + \pi)| \tag{2.72}$$

$$|H_0(\omega)|^2 + |H_1(\omega)|^2 = 2 \tag{2.73}$$

2.3.3 *Examples*

Coming back to the multiresolution equation, let us comment some solutions.

2.3.3.1 For M = 2

The necessary conditions imply that:

$$h_0(0) + h_0(1) = \sqrt{2} \tag{2.74}$$

$$h_0(0) = h_0(1) \tag{2.75}$$

Fig. 2.26 Haar scaling functions

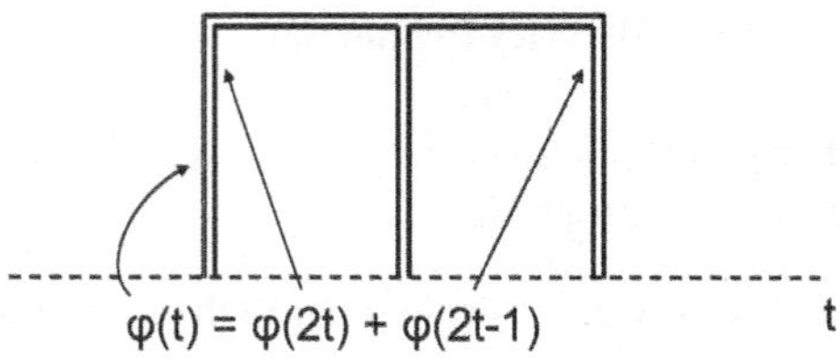

Then:

$$h_0(0) = \frac{1}{\sqrt{2}}, \ h_0(1) = \frac{1}{\sqrt{2}} \tag{2.76}$$

and the solution is given by the Haar scaling function.

Figure 2.26 shows how $\varphi(t)$ can be obtained with $\varphi(2t)$ and $\varphi(2t-1)$ according with Eq. (2.37) and the coefficients (2.76). Notice that $\varphi(2t)$ and $\varphi(2t-1)$ (or in general $\varphi(2t-k)$) are orthogonal since they do not overlap.

The Haar wavelets are obtained with Eq. (2.61) using:

$$h_1(0) = \frac{1}{\sqrt{2}}, \ h_1(1) = -\frac{1}{\sqrt{2}} \tag{2.77}$$

The values of $h_1(0)$ and the $h_1(1)$ are easily derived from orthogonality conditions. Figure 2.27 shows how $\psi(t)$ can be obtained with $\varphi(2t)$ and $\varphi(2t-1)$ according with Eq. (2.61) and the coefficients (2.77).

The Haar wavelets have the following properties:

- Orthogonal
- Compact time-domain support
- The scaling function is symmetric
- The wavelet function is anti-symmetric
- The wavelet function has only one vanishing moment

Fig. 2.27 Haar wavelet and scaling functions

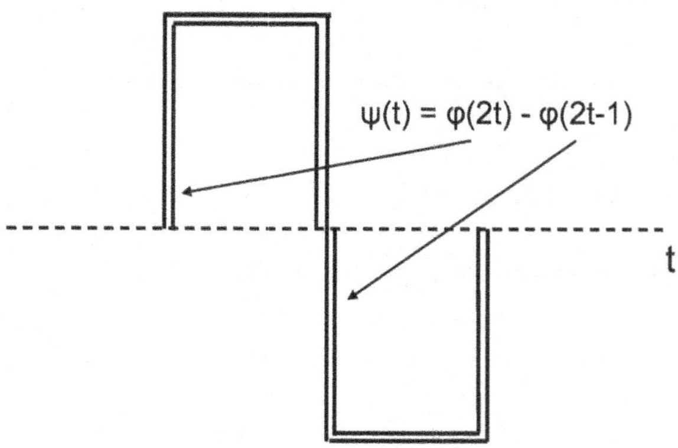

Actually, the Haar wavelet is the only one having the first three properties.

2.3.3.2 For M = 3

The necessary conditions imply that:

$$h_0(0) + h_0(1) + h_0(2) = \sqrt{2} \tag{2.78}$$

$$h_0(0) + h_0(2) = h_0(1) \tag{2.79}$$

A solution is provided by the triangle scaling function, with the following values:

$$h_0(0) = \frac{1}{2\sqrt{2}}, \; h_0(1) = \frac{1}{\sqrt{2}} \; h_0(2) = \frac{1}{2\sqrt{2}} \tag{2.80}$$

Figure 2.28 shows how $\varphi(t)$ can be obtained with $\varphi(2t)$, $\varphi(2t-1)$, and $\varphi(2t-2)$ according with Eq. (2.37) and the coefficients (2.80). Notice that $\varphi(2t)$ and $\varphi(2t-1)$ (or in general $\varphi(2t-k)$) are *not* orthogonal.

The triangle wavelets are obtained with Eq. (2.61) using:

$$h_1(0) = -\frac{1}{2\sqrt{2}}, \; h_1(1) = \frac{1}{\sqrt{2}}, \; h_1(2) = -\frac{1}{2\sqrt{2}} \tag{2.81}$$

As in the previous example, Fig. 2.29 shows how $\psi(t)$ can be obtained with $\varphi(2t)$, $\varphi(2t-1)$, and $\varphi(2t-2)$ according with Eq. (2.61) and the coefficients of the triangle wavelet (2.81).

Triangle scaling functions have compact support and are symmetric, but they are not orthogonal.

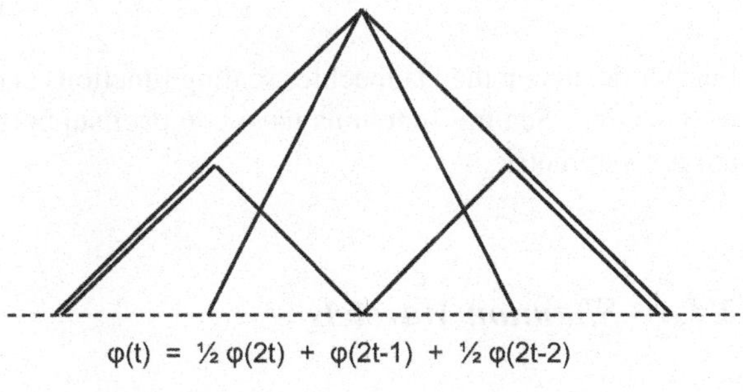

Fig. 2.28 Triangle scaling functions

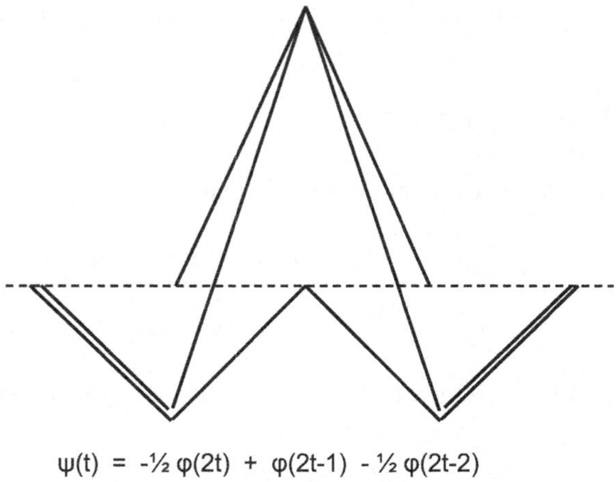

Fig. 2.29 Triangle wavelet and scaling functions

2.3.3.3 For M = 4

The necessary conditions, for the case that integer translates of φ(t) are orthogonal, imply that:

$$h_0(0) + h_0(1) + h_0(2) + h_0(3) = \sqrt{2} \tag{2.82}$$

$$h_0(0) + h_0(2) = h_0(1) + h_0(3) \tag{2.83}$$

$$h_0^2(0) + h_0^2(1) + h_0^2(2) + h_0^2(3) = 1 \tag{2.84}$$

A solution was found by Daubechies, with the following values

$$\begin{aligned} h_0(0) &= \frac{1+\sqrt{3}}{4\sqrt{2}}, \; h_0(1) = \frac{3+\sqrt{3}}{4\sqrt{2}}, \\ h_0(2) &= \frac{3-\sqrt{3}}{4\sqrt{2}}, \; h_0(3) = \frac{1-\sqrt{3}}{4\sqrt{2}} \end{aligned} \tag{2.85}$$

Further details on the Daubechies scaling functions and wavelets will be given in the next section. Scaling functions have compact support and are orthogonal, but they are not symmetric.

2.3.4 *Shannon Wavelets*

The Shannon scaling function has a well-known expression, related to the frequency response of a brickwall filter:

$$\varphi(t) = sinc\,(\pi\, t) = \frac{\sin\,(\pi\, t)}{\pi\, t} \tag{2.86}$$

with value 1 for $t = 0$.

The corresponding LP(z) coefficients are the following:

$$h_0(0) = \frac{1}{2} \tag{2.87}$$

$$h_0(2n) = 0,\ n \neq 0 \tag{2.88}$$

$$h_0(2n+1) = \frac{(-1)^n \sqrt{2}}{(2n+1)\,\pi},\ n \neq 0 \tag{2.89}$$

Cardinal functions have zero value at integer t values (except at $t = 0$). The Shannon scaling function $\varphi(t)$ is a cardinal function and this makes it useful for interpolation (actually it is the function used by the Shannon sampling theorem to reconstruct a signal from its samples).

- The Shannon scaling functions are orthogonal
- The Shannon wavelet is:

$$\psi(t) = 2\,\varphi(2t) - \varphi(t) = \frac{\sin(2\pi\, t) - \sin(\pi\, t)}{\pi\, t} \tag{2.90}$$

- The Shannon wavelet has infinite number of vanishing moments

Both $\varphi(t)$ and $\psi(t)$ are symmetric, with infinite time support and slow decay (Fig. 2.30).

Program 2.14 Display of Shannon scaling function and wavelet

```
% Display of Shannon scaling function and wavelet
t=-10:0.01:10; %time vector
phi=sinc(t) ; %the scaling function
psi=(2*sinc(2*t))-sinc(t); %the wavelet
figure(1)
subplot(2,1,1)
plot(t,phi,'k');
axis([-10 10 -0.4 1.2]);
xlabel('t')
title('Shannon scaling function');
subplot(2,1,2)
plot(t,psi,'k');
axis([-10 10 -1.2 1.2]);
xlabel('t')
title('Shannon wavelet');
```

The Haar and the Shannon scaling functions are Fourier duals. The Haar wavelet system is good for time localization, but not for frequency localization. The Shannon wavelet system is the opposite.

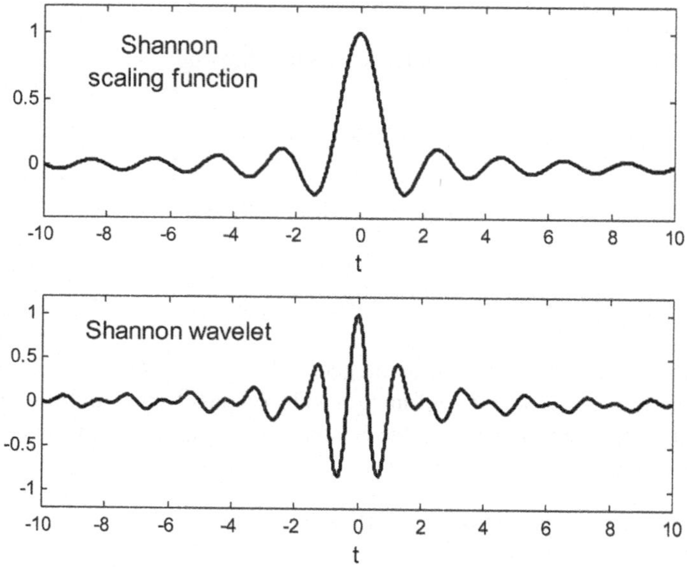

Fig. 2.30 Shannon scaling function and wavelet

2.3.5 *Splines*

Now it is convenient to introduce splines. They are important in signal processing, for instance for interpolation applications, and in the context of wavelets.

Suppose several consecutive times:

$$a = t_0 < t_1 < t_2 < \ldots < t_{k-1} = b \tag{2.91}$$

A spline $l(t)$ is a piecewise polynomial function, such that:

$$l(t) = P_0(t), \quad t_0 \leq t < t_1, \tag{2.92}$$

$$l(t) = P_1(t), \quad t_1 \leq t < t_2, \tag{2.93}$$

$$- -$$

$$l(t) = P_{k-2}(t), \quad t_{k-2} \leq t < t_{k-1}, \tag{2.94}$$

where $P_i(t)$ are polynomials.

The points t_i are called '*knots*'.

There are many types of splines. For instance, if the knots are equidistant the spline is '*uniform*', otherwise is '*non-uniform*'.

If all $P_i(t)$ have degree at most n, then the spline is of degree n.

For a given knot t_i, $P_{i-1}(t)$ and $P_i(t)$ may share a certain number r_i of common derivatives, which indicates how smooth is the spline at t_i.

'*Natural*' splines have zero second derivatives at a and b. '*Interpolating*' splines should have a given set of data values.

2.3.5.1 B-Splines

B-splines (basis splines) are bell-shaped functions generated with the (n + 1)-fold convolution of a rectangular pulse β^0:

$$\beta^0(t) = \begin{cases} 1, \ -1/2 < t < 1/2 \\ 1/2, \ |t| = 1/2 \\ 0, \ otherwise \end{cases} \tag{2.95}$$

$$\beta^n(t) = \underbrace{\beta^0 * \beta^0 * \ldots * \beta^0(t)}_{n+1 \ times} \tag{2.96}$$

where the asterisk (*) means convolution.

The Fourier transform of $\beta^n(t)$ is:

$$\mathrm{B}^n(\omega) = \left(\frac{\sin(\omega/2)}{\omega/2}\right)^{n+1} \tag{2.97}$$

B-splines satisfy the MAE equation, so they are scaling functions. In particular:

- For n = 1:

$$\beta^1(t) = \frac{1}{2}\left(\beta^1(2t+1) + 2\beta^1(2t) + \beta^1(2t-1)\right) \tag{2.98}$$

That corresponds to:

$$h_0(0) = \frac{1}{2\sqrt{2}},\ h_0(1) = \frac{1}{\sqrt{2}},\ h_0(2) = \frac{1}{2\sqrt{2}} \tag{2.99}$$

The B-spline of degree 1 is the triangle scaling function seen before.

- For n = 2:

$$\beta^1(t) = \frac{1}{4}\left(\beta^2(2t+1) + 3\beta^2(2t) + 3\beta^2(2t-1) + \beta^2(2t-2)\right) \tag{2.100}$$

That corresponds to:

$$h_0(0) = \frac{1}{4\sqrt{2}},\ h_0(1) = \frac{3}{4\sqrt{2}},\ h_0(2) = \frac{3}{4\sqrt{2}},\ h_0(3) = \frac{1}{4\sqrt{2}} \tag{2.101}$$

- For n = 3 we have the popular cubic B-spline, with:

$$\begin{array}{l} h_0(0) = \frac{1}{16\sqrt{2}},\ h_0(1) = \frac{1}{4\sqrt{2}}, \\ h_0(2) = \frac{3}{8\sqrt{2}},\ h_0(3) = \frac{1}{4\sqrt{2}},\ h_0(4) = \frac{1}{16\sqrt{2}} \end{array} \tag{2.102}$$

- In general:

$$\beta^n(t) = \frac{1}{2^n} \sum_{j=0}^{n+1} \binom{n+1}{j} \beta^n(2t + \frac{n}{2} - j),\ \ n\ even \tag{2.103}$$

$$\beta^n(t) = \frac{1}{2^n} \sum_{j=0}^{n+1} \binom{n+1}{j} \beta^n(2t + \frac{n+1}{2} - j),\ \ n\ odd \tag{2.104}$$

The B-splines are *not* orthogonal over integer translations.

Schoenberg has shown, in 1946, that all uniform splines with unit spacing are uniquely characterized in terms of a B-spline expression:

$$f(t) = \sum_l c(l)\, \beta^n(t - l) \tag{2.105}$$

with l an integer and $\beta^n(t)$ the central B-spline of degree n.

Influential authors prefer a slightly different definition of B-splines, being uniform splines starting from:

$$b_i^0(t) = \begin{cases} 1, & t_i \leq t < t_{i+1} \\ 0, & otherwise \end{cases} \tag{2.106}$$

and then the others can be obtained with a recursive relation:

$$b_i^n(t) = \frac{t - t_i}{t_{i+n} - t_i}\, b_{i,n-1}(t) + \frac{t_{i+n+1} - t}{t_{i+n+1} - t_{i+1}}\, b_{i+1,n-1}(t) \tag{2.107}$$

Based on this recursion, the Program 2.15 computes the B-splines corresponding to $i = 0$ with degrees n = *0 to 5*. Figure 2.31 shows the B-splines. Notice how the support of the B-splines grows as degree increases (actually this support is $n + 1$). For a B-spline of degree n, the program only has to consider the previously computed B-spline of degree $n - 1$, and a right-shifted version (by 1 second) of this B-spline of degree $n - 1$.

The difference between the b and the β versions of B-splines is that the centre of β versions is the origin, and the b versions are shifted to the right.

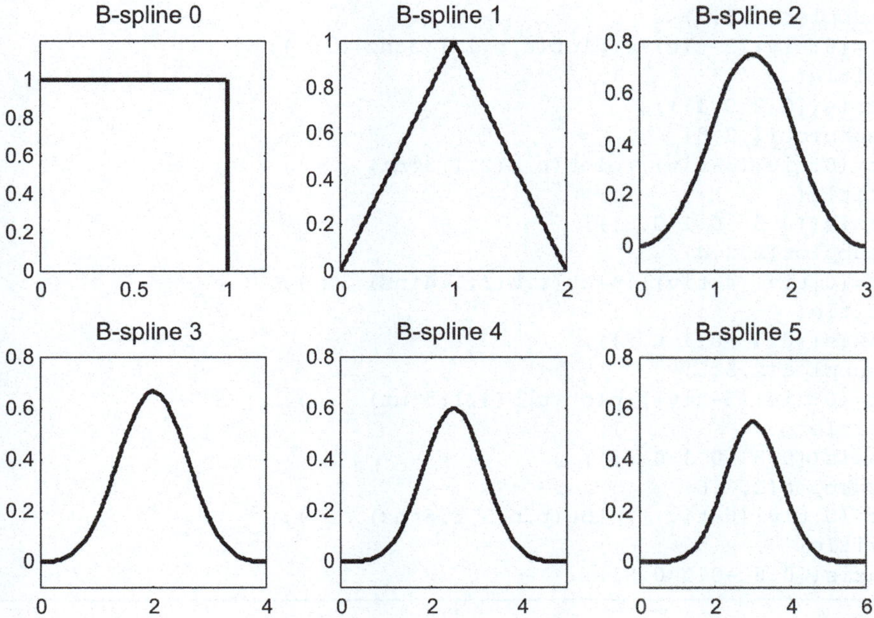

Fig. 2.31 B-splines of degrees 0 to 5

Program 2.15 Display of B-splines

```
% Display of B-splines
%space for splines
nz=100; tiv=1/nz;
b=zeros(4,5*nz); %space for splines
% the box spline (n=0)
b0=ones(1,nz);
% spline n=1
N=1;
b1L=(0:tiv:(N-tiv)).*b0; b1D=(N:-tiv:tiv).*b0;
b(N,1:((N+1)*nz))=[b1L b1D];
% splines 2..5
for N=2:5,
bL=(0:tiv:(N-tiv)).*b(N-1,1:N*nz);
bD=(N:-tiv:tiv).*b(N-1,1:N*nz);
b(N,1:((N+1)*nz))=(1/N)*[bL(1:nz) bL((nz+1):N*nz)+...
bD(1:((N-1)*nz)) bD((((N-1)*nz)+1):N*nz)];
end;
figure(1)
subplot(2,3,1)
t=(0:tiv:(1-tiv)); plot(t,b0,'k',[1 1],[1 0],'k');
title('B-spline 0');
axis([0 1.2 0 1.2]);
```

```
subplot(2,3,2)
t=(0:tiv:(2-tiv)); plot(t,b(1,1:2*nz),'k');
title('B-spline 1');
axis([0 2 0 1]);
subplot(2,3,3)
t=(0:tiv:(3-tiv));plot(t,b(2,1:3*nz),'k');
title('B-spline 2');
axis([0 3 -0.1 0.8]);
subplot(2,3,4)
t=(0:tiv:(4-tiv));plot(t,b(3,1:4*nz),'k');
title('B-spline 3');
axis([0 4 -0.1 0.8]);
subplot(2,3,5)
t=(0:tiv:(5-tiv));plot(t,b(4,1:5*nz),'k');
title('B-spline 4');
axis([0 5 -0.1 0.8]);
subplot(2,3,6)
t=(0:tiv:(6-tiv));plot(t,b(5,1:6*nz),'k');
title('B-spline 5');
axis([0 6 -0.1 0.8]);
```

2.4 Orthogonal Wavelets

Orthogonal scaling functions can be obtained by design, or by orthogonalization of non-orthogonal scaling functions.

Given a non-orthogonal scaling function, with Fourier transform $\Phi(\omega)$ a orthogonal scaling function with Fourier transform $\Phi_{on}(\omega)$ can be obtained as follows:

$$\Phi_{on}(\omega) = \frac{\Phi(\omega)}{\sqrt{E(\omega)}} \tag{2.108}$$

with:

$$E(\omega) = \sum_k |\Phi(\omega + 2\pi k)|^2 \tag{2.109}$$

where $E(\omega)$ is an Euler-Frobenius polynomial.

To see that $\Phi_{on}(\omega)$ is orthogonal, recall from Sect. 2.3.2 the necessary condition given in frequency domain, that in this case reads as:

$$\sum_n |\Phi_{on}(\omega + 2\pi k|^2 = 1 \tag{2.110}$$

A version of the frequency domain version of the MAE, for the orthogonal scaling function, is:

$$\Phi_{on}(\omega) \;=\; \frac{1}{\sqrt{2}} H_0(\frac{\omega}{2})\, \Phi_{on}(\frac{\omega}{2}) \tag{2.111}$$

Then, the corresponding LP filter is:

$$H_0(\omega) \;=\; \sqrt{2}\, \frac{\Phi_{on}(2\omega)}{\Phi_{on}(\omega)} \tag{2.112}$$

Being filters orthogonal, then:

$$H_1(\omega) \;=\; -\, e^{-j\,\omega}\, H_0^*(e^{j(\omega+\pi)}) \tag{2.113}$$

And, by a frequency domain version of (2.61):

$$\Psi(\omega) \;=\; \frac{1}{\sqrt{2}} H_1(\frac{\omega}{2})\, \Phi_{on}(\frac{\omega}{2}) \tag{2.114}$$

The coefficients of the LP or HP filters can be obtained with:

$$h_k \;=\; \sqrt{2}\, \frac{1}{2\pi} \int_0^{2\pi} H(\omega)\, e^{j\,k\,\omega} d\omega \tag{2.115}$$

2.4.1 Meyer Wavelet

The Meyer wavelet was one of the first wavelets, mid 1980s. Meyer wavelet can be viewed as an improvement with respect to the Shannon wavelet. Instead of a brickwall shape of $\Phi(\omega)$ with vertical transitions, discontinuities and transitions are smoothed to improve the time-domain localization.

The scaling function proposed by Meyer is, in the frequency domain:

$$\Phi(\omega) \;=\; \begin{cases} 1, & |\omega| \,\leq\, \frac{2\pi}{3} \\ \cos\,[\,\frac{\pi}{2}\, \upsilon\,(\frac{3}{2\pi}\,|\omega| \,-\, 1)\,], & \frac{2\pi}{3} \leq |\omega| \,\leq\, \frac{4\pi}{3} \\ 0, & otherwise \end{cases} \tag{2.116}$$

where $\upsilon(x)$ is a smooth polynomial interpolating function that:

$$\upsilon(x) \;=\; \begin{cases} 0, \; if \;\, x \leq 0 \\ 1, \; if \;\, x \geq 1 \end{cases}$$
$$\upsilon(x) \,+\, \upsilon(1 \,-\, x) \;=\; 1 \tag{2.117}$$

For example:

$$\upsilon(x) \;=\; x^4\,(35 \;-\; 84\,x \,+\, 70\,x^2 \;-20\,x^3), \;\; with \; x \in \,[0,\, 1] \tag{2.118}$$

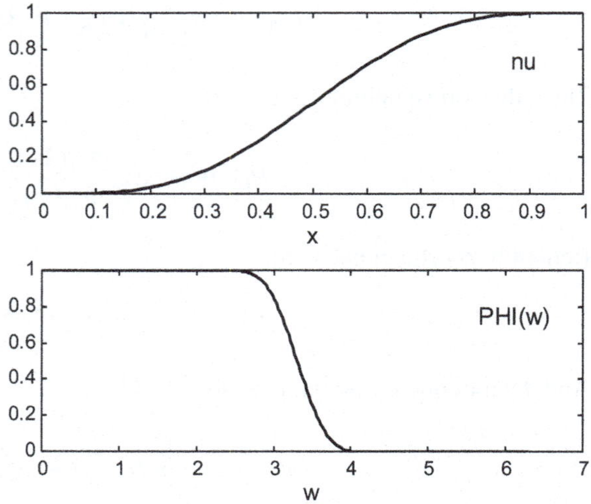

Fig. 2.32 The υ(x) function and the Meyer $\Phi(\omega)$

The scaling functions $\varphi(t - k)$ form an orthonormal basis.

The frequency response of the corresponding LP(z) filter is:

$$H_0(\omega) = \sum_k \Phi\,(2\,(\omega + 2\pi\,k)) \tag{2.119}$$

Figure 2.32 depicts $\upsilon(x)$ and (half) $\Phi(\omega)$ as given by Eqs. (2.117) and (2.116). The figure has been generated with the Program 2.16. Notice the difference with respect to a Sannon's $\Phi(\omega)$ brickwall shape.

Program 2.16 Display of Meyer nu function and 1/2 PHI(w) (right side)

```
% Display of Meyer nu function and 1/2 PHI(w) (right side)
Np=120; %samples per pi
w=0:(pi/Np):(2*pi); %frequencies
L=length(w);
n23=1+((2*Np)/3); %samples for 0..2/3 pi
n43=1+((4*Np)/3); %samples for 0..4/3 pi
wc=w(n23:n43); %central zone
x=((3/(2*pi))*abs(wc))-1;
nu=35-(84*x)+(70*(x.^2))-(20*(x.^3));
nu=(x.^4).*nu;
PHI=zeros(1,L);
PHI(1:(n23-1))=1;
PHI(n23:n43)=cos((pi/2)*nu);
figure(1)
subplot(2,1,1)
plot(x,nu,'k');
xlabel('x'); title('nu');
subplot(2,1,2)
plot(w,PHI,'k');
xlabel('w'); title('PHI(w)');
```

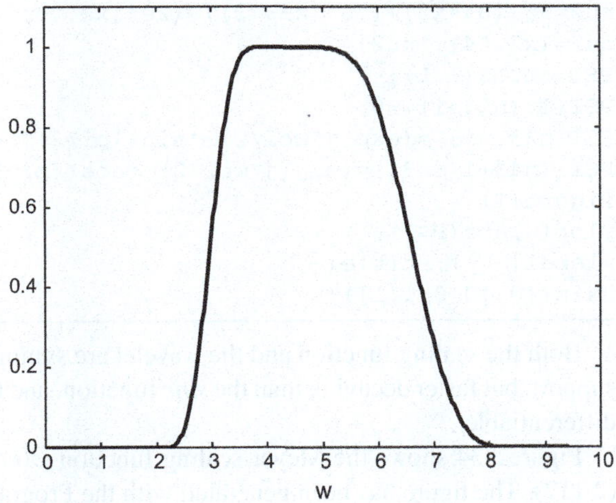

Fig. 2.33 The Meyer $\Psi(\omega)$

The wavelet is, in the frequency domain:

$$\Psi(\omega) = e^{j\omega/2}\,(\Phi(\omega - 2\pi) + \Phi(\omega + 2\pi))\,\Phi(\frac{\omega}{2}) \tag{2.120}$$

That is:

$$\Psi(\omega) = \begin{cases} e^{j(\omega/2)} \sin\left[\frac{\pi}{2}\, v\,(\frac{3}{2\pi}\,|\omega| - 1)\right], & \frac{2\pi}{3} \le |\omega| \le \frac{4\pi}{3} \\ e^{j(\omega/2)} \cos\left[\frac{\pi}{2}\, v\,(\frac{3}{4\pi}\,|\omega| - 1)\right], & \frac{4\pi}{3} \le |\omega| \le \frac{8\pi}{3} \\ 0, & otherwise \end{cases} \tag{2.121}$$

Figure 2.33, which has been generated with the Program 2.17, depicts the $\Psi(\omega)$ corresponding to $v(x)$ as in (2.117). Notice how we extended the range of frequencies above 2π. The plot shows clearly the pass-band nature of the wavelet.

Program 2.17 Display of Meyer 1/2 PSI(w) (right side)

```
% Display of Meyer 1/2 PSI(w) (right side)
Np=120; %samples per pi
w=0:(pi/Np):(3*pi); %frequencies
L=length(w);
n23=1+((2*Np)/3); %samples for 0..2/3 pi
n43=1+((4*Np)/3); %samples for 0..4/3 pi
n83=1+((8*Np)/3); %samples for 0..8/3 pi
wc1=w(n23:n43); %zone 1
wc2=w((n43+1):n83); %zone 2
x1=((3/(2*pi))*abs(wc1))-1;
x2=((3/(4*pi))*abs(wc2))-1;
nu1=35-(84*x1)+(70*(x1.^2))-(20*(x1.^3));
nu1=(x1.^4).*nu1;
```

```
nu2=35-(84*x2)+(70*(x2.^2))-(20*(x2.^3));
nu2=(x2.^4).*nu2;
PSI=zeros(1,L);
PSI(1:(n23-1))=0;
PSI(n23:n43)=(exp(j*wc1/2).*sin((pi/2)*nu1));
PSI((n43+1):n83)=(exp(j*wc2/2).*cos((pi/2)*nu2));
figure(1)
plot(w,abs(PSI),'k');
xlabel('w'); title('PSI(w)');
axis([0 10 0 1.1]);
```

Both the scaling function and the wavelet are symmetric. They have infinite time support, but faster decaying than the *sinc* function, and they are infinitely many times differentiable.

Figure 2.34 shows the Meyer scaling function $\varphi(t)$ corresponding to $v(x)$ as in (2.117). The figure has been generated with the Program 2.18.

Program 2.18 Display of Meyer scaling function phi(t)

```
% Display of Meyer scaling function phi(t)
Np=120; %samples per pi
Nc=18; %number of data cycles
%frequencies extended for better time resolution:
w=0:(pi/Np):(Nc*pi);
L=length(w);
n23=1+((2*Np)/3); %samples for 0..2/3 pi
n43=1+((4*Np)/3); %samples for 0..4/3 pi
wc=w(n23:n43); %central zone
x=((3/(2*pi))*abs(wc))-1;
nu=35-(84*x)+(70*(x.^2))-(20*(x.^3));
nu=(x.^4).*nu;
PHI=zeros(1,L);
PHI(1:(n23-1))=1;
PHI(n23:n43)=cos((pi/2)*nu);
yphi=ifftshift(ifft(PHI));
ct=L/Np; %scaling according with sampling rate
phi=ct*yphi;
figure(1)
tiv=100/(Nc*Np);
tx=(-50):tiv:(50);
t=tx*(Np/60);
Mt=floor(L/2); My=1+(Mt); Mw=4*(Nc+1);
plot(t((Mt-Mw):(Mt+Mw)),real(phi((My-Mw):(My+Mw))),'k');
axis([-7 7 -0.3 1.2]);
xlabel('t'); title('phi(t)');
```

Figure 2.35 shows the Meyer wavelet $\psi(t)$ corresponding to $v(x)$ as in (2.117). The figure has been generated with the Program 2.19.

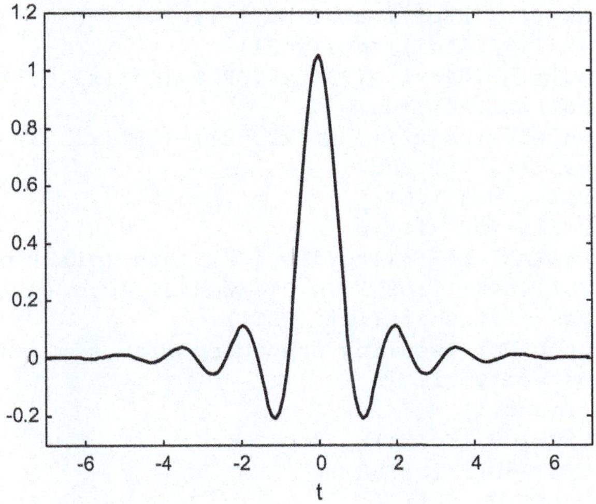

Fig. 2.34 The Meyer scaling function

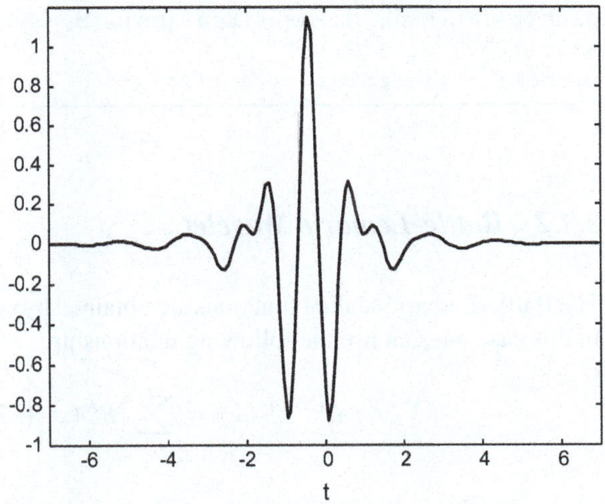

Fig. 2.35 The Meyer wavelet

Program 2.19 Display of Meyer wavelet psi(t)

```
% Display of Meyer wavelet psi(t)
Np=120; %samples per pi
Nc=19; %number of data cycles
%frequencies extended for better time resolution:
w=0:(pi/Np):(Nc*pi);
L=length(w);
n23=1+((2*Np)/3); %samples for 0..2/3 pi
n43=1+((4*Np)/3); %samples for 0..4/3 pi
n83=1+((8*Np)/3); %samples for 0..8/3 pi
wc1=w(n23:n43); %zone 1
wc2=w((n43+1):n83); %zone 2
```

```
x1=((3/(2*pi))*abs(wc1))-1;
x2=((3/(4*pi))*abs(wc2))-1;
nu1=35-(84*x1)+(70*(x1.^2))-(20*(x1.^3));
nu1=(x1.^4).*nu1;
nu2=35-(84*x2)+(70*(x2.^2))-(20*(x2.^3));
nu2=(x2.^4).*nu2;
PSI=zeros(1,L);
PSI(1:(n23-1))=0;
PSI(n23:n43)=(exp(j*wc1/2).*sin((pi/2)*nu1));
PSI((n43+1):n83)=(exp(j*wc2/2).*cos((pi/2)*nu2));
ypsi=ifftshift(ifft(PSI));
ct=L/Np; %scaling according with sampling rate
psi=ct*ypsi;
figure(1)
tiv=100/(Nc*Np);
tx=(-50):tiv:(50);
t=tx*(Np/60);
Mt=floor(L/2); My=1+(Mt); Mw=4*(Nc+1);
plot(t((Mt-Mw):(Mt+Mw)),real(psi((My-Mw):(My+Mw))),'k');
axis([-7 7 -1 1.2]);
xlabel('t'); title('psi(t)');
```

2.4.2 Battle-Lemarié Wavelet

The Battle-Lemarié scaling functions are obtained by orthogonalization of B-splines. In this case one can use the following relationship:

$$B^{(2N+1)}(\omega) = \sum_k \left|B^N(\omega + 2k\pi\right|^2 \tag{2.122}$$

where $B^{(2N+1)}$ is the DFT of the sampled version of the continuous time $b^{(2N+1)}$ B-spline.

The orthogonal scaling function is obtained with:

$$\Phi_{on}(\omega) = \frac{B^N(\omega)}{\sqrt{B^{(2N+1)}(\omega)}} \tag{2.123}$$

Using the MAE equation in the frequency domain, we can write:

$$\begin{aligned} H_0(\omega) &= \sqrt{2}\,\frac{\Phi_{on}(2\omega)}{\Phi_{on}(\omega)} = \sqrt{2}\left(\frac{\sin\omega}{\omega}\;\frac{(\omega/2)}{\sin(\omega/2)}\right)^{N+1} \frac{\sqrt{B^{(2N+1)}(\omega)}}{\sqrt{B^{(2N+1)}(2\omega)}} = \\ &= \sqrt{2}\,(\cos(\omega/2))^{N+1}\,\frac{\sqrt{B^{(2N+1)}(\omega)}}{\sqrt{B^{(2N+1)}(2\omega)}} \end{aligned} \tag{2.124}$$

The Battle-Lemarié scaling functions and the corresponding wavelets have infinite time-domain support, but decay exponentially fast.

For example, in the case of the triangular B-spline $B^1(\omega)$:

$$B^{(3)}(\omega) = \frac{2}{3} + \frac{1}{3} \cos(\omega) = 1 - \frac{2}{3} \sin^2(\frac{\omega}{2}) = \frac{1}{3} (1 + 2 \cos^2(\frac{\omega}{2})) \quad (2.125)$$

Denote:

$$V = \frac{1}{3} (1 + 2 \cos^2(\frac{\omega}{2})) \quad (2.126)$$

Then:

$$\Phi_{on}(\omega) = \frac{\sin^2(\omega/2)}{(\omega/2)^2 \sqrt{V}} = \sqrt{3} \frac{4 \sin^2(\omega/2)}{\omega^2 \sqrt{1 + 2 \cos^2(\omega/2)}} \quad (2.127)$$

And,

$$H_0(\omega) = \sqrt{2} \frac{\cos^2(\omega/2) \sqrt{V}}{\sqrt{1 - (2/3) \sin^2(\omega)}} \quad (2.128)$$

$$H_1(\omega) = -e^{-j\omega} H_0(\omega + \pi) \quad (2.129)$$

$$\Psi(\omega) = -e^{-j(\omega/2)} \frac{\sin^4(\omega/4)}{(\omega/4)^2 \sqrt{V}} \sqrt{\frac{1 - (2/3) \cos^2(\omega/4)}{1 - (2/3) \sin^2(\omega/4)}} \quad (2.130)$$

Figure 2.36 shows V.

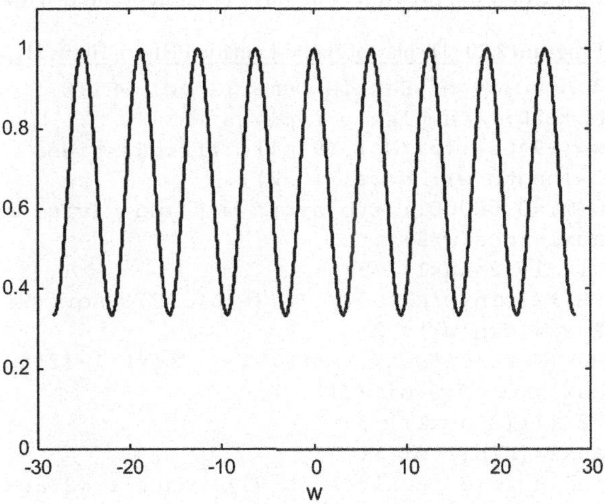

Fig. 2.36 The values of V along frequency

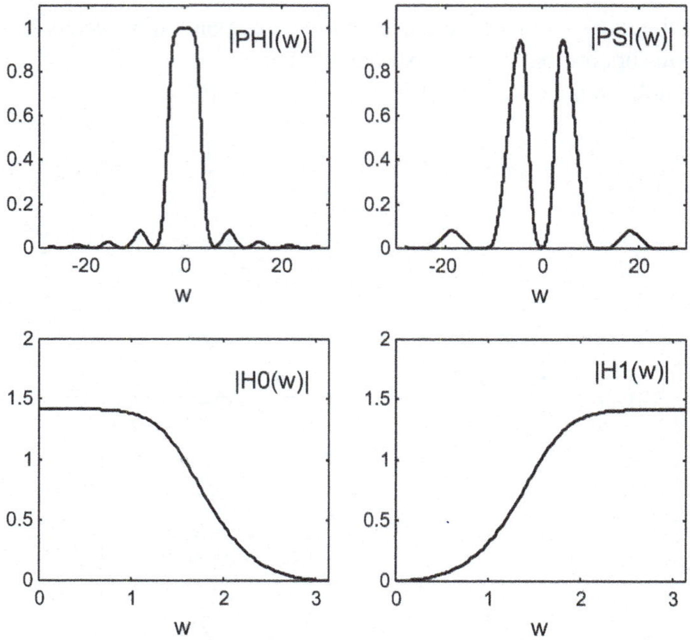

Fig. 2.37 The Battle-Lemarié first order scaling function $\Phi(\omega)$, wavelet $\Psi(\omega)$, and magnitude frequency response of filters H_0 and H_1

Figure 2.37 shows in frequency domain the Battle-Lemarié scaling function, $H_0(\omega)$, $H_1(\omega)$, and wavelet corresponding to the triangular B-spline $B^1(\omega)$. This figure and the previous one have been generated with the Program 2.20.

Program 2.20 Display of Battle-Lemarié PHI(w), H0(w, H1(w) and PSI(w)

```
% Display of Battle-Lemari\'{e} PHI(w), H0(w), H1(w) and PSI(w)
Np=120; %samples per pi
w=(-9*pi):(pi/Np):(9*pi); %frequencies
L=length(w); M=ceil(L/2);
w(M)=0.000001; %to avoid w=0 and division by cero
aux1=(cos(w/2)).^2;
V1=(1+(2*aux1))/3;
PHI=((sin(w/2)).^2)./(((w/2).^2).*sqrt(V1));
aux2=(sin(w)).^2;
H0=(sqrt(2)*aux1.*sqrt(V1))./sqrt(1-(2*aux2/3));
aux1=(cos((w+pi)/2)).^2;
V2=(1+(2*aux1))/3;
aux2=(sin(w+pi)).^2;
aux=sqrt(2)*aux1.*sqrt(V2)./sqrt(1-(2*aux2/3));
H1=-(exp(-j*w)).*aux;
aux0=(exp(-j*(w/2))).*((sin(w/4)).^4)./(((w/4).^2).*sqrt(V1));
aux1=(1-((2/3)*((cos(w/4)).^2)))./(1-((2/3)*((sin(w/4)).^2)));
```

```
PSI=aux0.*sqrt(aux1);
figure(1)
plot(w,V1,'k');
xlabel('w'); title('V');
axis([-30 30 0 1.1]);
figure(2)
subplot(2,2,1)
plot(w,abs(PHI),'k');
xlabel('w'); title('|PHI(w)|');
axis([-30 30 0 1.1]);
subplot(2,2,2)
plot(w,abs(PSI),'k');
xlabel('w'); title('|PSI(w)|');
axis([-30 30 0 1.1]);
subplot(2,2,3)
plot(w(M:(M+Np)),abs(H0(M:(M+Np))),'k');
xlabel('w'); title('|H0(w)|');
axis([0 pi 0 2]);
subplot(2,2,4)
plot(w(M:(M+Np)),abs(H1(M:(M+Np))),'k');
xlabel('w'); title('|H1(w)|');
axis([0 pi 0 2]);
```

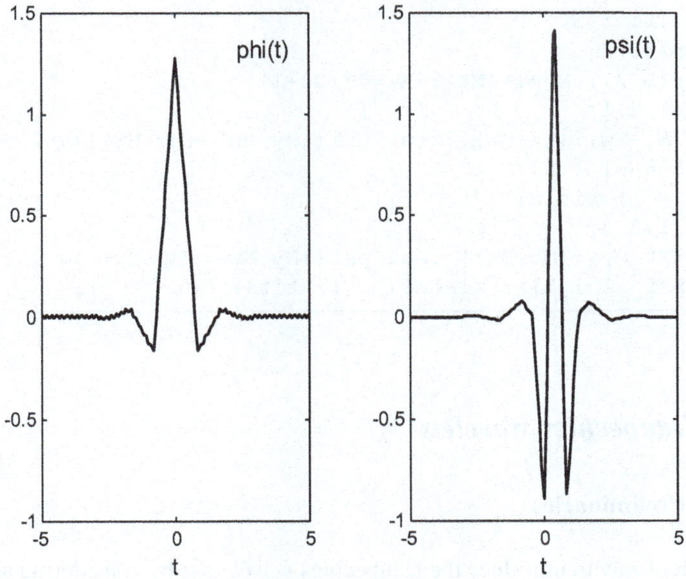

Fig. 2.38 The Battle-Lemarié first order scaling function φ(t)

Figure 2.38 shows in time domain the Battle-Lemarié scaling function and wavelet corresponding to the triangular B-spline $B^1(\omega)$. The figure has been generated with the Program 2.21.

Program 2.21 Display of Battle-Lemarié phi(t), and psi(t)

```
% Display of Battle-Lemari\'{e} phi(t), and psi(t)
Np=120; %samples per pi
Nc=18; %number of data cycles
wiv=pi/Np; %frequency interval
%frequencies extended for better time resolution:
w=wiv:wiv:(Nc*pi)+wiv;
L=length(w);M=ceil(L/2);
w(M)=0.000001; %to avoid w=0 and division by cero
V=(1+(2*(cos(w/2)).^2))/3;
PHI=((sin(w/2)).^2)./(((w/2).^2).*sqrt(V));
aux0=(exp(-j*(w/2))).*((sin(w/4)).^4)./(((w/4).^2).*sqrt(V));
aux1=(1-((2/3)*((cos(w/4)).^2)))./(1-((2/3)*((sin(w/4)).^2)));
PSI=aux0.*sqrt(aux1);
yphi=ifftshift(ifft(PHI));
ct=L/Np; %scaling according with sampling rate
phi=ct*yphi;
ypsi=ifftshift(ifft(PSI));
psi=ct*ypsi;
figure(1)
tiv=100/(Nc*Np);
tx=(-50):tiv:(50);
t=tx*(Np/60);
Mt=floor(L/2); My=1+(Mt); Mw=4*(Nc+1);
subplot(1,2,1)
plot(t((Mt-Mw):(Mt+Mw)),real(phi((My-Mw):(My+Mw))),'k');
axis([-5 5 -1 1.5]);
xlabel('t'); title('phi(t)');
subplot(1,2,2)
plot(t((Mt-Mw):(Mt+Mw)),real(psi((My-Mw):(My+Mw))),'k');
axis([-5 5 -1 1.5]); xlabel('t'); title('psi(t)');
```

2.4.3 Daubechies Wavelets

2.4.3.1 Preliminaries

A convenient way to introduce the Daubechies wavelets is by considering an orthogonal scaling function and its corresponding filter LP(z). This filter is said to be *K-regular* if $H_0(z)$ has K zeros at $z = exp(j\pi)$, so:

$$H_0(z) = \left(\frac{1+z^{-1}}{2}\right)^K Q(z) \tag{2.131}$$

where $Q(z)$ has no poles or zeros at $z = exp(j\pi)$.

Suppose that $H_0(z)$ is polynomial with degree $N-1$. It can be shown that:

$$1 \leq K \leq \frac{N}{2} \tag{2.132}$$

The Daubechies design aims at maximum regularity. To see the implications consider the moments of $\psi(t)$ and $\varphi(t)$:

$$m_0(k) = \int t^k \varphi(t)\, dt \tag{2.133}$$

$$m_1(k) = \int t^k \psi(t)\, dt \tag{2.134}$$

and the discrete moments:

$$\mu_0(k) = \sum_n n^k h_0(n) \tag{2.135}$$

$$\mu_1(k) = \sum_n n^k h_1(n) \tag{2.136}$$

It can be shown that the filter $H_0(z)$ is *K-regular* if and only if the following equivalent statements are true:

$$\forall \mu_1(k) = 0,\ k = 0, 1, 2, \ldots, K-1 \tag{2.137}$$

$$\forall m_1(k) = 0,\ k = 0, 1, 2, \ldots, K-1 \tag{2.138}$$

Recall from Sect. 2.3 that one of the necessary conditions for $\varphi(t)$ to be a solution of the MAE, if the integer translates of $\varphi(t)$ are orthogonal, is that:

$$|H_0(\omega)|^2 + |H_0(\omega+\pi)|^2 = 2 \tag{2.139}$$

2.4.3.2 The Daubechies Wavelets

Daubechies wavelets are an important alternative for many practical applications. These wavelets allow for localization in both time and frequency.

Daubechies obtained orthonormal wavelets with compact support and the maximum number of vanishing moments, with:

$$H_0(\omega) = \sqrt{2}\left(\frac{1+e^{-j\omega}}{2}\right)^K D(\omega), \quad K \leq \frac{N}{2} \tag{2.140}$$

($D(\omega)$ is a trigonometric polynomial).

The filter must satisfy the orthogonality condition (2.139).

Let us work on one of the squared terms that appear in the orthogonality condition:

$$\begin{aligned} |H_0(\omega)|^2 &= 2\left|\tfrac{1+e^{-j\omega}}{2}\right|^{2K} |D(\omega)|^2 = 2\left|\cos^2(\tfrac{\omega}{2})\right|^K |D(\omega)|^2 = \\ &= 2\left|\cos^2(\tfrac{\omega}{2})\right|^K P(\sin^2(\tfrac{\omega}{2})) = 2(1-y)^K P(y) \end{aligned} \tag{2.141}$$

(change of variable: $y = \sin^2(\omega/2)$)

Therefore the orthogonality condition (2.139) can be written as:

$$(1-y)^K P(y) + y^K P(1-y) = 1 \tag{2.142}$$

which is a Bezout's identity. In this case there are explicit solutions. If we set $K = N/2$, the solution is:

$$P(y) = \sum_{l=0}^{K-1} \binom{K-1+l}{l} y^l \tag{2.143}$$

If we want $K < N/2$, then the solution is:

$$P(y) = \sum_{l=0}^{K-1} \binom{K-1+l}{l} y^l + y^K R(\frac{1}{2} - y) \tag{2.144}$$

where $R(y)$ is an odd polynomial such that:

$$P(y) \geq 0 \; for \; 0 \leq y \leq 1 \tag{2.145}$$

Here are some examples of $P(y)$ for several values of $K = N/2$:

$$\begin{aligned} &K = 2, \; P(y) = 1 + 2y \\ &K = 3, \; P(y) = 1 + 3y + 6y^2 \\ &K = 4, \; P(y) = 1 + 4y + 10y^2 + 20y^3 \\ &K = 5, \; P(y) = 1 + 5y + 15y^2 + 35y^3 + 70y^4 \end{aligned} \tag{2.146}$$

Once $|H_0(\omega)|^2$ is obtained, its "square root" can be computed via factorization of $P(y)$. It is convenient for this purpose to use $z = exp(j\omega)$. The target is to get a factorization in the form $P(z) = L(z)\,L(z^{-1})$. First notice that:

$$y = \sin^2(\frac{\omega}{2}) = \frac{1}{2} - \frac{z + z^{-1}}{4} = \frac{2 - (z + z^{-1})}{4} \tag{2.147}$$

Introduce an auxiliary polynomial $Q(z)$ such that:

$$P(\frac{2-(z+z^{-1})}{4}) = z^{-(K-1)} Q(z) \tag{2.148}$$

Now compute the roots a_m of the polynomial $Q(z)$. Then:

$$L(z)\,L(z^{-1}) = r^2 \prod_{m=1}^{K-1} (1 - a_m z)\,(1 - a_m^{-1} z^{-1}) = z^{-(K-1)}\, Q(z) \tag{2.149}$$

with:

$$r = 1/(\prod_{m=1}^{K-1} (1 - a_m)) \tag{2.150}$$

Factorization becomes easy with the help of $Q(z)$. For example, in the case of $K = 2$:

$$\begin{aligned} &P(y) = 1 + 2y \\ &P(\tfrac{2-(z+z^{-1})}{4}) = 1 + 2\,(\tfrac{2-(z+z^{-1})}{4}) = \tfrac{4-(z+z^{-1})}{2} = z^{-1}\, Q(z) \end{aligned} \tag{2.151}$$

$$Q(z) = \frac{1}{2}(-z^2 + 4z - 1) \tag{2.152}$$

The roots of $Q(z)$ are $2 + \sqrt{3},\ 2 - \sqrt{3}$. Let us choose the root inside the unit circle ($|a_m| < 1$) for L(z). Therefore:

$$\begin{aligned} L(z) &= r\,(1 - a_1 z) = \tfrac{1}{1-a_1}(1 - a_1 z) = \tfrac{1}{\sqrt{3}-1}(1 - (2 - \sqrt{3})\,z) = \\ &= \tfrac{\sqrt{3}+1}{2}(1 - (2 - \sqrt{3})\,z) = \tfrac{\sqrt{3}+1}{2} - \tfrac{\sqrt{3}-1}{2} z \end{aligned} \tag{2.153}$$

Likewise:

$$L(z^{-1}) = \frac{\sqrt{3}+1}{2} - \frac{\sqrt{3}-1}{2} z^{-1} \tag{2.154}$$

with it the desired result is obtained:

$$\begin{aligned} H_0(z) &= \sqrt{2}\,\left(\tfrac{1+z}{2}\right)^2 L(z) = \tfrac{\sqrt{2}}{4}(1 + 2z + z^2)\left(\tfrac{\sqrt{3}+1}{2} - \tfrac{\sqrt{3}-1}{2} z\right) = \\ &= \tfrac{1}{4\sqrt{2}}(1 + \sqrt{3} + (3 + \sqrt{3})\,z + (3 - \sqrt{3})\,z^2 + (1 - \sqrt{3})\,z^3) \end{aligned} \tag{2.155}$$

(this gives the filter coefficients already obtained in Sect. 2.3.3. for $M = 4$).

Other Daubechies orthonormal wavelets can be obtained for higher values of K, following the same factorization procedure, which gives minimal phase H_0 filters (since roots inside the unit circle were chosen for $L(z)$). A colateral result of the minimal phase is the marked asymmetry of the scaling functions and wavelets. This asymmetry denotes that the filter H_0 is non-linear phase.

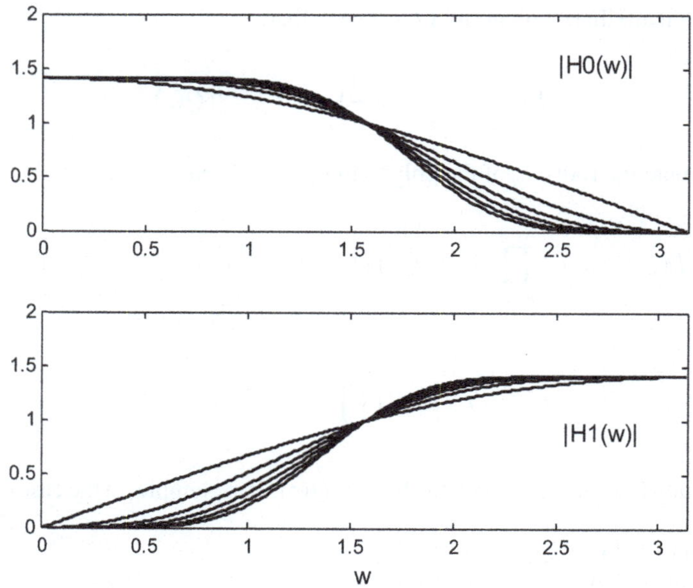

Fig. 2.39 Frequency magnitude responses of H_0 and H_1 Daubechies filters for N = 2, 4, 6, … 12

Other factorizations are possible. For instance a maximal phase factorization, choosing the roots outside the unit circle for $L(z)$. Or for instance mixed solutions, taking roots inside the unit circle and others outside the unit circle. Not always a factorization leads to an orthonormal wavelet basis; but it is sufficient that $H_0(\omega)$ has no zeros in the band $[0.\pi/2]$.

Daubechies has shown that, except by the Haar basis, there is no factorization yielding symmetric scaling functions and wavelets (for orthonormal wavelets having compact support).

Figure 2.39 shows the frequency magnitude responses of the filters H_0 and H_1 for $N = 2, 4, 6 \ldots 12$ and $K = N/2$. The figure has been generated with the Program 2.22, which contains interesting code.

Program 2.22 Compute Daubechies filters for N = 2, 4, 6 … 12

```
% Compute Daubechies filters for N=2,4,6...12
w=0:(2*pi/511):pi;
for K=1:6,
a=1; p=1; q=1;
h=[1 1];
M=2*K; %length of the filter
% the h0(n) coefficients
for nn=1:K-1,
h=conv(h,[1,1]);
a=-a*0.25*(nn+K-1)/nn;
p=conv(p,[1,-2,1]);
```

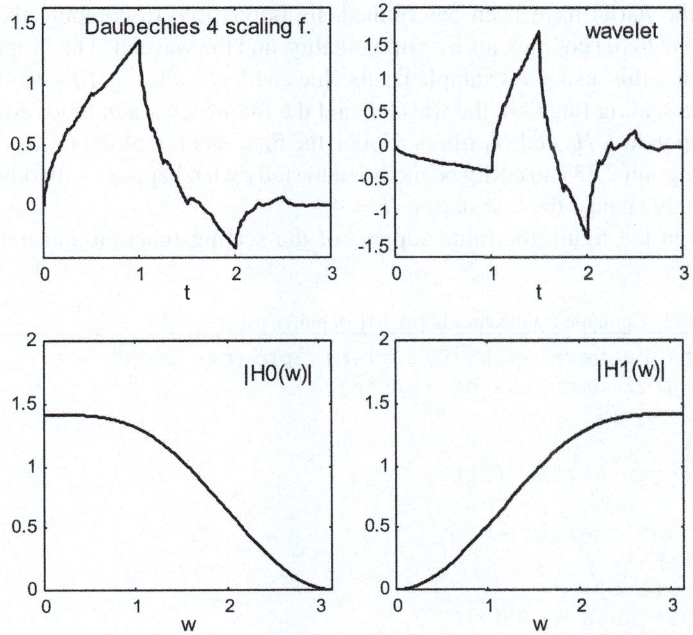

Fig. 2.40 Daubechies N = 4 scaling function, wavelet, and frequency magnitude responses of H_0 and H_1 filters

```
q=[0 q 0] + a*p;
end;
q=sort(roots(q));
aux=real(poly(q(1:K-1)));
h=conv(h,aux);
h0=(h*sqrt(2))/sum(h); %normalization
H0=fft(h0,512); %LP filter frequency response
%the h1(n) coefficients
h1=fliplr(h0); h1(1:2:end)=-h1(1:2:end);
H1=fft(h1,512); %HP filter frequency response
%display
subplot(2,1,1)
plot(w,abs(H0(1:256)),'k'); hold on;
axis([0 pi 0 2]);
title('|H0(w)|');
subplot(2,1,2)
plot(w,abs(H1(1:256)),'k'); hold on;
axis([0 pi 0 2]);
title('|H1(w)|'); xlabel('w');
end;
```

Once the $h_0(n)$ have been determined, its is possible to compute the scaling function, the $h_1(n)$ coefficients by orthogonality, and the wavelet. The Program 2.25 demonstrates this, using as example Daubechies with $N = 4$. Fig. 10.4.12 shows the results: the scaling function, the wavelet, and the frequency magnitude responses of the corresponding H_0 and H_1 filters. Notice the flatness of the filters.

The Program 2.23 can easily be modified to study what happens with other values of N. Simply change the line stating $N = 4$.

Notice in the figure the finite support of the scaling function and the wavelet (Fig. 2.40).

Program 2.23 Compute Daubechies h0(n), h1(n), phi(t), psi(t)

```
% Compute Daubechies h0(n), h1(n), phi(t), psi(t)
% for M=4 (length of the filter)
a=1; p=1; q=1;
h=[1 1];
M=4; %length of the filter
K=M/2;
% the h0(n) coefficients
for nn=1:K-1,
h=conv(h,[1,1]);
a=-a*0.25*(nn+K-1)/nn;
p=conv(p,[1,-2,1]);
q=[0 q 0] + a*p;
end;
q=sort(roots(q));
aux=real(poly(q(1:K-1)));
h=conv(h,aux);
h0=(h*sqrt(2))/sum(h); %normalization
H0=fft(h0,512); %LP filter frequency response
%the scaling function phi(t), using cascade algorithm
Ns=128; %number of fi samples
hN=sqrt(2)*h0;
phi=[ones(1,3*M*Ns),0]/(3*M); %initial iteration
%upsample hN, inserting Ns-1 zeros between samples
hup=[hN;zeros(Ns-1,M)];
hup=hup(1:(Ns*M));
%iteration
for nn=0:12,
aux=conv(hup,phi);
phi=aux(1:2:length(aux)); %downsampling by 2
end
%the h1(n) coefficients
h1=fliplr(h0); h1(1:2:end)=-h1(1:2:end);
H1=fft(h1,512); %HP filter frequency response
%the wavelet psi(t), using definition
%upsample by K
hN=-sqrt(2)*h1;
h1up=[hN;zeros(Ns-1,M)];
```

```
h1up=h1up(1:Ns*M-1);
%downsample by 2
aux=conv(h1up,phi);
psi=aux(1:2:length(aux));
%display
subplot(2,2,1)
phi=phi(1:(M-1)*Ns); %the supported part
t=(1:length(phi))/Ns;
plot(t,phi,'k'); %plots the scaling function
axis([0 max(t) 1.2*min(phi) 1.2*max(phi)]);
title('Daubechies 4 scaling f.'); xlabel('t');
subplot(2,2,2)
psi=psi(1:(M-1)*Ns); %the supported part
plot(t,psi,'k'); %plots the wavelet
axis([0 max(t) 1.2*min(psi) 1.2*max(psi)]);
title('wavelet'); xlabel('t');
w=0:(2*pi/511):pi;
subplot(2,2,3)
plot(w,abs(H0(1:256)),'k');
axis([0 pi 0 2]);
title('|H0(w)|');xlabel('w');
subplot(2,2,4)
plot(w,abs(H1(1:256)),'k');
axis([0 pi 0 2]);
title('|H1(w)|');xlabel('w');
```

The code of the previous programs has been re-used to study the changes on scaling functions and wavelets as N grows up. Figure 2.41 shows the scaling function (left) and the wavelet (right) for $N = 4, 8, 12, \ldots 26$. The figure has been generated with the Program A.4 that has been included in Appendix A.

It is interesting to note that Daubechies obtains the Haar wavelet for $N = 2$. This is not shown in the figure, but it should be easy for the reader to plot this case.

Notice again that for $N > 2$ the Daubechies scaling function and wavelet are non-symmetrical.

2.4.3.3 Symlets

Symmlets are the result of factorizations leading to the least asymmetric scaling function and wavelet. In other words, one tries to get as linear phase as possible.

Each of the roots chosen for $L(z)$ has a corresponding phase contribution to the phase θ of $H_0(\omega)$. A measure of the nonlinear part of θ was defined by Daubechies as follows:

$$\eta(\omega) \;=\; \theta(\omega) \;-\; \frac{\omega}{2\pi}\theta(2\pi) \tag{2.156}$$

The idea is then to choose the roots minimizing $\eta(\omega)$.

Next table gives the coefficients of symlets 2, 3 and 4.

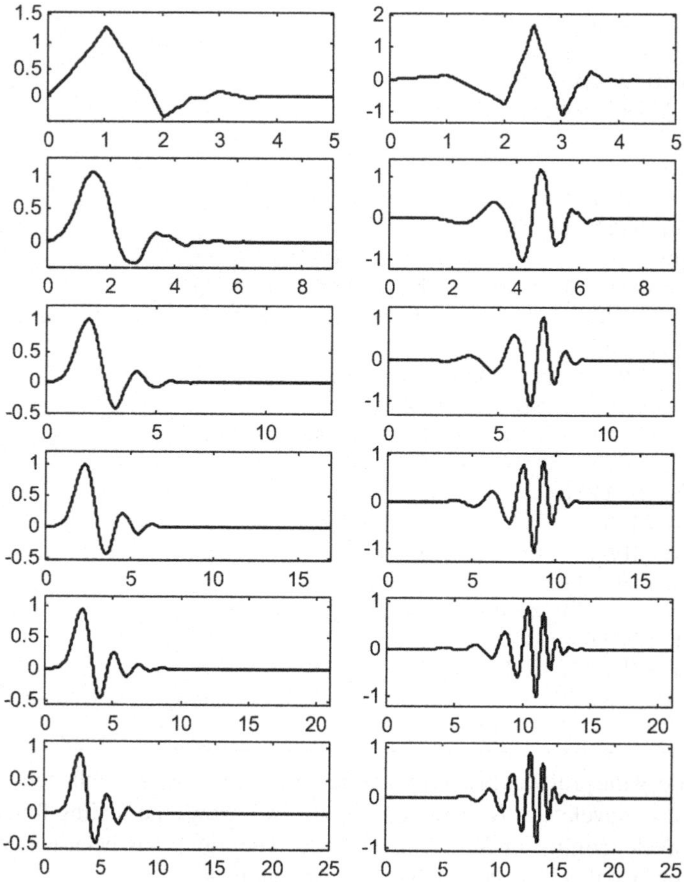

Fig. 2.41 Daubechies scaling function (*left*) and wavelet (*right*) for N = 4, 8, 12, ... 26

h_0 (symlet 2)	h_0 (symlet 3)	h_0 (symlet 4)
0.482962913145	0.332670552951	0.032223100604
0.836516303737	0.806891509313	−0.012603967262
0.224143868042	0.459877502119	−0.099219543577
−0.129409522551	−0.135011020010	0.297857795606
	−0.085441273882	0.803738751807
	0.035226291882	0.497618667633
		−0.029635527646
		−0.075765714789

A more complete table, with a different normalization of coefficients, is given in [25], (p. 198).

Figure 2.42 shows the symlet 4 scaling function and wavelet. The figure has been generated with Program 2.24.

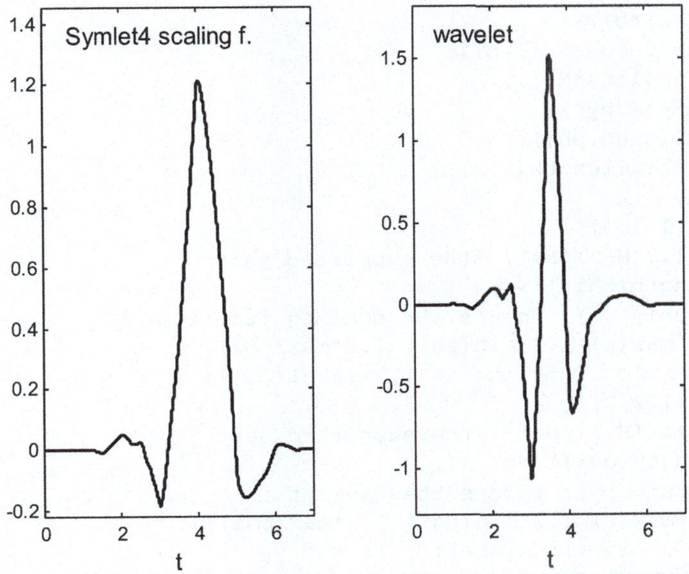

Fig. 2.42 Symlet 4 scaling function (*left*) and wavelet (*right*)

Program 2.24 Symlet4 phi(t), psi(t)

```
% Symlet4 phi(t), psi(t)
%coefficients
h0=[0.032223100604,-0.012603967262,-0.099219543577...
,0.297857795606,0.803738751807,0.497618667633...
,-0.029635527646,-0.075765714789];
Ns=128; %number of function samples
%scaling function
%using cascade algorithm
M=length(h0);
hN=sqrt(2)*h0;
phi=[ones(1,3*M*Ns),0]/(3*M); %initial iteration
%upsample hN, inserting Ns-1 zeros between samples
hup=[hN;zeros(Ns-1,M)];
hup=hup(1:(Ns*M));
%iteration
for nn=0:12,
aux=conv(hup,phi);
phi=aux(1:2:length(aux)); %downsampling by 2
end
%wavelet
%the h1(n) coefficients
h1=fliplr(h0); h1(1:2:end)=-h1(1:2:end);
%the wavelet psi(t), using definition
%upsample
```

```
hN=sqrt(2)*h1;
h1up=[hN;zeros(Ns-1,M)];
h1up=h1up(1:Ns*M-1);
%downsample by 2
aux=conv(h1up,phi);
psi=aux(1:2:length(aux));
%display
subplot(1,2,1)
phi=phi(1:(M-1)*Ns); %the supported part
t=(1:length(phi))/Ns;
plot(t,phi,'k'); %plots the scaling function
axis([0 max(t) 1.2*min(phi) 1.2*max(phi)]);
title('Symlet4 scaling f.'); xlabel('t');
subplot(1,2,2)
psi=psi(1:(M-1)*Ns); %the supported part
t=(1:length(psi))/Ns;
plot(t,psi,'k'); %plots the wavelet
axis([0 max(t) 1.2*min(psi) 1.2*max(psi)]);
title('wavelet'); xlabel('t');
```

2.4.3.4 Coiflets

In 1989 R. Coifman suggested to Daubechies to construct wavelet bases with vanishing moments not only for the wavelet, but also for the scaling function. That is:

$$\forall m_0(k) = 0, \; k = 0, 1, 2, \ldots, L-1 \tag{2.157}$$

$$\forall m_1(k) = 0, \; k = 0, 1, 2, \ldots, L-1 \tag{2.158}$$

where L is the order of the *coiflet.*

According with [25], in order to have the specified wavelet vanishing moments, $H_0(\omega)$ should be:

$$H_0(\omega) = \sqrt{2}\left(\frac{1+e^{-j\omega}}{2}\right)^L D(\omega) \tag{2.159}$$

In addition, to have the specified scaling function vanishing moments, $H_0(\omega)$ should be:

$$H_0(\omega) = \sqrt{2}\left(1 + (1-e^{-j\omega})^L\right) E(\omega) \tag{2.160}$$

($D(\omega)$ and $E(\omega)$ are trigonometric polynomials).

Suppose that L is even, so $L = 2K$.

Since

$$\begin{array}{l} \left(\frac{1+e^{-j\omega}}{2}\right)^{2K} = e^{-j\omega K}(\cos^2(\omega/2))^K, \\ (1-e^{-j\omega})^{2K} = e^{-j\omega K}(2j\,\sin^2(\omega/2))^{2K} \end{array} \tag{2.161}$$

Then, to satisfy (2.159) and (2.160), one has to find two trigonometric polynomials, $P_1(\omega)$, $P_2(\omega)$, such that:

$$(\cos^2(\omega/2))^K \, P_1(\omega) \; = \; 1 + (\sin^2(\omega/2))^K \, P_2(\omega) \tag{2.162}$$

This is again the Bezout equation. The solution for $P_1(\omega)$ is:

$$P_1(\omega) \; = \; \sum_{l=0}^{K-1} \binom{K-1+l}{l} (\sin^2(\omega/2))^l \; + \; (\sin^2(\omega/2))^K f(\omega) \tag{2.163}$$

In this expression a $f(\omega)$ must be found to satisfy the orthogonality condition. An approach for this is to take:

$$f(\omega) \; = \; \sum_{n=0}^{L-1} f_n e^{-jn\omega} \tag{2.164}$$

Then a system of K quadratic equations can be set.

Next table gives the coefficients of the coiflets for $K = 1$ and $K = 2$.

$h_0/\sqrt{2}$ (for K = 1)	$h_0/\sqrt{2}$ (for K = 2)
−0.051429728471	0.011587596739
0.238929728471	−0.029320137980
0.602859456942	−0.047639590310
0.272140543058	0.273021046535
−0.051429972847	0.574682393857
−0.011070271529	0.294867193696
	−0.054085607092
	−0.042026480461
	0.016744410163
	0.003967883613
	−0.001289203356
	−0.000509505399

A more complete table can be found in [25] (p. 261).

Figure 2.43 shows the Coiflet1 scaling function and wavelet. The figure has been generated with Program A.5 which is similar to the previous program, the only change is in the coefficients. This program has been included in Appendix A.

2.4.3.5 Example of Signal Analysis and Recovery with Daubechies 4

To conclude this section it seems rewarding to see an example of wavelet decomposition of a signal (analysis), and then a recovery of the signal from the wavelets (synthesis). We use for this example the Daubechies 4 wavelets.

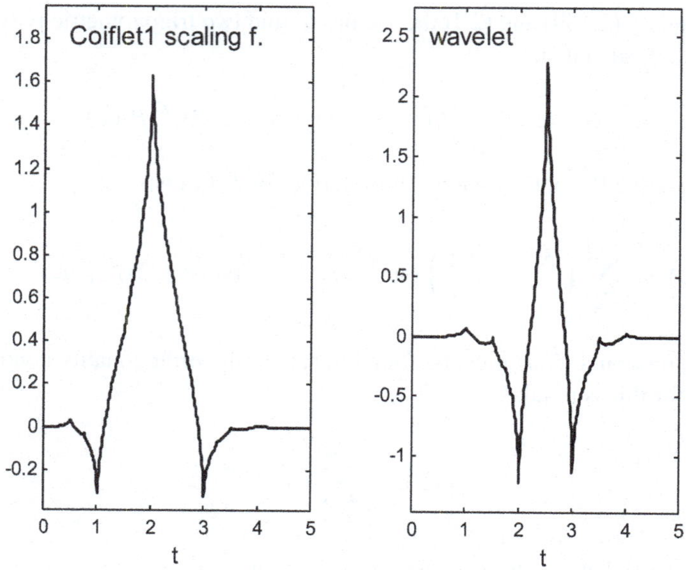

Fig. 2.43 Coiflet1 scaling function (*left*) and wavelet (*right*)

The signal selected for this example is a pattern visual evoked potential (see the web page of Rodrigo Quian Quiroga). It was recorded with a left occipital electrode.

The scalogram obtained with the wavelet analysis is shown in Fig. 2.44. Program 2.25 has been used for the analysis and graphical representation.

Program 2.25 Visual Evoked Potential analysis

```
% Visual Evoked Potential analysis
% Daubechies 4 Wavelet
% Plot of signal and scalogram
%The EVP signal
fs=250; %samplig frequency in Hz
tiv=1/fs; %time interval between samples
Ts=tiv; %sampling period
%read signal file
fer=0;
while fer==0,
fid2=fopen('EVP_short.txt','r');
if fid2==-1, disp('read error')
else sg=fscanf(fid2,'%f \r\n'); fer=1;
end;
end;
fclose('all');
Nss=length(sg); %number of signal samples
duy=(Nss-1)*tiv; %duration of signal
tss=0:tiv:duy; %time intervals set
```

```
%analysis of the signal with wavelets
y=sg;
%scaling filter
hden=4*sqrt(2); %coeff. denominator
hsq=sqrt(3); %factor
%Daubechies 4:
h=[(1+hsq)/hden, (3+hsq)/hden, (3-hsq)/hden, (1-hsq)/hden];
N=length(h);
K=9; %number of scales (512=2^9)
dd=zeros(K,Nss/2); %space for coefficients
a=y';
aux=0;
h0=fliplr(h);
h1=h; h1(1:2:N)=-h1(1:2:N);
%wavelet calculus using filters
NN=Nss;
for n=1:K,
L=length(a);
a=[a(mod((-(N-1):-1),L)+1) a];
d=conv(a,h1);
d=d(N:2:(N+L-2));
a=conv(a,h0);
a=a(N:2:(N+L-2));
aux=[d,aux];
dd(K+1-n,1:NN/2)=d;
NN=NN/2;
end;
wty=[a,aux(1:end-1)];
%preparing for scalogram
S=zeros(K,Nss); %space for S(j,k) scalogram coefficients
for n=1:K,
q=2^(n-1); L=Nss/q;
for m=1:q,
R=(1+(L*(m-1))):(L*m); %index range
S(n,R)=dd(n,m);
end;
end;
%figure
subplot('position',[0.04 0.77 0.92 0.18])
plot(y);
axis([0 512 1.2*min(y) 1.2*max(y)]);
title('signal');
subplot('position',[0.04 0.05 0.92 0.6])
imagesc(S); colormap('bone');
title('Scalogram of Daubechies w.t. Evoked Potential signal');
h=gca; set(h,'YDir','normal');
```

Program 2.26 does the recovery of the evoked potential signal, with Daubechies 4 wavelet synthesis. Figure 2.45 shows the recovered signal.

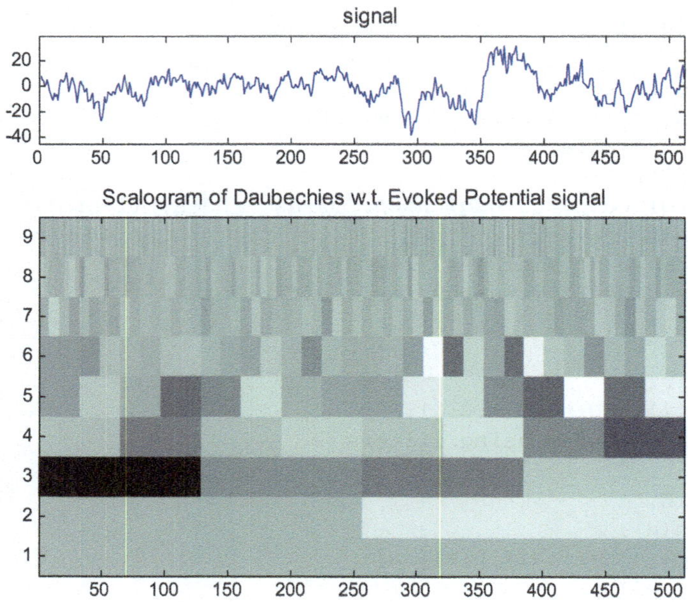

Fig. 2.44 Scalogram of Evoked Potential signal using Daubechies 4

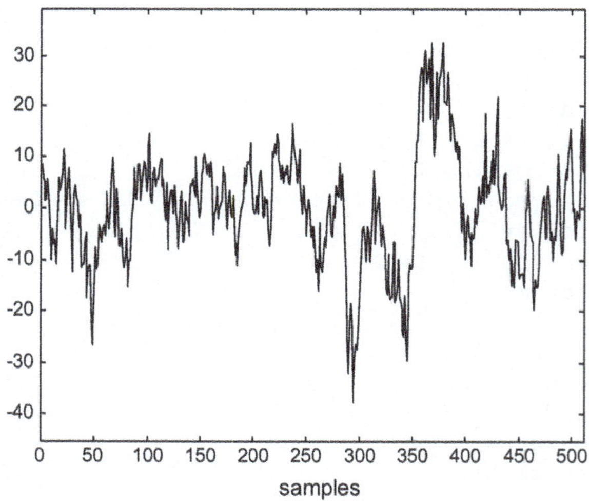

Fig. 2.45 Recovered Evoked Potential signal

Program 2.26 Inverse of the Daubechies DWT of Evoked Potential signal

```
% inverse of the Daubechies DWT
% Visual Evoked Potential signal
L=length(wty); %length of the DWT
%scaling filter
hden=4*sqrt(2); %coeff. denominator
```

```
hsq=sqrt(3); %factor
%Daubechies 4:
h=[(1+hsq)/hden, (3+hsq)/hden, (3-hsq)/hden, (1-hsq)/hden];
hN=h;
N=length(hN);
K=9; %number of scales
aux=0;
h0=hN;
h1=fliplr(hN); h1(2:2:N)=-h1(2:2:N);
Ln=1;
a=wty(1);
for n=1:K,
aux= 1+mod(0:N/2-1,Ln);
d=wty(Ln+1:2*Ln);
ax(1:2:2*Ln+N)= [a a(1,aux)];
dx(1:2:2*Ln+N)= [d d(1,aux)];
a=conv(ax,h0)+ conv(dx,h1);
a=a(N:(N+2*Ln-1));
Ln=2*Ln;
end;
figure(1)
plot(a,'k');
axis([0 512 1.2*min(a) 1.2*max(a)]);
xlabel('samples');
title('the recovered signal');
```

A first intent of comparing the original evoked potential signal, and the recovered signal, was to plot one on top the other. But both signals are practically identical, so

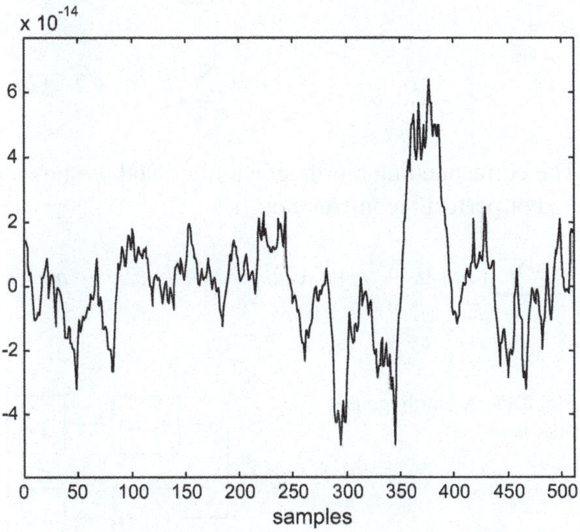

Fig. 2.46 Difference between original and recovered signals

it is not possible to discern one from the other. What has been done is to subtract both signals and depict—Fig. 2.46—the result. Notice that the scale of the vertical axis is 10^{-14}.

2.5 Biorthogonal Wavelets

The use of orthogonal wavelet systems cause limitations in design freedom, and prevent linear phase analysis and synthesis filter banks. Biorthogonal wavelet systems offer greater flexibility.

In a biorthogonal system, a primal scaling function $\varphi(t)$ and a dual scaling function $\tilde{\varphi}(t)$ are defined, such that:

$$\varphi(t) = \sum_n h_0(n)\,\sqrt{2}\,\varphi(2t - n) \tag{2.165}$$

$$\tilde{\varphi}(t) = \sum_n \tilde{h}_0(n)\,\sqrt{2}\,\tilde{\varphi}(2t - n) \tag{2.166}$$

In order to $\varphi(t)$ and $\tilde{\varphi}(t)$ to exist:

$$\sum_n h_0(n) = \sum_n \tilde{h}_0(n) = \sqrt{2} \tag{2.167}$$

A primal and a dual wavelet are also defined, such that:

$$\psi(t) = \sum_n h_1(n)\,\sqrt{2}\,\varphi(2t - n) \tag{2.168}$$

$$\tilde{\psi}(t) = \sum_n \tilde{h}_1(n)\,\sqrt{2}\,\tilde{\varphi}(2t - n) \tag{2.169}$$

The corresponding biorthogonal filter bank is shown in Fig. 2.47.

For perfect reconstruction:

$$\sum_k \left(h_0(2k - m)\,\tilde{h}_0(2k - n) + h_1(2k - m)\,\tilde{h}_1(2k - n) \right) = \delta_{m,n} \tag{2.170}$$

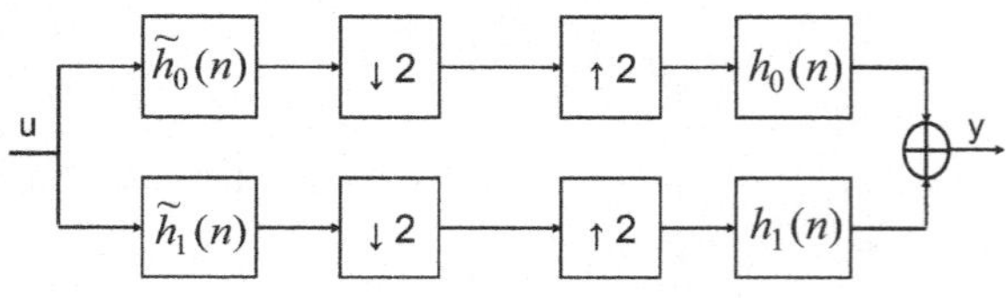

Fig. 2.47 A biorthogonal filter bank

where m, n, are integers.

In order for this condition to hold, the four filters have to be:

$$\tilde{h}_1(n) = \pm(-1)^n \, h_0\,(L - n) \tag{2.171}$$

$$\tilde{h}_0(n) = \pm(-1)^n \, h_1\,(L - n) \tag{2.172}$$

where L is an arbitrary odd integer.

Using these relationships, the condition (2.170) becomes:

$$\sum_n \tilde{h}_0(n)\; h_0(n + 2k) = \delta_k \tag{2.173}$$

Primal and dual baby families are also defined:

$$\varphi_{j,k}(t) = \sqrt{2^j}\varphi(2^j t \; - \; k) \tag{2.174}$$

$$\tilde{\varphi}_{j,k}(t) = \sqrt{2^j}\tilde{\varphi}(2^j t \; - \; k) \tag{2.175}$$

$$\psi_{j,k}(t) = \sqrt{2^j}\psi(2^j t \; - \; k) \tag{2.176}$$

$$\tilde{\psi}_{j,k}(t) = \sqrt{2^j}\tilde{\psi}(2^j t \; - \; k) \tag{2.177}$$

If Eq. (2.170) is satisfied and some other conditions are satisfied by the primal and dual scaling functions, the primal baby wavelets constitute a frame in L^2, and the dual baby wavelets constitute the dual frame if and only if:

$$\int \varphi(t)\tilde{\varphi}(t - k)dt = \delta_k \tag{2.178}$$

Then the primal and dual baby wavelets constitute two Riesz bases, with:

$$< \psi_{j,k}, \; \tilde{\psi}_{j',k'} >= \delta_{j,j'} \; \delta_{k,k'} \tag{2.179}$$

Concerning multiresolution formulations, we have:

$$V_j \subset V_{j+1} \,, \;\; \tilde{V}_j \subset \tilde{V}_{j+1} \tag{2.180}$$

$$V_j \perp \tilde{W}_j \,, \;\; \tilde{V}_j \perp W_j \tag{2.181}$$

The coefficients corresponding to the expansion of a signal $y(t)$ in terms of the primal baby scaling functions and wavelets, can be obtained with:

$$a_{j,k} \; =< \; y, \; \tilde{\varphi}_{j,k} \; > \tag{2.182}$$

$$d_{j,k} = \langle y, \tilde{\psi}_{j,k} \rangle \tag{2.183}$$

And:

$$a_{j,k} = \sum_m \tilde{h}_0(m - 2k)\, a_{j+1,m} \tag{2.184}$$

$$d_{j,k} = \sum_m \tilde{h}_1(m - 2k)\, a_{j+1,m} \tag{2.185}$$

All filters can be symmetric (linear phase). This is a great advantage of biorthogonal wavelets.

2.5.1 Daubechies Approach

Orthogonal wavelets can be considered as a particular case of biorthogonal wavelets. In fact the approach of *K-regular* filters can be used to obtain orthogonal or biorthogonal wavelets.

In orthogonal wavelets we needed to get the "square root" of $|H_0(\omega)|^2$, in the case of biorthogonal wavelets we only need a factorization into two filters (the other two filters are obtained according with (2.171) and (2.172)).

As required for perfect reconstruction in a two-filter bank, the distortion term (see Sect. 1.4.1.) must be in this case the following:

$$\tilde{H}_0(\omega)\, H_0(\omega)^* \cdot \tilde{H}_0(\omega + \pi)\, H_0(\omega + \pi)^* = 2 \tag{2.186}$$

(this also corresponds to condition (2.173)).

Symmetry of filters can be obtained if one chooses $H_0(\omega),\ \tilde{H}_0(\omega)$ of the form;

$$\sqrt{2}(\cos{(\omega/2)})^L\, q_0(\cos{\omega}) \tag{2.187}$$

for L even.
or

$$\sqrt{2}\, e^{-j\omega/2}\, (\cos{(\omega/2)})^L\, q_0(\cos{\omega}) \tag{2.188}$$

for L odd.

Where $q()$ is a polynomial. Substitution into (2.186) gives:

$$\begin{aligned}&(\cos(\omega/2))^K\, \tilde{q}_0(\cos{\omega})\, q_0(\cos{\omega})^* + \\ &+ (\sin(\omega/2))^K\, \tilde{q}_0(-\cos{\omega})\, q_0(-\cos{\omega})^* = 1\end{aligned} \tag{2.189}$$

where $K = L0 + L1$ for $L0$ even, or $K = L0 + L1 + 1$ for $L0$ odd. The integers $L0$ and $L1$ correspond to $\tilde{H}_0(\omega)$ and $H_0(\omega)$.

Now, if you define:

$$P(\sin^2(\omega/2)) = \tilde{q}_0(\cos\omega) \cdot q_0(\cos\omega)^* \tag{2.190}$$

Then:

$$(1-y)^K P(y) + y^K P(1-y) = 1 \tag{2.191}$$

and so, the solution is once again:

$$P(y) = \sum_{l=0}^{K-1} \binom{K-1+l}{l} y^l + y^K R(\frac{1}{2} - y) \tag{2.192}$$

Daubechies [25] obtained several families of biorthogonal wavelets, using different choices for $R()$ and for the factorization of $P()$.

A first alternative is to choose $R() = 0$. In this case, the design focuses on the summatorial term. Several families of biorthogonal wavelets have been obtained, including the spline biorthogonal wavelets.

2.5.1.1 Spline Biorthogonal Wavelets

From the previous equations, and taking $R() = 0$, a family of spline biorthogonal wavelets is directly obtained as follows.

- For odd-length filters, the analysis and synthesis low-pass filters are:

$$\tilde{H}_0(\omega) = \sqrt{2}\,(\cos^2(\omega/2))^{L0} \sum_{n=0}^{K-1} \binom{K+n-1}{n} (\sin^2(\omega/2))^n \tag{2.193}$$

$$H_0(\omega) = \sqrt{2}\,(\cos^2(\omega/2))^{L1} \tag{2.194}$$

  with $K = L0 + L1$.

- For even-length filters, the analysis and synthesis low-pass filters are:

$$\tilde{H}_0(\omega) = \sqrt{2}\,e^{-j\omega/2}(\cos^2(\omega/2))^{L0} \sum_{n=0}^{K-1} \binom{K+n}{n} (\sin^2(\omega/2))^n \tag{2.195}$$

$$H_0(\omega) = \sqrt{2}\,e^{-j\omega/2}(\cos^2(\omega/2))^{L1} \tag{2.196}$$

  with $K = L0 + L1 + 1$.

Here is a table with the coefficients of some members of the Cohen-Daubechies-Feauveau family of biorthogonal spline wavelets [20]:

A more complete table is given in [25], (p. 277), and in [20].

$h_0/\sqrt{2}$	$\tilde{h}_0/\sqrt{2}$
1/2, 1/2	1/2, 1/2
1/2, 1/2	−1/16, 1/16, 1/2, 1/2, 1/16, −1/16
1/4, 1/2, 1/4	−1/8, 1/4, 3/4, 1/4, −1/8
1/4, 1/2, 1/4	3/128, −3/64, −1/8, 19/64, 45/65, 19/64, −1/8, −3/64, 3/128
1/8, 3/8, 3/8,1/8	−1/4, 3/4, 3/4, −1/4
1/8, 3/8, 3/8,1/8	3/64, −9/64, −7/64, 45/64, 45/64, −7/64, −9/64, 3/64
1/8, 3/8, 3/8,1/8	−5/512, 15/512, 19/512, −97/512, −13/256, 175/256, 175/256, −13/256, −97/512, 19/512, 15/512, −5/512

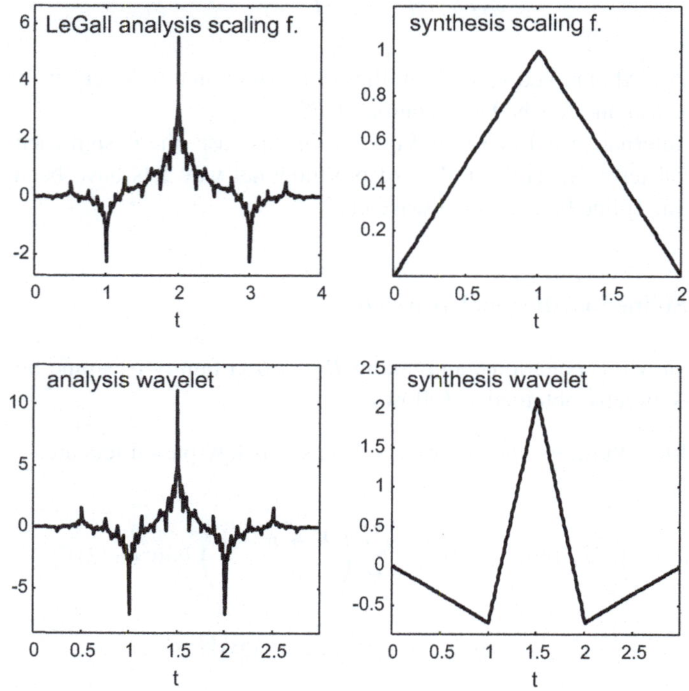

Fig. 2.48 LeGall scaling functions and wavelets

The third row of coefficients corresponds to the LeGall 5/3 wavelet [75]. Figure 2.48 shows the scaling functions and the wavelets of LeGall 5/3. The figure has been generated with Program 2.27.

Program 2.27 LeGall phi(t), psi(t)

```
% LeGall phi(t), psi(t)
%coefficients
ah0=[-1/8, 1/4, 3/4, 1/4, -1/8];
sh0=[1/4, 1/2, 1/4];
Ns=128; %number of function samples
%analysis scaling function
%using cascade algorithm
```

```
Ma=length(ah0);
ah0=2*ah0/sum(ah0); %normalization
aphi=[ones(1,3*Ma*Ns),0]/(3*Ma); %initial iteration
%upsample hN, inserting Ns-1 zeros between samples
hup=[ah0;zeros(Ns-1,Ma)];
hup=hup(1:(Ns*Ma));
%iteration
for nn=0:12,
aux=conv(hup,aphi);
aphi=aux(1:2:length(aux)); %downsampling by 2
end
%synthesis scaling function
%using cascade algorithm
Ms=length(sh0);
sh0=2*sh0/sum(sh0); %normalization
sphi=[ones(1,3*Ms*Ns),0]/(3*Ms); %initial iteration
%upsample hN, inserting Ns-1 zeros between samples
hup=[sh0;zeros(Ns-1,Ms)];
hup=hup(1:(Ns*Ms));
%iteration
for nn=0:12,
aux=conv(hup,sphi);
sphi=aux(1:2:length(aux)); %downsampling by 2
end
%analysis wavelet
%the ah1(n) coefficients
ah1=fliplr(sh0); ah1(1:2:end)=-ah1(1:2:end);
%the wavelet psi(t), using definition
%upsample
hN=sqrt(2)*ah1;
h1up=[hN;zeros(Ns-1,Ms)];
h1up=h1up(1:Ns*Ms-1);
%downsample by 2
aux=conv(h1up,aphi);
apsi=aux(1:2:length(aux));
%synthesis wavelet
%the sh1(n) coefficients
sh1=fliplr(ah0); sh1(1:2:end)=-sh1(1:2:end);
%the wavelet psi(t), using definition
%upsample
hN=-sqrt(2)*sh1;
h1up=[hN;zeros(Ns-1,Ma)];
h1up=h1up(1:Ns*Ma-1);
%downsample by 2
aux=conv(h1up,sphi);
spsi=aux(1:2:length(aux));
%display
subplot(2,2,1)
aphi=aphi(1:(Ma-1)*Ns); %the supported part
```

```
t=(1:length(aphi))/Ns;
plot(t,aphi,'k'); %plots the scaling function
axis([0 max(t) 1.2*min(aphi) 1.2*max(aphi)]);
title('LeGall analysis scaling f.'); xlabel('t');
subplot(2,2,3)
su=round(0.75*length(apsi));
t=(1:su)/Ns;
plot(t,apsi(1:su),'k'); %plots the wavelet
axis([0 max(t) 1.2*min(apsi) 1.2*max(apsi)]);
title('analysis wavelet'); xlabel('t');
subplot(2,2,2)
sphi=sphi(1:(Ms-1)*Ns); %the supported part
t=(1:length(sphi))/Ns;
plot(t,sphi,'k'); %plots the scaling function
axis([0 max(t) 1.2*min(sphi) 1.2*max(sphi)]);
title('synthesis scaling f.'); xlabel('t');
subplot(2,2,4)
su=round(0.75*length(spsi));
t=(1:su)/Ns;
plot(t,spsi(1:su),'k'); %plots the wavelet
axis([0 max(t) 1.2*min(spsi) 1.2*max(spsi)]);
title('synthesis wavelet'); xlabel('t');
```

2.5.1.2 Filters with Nearly Same Lengths

Notice that in the spline biorthogonal wavelets the part with the summatorial is entirely assigned to one of the filters, thus causing disparate filter lengths.

The filter lengths could be more balanced if $P(y)$ was factorized into two polynomials with similar lengths. Daubechies proposed to write $P(y)$ as a product of first and second order polynomial factors, based on real and complex zeros, and then build the two polynomials with these factors in appropriated form.

There is a table in [25], (p. 279) with some biorthogonal wavelets constructed in this way. One of these wavelets is the popular CDF 9/7, which has the following filter coefficients:

$h_0/\sqrt{2}$	$\tilde{h}_0/\sqrt{2}$
0.026748757411	−0.045635881557
−0.016864118443	−0.028771763114
−0.078223266529	0.295635881557
0.266864118443	0.557543526229
0.602949018236	0.295635881557
0.266864118443	−0.028771763114
−0.078223266529	−0.045635881557
−0.016864118443	
0.026748757411	

A main reason for the CDF 9/7 to be popular is that it has been adopted by the FBI for fingerprint image compression [13]. Also, the JPEG 2000 compression standard uses CDF 9/7, and LeGall 5/3 [64, 75].

Figure 2.49 shows the scaling functions and the wavelets of cdf 9/7. The figure has been generated with Program A.6 that is very similar to Program 2.27, with only changes of coefficients. Anyway, the Program A.6 has been included in Appendix A.

To complete the view of the CDF 9/7 filter bank, Fig. 2.50 shows the frequency responses of the four filters. The figure has been generated with Program 2.28.

Program 2.28 CDF 9/7 frequency response of filters

```
% CDF 9/7 frequency response of filters
%coefficients
ah0=[-0.045635881557,-0.028771763114,0.295635881557...
,0.557543526229,0.295635881557,-0.028771763114...
,-0.045635881557];
sh0=[0.026748757411,-0.016864118443,-0.078223266529...
,0.266864118443,0.602949018236,0.266864118443...
,-0.078223266529,-0.016864118443,0.026748757411];
ah0n=(ah0*sqrt(2))/sum(ah0); %normalization
sh0n=(sh0*sqrt(2))/sum(sh0); %normalization
aH0=fft(ah0n,512); %frequency response
sH0=fft(sh0n,512); %frequency response
%the ah1(n) coefficients
ah1=fliplr(sh0n); ah1(1:2:end)=-ah1(1:2:end);
%the sh1(n) coefficients
sh1=fliplr(ah0n); sh1(1:2:end)=-sh1(1:2:end);
aH1=fft(ah1,512); %frequency response
sH1=fft(sh1,512); %frequency response
%display
w=0:(2*pi/511):pi;
subplot(2,2,1)
plot(w,abs(aH0(1:256)),'k');
axis([0 pi 0 2]);
title('|aH0(w)|');xlabel('w');
subplot(2,2,2)
plot(w,abs(aH1(1:256)),'k');
axis([0 pi 0 2]);
title('|aH1(w)|');xlabel('w');
subplot(2,2,3)
plot(w,abs(sH0(1:256)),'k');
axis([0 pi 0 2]);
title('|sH0(w)|');xlabel('w');
subplot(2,2,4)
plot(w,abs(sH1(1:256)),'k');
axis([0 pi 0 2]);
title('|sH1(w)|');xlabel('w');
```

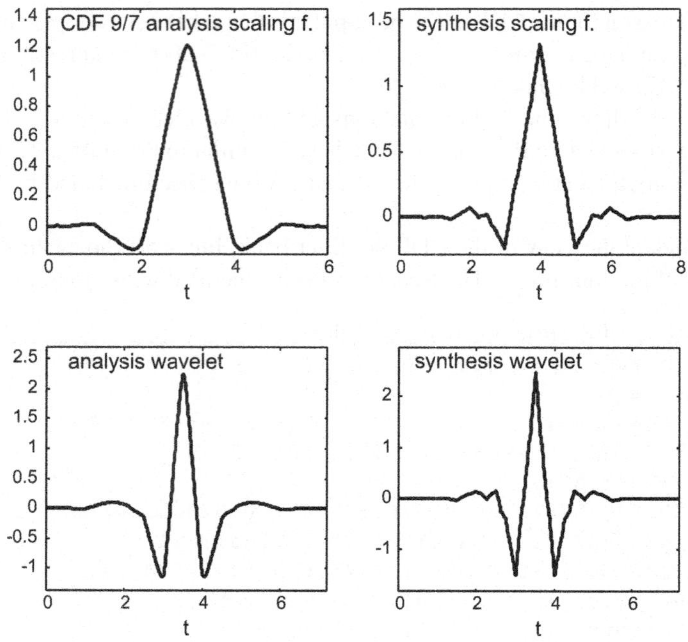

Fig. 2.49 CDF 9/7 scaling functions and wavelets

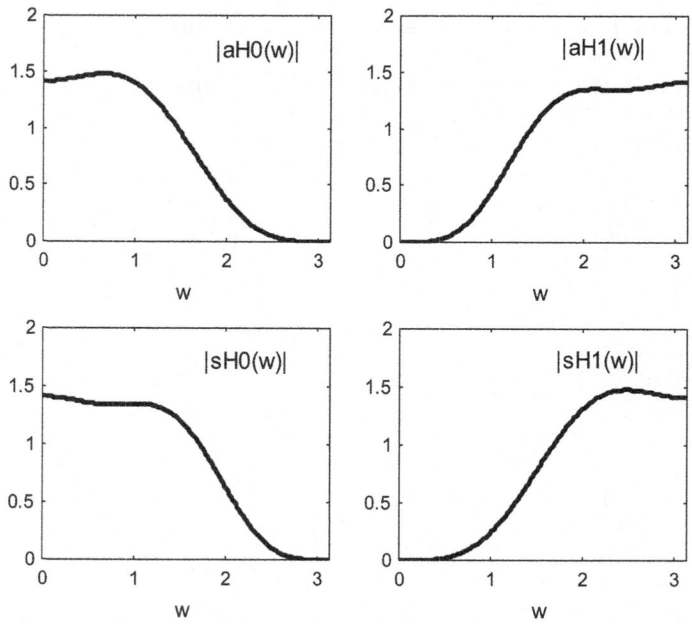

Fig. 2.50 Frequency responses of the CDF 9/7 filter banks

2.5.1.3 Almost Orthonormal Filters

Following the work of Daubechies in [25], it was considered that, based on a Laplacian pyramid filter [15], an almost orthonormal biorthogonal wavelet could be found. We will give more details of this type of filters in a next chapter.

The analysis filter was chosen to be a Laplacian pyramid filter:

$$\tilde{H}_0(\omega) = \sqrt{2}\,(\cos^2(\omega/2))\,(1 + \frac{4}{5}\sin^2(\omega/2)) \tag{2.197}$$

Then, using (2.186):

$$H_0(\omega) = \sqrt{2}\,(\cos^2(\omega/2))\,(1 + \frac{6}{5}\sin^2(\omega/2) - \frac{24}{35}\sin^4(\omega/2)) \tag{2.198}$$

The result is a biorthogonal wavelet that is very close to one of the orthonormal coiflets.

A second initiative of [25] was to drive near orthonormal wavelets using optimization procedures. A proximity criterion was defined, and some requisites were added, like to have rational filter coefficients. The analysis filter was chosen to be:

$$\begin{aligned}\tilde{H}_0(\omega) &= \sqrt{2}\,(\cos^2(\omega/2))^{2K} \cdot \\ &\cdot \left[\sum_{n=0}^{K-1}\binom{K+k-1}{k}(\sin^2(\omega/2))^{2k} + a\cdot\,(\sin^2(\omega/2))^{2K}\right]\end{aligned} \tag{2.199}$$

This expression has one parameter, a, to be optimized. Once its optimal value is found, the corresponding synthesis filter is obtained using (2.186) (actually it is similar to $\tilde{H}_0(\omega)$).

2.5.2 *More Ways to Find Biorthogonal Wavelets*

2.5.2.1 Still More About Coifman Biorthogonal Wavelets

There is more to say about Coifman and biorthogonal wavelets. In the extensive work on coiflet-type wavelets presented in the Thesis [78], there are two chapters on biorthogonal wavelets that detail its construction using Vandermonde matrices. The results are similar to those presented in [70]. More recently, [36] completed the results.

According to [36], if $1 \leq a \leq N$, the Coifman biorthogonal wavelet system of order N with minimum length $2N + 1$ synthesis filter, is obtained as follows:

- The nonzero coefficients of the synthesis filter are given by:

$$h_0 = \frac{1}{\sqrt{2}}, \quad h_{1-2a+2k} = \frac{\prod_{m=0,\, m\neq k}^{N} (2(m-a)+1)}{2^N \sqrt{2} \prod_{m=0, m\neq k}^{N} (m-k)}, \quad k = 0, \ldots, N \tag{2.200}$$

where a is an integer.

- The coefficients of the analysis filter are calculated in recursively form:

$$\tilde{h}_{2k} = \sqrt{2} \left[\delta_{k0} - \sum_n h_{1+2n} h_{1+2(n-k)} \right], \quad \tilde{h}_{2k+1} = h_{2k+1} \tag{2.201}$$

With this procedure, all the list of biorthogonal Coifman wavelet systems given in [70] can be obtained, and other new examples can be found.

An important aspect of the results attained by [36, 70, 78] is that all the filter coefficients are in the form of fractions of integers.

2.5.2.2 Filter Banks and Biorthogonal Wavelets

Clearly, wavelets are linked to perfect reconstruction filter banks. Actually, there are books that arrive to the construction of orthogonal and biorthogonal wavelets in the framework of filter banks. For instance, [77] presents in this way a fairly complete treatement of orthogonal, biorthogonal, etc. wavelets, and other related topics. It extends the scenario by considering the use of IIR filters, instead of the FIR filters. As one interesting example, a wavelet system based on the Butterworth filter is discussed. With IIR filters, orthogonal symmetric wavelets (not Haar) can be designed.

2.5.2.3 Getting Rational Coefficients

In most of the tables of coefficients already shown, the coefficients are irrational. This is not convenient for the use of digital processors. The research has proposed some solutions, for instance the procedure of [68]. This article belongs to the filter bank context. It starts with analysis H_0 and synthesis F_0 low-pass filters, and high-pass filters given by:

$$H_1(z) = z^{-1} F_0(-z); \quad F_1(z) = z\, H_0(-z) \tag{2.202}$$

The product of the low-pass filters $D(z) = H_0(z)\, F_0(z)$ must satisfy the PR condition, $D(z) + D(-z) = 2$. Many wavelet systems have been obtained by factorizing the *Lagrange half-band filter (LHBF):*

$$D(z) = L_K(z) = z^K \left(\frac{1+z^{-1}}{2}\right)^{2K} R_K(z) \tag{2.203}$$

where:

$$R_K(z) = \sum_{n=0}^{K-1} \binom{K+n-1}{n} \left(\frac{2-(z+z^{-1})}{4}\right)^n \tag{2.204}$$

The transition bandwidth of the LHBF is proportional to $\sqrt{N}$, where $N = 2K - 1$. The LHBF is a maximally flat filter. The LHBF is linked to Lagrange interpolation formulae. Obviously, the orthogonal and biorthogonal wavelets of Daubechies are related to factorizations of the LHBF. Using the change of variable: $x = \cos(\omega) = (1/2)(z + z^1)$, the LHBF can be written as follows:

$$L'_K(x) = (x+1)^K \, R'_K(x) \tag{2.205}$$

The presence of irrational coefficients is due to $R'_K(x)$. The idea of [68] is to allow one degree of freedom by using:

$$L_K^M(x) = (x+1)^{K-1} (x+\alpha) \, R_K^M(x) \tag{2.206}$$

and then try to get rational coefficients. For instance, let us factorize the K = 4 LHBF:

$$H_0 = k_1 (x+1)(x^3 + A x^2 + Bx + C) \tag{2.207}$$

$$F_0 = k_2 (x+1)^2 (x+\alpha) \tag{2.208}$$

(k_1, k_2 are normalizing constants)

Then, one gets the product $D(x) = H_0(x) \, F_0(x)$ and applies the PR condition. The terms in x^2, x^4, x^6, must be zero. Three equations are obtained, and finally:

$$A = -(3+\alpha) \tag{2.209}$$

$$B = \frac{9\alpha^3 + 35\alpha^2 + 45\alpha + 24}{3\alpha^2 + 9\alpha + 8} \tag{2.210}$$

$$C = -\frac{8(1+\alpha)^3}{3\alpha^2 + 9\alpha + 8} \tag{2.211}$$

Setting $\alpha = -1.6848$ will give the CDF 9/7. But with values such −3/2, or −5/3, or −8/5, or −2, biorthogonal filters can be obtained with all coefficients rational. In particular, with $\alpha = -5/3$ the corresponding filter bank has responses close to the CDF 9/7.

2.5.2.4 Getting Families by Using One Parameter

In the previous paragraphs the introduction of one parameter, α, has proved to be useful. This idea could be extended for the parameterized construction of many families of biorthogonal wavelets. The contribution of [43] proposes to use the following low-pass analysis filter, which has $2l$ vanishing moments (case I), or $2l + 1$ vanishing moments (case II):

- (case I):

$$H_0(\omega) = \cos^{2k}(\omega/2)\,(\alpha + (1-\alpha)\cos\omega) \tag{2.212}$$

- (case II):

$$H_0(\omega) = e^{-j\omega/2}\cos^{2k+1}(\omega/2)\,(\alpha + (1-\alpha)\cos\omega) \tag{2.213}$$

The dual low-pass filter is found to be:

- (case I):

$$F_0(\omega) = \cos^{2m}(\omega/2)\;Q(\sin^2(\omega/2)) \tag{2.214}$$

- (case II):

$$F_0(\omega) = e^{-j\omega/2}\cos^{2m+1}(\omega/2)\;Q(\sin^2(\omega/2)) \tag{2.215}$$

The filter has m vanishing moments.

Now, it is shown that, if we define:

$$Q() = \sum_{n=0}^{L} q_n (\sin^2(\omega/2))^n \tag{2.216}$$

Then the filter bank satisfies the PR condition if:

$$q_n = \sum_{k=0}^{L} \binom{L+n-k-1}{L-1} (2\,(1-\alpha))^k, \quad n = 0,\,1,\,\ldots,\,L-1 \tag{2.217}$$

and:

$$q_L = \frac{1}{2\alpha}\left\{\sum_{k=0}^{L} \binom{2L-k-1}{L-1} (2\,(1-\alpha))^k + (1-2\alpha)\sum_{n=0}^{L-1} q_n\right\} \tag{2.218}$$

where $L = l + m$ in case I, or $L = l + m + 1$ in case II.

Once this parameterization is created, [43] constructs ten families of biorthogonal wavelets. The CDF 9/7 is obtained with $\alpha = 2.460348$. Families with rational coeffcients are found, and also families with binary coefficients.

2.5.2.5 Using Bernstein Polynomials

Given a function $f(x)$ on the interval [0,1], the *Nth* order Bernstein polynomial approximation to $f(x)$ is:

$$B_N(f, x) = \sum_{k=0}^{N} f(k/N) \cdot \binom{N}{k} x^k (1-x)^{N-k} \tag{2.219}$$

Notice that the approximation uses only $(N+1)$ samples of $f(x)$ at equally spaced values of x. The approximation has nice mathematical properties.

Now suppose that $f(x)$ express the frequency response of a desired low-pass filter. The description of that response could be done as follows:

$$f\left(\frac{k}{2N-1}\right) = \begin{cases} 1\, k = 0 \\ 1 - \alpha_k\, 1 \leq k \leq N-1 \\ \alpha_k\, N \leq k \leq 2(N-1) \\ 0\, k = 2N-1 \end{cases} \tag{2.220}$$

where $0 \leq \alpha_k < 0.5$, and $\alpha_k = \alpha_{2N-k-1}$.

The approximation by Bernstein polynomial would be:

$$\begin{aligned} B_{2N-1}(f, x) &= \sum_{k=0}^{N-1} \binom{2N-1}{k} x^k (1-x)^{2N-1-k} - \\ &- \sum_{k=1}^{N-1} \alpha_k \binom{2N-1}{k} x^k (1-x)^{2N-1-k} + \\ &+ \sum_{k=N}^{2(N-1)} \alpha_k \binom{2N-1}{k} x^k (1-x)^{2N-1-k} \end{aligned} \tag{2.221}$$

By substituting $x = -(1/4)\, z\, (1 - z^{-1})^2$ a half band filter $R(z)$ is obtained. A PR filter bank can be easily found by factorization.

It has been shown (see for instance [79]) that the number of zeros in the Bernstein coefficients (that is, the α_k) determines the vanishing moments of wavelets. Also, the number of ones in Bernstein coefficients determines the filter flatness.

This method has been introduced by [16]. It would be convenient for the reader to see [12] in order to know more about Bernstein polynomials. The article [22] extends the application of the method, designing orthonormal and biorthogonal wavelet filters. Relevant theoretical aspects are covered in [79]. Indeed, the frequency response of $R(z)$ should be nonnegative. This is ensured when using $0 \leq \alpha_k < 0.5$. If you try other values, nonnegativity is not guaranteed. The research has proposed some alternatives to overcome this problem [69, 80, 81].

Once the way is open for the use of interpolation functions, new proposals could be expected. For instance, using Lagrange interpolation, etc.

2.5.2.6 Factorization of a General Half-Band Filter

A more radical proposal is to design the biorthogonal filter bank starting from a general half-band filter, to be factorized. The target is to have more control on the frequency response of the filters.

An example of this approach is [53]. The general half-band filter would be:

$$\begin{aligned} D(z) = a_0 + a_2 z^{-2} + \cdots + a_{(K/2)-1}\, z^{-(K/2)-1} + z^{-K/2} \\ + a_{(K/2)-1}\, z^{-(K/2)+1} + \cdots + a_0 z^{-K} \end{aligned} \tag{2.222}$$

A *Mth* order flatness is imposed:

$$\frac{d^i}{d\omega^i} D(\omega)|_{\omega=\pi} = 0, \quad i = 0,\, 1,\, \ldots, M \tag{2.223}$$

with $M < ((K/2) + 1)$. We do not impose the maximum number of zeros (which would mean maximum flatness). In this way, some design freedom is gained. After imposing the specified flatness, $D(z)$ can be expressed in terms of some independent parameters.

Then, D(z) is factorized into $H_0(z)$ and $F_0(z)$. An optimization criterion is established, considering the desired frequency response of the two filters (like, for instance, not to be far from 1 in the pass band). Finally, a standard optimization method is applied to get the parameters that optimize the criterion.

2.6 Continuous Wavelets

Some examples of continuous wavelets have already been considered in this chapter, like for instance the Shannon wavelet. Now, let us see other instances of this type of wavelets.

An important advantage of the next wavelets is that they have explicit definition formulae.

2.6.1 The Mexican Hat Wavelet

The Mexican hat wavelet, also denoted as Ricker wavelet, is proportional to the second derivative of the Gaussian density function:

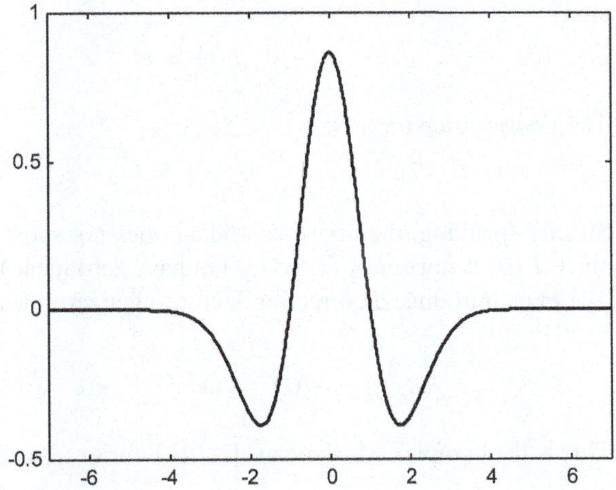

Fig. 2.51 Mexican hat wavelet

$$\psi(t) = (\frac{2}{\sqrt{3}} \cdot \frac{1}{\sqrt[4]{\pi}})(1 - t^2)\, e^{-\frac{t^2}{2}} \qquad (2.224)$$

Figure 2.51 shows the Mexican hat wavelet. The figure has been generated with the Program 2.29.

Program 2.29 Display of Mexican hat wavelet

```
% Display of Mexican hat wavelet
t=-7:0.01:7;
C=2/(sqrt(3)*sqrt(sqrt(pi)));
psi=C*(1-t.^2).*exp(-0.5*t.^2);
plot(t,psi,'k'); %plots the wavelet
axis([-7 7 -0.5 1]);
title('Mexican hat wavelet');
```

There is a complex Mexican hat wavelet, proposed by Add. Its Fourier transform is:

$$\Psi(\omega) = (\frac{2\sqrt{2}}{\sqrt{3}} \cdot \frac{1}{\sqrt[4]{\pi}})\, \omega^2 e^{-\frac{\omega^2}{2}} \qquad (2.225)$$

for $\omega \geq 0$; and $\Psi(\omega) = 0$ otherwise.

The Mexican hat wavelet is a special case of *Hermitian wavelets*, which are derivatives of a Gaussian.

2.6.2 The Morlet Wavelet

Consider as wavelet candidate a sine wave multiplied by a Gaussian envelope (this is similar to the complex Gaussian modulated pulse (GMP) used in the Gabor expansion):

$$f(t) = e^{j\,\omega_0\,t} e^{-\frac{t^2}{2}} \tag{2.226}$$

The Fourier transform is:

$$F(\omega) = e^{-\frac{(\omega-\omega_0)^2}{2}} \tag{2.227}$$

Strictly speaking, the above candidate does not satisfy the admissibility condition, since $F(0)$ is not zero ($f(t)$ does not have zero mean).

Let us introduce a correction factor to get zero mean:

$$\psi(t) = (e^{-j\,\omega_0\,t} - e^{-\frac{\omega_0^2}{2}})\,e^{-\frac{t^2}{2}} \tag{2.228}$$

This is the complex Morlet wavelet. Its Fourier transform is:

$$\Psi(\omega) = e^{-\frac{(\omega-\omega_0)^2}{2}} - e^{-\frac{\omega_0^2}{2}}\,e^{-\frac{\omega^2}{2}} \tag{2.229}$$

Since the Morlet wavelet is complex, the wavelet transform of a signal is plotted as in Bode diagrams, magnitude and phase. The frequency ω_0 is usually chosen equal to five, to make the ratio of main peak and neighbour peak of the wavelet be near 2. For $\omega_0 = 5$ the correction factor is very small and can be neglected. Some authors take the real part, and say that the Morlet wavelet is:

$$\psi(t) = C\,e^{-t^2/2}\cos(5t) \tag{2.230}$$

Figure 2.52 shows the magnitude, real part and imaginary part of the complex Morlet wavelet. The figure has been generated with the Program 2.30.

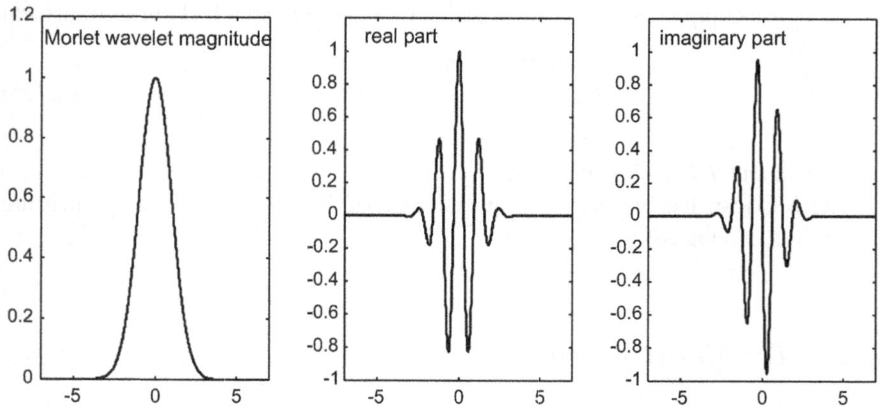

Fig. 2.52 The complex Morlet wavelet

Program 2.30 Display of complex Morlet wavelet

```
% Display of complex Morlet wavelet
t=-7:0.01:7;
w0=5;
cc=exp(-0.5*(w0^2));
aux=exp(-j*w0*t)-cc;
psi=aux.*exp(-0.5*t.^2);
%plots the wavelet
subplot(1,3,1)
plot(t,abs(psi),'k'); %magnitude
axis([-7 7 0 1.2]);
title('Morlet wavelet magnitude');
subplot(1,3,2)
plot(t,real(psi),'k'); %magnitude
axis([-7 7 -1 1.2]);
title('real part');
subplot(1,3,3)
plot(t,imag(psi),'k'); %magnitude
axis([-7 7 -1 1.2]);
title('imaginary part');
```

2.6.3 Complex B-Spline Wavelets

Complex B-spline wavelets are obtained with the following formula:

$$\psi(t) = \sqrt{\frac{\omega_b}{2\pi\, m}} \left(\frac{\sin(\frac{\omega_b t}{2m})}{(\frac{\omega_b t}{2m})} \right)^m e^{j\omega_c t} \tag{2.231}$$

where m is the order of the wavelet, ω_c is the central frequency of the wavelet and ω_b is a window parameter.

For $m = 1$ we obtain the Shannon wavelet.

Figure 2.53 shows the magnitude, real part and imaginary part of the complex B-spline wavelet. The figure has been generated with the Program 2.31.

Program 2.31 Display of complex B-spline wavelet

```
% Display of complex B-spline wavelet
t=-9:0.01:9;
m=2; wc=8; wb=5;
cc=sqrt(wb/(2*pi*m));
aa=(wb*t)/(2*m); aux=sin(aa)./aa;
psi=cc*(aux.^m).*exp(j*wc*t);
%plots the wavelet
subplot(1,3,1)
plot(t,abs(psi),'k'); %magnitude
```

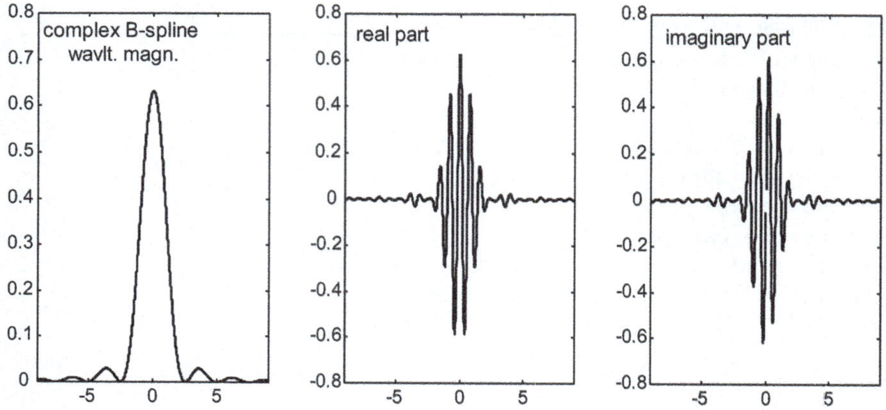

Fig. 2.53 The complex B-spline wavelet

```
axis([-9 9 0 0.8]);
title('complex B-spline wavlt. magn.');
subplot(1,3,2)
plot(t,real(psi),'k'); %magnitude
axis([-9 9 -0.8 0.8]);
title('real part');
subplot(1,3,3)
plot(t,imag(psi),'k'); %magnitude
axis([-9 9 -0.8 0.8]);
title('imaginary part');
```

2.7 Continuous Wavelet Transform (CWT)

The continuous Wavelet transform (CWT) is:

$$W_y(\tau, s) = \langle y(t), \psi(t) \rangle = \int_{-\infty}^{\infty} y(\tau) \frac{1}{\sqrt{|s|}} \psi^*(\frac{t-\tau}{s})\, dt \tag{2.232}$$

The original signal can be reconstructed with the inverse transform:

$$y(t) = \frac{1}{C_\psi} \int_{-\infty}^{\infty} \int_{-\infty}^{\infty} W_y(\tau, s) \frac{1}{\sqrt{|s|}} \psi(\frac{t-\tau}{s})\, d\tau \frac{ds}{s^2} \tag{2.233}$$

where:

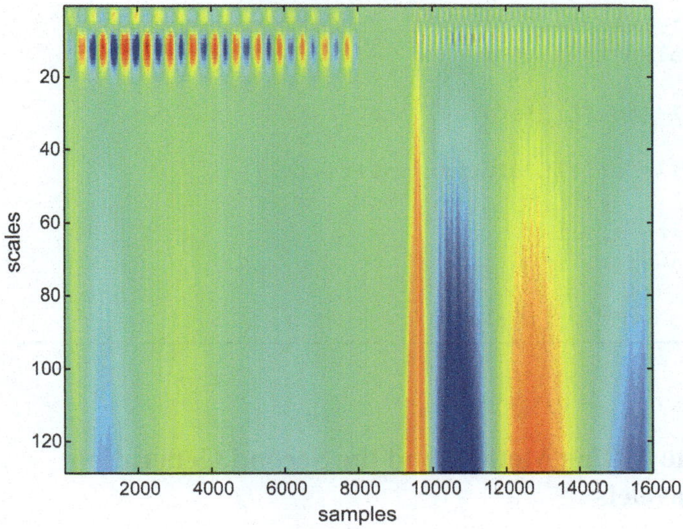

Fig. 2.54 Scalogram of doorbell using mexican hat

$$C\psi = \frac{1}{2\pi}\int_{-\infty}^{\infty} \frac{|\Psi(\omega)|^2}{|\omega|}\, d\omega \tag{2.234}$$

and Ψ is the Fourier transform of ψ.

See [3] for an extensive introduction of CWT.

Since MATLAB has optimized so much the speed of the Fourier transform, it is convenient to implement the CWT in the Fourier domain. This is demonstrated in the next example, where a typical DIIN-DOON doorbell is analyzed.

Figure 2.54 shows the analysis result in the form of a continuous scalogram. This figure has been generated with the Program 2.32. The Mexican hat wavelet has been chosen for this example; the reader is invited to use any other wavelet.

The execution of the program requires about one minute.

Program 2.32 Continuous wavelet analysis with Mexican hat

```
%Continuous wavelet analysis with Mexican hat
%harp signal
[y1,fs1]=wavread('doorbell1.wav'); %read wav file
y1=y1(1:16000); %select part of the signal
Nss=length(y1);
soundsc(y1,fs1); %hear wav
t=(-Nss/2):((Nss/2)-1); %normalized time intervals set
C=2/(sqrt(3)*sqrt(sqrt(pi))); %Mexican hat constant
NS=128; %number of scales
CC=zeros(NS,Nss); %space for wavelet coeffs.
for ee=1:NS,
```

```
s=(ee*1); %scales
%the scaled Mexican hat
ts=t/s;
psi=C*(1-(ts.^2).*exp(-0.5*ts.^2));
%CWT
CC(ee,:)=abs(ifft(fft(psi).*fft(y1')));
end
figure(1)
imagesc(CC);
title('Scalogram of doorbell');
xlabel('samples'); ylabel('scales');
```

2.8 The Lifting Method and the Second Generation Wavelets

The lifting method is a recent alternative for the construction of wavelet bases, or for the computation of wavelet transforms. It constitutes the basis for the development of the so-called second generation wavelets [35, 67].

Let us first introduce the main ideas. Consider a signal $y(k)$. Let us split it into two sets, one is formed by the even indexed samples and the other by the odd indexed samples:

$$\bar{y}_e = y(2k), \textit{ even indexed samples} \tag{2.235}$$

$$\bar{y}_o = y(2k+1), \textit{ odd indexed samples} \tag{2.236}$$

Notice that these are the polyphase components (see Sect. 1.3.1.).

Usually both sets are correlated. In fact the even samples could be used to predict the intermediate odd samples, or vice versa.

Suppose the even samples are used, with an operator $P()$, to predict the odd samples. There would be differences (details) between the actual and the predicted values:

$$\bar{d} = \bar{y}_o - P(\bar{y}_e) \tag{2.237}$$

Given the details and the even samples, it is always possible to recover the odd samples:

$$\bar{y}_o = P(\bar{y}_e) + \bar{d} \tag{2.238}$$

The operation of computing a prediction and recording the details is a 'lifting step'.

A simple predictor is the average of the even neighbours:

$$d_k = y_{2k+1} - \frac{(y_{2k} + y_{2k+2})}{2} \tag{2.239}$$

A second lifting step, which obtains smoothed data, could be the following:

$$\bar{a} = \bar{y}_e + U(\bar{d}) \tag{2.240}$$

where $U()$ is an update operator.

The original values can always be recovered with:

$$\bar{y}_e = \bar{a} - U(\bar{d}) \tag{2.241}$$

An update example, which is able to keep the moving average of the signal, is:

$$a_k = y_{2k} + \frac{(d_{k-1} + d_k)}{4} \tag{2.242}$$

Figure 2.55 shows a conceptual diagram of the complete scheme.

It is clear that the scheme is invertible, so it can be used for a perfect reconstruction filter bank. Let us specify more about this filter. The separation of even and odd samples is done with the so-called Lazy wavelet, as shown in the next Fig. 2.56.

The combination of Lazy wavelet, and the predict and update lifting steps, yields the PR filter bank represented in Fig. 2.57.

The filter bank just represented can be also seen as a polyphase transform. Let us translate the filtering steps to matrix algebra. To begin with, the next expression corresponds to Fig. 2.58.

$$\begin{pmatrix} x_e \\ x_o \end{pmatrix} = \begin{pmatrix} 1 & 0 \\ -P & 1 \end{pmatrix} \begin{pmatrix} y_e \\ y_o \end{pmatrix} \tag{2.243}$$

And the next expression corresponds to Fig. 2.59.

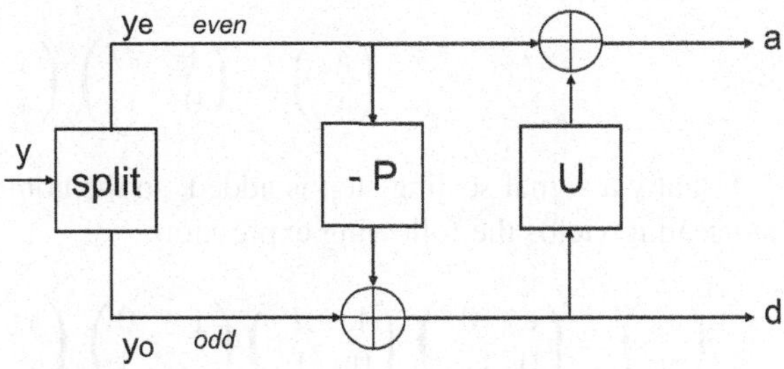

Fig. 2.55 Conceptual diagram of the lifting method

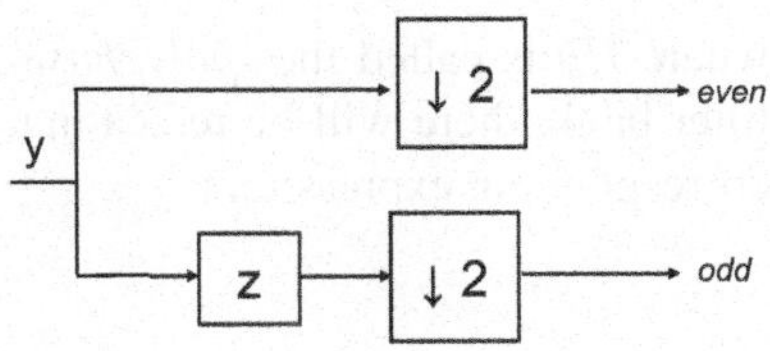

Fig. 2.56 The Lazy wavelet filter bank

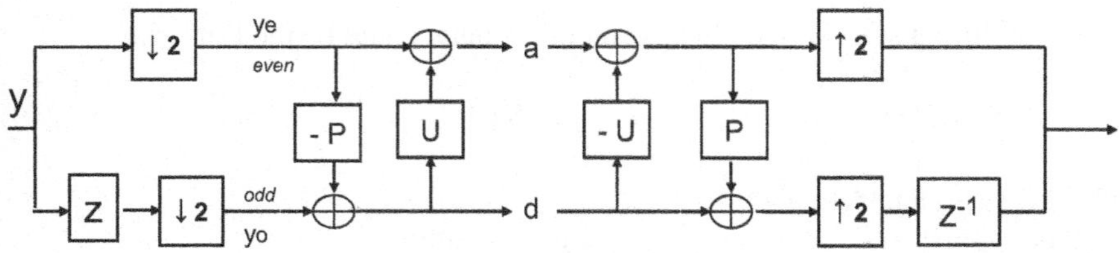

Fig. 2.57 The complete filter bank

Fig. 2.58 A prediction branch

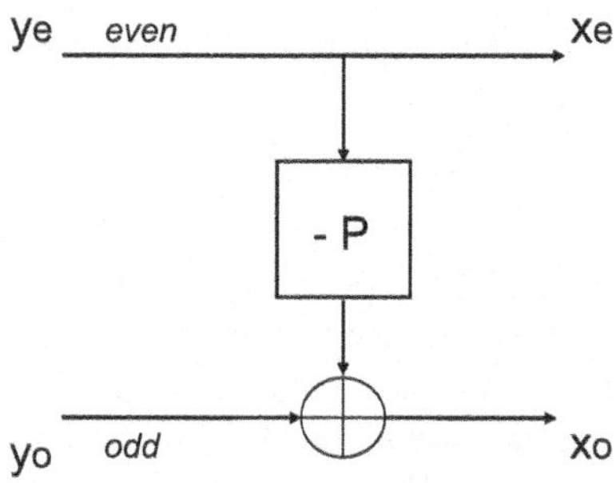

Fig. 2.59 An update branch

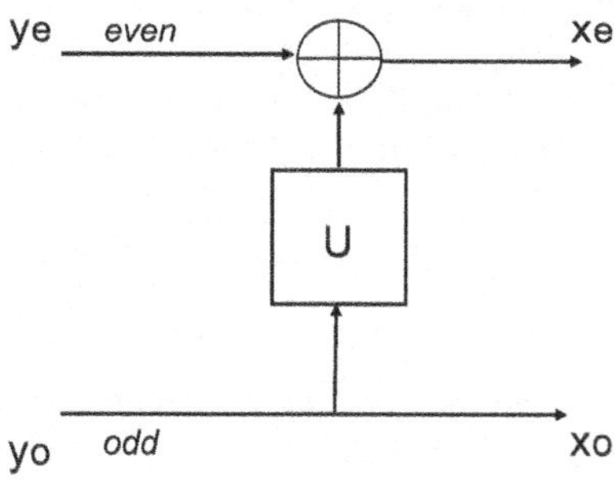

$$\begin{pmatrix} x_e \\ x_o \end{pmatrix} = \begin{pmatrix} 1 & U \\ 0 & 1 \end{pmatrix} \begin{pmatrix} y_e \\ y_o \end{pmatrix} \tag{2.244}$$

Usually a signal scaling step is added, so the combination of prediction, update and scaling yields the following expression:

$$\begin{pmatrix} x_e \\ x_o \end{pmatrix} = \begin{pmatrix} c & 0 \\ 0 & 1/c \end{pmatrix} \begin{pmatrix} 1 & U \\ 0 & 1 \end{pmatrix} \begin{pmatrix} 1 & 0 \\ -P & 1 \end{pmatrix} \begin{pmatrix} y_e \\ y_o \end{pmatrix} = H_p \begin{pmatrix} y_e \\ y_o \end{pmatrix} \tag{2.245}$$

where *Hp* is called the '*polyphase matrix*'. Likewise, in the synthesis part of the filter bank, there will be re-scaling, and the inverse of update and prediction. The corresponding expression is:

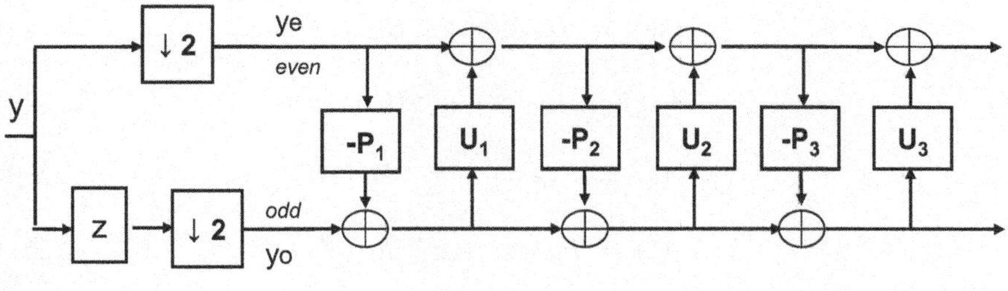

Fig. 2.60 Lifting steps

$$\begin{pmatrix} y_e \\ y_o \end{pmatrix} = \begin{pmatrix} 1 & 0 \\ P & 1 \end{pmatrix} \begin{pmatrix} 1 & -U \\ 0 & 1 \end{pmatrix} \begin{pmatrix} 1/c & 0 \\ 0 & c \end{pmatrix} \begin{pmatrix} x_e \\ x_o \end{pmatrix} = F_p \begin{pmatrix} x_e \\ x_o \end{pmatrix} \tag{2.246}$$

It is easy to check that he product of the *Fp* and *Hp* matrices is the identity matrix. Also *det (Hp) = det (Fp) = 1.*

In general it is possible to continue with the chain of lifting steps, as illustrated in Fig. 2.60:

Again, the chain of lifting steps can be represented with a polyphase matrix *Hp*, so the analysis part of the filter bank can be drawn as in Fig. 2.61 (similarly, *mutatis mutandis*, with the synthesis part and the matrix *Fp*):

Let us write the polyphase matrix as:

$$H_p(z^2) = \begin{pmatrix} H_{00}(z^2) & H_{01}(z^2) \\ H_{10}(z^2) & H_{11}(z^2) \end{pmatrix} \tag{2.247}$$

It is possible to relate the filters with the lifting branches (*FLB*) and the filter banks with parallel structure (*FPS*) (as depicted in Fig. 2.62).

It can be shown that both filters, *FLB* and *FPS*, are equal when:

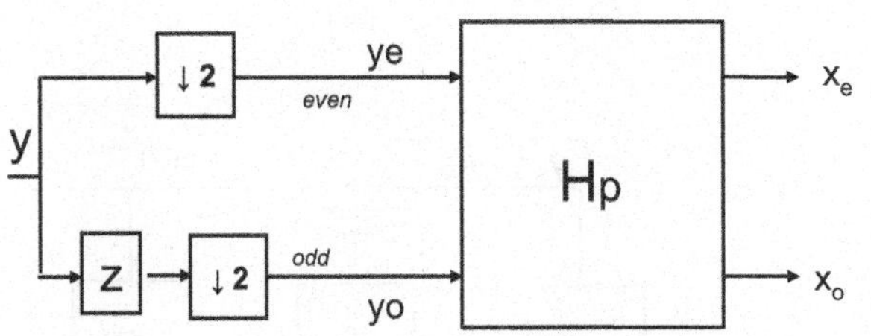

Fig. 2.61 Analysis part of the filter

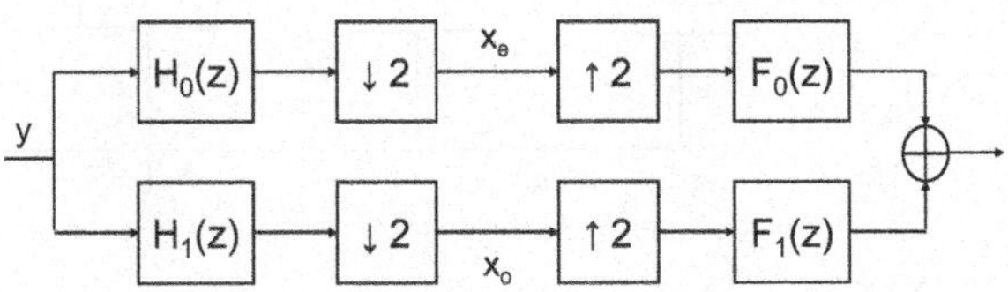

Fig. 2.62 A FPS filter bank

$$H_0(z) = H_{00}(z^2) + z\, H_{01}(z^2) \tag{2.248}$$

$$H_1(z) = H_{10}(z^2) + z\, H_{11}(z^2) \tag{2.249}$$

$$F_0(z) = F_{00}(z^2) + z^{-1}\, F_{01}(z^2) \tag{2.250}$$

$$F_1(z) = F_{10}(z^2) + z^{-1}\, F_{11}(z^2) \tag{2.251}$$

In consequence the wavelet transform can be given in three equivalent forms: lifting steps, polyphase matrix, or *FPS* filter bank. If a perfect reconstruction (PR) filter can be represented by a polyphase matrix *Hp*, then it can be implemented with lifting steps, which are obtained by factorization of H into triangle matrices.

The filter pair $(H_0(z), H_1(z))$ is *complementary* if the corresponding polyphase matrix *Hp* has *det(Hp) = 1.*

If $(H_0(z), H_1(z))$ is complementary, $(F_0(z), F_1(z))$ is also complementary.

2.8.1 Example

Let us take as example the Daubechies wavelet introduced in Sect. 2.4.3.2. Figure 2.63 shows how it is implemented using lifting steps.

The lifting equations are the following:

$$a_k^{(1)} = y_{2k} + \sqrt{3}\, y_{2k+1} \tag{2.252}$$

$$d_k^{(1)} = y_{2k+1} - \frac{\sqrt{3}}{4}\, a_k^{(1)} - \frac{(\sqrt{3}-2)}{4}\, a_{k-1}^{(1)} \tag{2.253}$$

$$a_k^{(2)} = a_k^{(1)} - d_{k+1}^{(1)} \tag{2.254}$$

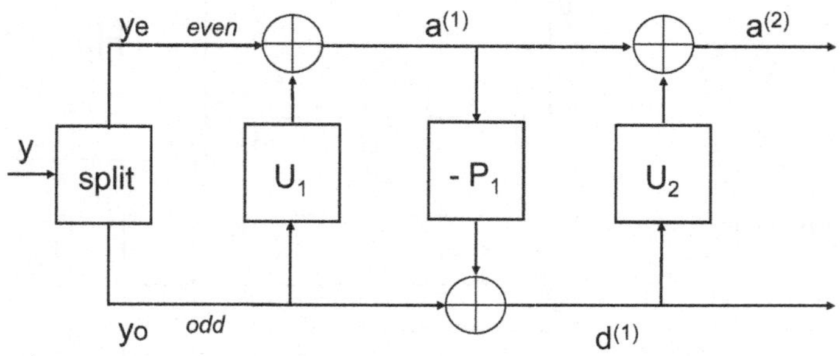

Fig. 2.63 Daubechies 4 wavelet lifting steps

Then, the polyphase matrix is:

$$H_p(z) = \begin{pmatrix} \frac{\sqrt{3}-1}{\sqrt{2}} & 0 \\ 0 & \frac{\sqrt{3}+1}{\sqrt{2}} \end{pmatrix} \begin{pmatrix} 1 & -z \\ 0 & 1 \end{pmatrix} \begin{pmatrix} 1 & 0 \\ -\frac{\sqrt{3}}{4} - \frac{\sqrt{3}-2}{4} z^{-1} & 1 \end{pmatrix} \begin{pmatrix} 1 & \sqrt{3} \\ 0 & 1 \end{pmatrix} \tag{2.255}$$

(with scaling $c = \frac{\sqrt{3}-1}{\sqrt{2}}$)

Develop the equation:

$$H_p(z) = \begin{pmatrix} \frac{3-\sqrt{3}}{4\sqrt{2}} z + \frac{1+\sqrt{3}}{4\sqrt{2}} & \frac{1-\sqrt{3}}{4\sqrt{2}} z + \frac{3+\sqrt{3}}{4\sqrt{2}} \\ -\frac{(3+\sqrt{3})}{4\sqrt{2}} - \frac{(1-\sqrt{3})}{4\sqrt{2}} z^{-1} & \frac{1+\sqrt{3}}{4\sqrt{2}} + \frac{3-\sqrt{3}}{4\sqrt{2}} z^{-1} \end{pmatrix} =$$

$$= \begin{pmatrix} H_{00}(z^2) & H_{01}(z^2) \\ H_{10}(z^2) & H_{11}(z^2) \end{pmatrix} \tag{2.256}$$

In particular:

$$\begin{aligned} H_0(z) &= H_{00}(z^2) + z\, H_{01}(z^2) = \\ &= \frac{3-\sqrt{3}}{4\sqrt{2}} z^2 + \frac{1+\sqrt{3}}{4\sqrt{2}} + \frac{1-\sqrt{3}}{4\sqrt{2}} z^3 + \frac{3+\sqrt{3}}{4\sqrt{2}} z = \\ &= \frac{1+\sqrt{3}}{4\sqrt{2}} + \frac{3+\sqrt{3}}{4\sqrt{2}} z + \frac{3-\sqrt{3}}{4\sqrt{2}} z^2 + \frac{1-\sqrt{3}}{4\sqrt{2}} z^3 \end{aligned} \tag{2.257}$$

$$\begin{aligned} H_1(z) &= H_{10}(z^2) + z\, H_{11}(z^2) = \\ &= -\frac{3+\sqrt{3}}{4\sqrt{2}} - \frac{1-\sqrt{3}}{4\sqrt{2}} z^{-2} + \frac{1+\sqrt{3}}{4\sqrt{2}} z + \frac{3-\sqrt{3}}{4\sqrt{2}} z^{-1} = \\ &= -\frac{1-\sqrt{3}}{4\sqrt{2}} z^{-2} + \frac{3-\sqrt{3}}{4\sqrt{2}} z^{-1} - \frac{3+\sqrt{3}}{4\sqrt{2}} + \frac{1+\sqrt{3}}{4\sqrt{2}} z \end{aligned} \tag{2.258}$$

Note that the coefficients in the expression of $H_0(z)$ are:

$$\begin{aligned} h_0(0) &= \frac{1+\sqrt{3}}{4\sqrt{2}},\ h_0(1) = \frac{3+\sqrt{3}}{4\sqrt{2}}, \\ h_0(2) &= \frac{3-\sqrt{3}}{4\sqrt{2}},\ h_0(3) = \frac{1-\sqrt{3}}{4\sqrt{2}} \end{aligned} \tag{2.259}$$

(these values already given in Sect. 2.3.3).

A segment of a long sonar signal has been chosen to present now an example of wavelet analysis using lifting for Daubechies 4. Program 2.33 shows a MATLAB implementation of the lifting scheme, based on [35]. The result is depicted in Fig. 2.64.

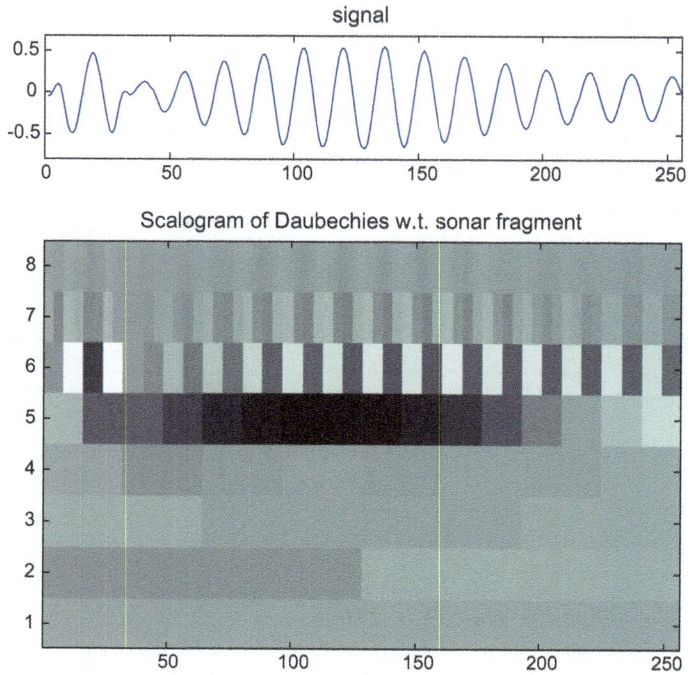

Fig. 2.64 Scalogram of sonar fragment using lifting Daubechies 4

Program 2.33 Lifting example: analysis of sonar signal

```
% Lifting example
% analysis of sonar signal
% Daubechies 4 Wavelet
% Plot of signal and scalogram
%The sonar signal
[y1,fs1]=wavread('sonar1.wav'); %read wav file
ta=7815; tb=ta+255;
sg=y1(ta:tb);
Nss=length(sg);%number of signal samples
tiv=1/fs1;
duy=(Nss-1)*tiv; %duration of signal
tss=0:tiv:duy; %time intervals set
Ts=tiv; %sampling period
%analysis of the signal with wavelets
y=sg;
a=y';
aux=0; cq=sqrt(3);
K=8; %number of scales (256=2^8)
%wavelet calculus using lifting
NN=Nss;
```

```
for n=1:K,
L=length(a); L2=L/2;
a1=a(1:2:L-1)+ (cq*a(2:2:L));
d1=a(2:2:L)-((cq/4)*a1)-(((cq-2)/4)*[a1(L2) a1(1:(L2-1))]);
a2=a1-[d1(2:L2) d1(1)];
a=((cq-1)/sqrt(2))*a2;
d=((cq+1)/sqrt(2))*d1;
aux=[d,aux];
dd(K+1-n,1:NN/2)=d;
NN=NN/2;
end;
wty=[a,aux(1:(end-1))];
%preparing for scalogram
S=zeros(K,Nss); %space for S(j,k) scalogram coefficients
for n=1:K,
q=2^(n-1); L=Nss/q;
for m=1:q,
R=(1+(L*(m-1))):(L*m); %index range
S(n,R)=dd(n,m);
end;
end;
%figure
subplot('position',[0.04 0.77 0.92 0.18])
plot(y);
axis([0 256 1.2*min(y) 1.2*max(y)]);
title('signal');
subplot('position',[0.04 0.05 0.92 0.6])
imagesc(S); colormap('bone');
title('Scalogram of Daubechies w.t. sonar fragment');
h=gca; set(h,'YDir','normal');
```

To complete the example, the original signal is recovered by the Program 2.34. It is clearly seen in this program how easy is to invert the lifting process, re-using and re-ordering the lifting equations. Figure 2.65 shows the recovered sonar signal, which is practically identical to the original signal.

Program 2.34 Inverse of the Lifting

```
% inverse of the Lifting
% Sonar signal
% Daubechies 4 wavelet
% use after lifting analysis (it needs wty)
L=length(wty); %length of the DWT
sg=zeros(1,L);
K=8; %number of scales
cq=sqrt(3);
a=wty(1);
for n=1:K,
wd=2^(n-1);
bg=1+wd; fl=bg+wd-1;
```

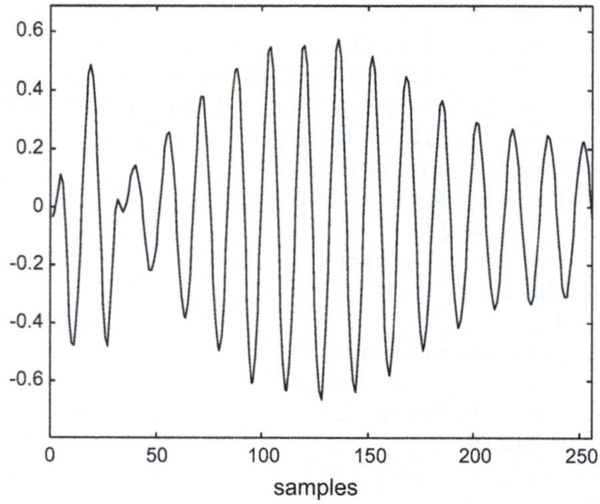

Fig. 2.65 Recovered sonar fragment using lifting Daubechies 4

```
d=wty(bg:fl);
d1=d/((cq+1)/sqrt(2));
a2=a/((cq-1)/sqrt(2));
a1=a2+[d1(2:end) d1(1)];
sg(2:2:(2^n))=d1+((cq/4)*a1)+(((cq-2)/4)*[a1(end) a1(1:end-1)]);
sg(1:2:(2^n)-1)=a1-(cq*sg(2:2:(2^n)));
a=sg(1:(2^n));
end;
figure(1)
plot(sg,'k');
axis([0 256 1.2*min(sg) 1.2*max(sg)]);
xlabel('samples');
title('the recovered signal');
```

2.8.2 Decomposition into Lifting Steps

By writing any FIR wavelet or PR filter bank in the polyphase form, it can be confirmed that in every case it can be decomposed into lifting steps (this is an indication of the power of the lifting method) [23, 27, 46, 76].

The decomposition into lifting steps can be done by applying the Euclidean algorithm on Laurent polynomials.

A Laurent polynomial is an expression of the form:

$$p(z) = \sum_{k=a}^{b} c_k \, z^{-k} \tag{2.260}$$

The degree of a Laurent polynomial is $b - a$. Thus the following z-transform:

$$h(z) = \sum_{k=a}^{b} h_k \, z^{-k} \tag{2.261}$$

is a Laurent polynomial.

The Euclidean algorithm is a method for finding the greatest common divisor (gcd) of two integers. Suppose two integers a and b, with $|a| > |b|$ and $b \neq 0$. By division one can obtain a quotient q_1 and a remainder r_1 (if $r_1 = 0$, then *gcd(a,b)* = *b*):

$$a = b\, q_1 + r_1 \tag{2.262}$$

Rewriting this:

$$r_1 = a - b\, q_1 \tag{2.263}$$

It is evident that any common factor of a and b is a factor of r_1, and it is also a common factor of r_1 and b: $gcd(a, b) = gcd\,(b, r_1)$

Let us continue with the division:

$$b = r_1\, q_2 + r_2 \tag{2.264}$$

Its is clear that $gcd\,(b, r_1) = gcd\,(r_1, r_2)$.

Iterating the process, a $r_i = 0$ will be obtained, and:

$$r_{i-2} = r_{i-1}\, q_i \tag{2.265}$$

So $gcd\,(a, b) = gcd\,(r_{i-2}, r_{i-1}) = r_{i-1}$.

For example, the gcd of 182 and 34:

$$\begin{aligned} 182 &= 34 \cdot [5] + 12 \\ 34 &= 12 \cdot [2] + 10 \\ 12 &= 10 \cdot [1] + 2 \\ 10 &= 2 \cdot [5] \end{aligned} \tag{2.266}$$

The gcd is 2.

The procedure can be implemented in the following computational recursive form:

$$\begin{aligned} q_{i+1}, r_{i+1} &\leftarrow a_i \,/\, b_i \\ a_{i+1} &\leftarrow b_i \\ b_{i+1} &\leftarrow r_{i+1} \end{aligned} \tag{2.267}$$

For Laurent polynomials division with remainder is possible, so given two non-zero Laurent polynomials a(z), with degree n, and b(z), with degree m (n $\geq$ m), we can write:

$$a(z) = b(z)\,q(z) + r(z) \tag{2.268}$$

where q(z) and r(z) are also Laurent polynomials. Notice that the division is not necessarily unique.

The Euclidean algorithm can be applied to Laurent polynomials, starting from $a_0(z) = a(z)$ and $b_0(z) = b(z)$, $i = 0$, and according with the following iteration:

$$\begin{aligned} &q_{i+1}(z),\ r_{i+1}(z) \leftarrow a_i(z)\,/\,b_i(z) \\ &a_{i+1}(z) \leftarrow b_i(z) \\ &b_{i+1}(z) \leftarrow r_{i+1}(z) = a_i(z) - b_i(z)\,q_{i+1}(z) \end{aligned} \tag{2.269}$$

which in matrix form is:

$$\begin{bmatrix} a_{i+1}(z) \\ b_{i+1}(z) \end{bmatrix} = \begin{pmatrix} 0 & 1 \\ 1 & -q_{i+1}(z) \end{pmatrix} \begin{bmatrix} a_i(z) \\ b_i(z) \end{bmatrix} \tag{2.270}$$

After N iterations we obtain a $b_N(z) = 0$, so $a_N(z)$ is the gcd of $a(z)$ and $b(z)$. Therefore:

$$\begin{bmatrix} a_N(z) \\ 0 \end{bmatrix} = \prod_{i=N}^{1} \begin{pmatrix} 0 & 1 \\ 1 & -q_i(z) \end{pmatrix} \begin{bmatrix} a(z) \\ b(z) \end{bmatrix} \tag{2.271}$$

With matrix inversion and side exchanging:

$$\begin{bmatrix} a(z) \\ b(z) \end{bmatrix} = \prod_{i=1}^{N} \begin{pmatrix} q_i(z) & 1 \\ 1 & 0 \end{pmatrix} \begin{bmatrix} a_N(z) \\ 0 \end{bmatrix} \tag{2.272}$$

Let us apply the Euclidean algorithm to our lifting factorization problem. Now take a complementary filter pair and substitute $a(z) = H_{00}(z)$, $b(z) = H_{01}(z)$:

$$\begin{bmatrix} H_{00}(z) \\ H_{01}(z) \end{bmatrix} = \prod_{i=1}^{N} \begin{pmatrix} q_i(z) & 1 \\ 1 & 0 \end{pmatrix} \begin{bmatrix} K \\ 0 \end{bmatrix} \tag{2.273}$$

The constant K appears because $det(Hp) = 1$:

$$H_{00}(z)\,H_{11}(z) - H_{01}(z)\,H_{10}(z) = 1 \tag{2.274}$$

so the gcd of $H_{00}(z)$ and $H_{01}(z)$ must be a monomial ($K z^n$), and the quotients can be chosen to have $n = 0$.

To complete a complementary filter pair, ensuring a determinant 1, we can use:

$$\begin{pmatrix} H_{00}(z) & H'_{10}(z) \\ H_{01}(z) & H'_{11}(z) \end{pmatrix} = \prod_{i=1}^{N} \begin{pmatrix} q_i(z) & 1 \\ 1 & 0 \end{pmatrix} \begin{pmatrix} K & 0 \\ 0 & 1/K \end{pmatrix} \tag{2.275}$$

Considering that:

$$\begin{pmatrix} q_i(z) & 1 \\ 1 & 0 \end{pmatrix} = \begin{pmatrix} 1 & q_i(z) \\ 0 & 1 \end{pmatrix} \begin{pmatrix} 0 & 1 \\ 1 & 0 \end{pmatrix} = \begin{pmatrix} 0 & 1 \\ 1 & 0 \end{pmatrix} \begin{pmatrix} 1 & 0 \\ q_i(z) & 1 \end{pmatrix} \tag{2.276}$$

and transposing both sides:

$$\begin{pmatrix} H_{00}(z) & H_{01}(z) \\ H'_{10}(z) & H'_{11}(z) \end{pmatrix} = \begin{pmatrix} K & 0 \\ 0 & 1/K \end{pmatrix} \prod_{i=N/2}^{1} \begin{pmatrix} 1 & q_{2i}(z) \\ 0 & 1 \end{pmatrix} \begin{pmatrix} 1 & 0 \\ q_{2i-1}(z) & 1 \end{pmatrix} \tag{2.277}$$

The original filter with polyphase matrix *Hp(z)* can be obtained with:

$$H_p(z) = \begin{pmatrix} 1 & 0 \\ -K^2 t(z) & 1 \end{pmatrix} \begin{pmatrix} K & 0 \\ 0 & 1/K \end{pmatrix} \prod_{i=N/2}^{1} \begin{pmatrix} 1 & q_{2i}(z) \\ 0 & 1 \end{pmatrix} \begin{pmatrix} 1 & 0 \\ q_{2i-1}(z) & 1 \end{pmatrix} \tag{2.278}$$

where $t(z)$ is a lifting term, which is found by comparison between both sides of the equation.

The expression above is a factorization of the filter into lifting steps.

2.8.3 Examples

2.8.3.1 A Filter Branch is Half-Band

Suppose that in the complementary pair, $H_0(z)$ is half-band. That means that $H_{00}(z) = 1$ and so the polyphase matrix is:

$$H_p(z) = \begin{pmatrix} 1 & H_{01}(z) \\ H_{10}(z) & 1 + H_{01}(z)\, H_{10}(z) \end{pmatrix} \tag{2.279}$$

(since *det(Hp(z)) = 1*)

This polyphase matrix can be easily factorized:

$$H_p(z) = \begin{pmatrix} 1 & 0 \\ H_{10}(z) & 1 \end{pmatrix} \begin{pmatrix} 1 & H_{01}(z) \\ 0 & 1 \end{pmatrix} \tag{2.280}$$

This expression is connected with a family of symmetric biorthogonal wavelets with Deslauriers-Dubuc scaling functions.

2.8.3.2 The (9–7) Filter Pair

The (9–7) filter pair is frequently being used in important applications. The analysis filter has 9 coefficients, and the synthesis filter has 7 coefficients. Thanks to the work of Daubechies and Sweldens, the following factorization has been obtained:

$$H_p(z) = \begin{pmatrix} 1 & \alpha(1+z^{-1}) \\ 0 & 1 \end{pmatrix} \begin{pmatrix} 1 & 0 \\ \beta(1+z) & 1 \end{pmatrix} \begin{pmatrix} 1 & \gamma(1+z^{-1}) \\ 0 & 1 \end{pmatrix} \cdot \cdot \begin{pmatrix} 1 & 0 \\ \delta(1+z) & 1 \end{pmatrix} \begin{pmatrix} \varphi & 0 \\ 0 & 1/\varphi \end{pmatrix} \tag{2.281}$$

with: $\alpha = -1.58613$, ß $= -0.05298$, $\gamma = 0.88298$, $\delta = 0.4435$, $\varphi = 1.1496$

Then, the lifting implementation of the filter is:

$$a_k^{(0)} = y_{2k} \tag{2.282}$$

$$d_k^{(0)} = y_{2k+1} \tag{2.283}$$

$$d_k^{(1)} = d_k^{(0)} + \alpha\,(a_k^{(0)} + a_{k+1}^{(0)}) \tag{2.284}$$

$$a_k^{(1)} = a_k^{(0)} + \beta\,(d_k^{(1)} + d_{k-1}^{(1)}) \tag{2.285}$$

$$d_k^{(2)} = d_k^{(1)} + \gamma\,(a_k^{(1)} + a_{k+1}^{(1)}) \tag{2.286}$$

$$a_k^{(2)} = a_k^{(1)} + \delta(d_k^{(2)} + d_{k-1}^{(2)}) \tag{2.287}$$

$$a_k = \varphi\, a_k^{(2)} \tag{2.288}$$

$$d_k = {}^1\!/_{\varphi} d_k^{(2)} \tag{2.289}$$

Presuming that the reader would be tempted to check the CDF 9/7, the same example as before (the analysis and recovery of the sonar signal) is again selected for an exercise of CDF 9/7 coding in MATLAB.

Program 2.35 does the wavelet analysis of the sonar signal using CDF 9/7. The result is shown in Fig. 2.66.

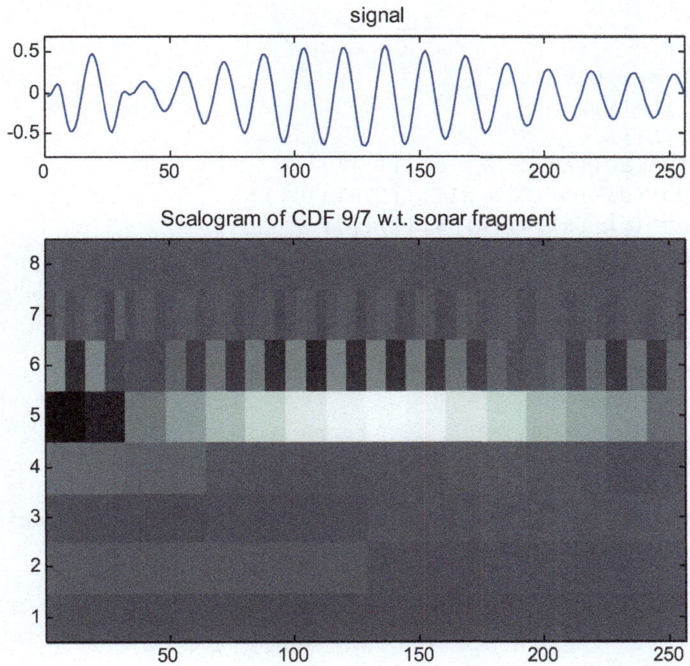

Fig. 2.66 Scalogram of sonar fragment using lifting CDF 9/7

Program 2.35 Lifting example (sonar signal), CDF 9/7

```
% Lifting example
% analysis of sonar signal
% CDF 9/7 Wavelet
% Plot of signal and scalogram
%The sonar signal
[y1,fs1]=wavread('sonar1.wav'); %read wav file
ta=7815; tb=ta+255;
sg=y1(ta:tb);
Nss=length(sg);%number of signal samples
tiv=1/fs1;
duy=(Nss-1)*tiv; %duration of signal
tss=0:tiv:duy; %time intervals set
Ts=tiv; %sampling period
%analysis of the signal with wavelets
y=sg;
a=y';
aux=0;
%CDF coeffs.
pa=-1.58613; pb=-0.05298; pg=0.88298; pd=0.4435; pp=1.1496;
K=8; %number of scales (256=2^8)
%wavelet calculus using lifting
```

```
NN=Nss;
for n=1:K,
L=length(a); L2=L/2;
a0=a(1:2:L-1);
d0=a(2:2:L);
d1=d0+(pa*(a0+[a0(2:L2) a0(1)]));
a1=a0+(pb*(d1+[d1(L2) d1(1:(L2-1))]));
d2=d1+(pg*(a1+[a1(2:L2) a1(1)]));
a2=a1+(pd*(d2+[d2(L2) d2(1:(L2-1))]));
a=pp*a2;
d=d2/pp;
aux=[d,aux];
dd(K+1-n,1:NN/2)=d;
NN=NN/2;
end;
wty=[a,aux(1:(end-1))];
%preparing for scalogram
S=zeros(K,Nss); %space for S(j,k) scalogram coefficients
for n=1:K,
q=2^(n-1); L=Nss/q;
for m=1:q,
R=(1+(L*(m-1))):(L*m); %index range
S(n,R)=dd(n,m);
end;
end;
%figure
subplot('position',[0.04 0.77 0.92 0.18])
plot(y);
axis([0 256 1.2*min(y) 1.2*max(y)]);
title('signal');
subplot('position',[0.04 0.05 0.92 0.6])
imagesc(S); colormap('bone');
title('Scalogram of CDF 9/7 w.t. sonar fragment');
h=gca; set(h,'YDir','normal');
```

The signal recovery is done with the Program 2.36, which inverts the application of the lifting equations. The recovered signal is shown in Fig. 2.67; it is close to the original signal.

Program 2.36 Inverse of the Lifting, CDF 9/7

```
% inverse of the Lifting
% Sonar signal
% CDF 9/7 wavelet
% use after lifting analysis (it needs wty)
L=length(wty); %length of the DWT
sg=zeros(1,L);
%CDF coeffs.
pa=-1.58613; pb=-0.05298; pg=0.88298; pd=0.4435; pp=1.1496;
K=8; %number of scales
```

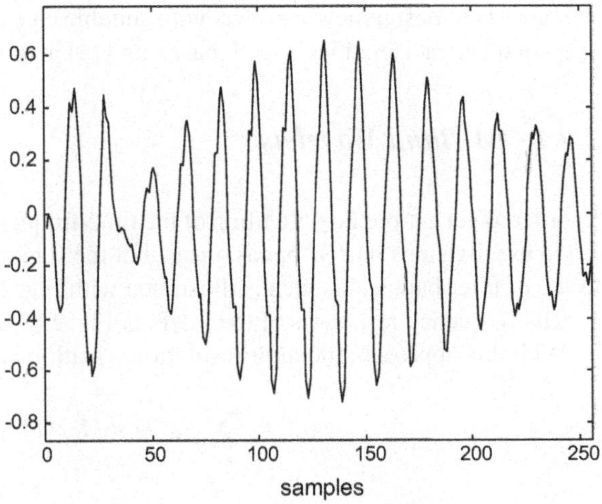

Fig. 2.67 Recovered sonar fragment using lifting CDF 9/7

```
cq=sqrt(3);
a=wty(1);
for n=1:K,
wd=2^(n-1);
bg=1+wd; fl=bg+wd-1;
d=wty(bg:fl);
d2=d*pp;
a2=a/pp;
a1=a2-(pd*(d2+[d2(end) d2(1:(end-1))]));
d1=d2-(pg*(a1+[a1(2:end) a1(1)]));
a0=a1-(pb*(d1+[d1(end) d1(1:(end-1))]));
d0=d1-(pa*(a0+[a0(2:end) a0(1)]));
sg(2:2:(2^n))=d0;
sg(1:2:(2^n)-1)=a0;
a=sg(1:(2^n));
end;
figure(1)
plot(sg,'k');
axis([0 256 1.2*min(sg) 1.2*max(sg)]);
xlabel('samples');
title('the recovered signal');
```

2.9 More Analysis Flexibility

There are applications that require more processing flexibility. This in particular concerns to images, as it will be clearly seen in the next chapter. In general two approaches have been proposed. One is to extend the architecture of filter banks, and the other is to design new wavelets with suitable capabilities. In this section the first approach is concisely introduced, based on [14] and other publications.

2.9.1 M-Band Wavelets

In order to get a more flexible tiling of the time-frequency plane it has been proposed [7] to use divisions into M bands, instead of the standard division into two bands. In terms of filter banks, this idea is illustrated with Fig. 2.68 for the case $M = 4$.

The frequency responses of the filters in Fig. 2.68 are depicted in Fig. 2.69.

With this approach, the multiresolution equation is the following:

$$\varphi(t) = \sum_n h_0(n)\,\sqrt{M}\,\varphi(Mt - n) \tag{2.290}$$

The frequency domain version of this MAE is:

$$\Phi(\omega) = \frac{1}{\sqrt{M}} H_0(\frac{\omega}{M})\,\Phi(\frac{\omega}{M}) \tag{2.291}$$

which after iteration becomes:

$$\Phi(\omega) = \prod_{k=1}^{\infty} [\frac{1}{\sqrt{M}} H_0(\frac{\omega}{M^k})]\;\Phi(0) \tag{2.292}$$

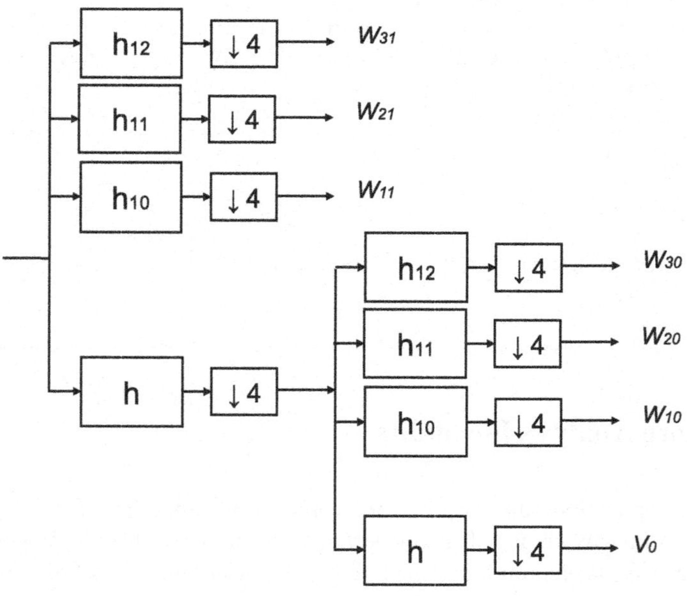

Fig. 2.68 4-band wavelet system

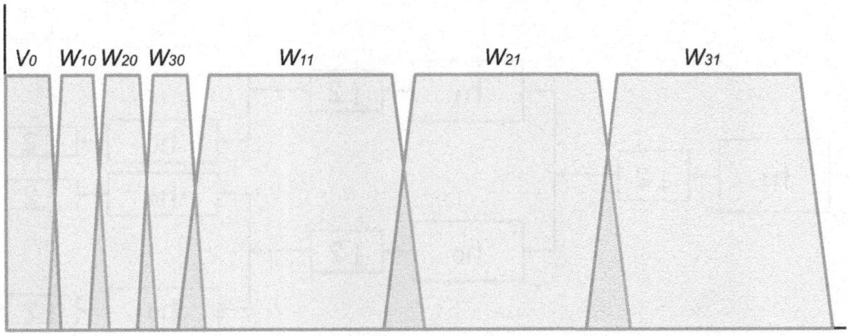

Fig. 2.69 Frequency bands of 4-band wavelet system

The computation of filter coefficients could be involved, especially for odd values of M. The article [41] shows an algebraic computation for orthonormal M-band wavelets, with several examples. In [39] the M-band wavelets are applied to audio signals; a case of adaptation to musical scales is considered. There are versions for image texture studies.

2.9.2 *Wavelet Packets*

In 1992, in a frequently cited article [21], Coifman proposed the '*wavelet packet*' system. Figure 2.70 depicts the idea in terms of filter bank. Like for the low frequencies, there are filters dividing into bands the high frequencies. This allows for a better analysis of high frequencies.

When wavelet packets were introduced, it was noticed in real applications that the general architecture could be simplified and optimized for each particular case. The signal power is distributed among branches. It is not practical to have branches with negligible signal energy flow.

The issue can be regarded in terms of assignments of frequency bands to signal characteristics.

The research has formulated this problem as finding a *best basis*. An objective function should be defined, and then an optimization algorithm should be applied. For instance, in [21] an additive cost function was suggested; it is an entropy-based cost function that is minimized if the energy is concentrated in a few nodes. A better basis employs fewer components to represent a signal.

Supposing that a signal has been analyzed with a complete structure, then a pruning process can be applied. Start with terminal nodes, and prune bottom-up. Consider the following example of node costs (only the two last levels of a structure):

0.23		0.77	
0.18	0.19	0.4	0.2

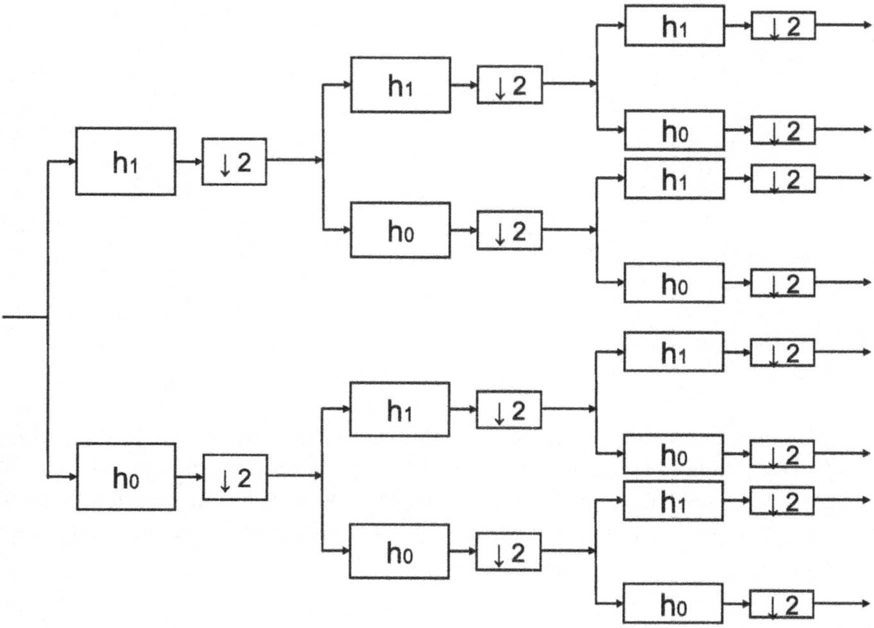

Fig. 2.70 A wavelet packet example

If the sum of costs of children (S) is greater than the parent's cost, the children are pruned. Otherwise, the parent's cost is replaced with S. Therefore, after the first step of pruning the two last levels become:

0.23	**0.6**
|0.18| 0.19 |	|**0.4**| **0.2**|

Bold numbers correspond to selected members of the best basis. The algorithm continues towards the top of the hierarchy.

Several cost functions have been proposed based on thresholds, the number of bits to represent the signal, and other measures.

2.9.3 *Multiwavelets*

Whereas wavelets have an associated scaling function and wavelet function, *multi-wavelets* have two or more scaling and wavelet functions. In this way more analysis freedom—with good properties—is allowed.

The set of scaling functions can be written in vector notation as follows:

$$\Phi(t) = [\varphi_1(t),\ \varphi_2(t), \ldots,\ \varphi_R(t)]^T \tag{2.293}$$

and the set of wavelet functions

$$\Psi(t) = [\psi_1(t),\ \psi_2(t), \ldots,\ \psi_R(t)]^T \tag{2.294}$$

The multiresolution equation is the following:

$$\Phi(t) = \sqrt{2} \sum_n H(n)\, \Phi(2t - n) \tag{2.295}$$

where $H(n)$ is a $R \times R$ matrix.

Likewise, the construction of wavelets is:

$$\Psi(t) = \sqrt{2} \sum_n G(n)\, \Phi(2t - n) \tag{2.296}$$

where $G(n)$ is another $R \times R$ matrix.

2.9.3.1 A Simple Example

Most published multiwavelet systems include just two scaling functions. For example [8] one Haar scaling function as $\varphi_1(t)$ and a second scaling function such as:

$$\varphi_2(t) = \frac{\sqrt{3}}{2}\varphi_1(2t) + \frac{1}{2}\varphi_2(2t) - \frac{\sqrt{3}}{2}\varphi_1(2t-1) + \frac{1}{2}\varphi_2(2t-1) \tag{2.297}$$

Fig. 2.71 shows $\varphi_2(t)$:

The MAE for this case can be written as follows:

$$\begin{bmatrix} \varphi_1(t) \\ \varphi_2(t) \end{bmatrix} = \begin{pmatrix} 1 & 0 \\ \frac{\sqrt{3}}{2} & \frac{1}{2} \end{pmatrix} \begin{bmatrix} \varphi_1(2t) \\ \varphi_2(2t) \end{bmatrix} + \begin{pmatrix} 1 & 0 \\ -\frac{\sqrt{3}}{2} & \frac{1}{2} \end{pmatrix} \begin{bmatrix} \varphi_1(2t-1) \\ \varphi_2(2t-1) \end{bmatrix} \tag{2.298}$$

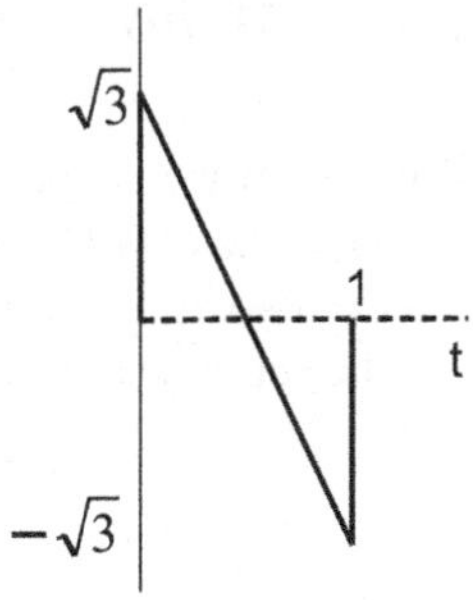

Fig. 2.71 The second scaling function $\varphi_2(t)$

2.9.3.2 The Geronimo, Hardin, and Massopust Multiwavelet System

The dilation and wavelet equations of the multiwavelet system proposed by Geronimo, Hardin and Massopust (see [65] and references therein), have the following expressions:

$$\begin{aligned}\Phi(t) &= \begin{bmatrix}\varphi_1(t)\\ \varphi_2(t)\end{bmatrix} = H(0)\,\Phi(2t) + H(1)\,\Phi(2t-1)\\ &+ H(2)\;\Phi(2t-2) + H(3)\;\Phi(2t-3)\end{aligned} \tag{2.299}$$

with:

$$H(0) = \begin{pmatrix}\frac{3}{5} & \frac{4\sqrt{2}}{5}\\ -\frac{1}{10\sqrt{2}} & -\frac{3}{10}\end{pmatrix};\quad H(1) = \begin{pmatrix}\frac{3}{5} & 0\\ \frac{9}{10\sqrt{2}} & 1\end{pmatrix} \tag{2.300}$$

$$H(2) = \begin{pmatrix}0 & 0\\ \frac{9}{10\sqrt{2}} & -\frac{3}{10}\end{pmatrix};\quad H(3) = \begin{pmatrix}0 & 0\\ -\frac{1}{10\sqrt{2}} & 0\end{pmatrix} \tag{2.301}$$

and,

$$\begin{aligned}\Psi(t) &= \begin{bmatrix}\psi_1(t)\\ \psi_2(t)\end{bmatrix} = G(0)\,\Phi(2t) + G(1)\,\Phi(2t-1)\\ &+ G(2)\;\Phi(2t-2) + G(3)\;\Phi(2t-3)\end{aligned} \tag{2.302}$$

with:

$$G(0) = \frac{1}{10}\begin{pmatrix}-\frac{1}{\sqrt{2}} & -3\\ 1 & -3\sqrt{2}\end{pmatrix};\quad G(1) = \frac{1}{10}\begin{pmatrix}\frac{9}{\sqrt{2}} & -10\\ -9 & 0\end{pmatrix} \tag{2.303}$$

$$G(2) = \frac{1}{10}\begin{pmatrix}\frac{9}{\sqrt{2}} & -3\\ 9 & -3\sqrt{2}\end{pmatrix};\quad G(3) = \frac{1}{10}\begin{pmatrix}-\frac{1}{\sqrt{2}} & 0\\ -1 & 0\end{pmatrix} \tag{2.304}$$

This multiwavelet system has notable properties: the scaling functions have short support, both scaling functions are symmetric, the wavelets form a symmetric/antisymmetric pair, and all integer translates of the scaling functions are orthogonal.

2.9.3.3 Other Examples

Let us include two more examples. The first is the symmetric pair determined by the following three coefficients [65]:

$$H(0) = \begin{pmatrix} 0 & \frac{2+\sqrt{7}}{4} \\ 0 & \frac{2-\sqrt{7}}{4} \end{pmatrix}; \quad H(1) = \begin{pmatrix} \frac{3}{4} & \frac{1}{4} \\ \frac{1}{4} & \frac{3}{4} \end{pmatrix} \tag{2.305}$$

$$H(2) = \begin{pmatrix} \frac{2-\sqrt{7}}{4} & 0 \\ \frac{2+\sqrt{7}}{4} & 0 \end{pmatrix} \tag{2.306}$$

The other example is the multiwavelet system proposed by Chui-Lian (see [65] and references therein):

$$H(0) = \begin{pmatrix} \frac{1}{2} & -\frac{1}{2} \\ \frac{\sqrt{7}}{4} & -\frac{\sqrt{7}}{4} \end{pmatrix}; \quad H(1) = \begin{pmatrix} 2 & 0 \\ 0 & 1 \end{pmatrix} \tag{2.307}$$

$$H(2) = \begin{pmatrix} \frac{1}{2} & \frac{1}{2} \\ -\frac{\sqrt{7}}{4} & -\frac{\sqrt{7}}{4} \end{pmatrix} \tag{2.308}$$

2.10 Experiments

Next experiments consider the most popular applications of wavelets, which are signal analysis, denoising, and compression.

2.10.1 ECG Analysis Using the Morlet Wavelet

One of the most important fields of wavelet application is medicine. In this experiment a fragment of a normal electrocardiogram (ECG) signal has been selected. The fragment is shown in the upper plot in Fig. 2.72. It includes two consecutive hearthbeats.

The scalograms corresponding to continuous wavelet analysis are very expressive regarding local frequency contents. In this experiment the Morlet wavelet has been

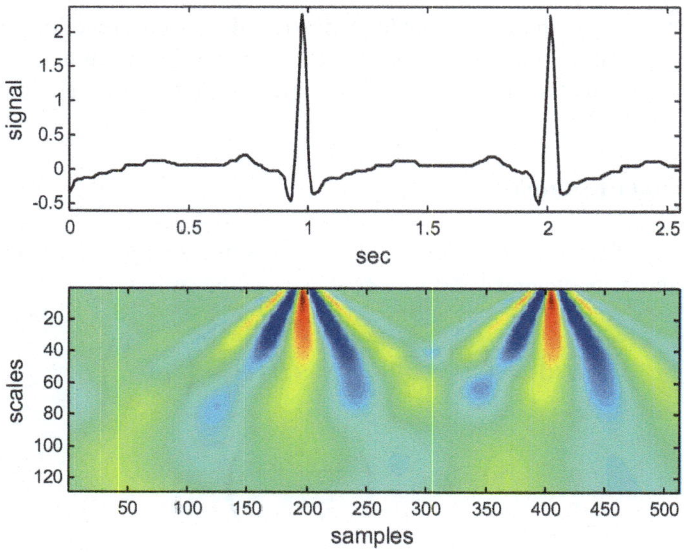

Fig. 2.72 Signal analysis using the Morlet wavelet

used. The scalogram of the two heartbeats is shown in Fig. 2.72. The pictured signal energies at each time and scale correspond well to the peaks and the valleys of hearthbeats. The figure has been obtained with the Program 2.37.

Notice important aspects of the Program 2.37. Zero padding is applied for both extremes of the signal. The program has a first vectorized part to compute a large wavelet tensor. Then, the program continues with the CWT as a multiplication of signal and tensor. The execution of the program may take around one or two minutes. Notice that the result is a large matrix, corresponding to the scalogram pixels.

Program 2.37 ECG analysis by continuous Morlet wavelet transform

```
% ECG analysis by continuous wavelet transform
% Morlet Wavelet
% Plot of signal and scalogram
clear all;
disp('please wait'); %ask for patience
%The ECG signal
fs=200; %samplig frequency in Hz
tiv=1/fs; %time interval between samples
Ts=tiv; %sampling period
%read signal file
fer=0;
while fer==0,
fid2=fopen('ECG_short.txt','r');
if fid2==-1, disp('read error')
else sg=fscanf(fid2,'%f \r\n'); fer=1;
end;
```

```
end;
fclose('all');
Nss=length(sg); %number of signal samples
duy=(Nss-1)*tiv; %duration of signal
tss=0:tiv:duy; %time intervals set
y=[sg(1)*ones(1,128) sg' sg(end)*ones(1,128)]; %padding
y=y-mean(y); %zero-mean signal
ND=length(y); %number of data
NS=128; %number of scales
CC=zeros(NS,ND); %for output (coeffs)
% Pre-compute wavelet tensor------------
PSIa=zeros(NS,ND,ND); %array
nn=1:ND;
t=Ts*(nn-1);
for ee=1:NS,
s=(ee*0.006)+0.05; %scales
for rr=1:ND, %delays
a=Ts*(rr-1);
val=0;
%vectorized part (t)
x=(t-a)/s;
PSIa(ee,rr,:)=(1/sqrt(s))*(exp(-(x.^2)/2).*cos(5*x));
end;
end;
disp('wavelet tensor is now ready')
%CWT------------------------
nd=1:ND;
for ne=1:NS,
aux=squeeze(PSIa(ne,nd,:));
val=(y)*aux;
CC(ne,nd)=val;
end;
%display--------------------
figure (1)
subplot(2,1,1)
plot(tss,y(129:(128+Nss)),'k');
axis([0 (Nss-1)*tiv min(y)-0.1 max(y)+0.1]);
xlabel('sec'); ylabel('signal');
title('wavelet analysis');
subplot(2,1,2)
imagesc(CC(:,129:(128+Nss)));
colormap('jet');
xlabel('samples'); ylabel('scales');
```

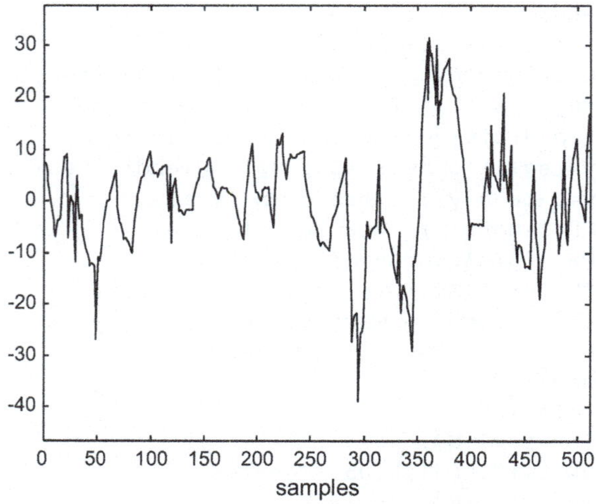

Fig. 2.73 Denoised Evoked Potential signal

2.10.2 *Signal Denoising*

Once a signal is decomposed into wavelet coefficients at several scales, there is the opportunity of attenuating or deleting high-frequency components often related to noise.

The idea is to choose a value, a threshold, and modify wavelet coefficients being smaller than the threshold. Two alternatives have been proposed: hard and soft thresholding. Both substitute by zeros the smaller coefficients. Hard thresholding keeps untouched the rest of coefficients. Soft thresholding substracts the threshold to the rest of coefficients.

The thresholding is usually applied to the highest scales. However there are alternatives: global, or scale level by scale level.

Of course there is a key issue: what threshold value should be chosen. The research has proposed several ways to take into account estimated noise.

Another issue is strategic: to keep a constant threshold, or to change it along the signal. Again, the research proposes some alternatives, being adaptive or not, local in time or by blocks, etc.

Opportune references on signal denoising are [29, 56].

Our experiment is to denoise the evoked potential signal that has been studied in Sect. 2.4.3. A hard thresholding has been applied using the Program 2.38. The result is shown in Fig. 2.73.

Program 2.38 Denoising example with Daubechies DWT

```
% Denoising example
% with inverse of the Daubechies DWT
% Visual Evoked Potential signal
% Use after analysis routine (to get wty)
L=length(wty); %length of the DWT
```

```
%scaling filter
hden=4*sqrt(2); %coeff. denominator
hsq=sqrt(3); %factor
%Daubechies 4:
h=[(1+hsq)/hden, (3+hsq)/hden, (3-hsq)/hden, (1-hsq)/hden];
hN=h;
N=length(hN);
K=9; %number of scales
aux=0;
h0=hN;
h1=fliplr(hN); h1(2:2:N)=-h1(2:2:N);
Ln=1;
a=wty(1);
th=8; %noise threshold
for n=1:K,
aux= 1+mod(0:N/2-1,Ln);
d=wty(Ln+1:2*Ln);
%denoising at the higher scales
if n>K-6,
mm=length(d);
for m=1:mm,
if abs(d(m))<th,
d(m)=0;
end;
end;
end;
ax(1:2:2*Ln+N)= [a a(1,aux)];
dx(1:2:2*Ln+N)= [d d(1,aux)];
a=conv(ax,h0)+ conv(dx,h1);
a=a(N:(N+2*Ln-1));
Ln=2*Ln;
end;
figure(1)
plot(a,'k');
axis([0 512 1.2*min(a) 1.2*max(a)]);
xlabel('samples');
title('the denoised signal');
```

2.10.3 Compression

It has been noticed in many cases that if you delete the wavelet coefficients of the highest scales, and then you recover the signal, this recovered signal has acceptable characteristics. Even, it may happen that this signal has less noise than the original. Therefore, if you have say a 10 Mb signal and you apply wavelet analysis using 12 scale levels, it can be reduced to 5 Mb just by deleting the 12th scale coefficients;

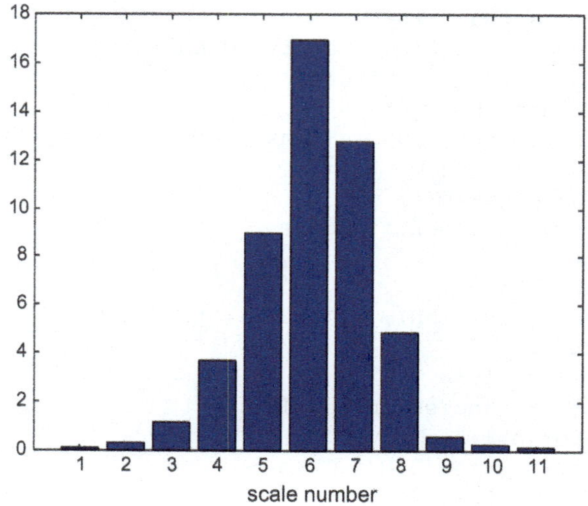

Fig. 2.74 Energy of the signal to be compressed

or even smaller sizes if you continue deleting coefficients of 11th, 10th, etc., scale levels. Obviously you loose signal details: it is lossy compression.

A piece of Bizet music has been chosen for our signal comprssion experiment. Since the signal is long in terms of number of samples, the signal is divided into segments, and a Haar wavelet analysis has been repeatedly applied.

The idea has been to obtain with the wavelet analysis a set of vectors *wty(n,:)* containing the wavelet coefficients of each signal segment. This is an uncompressed representation of the original signal. Then, the upper half part of each *wty(n,:) (with* $n = 1, 2 \ldots$*)* is eliminated. Hence, a 50 % signal compression is achieved.

Program 2.39 implements this simple procedure. The result of compression is a set of vectors *cwty(n,:)*. This is supposed to be stored on a file for further use.

Then, after compression a signal is recovered by the second part of the program. Here the simple idea is to append zeros in substitution of the deleted wavelet coefficients, getting a set of vectors *rwty(n,:)*, and then apply Haar synthesis to obtain an uncompressed signal.

When running the Program 2.39 the user can hear the original music, and then, after some moments, hear the recovered uncompressed signal for comparison. The program also generates the Fig. 2.74 that shows the energy distribution of the music signal among scale levels. The reason for the compressing procedure to be acceptable is that it only touches the highest scale, where there is not much signal energy. The figure also suggests that more compressing could be tried.

Program 2.39 Audio compression example with Haar wavelet

```
% Audio compression example
% analysis of music signal
% Haar Wavelet
% Hear original and compressed signals
```

```
%The music signal
[y,fs]=wavread('Bizet1.wav'); %read wav file
y1=y(:,1); %mono channel
y1=y1-mean(y1); %zero mean
Nss=300*2048; %number of signal samples
sg=y1(1:Nss);
tiv=1/fs;
disp('the original music');
soundsc(sg,fs);
pause(16);
disp('now, wavelets in action');
%analysis of the signal with wavelets
%divide the signal into 230 segments
%apply wavelet analysis to each segment
K=11; %number of scales (2048=2^11)
cq=sqrt(3);
wty=zeros(300,2048);
En=zeros(1,K); %for energy measurement
for nst=1:300,
bg=((nst-1)*2048)+1;
y=sg(bg:(bg+2047)); %music segment
%Haar wavelet
NN=2048;
for n=1:K,
aux1= y(1:2:NN-1) + y(2:2:NN);
aux2= y(1:2:NN-1) - y(2:2:NN);
y(1:NN)=[aux1,aux2]/sqrt(2);
En(n)=En(n)+sum(y((NN/2)+1:NN).^2);
NN=NN/2;
end;
wty(nst,:)=y';
end;
%-------------------------------------------
%compress the signal by deleting highest scale data
aux=zeros(300,1024);
cwty=zeros(300,1024); %50% smaller
for nn=1:300,
aux(nn,:)=wty(nn,1:(2^10)); %delete the upper half of wty
end;
cwty=aux; %compressed audio (to be stored)
%-------------------------------------------
% read compressed audio and recover the signal
aux=zeros(300,1024);
rwty=zeros(300,2048);
%append zeros for the upper half of rwty
for nn=1:300,
rwty(nn,:)=[cwty(nn,:) aux(nn,:)];
end;
%---------------
```

```
%conventional signal recovery
ry=zeros(1,Nss);
J=K+1;
a=zeros(J,(2^K)); %space for a(j,k) coefficients
%signal recovering (wavelet synthesis)
for nst=1:300,
m=1;
z=rwty(nst,:);
a(1,1)=z(1);
for n=1:K,
a(n+1,1:2:(2*m-1))=(a(n,1:m)+z((1+m):(2*m)))/sqrt(2);
a(n+1,2:2:(2*m))=(a(n,1:m)-z((1+m):(2*m)))/sqrt(2);
m=m*2;
end;
bg=((nst-1)*2048)+1;
ry(bg:(bg+2047))=a(J,:); %signal recovery
end;
disp('and the decompressed music');
soundsc(ry,fs);
%display of signal energy vs scale------------------------
figure(1)
bar(En);
title('signal energy vs. scale');
xlabel('scale number');
```

2.11 Applications

The area of wavelet applications is continuously expanding. Possibly, two main application directions could be identified. One is linked to signal/data compression, and denoising. The other comes from the origins of wavelets: the analysis of signals/data.

In this section a short overview of different fields of application will be done. Most cited references are intended as starting points for the reader, and so they include abundant bibliographies.

A review of emerging applications of wavelets is [4]. The book [37] includes a set of case studies. There are several scientific journals focusing on wavelets, like the International Journal of Wavelets, Multiresolution and Information Processing, the Journal of Wavelet Theory and Applications, and others.

2.11.1 Earth Sciences

During the last 1970s, Morlet introduced what he called “wavelets of constant shapes”. Mr. Morlet was a geophysical engineer at an oil company. Daubechies, in her view of the wavelet history [26], tells that, soon after, the research adopted the term *wavelet* letting implicit the mention of constant shape.

Hence, the origin of wavelets is in geophysical signal analysis. It is not strange that now there is an extensive literature on geophysical wavelet applications.

Seismic signals are of evident interest for at least two reasons. One is earthquakes, and the other is detection and characterization of hydrocarbon reservoirs, minerals, etc.

With respect to earthquakes, the article [61] provides pertinent information and references.

And with respect to natural resources the key topic is seismic inversion. The web page of the Wavelet Seismic Inversion Lab, and the publications of the Society of Exploration Geophysicists offer a good ingress on this field. In [62] the use of wavelets and the like on the sphere, with application to geophysical data, is treated; see also [63] about the use of continuous wavelets.

There are many other branches of Earth sciences interested on the use of wavelets. This is revealed by research papers like [19] on tsunami warning, [32] on ocean waves, [28] on atmospheric sciences and space, etc. The article [34] presents an overview of wavelet applications in earthquake, wind and ocean engineering,

There are some books on wavelets and geophysics, like [33, 38]. Denoising of geophysical data is considered in a number of papers, see [71] and references therein.

2.11.2 Medicine, Biology

The denoising and compressing capabilities of wavelets have found great acceptance for biomedical applications. In addition, the research has proposed several applications of wavelets for the signal/data processing and analysis in this area. There are some books about the use of wavelets in medicine and biology [6, 18]. Other books include chapters on wavelet biomedical applications [51, 52, 58]. The book [11] deals with wavelet applications in biology and geosciences. Articles with reviews of this topic are [1, 5, 74].

A lot of research effort has been directed to image processing. This aspect will be covered in the next chapter.

Regarding physiological signals, a main source of information is the web page Physionet

The application of wavelet techniques in electrocardiograms (ECG) is reviewed in [50]. The paper [30] deals with sleep stage classification on the basis of electroencephalogram. As an aspect of robotics, the recognition of emotions is treated in the Thesis [10].

The paper [42] presents the state of the art in bioinformatics.

Since natural sounds, like bird singing or other animal sounds, are transients, it is pertinent to use wavelets for their study; [40] is a frequently cited article about it.

2.11.3 Chemical

According with the perspective given in [59] there is an increasing number of wavelet applications in chemistry. In particular for the study of spectra, chromatography, or in case of oscillatory signals.

Some examples are [31], which describes the applications of wavelets to analytical chemistry for denoising, compression, etc. The article [60] on the study of pressure in a fluizided bed. And [17] on Raman spectrography, and [57] on spectral libraries and remote environmental sensing.

2.11.4 Industrial

In the reviews [72, 73] of industrial applications of wavelets, the authors cite applications related to acoustical signal processing, power production and transport, power electronics, non-destructive testing, chemical processes, stochastic signal analysis, image compression, satellite imaginery, machine vision, bioinformatics, and flow analysis.

Vibration and mechanical damage is an obvious field of application for wavelets. Two illustrative papers concerned with this topic are [66] on structural damage, and [9] on bridge integrity assessment.

Another kind of application is machine condition monitoring. The article [54] presents a review of this topic, with extensive bibliography.

In [2] the authors deal with crack propagation in rotating shafts. The article of [55] uses wavelet analysis of acoustic emissions for material characterization. There is a number of papers on wavelets and materials fatigue, considering cases related to railways, ships, airplanes, etc.

2.12 The MATLAB Wavelet Toolbox

With the help of the MATLAB Wavelet Toolbox, wavelets can be applied and studied in a fast and easy way. This Toolbox has extensive and well organized documentation.

The purpose of this section is a quick introduction to the Toolbox.

Since the Toolbox includes many wavelets, the user can recall names and short descriptions using the functions *waveinfo()* and *waveletfamilies()*.

With *wavedemo()*, the user can get first impressions of wavelet possibilities.

An important general tool is *wavemenu()*. It is an entrance to the graphical tools. The following screen appears (Fig. 2.75):

The Toolbox provides functions for 1-D continuous and discrete wavelets, and 2-D discrete wavelets. In addition the Toolbox includes wavelet packets, lifting wavelets, multiwavelets, denoising and compression and other wavelet applications.

Fig. 2.75 Screen corresponding to *wavemenu()*

2.12.1 1-D Continuous Wavelet

The main function for 1-D continuous wavelets is *cwt()*. The scalogram can be obtained in two ways: one by using *wscalogram()* after *cwt()*, or directly from *cwt()* with '*plot*' option.

If you use the [Continuous Wavelet 1-D] option after *wavemenu()*, a graphical tool opens. You import a file with the signal to analyze (for instance in *.wav* format), then choose a wavelet, and then select display options (colors, subplots, etc.). The following screen (Fig. 2.76) has been obtained using the Morlet wavelet for the analysis of the cow moaning already studied in Chap. 6 with the spectrogram. The central plot is the scalogram.

2.12.2 1-D Discrete Wavelet

The main functions for 1-D discrete wavelets are is *dwt()*and *idwt()* for single-level transform, and *wavedec()* and *waverec()* for multiple level transform.

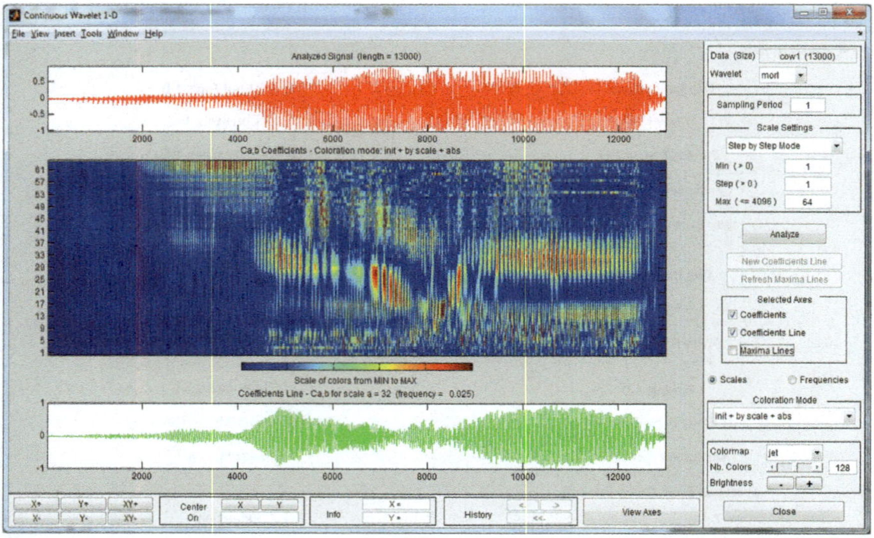

Fig. 2.76 Screen corresponding to cow moaning

If you use the [Wavelet 1-D] option after *wavemenu()*, then a screen similar to the previous example opens. Figure 2.77 shows the result of analyzing a duck quack. The subplots show the decomposition into scales using the Haar wavelet (the numbering of scales is the opposite to the numbering in this chapter).

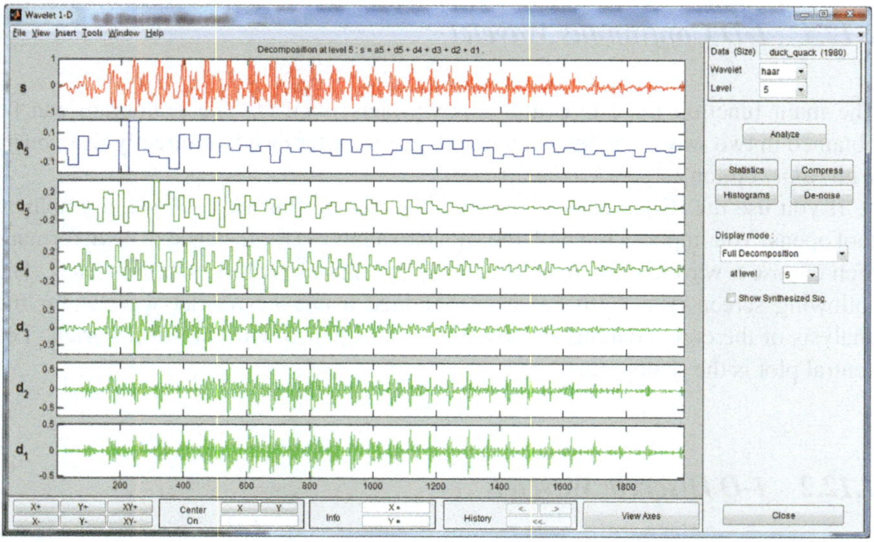

Fig. 2.77 Screen corresponding to duck quack

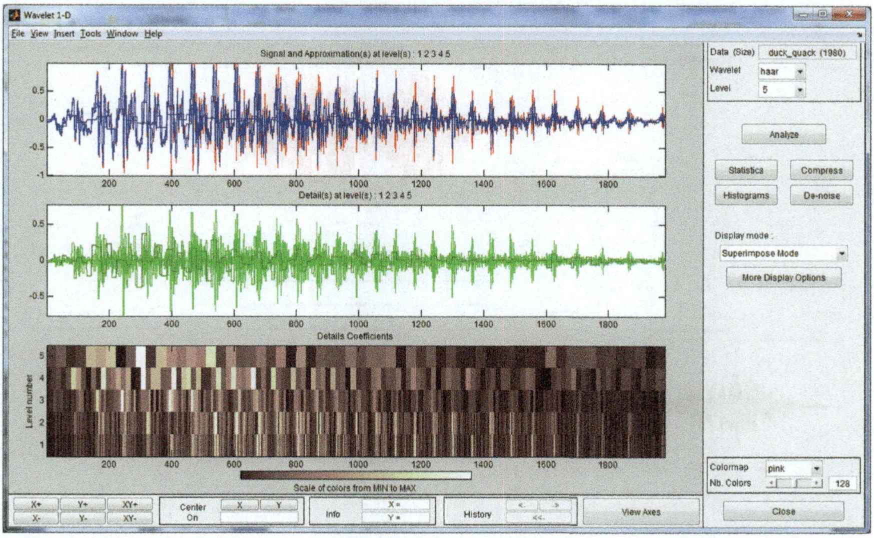

Fig. 2.78 Another screen corresponding to duck quack

Another visualization option leads to the screen shown in Fig. 2.78. It includes an approximation to the signal and the scalogram.

2.12.3 *Wavelet Packets*

The main functions for 1-D wavelet packets are *wpdec()* and *wprec()*. To find best basis one uses *bestlevt()* and *besttree()*.

If you use the [Wavelet Packet 1-D] option after *wavemenu()*, then a screen opens. This is a different type of screen, specific for wavelet packets. Figure 2.79 shows the screen when a harp signal was analyzed with Haar wavelet. The left plot shows the structure of the packet decomposition. If you click on one of the nodes, a plot appears at bottom left with the part of the analized signal that crosses this node. Another plot, bottom right, shows the coefficients at the terminal nodes. Taken as a whole, the screen is a tool that helps to find a best basis.

2.12.4 *Lifting*

The main functions for 1-D lifting wavelet transforms are *lwt()* and *ilwt()*. Information about lifting schemes is given by *lsinfo()* and *displs()*, the lifting scheme for

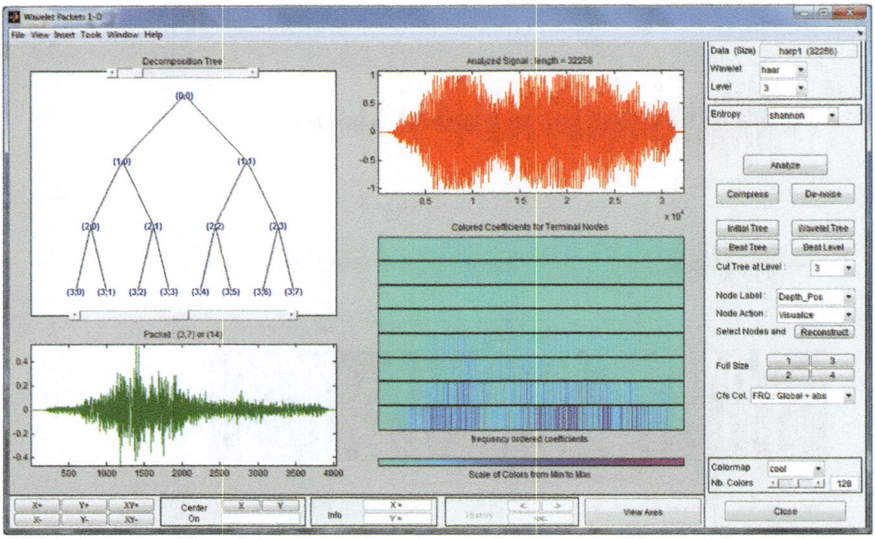

Fig. 2.79 Screen corresponding to harp and wavelet packet analysis

a particular wavelet is given by *liftwave()*, the names of wavelets is provided by *wavenames()*. The lifting schemes can be translated to filter banks using *ls2filt()*.

2.12.4.1 Multiwavelets (multisignal)

The main functions for 1-D multisignal wavelet transforms are *mdwtdec()* and *mdtrec()*. There are functions for denoising and compression using multisignal transform.

2.13 Resources

2.13.1 MATLAB

Usually the toolboxes include tables of coefficients for the different wavelet families.

2.13.1.1 Toolboxes

- WaveLab:
  http://www-stat.stanford.edu/~wavelab/
- Uvi-Wave Wavelet Toolbox:
  http://www.gts.tsc.uvigo.es/~wavelets/

- Rice Wavelet Toolbox:
  http://dsp.rice.edu/software/rice-wavelet-toolbox
- UST Wavelets:
  http://cam.mathlab.stthomas.edu/wavelets/packages.php
- WavBox:
  http://www.toolsmiths.com/wavelet/WavBox
- WAVOS Toolkit:
  http://sourceforge.net/projects/wavos/files/
- WMTSA Wavelet Toolkit:
  http://www.atmos.washington.edu/~wmtsa/
- Wavekit Toolkit:
  http://www.math.rutgers.edu/ojanen/wavekit/

2.13.1.2 Matlab Code

- Mathworks file exchange:
  http://www.mathworks.com/matlabcentral/fileexchange/5104-toolbox-wavelets
- Gabriel Peyre:
  http://www.mathworks.es/matlabcentral/fileexchange/authors/14044
- Ripples in Mathematics:
  http://www.control.auc.dk/alc/ripples.html
- Matthew Roughan:
  http://www.maths.adelaide.edu.au/matthew.roughan/
- Pascal Getreuer:
  http://www.getreuer.info/
  (see also): http://www.mathworks.com/matlabcentral/fileexchange/authors/14582
- Brooklin Poly:
  http://eeweb.poly.edu/iselesni/WaveletSoftware/standard1D.html
- Joseph Salmon:
  http://josephsalmon.eu/enseignement/M1/ondelettes.sci

2.13.2 Internet

2.13.2.1 Web Sites

- Wavelet.org:
  http://www.wavelet.org/
- Wim Sweldens'Homepage:
  www.cm.bell-labs.com/who/wim/
- Jacket's Wavelets:
  http://gtwavelet.bme.gatech.edu/

- Rodrigo Quian, Evoked Potential data:
  http://www2.le.ac.uk/departments/engineering/research/bioengineering/neuroengineering-lab/software/
- The Wavelet Seismic Inversion Lab:
  http://timna.mines.edu/~zmeng/waveletlab/waveletlab.html
- Physionet:
  www.physionet.org
- The BioSig Project:
  http://biosig.sourceforge.net/
- JPEG Homepage:
  http://www.jpeg.org/jpeg/index.html

2.13.2.2 Link Lists

- Wavelet Software:
  http://www.amara.com/current/wavesoft.html
- SIVA links:
  http://www.laurent-duval.eu/siva-signal-image-links.html
- Baum links:
  http://stommel.tamu.edu/~baum/wavelets.html

References

1. P.S. Addison, J. Walker, R.C. Guido, Time-frequency analysis of biosignals. IEEE Eng. Med. Biol. Mgz. pp. 14–29 (2009)
2. S.A. Adewuai, B.O. Al-Bedoor, Wavelet analysis of vibration signals of an overhang rotor with a propagating transverse crack. J. Sound Vibr. **246**(5), 777–793 (2001)
3. L. Aguiar-Conraria, M.J. Soares, The continuous wavelet transform: Moving beyond uni-and bivariate analysis. J. Econ. Surv. **28**(2), 344–375 (2014)
4. A.N. Akansu, W.A. Serdijn, I.W. Selesnick, Emerging applications of wavelets: A review. Phys. Commun. **3**, 1–18 (2010)
5. M. Akay, Wavelet applications in medicine. IEEE Spectrum **34**(5), 50–56 (1997)
6. A. Aldroubi, M. Unser, *Wavelets in Medicine and Biology* (CRC Press, 1996)
7. O. Alkin, H. Caglar, Design of efficient M-band coders with linear-phase and perfect-reconstruction properties. IEEE Trans. Sign. Process. **43**(7), 1579–1590 (1995)
8. B.K. Alpert, A class of bases in L2 for the sparse representation of integral operators. SIAM J. Math. Anal. **24**(1), 246–262 (1993)
9. A. Alvandi, J. Bastien, E. Gregoire, M. Jolin, Bridge integrity assessment by continuous wavelet transforms. Intl. J. Struct. Stab. Dyn. **9**(11) (2009)
10. V. Aniket, *Biosignal Processing Challenges in Emotion Recognition for Adaptive Learning*. PhD thesis (Univ. Central Florida, 2010)
11. D. Balenau, *Wavelet Transforms and Their Recent Applications in Biology and Geoscience* (InTech., 2012)
12. R.P. Boyer, Generalized Bernstein polynomials and symmetric functions. Adv. Appl. Math. **28**, 17–39 (2002)

13. J. Bradley, C. Brislawn, T. Hopper, The FBI wavelet/scalar quantization standard for gray-scale fingerprint image compression, in *SPIE v.1961: Visual Image Processing* (1993), pp. 293–304
14. C.S. Burrus, R.A. Gopinath, H. Guo, *Wavelets and Wavelet Transforms* (Prentice-Hall, 1998)
15. P. Burt, E. Adelson, The Laplacian pyramid as a compact image code. IEEE Trans. Commun. **31**, 482–540 (1983)
16. H. Caglar, A.N. Akansu, A generalized parametric PR-QMF design technique based on Bernstein polynomial approximation. IEEE Trans. Sign. Process. **41**(7), 2314–2321 (1993)
17. T.T. Cai, D. Zhang, D. Ben-Amotz, Enhanced chemical classification of Raman images using multiresolution wavelet transformation. Appl. Spectrosc. **55**(9), 1124–1130 (2001)
18. R. Capobianco, *Emergent Applications of Fractals and Wavelets in Biology and Biomedicine* (Elsevier, 2009)
19. A. Chamoli, V.S. Rani, K. Srivastava, D. Srinagesh, V.P. Dimri, Wavelet analysis of the seismograms for tsunami warning. Nonlinear Process. Geophys. **17**, 569–574 (2010)
20. A. Cohen, I. Daubechies, J.C. Feauveau, Biorthogonal bases of compactly supported wavelets. Commun. Pure Appl. Math. **45**, 485–560 (1992)
21. R.R. Coifman, M.V. Wickerhauser, Entropy-based algorithms for best basis selection. IEEE Trans. Inform. Theory **38**(2), 713–718 (1992)
22. T. Cooklev, A. Nishihara, M. Sablatash, Regular orthonormal and biorthogonal wavelet filters. Sign. Process. **57**, 121–137 (1997)
23. A. Cour-Harbo, A. Jensen. Wavelets and the lifting scheme, in *Encyclopedia of Complexity and Systems Science*, ed. by R.A. Meyers (Springer, 2009), pp. 10007–10031
24. I. Daubechies, The wavelet transform, time-frequency localization and signal analysis. IEEE Trans. Inform. Theory **36**(5), 961–1005 (1990)
25. I. Daubechies, *Ten Lectures on Wavelets* (SIAM, Philadelphia, 1992)
26. I. Daubechies, Where do wavelets come from?- A personal point of view. Proc. IEEE **84**(4), 510–513 (1996)
27. I. Daubechies, W. Sweldens, Factoring wavelet and subband transforms into lifting steps. Technical report, TechnicalBell Laboratories, Lucent Technologies (1996)
28. M.O. Domingues, O. Jr. Mendes, A. Mendes da Costa, On wavelet techniques in atmospheric sciences. Adv. Space Res. **35**, 831–842 (2005)
29. D.L. Donoho, Denoising by soft-thresholding. IEEE Trans. Inform. Theory **41**(3), 613–627 (1995)
30. F. Ebrahimi, M. Mikaeili, E. Estrada, H. Nazeran, Automatic sleep stage classification based on EEG signals by using neural networks and wavelet packet coefficients. Proc. IEEE Int. Conf. EMBS 1151–1154 (2008)
31. F. Ehrentreich, Wavelet transform applications in analytical chemistry. Anal. Bioanal. Chem. **372**(1), 115–121 (2002)
32. M. Elsayed, An overview of wavelet analysis and its application to ocean wind waves. J. Coast. Res. **26**(3), 535–540 (2010)
33. Foufola-Georgiou, E., P. Kumar, *Wavelets in Geophysics* (Academic Press, 1994)
34. K. Gurley, A. Kareem, Applications of wavelet transforms in earthquakes, wind and ocean engineering. Eng. Struct. **21**, 149–167 (1999)
35. A. Jensen, A. la Cour-Harbo, *Ripples in Mathematics* (Springer, 2001)
36. Z. Jiang, X. Guo, A note on the extension of a family of biorthogonal Coifman wavelet systems. The ANZIAM J. **46**, 111–120 (2004)
37. M. Kobayashi, Wavelets and their applications: Case studies. Technical report, IBM Tokyo Research Lab (1998)
38. P. Kumar, Wavelet analysis for geophysical applications. Rev. Geophys. **35**(4), 385–412 (1997)
39. F. Kurth, M. Clausen, Filter bank tree and M-band wavelet packet algorithms in audio signal processing. IEEE Trans. Sign. Process. **47**(2), 549–554 (1999)
40. M.S. Lewicki, Efficient coding of natural sounds. Nat. Neurosci. **5**(4), 356–363 (2002)
41. T. Lin, S. Xu, Q. Shi, P. Hao, An algebraic construction of orthonormal M-band wavelets with perfect reconstruction. Appl. Math. Comput. **172**, 717–730 (2006)

42. P. Lio, Wavelets in bioinformatics and computational biology: State of art and perspectives. Bioinform. Rev. **19**(1), 2–9 (2003)
43. Z. Liu, N. Zheng, Parametrization construction of biorthogonal wavelet filter banks for image coding. Sign. Image Video Process. **1**, 63–76 (2007)
44. S. Mallat, *A Wavelet Tour of Signal Processing: The Sparse Way* (Academic Press, 2008)
45. S.G. Mallat, A theory for multiresolution signal decomposition: The wavelet representation. IEEE Trans. Pattern Anal. Mach. Intell. **11**(7), 674–693 (1989)
46. M. Maslen, P. Abbott, Automation of the lifting factorization of wavelet transforms. Comput. Phys. Commun. **127**, 309–326 (2000)
47. Y. Meyer. *Principe D'incertitude, Bases Hilbertiennes Et Algebres D'operateurs* (1985). Seminaire Bourbaki, n. 662. http://archive.numdam.org/article/SB_1985-1986_28_209_0.pdf
48. M. Misiti, Y. Misiti, G. Oppenheim, J.M. Poggi, *Wavelets and Their Applications* (ISTE, London, 20070
49. J. Morlet, A. Grossman, Decomposition of Hardy functions into square integrable wavelets of constant shape. SIAM J. Math. Anal. **15**, 723–736 (1984)
50. H. Nagendra, S. Mukherjee, V. Kumar, Application of wavelet techniques in ECG signal processing: An overview. Int. J. Eng. Sci. Technol. **3**(10), 7432–7443 (2011)
51. A. Nait-Ali, *Advanced Biosignal Processing* (Springer, 2009)
52. K. Najarian, R. Splinter, *Biomedical Signal and Image Processing* (CRC Press, 2012)
53. B.D. Patil, P.G. Patwardhan, V.M. Gadre, On the design of FIR wavelet filter banks using factorization of a halfband polynomial. IEEE Sign. Process. Lett. **15**, 485–488 (2008)
54. Z.K. Peng, F.L. Chu, Application of the wavelet transform in machine condition monitoring and fault diagnostics: A review with bibliography. Mech. Syst. Sign. Process. **18**, 199–221 (2004)
55. G. Qi, Wavelet-based AE characterization of composite materials. NDT E Int. **33**(3), 133–144 (2000)
56. M. Raphan, E.P. Simoncelli, Optimal denoising in redundant representations. IEEE Trans. Image Process. **17**(8), 1342–1352 (2008)
57. B. Rivard, J. Feng, A. Gallie, A. Sanchez-Azofeifa, Continuous wavelets for the improved use of spectral libraries and hyperspectral data. Remote Sens. Environ. **112**, 2850–2862 (2008)
58. J.L. Semmlow, *Biosignal and Biomedical Image Processing* (CRC Press, 2008)
59. X.G. Shao, A.K. Leung, F.T. Chau, Wavelet: A new trend in chemistry. Acc. Chem. Res **36**(4), 276–283 (2003)
60. M.C. Shou, L.P. Leu, Energy of power spectral density function and wavelet analysis of absolute pressure fluctuation measurements in fluidized beds. Chem. Eng. Res. Des. **83**(5), 478–491 (2005)
61. F.J. Simons, B.D.E. Dando, R.M. Allen, Automatic detection and rapid determination of earthquake magnitude by wavelet multiscale analysis of the primary arrival. EarthPlanet. Sci. Lett. **250**, 214–223 (2006)
62. F.J. Simons, I. Loris, E. Brevdo, I. Daubechies, Wavelets and wavelet-like transforms on the sphere and their application to geophysical data inversion. Proc. SPIE **8138**, 1–15 (2011)
63. S. Sinha, P.S. Routh, P.D. Anno, J.P. Castagna, Spectral decomposition of seismic data with continuous-wavelet transform. Geophysics **70**, 19–25 (2005)
64. A.N. Skodras, C.A. Christopoulos, T. Ebrahimi, JPEG2000: The upcoming still image compression standard. Pattern Recogn. Lett. **22**(12), 1337–1345 (2001)
65. V. Strela, P.N. Heller, G. Strang, P. Topiwala, C. Heil, The application of multiwavelet filterbanks to image processing. IEEE T. Image Proc. **8**(4), 548–563 (1999)
66. Z. Sun, C.C. Chang, Structural damage assessment based on wavelet packet transform. J. Struct. Eng. **128**(10), 1354–1361 (2002)
67. W. Sweldens, The lifting scheme: A construction of second generation wavelets. SIAM. J. Math. Anal. **29**(2), 511–546 (1997)
68. D.B.H. Tay, Rationalizing the coefficients of popular biorthogonal wavelet filters. IEEE Trans. Circ. Syst. Video Technol. **10**(6), 998–1005 (2000)

69. D.B.H. Tay, M. Palaniswami, A novel approach to the design of the class of triplet halfband filterbanks. IEEE Trans. Circ. Syst.-II: Express Briefs **51**(7), 378–383 (2004)
70. J. Tian, R.O. Wells Jr, Vanishing moments and biorthogonal wavelet systems, in *Mathematics in Signal Processing IV*, ed. by Mc.Whirter (Oxford University Press, 1997)
71. A.C. To, J.R. Moore, S.D. Glaser, Wavelet denoising techniques with applications to experimental geophysical data. Sign. Process. **89**, 144–160 (2009)
72. F. Truchelet, O. Laligant, Wavelets in industrial applications: A review, in *Proceedings SPIE*, vol. 5607 (2004), pp. 1–14
73. F. Truchelet, O. Laligant, Review of industrial applications of wavelet and multiresolution-based signal and image processing. J. Electron. Imaging **17**(3) (2008)
74. M. Unser, A. Aldroubi, A review of wavelets in biomedical applications. Proc. IEEE **84**(4), 626–638 (1996)
75. M. Unser, T. Blu, Mathematical properties of the JPEG2000 wavelet filters. IEEE Trans. Image Process. **12**(9), 1080–1090 (2003)
76. G. Uytterhoeven, D. Roose, A. Bultheel, Wavelet transforms using the lifting scheme. Technical report, Katholieke Universiteit Leuven, 1997. ITA-Wavelets-WP.1.1
77. M. Vetterli, J. Kovacevic. *Wavelets and Subband Coding* (Prentice Hall, 1995)
78. D. Wei. *Coiflet-type Wavelets: Theory, Design, and Applications*. PhD thesis, University of Texas at Austin (1998)
79. X. Yang, Y. Shi, B. Yang, General framework of the construction of biorthogonal wavelets based on Bernstein bases: Theory analysis and application in image compression. IET Comput. Vision **5**(1), 50–67 (2011)
80. R. Yu, A. Baradarani, Design of halfband filters for ortogonal wavelets via sum of squares decomposition. IEEE Sign. Process. **15**, 437–440 (2008)
81. X. Zhang, Design of FIR halfband filters for orthonormal wavelets using Remez exchange algorithm. IEEE Sign. Proces. Lett. **16**(9), 814–817 (2009)

Chapter 3
Image and 2D Signal Processing

3.1 Introduction

While many animals depend on audition, olfaction, or pressure sensing, for their lives, in the case of humans vision is of utmost importance.

One of the characteristic advances of modern times is the digitalization of images. Several signal processing methods are nowadays used by common people, and taken as granted; for example, photograph and video encoding/compression, visual effects, image stabilization, face recognition, etc.

This chapter is an introduction to the topic of image processing. In its first sections, the interest is centred on image filtering, improvement, certain modifications, and emphasizing of edges and borders. Then, the chapter deals with important techniques of medical diagnosis related to computational tomography. Finally, the chapter introduces filter banks for bi-dimensional processing.

It should be said that images are not the only 2D scenario to be considered. In fact, networked sensor systems are of increasing interest, for a number of applications related to seismi activity, mineral prospection, weather prediction, tsunami prevention, etc.

In this chapter we used some of the specialized functions provided by the MATLAB Image Processing Toolbox.

It would be recommended for the reader to further explore, with other types of images and targets, the potentialities of the methods introduced in this chapter

3.2 Image Files and Display

Image capture devices usually give the results as arrays of values. The elements of these arrays are called *pixels*.

There are four basic types of images: binary, gray scale, true color, and indexed. In the case of binary images, each pixel has only two possible values, corresponding

© Springer Science+Business Media Singapore 2017

J.M. Giron-Sierra, *Digital Signal Processing with Matlab Examples, Volume 2*,
Signals and Communication Technology, DOI 10.1007/978-981-10-2537-2_3

to black or white. In the case of gray scale, each pixel has a value that corresponds to a shade of gray. A frequent gray scale representation uses integer values between 0, black, and 255, white: with this representation each pixel uses a byte (8 bits).

RGB, red-blue-green, is a common way to deal with colors. Extending the idea of gray scale representation, the amount of red can be specified with integers between 0 and 255, and so with blue and with green; and then each pixel uses 3 bytes (24 bits).

In many cases it is possible to use less than 16 million colors for a particular image. A color map or palette, a list of colors, can be defined, and each pixel can just point to the pertinent entry of the list: in other words, pixels are indexes to the list. For instance, the list could contain 256 colors, so each pixel uses a byte. This is the indexed image type, which tries to reduce the number of bytes for image representation.

3.2.1 Image Files

Images can be saved on a variety of file formats. The MATLAB Image Processing Toolbox (IPT) provides functions to read and write in a number of well-know file formats, which are briefly listed below:

- TIFF: Tagged Image File Format. It can be used for binary, grey scale,true color and indexed images. It supports several compression methods.
- JPEG: Joint Photographics Experts Group compression method.
- BMP: Microsoft Bitmap.
- HDF: Hierarchical Data Format. Used for scientific images.
- XWD: X Window Dump, in UNIX systems.
- Old ones, like PCX, used in MS-DOS Paintbrush and many others, or GIF, Graphics Interchange Format. An improved replacement of GIF is PNG, which supports gray scale, true color and indexed images.

IPT can also handle the ICO format, for Ms-Windows icons, and the CUR format for the Ms-Windows mouse cursor.

The compression methods used by GIF and PNG do not loose information. On the contrary, JPEG uses a lossy compression.

3.2.2 Image Display with MATLAB

Let us display an image, taking it from a TIFF file. This is done in the Program 3.1 using two IPT functions: *imread()* and *imshow()*. In the first line of the program the file is read, and the image is saved in a matrix, named keaton. The third line of the program displays the image matrix. Also, in this line, the function *pixval* is invoked to get information on the numeric value of any pixel in the image. Figure 3.1 shows the result of this short program.

Fig. 3.1 Image display

Program 3.1 Display a gray scale picture

```
% Display gray scale picture
%read the image file into a matrix:
keaton=imread('keaton1bw.tif');
figure(1)
imshow(keaton); pixval on; %display the photo
```

Figure 3.1 shows a gray scale image, as it was stored in the TIFF file that has been read. The line below the image is generated by *pixval*, you can move the mouse on the image, and see the integer value, between 0 and 255, corresponding to the image pixels. Notice that the (1, 1) coordinates correspond to the top left corner.

If you write in the MATLAB command window *size(keaton)* you will obtain 368 461, meaning that the matrix keaton has 368 rows and 461 columns.

In a similar way, color images can be read and displayed. The last section of this chapter will deal with color images in particular.

MATLAB can handle several data types. One of these data types is *uint8,* unsigned integer, with integer values between 0 and 255. The matrix keaton has unsigned integer data. Many SPT functions use another data type, the *double* data type: double precision real numbers. The function *imshow()* can display a matrix of double precision real numbers with values between 0 and 1.

The function *im2double()* converts *unit8* data, values between 0 and 255, to *double* data, values between 0 and 1. The function *im2uint8()* does the reverse.

Arithmetic operations are not permitted with unsigned integer data; therefore you should convert to *double* in order to apply mathematics to images.

3.3 Basic Image Analysis and Filtering

Given a gray scale image, it is interesting to analyze its contents in terms of amounts of white, black and shades of grey. On the basis of this analysis the image can be processed to modify these amounts.

3.3.1 Histograms

The histogram is a bar graph showing the number of pixels with value 0, with value 1, with value 2, etc. until 255.

Program 3.2 uses the IPT function *imhist()* to compute and show the histogram of the image matrix keaton. Figure 3.2 shows the result.

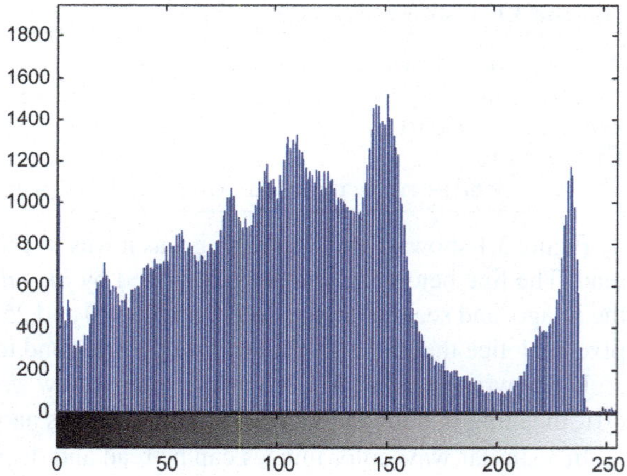

Fig. 3.2 Image histogram

Program 3.2 Histogram of the gray scale picture

```
% Histogram of gray scale picture
%read the image file into a matrix:
keaton=imread('keaton1bw.tif');
figure(1)
imhist(keaton); %plots histogram
title('image histogram');
```

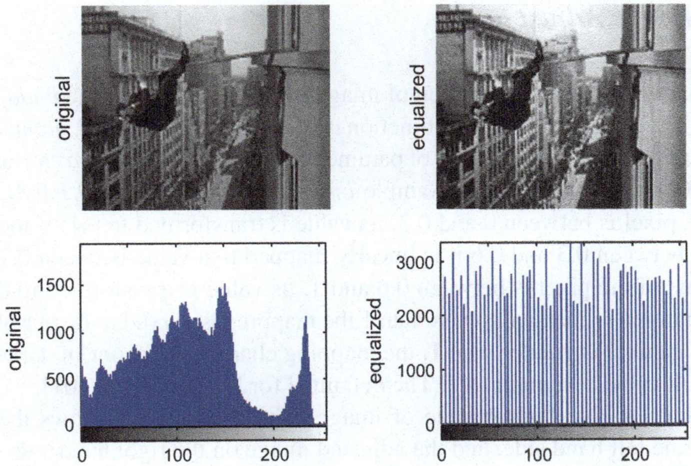

Fig. 3.3 Histogram equalization

3.3.2 *Histogram Equalization*

IPT offers the function *histeq()* to obtain from an image another image with equalized histogram.

Program 3.3 gives an example of histogram equalization, using the keaton image. Figure 3.3, which is obtained with this program, shows the original image and histogram on the left hand side, and the equalized image and its histogram on the right hand side.

Program 3.3 Histogram equalization of the gray scale picture

```
% Histogram equalization of gray scale picture
%read the image file into a matrix
keaton=imread('keaton1bw.tif');
hqk=histeq(keaton); %histogram equalization
figure(1)
subplot(2,2,1)
imshow(keaton); %plots original picture
title('histogram equalization');
ylabel('original');
subplot(2,2,2)
imshow(hqk); %plots equalized picture
ylabel('equalized');
subplot(2,2,3)
imhist(keaton); %plots histogram original picture
ylabel('original');
subplot(2,2,4)
imhist(hqk); %plots histogram equalized picture
ylabel('equalized');
```

3.3.3 Image Adjust

An alternative for the manipulation of image contrast is given by the *imadjust()* IPT function. The general form of this function is *imadjust(im,[a b],[c d], gamma)*, where *im* is the image matrix. The values of parameters *a, b, c, d,* and *gamma* can take values between 0 and 1. Suppose the example *imadjust(im, [0.3 0.6], [0.1 0.8],1)*, if the value of a pixel is between 0 and 0.3, its value is transformed to 0.1; if the value of a pixel is between 0.3 and 0.6 it is linearly mapped to a value between 0.1 and 0.8; if the value of the pixel is between 0.6 and 1, its value is transformed to 0.8. If the value of gamma is changed to less than 1 the mapping is curved in favour of brighter values; if gamma is greater than 1, the mapping changes in favour of darker values. The default value of gamma is 1. The default [] for [*ab*] or [*cd*] is [01].

Program 3.4 gives an example of image adjust. Figure 3.4 shows the original image on the left hand side, and the adjusted image on the right hand side.

Fig. 3.4 Image adjust

Program 3.4 Adjust gray scale picture

```
% Adjust gray scale picture
%read the image file into a matrix
keaton=imread('keaton1bw.tif');
adk=imadjust(keaton,[0.1 0.8],[]); %image adjust
figure(1)
subplot(1,2,1)
imshow(keaton); %plots original picture
title('image adjust');
ylabel('original');
subplot(1,2,2)
imshow(adk); %plots adjusted picture
ylabel('adjusted');
```

3.3.4 2D Filtering with Neighbours

Given an image, edges and other significant changes of grey shades are high frequency components. Regions with constant or almost constant grey value are low frequency components. High pass filters tend to highlight edges. Low pass filter tend to smooth edges and give blurred images.

A simple scheme to obtain a low pass filter is to take each pixel and several neighbours symmetrically around, compute the average of pixel values, and substitute the original pixel value by the average.

For instance, let us select pixel $p(81, 31)$, and take a 3×3 set of pixels, with the pixel of interest at the centre:

p(80,30) p(80,31) p(80,32)
p(81,30) p(81,31) p(81,32)
p(82,30) p(82,31) p(82,32)

Now, the average of the nine pixel values is computed, and given to $p(81, 31)$.

Continuing with the example, the 2D linear filtering that has been applied can be written as:

$$p(x, y) = \sum_{i,j} w(i, j)\, p(x + i, y + j) \tag{3.1}$$

with *i=-1,0,1, j=-1,0,1* and $w(i, j) = 1$.

The matrix of weights $w(i, j)$ can be described as a 2D 3×3 'mask'. And the filtering process is a 2D convolution of the image and the mask.

In other words, to filter the complete image the mask is moved on each of the image pixels $p(x, y)$, and the convolution is computed for each $p(x, y)$.

The simplest mask is made with ones, for averaging. Other masks can be designed for high-pass filtering to highlight edges, or any other purposes.

Program 3.5 provides an example of 2D low-pass filtering. The program uses the *filter2()* MATLAB function for 2D filtering. The *fspecial()* IPT function offer a set of *filter molecules* (masks) for different purposes, including high or low pass filtering. In the Program 3.5, averaging with a 5×5 set of pixels is the selected filter molecule. Figure 3.5 compares the original picture, left hand side, with the filtered picture, right hand side. This last image looks blurred.

Notice in the Program 3.5 that *fk*, the output of *filter2()*, is a set of real values (with decimals) between 0 and 255. These values are converted to unsigned integer data in the line with *uint8(round())*.

Program 3.5 Result of average filter

```
% Display filtered gray scale picture
% average filter
%read the image file into a matrix:
keaton2=imread('keaton2bw.tif');
fil=fspecial('average',[5,5]); %filter molecule
fk=filter2(fil,keaton2);
bfk=uint8(round(fk)); %convert to unsigned 8-bit
```

```
figure(1)
subplot(1,2,1)
imshow(keaton2);
title('average filtering');
ylabel('original')
subplot(1,2,2)
imshow(bfk); %display the filtered image
ylabel('filtered')
```

Program 3.6 gives an example of 2D high pass filtering. It uses the *laplacian* filter molecule for approximate spatial derivative. Figure 3.6 compares the original picture, left hand side, with the filtered picture, right hand side. The result of the 2D filter has been further adjusted for contrast increase. Borders are clearly highlighted.

Program 3.6 Result of Laplacian filter

```
% Display filtered gray scale picture
% Laplacian filter
%read the image file into a matrix:
keaton2=imread('keaton2bw.tif');
fil=fspecial('laplacian'); %filter molecule
fk=filter2(fil,keaton2);
bfk=uint8(round(abs(fk))); %convert to unsigned 8-bit
abfk=imadjust(bfk,[0.1 0.7],[]); %image adjust
figure(1)
subplot(1,2,1)
imshow(keaton2);
title('laplacian filtering');
ylabel('original')
subplot(1,2,2)
imshow(abfk); %display the filtered image
ylabel('filtered')
```

Fig. 3.5 Average filtering

Fig. 3.6 Laplacian filtering

3.3.5 *Gaussian 2D Filters*

A particular case of filtering with neighbours is Gaussian filters. These filters are frequently used due to its good properties and easy implementation. The mask is a discretized bidimensional Gaussian curve, with its typical bell shape.

Figure 3.7, which has been generated by the simple Program 3.7, shows an example of Gaussian mask, produced by the *fspecial()* function. As it has been specified, the mask is a 20 × 20 matrix. In this example, the value given for the last parameter is 4, to get a not very steep 'mountain'; smaller values raise the peak and narrow the mountain.

Program 3.7 Display Gaussian mask

```
% Display gaussian mask
fil=fspecial('gaussian',[20 20],4);
surf(fil);
title('gaussian mask');
```

Program 3.8 applies the Gaussian mask for low-pass image filtering. Figure 3.8 shows the result. As expected, the filtered picture looks blurred.

Fig. 3.7 A Gaussian mask

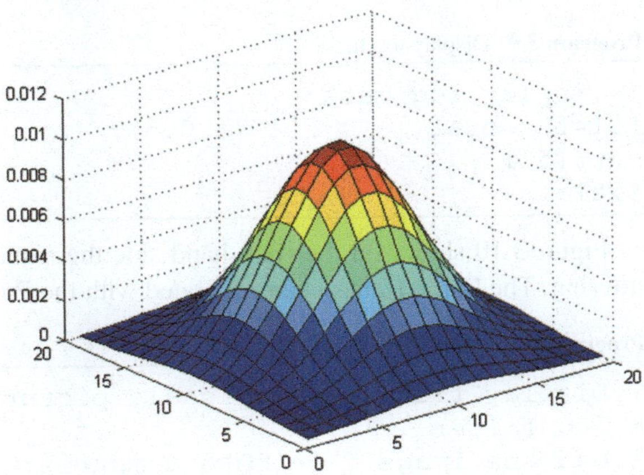

Fig. 3.8 Gaussian filtering

Program 3.8 Result of Gaussian filter

```
% Display filtered gray scale picture
% Gaussian filter
%read the image file into a matrix:
keaton2=imread('keaton2bw.tif');
fil=fspecial('gaussian',[20 20],4); %filter molecule
fk=filter2(fil,keaton2);
bfk=uint8(round(abs(fk))); %convert to unsigned 8-bit
abfk=imadjust(bfk,[0.1 0.7],[]); %image adjust
figure(1)
subplot(1,2,1)
imshow(keaton2);
title('gaussian filtering');
h=gca;ht=get(h,'Title'); set(ht,'FontSize',14);
ylabel('original')
subplot(1,2,2)
imshow(abfk); %display the filtered image
ylabel('filtered')
```

High pass filtering can be done using the derivative of the gaussian mask. This is done with the choice '*log*' in the *fspecial()* function. Figure 3.9, which has been generated with the Program 3.9, shows an example of this mask.

Program 3.9 Display log mask

```
% Display log mask
fil=fspecial('log',[7 7],0.5);
surf(fil);
title('log mask');
```

Figure 3.10 shows on the right hand side the results of the Gaussian high pass filtering. The Fig. 3.10 has been generated with the Program 3.10.

Program 3.10 Result of log filter

```
% Display filtered gray scale picture
% log filter
%read the image file into a matrix:
```

```
keaton2=imread('keaton2bw.tif');
fil=fspecial('log',[7 7],0.5); %filter molecule
fk=filter2(fil,keaton2);
bfk=uint8(round(abs(fk))); %convert to unsigned 8-bit
abfk=imadjust(bfk,[0.1 0.7],[]); %image adjust
figure(1)
subplot(1,2,1)
imshow(keaton2);
title('log filtering');
ylabel('original')
subplot(1,2,2)
imshow(abfk); %display the filtered image
ylabel('filtered')
```

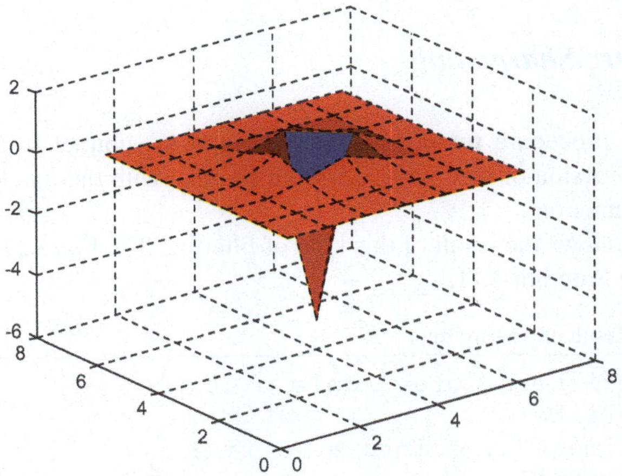

Fig. 3.9 A 'log' mask

Fig. 3.10 Log filtering

Fig. 3.11 Unsharp filtering

3.3.6 *Picture Sharpening*

An interesting procedure for picture sharpening is to subtract from the image a scaled blurred version of the image. This can be done with the *'unsharp'* choice in the *fspecial()* function.

Figure 3.11 shows the result of this kind of filtering. The Fig. 3.11 has been generated with the Program 3.11.

Program 3.11 Result of unsharp filter

```
% Display filtered gray scale picture
% unsharp filter
%read the image file into a matrix:
keaton=imread('keaton2bw.tif');
fil=fspecial('unsharp'); %filter molecule
fk=filter2(fil,keaton);
bfk=uint8(round(abs(fk))); %convert to unsigned 8-bit
figure(1)
imshow(bfk); %display the filtered image
title('unsharp filtering');
```

3.4 2D Fourier Transform

The 2D DFT of a matrix is another matrix with the same size. This transform can be obtained in two steps: first one computes the DFT of the rows, obtaining an intermediate matrix, and second one computes the DFT of the columns of the intermediate matrix.

Suppose you have a picture with some added noise or mark, so you have:

$$im = p + n \tag{3.2}$$

where p is the original picture and n the added noise.

Then the 2D DFT will be:

$$F(im) = F(p) + F(n) \tag{3.3}$$

This expression shows a good opportunity to remove the noise.

The filtering with neighbours, described in the section before, is based on the convolution of the image *im* and a mask *mk*. The corresponding computational load can be large. The 2D DFT offers an alternative, since the DFT of *im*mk* (the asterisk means convolution) is the product *F(im) F(mk)*. The idea is to compute this product and then use the inverse 2D DFT to get the filtered picture.

MATLAB provides the *fft2()* function for the 2D DFT.

The following parts of this section introduce step by step the use of the 2D Fourier transform for image filtering.

3.4.1 2D Fourier Transform of Edges

Recall that image edges correspond to high frequency components. The Fourier transform takes you to the frequency domain. It is interesting to see the 2D Fourier transform of simple images with evident edges.

A very simple case is presented in the left hand side of Fig. 3.12. There is a neat vertical border between white and black. The right hand side of the Fig. 3.12 shows the corresponding 2D DFT, which is a white horizontal line on black background. This line depicts the high frequencies due to the border. The Fig. 3.12 has been generated with the simple Program 3.12.

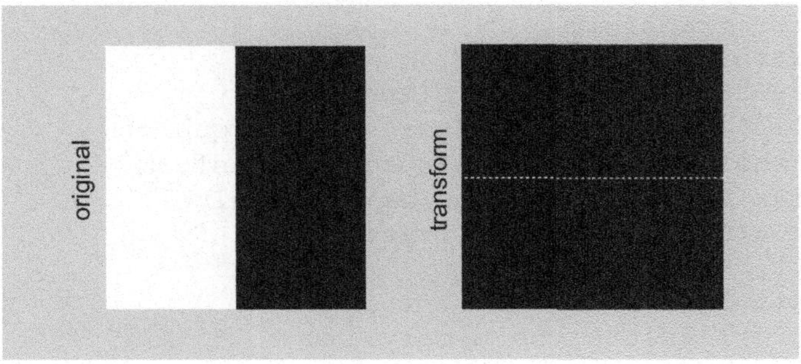

Fig. 3.12 Fourier transform of an edge

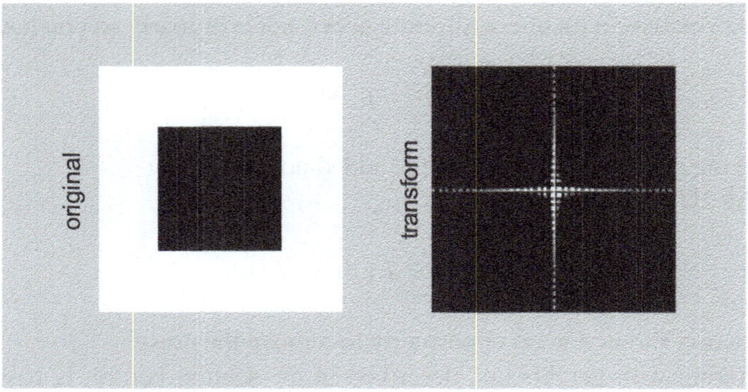

Fig. 3.13 Fourier transform of a square

Program 3.12 Fourier transform of an edge

```
% Fourier transform
% a simple edge
fg=[ones(256,128) zeros(256,128)]; %simple edge
Ffg=fftshift(fft2(fg)); %Fourier transform
M=max(max(Ffg)); % the one maximum value
sFfg=(256*Ffg/M); %normalization
figure(1)
subplot(1,2,1)
imshow(fg); %plots the binary image with a simple edge
title('Fourier transform of an edge');
ylabel('original');
subplot(1,2,2)
imshow(abs(sFfg)); %plots the Fourier transform
ylabel('transform');
```

The next experiment is to obtain the 2D DFT of a black square on white background. This is done with the Program 3.13, and the results are shown in Fig. 3.13.

Program 3.13 Fourier transform of a square

```
% Fourier transform
% a square
fg=ones(256,256); %white plane
fg(64:192, 64:192)=0; %insert black square
Ffg=fftshift(fft2(fg)); %Fourier transform
M=max(max(Ffg)); % the one maximum value
sFfg=(256*Ffg/M); %normalization
figure(1)
subplot(1,2,1)
imshow(fg); %plots the binary image
title('Fourier transform of a square');
ylabel('original');
```

```
subplot(1,2,2)
imshow(abs(sFfg)); %plots the Fourier transform
ylabel('transform');
```

As a final experiment of this series, Program 114.3 obtains the 2D DFT of a rhombus. The result is shown in Fig. 3.14.

Program 3.14 Fourier transform of a rhombus

```
% Fourier transform
% rhombus
fg=ones(256,256); %white plane
for n=1:64,
fgx=128+(-n:n); fgy=64+n;
fg(fgx,fgy)=0; %one triangle
fg(fgx,256-fgy)=0; %the other triangle
end
Ffg=fftshift(fft2(fg)); %Fourier transform
M=max(max(Ffg)); % the one maximum value
sFfg=(256*Ffg/M); %normalization
figure(1)
subplot(1,2,1)
imshow(fg); %plots the binary image
title('Fourier transform of a rhombus');
ylabel('original');
subplot(1,2,2)
imshow(abs(sFfg)); %plots the Fourier transform
ylabel('transform');
```

3.4.2 2D Fourier Transform of a Picture

Now, let us take the 2D DFT transform of a picture, and then take the inverse transform to recover the picture. Program 3.15 applies these steps, and generates Fig. 3.15. The

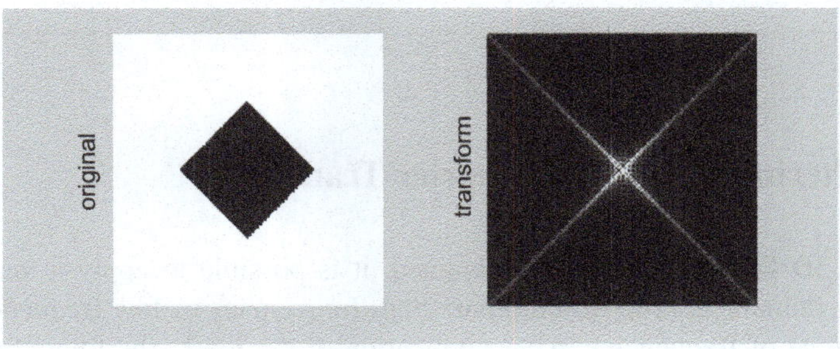

Fig. 3.14 Fourier transform of a rhombus

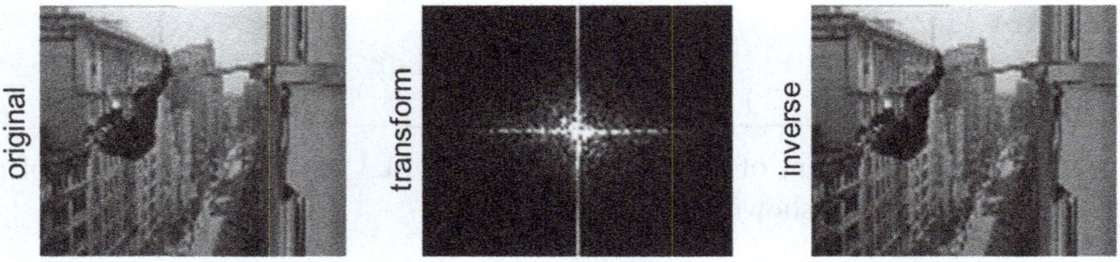

Fig. 3.15 Fourier transform and its inverse

left hand side of the Fig. 3.15 shows the original picture, and the right hand side shows the recovered picture.

The most interesting thing is the 2D transform itself, which is shown at the middle of the Fig. 3.15, looking like a cross. The lowest frequencies of the picture are depicted at the centre of the cross; this fact can be exploited for filtering, as will be described in the next section.

Program 3.15 Fourier transform of an image, and its inverse

```
% Fourier transform and inverse
%read the image file into a matrix:
keaton=imread('keaton1bw.tif');
FKe=fftshift(fft2(keaton)); %Fourier transform
M=max(max(FKe)); % the one maximum value
sFKe=(256*FKe/M); %normalization
IKe=ifft2(FKe); %inverse transform
uIKe=uint8(abs(IKe)); %convert to unsigned 8-bit
figure(1)
subplot(1,3,1)
imshow(keaton); %display the photo
title('Fourier transform');
ylabel('original');
subplot(1,3,2)
imshow(abs(sFKe)); %Fourier transform of the photo
ylabel('transform');
subplot(1,3,3)
imshow(uIKe); %inverse of the Fourier transform
ylabel('inverse');
```

3.5 Filtering with the 2D Fourier Transform

Once the 2D DFT of a picture is obtained, it is possible to apply a mask in the frequency domain in order to eliminate high frequencies or low frequencies, or a desired range of frequencies. After application of the mask, the filtered picture is obtained by inverse 2D DFT.

The objective of this section is to describe this filtering technique, starting with the simplest filters and then introducing some alternatives.

3.5.1 Basic Low Pass and High Pass Filtering using 2D DFT

The simplest mask in the frequency domain is a circle. A white circle on black background can be used for low pass filtering. A black circle on white background can be used for high pass filtering

Program 3.16 shows how to apply the low pass filtering. After reading a file with the picture to be filtered, the program prepares an adequate circle, denoted as *circ*. Then there are three lines with the filtering process: first the 2D DFT transform is applied to obtain *FKe*; then the mask (the circle) is applied to obtain *FKefil*; and finally the inverse 2D DFT is applied to get the filtered picture *Ike*. The rest of the program is devoted to display two figures. Figure 3.16 shows the original picture and the circle. Figure 3.17 shows the result: the filtered picture.

Fig. 3.16 Original picture and circle filter

Fig. 3.17 Filtered image

Program 3.16 Fourier low-pass filtering

```
% Filtering the Fourier transform and then inverse
% low-pass circle filter
%read the image file into a matrix:
keaton=imread('keaton1bw.tif');
[ly,lx]=size(keaton); %determine image size
dx=(0:(lx-1))-round(lx/2); dy=(0:(ly-1))-round(ly/2);
%a grid with the same size as image, with center
(0,0):
[x,y]=meshgrid(dx,dy);
R=sqrt((x.^2)+(y.^2));
circ=(R <30); %circle (low-pass filter)
FKe=fftshift(fft2(keaton)); %Fourier transform
FKefil=FKe.*circ; %filtering in the frequency domain
IKe=ifft2(FKefil); %inverse transform
uIKe=uint8(abs(IKe)); %convert to unsigned 8-bit
figure(1)
subplot(1,2,1)
imshow(keaton); %display the original photo
title('Fourier circle low-pass filtering');
ylabel('original');
subplot(1,2,2)
imshow(circ); %the filter
ylabel('the filter');
figure(2)
imshow(uIKe); %the filtered image
title('filtered image');
```

Program 3.17 shows how to apply a circle for high pass filtering. The program is very similar to Program 1.5.1. Figure 3.18 shows the original picture and the circle. Figure 3.19 shows the result of the filtering process.

Fig. 3.18 Original picture and circle filter

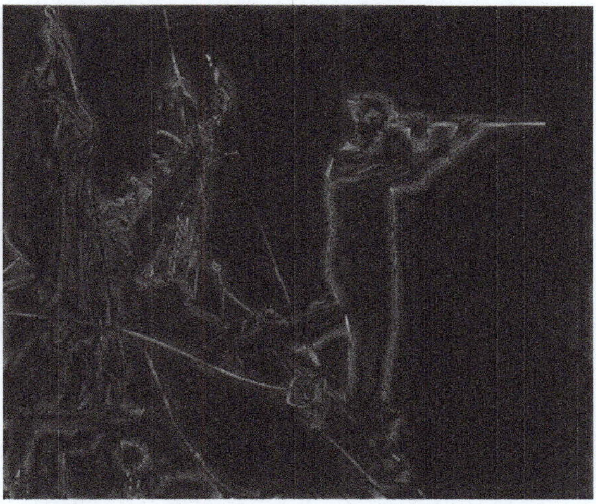

Fig. 3.19 Filtered image

Program 3.17 Fourier high-pass filtering

```
% Filtering the Fourier transform and then inverse
% high-pass circle filter
%read the image file into a matrix:
keaton=imread('keaton2bw.tif');
[ly,lx]=size(keaton); %determine image size
dx=(0:(lx-1))-round(lx/2); dy=(0:(ly-1))-round(ly/2);
%a grid with the same size as image, with center
(0,0):
[x,y]=meshgrid(dx,dy);
R=sqrt((x.^2)+(y.^2));
circ=(R >16); %circle (high-pass filter)
FKe=fftshift(fft2(keaton)); %Fourier transform
FKefil=FKe.*circ; %filtering in the frequency domain
IKe=ifft2(FKefil); %inverse transform
uIKe=uint8(abs(IKe)); %convert to unsigned 8-bit
figure(1)
subplot(1,2,1)
imshow(keaton); %display the photo
title('Fourier circle high-pass filtering');
ylabel('original');
subplot(1,2,2)
imshow(circ); %the filter
ylabel('the filter');
figure(2)
imshow(uIKe);
title('filtered image');
```

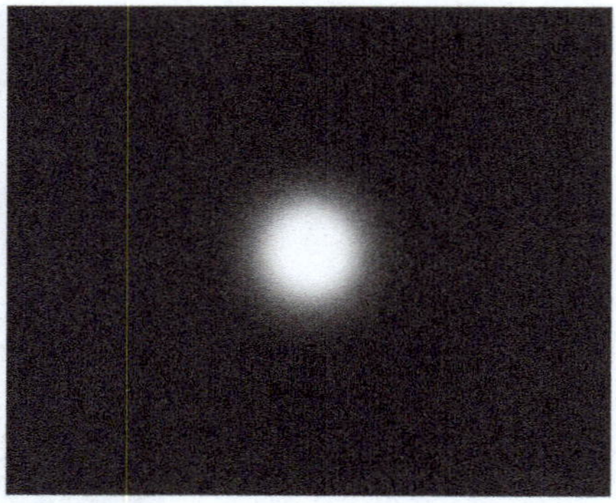

Fig. 3.20 Butterworth mask

3.5.2 *Other Low Pass Filters Using 2D DFT*

The abrupt border of the circle mask induce undesirable effects, such as the ringing that can be noticed in Fig. 3.17. There are alternative masks with smoother borders that can be used for better filtering. Let us take two notable examples.

3.5.2.1 Butterworth

A spatial low pass Butterworth filter can be obtained based on the following mask:

$$mk(r) = \frac{1}{1 + (r/R)^{2n}} \tag{3.4}$$

where r is distance to the centre, R is a radius, and n is the order of the filter.

Program 3.18 provides an example of this type of filtering, with a Butterworth mask of order 3. Figures 3.20 and 3.21 provide views of this mask. Figure 3.22 shows the filtered picture, in which there is no ringing.

Program 3.18 Fourier Butterworth low-pass filtering

```
% Filtering the Fourier transform and then inverse
% low-pass Butterworth filter
%read the image file into a matrix:
keaton=imread('keaton1bw.tif');
[ly,lx]=size(keaton); %determine image size
dx=(0:(lx-1))-round(lx/2); dy=(0:(ly-1))-round(ly/2);
%a grid with the same size as image, with center
(0,0):
[x,y]=meshgrid(dx,dy);
r=sqrt((x.^2)+(y.^2));
M=40; nf=3; %filter specification
fbut=1./(1+((r/M).^(2*nf))); %Butterworth filter
FKe=fftshift(fft2(keaton)); %Fourier transform
```

```
FKefil=FKe.*fbut; %filtering in the frequency domain
IKe=ifft2(FKefil); %inverse transform
uIKe=uint8(abs(IKe)); %convert to unsigned 8-bit
figure(1)
imshow(fbut); %the filter
title('Fourier Butterworth low-pass filtering');
ylabel('the filter');
figure(2)
mesh(x,y,fbut); %3D view of the filter
title('3D view of the filter');
figure(3)
imshow(uIKe); %display the photo
title('filtered image');
```

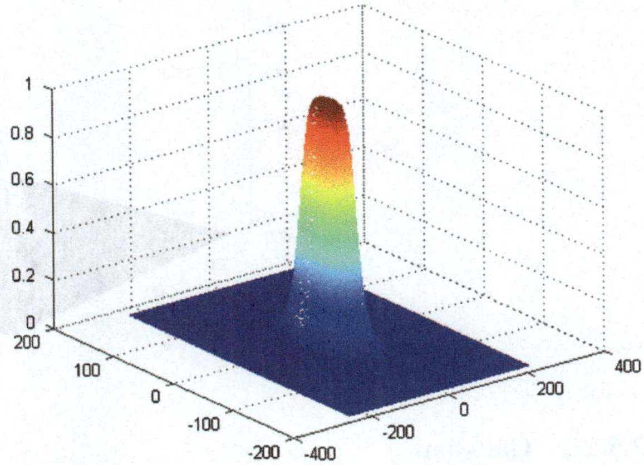

Fig. 3.21 3D view of the mask

Fig. 3.22 Filtered image

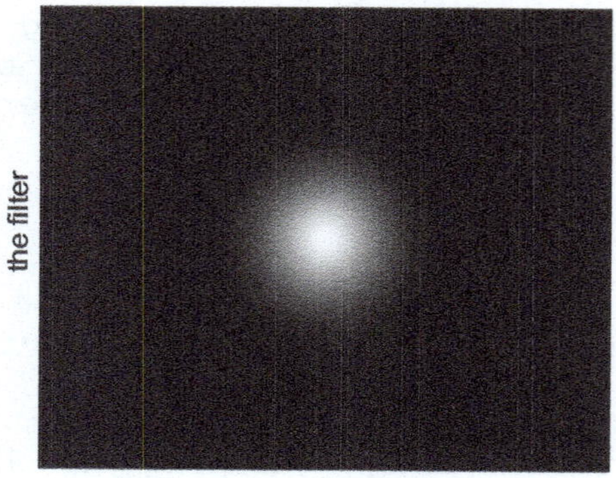

Fig. 3.23 Gaussian mask

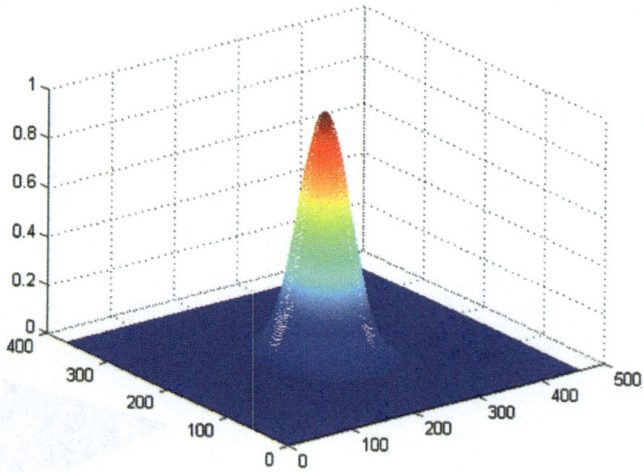

Fig. 3.24 3D view of the mask

3.5.2.2 Gaussian

The Gaussian filter obtained with *fspecial('gaussian',etc)* can be used as mask in the frequency domain. Program 3.19 gives an example of how to use this technique. The function *mat2gray()* scales to real values between 0 and 1. Figures 3.23 and 3.24 show the Gaussian mask. Figure 3.25 shows the filtered picture, which also has no ringing.

Program 3.19 Fourier Gaussian low-pass filtering

```
% Filtering the Fourier transform and then inverse
% low-pass gaussian filter
%read the image file into a matrix:
keaton=imread('keaton1bw.tif');
[ly,lx]=size(keaton); %determine image size
yo=ceil(ly/2); xo=ceil(lx/2); %figure \ref{fig:3.center
xgfil=zeros(ly,lx); %filter initialization
if ly >lx rcir=xo-1; fxo=xo; fyo=xo;
else rcir=yo-1; fyo=yo; fxo=yo;
end; %radius and center of the filter
```

```
sigma=30; %filter bandwidth parameter
gfil=fspecial('gaussian',1+(2*rcir),sigma); %gaussian filter
ngfil=mat2gray(gfil); %normalization
spnx=[(xo-rcir):(xo+rcir)]; %set of indexes
spny=[(yo-rcir):(yo+rcir)]; %set of indexes
spn=[(fyo-rcir):(fyo+rcir)]; %set of indexes
xgfil(spny,spnx)=ngfil(spn,spn);
FKe=fftshift(fft2(keaton)); %Fourier transform
FKefil=FKe.*xgfil; %filtering in the frequency domain
IKe=ifft2(FKefil); %inverse transform
uIKe=uint8(abs(IKe)); %convert to unsigned 8-bit
figure(1)
imshow(xgfil); %the filter
title('Fourier gaussian low-pass filtering');
ylabel('the filter');
figure(2)
mesh(xgfil); %3D view of the filter
title('3D view of the filter');
figure(3)
imshow(uIKe); %display the photo
title('filtered image');
```

3.5.3 *Other High-Pass Filters Using 2D DFT*

Inverted versions of the previous masks can be used for high pass filtering.

Fig. 3.25 Filtered image

3.5.3.1 Butterworth

The spatial high pass Butterworth filter can be obtained based on the following mask:

$$mk(r) = \frac{1}{1 + (R/r)^{2n}} \tag{3.5}$$

Program 3.20 applies this high pass filtering. Figures 3.26 and 3.27 present views of this mask. Figure 3.28 shows the filtered picture.

Fig. 3.26 Butterworth mask

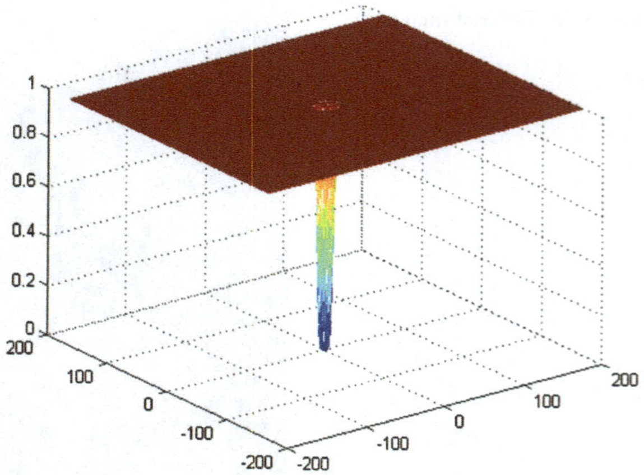

Fig. 3.27 3D view of the mask

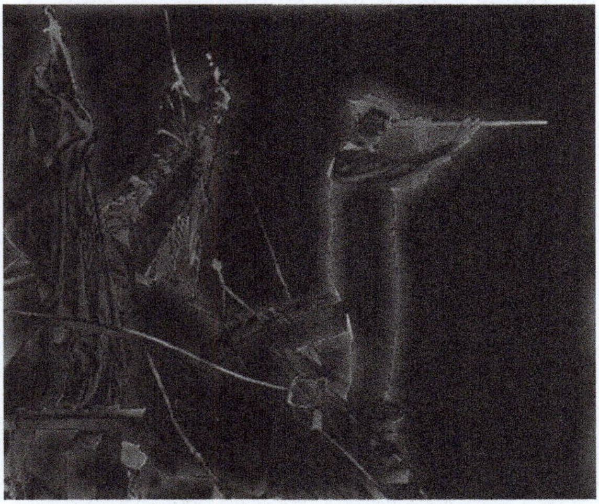

Fig. 3.28 Filtered image

Program 3.20 Fourier Butterworth high-pass filtering

```
% Filtering the Fourier transform and then inverse
% high-pass Butterworth filter
%read the image file into a matrix:
keaton=imread('keaton2bw.tif');
[ly,lx]=size(keaton); %determine image size
dx=(0:(lx-1))-round(lx/2); dy=(0:(ly-1))-round(ly/2);
%a grid with the same size as image, with center
(0,0):
[x,y]=meshgrid(dx,dy);
r=sqrt((x.^2)+(y.^2));
M=8; nf=3; %filter specification
fbut=1./(1+((r/M).^(2*nf))); %Butterworth filter
fhpbut=1-fbut; %high-pass filter
FKe=fftshift(fft2(keaton)); %Fourier transform
FKefil=FKe.*fhpbut; %filtering in the frequency domain
IKe=ifft2(FKefil); %inverse transform
uIKe=uint8(abs(IKe)); %convert to unsigned 8-bit
figure(1)
imshow(fhpbut); %the filter
title('Fourier Butterworth high-pass filtering');
ylabel('the filter');
figure(2)
mesh(x,y,fhpbut); %3D view of the filter
title('3D view of the filter');
figure(3)
imshow(uIKe); %display the photo
title('filtered image');
```

Fig. 3.29 Gaussian mask

3.5.3.2 Gaussian

Program 3.21 is almost identical to Program 3.19. The main difference is the line with *xgfil() = 1 – ngfil()*: here the circle is inverted. Figures 3.29 and 3.30 show views of the mask. Figure 3.31 displays the filtered picture.

Program 3.21 Fourier Gaussian high-pass filtering

```
% Filtering the Fourier transform and then inverse
% high-pass gaussian filter
%read the image file into a matrix:
keaton=imread('keaton2bw.tif');
[ly,lx]=size(keaton); %determine image size
yo=ceil(ly/2); xo=ceil(lx/2); %figure \ref{fig:3.center
xgfil=ones(ly,lx); %filter initialization
if ly >lx rcir=xo-1; fxo=xo; fyo=xo;
else rcir=yo-1; fyo=yo; fxo=yo;
end; %radius and center of the filter
sigma=10; %filter bandwidth parameter
gfil=fspecial('gaussian',1+(2*rcir),sigma); %gaussian filter
ngfil=mat2gray(gfil); %normalization
spnx=[(xo-rcir):(xo+rcir)]; %set of indexes
spny=[(yo-rcir):(yo+rcir)]; %set of indexes
spn=[(fyo-rcir):(fyo+rcir)]; %set of indexes
%put circular filter on image
rectangle:
xgfil(spny,spnx)=1-ngfil(spn,spn);
FKe=fftshift(fft2(keaton)); %Fourier transform
FKefil=FKe.*xgfil; %filtering in the frequency domain
IKe=ifft2(FKefil); %inverse transform
uIKe=uint8(abs(IKe)); %convert to unsigned 8-bit
figure(1)
imshow(xgfil); %the filter
title('Fourier gaussian high-pass filtering');
```

```
ylabel('the filter');
figure(2)
mesh(xgfil); %3D view of the filter
title('3D view of the filter');
figure(3)
imshow(uIKe); %the filtered picture
title('filtered image');
```

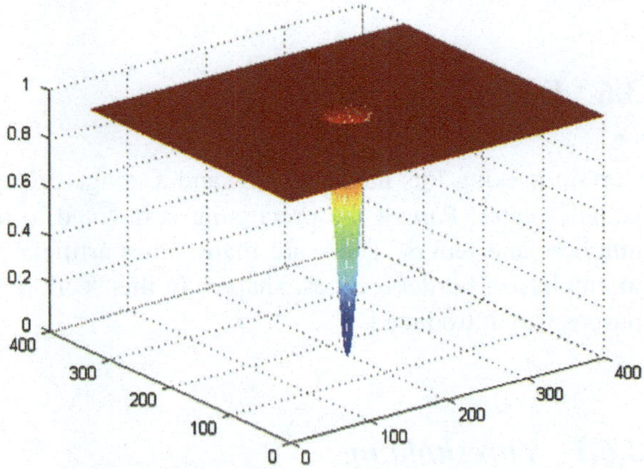

Fig. 3.30 3D view of the mask

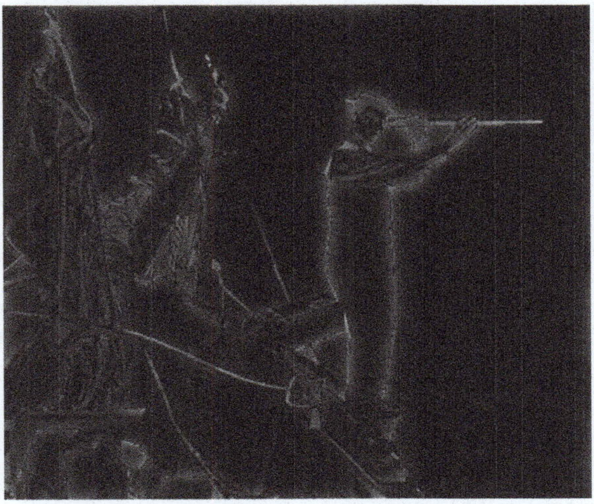

Fig. 3.31 Filtered image

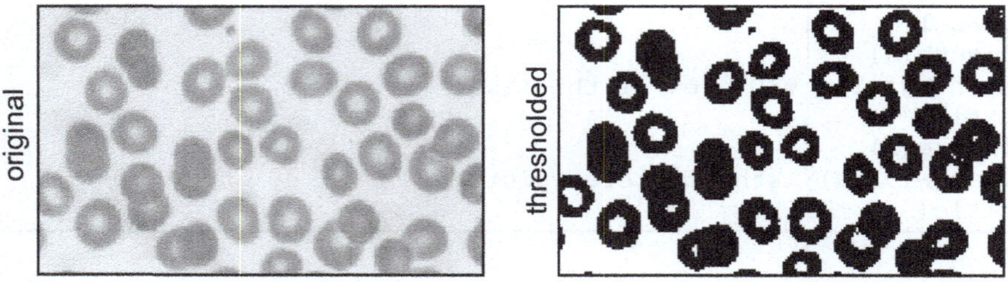

Fig. 3.32 Image thresholding

3.6 Edges

Certain parking lots have cameras and a image processing system that recognizes vehicle plates. Part of the processing is devoted to make stand out the shapes of numbers and letters. There are many other artificial vision applications that need to emphasize borders, edges, shapes. In this section a series of techniques for this purpose are introduced.

3.6.1 Thresholding

Sometimes it is adequate to apply thresholding in order to highlight image details. The idea is to guess a threshold value, say *Th*, and then display as white pixels with values above *Th*, and display the rest of the pixels in black. With MATLAB this technique can be applied quite easily.

An important example has been selected. A sample of cells, for instance blood cells, has been taken, and the number of cells must be counted. Thresholding comes to help counting.

Figure 3.32 shows on the left hand side the original image with the cells. Program 3.22 applies thresholding, in the line with *cells >180*, and obtains a picture, on the right hand side of Fig. 3.32, with better contrast and cell isolation for counting.

Program 3.22 Result of thresholding

```
% Display B&W thresholded picture
%read the image file into a matrix:
cells=imread('cells1bw.tif');
figure(1)
subplot(1,2,1)
imshow(cells); %display original
title('image threholding');
ylabel('original')
subplot(1,2,2)
imshow(cells >180); %display thresholded image
ylabel('thresholded');
```

Fig. 3.33 Original image

3.6.2 Edges

Since the detection of edges is important for several applications, there is a number of alternative procedures that have been proposed for this purpose.

The *edge()* IPT function offers six alternative methods. The zebra picture, as shown in Fig. 3.33, offers a good target to test the six methods. Program 3.23 generates two figures, Fig. 3.34 with the six results, and the other, Fig. 3.35, with one of these results corresponding to the Canny method. Notice the use of *montage()* to generate Fig. 3.34.

Program 3.23 Edges of a B&W picture

```
% Edges of B&W picture
%read the image file into a matrix:
zebra=imread('zebra1bw.tif');
[ly,lx]=size(zebra);
EZb=zeros(ly,lx,1,6);
figure(1)
imshow(zebra); %original picture
title('original picture');
figure(2)
EZb(:,:,1,1)=edge(zebra,'prewitt'); %Prewitt method
EZb(:,:,1,2)=edge(zebra,'roberts'); %Roberts method
EZb(:,:,1,3)=edge(zebra,'sobel'); %Sobel method
EZb(:,:,1,4)=edge(zebra,'log'); %Laplacian or Gaussian method
EZb(:,:,1,5)=edge(zebra,'zerocross'); %zero crossings method
EZb(:,:,1,6)=edge(zebra,'canny'); %Canny method
montage(EZb); %shows the 6 images
title('edge computation alternatives');
figure(3)
imshow(EZb(:,:,1,6)); %one of the results
title('result of the Canny method');
```

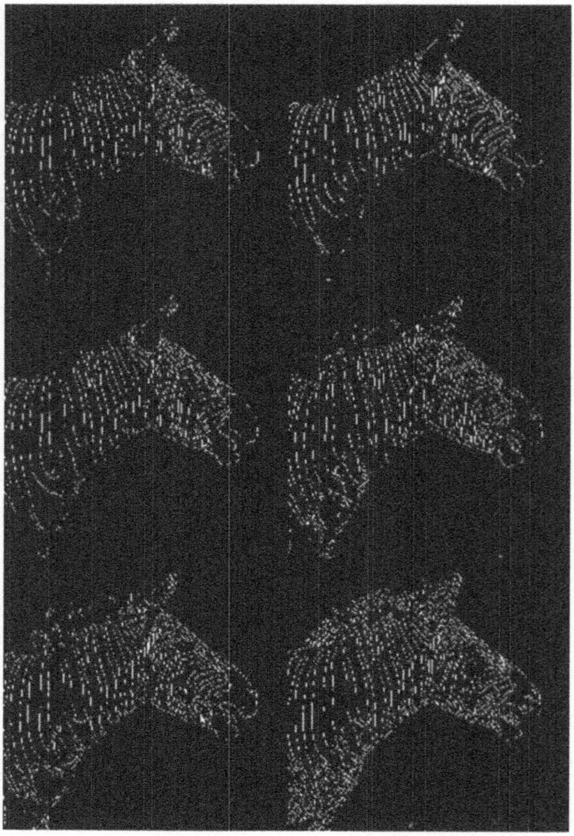

Fig. 3.34 Six results

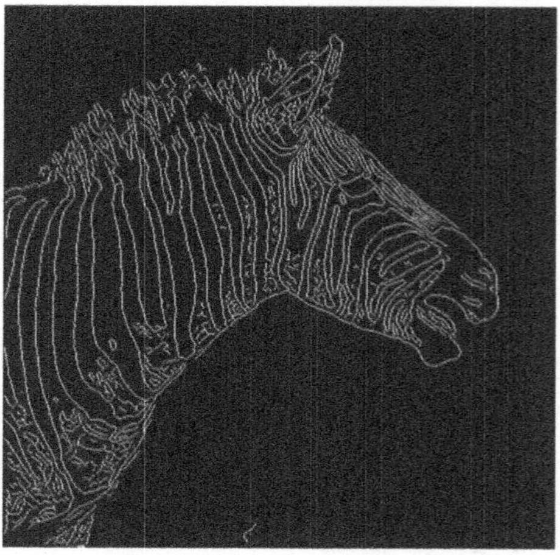

Fig. 3.35 Result of the Canny method

3.7 Color Images

Color images are saved as three planes: three 2D matrices. The three matrices are given as one 3D matrix. For instance *Im(i,j,k)*, where $i = 1, 2, 3$, and j, k pointing to the rows and columns of the image.

The color image can be decomposed into the three planes in several ways. But before entering into programming details it is relevant to briefly look into the captivating world of colors. Internet offers a large mine of information on this topic. Likewise, there is an extensive scientific literature on color theory and representations; anyway, the reader may find useful to consult [22, 38] or the book [25].

The retina contains two types of photoreceptors, rods and cones. The rods are more numerous and are more sensitive than the cones. However, rods are not sensitive to color. The color is seen by the cones. There are three types of cones, L-cones, more sensitive to red light, M-cones, more sensitive to green light, and S-cones, more sensitive to blue light. There are more L-cones (64 %) than M-cones (32 %); S-cones are 2 %, and have the highest sensitivity.

Cones are for daylight; rods for night. Rods see blue, but do not see red; at twilight we tend to see green leaves brighter than red petals.

Many different colors can be obtained by combination of components, as painters do. Different sets—called '*color spaces*'-of three or four components have been proposed as a basis.

The *Commission Internacionale de l'Eclairage* (CIE) established in 1931, after a series of experiments, the CIE-XYZ standard, which uses three components X, Y and Z (the '*tristimulus*' coordinate system). The Y component means luminance. Three normalized variables, x, y and z, are derived from X, Y and Z.

$$x = \frac{X}{X+Y+Z} \tag{3.6}$$

$$y = \frac{Y}{X+Y+Z} \tag{3.7}$$

$$z = \frac{Z}{X+Y+Z} = 1 - x - y \tag{3.8}$$

The chromaticity of a color is specified by x and y. Figure 3.36 shows the CIE-31 chromaticity diagram, which is a most used standard. Monochromatic colors are located on the perimeter, and white light is at the center.

Professionals of illumination pay significant attention to white, and color temperature. The CIE has published a set of standard illuminants A,B,F. The illuminants series D correspond to daylight; D50 is horizon light, D55 is mid-morning light, etc. Figure 3.37 shows some of the illuminant white points, which are located on the black-body (or '*plankian*') curve.

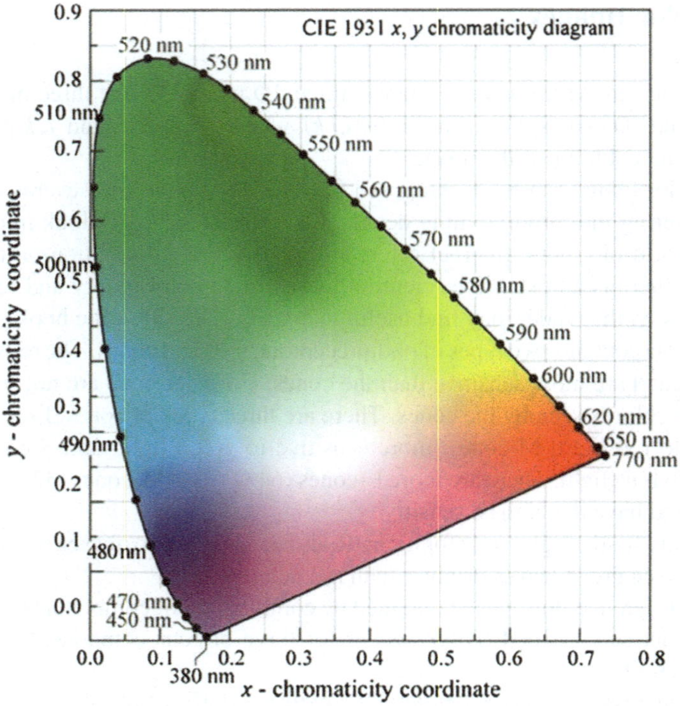

Fig. 3.36 The CIE 31 color space

One could list the most important color spaces under the following categories:

- Linear: RGB, CMYK
- Artistic: Munsell, HSV, HLS
- Perceptual: LUV, Lab
- Opponent: YIQ, YUV

Some of these models sound familiar, because of color TV or computer screens. For instance the RGB space, red-green-blue. Figure 3.38 shows a sketch of the RGB cube, with some additive color combinations.

The CMYK model, cyan-magenta-yellow-black, is more related to printing: it is better to directly use black ink, than a mix of three colors.

There are additive or subtractive color combinations. Figure 3.39 shows an example.

It is possible to convert RGB values to CIE-XYZ values to express the same color. For instance:

$$\begin{pmatrix} X \\ Y \\ Z \end{pmatrix} = \begin{pmatrix} 0.6326696 & 0.2045558 & 0.1269946 \\ 0.2284569 & 0.7373523 & 0.0341908 \\ 0.0000000 & 0.0095142 & 0.8156958 \end{pmatrix} \begin{pmatrix} R \\ G \\ B \end{pmatrix} \quad (3.9)$$

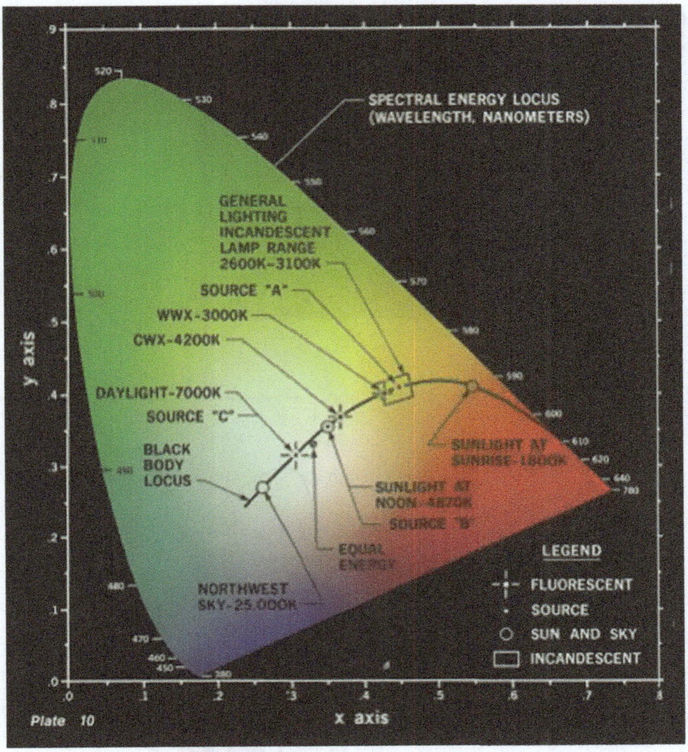

Fig. 3.37 Some illuminant white points

Fig. 3.38 The RGB cube

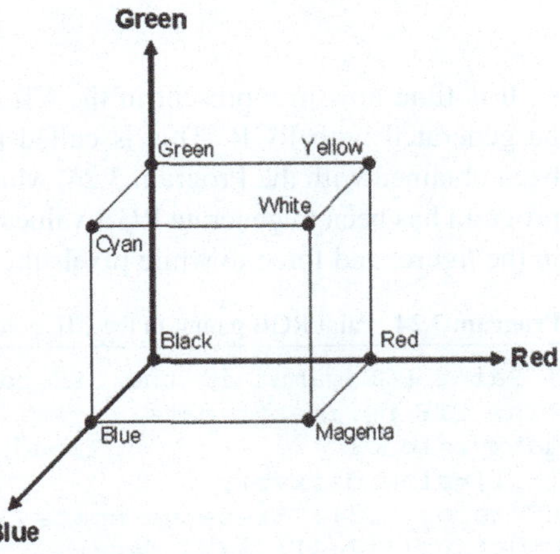

$$\begin{pmatrix} R \\ G \\ B \end{pmatrix} = \begin{pmatrix} 1.7552599 & -0.4836786 & -0.2530000 \\ -0.5441336 & 1.5068789 & 0.0215528 \\ 0.0063467 & -0.0175761 & 1.2256959 \end{pmatrix} \begin{pmatrix} X \\ Y \\ Z \end{pmatrix} \tag{3.10}$$

The matrices above correspond to using D50 as the white point.

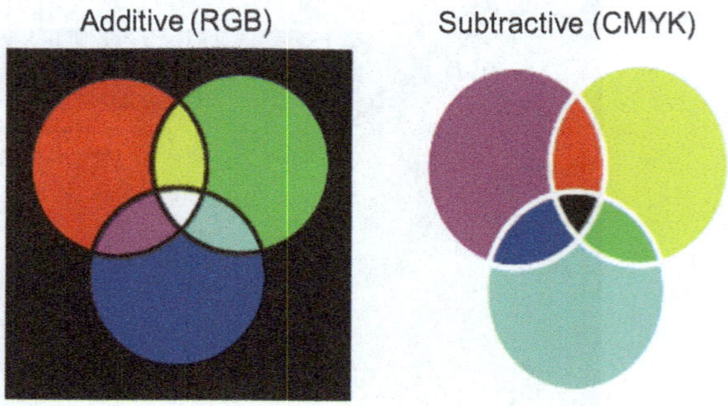

Fig. 3.39 Color combinations

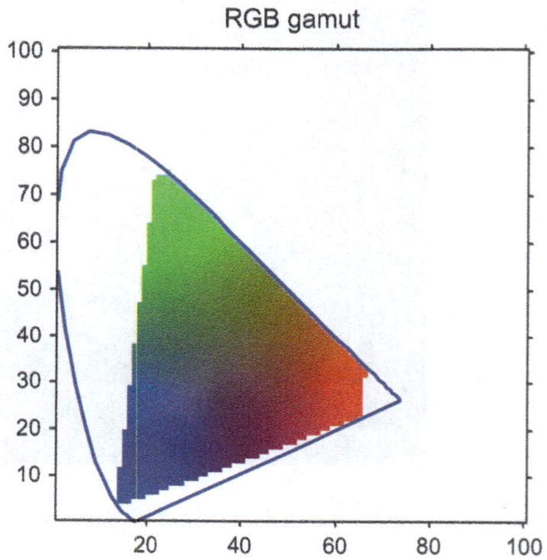

Fig. 3.40 The RGB gamut and the CIE-31 color space

It is time now to represent in the CIE-31 color space the set of colors that can be generated with RGB. This is called the RGB gamut. This representation has been obtained with the Program 3.24, which generates the Fig. 3.40. The idea of the program has been to generate RGB values corresponding to all the *100* × *100* pixels in the figure, and force to white pixels the invalid RGB values (negative or >1).

Program 3.24 Paint RGB gamut in the CIE color space

```
% Paint RGB gamut in the CIE color space
%The CIE perimeter data
datxyz=load('cie1.txt'); %load cie table
[N,l]=size(datxyz);
nC=zeros(N,3); %reserve space
C=datxyz(:,2:4); %XYZ data
for nn=1:N,
nC(nn,:)=C(nn,:)/sum(C(nn,:)); %normalize: xyz data
end;
%The XYZ to RGB conversion matrix
C2R=[1.7552599 -0.4836786 -0.2530;
-0.5441336 1.5068789 0.0215528;
```

```
0.0063467 -0.0175761 1.2256959];
gamut computation
G=zeros(100,100,3);
for ni=1:100,
for nj=1:100,
x=nj/100; y=ni/100; z=1-x-y;
G(ni,nj,:)=C2R*[x y z]';
% negative or  >1 values set to white
m=min(G(ni,nj,:)); M=max(G(ni,nj,:));
if (m <0 | M >1), G(ni,nj,:)=ones(1,3); end; %white
end;
end;
figure(1)
Cx=100*nC(:,1); Cy=100*nC(:,2);
imshow(G);
line([Cx' Cx(1)],[Cy' Cy(1)],'LineWidth',2);
axis square;
axis xy;
axis on;
title('RGB gamut');
```

Painters use *tints*, adding white to pure pigments (red, orange, yellow, blue, purple, etc.), *shades*, adding black, and *tones*, adding gray (a mix of white and black). The HSV color model uses terms like *hue*, for pure colors, *saturation*, for the amount of added white, and *value*, for the amount of added black. Other similar color models are HSI (hue, saturation, intensity), HSL (hue, saturation, luminosity), and HBL (hue, brightness, luminosity).

In MATLAB hue (H) is normalized from 0 to 1, according with the color wheel depicted in Fig. 3.41 (generated with the Program 3.25). Red corresponds to $H = 0$, green corresponds to $H = 1/4$, etc.

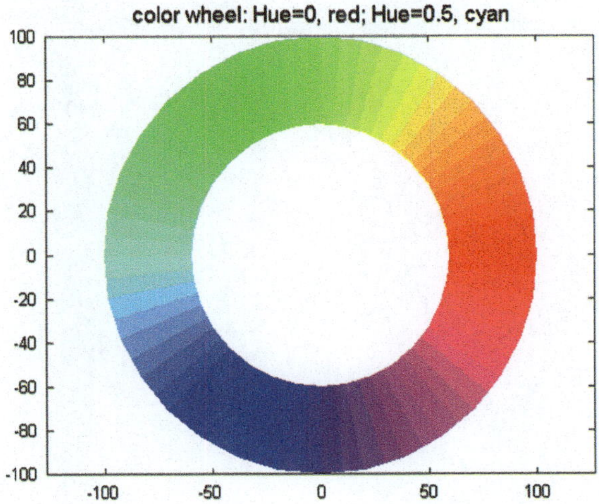

Fig. 3.41 Color wheel with Hue values

Program 3.25 Color wheel (hue)

```
% Color wheel (hue)
N=32; %number of segments
R=(60:10:100)'; %radius zone
phi=2*pi*(0:N)/N;
zz=ones(length(R),1);
x=R*cos(phi);
y=R*sin(phi);
c=zz*(phi/(2*pi));
figure(1)
colormap('HSV');
pcolor(x,y,c);
shading interp;
axis equal;
title('color wheel: Hue=0, red; Hue=0.5, cyan');
```

The Munsell color system uses Hue = R, YR, Y, GY, G, BG, B, PB, P, RP, each subdivided into 10; Value, from 0 to 10; Chroma (Saturation), from 0 to 20. There are books of Munsell colors.

Typically, the HSV color space is described with a hexagon or a cone. Figure 3.42 shows a diagram of HSV coordinates. In MATLAB, saturation and values are normalized from 0 to 1.

Program 3.26 can be used to generate in 3D an HSV color cone (Fig. 3.43).

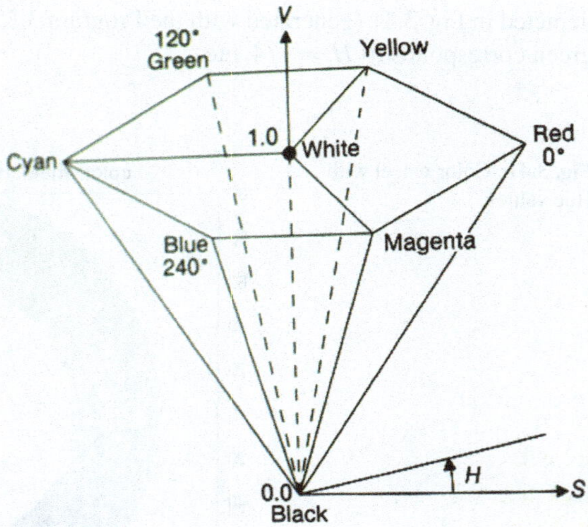

Fig. 3.42 The HSV coordinates

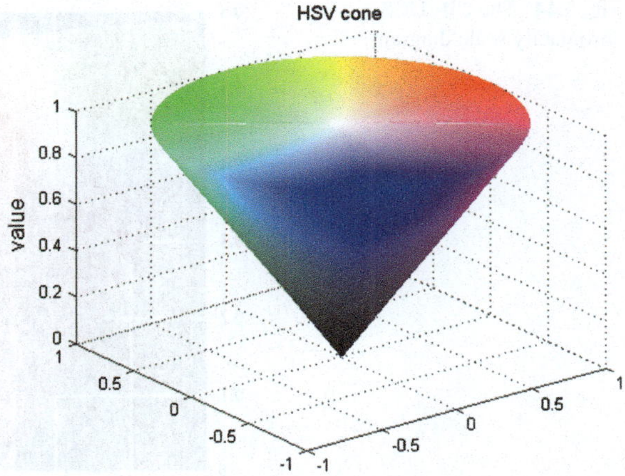

Fig. 3.43 The HSV cone

Program 3.26 HSV cone

```
% HSV cone
%make an HSV image
Nh=1000; Ns=500; Nv=500;
H=repmat(linspace(0,1,Nh),Nh,1);
S=repmat([linspace(0,1,Ns) linspace(1,0,Ns)].',1,2*Ns);
V=repmat([ones(1,Nv) linspace(1,0,Nv)].',1,2*Nv);
Ihsv=cat(3,H,S,V);
%convert to RGB
Irgb=hsv2rgb(Ihsv);
%make the cone
phi=linspace(0,2*pi,1000);
zz=zeros(1,1000);
X=[zz; cos(phi); zz];
Y=[zz; sin(phi); zz];
Z=[1.*ones(2,1000); zz];
figure(1)
surf(X,Y,Z,Irgb,'FaceColor','texturemap','EdgeColor','none');
axis([ -1 1 -1 1 0 1 ]);
title('HSV cone');
zlabel('value');
```

Both LUV and Lab color models are perceptually uniform, which means that two colors that are equally distant in the color space are equally distant perceptually. Both color models were derived from the CIE 31 model. A uniform chromacity scale (UCS) diagram was proposed by CIE; the Y lightness scale is replaced by a new scale L, and the other two components are U and V for CIE LUV, and *a* and *b* for CIE Lab. These components are non-linearly derived from CIE 31. A first UCS diagram was proposed by CIE in 1960, and then it was improved in 1976. Figure 3.44 shows the new UCS diagram.

In the CIE Lab model, L is luminance, positive values of *a* correspond to red and negative values to green, positive values of *b* correspond to blue and negative to yellow. Blue and yellow are opponent colors, as well as red with respect to green.

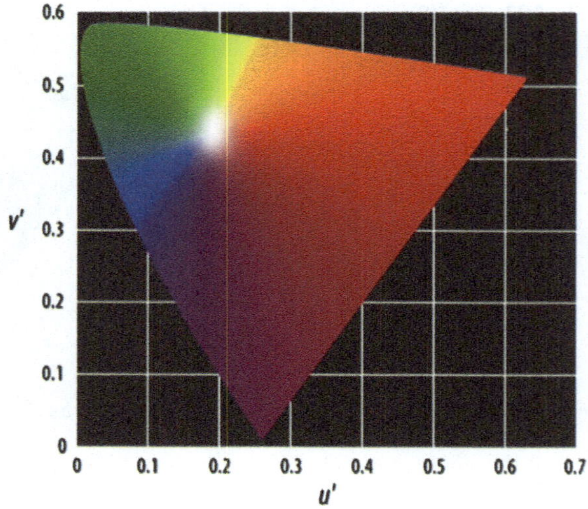

Fig. 3.44 The CIE UCS chromacity scale diagram

YIQ and YUV also use opponent colors. Both color spaces are utilized in color TV. YIQ is used in America (NTSC), and YUV in Europe (PAL). The component Y is luminance. An advantage of Y is that it is sufficient for black and white TV, therefore color transmissions are compatible with old TV sets.

In the case of YIQ, I stands for *in-phase*, and Q stands for *quadrature*. In order to consider that our sensitivity to yellow-blue (I) changes is different than to green-red (Q) changes, more NTSC TV bandwidth is assigned to I and lesss to Q. In the case of YUV, both U and V contain yellow-blue information, so both components are assigned the same bandwidth.

Other color spaces are LCh (C, chromacity; h, hue), and YCrCb (Cr, red/yellow; Cb, blue/yellow).

Fig. 3.45 The IQ color plane for Y = 1

Figure 3.45 shows the IQ color plane for Y = 1. It has been generated with the Program 3.27: the program uses the function *ntsc2rgb( ),* which is one of the functions provided by the MATLAB IPT for conversions between different color spaces.

Program 3.27 Paint YIQ color space

```
% Paint YIQ color space
%create an IQ color plane
nn=linspace(-1,1,200);
mm=fliplr(nn);
mY=ones(200,200);
mI=repmat(nn,[200,1]);
mQ=repmat(mm',[1,200]);
%prepare image
Iiq=zeros(200,200,3);
Iiq(:,:,1)=mY; Iiq(:,:,2)=mI; Iiq(:,:,3)=mQ;
%convert to RGB
Irgb=ntsc2rgb(Iiq);
%display
figure(1)
imshow(Irgb);
title('I-Q color space, Y=1');
xlabel('I'); ylabel('Q');
```

There are many possible conversions between color spaces, and the MATLAB IPT does not cover that all. Anyway, it is easy to obtain from Internet matrices for mathematical conversion between color components, whenever this is possible (some conversions require nonlinear equations). Based on one of these matrices, a picture of the UV color plane, depicted in Fig. 3.46 is obtained with the simple Program 3.28. In the 8-bit YUV space, the range of components is Y = 16..230, U = 0..255, V = 0..255 (there are YUV versions were U and V should be 16..230).

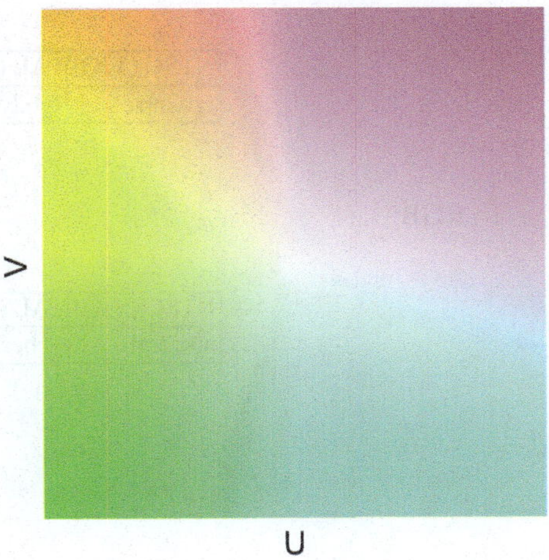

Fig. 3.46 The UV color plane for Y = 230

Program 3.28 Paint YUV color space

```
% Paint YUV color space
%YUV to RGB conversion matrix
U2R=[298 0 409;
298 -100 -208;
298 516 0];
U2R=U2R/256;
Y=230; %maximum
C=Y-16;
Irgb=zeros(256,256,3);
for nV=1:256,
V=nV-1;
for nU=1:256,
U=nU-1;
D=U-128;
E=V-128;
aux0=[C;D;E]; %data vector
offs=[0.5;0.5;0.5]; %offset
aux1=(U2R*aux0)+offs; %convert pixel to RGB
aux1=uint8(aux1); %clip values into 0..255
Irgb(257-nV,nU,:)=aux1; %put pixel in image
end;
end;
Irgb=Irgb/256; %0..1 values range
%display
figure \ref{fig:3.(1)
imshow(Irgb);
title('U-V color space, Y=230');
xlabel('U'); ylabel('V');
```

The set of IPT MATLAB functions for conversions between color spaces is the following:

From RGB to:

NTSC (YIQ)	PAL (YCbCr)	HSV
rgb2ntsc	rgb2ycbcr	rgb2hsv

To RGB:

NTSC (YIQ)	PAL (YCbCr)	HSV
ntsc2rgb	ycbcr2rgb	hsv2rgb

Standard illuminants data are given by *whitepoint( )*.

It is interesting to remark that many webcams use YUV. Actually, there are many versions of YUV, however in most cases it is like YcbCr.

3.7.1 RGB Example

The three planes of the image representation in MATLAB (three matrices) correspond to red, green and blue (RGB).

Program 3.29 reads an RGB image, a colourful parrot, and shows it in Fig. 3.47. The program also shows the three gray scale planes, R, G and B, which are used to compose the colors (Fig. 3.48).

Fig. 3.47 Color picture

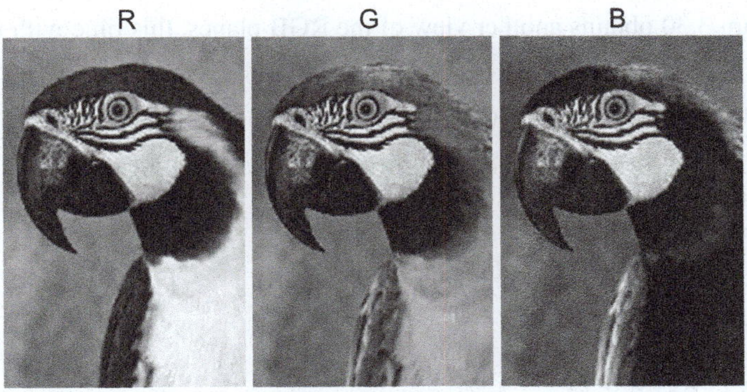

Fig. 3.48 RGB components of the picture

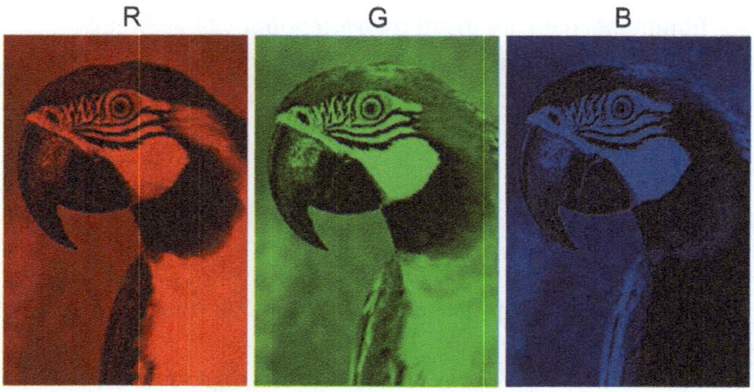

Fig. 3.49 RGB components of the picture

Program 3.29 Display RGB planes of Parrot picture

```
% Display RGB planes of Parrot picture
%read the image file into a matrix:
parrot=imread('parrot1.jpg');
figure(1)
imshow(parrot); %display color picture
title('original');
figure(2)
subplot(1,3,1);
imshow(parrot(:,:,1)); %display R
title('R');
subplot(1,3,2);
imshow(parrot(:,:,2)); %display G
title('G');
subplot(1,3,3);
imshow(parrot(:,:,3)); %display B
title('B');
```

Program 3.30 obtains another view of the RGB planes, this time with the corresponding colors (Fig. 3.49).

Program 3.30 Display Coloured RGB planes of Parrot picture

```
% Display Coloured RGB planes of Parrot picture
%read the image file into a matrix:
parrot=imread('parrot1.jpg');
[M,N,P]=size(parrot);
figure(1)
subplot(1,3,1);
A=uint8(zeros(M,N,3));
A(:,:,1)=parrot(:,:,1);
imshow(A); %display R
subplot(1,3,2);
```

```
A=uint8(zeros(M,N,3));
A(:,:,2)=parrot(:,:,2);
imshow(A); %display G
subplot(1,3,3);
A=uint8(zeros(M,N,3));
A(:,:,3)=parrot(:,:,3);
imshow(A); %display B
```

3.7.2 HSV Example

Another decomposition into three is hue, saturation and value (HSV).

Program 3.31 reads the RGB parrot image, and then convert it to HSV. Then the program shows in Fig. 3.50 the HSV gray scale planes of the picture.

Program 3.31 Convert Parrot picture to HSV

```
% Convert Parrot picture to HSV
%read the image file into a matrix:
parrot=imread('parrot1.jpg');
parrothsv=rgb2hsv(parrot);
figure(1)
subplot(1,3,1);
imshow(parrothsv(:,:,1));
title('hue');
subplot(1,3,2);
imshow(parrothsv(:,:,2));
title('saturation');
subplot(1,3,3);
imshow(parrothsv(:,:,3));
title('value');
```

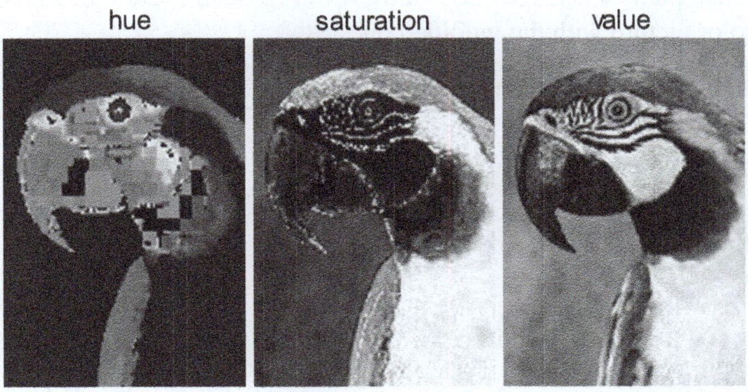

Fig. 3.50 HSV components of the picture

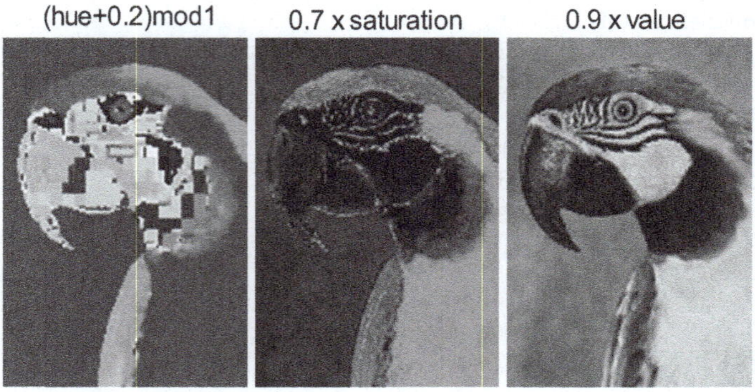

Fig. 3.51 Changing the HSV components

Fig. 3.52 Comparison of original and modified picture

Once in HSV format, it is interesting to play with changes on each of the three planes, and see what happens with the picture. Program 3.32 gives an example: the values in each plane is changed according with Fig. 3.51. Figure 3.52 compares the original color picture with the modified one.

Program 3.32 Modifiy HSV Parrot picture

```
% Modifiy HSV Parrot picture
%read the image file into a matrix:
parrot=imread('parrot1.jpg');
parrothsv=rgb2hsv(parrot);
%change of Hue
aux=parrothsv(:,:,1);
aux=aux+0.2; %hue shift
aux=mod(aux,1); %use reminder if  >1
newparrot(:,:,1)=aux; %change of H
newparrot(:,:,2)=0.7*(parrothsv(:,:,2)); %change of S
```

```
newparrot(:,:,3)=0.9*(parrothsv(:,:,3)); %change of V
figure(1)
subplot(1,3,1);
imshow(newparrot(:,:,1));
title('(hue+0.2)mod1');
subplot(1,3,2);
imshow(newparrot(:,:,2));
title('0.7 x saturation');
subplot(1,3,3);
imshow(newparrot(:,:,3));
title('0.9 x value');
figure(2)
subplot(1,2,1)
imshow(parrot); %original
title('original');
h=gca;ht=get(h,'Title'); set(ht,'FontSize',14);
subplot(1,2,2)
newparrotrgb=hsv2rgb(newparrot); %to RGB
imshow(newparrotrgb);
title('modified');
h=gca;ht=get(h,'Title'); set(ht,'FontSize',14);
```

3.7.3 YIQ Example

It is also possible a decomposition into intensity (Y), x-color(I) and y-color (Q).

Program 3.33 reads the RGB parrot image, and then convert it to YIQ. Then the program shows in Fig. 3.53 the YIQ planes of the picture.

Fig. 3.53 YIQ component

Fig. 3.54 Comparison of original and modified picture

Program 3.33 Convert Parrot picture to YIQ with I and Q coloured planes

```
% Convert Parrot picture to YIQ with I and Q coloured planes
%read the image file into a matrix:
parrot=imread('parrot1.jpg');
parrotyiq=rgb2ntsc(parrot);
[M,N,P]=size(parrot);
figure(1)
subplot(1,3,1);
imshow(parrotyiq(:,:,1));
title('Y (intensity)');
subplot(1,3,2);
A=zeros(M,N,3);
A(:,:,1)=parrotyiq(:,:,1);
A(:,:,2)=parrotyiq(:,:,2);
Ii=ntsc2rgb(A);
imshow(Ii);
title('I-color');
subplot(1,3,3);
A=zeros(M,N,3);
A(:,:,1)=parrotyiq(:,:,1);
A(:,:,3)=parrotyiq(:,:,3);
Iq=ntsc2rgb(A);
imshow(Iq);
title('Q-color');
```

Like in HSV, it is interesting to play with changes on each of the YIQ three planes. Program 3.34 includes modifications of the values in each YIQ plane. Figure 3.54 compares the original color picture with the modified one.

Program 3.34 Modifiy YIQ Parrot picture

```
% Modifiy YIQ Parrot picture
%read the image file into a matrix:
parrot=imread('parrot1.jpg');
parrotyiq=rgb2ntsc(parrot);
```

```
newparrot(:,:,1)=0.8*(parrotyiq(:,:,1)); %change of Y
newparrot(:,:,2)=1.5*(parrotyiq(:,:,2)); %change of I
newparrot(:,:,3)=0.3*(parrotyiq(:,:,3)); %change of Q
figure(1)
subplot(1,2,1)
imshow(parrot); %original
title('original');
subplot(1,2,2)
newparrotrgb=ntsc2rgb(newparrot); %to RGB
imshow(newparrotrgb);
title('modified');
```

3.7.4 *Indexed Images*

One of the image representation formats that MATLAB IPT can use is indexed images. A color map—which is a sort of color catalog-, is used, with the colors numbered. An indexed image *I* is a *MxN* matrix, with entries being color numbers (indices). This image representation is compact, saving a lot of memory.

RGB images can be converted to indexed images using the *rgb2ind( )* function; it removes hue and saturation, retaining luminance. Likewise, gray scale images can be converted to indexed images using *gray2ind( )*. The reverse conversions are *ind2rgb( )* and *ind2gray( )*.

Further representation compaction could be obtained by using *imapprox()*, which attempts to approximate the indexed image with fewer colors.

Program 3.35 reads an RGB image, and converts it to an indexed image. The result is depicted in Fig. 3.55 The program also gives information on sizes of the RGB and the indexed representations.

Program 3.35 Convert Parrot to indexed image

```
% Convert Parrot to indexed image
%read the image file into a matrix:
parrot=imread('parrot1.jpg');
[Ip,Cmap]=rgb2ind(parrot,64); %make a colormap with 64 entries
figure(1)
imshow(Ip,Cmap); %display color picture
title('indexed image');
disp('parrot size')
size(parrot)
disp('indexed parrot size')
size(Ip)
disp('colormap size')
size(Cmap)
```

Fig. 3.55 Indexed color picture

3.8 Hough Transform and Radon Transform

This section is devoted to transforms that use '*sinograms*' for certain purposes, which will be illustrated with some examples.

The Radon transform, in particular, is very useful in the medical context. For example, it is used in Computerized Tomography (CT).

3.8.1 The Sinogram

Let us select a point (x_0, y_0) on the $x - y$ plane. Corresponding to this point, a curve can be obtained on the plane $P - \alpha$, using the following expression:

$$P = x_0 \cos(\alpha) + y_0 \sin(\alpha), \quad \forall \alpha \in [0, 2\pi] \tag{3.11}$$

(in practice, the expression is computed for a set of α angle values)

If one represents the curve $P(\alpha)$, one obtain the *sinogram*. Figure 3.56 shows an example (corresponding to the point (5, 4)).

Program 3.36 Sinogram for one point

```
% Sinogram for one point
%the point
x0=5; y0=4;
%sinogram preparation
NP=360; %number of points
alpha=1:1:NP;
alpha=alpha.*pi/180; %angle in rads
```

```
P=zeros(1,NP); %space for projections
for nn=1:NP,
P(nn)=(x0*cos(alpha(nn)))+(y0*sin(alpha(nn)));
end;
figure(1)
plot(alpha(:),P(:),'k');
axis([0 2*pi 1.2*min(P) 1.2*max(P)]);
title('sinogram for one point');
xlabel('angle in degrees');
ylabel('Projection');
```

Coming to the $x - y$ plane, there is a radius from the origin to our point (x_0, y_0), and there is a perpendicular to the radius at (x_0, y_0). This is depicted in Fig. 3.57.

If one selects three points, and represents the corresponding sinograms (for an angle range between 0 and 270°, for example), three curves are obtained, as depicted in Fig. 3.58.

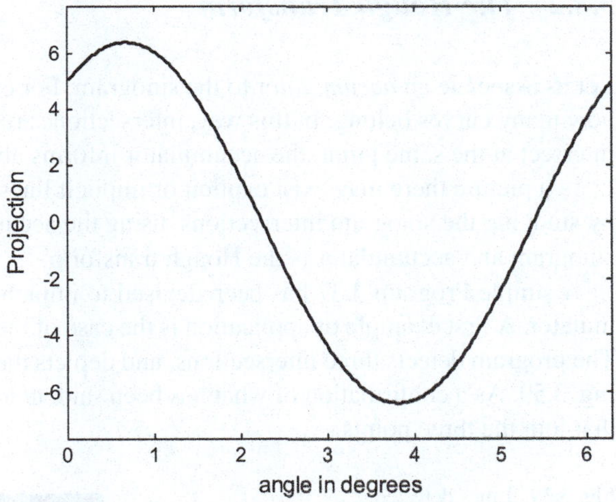

Fig. 3.56 An example of sinogram

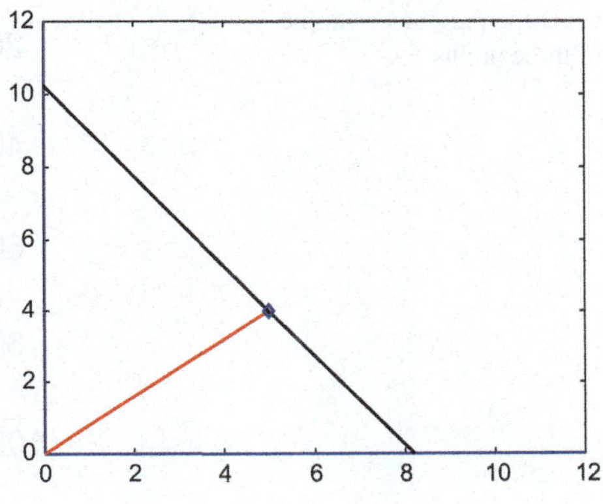

Fig. 3.57 The point (x_0, y_0) *on the x-y plane, radius and perpendicular line*

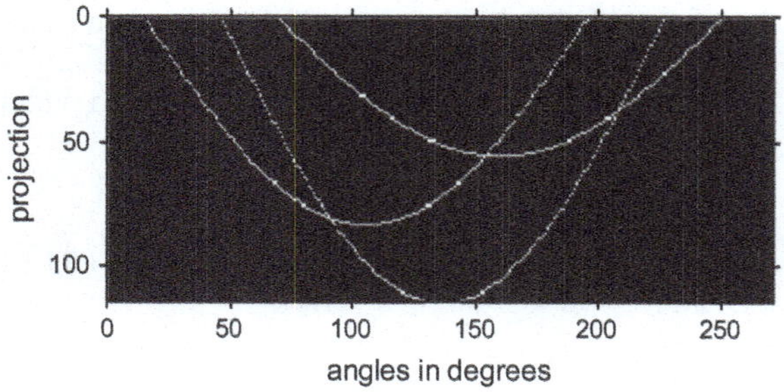

Fig. 3.58 Sinogram corresponding to three points

Notice that there are three intersections of the curves in Fig. 3.58. This is an indication that three lines can be found, joining the three points: this is the basis of the Hough transform.

3.8.2 The Hough Transform

Let us associate an *accumulator* to the sinogram. For every image pixel, it counts to how many curves belong. In this way, intersections are detected. Several curves can intersect at the same point, the accumulator informs about it.

In a picture there may exist explicit or implicit lines. These lines can be detected by studying the sinogram intersections, using the accumulator. This combination of sinogram and accumulator is the Hough transform.

A simple Program 3.37 has been devised to implement the sinogram with accumulator. A first example of application is the case of the three points just introduced. The program detects three intersections, and depicts the three corresponding lines in Fig. 3.59. As a confirmation of what has been said, note that the three lines are those that join the three points.

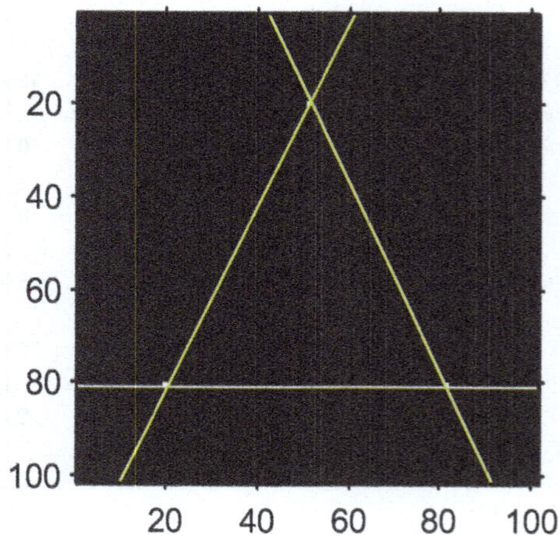

Fig. 3.59 Lines detected with the Hough transform, using the previous example of three points

Program 3.37 Hough projection basic example

```
% Hough projection basic example
clear all;
%simple image with 3 points
A=zeros(101,101);
A(20,51)=1; A(80,20)=1; A(80,81)=1;
%prepare for Hough transform
[x,y]=find(A); %indices of non-zero pixels
Nx=length(x); %number of points
%compute projections
ang=([-90:180]*pi)/180; %set of angles
Na=length(ang); %number of angles
P=floor(x*cos(ang)+y*sin(ang)); %integer projections
Pmax=max(max(P)); %maximum P
%accumulate
AC=zeros(Pmax+1,Na); %accumulator
for nj=1:Na, %angles
for ni=1:Nx, %points
aux=P(ni,nj);
if aux >=0,
AC(1+aux,nj)=1+AC(1+aux,nj);
end;
end;
end;
figure(1)
iptsetpref('ImshowAxesVisible','on');
imshow(AC);
axis([0 Na 0 Pmax+1]);
title('sinogram of three points');
xlabel('angles in degrees'); ylabel('projection');
%---------------------------------
% plot detected lines on photograph
M=max(max(AC)); %accumulator maximum
[dP,dphi]=find(AC==M);
Lm=length(dP); %number of maximum fits
dang=((181-dphi)*pi)/180; %from degrees to radians
figure(2)
imshow(A); hold on; %the photo
title('image and detected lines');
%add lines
[X,Y]=size(A);
for nn=1:Lm,
S=sin(dang(nn)); C=cos(dang(nn));
rr=dP(nn);
if S==0,
line([rr,rr], [0,Y], 'Color','Cyan');
else
yrr=(rr-(Y*C))/S;
line([0,Y], [rr/S,yrr],'Color','Yellow');
```

```
end;
end
iptsetpref('ImshowAxesVisible','off'); %back to default
```

Now, let us study a photograph, and try to detect lines in the photograph. Figure 3.60 presents an interesting example to be studied.

Figure 3.61 shows the sinogram corresponding to the chosen photograph. There are many intersections.

The main body of Program 3.38 is similar to the previous program: it implements the sinogram with accumulator. The program displays three Figs. 3.60, 3.61 and 3.62.

The Fig. 3.62 displays the photograph and draws over it the principal lines detected with the accumulator.

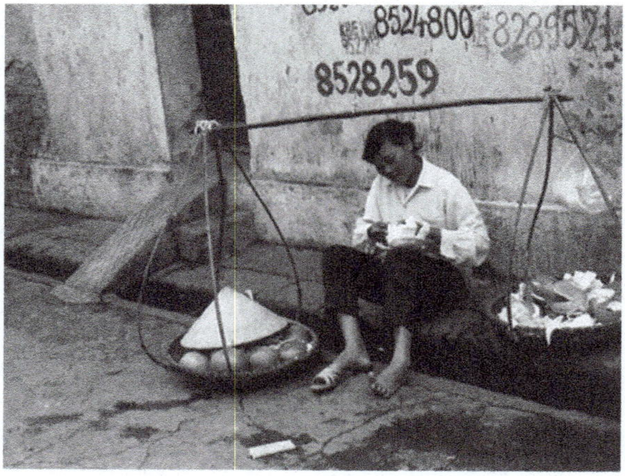

Fig. 3.60 A photograph to be studied

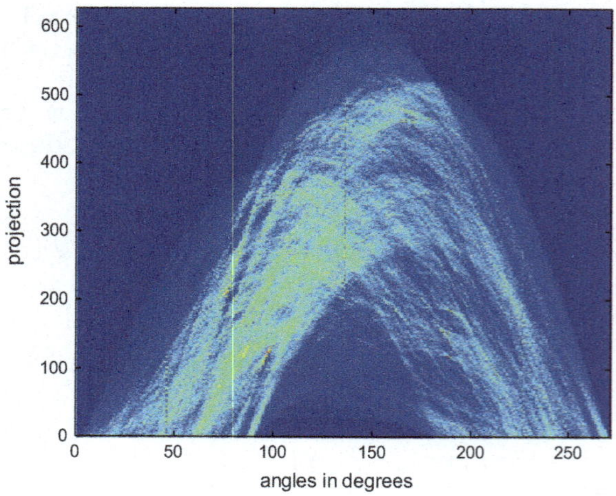

Fig. 3.61 Sinogram of the previous photograph

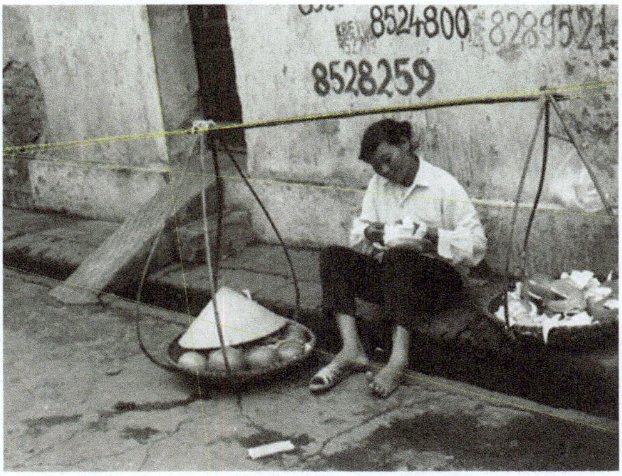

Fig. 3.62 Image and detected lines

Program 3.38 Hough photo projection example

```
% Hough photo projection example
clear all;
%read the image file into a matrix:
phot=imread('Viet1.jpg');
phot1=phot(:,:,1);
pho=phot1(40:430,70:590); %limit the size
A=edge(pho,'canny'); %get edges
%prepare for Hough transform
[x,y]=find(A); %indices of non-zero pixels
Nx=length(x); %number of points
%compute projections
ang=([-90:180]*pi)/180; %set of angles
Na=length(ang); %number of angles
P=floor(x*cos(ang)+y*sin(ang)); %integer projections
Pmax=max(max(P)); %maximum P
%accumulate
AC=zeros(Pmax+1,Na); %accumulator
for nj=1:Na, %angles
for ni=1:Nx, %points
aux=P(ni,nj);
if aux >=0,
AC(1+aux,nj)=1+AC(1+aux,nj);
end;
end;
end;
figure(1)
imagesc(AC);
axis([0 Na 0 Pmax+1]);
h=gca; set(h,'YDir','normal');
title('Hough transform: sinogram');
```

```
xlabel('angles in degrees'); ylabel('projection');
%----------------------------------------
% plot detected lines on photograph
M=max(max(AC)); %accumulator maximum
[dP,dphi]=find((AC <=M-44) & (AC >M-46));
Lm=length(dP); %number of maximum fits
dang=((181-dphi)*pi)/180; %from degrees to radians
figure(2)
imshow(pho); hold on; %the photo
title('original photo');
figure(3)
imshow(pho); hold on; %the photo
title('image and detected lines');
%add lines
[X,Y]=size(pho);
for nn=1:Lm,
S=sin(dang(nn)); C=cos(dang(nn));
rr=dP(nn);
if S==0,
line([rr,rr], [0,Y], 'Color','Cyan');
else
yrr=(rr-(Y*C))/S;
line([0,Y], [rr/S,yrr],'Color','Yellow');
end;
end
```

It is possible to modify the Hough transform, so it can be used to detect other shapes, like circles, ellipses, etc. [18, 35].

3.8.3 *The Radon Transform, and Computerized Tomography*

Computerized tomography is an important diagnostic method, being applied in medicine from years ago. This subsection is devoted to introduce the fundamental ideas of the method. Most of the mathematical treatment is based on [16]; more mathematical details may be found in [11, 13, 39].

A basic device could consist of a X-ray source drawing a narrow beam through the patient, and a X-ray detector on the opposite side. Both the source and the detector turn around the patient. The result obtained is an accumulated projection. The aim of the system is to obtain an image that reproduces what is inside the patient. Actually the data obtained correspond to a 'slice' of the patient, according to a virtual cutting plane perpendicular to the patient. To obtain 3D data it suffices with moving the table with the patient while taking successive data planes.

Figure 3.63 depicts the concept, and shows a photograph of a medical device for computerized tomography.

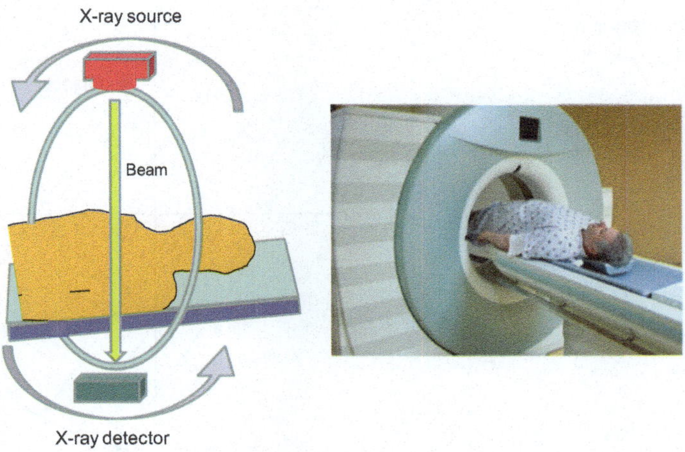

Fig. 3.63 Concept of computerized tomography

Let us study the physics of the basic device. Figure 3.64 shows a X.ray crossing an object. This ray will be represented with the pair (r, θ).

Denote $f(x, y)$ the density in any place of the object. Suppose that the energy of the ray was E_0 at the source. If there was a detector at the end of the X-ray, it would measure an energy E_1. Denote $p = \ln(E_0/E_1)$. Then:

$$p = \int_L f(x, y)\, ds \tag{3.12}$$

where the integral is computed along the line L.

Fig. 3.64 Representation of X-rays

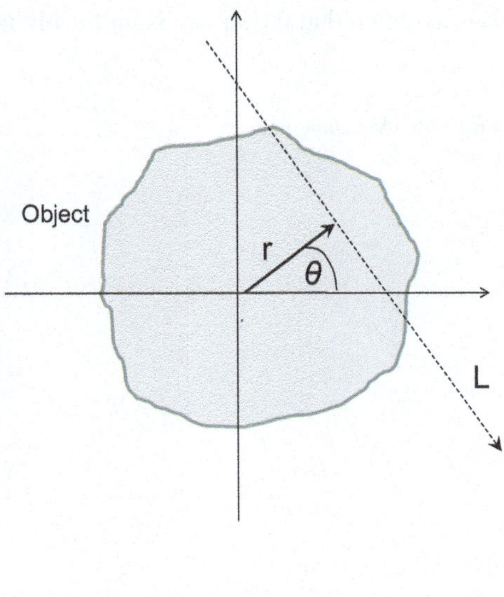

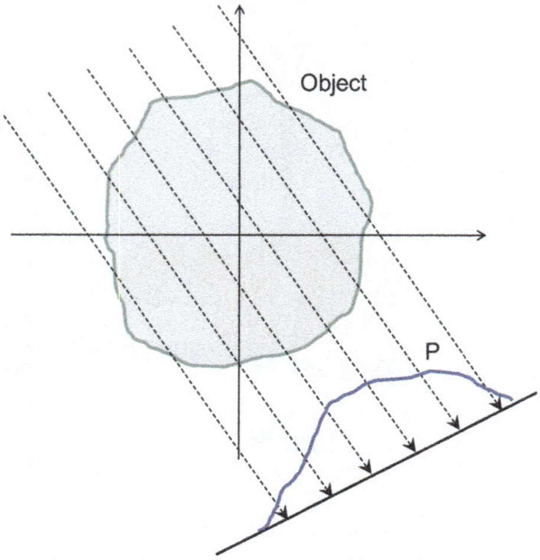

Fig. 3.65 Projection of X-rays

Consider now that there are many parallel X-rays, as sketched in Fig. 3.65 (where a projection, corresponding to a certain angle θ has been pictured).

The projection, for any angle θ can be computed with the following expression, which represents all line integrals of the density function:

$$p(r, \theta) = \int_{-\infty}^{\infty} \int_{-\infty}^{\infty} f(x, y)\, \delta(x \cos\theta + y \sin\theta - r)\, dx\, dy \tag{3.13}$$

where $\delta(x)$ is the Dirac delta.

The right hand side of the equation above is the *Radon transform*: $[\Re f](r, \theta)$.

The result of a Radon transform can be visualized with a sinogram. As a simple example, let us take the Radon transform of a square, as the one depicted in Fig. 3.66. It is assumed that the square is uniformly filled.

Fig. 3.66 A rectangular object

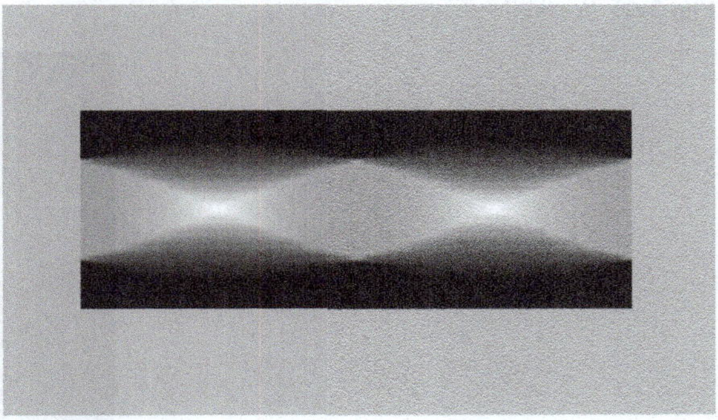

Fig. 3.67 Radon transform of the square (sinogram)

The Radon transform of the square has been computed with the Program 3.39. Actually it is possible to use code similar to Program 3.38, but it is also the occasion to explore other coding alternatives, using in this case the *imrotate()* MATLAB IPT function. The result of the Radon transform is shown in Fig. 3.67.

Program 3.39 Radon transform of a rectangle

```
% Radon transform of a rectangle
%padded rectangle
B=ones(32,32);
R=zeros(64,64);
R(17:48,17:48)=B;
theta=0:1:180; %set of angles
N=length(theta);
PR=zeros(64,181);
%projections
for nn=1:N,
aux=imrotate(R,theta(nn),'bilinear','crop');
PR(:,nn)=sum(aux)';
end
nPR=PR/max(max(PR)); %normalize for display
figure(1)
imshow(R);
title('rectangular image');
figure(2)
imshow(nPR);
title('Radon transform (sinogram)');
```

It is usual, for biomedical computing exercises, to use *"phantoms"* that simulate certain types of medical images obtained with X-rays or whatever. Figure 3.68 shows an example.

Figure 3.69 shows the sinogram of the corresponding Radon transform.

The phantom has been included at the beginning of Program 3.39, instead of the code to generate the rectangle. The corresponding fragment is the following

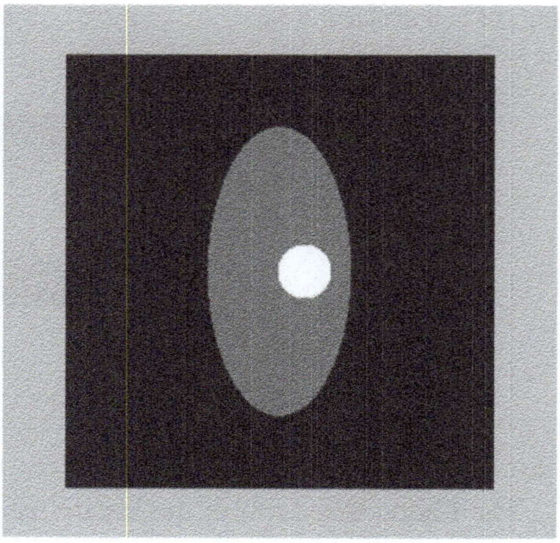

Fig. 3.68 A simple phantom

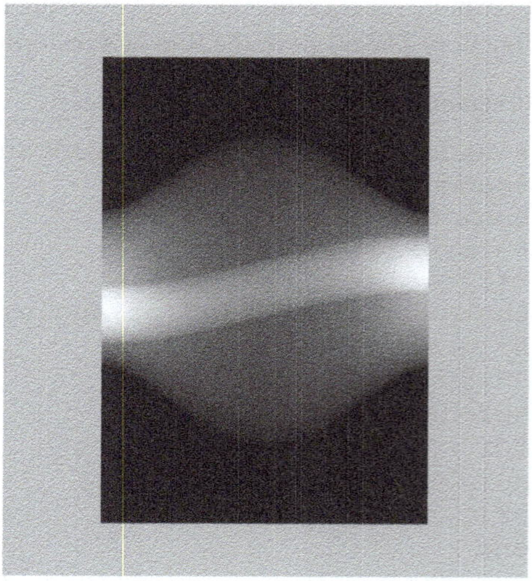

Fig. 3.69 Radon transform of the phantom (sinogram)

Fragment 3.40 A phantom

```
%a "phantom"
x=-127:1:128; y=-127:1:128;
[X,Y]=meshgrid(x,y);
z1=(4*(X.*X)+(Y.*Y)) <(128/1.5)^2;
aX=X-15; aY=Y;
z2=((aX.*aX)+(aY.*aY)) <(128/8)^2;
R=(0.9*z2)+(0.5*z1);
```

What one wants is to determine $f(x, y)$, on the basis of projections. In other words, the problem is to compute the inverse Radon transform:

$$f(x, y) = [\Re^{-1} p](x, y) \tag{3.14}$$

In practical terms, there are three ways of attack for this problem.

3.8.3.1 Algebraic Reconstruction Method (ARM)

A set of *basis images* is taken into consideration:

$$b_j(x, y) = \begin{cases} 1 \ \ if \ f(x, y) == pixel(j) \\ 0 \ \ else \end{cases} \tag{3.15}$$

The function $f(x, y)$ is regarded as an image with N pixels. This image is put in function of the basis:

$$f(x, y) = \sum_{j=1}^{N} c_j b_j(x, y) \tag{3.16}$$

Now, the problem is to determine the coefficients c_j. One departs from the information given by the projections p_i. Define a projection operator R_i as follows:

$$R_i f = \int_{Li} f(x, y)\, ds = p_i \tag{3.17}$$

Then:

$$R_i f = \sum_{j=1}^{N} c_j R_i b_j(x, y) \tag{3.18}$$

where $b_j(x, y)$ are known. Therefore it is possible to compute:

$$a_{ij} = R_i b_j(x, y) \tag{3.19}$$

The coefficients a_{ij} can be grouped in a matrix A. Likewise, the projections p_i form a vector $\mathbf{p}$, and the coefficients c_j form a vector $\mathbf{c}$. The following system of equations can be written:

$$A\,\mathbf{c} = \mathbf{p} \tag{3.20}$$

In general there are more equations than unknowns and there is no solution of the system. An approximation can be obtained by the transforming the problem into an optimization one: to find $\mathbf{c}$ such as:

$$\min \|A\mathbf{c} - \mathbf{p}\| \tag{3.21}$$

Usually this is a ill posed problem. To alleviate numerical difficulties, *regularization* may be used:

$$\min (\|A\mathbf{c} - \mathbf{p}\|^2 + \gamma \|\mathbf{c}\|^2) \tag{3.22}$$

where γ is a constant.

The topic of inverse problems is treated with more detail in the chapter on adaptive filters.

3.8.3.2 Filtered Back-Projection Method (FBM)

For any line L defined by a pair of coordinates (r, θ), the corresponding projection is:

$$p(r, \theta) = \int_L f(x, y)\, ds \tag{3.23}$$

The Fourier transform with respect to r is the following:

$$P(\omega, \theta) = \int_{-\infty}^{\infty} p(r, \theta)\, e^{-j 2\pi \omega r}\, dr \tag{3.24}$$

It can be deduced that:

$$f(x, y) = \int_0^{\pi} \int_{-\infty}^{\infty} P(\omega, \theta)\, |\omega| e^{j 2\pi \omega r}\, d\omega\, d\theta \tag{3.25}$$

where $r = x \cos\theta + y \sin\theta$.

Therefore, one can proceed as follows: (1) Fourier transform with respect to r, (2) multiplication by $|\omega|$, (3) inverse Fourier transform, (4) integral with respect to θ.

The step (4) is called back-projection. Denote as $q(r, \theta)$ the result of step (3); the back-projection can be approximated as follows:

$$f(x, y) = \int_0^{\pi} q(x \cos\theta + y \sin\theta,\, \theta)\, d\theta \approx \frac{\pi}{\Delta\theta} \sum_j q(x \cos\theta_j + y \sin\theta_j,\, \theta_j) \tag{3.26}$$

The step (2) can be done by filtering in the Fourier domain, or by convolution in the projection domain. The basic filter for this is called '*ramp filter*' or '*Ram-Lak*' filter.

Program 3.41 implements the recovery of the square (Fig. 3.66), after Radon transform and then backprojection (using ramp filter). The result is shown in Fig. 3.70.

Fig. 3.70 Example of backprojection: the square is recovered

Program 3.41 Inverse Radon transform with filter

```
% Inverse Radon transform with filter
% for the rectangle example
%The Radon transform
%padded rectangle
B=ones(32,32);
R=zeros(64,64);
R(17:48,17:48)=B;
theta=0:1:180; %set of angles
N=length(theta);
PR=zeros(64,181);
%projections
for nn=1:N,
aux=imrotate(R,theta(nn),'bilinear','crop');
PR(:,nn)=sum(aux)';
end
%ramp filter
rampf=[2*[0:31,32:-1:1]'/64];
%Fourier domain filtering
FPR=fft(PR,64);
for nn=1:N,
FPRf(:,nn)=FPR(:,nn).*rampf;
end;
%inverse Fourier
PR=real(ifft(FPRf));
%Backprojection
aux=zeros(64,64);
IR=aux;
for nn=1:N,
for np=1:64,
aux(:,np)=PR(:,nn);
end;
IR=IR+imrotate(aux,theta(nn),'bilinear','crop');
end;
```

```
nIR=IR/max(max(IR)); %normalize
figure(1)
imshow(nIR);
title('Recovered rectangle, with ramp filter');
```

Had the filter not been used, the result of backprojection would have been a blurred recovery, as shown in Fig. 3.71. This Fig. 3.71 has been obtained by deactivating the filtering sentences in Program 3.41.

In order to improve the recovery, some other filters could be tried. A typical approach is to multiply the ramp filter by a sinc function (this is the Shepp-Logan filter), or by a cosine function, or by a Hamming or Hanning window.

3.8.3.3 Direct Fourier Method (DFM)

The *projection slice theorem* states that:

$$P(\omega, \theta) = F(\omega \cos\theta, \omega \sin\theta) \tag{3.27}$$

where $F()$ denotes the Fourier transform of $f(x, y)$.

At first sight it seems that our problem can be directly solved by inverse 2-D Fourier transform. However there is a serious difficulty: the data are given in polar coordinates, but the desired result should be in Cartesian coordinates. Interpolation in the Fourier domain may cause important image recovery errors, so there is intense research on this aspect. A next section in this chapter is devoted to the Fourier transform for nonequispaced data.

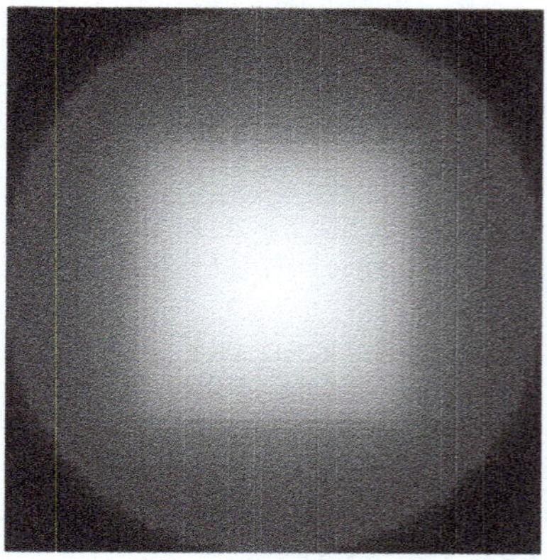

Fig. 3.71 Example of unfiltered backprojection

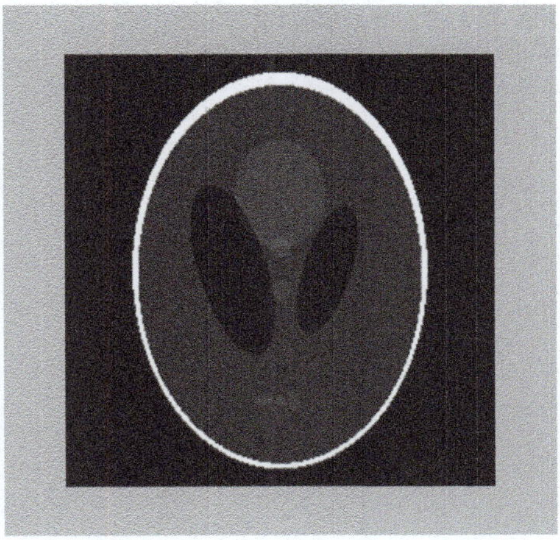

Fig. 3.72 A Shepp-Logan phantom

3.8.4 IPT Functions for the Radon Transform

The MATLAB Image Processing Toolbox offers the *radon( )* and *iradon( )* functions, and the *phantom( )* function. A typical use of *phantom( )* is for generating the '*Shepp-Logan*' phantom. With these functions the coding of Radon applications is considerably simplified. To demonstrate it, a program has been developed (Program 3.42) that generates the phantom and then computes the Radon transform of it; the program presents two figures.

Figure 3.72 shows the Shepp-Logan phantom. Using the parameters of the function *phantom( )*, it has been specified to use a modified Shepp-Logan phantom that has better contrast.

Figure 3.73 shows the Radon transform of the Shepp-Logan phantom.

Program 3.42 Radon transform of a phantom

```
% Radon transform of a phantom
P=phantom('Modified Shepp-Logan',256);
theta=0:1:180;
PR=radon(P,theta); %the Radon transform
nPR=PR/max(max(PR)); %normalization
figure(1)
imshow(P);
title('the Shepp-Logan phantom');
figure(2)
imshow(nPR);
title('Radon transform (sinogram)');
```

To complete the example, Program 3.43 applies the inverse Radom transform, using *iradon( )*, to recover the phantom. Figure 3.74 shows the good result.

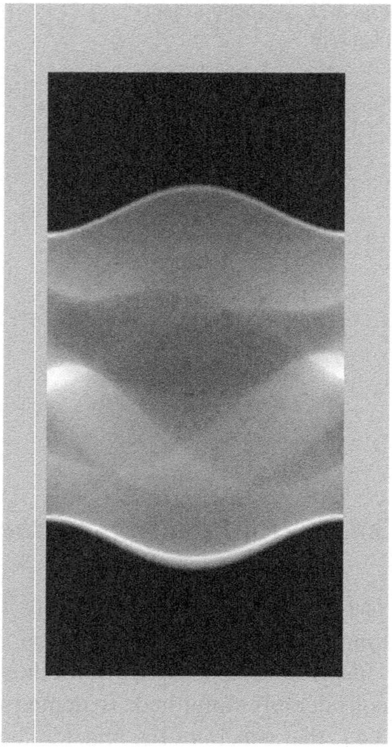

Fig. 3.73 Radon transform of the phantom (sinogram)

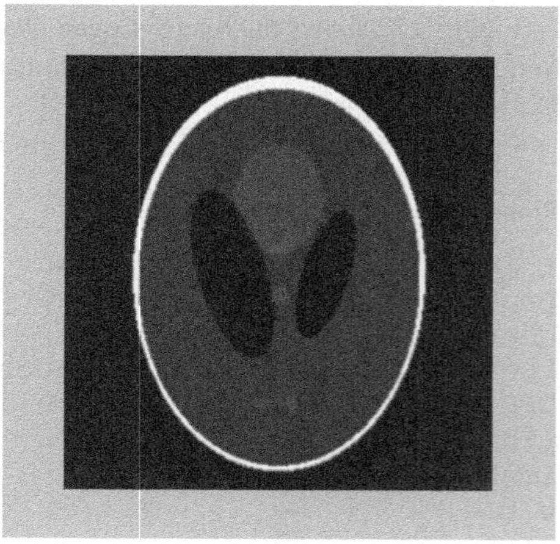

Fig. 3.74 Recovery of the phantom by Inverse Radon transform

Program 3.43 Inverse Radon transform

```
% Inverse Radon transform
% phantom data
% The Radon transform of the phantom
P=phantom('Modified Shepp-Logan',256);
theta=0:1:180;
PR=radon(P,theta); %the Radon transform
% The inverse transform
IR=iradon(PR,theta);
figure(1)
imshow(IR);
title('the recovered phantom');
```

Before closing this section it is opportune to remark that there are many extensions worth to mention, and that remains as further work for the reader. For instance the use of kernels in the Hough transform, or the treatement on non parallel rays in computer assisted tomographic techniques for medical applications. By the end of this chapter, pertinent references and sources of information are included.

3.9 Filter Banks and Images

This section is intended just as an introduction to some fundamental aspects concerning the use of filter banks for bi-dimensional applications. Further reading, with relevant links, could be based on [8, 19, 41].

Of course, there are obvious links between wavelets and filter banks. Actually, there are some new developments that try to create wavelets well adapted to the analysis of images, being able to capture directional edges and details (this is the case with textures). Part of these developments relies on the use of directional filters. One of the targets of this section is to show what directional filters are.

3.9.1 Initial Overview

One of the representative structures of filter banks, as it was studied in a previous chapter, is shown in Fig. 3.75.

There are three elements of interest in the filter bank: the signals (samples), the filters, and the subsampling. It is now to have a first glance of what happens with these elements when playing with images.

When you take a picture with a digital camera, the real image is translated to pixels. This is two-dimensional sampling. It may happen that the camera has not sufficient pixels to capture the desired details, and problems may arise. For example, if you take a photograph of a brick wall, with many visible bricks, Moiré curves (concentric circles) may appear. This corresponds to aliasing.

In general, the image sensor is a rectangular array of punctual sensors. To formally deal with such 2D structure, it is common to use the concept of **lattice**. Here, one could recall aspects of crystallography and solid state physics, where several types of lattices are studied: triangular , rectangular, hexagonal, etc.

Once an image has been translated to a rectangular lattice, it is possible to extract some of the pixels, according with any established rule. For instance, one could take pixels (1, 3, 5, …127) from even rows and even rows. Another possible scheme could be take pixels (1, 3, 5, …, 127) from odd rows, and pixels (2, 4, 6, …, 128) from even rows. Both are different types of subsampling. The first generates another rectangular lattice, and the second a *'quincunx'* lattice.

Figure 3.76 shows the initial lattice, the subsampled rectangular lattice, and the subsampled quincunx lattice. Selected samples are painted in black, rejected samples in white.

Indeed, many other subsampling schemes could be applied, resulting in different types of lattices.

In previous pages, some 2D filters have been introduced. These were more oriented to blurring or edge visualization. However, if one looks at the literature on multi-dimensional filters, a large repertory of filter archetypes appears on scene. Some of them are represented in Fig. 3.77 on the 2D Fourier plane, dark colour corresponds to pass, and white colour corresponds to reject.

Before coming to the design of filter banks, it is convenient to formalize a number of concepts. This subsection is guided by [9, 27].

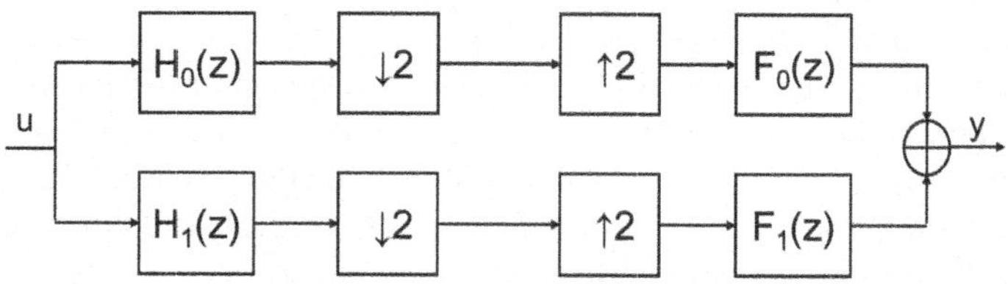

Fig. 3.75 A typical filter bank

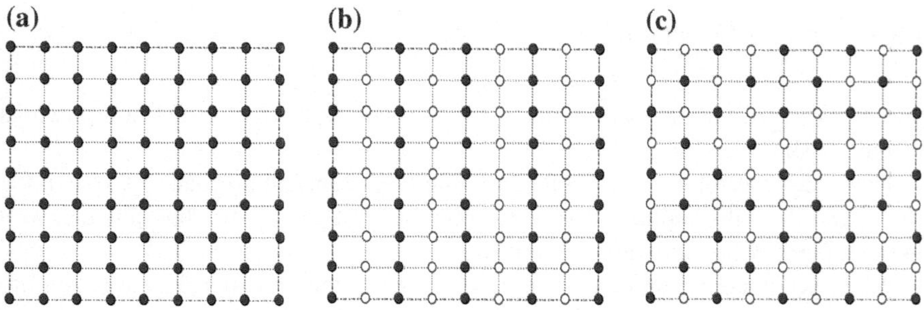

Fig. 3.76 **a** Initial lattice, **b** subsampled rectangular lattice, **c** subsampled quincunx lattice

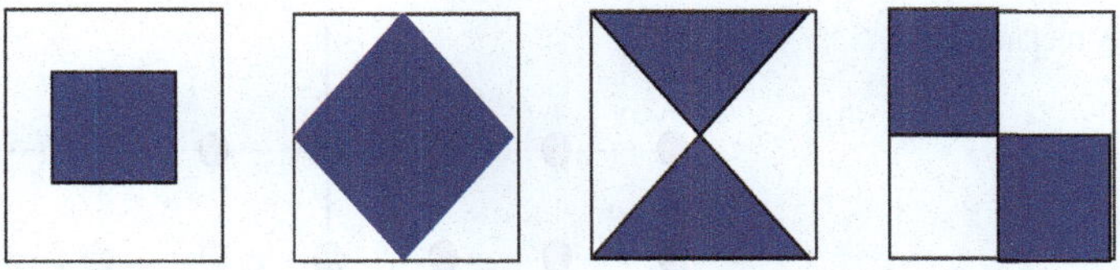

Fig. 3.77 Some filter archetypes (supports on 2D Fourier plane)

3.9.1.1 Lattices

Let $\Re^m$ be the m-dimensional Euclidean space. A lattice in $\Re^m$ is the set:

$$LAT(\mathbf{b}_1, .., \mathbf{b}_n) = \left\{ \sum_i x_i \mathbf{b}_i; \ x_i \text{ integer} \right\} \tag{3.28}$$

of all integer combinations of linearly independent vectors $\mathbf{b}_1$, ..., $\mathbf{b}_n$ in $\Re^m$, with $m \geq n$. The set $\mathbf{b}_1$, ..., $\mathbf{b}_n$ is called a lattice basis, and can be represented as a matrix (each column a vector). For instance, the bi-dimensional quincunx lattice depicted in Fig. 3.78 is generated by the following basis:

$$Q_1 = \begin{pmatrix} 1 & 1 \\ -1 & 1 \end{pmatrix} \tag{3.29}$$

Fig. 3.78 A quincunx lattice

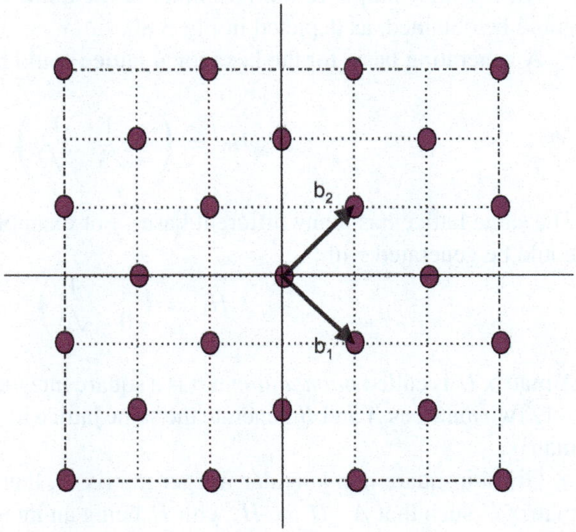

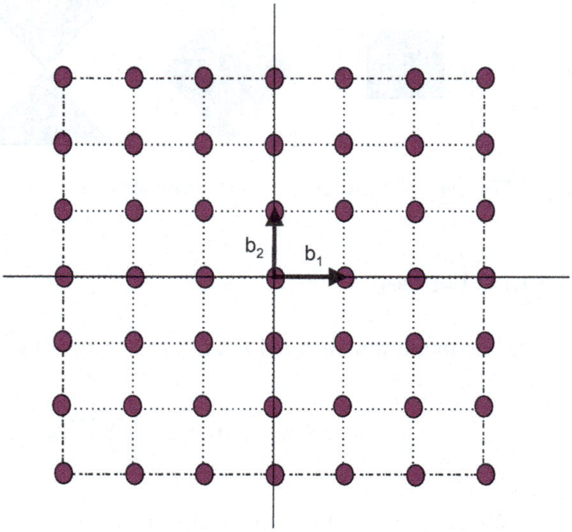

Fig. 3.79 A rectangular lattice

Figure 3.79 shows another example, a rectangular lattice, which is generated with the basis:

$$I = \begin{pmatrix} 1 & 0 \\ 0 & 1 \end{pmatrix} \tag{3.30}$$

Notice that the quincunx lattice could be regarded as a regular lattice rotated 45° with some re-dimensioning.

Also, with a simple re-dimensioning of the quincunx lattice, a hexagonal lattice could be obtained, as depicted in Fig. 3.80.

A generating basis for the hexagonal lattice could be:

$$H_0 = \begin{pmatrix} 1 & 1 \\ -\sqrt{3} & \sqrt{3} \end{pmatrix} \tag{3.31}$$

The same lattice has many different bases. For example, the same hexagonal lattice could be generated with:

$$H_1 = \begin{pmatrix} 2 & 1 \\ 0 & \sqrt{3} \end{pmatrix} \tag{3.32}$$

A matrix U is called *unimodular* if it is a square integer matrix with determinant 1 or -1. Two matrices A and B generate the same lattice iif $A = BU$, with U a unimodular matrix.

Given a square non-singular integer matrix A, there exists a unique unimodular matrix U such that $A \cdot U = H$, with H being an integer matrix in *Hermite normal form*. This form is upper triangular, with $h_{ii} > 0$, and $0 \leq h_{ij} < h_{ii}$ for $1 \leq i < j$.

There are only three 2×2 Hermite normal matrices with determinant 2:

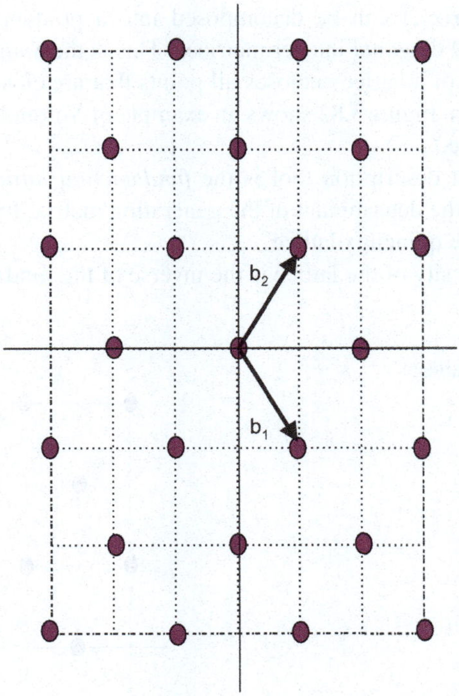

Fig. 3.80 A hexagonal lattice

$$\begin{pmatrix} 1 & 0 \\ 0 & 2 \end{pmatrix}, \begin{pmatrix} 2 & 0 \\ 0 & 1 \end{pmatrix}, \begin{pmatrix} 2 & 1 \\ 0 & 1 \end{pmatrix} \quad (3.33)$$

The first matrix generates a vertical rectangular lattice, the second a horizontal rectangular lattice, and the third a quincunx lattice. Figure 3.81 shows the two variants of the rectangular lattices.

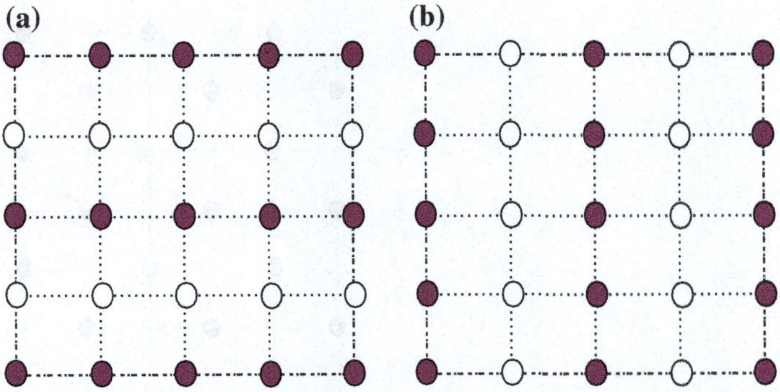

Fig. 3.81 **a** Vertical rectangular lattice, **b** horizontal rectangular lattice

Any integer matrix A can be decomposed into a product UDV, with U and V unimodular, and D diagonal integer matrices. This is the *Smith normal form*.

The Voronoi cell of a lattice encloses all points that are closer to the origin than to other lattice points. Figure 3.82 shows an example of Voronoi cell, corresponding to a hexagonal lattice.

Another pertinent description tool is the *fundamental lattice cell*. The area of this cell is given by the determinant of the generating matrix. Figure 3.83 shows the fundamental cell of a quincunx lattice.

The sampling density of the lattice is the inverse of the fundamental cell area:

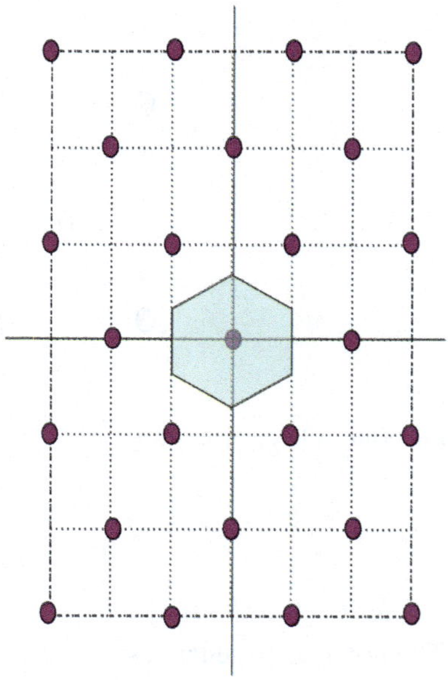

Fig. 3.82 Hexagonal lattice: Voronoi cell

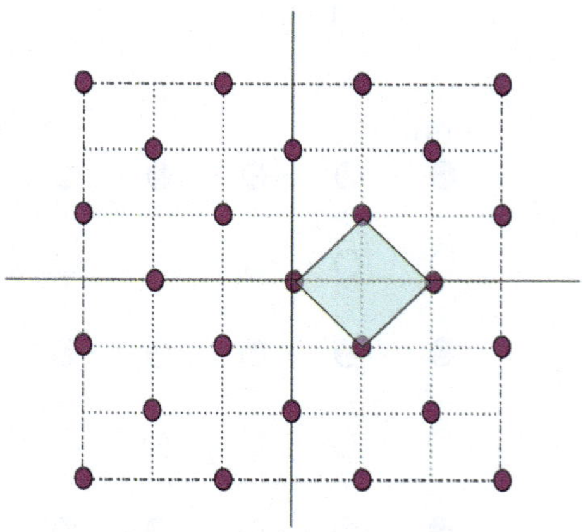

Fig. 3.83 Quincunx lattice: fundamental cell

$$density = \frac{1}{|\det(S)|} \tag{3.34}$$

where S is the generating matrix of the lattice.

The *reciprocal lattice* is a lattice generated by $(S^T)^{-1}$.

More details on lattices can be found in [7, 34].

3.9.1.2 2-D Discrete Signals

A 2-D discrete signal is a function defined over the set of ordered pairs of integers:

$$x = \{x(n_1, n_2);\ n_1, n_2, \text{integers}\} \tag{3.35}$$

Hence, $x(n_1, n_2)$ is the sample of the signal at the point (n_1, n_2). The samples could come in vector (sequence) or matrix (array) format.

An interesting example is the exponential sequence:

$$x = a^{n_1}\, b^{n_2} \tag{3.36}$$

where a and b are complex numbers.

Any 2-D signal that can be written as follows:

$$x\,(n_1,\, n_2) = x_1(n_1)\, \cdot x_2(n_2) \tag{3.37}$$

is a *separable* 2-D signal. An example of this is the exponential sequence.

Given a continuous 2-D signal x_c, a rectangular sampling could be applied to obtain a discrete 2-D signal. Suppose that the continuous signal is bandlimited; then, its support in the 2-D Fourier domain is a circle C, with radius Ω_m. In order to be able to recover x_c from its samples, the sampling (spatial) frequency must be at least twice Ω_m. The support of the sampled signal is, in the 2-D Fourier domain, a set of circles C, placed on the reciprocal lattice.

The spatial sampling frequency is given by the sampling density. Figure 3.84 shows in the 2-D Fourier domain the effect of a good sampling density. The circles C do not overlap.

Although rectangular sampling is very common, there are other periodic sampling patterns being used in practical applications. The most efficient sampling is provided by hexagonal sampling. More efficiency means that signals are represented with fewer sampling points than any other sampling pattern. In particular, the optimal hexagonal lattice has 90.7 % efficiency, while the rectangular lattice has 78.5 % efficiency. This fact can be illustrated considering the Voronoi cell of the reciprocal lattice, and how the circular support of the sampled signal fits inside the cell (see Fig. 3.85).

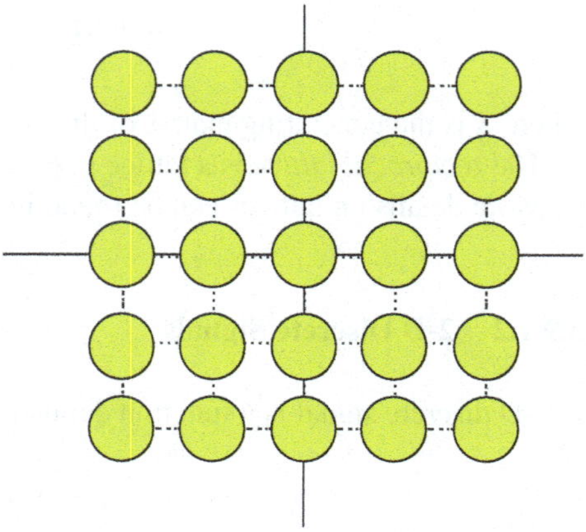

Fig. 3.84 The support, in the 2-D Fourier domain, of the sampled signal

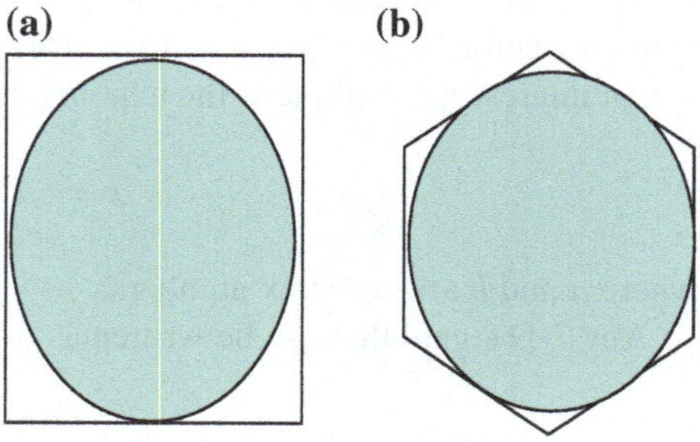

Fig. 3.85 Voronoi cell of the reciprocal lattice and in circle support: a rectangular lattice, b hexagonal lattice

There are experimental scenarios where a set of sensors is placed covering a certain 2-D area. For instance, for seismic studies, or using hydrophones on the sea. Hexagonal sampling is, in such cases, preferred.

3.9.1.3 From Discrete to Discrete

Important operations in the context of 2-D discrete signals are downsampling, and upsampling. These operations can be represented with a square non-singular integer matrix M.

Denote $\mathbf{n} = (n_1, n_2)$. In the case of downsampling, the input $x(\mathbf{n})$ and the output $x_d(\mathbf{n})$ are related by:

$$x_d\,(\mathbf{n}) = x(M \cdot \mathbf{n}) \tag{3.38}$$

In the case of upsampling, the input $x(\mathbf{n})$ and the output $x_u(\mathbf{n})$ are related by:

$$x_u\,(\mathbf{n}) = \begin{cases} x(\mathbf{k}), & if\ \mathbf{n} = M\,\mathbf{k} \\ 0, & otherwise \end{cases} \tag{3.39}$$

where $\mathbf{k}$ is an integer vector.

Fig. 3.86 A typical photograph, chosen for our experiments

Suppose that M is unimodular. In this case sampling does not reject any input sample, it only rearranges the samples. It is usually called a *resampling* operation. Another important fact, when M is unimodular, is that upsampling by M is equivalent to downsampling by M^{-1}, and vice versa.

Interesting examples of unimodular matrices are the following:

$$\begin{aligned} R_0 &= \begin{pmatrix} 1 & 1 \\ 0 & 1 \end{pmatrix}, \; R_1 = \begin{pmatrix} 1 & -1 \\ 0 & 1 \end{pmatrix}, \\ R_2 &= \begin{pmatrix} 1 & 0 \\ 1 & 1 \end{pmatrix}, \; R_3 = \begin{pmatrix} 1 & 0 \\ -1 & 1 \end{pmatrix}, \end{aligned} \tag{3.40}$$

Notice that $R_0\ R_1\ =\ R_2\ R_3\ =\ I$. Then, for example, upsampling by R_0 is equivalent to downsampling by R_1. Any of these four matrices can be used for *shearing* operations.

Let us do an experiment. Figure 3.86 shows a typical photograph, which will be subject to downsampling.

Now, let us apply R_0 for resampling. The effect—a 45° image shearing—is shown in Fig. 3.87.

The resampling has been done with the Program 3.44.

Program 3.44 Image downsampling (causing shearing) example

```
% Image downsampling (causing shearing) example
%read the 256x256 image file into a matrix:
phot=imread('London1.tif');
N=256;
aphot=uint8(255*ones(N,3*N)); %space for padded image
subphot=uint8(255*ones(N,2*N)); %space for sheared image
aphot(1:N,N+1:2*N)=phot(:,:); %photo padding
M=[1 1;0 1]; %downsampling matrix
disp('please wait');
```

```
for nr=1:N, %rows
for nc=1:2*N, %columns
p=M*[nc;nr]; %select a sample
subphot(nr,nc)=aphot(p(2),p(1));
end;
end;
figure(1)
imshow(subphot);
```

Coming back to the original photograph, if one applies a downsampling by Q_1, which is a quincunx downsampling, one obtains a 45° rotation, as depicted in Fig. 3.88. The original photograph size is 256×256; the downsampled image fits in a 256×256 vertical rectangle, so the rotated photograph is smaller (half the area).

Fig. 3.87 Image shearing obtained with resampling

Fig. 3.88 Image rotation and re-sizing obtained with Q_1 downsampling

The downsampling has been done with the Program 3.45.

Program 3.45 Image downsampling (quincunx) example

```
% Image downsampling (quincunx) example
%read the 256x256 image file into a matrix:
phot=imread('London1.tif');
N=256; L=N/2;
aphot=uint8(255*ones(3*N,3*N)); %space for padded image
subphot=uint8(255*ones(N,N)); %space for subsampled image
aphot(N+1:2*N,N+1:2*N)=phot(:,:); %photo padding
M=[1 1;-1 1]; %downsampling matrix
disp('please wait');
for nr=1:N, %rows
for nc=1:N, %columns
p=M*[nc;nr]; %select a sample
subphot(nr,nc)=aphot(N+L+p(2),L+p(1));
end;
end;
figure(1)
imshow(subphot);
```

The quincunx generating matrices:

$$Q_0 = \begin{pmatrix} 1 & -1 \\ 1 & 1 \end{pmatrix}; \quad Q_1 = \begin{pmatrix} 1 & 1 \\ -1 & 1 \end{pmatrix} \tag{3.41}$$

can be written in the Smith normal form as follows:

$$Q_0 = R_1 \, D_1 \, R_2 = R_2 \, D_2 \, R_1 \tag{3.42}$$

$$Q_1 = R_0 \, D_1 \, R_3 = R_3 \, D_2 \, R_0 \tag{3.43}$$

where:

$$D_1 = \begin{pmatrix} 2 & 0 \\ 0 & 1 \end{pmatrix}; \quad D_2 = \begin{pmatrix} 1 & 0 \\ 0 & 2 \end{pmatrix} \tag{3.44}$$

3.9.1.4 Filters

The response of a linear shift-invariant system can be computed with a 2-D convolution product:

$$y(n_1, n_2) = \sum_{k_1=-\infty}^{\infty} \sum_{k_2=-\infty}^{\infty} u(k_1, k_2) \, h(n_1 - k_1, \, n_2 - k_2) \tag{3.45}$$

where $u()$ is the input, $y()$ is the output, and $h()$ is the impulse response. Shift-invariant systems are systems such that the response corresponding to $u(n_1 - l_1, n_2 - l_2)$, is $y(n_1 - l_1, n_2 - l_2)$ for all inputs $u()$ and shifts $(l_1, \, l_2)$.

A *separable* system is a system with separable impulse response, that is:

$$h(n_1, n_2) = h(n_1)\, h(n_2) \tag{3.46}$$

The frequency response of a linear filter is:

$$H(\omega_1, \omega_2) = \sum_{n_1=-\infty}^{\infty} \sum_{n_2=-\infty}^{\infty} h(n_1, n_2) \cdot e^{-j(\omega_1 n_1 + \omega_2 n_2)} \tag{3.47}$$

For example, suppose that the impulse response of the filter is the case depicted in Fig. 3.89.

Then, the frequency response of this filter is:

$$\begin{aligned} H(\omega_1, \omega_2) &= e^{j\omega_1} + e^{-j\omega_1} + e^{j\omega_2} + e^{-j\omega_2} = \\ &2(\cos\omega_1 + \cos\omega_2) \end{aligned} \tag{3.48}$$

The inverse 2-D Fourier transform can be used to compute the impulse response corresponding to desired frequency responses. For example, consider the frequency response depicted in Fig. 3.90.

The inverse Fourier transform is:

$$\begin{aligned} h(n_1, n_2) &= \frac{1}{4\pi^2} \int_{-a}^{a} \int_{-b}^{b} e^{j(\omega_1 n_1 + \omega_2 n_2)}\, d\omega_1\, d\omega_2 = \\ &= \left(\frac{1}{2\pi} \int_{-a}^{a} e^{j\omega_1 n_1}\, d\omega_1\right) \left(\frac{1}{2\pi} \int_{-b}^{b} e^{j\omega_2 n_2}\, d\omega_2\right) = \frac{\sin(a\, n_1)}{\pi\, n_1} \cdot \frac{\sin(b\, n_2)}{\pi\, n_2} \end{aligned} \tag{3.49}$$

It has been easy to calculate this inverse Fourier transform, since the filter is separable. Figure 3.91 shows a top view of the impulse response corresponding to the rectangular filter in Fig. 3.90.

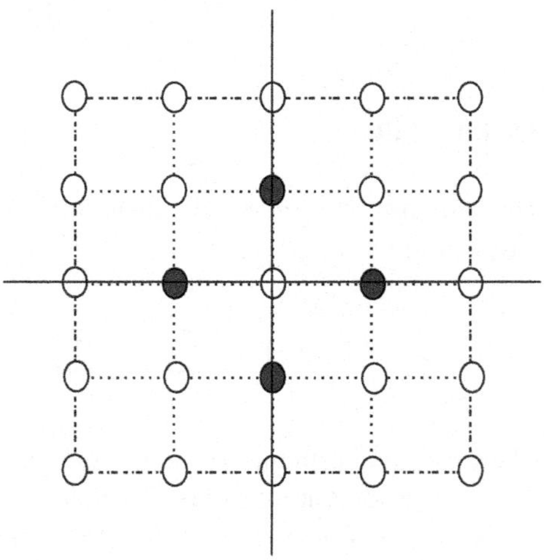

Fig. 3.89 Example of impulse response of a 2-D filter

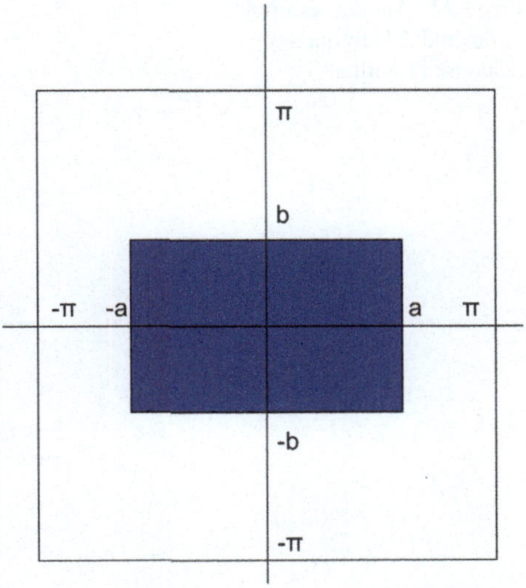

Fig. 3.90 Example of desired 2-D frequency response of a filter

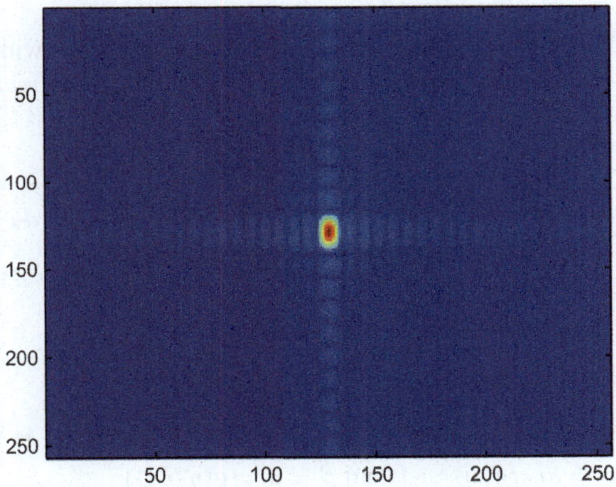

Fig. 3.91 Impulse response corresponding to the rectangular filter

Program 3.46 Impulse response of the rectangular filter

```
% Impulse response of the rectangular filter
%the rectangular filter in 2-D Fourier domain
a=20; b=10; %rectangle half sides
Ffilt=zeros(256,256);
Ffilt((129-b):(128+b),(129-a):(128+a))=1;
%inverse transform
hfilt=ifftshift(ifft2(Ffilt));
ah=abs(hfilt);
figure(1)
imagesc(ah);
title('Impulse response');
```

Fig. 3.92 Another example of desired 2-D frequency response of a filter

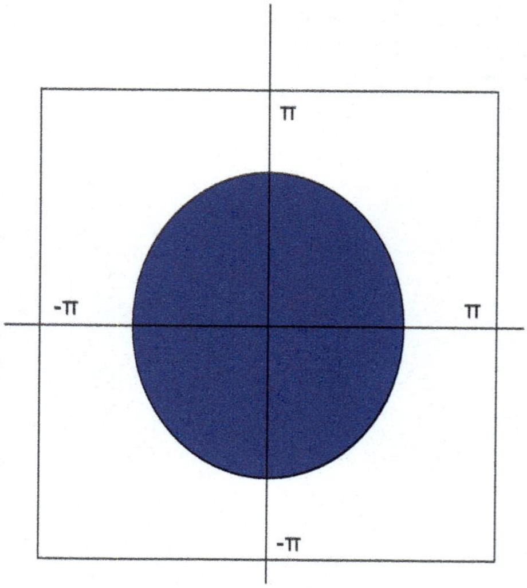

Consider another example. It is a circular filter, with frequency response as represented in Fig. 3.92. *Separable filters can only have rectangular responses*; therefore, the circular filter is nonseparable.

The inverse Fourier transform is:

$$\begin{aligned} h(n_1, n_2) &= \frac{1}{4\pi^2} \int_{-a}^{a} \int_{-b}^{b} e^{j(\omega_1 n_1 + \omega_2 n_2)}\, d\omega_1\, d\omega_2 = \\ &= \frac{1}{4\pi^2} \int_0^R \int_0^{2\pi} \omega \exp\left[j\omega\sqrt{n_1^2 + n_2^2} \cos(\beta - \phi) \right] d\phi\, d\omega = \\ &= \frac{R}{2\pi} \cdot \frac{J_1(R\sqrt{n_1^2+n_2^2})}{\sqrt{n_1^2+n_2^2}} \end{aligned} \tag{3.50}$$

where $J_1()$ is the Bessel function of order 1, R is the radius, $\omega = \sqrt{\omega_1^2 + \omega_2^2}$, $\phi = arctg(\omega_2/\omega_1)$, and $\beta = arctg(n_2/n_1)$.

Figure 3.93 shows a 3-D view of the impulse response corresponding to the circular filter. The circle has been narrowed perhaps too much, but for good illustrative reasons: the 3-D view clearly shows the undulations of the impulse response.

Program 3.47 Impulse response of the circular filter

```
% Impulse response of the circular filter
%the circular filter in 2-D Fourier domain
R=4; %radius
rs=R^2;
Ffilt=zeros(256,256);
for nx=1:256,
for ny=1:256,
pr=((nx-128)^2)+((ny-128)^2);
if pr <=rs,
```

```
Ffilt(nx,ny)=1;
end;
end;
end;
%inverse transform
hfilt=ifftshift(ifft2(Ffilt));
ah=abs(hfilt);
figure(1)
mesh(ah);
title('Impulse response');
```

3.9.1.5 Filter Banks

Figure 3.94 shows the Noble identities when using the downsampling or upsampling matrix M.

An important tool for analytical purposes is the polyphase representation. The polyphase components with respect to the matrix M are defined as follows:

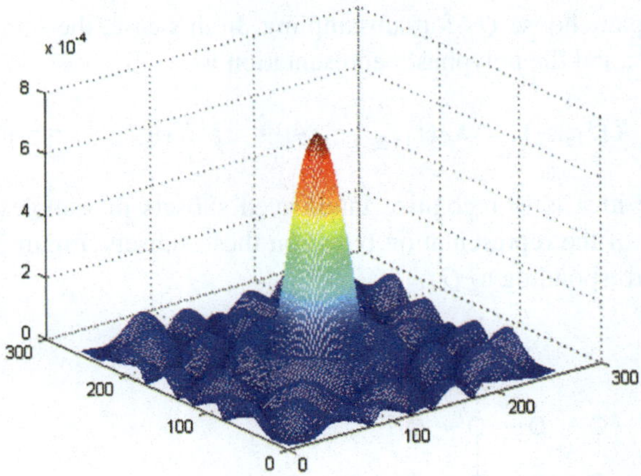

Fig. 3.93 Impulse response corresponding to the circular filter

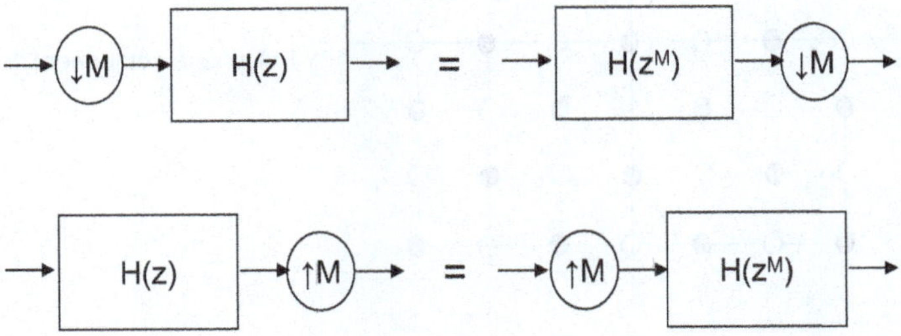

Fig. 3.94 Noble identities

$$x_i(\mathbf{n}) = x(M\,\mathbf{n} + \mathbf{l}_i) \tag{3.51}$$

The set of integer vectors $\mathbf{l}_i$ has K members, where $K = det(M)$. This set is contained in the fundamental cell of the lattice generated by M.

In the z-domain, the polyphase representation is:

$$X(\mathbf{z}) = \sum_{i=0}^{K-1} (\mathbf{z})^{-\mathbf{l}_i} X_i(\mathbf{z}^M) \tag{3.52}$$

For example, suppose that:

$$M = \begin{pmatrix} 2 & 0 \\ 0 & 2 \end{pmatrix} \tag{3.53}$$

In this case, the set of vectors $\mathbf{l}_i$ is $\{(0,0)^T, (1,0)^T, (0,1)^T, (1,1)^T\}$, and the z-domain representation can be written as follows:

$$\begin{aligned} X(z_1, z_2) = X_0(z_1^2, z_2^2) + z_1^{-1} X_1(z_1^2, z_2^2) + \\ + z_2^{-1} X_2(z_1^2, z_2^2) + z_1^{-1} z_2^{-1} X_3(z_1^2, z_2^2) \end{aligned} \tag{3.54}$$

Another example: choose Q_1 for subsampling. In this case, the set of vectors $\mathbf{l}_i$ is: $(0,0)^T$, $(1,0)^T$; and the polyphase representation is:

$$X(z_1, z_2) = X_0(z_1 z_2^{-1}, z_1 z_2) + z_1^{-1} X_1(z_1 z_2^{-1}, z_1 z_2) \tag{3.55}$$

In general, the idea is to recognize interleaved subsets of samples, and obtain a decomposition of the representation based on these subsets. Figure 3.95 shows the two subsets corresponding to Q_1.

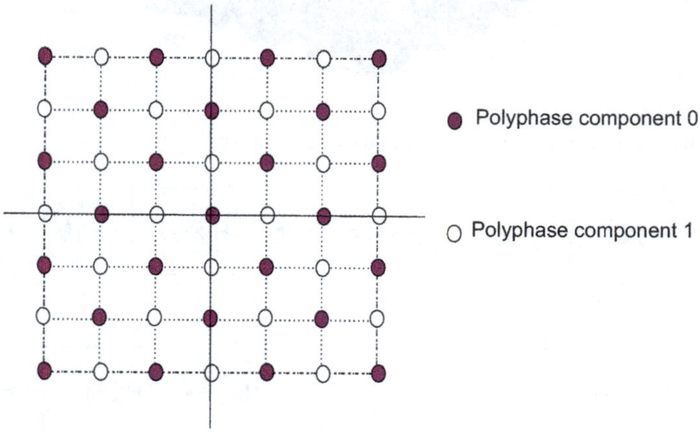

Fig. 3.95 Polyphase components for the quincunx case

In the next paragraphs, the target is to analyse filter banks composed by several branches, according with typical structures. Let us begin by looking at the filter branch depicted in Fig. 3.96.

Using polyphase representation, the output of the filter branch is:

$$\begin{aligned} y(\mathbf{n}) &= \sum_{\mathbf{m}} u(\mathbf{m})\, h(M\,\mathbf{n} - \mathbf{m}) = \\ &= \sum_{i=0}^{K-1} \sum_{\mathbf{p}} u(M\,\mathbf{p} + \mathbf{l}_i)\, h(M\,\mathbf{n} - M\,\mathbf{p} - \mathbf{l}_i) = \\ &= \sum_{i=0}^{K-1} \sum_{\mathbf{p}} u_i(\mathbf{p})\, h_i(\mathbf{n} - \mathbf{p}) \end{aligned} \tag{3.56}$$

This is equivalent in the z-domain to:

$$Y(\mathbf{z}) = \sum_{i=0}^{K-1} H_i(\mathbf{z})\, U_i(\mathbf{z}) = \mathbf{H}(\mathbf{z})\, \mathbf{u}(\mathbf{z}) \tag{3.57}$$

where:

$$\mathbf{u}(\mathbf{z}) = (U_0(\mathbf{z}), U_1(\mathbf{z}), \ldots, U_{K-1}(\mathbf{z}))^T \tag{3.58}$$

$$\mathbf{H}(\mathbf{z}) = (H_0(\mathbf{z}), H_1(\mathbf{z}), \ldots, H_{K-1}(\mathbf{z})) \tag{3.59}$$

In the case of the filter branch depicted in Fig. 3.97.

The output of this branch is:

$$\mathbf{r}(\mathbf{z}) == \mathbf{G}(\mathbf{z})\, Y(\mathbf{z}) \tag{3.60}$$

where:

$$\mathbf{G}(\mathbf{z}) = (G_0(\mathbf{z}), G_1(\mathbf{z}), \ldots, G_{K-1}(\mathbf{z}))^T \tag{3.61}$$

$$\mathbf{r}(\mathbf{z}) = (R_0(\mathbf{z}), R_1(\mathbf{z}), \ldots, R_{K-1}(\mathbf{z}))^T \tag{3.62}$$

Consider now a general filter structure, with an analysis part followed by a synthesis part (Fig. 3.98).

The equivalent filter bank, in the polyphase domain is shown in Fig. 3.99.

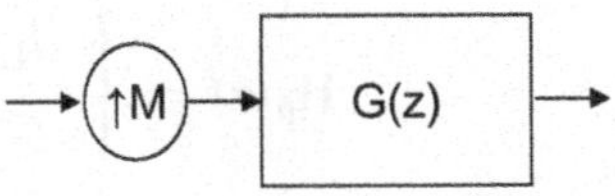

Fig. 3.96 A filter branch

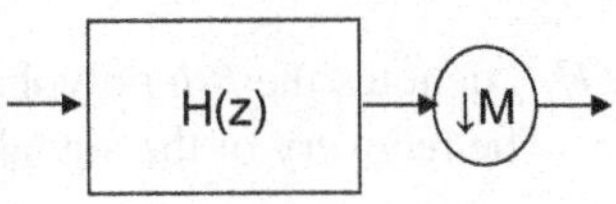

Fig. 3.97 Another filter branch

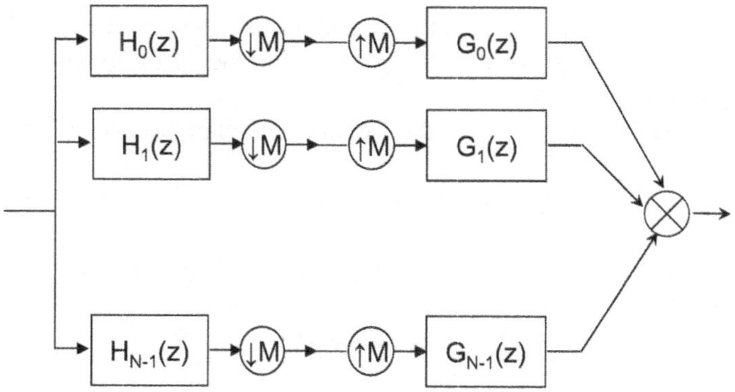

Fig. 3.98 A filter structure

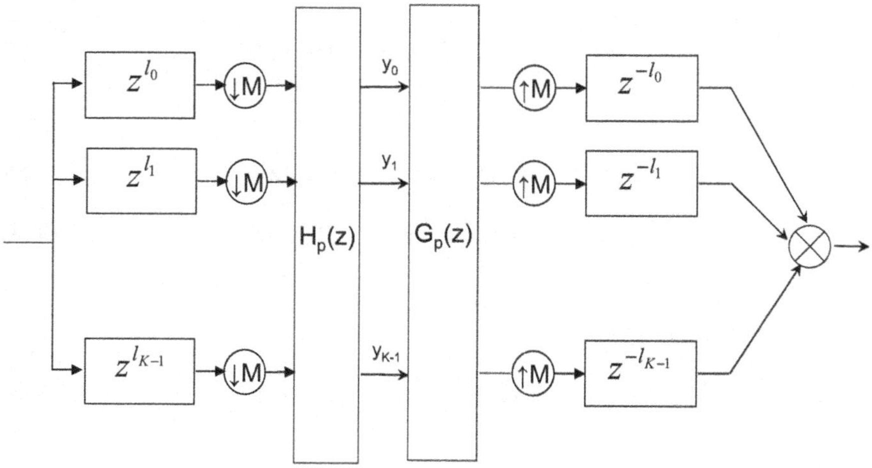

Fig. 3.99 The equivalent filter structure in the polyphase domain

The outputs of the analysis part can be obtained from:

$$\mathbf{y}(\mathbf{z}) = \mathbf{H_p}(\mathbf{z})\, \mathbf{u}(\mathbf{z}) \tag{3.63}$$

where:

$$\mathbf{H_p}(\mathbf{z}) = \begin{pmatrix} H_{0,0}(\mathbf{z}) & H_{0,1}(\mathbf{z}) & \ldots & H_{0,K-1}(\mathbf{z}) \\ H_{1,0}(\mathbf{z}) & H_{1,1}(\mathbf{z}) & \ldots & H_{1,K-1}(\mathbf{z}) \\ \vdots & \vdots & & \vdots \\ H_{N-1,0}(\mathbf{z}) & H_{N-1,1}(\mathbf{z}) & \ldots & H_{N-1,K-1}(\mathbf{z}) \end{pmatrix} \tag{3.64}$$

($H_{i,j}$ denotes the *j-th* polyphase component of the filter H_i)

The recovery of the signal by the synthesis part can be obtained with:

$$\hat{\mathbf{u}}(\mathbf{z}) = \mathbf{G_p}(\mathbf{z})\,\mathbf{y}(\mathbf{z}) \tag{3.65}$$

where:

$$\mathbf{G_p}(\mathbf{z}) = \begin{pmatrix} G_{0,0}(\mathbf{z}) & G_{1,0}(\mathbf{z}) & G_{N-1,0}(\mathbf{z}) \\ G_{0,1}(\mathbf{z}) & G_{1,1}(\mathbf{z}) & G_{N-1,1}(\mathbf{z}) \\ \vdots & \vdots & \vdots \\ G_{0,K-1}(\mathbf{z}) & G_{1,K-1}(\mathbf{z}) & G_{N-1,K-1}(\mathbf{z}) \end{pmatrix} \tag{3.66}$$

The filter bank has perfect reconstruction (PR), if:

$$\mathbf{G_p}(\mathbf{z})\,\mathbf{H_p}(\mathbf{z}) = I \tag{3.67}$$

If $K = N$ then the filter bank is critically sampled. In this case, $\mathbf{H_p}$ and $\mathbf{G_p}$ are square.

A critically sampled filter bank which has PR, is a biorthogonal filter bank. It is orthogonal if the analysis and synthesis filters are related as follows:

$$h_i(\mathbf{n}) = g_i^*(-\mathbf{n}) \tag{3.68}$$

Real-valued filter banks are orthogonal iif:

$$\mathbf{H_p}(\mathbf{z}) = \mathbf{G_P}(\mathbf{z}^{-1})^T \tag{3.69}$$

and:

$$\mathbf{G_P}(\mathbf{z})\,\mathbf{G_P}(\mathbf{z}^{-1})^T = \mathbf{G_P}(\mathbf{z}^{-1})^T\mathbf{G_P}(\mathbf{z}) = I \tag{3.70}$$

In the particular case of two-channel filter banks, the PR condition for the critically sampled filter bank is:

$$\begin{pmatrix} H_{0,0}(\mathbf{z}) & H_{0,1}(\mathbf{z}) \\ H_{1,0}(\mathbf{z}) & H_{1,1}(\mathbf{z}) \end{pmatrix} \begin{pmatrix} G_{0,0}(\mathbf{z}) & G_{0,1}(\mathbf{z}) \\ G_{1,0}(\mathbf{z}) & G_{1,1}(\mathbf{z}) \end{pmatrix} = I \tag{3.71}$$

This condition implies that:

$$H_{0,0}(\mathbf{z})\,G_{0,0}(\mathbf{z}) + H_{0,1}(\mathbf{z})\,G_{0,1}(\mathbf{z}) = 1 \tag{3.72}$$

If all filters are FIR, then $\det(\mathbf{G_p}(\mathbf{z})) = \alpha\,\mathbf{z^m}$, for some α real. It can be deduced that:

$$[H_{0,0}(\mathbf{z}) \;\; H_{1,1}(\mathbf{z})] = \alpha^{-1}\,\mathbf{z^{-m}}\,[-G_{0,1}(\mathbf{z}) \;\; G_{0,0}(\mathbf{z})] \tag{3.73}$$

$$[G_{1,0}(\mathbf{z}) \;\; G_{1,1}(\mathbf{z})] = \alpha\,\mathbf{z^m}\,[-H_{0,1}(\mathbf{z}) \;\; H_{0,0}(\mathbf{z})] \tag{3.74}$$

The design of two-channel FIR filter banks could proceed in two steps: first accomplish condition (3.72), and second fulfil conditions (3.73) and (3.74).

3.9.2 Design of 2D Filters

In previous sections the subject of 2-D filtering has been treated using neighbours, or using 2-D Fourier transform. There still remains an important topic to be considered: the design on the z-domain.

The most popular design approach is to design a 1-D filter prototype, and then use a mapping from 1-D to 2-D to obtain a 2-D filter with the desired characteristics.

Originally the mapping approach was proposed for FIR filters, see [33, 36] and references therein. Soon, it was also applied for IIR filters, in search of better efficiency [1, 31, 43].

This subsection starts with a summary of the McClelland transformation, which was introduced as soon as 1973. Other parts of the subsection consider generalizations of the transformation idea, and other design approaches. The subsection is based on [6, 15, 30, 41].

Diamond and fan filters are a frequent target of filter design. They can be easily transformed one into the other. Figure 3.100 shows typical supports of these filters in the 2-D Fourier domain:

Figure 3.101 shows two examples of quadrant filters. Some authors consider these filters as quadrant fan filters [29]. They can be easily obtained by transformation of the fan or the diamond filter.

Finally, Fig. 3.102 shows a general fan filter.

There are many other examples of 2-D filters. Being so, fan filters attract a lot of interest for directional filter applications.

Diamond and fan filters are nonseparable filters. Given a $N \times M$ 2-D filter, if it is separable it offers $N + M$ degrees of freedom for design; if it is nonseparable, it offers $N \times M$ degrees of freedom: much more flexibility.

In order to demonstrate the effect of the fan filter, a simple Program 3.48 has been created that applies this filter to the typical photograph used before. The filter has

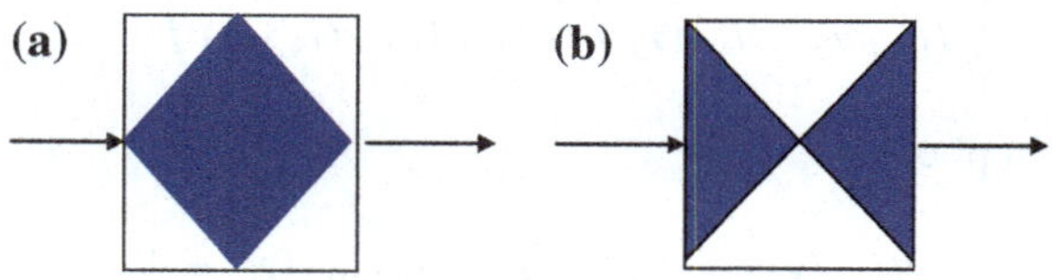

Fig. 3.100 Examples of **a** diamond filter, **b** fan filter

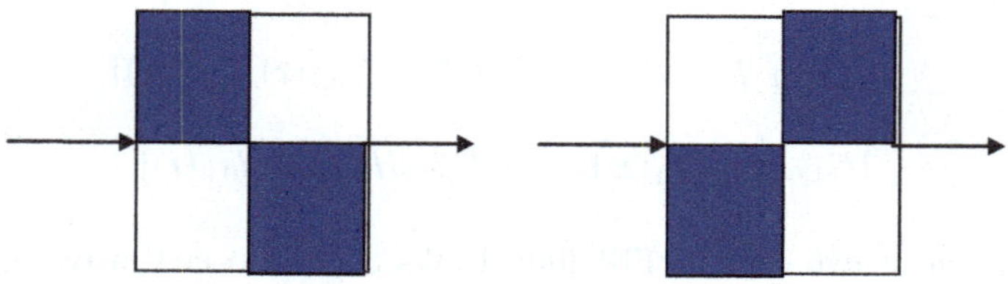

Fig. 3.101 Examples of quadrant filters

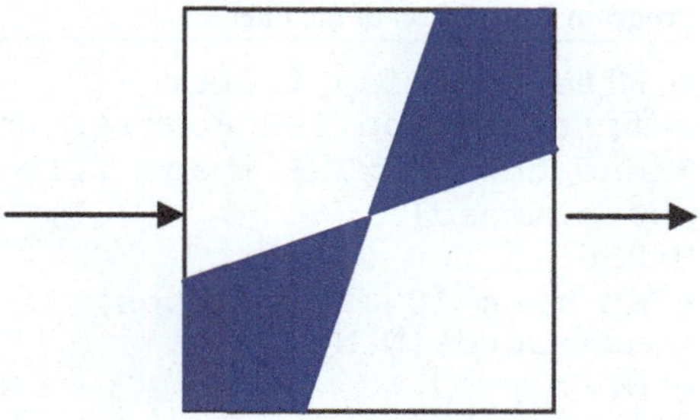

Fig. 3.102 A general fan filter

Fig. 3.103 The fan-shaped mask

been implemented using a fan shaped mask in the 2-D Fourier domain. Figure 3.103 shows the mask.

Figure 3.104 shows the filtered image.

Fig. 3.104 The filtered photograph

Program 3.48 Effect of fan filter

```
% Effect of fan filter
% filtering on 2-D Fourier domain
%read the 256x256 image file into a matrix:
phot=imread('London1.tif');
N=256;
%Fan mask (0 or 1 values)
fmask=zeros(N,N);
L=N/2;
for nh=1:L,
nv=1:nh;
fmask(L+nv,L+nh)=1;
end;
fmask((L+1):N,L:-1:1)=fmask((L+1):N,(L+1):N);
fmask(L:-1:1,L:-1:1)=fmask((L+1):N,(L+1):N);
fmask(L:-1:1,(L+1):N)=fmask((L+1):N,(L+1):N);
%filtering (Fourier domain)
Fph=fftshift(fft2(phot));
Fphfil=Fph.*fmask;
Iph=ifft2(Fphfil);
uIp=uint8(abs(Iph));
%display
figure(1)
ufmask=256*(1-fmask); %color inversion
imshow(ufmask);
title('Fan filter support');
figure(2)
imshow(uIp);
title('Filtered image');
```

3.9.2.1 The McClelland Transformation

The frequency response of a zero-phase symmetric filter can be written as follows:

$$H(\omega) = \sum_{k=-L}^{L} a(k)\,\cos(k\omega) \tag{3.75}$$

where $a(0) = h(0)$ and $a(k) = 2\,h(k)$ for $k \neq 0$.

One can replace $\cos(k\omega)$ by $T_k(\cos\omega)$, where $T_k()$ is the *n-th* Chebyshev polynomial. Hence:

$$H(\omega) = \sum_{k=-L}^{L} a(k)\,T_k(\cos\omega) \tag{3.76}$$

Now, the idea is to replace $\cos \omega$ by a zero-phase 2-D filter $F(\omega_1, \omega_2)$. In this way, a zero-phase 2-D filter is obtained:

$$H(\omega) = \sum_{k=-L}^{L} a(k)\, T_k(F(\omega_1, \omega_2)) \tag{3.77}$$

In the case of quincunx downsampling, it is usual to take:

$$F(\omega_1, \omega_2) = \frac{1}{2}(\cos \omega_1 + \cos \omega_2) \tag{3.78}$$

Supposing that the target is to design a two-channel PR filter bank, the suggested procedure would be: first, design $H_0(z)$ and $G_0(z)$ such that $D(z) = H_0(z)\, G_0(z)$ is halfband; second, apply the McClelland transformation to $H_0(z)$ and $G_0(z)$. This results in two 2-D filters that guarantee PR (the other two 2-D filters are easily derived).

3.9.2.2 Other Mapping Approaches

Similarly to the above transformation, one could take a zero-phase symmetric filter $H(z)$ and formulate it in function of $(z + z^{-1})$; that is:

$$H(z) = \tilde{H}(z + z^{-1}) = \tilde{H}(x). \tag{3.79}$$

Then, one applies the following transformation:

$$H(z_1, z_2) = \tilde{H}(F(z_1, z_2)) \tag{3.80}$$

An appropriate choice of $F(z_1, z_2)$ is:

$$F(z_1, z_2) = \frac{1}{2}(z_1 + z_1^{-1} + z_2 + z_2^{-1}) \tag{3.81}$$

With this approach, if $H(z)$ has a brickwall frequency response, then $H(z_1, z_2)$ has a diamond support (is a diamond filter).

A more general view of the mapping technique is the following. Consider the PR condition for the 1-D prototype filters:

$$H_0(z)\, G_0(z) + H_0(-z)\, G_0(-z) = 2 \tag{3.82}$$

A change of variables is applied $z \rightarrow F(z_1, z_2)$. The mapping $F(z_1, z_2)$ is chosen so that 2-D PR is accomplished:

$$H_0(z_1, z_2)\, G_0(z_1, z_2) + H_0(-z_1, s\, z_2)\, G_0(-z_1, s\, z_2) = 2 \tag{3.83}$$

where $s = 1$ if the sampling lattice is rectangular, or $s = -1$ if it is a quincunx lattice.

For the 2-D filter bank to be PR:

$$F(z_1, z_2) + F(-z_1, s\,z_2) = 0 \tag{3.84}$$

It is convenient to use for $F(z_1, z_2)$ the product of two 1-D filters.

3.9.2.3 An Example

An interesting example of mapping technique is provided in [26]. The proposed method starts from a given analog prototype filter, like, for instance $H_b(s) = \alpha/(s+\alpha+j\beta)$, or $H_r(s) = \alpha\,s/(s^2+\alpha\,s+\omega_0^2)$. The frequency response of the prototype filter is obtained with $s \to j\omega$. Then, the following mapping from 1-D to 2-D is proposed:

$$\omega \;\to\; f_\varphi(\omega_1, \omega_2) = \frac{a \cdot (\omega_1 \cos\varphi \,-\, \omega_2 \sin\varphi)}{(\omega_1 \cos\varphi \,+\, \omega_2 \sin\varphi)} \tag{3.85}$$

With this mapping a general fan filter is obtained, with an aperture angle θ and direction angle φ. The parameter a is the aperture coefficient, given by $a = 1/tg(\theta/2)$. By using $s_1 = j\omega_1$ and $s_2 = j\omega_2$, the mapping can be expressed as:

$$s \;\to\; f_\varphi(s_1, s_2) = \frac{ja \cdot (s_1 \cos\varphi \,-\, s_2 \sin\varphi)}{(s_1 \cos\varphi \,+\, s_2 \sin\varphi)} \tag{3.86}$$

The conventional way to obtain a discrete version of the transformed prototype would be to apply a bilinear transform. But this may cause distortion due to warping. To prevent this problem, a second transformation is applied:

$$\omega_1 \;\to\; 2\,arctg(\omega_1\,/2); \quad \omega_2 \;\to\; 2\,arctg(\omega_2\,/2) \tag{3.87}$$

It is possible to get a good approximation of $arctg()$:

$$arctg(\omega\,/2) \approx 0.4751\, \frac{\omega}{1+0.05\,\omega^2} \tag{3.88}$$

Combining all that, the proposed transformation is:

$$s \;\to\; f_\varphi(s_1, s_2) = \frac{ja \cdot (s_1\,(1-0.05\,s_2^2)\cos\varphi \,-\, s_2\,(1-0.05\,s_1^2)\sin\varphi)}{(s_1\,(1-0.05\,s_2^2)\cos\varphi \,+\, s_2\,(1-0.05\,s_1^2)\sin\varphi)} \tag{3.89}$$

Next, the discretized version is obtained with the bilinear transform:

$$s_1 = 2(z_1-1)/(z_1+1)\;,\;\; s_2 = 2(z_2-1)/(z_2+1) \tag{3.90}$$

The complete transformation can be written in matrix form:

$$s \;\rightarrow\; F_{\varphi}(z_1,\, z_2) \;=\; k\,\frac{[z_1^{-1}\ 1\ z_1]\,P\,[z_2^{-1}\ 1\ z_2]^T}{[z_1^{-1}\ 1\ z_1]\,Q\,[z_2^{-1}\ 1\ z_2]^T} \tag{3.91}$$

where:

$$P \;=\; \cos\varphi \begin{pmatrix} -1 & -3 & -1 \\ 0 & 0 & 0 \\ 1 & 3 & 1 \end{pmatrix} - \sin\varphi \begin{pmatrix} -1 & 0 & 1 \\ -3 & 0 & 3 \\ -1 & 0 & 1 \end{pmatrix} \tag{3.92}$$

$$Q \;=\; \sin\varphi \begin{pmatrix} -1 & -3 & -1 \\ 0 & 0 & 0 \\ 1 & 3 & 1 \end{pmatrix} + \cos\varphi \begin{pmatrix} -1 & 0 & 1 \\ -3 & 0 & 3 \\ -1 & 0 & 1 \end{pmatrix} \tag{3.93}$$

Let us apply the transformation to the simplest prototype, $H_b(s)$. The result is a 2-D filter with the following expression:

$$H_b(z_1, z_2) = \frac{(\mathbf{Z}_1\, B_b\, \mathbf{Z}_2^T)}{(\mathbf{Z}_1\, A_b\, \mathbf{Z}_2^T)} \tag{3.94}$$

where:

$$\mathbf{Z}_1 = [1\ \ z_1\ \ z_1^2];\ \ \mathbf{Z}_2 = [1\ \ z_2\ \ z_2^2] \tag{3.95}$$

and:

$$B_b \;=\; \alpha\, Q;\ \ A_b = \alpha\, Q \,+\, j\,(a\, P \,+\, \beta\, Q) \tag{3.96}$$

This example of transform has been implemented in a program, which has been included in Appendix A. Figure 3.105 shows the frequency response of the general fan filter, as specified in the program.

Fig. 3.105 The general fan filter frequency response

Fig. 3.106 The image before filtering

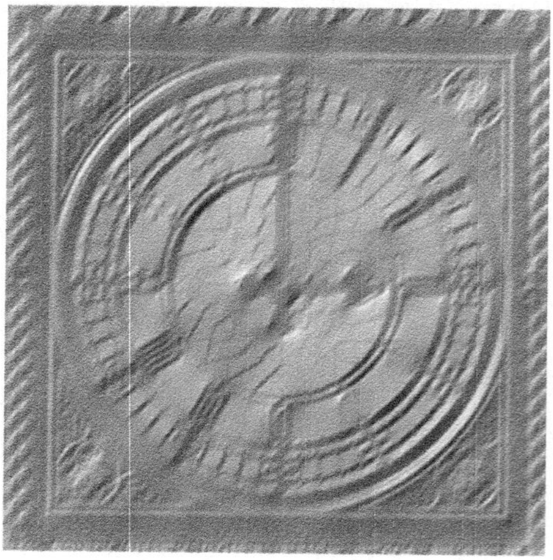

Fig. 3.107 The filtered photograph

The filter has been applied to the image depicted in Fig. 3.106. This Fig. 3.106 has been selected for better illustration of the directional effect of the filter.

Figure 3.107 shows the filtered image. Notice the blurring of details, except for minutes 5–10 (approx.) and other traits parallel to the white line of the filter frequency response.

3.9.2.4 Other Approaches

Based on the polyphase components of a 1D filter, and multipying these components, it is possible to build a multi-dimensional filter with separable polyphase components [41]. For example, suppose a 1D filter with two polyphase components, $H_0(z)$

and $H_1(z)$, so the filter is $H(z) = H_0(z^2) + z^{-1}H_1(z^2)$. The following separable polyphase components can be devised:

$$H_0(z_1, z_2) = H_0(z_1)H_0(z_2) \tag{3.97}$$

$$H_1(z_1, z_2) = H_1(z_1)H_1(z_2) \tag{3.98}$$

Now, by upsampling with respect to the quincunx lattice, the following 2D filter can be obtained:

$$H(z_1, z_2) = H_0(z_1 z_2)H_0(z_1 z_2^{-1}) + z_1^{-1}H_1(z_1 z_2)H_1(z_1 z_2^{-1}) \tag{3.99}$$

An example of this approach is included in [40] for high-definition television. In this example a very good 7 × 7 diamond filter is designed, departing from the following 1D filter:

$$H(z) = -1 + 9z^{-2} + 16\,z^{-3} + 9\,z^{-4} - z^{-6} \tag{3.100}$$

Since an orthogonal filter 1D bank is mapped into an orthogonal 2D bank iff the polyphase components of the 1D filter are allpass functions; it is remarked in [41] that in general PR is not achieved.

Another design approach is to use cascade structures. Cascades of orthogonal matrices and/or diagonal delay matrices yield orthogonal filter banks. For example [41], a separable design can be based on the following:

$$\mathbf{H_p}(z_1, z_2) = \left[\prod_{i=K-1}^{1} R_i D(z_1, z_2)\right] S_0 \tag{3.101}$$

where D is the matrix of delays (its diagonal is $(1\ z_1^{-1}\ z_2^{-1}\ (z_1 z_2)^{-1}\ \ldots)$), and both R_i and S_0 are scalar persymmetric matrices (that is: they are square and symmetric in the northeast to southwest diagonal). With R_i being also unitary, it is possible to obtain linear phase and orthogonal filters; in particular with the following specifications:

$$R_i = \frac{1}{2}\begin{pmatrix} I & \\ & J \end{pmatrix}\begin{pmatrix} I & I \\ I & -I \end{pmatrix}\begin{pmatrix} R_{2i} & \\ & R_{2i+1} \end{pmatrix}\begin{pmatrix} I & I \\ I & -I \end{pmatrix}\begin{pmatrix} I & \\ & J \end{pmatrix} \tag{3.102}$$

$$S_0 = \frac{1}{2}\begin{pmatrix} R_0 & \\ & R_1 \end{pmatrix}\begin{pmatrix} I & I \\ I & -I \end{pmatrix}\begin{pmatrix} I & \\ & J \end{pmatrix} \tag{3.103}$$

where R_{2i} and R_{2i+1} are 2 × 2 rotation matrices.

This 2D example is connected with linear phase and orthonormal wavelets.

Continuing with cascade structures, a second example [41] is the following:

$$\mathbf{H_p}(z_1, z_2) = \left[\prod_{i=K-1}^{1} R_{2i} \begin{pmatrix} 1 & 0 \\ 0 & z_2^{-1} \end{pmatrix} R_{1i} \begin{pmatrix} 1 & 0 \\ 0 & z_1^{-1} \end{pmatrix} \right] R_0 \tag{3.104}$$

(a quincunx cascade)

The matrices R_j have to be unitary for getting orthogonal filters. They have to be symmetric for achieving linear phase filters. See [21] for design examples. With some additional requirements, this cascade will correspond to Daubechies D2 wavelet.

3.10 Nonequispaced Data and the Fourier Transform

Some alarm systems should wait for a key event to occur, like for instance an earthquake, and in such case start the gathering of many samples along the event. This is an example of irregular sampling.

In certain important applications one has nonequispaced data. The Fourier transform can be adapted for this situation.

In this context, two main research paths could be identified:

1. Keep the polar *modus vivendi*
2. Consider the general case of nonequally spaced samples.

Let us introduce those two alternatives.

3.10.1 Fourier Transform Versions for the Polar Context

An application example where is natural to employ polar representations is nuclear magnetic resonance (NMR). It is found in [5] a polar Fourier transform (PFT), on the basis of the '*radial sampling*' proposed by [24]. One of the aspects discussed in [5] is that its is better a radial form of the PFT—evaluating first with respect to the radius and then with respect to the angle-than an azimuthal form—which follows the opposite order. An important comment of [5] is that in polar representations the sample density is larger near the center, while this density is uniform in a Cartesian grid. The Thesis [Bl] comprises an extensive study of radial sampling in NMR.

The *radial PFT* presented by [5] has the following expression:

$$F(X, Y) = \int_0^{2\pi} \int_0^{\infty} f(r, \theta)\, e^{-j\,2\pi\, r\, R'}\, r\, dr\, d\theta \tag{3.105}$$

where: $R' = X \cos\theta + Y \sin\theta$ and X, Y, are frequency domain coordinates.

A pseudo-polar Fourier transform (PPFT) has been introduced in [2]. In this article the question is how to obtain the PFT when the data form a regular Cartesian grid.

It is proposed to proceed along two steps: the first step is to apply a PPFT, using a pseudo-polar sampling; the second step is a conversion from PPFT to PFT. The first step involves the use of chirp-z transform, or, equivalently, the fractional Fourier transform (FRFT). The conversion from PPFT to PFT is described in detail: it is a matter of interpolations via 1D operations, with no accuracy loss.

Figure 3.108 shows a pseudo-polar sample grid. This grid was proposed for a direct and inverse Radon transform called '*Fast Slant-Stack*' (see references in [3]).

According with the comments of [4], the new PPFT has been well received. Since the research keeps moving, [17] has proposed an improvement: to use an octa-polar (OP) grid (Fig. 3.109), which is nearest to the polar coordinates than the pseudo-polar grid. The OP-FT seems to be promising.

Continuing with [4], this paper proposes a generalized pseudo-polar Fourier transform (GPPFT), on the basis of a GPP-grid, of which the PP-grid is a first order approximation and the OP-grid is a second order approximation.

There is a MATLAB Toolbox related to PPFT (see section on resources, near the end of the chapter). In addition, another Toolbox, called ShearLab, includes functions for PPFT.

3.10.2 Nonequispaced Fourier Transform

Now, let us consider the general case of nonequally (unequally) spaced data. A number of different names can be found related to the Fourier transform in this case. Some authors employ the term '*nonuniform Fourier transform*' (NUFT), other authors use '*nonequidistant*' or '*nonequispaced*' (NFT), and others use '*generalized*' or '*approximate for irregular spaced data*'. This topic is also related with '*gridding*'.

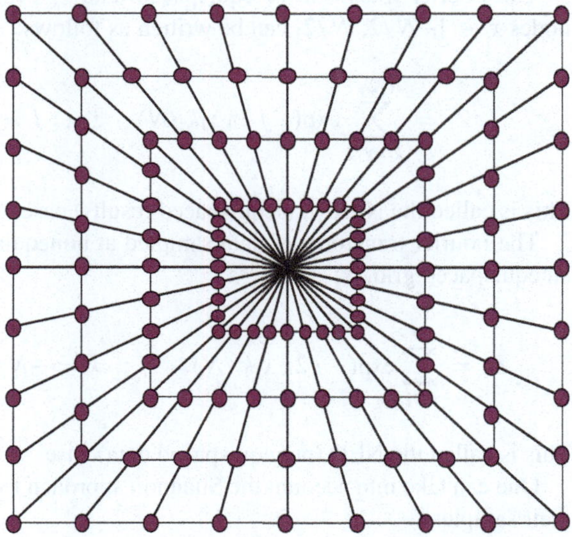

Fig. 3.108 Pseudo-polar grid

Fig. 3.109 Octa-polar grid

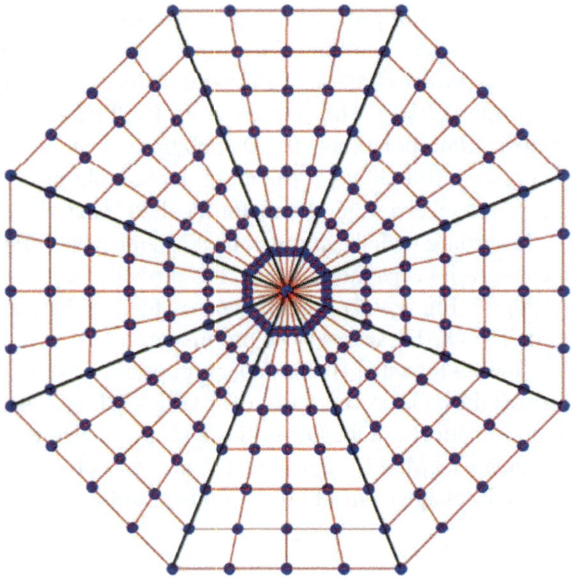

A first review of NFT was presented by [42] in 1998; also in that year, [37] obtained a unified approach for NFFT (nonequispaced *fast* Fourier transform). A major contribution in this field is the NFFT3 software library, in C. The documentation of this library, and related papers [20, 32], provide a good review of the methods that have been proposed for NFFT. Other relevant sources of information are [14] and several academic reports and dissertations [12, 23]. The NFFT· library includes a MATLAB interface. Let us introduce mathematical aspects of the topic, starting from basic concepts [14].

The Fourier transform of equispaced data z_k evaluated at nonequispaced grid nodes $x_l \in [-N/2,\ N/2]$ can be written as follows:

$$Z_l = \sum_{k=N/2}^{N/2-1} \exp(-j\,2\pi\,x_l k\,/N)\ \cdot z_k\ ;\quad l = 1,\,2,\ldots,M \tag{3.106}$$

This is called the NER (nonequispaced results) case.

The Fourier transform of data sampled at nonequispaced points x_l evaluated on an equispaced grid, is:

$$Z_k = \sum_{l=1}^{M} \exp(-j\,2\pi\,x_l k\,/N)\ \cdot z_l\ ;\quad k = -N/2,\ldots,N/2-1 \tag{3.107}$$

This is called the NED (nonequispaced data) case.

One can take into account the Shannon approach for the recovery of signals from their samples:

$$f(x) = \sum_m \text{sinc}(\pi(x - m))\, f(m) \tag{3.108}$$

(the bandwidth of $f(x)$ is $< \pi$) Our interest is centered on choosing $f(x) = \exp(-j\, x\, \xi)$, with $|\xi| < \pi$. Then:

$$\exp(-j\, x\, \xi) = \sum_m \text{sinc}(\pi(x - m))\, \exp(-j\, m\, \xi) \tag{3.109}$$

This interpolation is not convenient for fast computation, but provides a hint about what to do next. The idea is to use a better interpolation function (some authors refer to it as a window). Suppose that the bandwidth of $f(x)$ is $< \pi/c$, with $c > 1$. It can be shown that given that $0 < \pi/c < \beta$ and $\beta < \pi(2 - 1/c)$, and a function $\phi(x)$ with the following properties:

- Continuous and piecewise continuously differentiable in $[-\beta,\ \beta]$
- Vanishing outside $[-\beta,\ \beta]$
- Non-zero in $[-\pi/c,\ \pi/c]$

Then:

$$\exp(-j\, x\, \xi) = \frac{1}{\phi(\xi) \cdot \sqrt{2\pi}} \sum_m \phi(x - m)\, \exp(-j\, m\, \xi) \tag{3.110}$$

If one now takes $\xi = 2\pi\, k/cN$ and $x = c\, x_l$:

$$\begin{aligned} &\exp(-j\, 2\pi\, x_l\, k/N) = \\ &= \frac{1}{\phi(2\pi\, k/cN) \cdot \sqrt{2\pi}} \sum_m \phi(c\, x_l - m)\, \exp(-j\, 2\pi\, m\, k/cN) \end{aligned} \tag{3.111}$$

Therefore, for the NER case:

$$Z_l = \frac{1}{\sqrt{2\pi}} \sum_m \Phi(c\, x_l - m) \sum_{k=N/2}^{N/2-1} \exp(-j\, 2\pi\, m\, k\, /cN) \cdot \frac{z_k}{\phi(2\pi k/cN)} \tag{3.112}$$

with: $l = 1,\ 2,\ \ldots,\ M$

And for the NED case:

$$Z_l = \frac{1}{\phi(2\pi k/cN) \cdot \sqrt{2\pi}} \sum_{l=1}^{M} \sum_m z_l\, \Phi(c\, x_l - m)\, \exp(-j\, 2\pi\, mk/\, cN) \tag{3.113}$$

with: $k = -N/2,\ \ldots,\ N/2 - 1$

In both NER and NED expressions, $\Phi()$ means the equispaced FFT of $\phi()$ (with length cN).

It is convenient to use a $\phi()$ with compact support in $[-\beta,\ \beta]$ and $\Phi()$being mostly localized in some interval $[-K,\ K]$. The proper candidate for this would be a prolate spheroidal wave function. The literature has explored some types of windows (for instance, Kaiser-Bessel windows) for this application, [12, 20, 28].

One of the proposed generalizations of the Fourier transform, replaces its complex exponential basis functions with other functions, like for instance spherical harmonics [10]. This alternative has been included in the NFFT3 library. Another interesting initiative, for computation economy, is to use hyperbolic cross points with nonequispaced spatial nodes [12].

There is a web page about Fast Multipole Methods with links to NFFT3 (Chemnitz University), to the MATLAB Toolbox NUFFT (University of Michigan), and others (see web pages in the section on resources, near the end of the chapter). The MATLAB Central repository has a NUFFT-NFFT-USFFT Toolbox by M. Ferrara.

3.11 Experiments

3.11.1 Capturing Images with a Webcam

The MATLAB Toolbox for Image Acquisition provides a set of functions that can be used for the capture of videos or still images with a webcam, or other image acquisition systems.

Program 3.49 presents a simple example of image acquisition from a webcam. The user may opt for grayscale display by a simple modification of the program (which is included as a comment sentence). Figure 3.110 shows a photograph of my room at the University, captured with a webcam and using the program listed below.

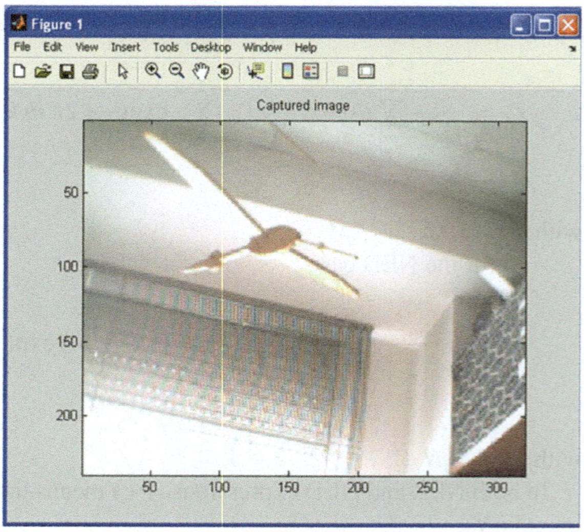

Fig. 3.110 Image captured with a webcam

Program 3.49 Webcam image acquisition

```
% Webcam image acquisition
% Preparation----------------------------------------------
Nfs=10; % set frame rate per second
hFigure=figure(1); %set figure \ref{fig:3.with handle
%set webcam driver
try
vid = videoinput('winvideo', 1);
catch
errordlg('No webcam available');
end
%set video parameters
set(vid,'FramesPerTrigger',1); %capture one frame each time
set(vid,'TriggerRepeat',Inf); %loop until stop
%set color acquisition
%set(vid,'ReturnedColorSpace','grayscale');
set(vid,'ReturnedColorSpace','rgb');
triggerconfig(vid, 'Manual');
% set timer (it uses the function ColorWdisp)
TimerData=timer('TimerFcn',
{@ColorWdisp,vid},'Period',1/Nfs,
'ExecutionMode','fixedRate','BusyMode','drop');
% Work------------------------------------------
% Start video and timer object
start(vid);
start(TimerData);
% Continue until the figure \ref{fig:3.is closed
uiwait(hFigure);
% Clean everything
stop(TimerData);
delete(TimerData);
stop(vid);
delete(vid);
clear functions;
```

Function ColorWdisp

```
% Function for frame display
% It is called by the timer
function ColorWdisp(obj, event,vid)
persistent IM;
persistent handlesRaw;
trigger(vid);
IM=getdata(vid,1,'uint8');
if isempty(handlesRaw) %if first frame, set figure
handlesRaw=imagesc(IM);
title('Captured image');
else
%updating
set(handlesRaw,'CData',IM);
end
```

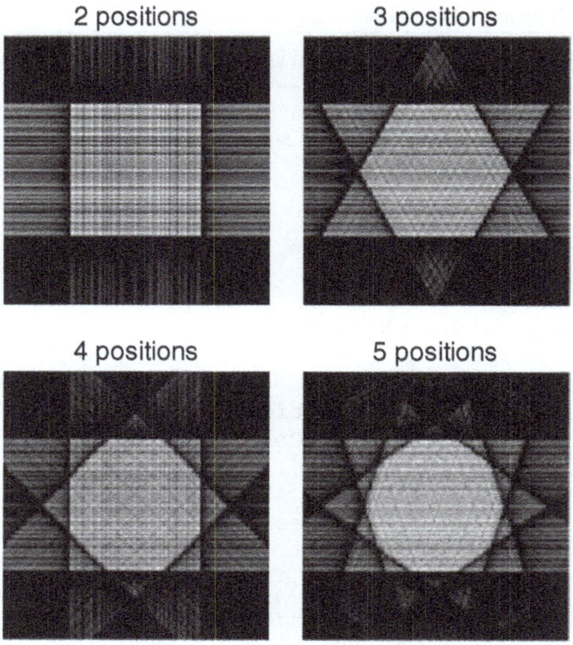

Fig. 3.111 Recovering the circle with backprojections

3.11.2 Backprojection Steps

It is illustrative to consider in more detail backprojection for inverse Radon transform. Actually, this method combines sets of rays, from several rotated directions. The idea of the proposed exercise is to visualize this mechanism using first 2 directions, then 3 directions, etc., and see the result. This result should be, when the number of backprojections is sufficient, the recovered image.

In this example, a filled circle has been chosen as example of the original image that, after Radon transform and then backprojection, should be recovered. The reader is invited to use other images, like a rectangle, a face, etc.

Program 3.50 obtains four images, which are depicted in Fig. 3.111. The program uses four times a function that is listed after the program. The images are reminiscent of a kaleidoscope.

Program 3.50 Inverse Radon transform with filter

```
% Inverse Radon transform with filter
% for circle
clear all;
%The Radon transform--------------
%padded circle
A=ones(64,64);B=A;
for nn=1:64;
for k=1:64,
```

```
m=(nn-32)^2+(k-32)^2;
if m >32^2, B(nn,k)=0; end;
end;
end;
R=zeros(128,128);
R(33:96,33:96)=B;
%projections
theta=0:1:180; %set of angles
N=length(theta);
PR=zeros(128,181);
for nn=1:N,
aux=imrotate(R,theta(nn),'bilinear','crop');
PR(:,nn)=sum(aux)';
end
%ramp filter
rampf=[2*[0:63,64:-1:1]'/128];
%Fourier domain filtering
FPR=fft(PR,128);
for nn=1:N,
FPRf(:,nn)=FPR(:,nn).*rampf;
end;
%inverse Fourier
PR=real(ifft(FPRf));
%Backprojections----------------------
figure(1)
subplot(2,2,1)
aux=zeros(128,128); IR=aux;
MA=90;
Rbak;
imshow(nIR);
title('2 positions');
subplot(2,2,2)
aux=zeros(128,128); IR=aux;
MA=60;
Rbak;
imshow(nIR);
title('3 positions');
subplot(2,2,3)
aux=zeros(128,128); IR=aux;
MA=45;
Rbak;
imshow(nIR);
title('4 positions');
subplot(2,2,4)
aux=zeros(128,128); IR=aux;
MA=36;
Rbak;
imshow(nIR);
title('5 positions');
```

Function Rbak.

```
%Backprojections-----------------------
aux=zeros(128,128); IR=aux;
for nn=1:MA:N,
for np=1:128,
aux(:,np)=PR(:,nn);
end;
IR=IR+imrotate(aux,theta(nn),'bilinear','crop');
end;
nIR=IR/max(max(IR)); %normalize
```

3.12 Resources

3.12.1 MATLAB

3.12.1.1 Toolboxes

- Image Reconstruction Toolbox (Including NUFFT):
  http://web.eecs.umich.edu/~fessler/code/mri.htm
- NFFT3:
  www-user.tu-chemnitz.de/~potts/nfft/
- FRIT Toolbox:
  http://www.ifp.illinois.edu/~minhdo/software/

3.12.1.2 Matlab Code

- Nonequispaced FFT; Mathworks file exchange:
  http://www.mathworks.com/matlabcentral/fileexchange/25135-nufft-nufft-usfft

3.12.2 Internet

There are several sources of technical information concerning color theory and applications, including presentations like the one by Idit Haran on color and the human response to light, or technical reports from Adobe.

The page of Michael Elad has links to several toolboxes, including PPFT.

References

1. R. Ansari, Efficient IIR and FIR fan filters. IEEE Trans. Circuits Syst. **34**(8), 941–945 (1987)
2. A. Averbuch, R.R. Coifman, D.L. Donoho, M. Elad, M. Israeli, Fast and accurate polar Fourier transform. Appl. Comput. Harmonic Anal. **21**, 145–167 (2006)
3. A. Averbuch, R.R. Coifman, D.L. Donoho, M. Israeli, J. Walden, Fast slant stack: a notion of radon transform for data in a cartesian grid which is rapidly computible, algebraically exact, geometrically faithful and invertible. Technical Report 11, Dept. Statistics, Stanford University, USA, 2001
4. N. Chou, J.A. Izatt, S. Farsiu, Generalized pseudo-polar Fourier grids and applications in registering ophthalmic optical coherence tomography images, in *Proceedings 43rd ASILOMAR Conference Signal, Systems and Computers*, pp. 807–811 (2009)
5. B.E. Coggins, P. Zhou, Polar Fourier transforms of radially sampled NMR data. J. Magn. Reson. **182**, 84–95 (2006)
6. A.L. Cunha, M.N. Do, Filter design for directional multiresolution decomposition, in *Proceedings SPIE*, vol. 5014 (2005)
7. H. Dammertz, A. Keller, S. Dammertz, Simulation of rank-1 lattices, in *Monte Carlo and Quasi-Monte Carlo Methods*, pp. 205–216 (Springer, 2008)
8. M.N. Do, Directional multiresolution image representation. Ph.D. thesis, Department of Communication Systems, Swiss Federal Institute of Technology Lausanne, 2001
9. M.N. Do, Y.M. Lu, Multidimensional filter banks and multiscale geometric representations. Found. Trends Sign. Process. **5**(3), 157–264 (2012)
10. J.R. Driscoll, D. Healy, Computing Fourier transforms and convolutions on the 2-sphere. Adv. Appl. Math. **15**, 202–250 (1994)
11. A. Faridani, E. Grinberg, E.T. Quinto, *Radon Transform and Tomograpy* (Amer. Mathematical Society, 2001)
12. M. Fenn, Fast Fourier transform at nonequispaced nodes and applications. Ph.D. thesis, Universität Manheim, Germany, 2005
13. J. Fessler, *Fessler Book Chapter on Tomography* (Course 516, University of Michigan, 2012). http://web.eecs.umich.edu/~fessler/course/516/l/c-tomo.pdf
14. K. Fourmont, Non-equispaced fast Fourier transforms with applicatios to tomography. J. Fourier Anal. Appl. **9**(5), 431–450 (2003)
15. Q. Gu, M.N.S. Swamy, On the design of a broad class of 2-D recusive digital filters with fan, diamong and elliptically-symmetric responses. IEEE Trans. Circuits Syst.-II **41**(9), 603–614 (1994)
16. B. Gustafsson, Mathematics for computer tomography. Phys. Scr. **T61**, 38–43 (1996)
17. O. Harari, A new nearly polar fft and analysis of Fourier-Radon relations in discrete spaces. Master's thesis, Ben-Gurion University of the Negev, Beersheba, 2007
18. AS. Hassanein, S. Mohammad, M. Sameer, M.E. Ragab, A survey on hough transform, theory, techniques and applications (2015). arXiv preprint arXiv:1502.02160
19. E.S. Karlsen, Multidimensional multirate sampling and seismic migration. Master's thesis, Dep. Earth Science, University of Bergen, Norway, 2010
20. J. Keiner, S. Kunis, D. Potts, Using nfft3—a software library for various nonequispaced fast Fourier transforms. ACM Trans. Math. Softw. **36**(4) (2009). art.no. 19
21. J. Kovaèeviæ, M. Vetterli, Design of multidimensional non-separable regular filter banks and wavelets, in *Proceedings IEEE International Conference Acoustics, Speech, and Signal Processing, ICASSP-92*, vol. 4, pp. 389–392 (1992)
22. T. Kratochvil, J. Melo. Utilization of MATLAB for TV colorimetry and color spaces analysis, in *Proceedings 14th Annual Conference Technical Computing*, pp. 53–60 (2006)
23. S. Kunis, Nonequispaced FFT- generalization and inversion. Ph.D. thesis, Institut für Mathematik, Universität zu Lübeck, Germany, 2006
24. E. Kupce, R. Freeman, Fast multidimensional NMR: radial sampling of evolution space. J. Magn. Reson. **173**, 317–321 (2005)

25. O. Marques, *Practical Image and Video Processing Using MATLAB* (J. Wiley, 2011)
26. R. Matei, Multi-directional filters designed from 1D prototypes, in *Proceedings IEEE 55th International Midwest Symposium on Circuits and Systems, (MWSCAS)*, pp. 864–867 (2012)
27. F. Nicolls, A. Wilkinson, *Multidimensional Digital Signal Processing* (Lecture of the EEE4001F course, University of Cape Town, South Africa, 2012). http://www.dip.ee.uct.ac.za/~nicolls/lectures/eee401f/01_mdsp_slides.pdf
28. J.P. Nilchian, M. Ward, C. Vonesch, M. Unser, Optimized Kaiser-Bessel window functions for computed tomography. IEEE Trans. Image Process. **24**(11), 3826–3833 (2015)
29. S.C. Pei, S.B. Jaw, Two-dimensional general fan-type FIR digital filter design. Sign. Process. **37**, 265–274 (1994)
30. Y. Pellin, P.P. Vaidyanathan, Theory and design of two-dimensional filter banks: a review. Multidimension. Syst. Signal Process. **7**, 263–330 (1996)
31. N.A. Pendergrass, S.K. Mitra, E.I. Jury, Spectral transformations for two-dimensional digital filters. IEEE Trans. Circuits Syst. **23**(1), 26–35 (1976)
32. D. Potts, G. Steidl, M. Tasche, Fast Fourier transforms for nonequispaced data: a tutorial, in *Modern Sampling Theory: Mathematics and Applications*, pp. 247–270 (Birkhauser, 2001)
33. E.Z. Psarakis, V.G. Mertzios, GPh Alexiou, Design of two-dimensional zero phase FIR fan filters via the McClelland transform. IEEE Trans. Circuits Syst. **37**(1), 10–16 (1990)
34. O. Regev, *Lattices in Computer Science: Introduction* (Lecture of the 0368.4282 course, Tel Aviv University, Israel, 2004). https://www.cims.nyu.edu/~regev/teaching/lattices_fall_2009/
35. M. Smereka, I. Dulêba, Circular object detection using a modified Hough transform. Int. J. Appl. Math. Comput. Sci. **18**(1), 85–91 (2008)
36. Y.S. Song, Y.H. Lee, Formulas for McClelland transform parameters in designing 2-D zero-phase FIR fan filters. IEEE Sign. Process. Lett. **3**(11), 291–293 (1996)
37. G. Steidl, A note on fast Fourier transforms for nonequispaced grids. Adv. Comput. Math. **9**, 337–353 (1998)
38. M. Tkalcic, J.F. Tasic, Colour spaces—perceptual, historical and applicational background, in *Proceedings IEEE Region 8 EUROCON, Computer as a Tool*, pp. 304–308 (2003)
39. P. Toft, The radon transform. Ph.D. thesis, Dept. Mathematical Modelling, Technical University of Denmark, 1996
40. M. Vetterli, E. Kovacevic, D.J. Legall, Perfect reconstruction filter banks for HDTV representation and coding. Sign. Process.: Image Commun. **2**, 349–363 (1990)
41. M. Vetterli, J. Kovacevic, *Wavelets and Subband Coding* (Prentice Hall, 1995)
42. A.F. Ware, Fast approximate Fourier transforms for irregulary spaced data. SIAM Rev. **40**, 838–856 (1998)
43. W.P. Zhu, S. Nakamura, An efficient approach for the synthesis of 2-D recursive fan filters using 1-D prototypes. IEEE Trans. Sign. Process. **44**(4), 979–983 (1996)

Chapter 4
Wavelet Variants for 2D Analysis

4.1 Introduction

This chapter is devoted to the analysis of 2D signals and/or images, with emphasis on the use of wavelet variants Many important applications are interested on such methods. In particular, the detection of edges, borders, contours, etc., is suitable for several purposes, like for instance spatially selective filtering. For example, in medical studies that want to have a clearer view of vessels against a noisy image background.

One of the topics covered in this chapter is the adaptation of 1D wavelets for the analysis of images. It can be clearly seen that the decomposition obtained with wavelets suggest simple methods for denoising, and for compression. Indeed, image compression is a major driving force in contemporary IT industrial and research activity.

The word "analysis", as defined in dictionaries, is related with decomposition for better knowledge. An image can be decomposed in several ways, like for instance according with a set of angular orientations, or into several frequency bands, etc. A relevant number of such decompositions will be described in this chapter.

Corresponding to the high interest of the topics contemplated in this chapter, many methods and techniques have been proposed, and in many cases this was accompanied by MATLAB toolboxes. One of the missions of this chapter has been to offer a panorama of analysis tool alternatives, together with references to software and documentation sources.

4.2 Laplacian Pyramid

Laplacian pyramids were introduced by [18]. Given an image *L0*, a low pass filter is applied, and a *L1** image is obtained. The image *L1** can be downsampled to obtain *L1*, with no information loss with respect to *L1**. The difference between images

© Springer Science+Business Media Singapore 2017

J.M. Giron-Sierra, *Digital Signal Processing with Matlab Examples, Volume 2*,
Signals and Communication Technology, DOI 10.1007/978-981-10-2537-2_4

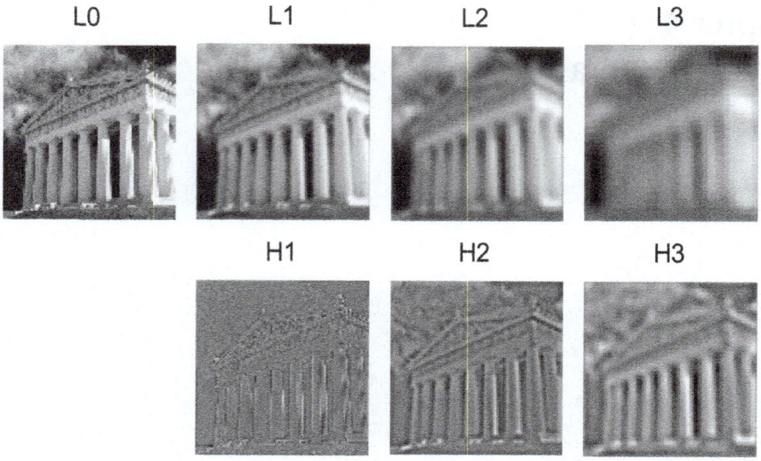

Fig. 4.1 Laplacian pyramid: a view of successive images

would be *H1* = *L0* – *L1**. *H1* is a bandpass image. For compression purposes, it is more convenient to encode *H1* and *L1*, rather than *L0*.

The Laplacian pyramid repeats the decomposition process. From *L0* it gets *H1* and *L1*, from *L1* it gets *H2* and *L2*, and so on.

Figure 4.1 shows a sequence of images corresponding to the Laplacian pyramid decomposition. Notice the effect of low pass filtering: more and more blurring. The images *H1*…*H3* mainly contain edges. The figure has been generated with the Program 4.1. The *cell()* MATLAB function is employed to devise an indexed set of matrices with different sizes.

Program 4.1 Laplacian pyramid

```
%Laplacian pyramid
%get an image
ufg=imread('Parth1.tif');
fg=double(ufg); %convert to float
fg=fg-mean(mean(fg)); %zero mean
[Nr,Nc]=size(fg);
nlevels=4; %number of pyramid levels
%preparing indexed matrix sets
PH=cell(nlevels,1); %for high pass
PL=cell(nlevels,1); %for low pass
PL(1)={fg}; PH(1)={ones(Nr,Nc)};
aux=fg;
for nn=2:nlevels,
  fil=fspecial('gaussian',[16,16],4);
  FL=filter2(fil,aux);
  dFL=FL(1:2:end,1:2:end); %subsampling
  FH=aux-FL;
  PL(nn)={dFL};
  PH(nn)={FH};
  aux=dFL;
```

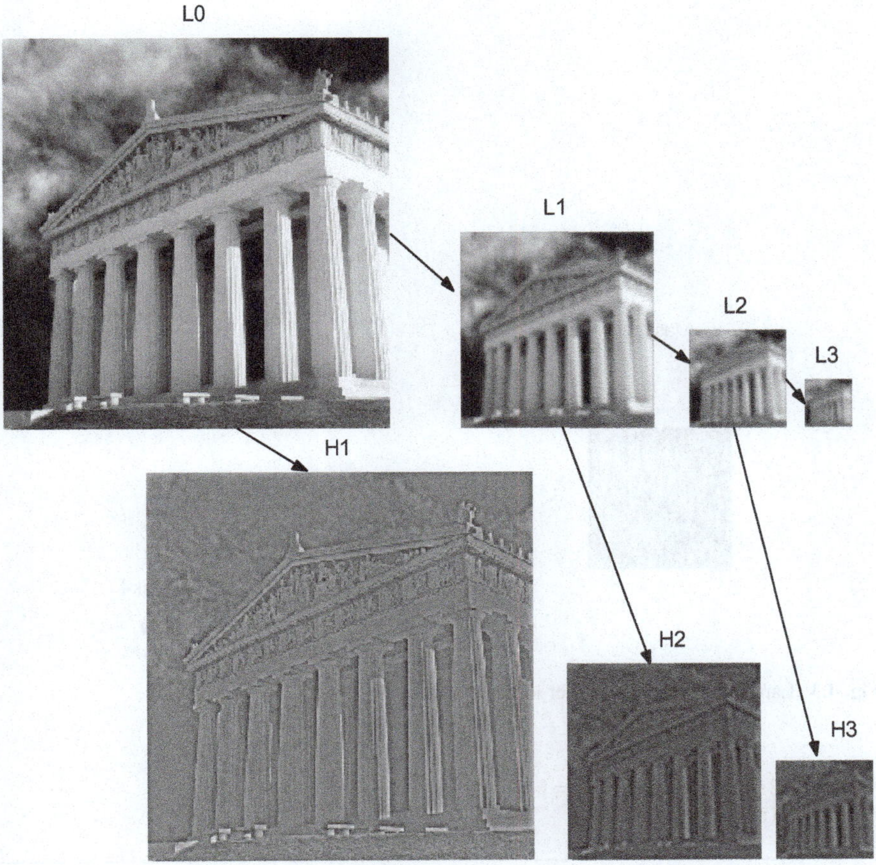

Fig. 4.2 Laplacian pyramid: sizes of images shrink along analysis

```
end;
%display (with conversion to imshow range)
for nn=1:nlevels
  subplot(2,nlevels,nn)
  aux=PL{nn};
  m=min(min(aux));
  M=max(max(aux));
  imshow((aux-m)/(M-m));
  s=num2str(nn-1); msg=['L',s];
  title(msg);
end;
for nn=2:nlevels
  subplot(2,nlevels,nlevels+nn)
  aux=PH{nn};
  m=min(min(aux));
  M=max(max(aux));
  imshow((aux-m)/(M-m));
```

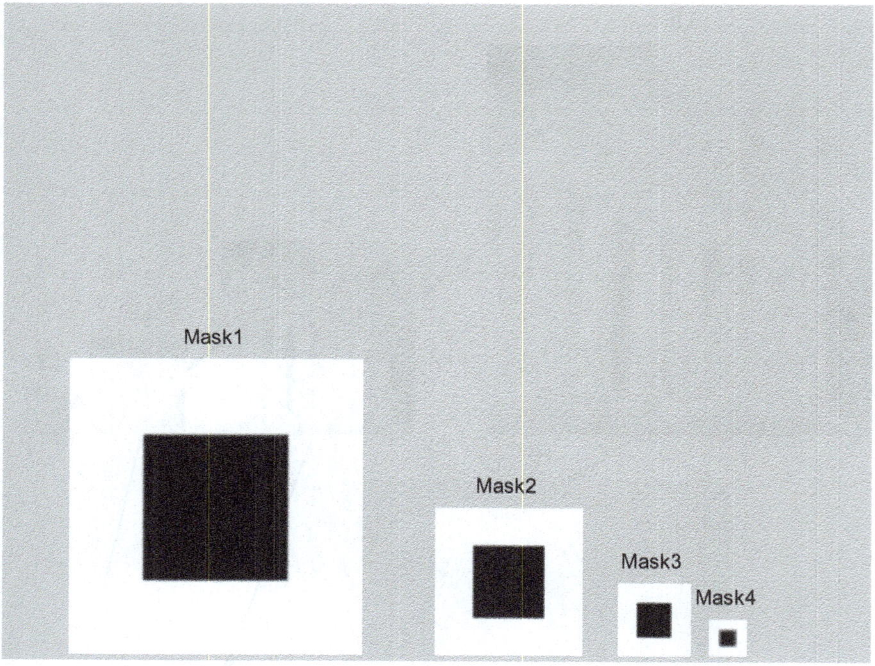

Fig. 4.3 Laplacian pyramid: Fourier low-pass masks

```
  s=num2str(nn-1); msg=['H',s];
  title(msg);
end;
```

To get more insight into the Laplacian pyramid, it is opportune to visualize the size of the images obtained along the decomposition process. Figure 4.2 shows a size comparison view. It has been generated with a program that has been included in Appendix A.

In order to explore alternatives, a second version of the Laplacian pyramid has been developed, this time on the basis of 2-D Fourier. In particular, the low-pass filter has been implemented using rectangular masks in the 2-D Fourier domain. The Program 4.2 implements the Laplacian pyramid, and displays two figures. Figure 4.3 shows the low-pass masks that have been applied. These masks have smoothed borders (see Program).

Program 4.2 Laplacian pyramid, Fourier masks

```
%Laplacian pyramid
%use of masks in the Fourier domain
%get an image
ufg=imread('Parth1.tif');
fg=double(ufg); %convert to float
fg=fg-mean(mean(fg)); %zero mean
```

```
[Nr,Nc]=size(fg);
knr=sqrt(Nr*Nc); %normalization constant
FIG=fftshift(fft2(ifftshift(fg)))/knr; %2D fft of the picture
nlevels=5; %number of pyramid levels
%preparing indexed matrix sets
PH=cell(nlevels,1); %for high pass
PL=cell(nlevels,1); %for low pass
MK=cell(nlevels,1); %for masks
PL(1)={fg}; PH(1)={ones(Nr,Nc)};
mkb1=[0.2 0.4 0.6 0.8]; mkb2=fliplr(mkb1); %mask borders
aux=FIG; mr=Nr; mc=Nc;
faux=fg;
for nn=2:nlevels,
%prepare a low-pass rectangular mask
sz1=zeros(1,mc/4); so1=ones(1,(mc/2)-8);
mk1=[sz1,mkb1,so1,mkb2,sz1]; %mask side 1
sz2=zeros(1,mr/4); so2=ones(1,(mr/2)-8);
mk2=[sz2,mkb1,so2,mkb2,sz2]; %mask side 2
mask=mk2'*mk1; %the low-pass mask
%low-pass filtering in 2D Fourier domain
FFL=aux.*mask; %image filtering (full size)
%extract internal rectangle (subsampling)
dFFL=FFL((1+(mr/4)):((3*mr)/4),(1+(mc/4)):((3*mc)/4));
aux=dFFL;
%back to image domain
FL=fftshift(ifft2(ifftshift(FFL)))*knr; %L*
dFL=fftshift(ifft2(ifftshift(dFFL)))*knr; %L
FH=faux-FL; %H
faux=dFL;
PL(nn)={real(dFL)};
PH(nn)={real(FH)};
MK(nn-1)={mask};
mc=mc/2; mr=mr/2;
end;
%display of masks
figure(1)
aw=0;
for nn=1:nlevels-1
  ax=2^(nn-1);
  subplot('Position',[0.02+aw,0.01,0.45/ax,0.45/ax])
  aux=MK{nn};
  imshow(1-aux);
  s=num2str(nn); msg=['Mask',s];
  title(msg);
  aw=aw+(0.45/ax);
end;
%display (with conversion to imshow range)
figure(2)
for nn=1:nlevels
  subplot(2,nlevels,nn)
  aux=PL{nn};
  m=min(min(aux));
  M=max(max(aux));
  imshow((aux-m)/(M-m));
  s=num2str(nn-1); msg=['L',s];
```

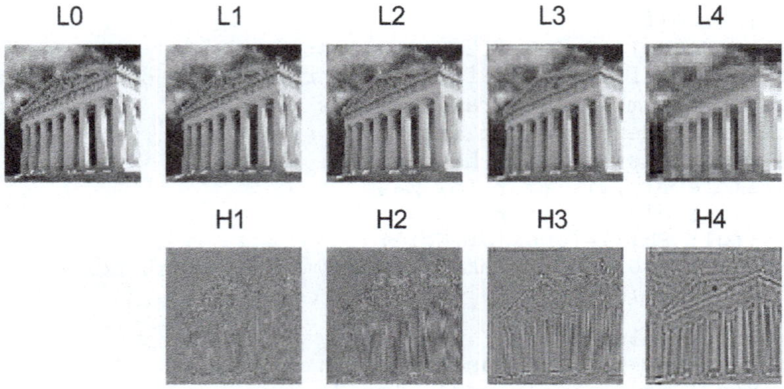

Fig. 4.4 Laplacian pyramid: a view of successive images

```
  title(msg);
end;
for nn=2:nlevels
  subplot(2,nlevels,nlevels+nn)
  aux=PH{nn};
  m=min(min(aux));
  M=max(max(aux));
  imshow((aux-m)/(M-m));
  s=num2str(nn-1); msg=['H',s];
  title(msg);
end;
```

Figure 4.4 shows the series of images, which have been re-dimensioned to be displayed with the same size.

An interesting aspect of the Program 4.2 is that subsampling is a simple matter: just to take the samples covered by the masks.

As it will be seen in the next section, there are wavelet variants—for instance, contourlets—that use Laplacian pyramids.

4.3 Steerable Filters and Pyramids

Consider the following scenario. This is a factory producing hammers. There is a conveyor system carrying the hammers, which have been deposited on the belt at random orientations. At the end of the conveyor a robot is in charge of picking hammers. A video system should tell to the robot what is the orientation of each arriving hammer.

There are image processing applications interested on orientation. This section deals with steerable filters. These filters can be oriented, so they may be called orientation-pass filters. In the scenario just described, a steerable filter with adequate orientation could match a hammer.

In 2D geometry a natural tool is the partial derivative. A very representative example of its use is the gradient, which in turn is a natural way for orientation studies. Therefore, it is not strange that the initial ideas about steerable filters were related to the use of partial derivatives.

Many examples of steerable functions are based in the use of Gaussians. It is opportune, before entering 2D scenarios, to visualize the derivative of the 1D Gaussian:

$$\frac{dg(x)}{dx} = \frac{d}{dx} e^{-x^2/2} = -x\, e^{-x^2/2} \tag{4.1}$$

where the normalization constant has not been used.

Figure 4.5 shows this derivative and the amplitude of its Fourier transform.

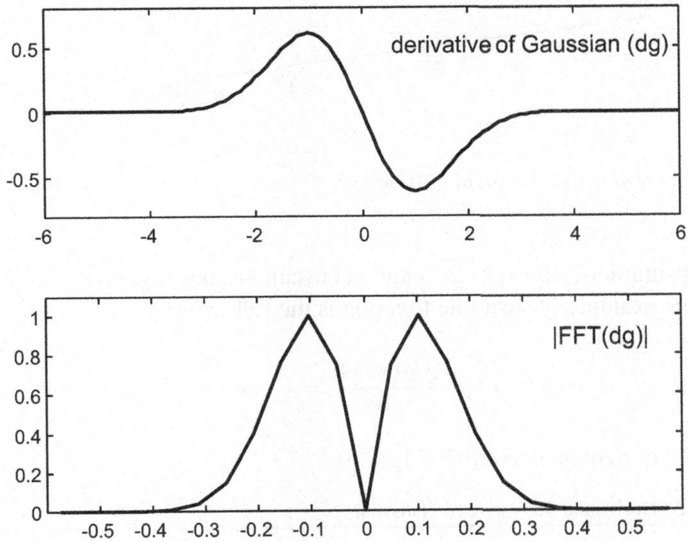

Fig. 4.5 The derivative of 1D Gaussian and its Fourier transform amplitude

4.3.1 *Steerable Filters*

The steerable filters were introduced by Freeman and Adelson [75] in 1991. Since then, it has received a lot of attention. The term '*steerable*' was coined in that article [75], and then it has been generalized so today one can speak of steerable functions.

Consider the following Gaussian function:

$$g(x, y) = e^{-(x^2+y^2)/2} \tag{4.2}$$

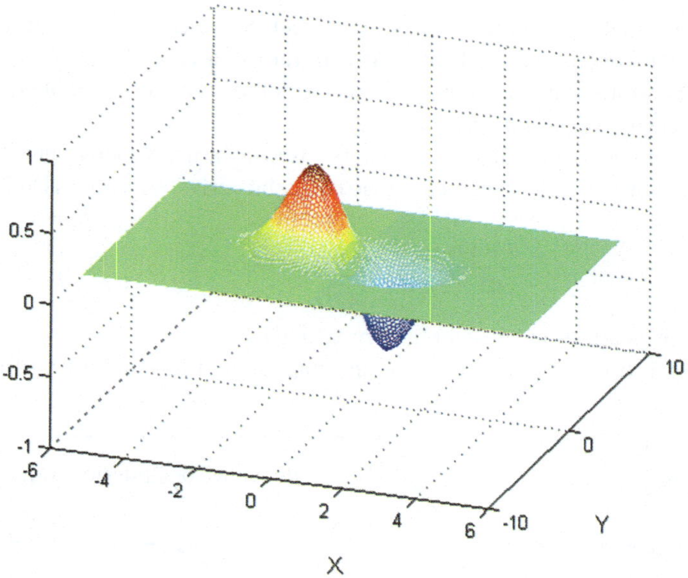

Fig. 4.6 The horizontal derivative of 2D Gaussian

where, for simplicity, the $1/\sqrt{2\pi}$ scaling constant has been ignored.

A simple example of steerable function is the following:

$$g_x(x,y) = \frac{\partial g(x,y)}{\partial x} = -x\, e^{-(x^2+y^2)/2} \tag{4.3}$$

which is the horizontal derivative (Fig. 4.6).

Program 4.3 Horizontal derivative of Gaussian

```
% Horizontal derivative of Gaussian
x=-6:0.1:6;
y=-6:0.1:6;
N=length(x);
z=zeros(N,N); %space for the function
for ny=1:N,
   for nx=1:N,
      aux=(x(nx)^2)+(y(ny)^2);
      z(ny,nx)=-x(nx)*exp(-aux/2);
   end;
end;
figure(1)
mesh(x,y,z);
view(20,30);
title('Horizontal derivative of Gaussian');
xlabel('X'); ylabel('Y');
```

The steerable function can be expressed in polar coordinates:

$$g_x(r,\theta) = -r\,e^{-r^2/2}\cos(\theta) \tag{4.4}$$

By the way, the expression above corresponds to a *polar-separable* function: it is the product of a radial component and an angular component.

Now let us rotate this function [87]. For instance, a $\pi/2$ rotation would yield:

$$g_x(r,\theta-\pi/2) = -r\,e^{-r^2/2}\cos(\theta-\pi/2) = -r\,e^{-r^2/2}\sin(\theta) \tag{4.5}$$

This result is coincident with the vertical derivative $\partial g(x,y)/\partial y$ in polar coordinates. This is an indication that the directional derivative at any orientation is a rotated copy of the steerable function.

Actually, if the steerable function is rotated to any angle α, the result would be:

$$\begin{aligned} g_\alpha(r,\theta-\alpha) &= -r\,e^{-r^2/2}\cos(\theta-\alpha) \\ &= -r\,e^{-r^2/2}(\cos(\theta)\cos(\alpha)+\sin(\theta)\sin(\alpha)) = \cos(\alpha)\,g_x+\sin(\alpha)\,g_y \end{aligned} \tag{4.6}$$

This expression can be written in vectorial format as follows:

$$g_\alpha(r,\theta-\alpha) = (\cos(\alpha),\sin(\alpha))\begin{pmatrix} g_x \\ g_y \end{pmatrix} = \mathbf{v}(\alpha)\cdot\mathbf{b} \tag{4.7}$$

Therefore any directional derivative of $g(x,y)$ can be obtained with a linear combination of the horizontal and vertical derivatives. These two derivatives constitute a *basis*; and $\cos(\alpha)$, $\sin(\alpha)$ are called *interpolation functions*.

To summarize, the rotated version of a steerable function is a rotated copy that can be obtained from a basis set.

Figure 4.7 shows (top views) an example of synthesis for -30*@°@* orientation.

Program 4.4 Using x-y basis

```
% Using x-y basis
x=-5:0.05:5;
y=-5:0.05:5;
N=length(x);
%horizontal
zx=zeros(N,N); %space for the function x
for ny=1:N,
   for nx=1:N,
      aux=(x(nx)^2)+(y(ny)^2);
      zx(ny,nx)=-x(nx)*exp(-aux/2);
   end;
end;
%vertical
zy=zeros(N,N); %space for the function y
for ny=1:N,
   for nx=1:N,
      aux=(x(nx)^2)+(y(ny)^2);
      zy(ny,nx)=-y(ny)*exp(-aux/2);
   end;
```

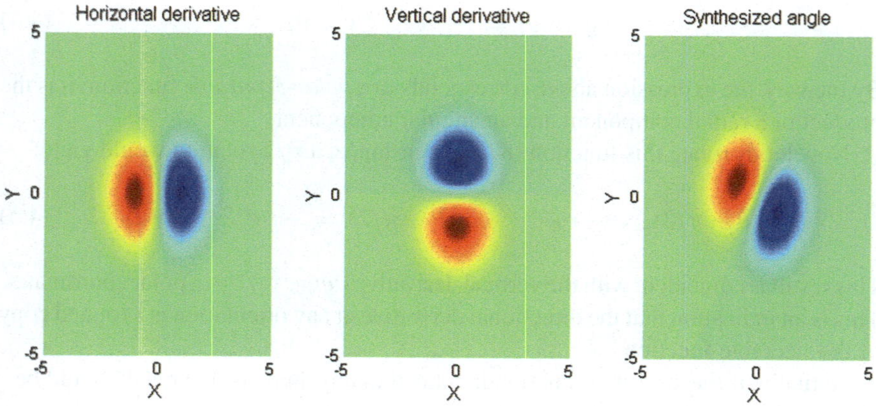

Fig. 4.7 Using two basis functions to generate a -30*@°@* oriented copy

```
end;
%with 30 deg}$
za=zeros(N,N); %space for the function za
alpha=-pi/6;
ca=cos(alpha); sa=sin(alpha);
for ny=1:N,
  for nx=1:N,
    za(ny,nx)=ca*zx(ny,nx)+sa*zy(ny,nx);
  end;
end;
figure(1)
subplot(1,3,1);
mesh(x,y,zx);
view(0,90);
title('Horizontal derivative');
xlabel('X'); ylabel('Y');
subplot(1,3,2);
mesh(x,y,zy);
view(0,90);
title('Vertical derivative');
xlabel('X'); ylabel('Y');
subplot(1,3,3);
mesh(x,y,za);
view(0,90);
title('Synthesized angle');
xlabel('X'); ylabel('Y');
```

The first-order directional derivative of a Gaussian is steerable with a basis of 2 functions. It can be shown that a nth-order directional derivative is steerable with a basis of $n + 1$ functions.

Directional derivatives are not the only case of steerable functions. Any polar-separable function with a band-limited angular component and arbitrary radial component can be steerable.

Apart from rotations, the concept of steerability has been extended to affine transformations and other spatial transformations.

4.3.1.1 Aspects of the Design and Use of Steerable Filters

Let us look in more detail at the results presented in [75]. The concept of steerable filter was introduced using the horizontal and the vertical first-order derivatives of a Gaussian as a basis. This corresponds to what has been already explained before. Then, the following question was considered: what are the conditions a function should obey for being steerable. In [75] words, a steerable function can be written as a linear sum of rotated versions of itself:

$$f_\alpha(x, y) = \sum_{j=1}^{M} k_j(\alpha)\, f_{\alpha_j}(x, y) \tag{4.8}$$

For an easier attack of the problem, the study focuses on functions that can be expanded in a Fourier series as follows:

$$f(r, \theta) = \sum_{k=-N}^{N} a_k(r) \cdot e^{jk\theta} \tag{4.9}$$

It was found that the function $f(r, \theta)$ is steerable, so it can be decomposed as in (4.8), iff the interpolation functions $k_k(\alpha)$ were the solutions of:

$$\begin{pmatrix} 1 \\ e^{j\alpha} \\ \vdots \\ e^{jN\alpha} \end{pmatrix} = \begin{pmatrix} 1 & 1 & \dots 1 \\ e^{j\alpha_1} & e^{j\alpha_2} & \dots e^{j\alpha_M} \\ \vdots & \vdots & \dots \vdots \\ e^{jN\alpha_1} & e^{jN\alpha_1} & \dots e^{jN\alpha_1} \end{pmatrix} \begin{pmatrix} k_1(\alpha) \\ k_2(\alpha) \\ \vdots \\ k_M(\alpha) \end{pmatrix} \tag{4.10}$$

If for any k, $a_k(r) = 0$, then the nth row of the left-hand side and of the matrix of the right-hand side should be removed.

Denote as T the number of nonzero coefficients $a_k(r)$ in (4.9). It was also found that the minimum number of basis functions is T. It is convenient to choose basis functions spaced equally in angle between 0 and π.

Notice that a 1D band-limited function can be represented by a finite number of samples (Shannon), which correspond to the number of Fourier terms.

All functions that are band-limited are steerable, provided a sufficient number of basis functions. For practical reasons, those steerable functions requiring a few basis functions are preferred.

A simple design example is the following. Choose the second derivative of a Gaussian, which is an even function. In this case the number of basis functions is three. Hence:

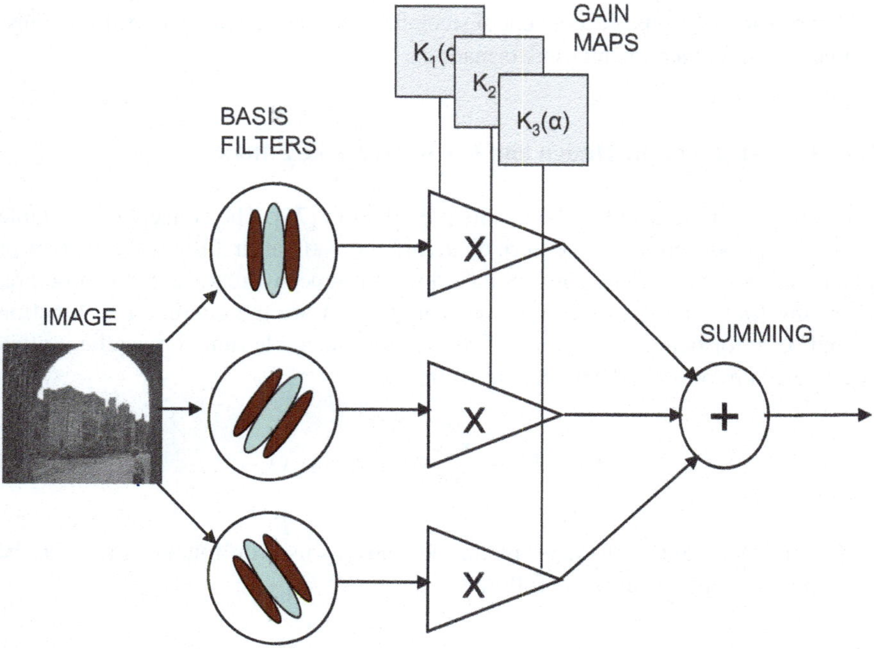

Fig. 4.8 Block diagram of a steerable filter system

$$\begin{pmatrix} 1 \\ e^{j2\alpha} \end{pmatrix} = \begin{pmatrix} 1 & 1 & \ldots 1 \\ e^{j2\alpha_1} & e^{j2\alpha_2} & \ldots e^{j2\alpha_M} \end{pmatrix} \begin{pmatrix} k_1(\alpha) \\ k_2(\alpha) \\ k_3(\alpha) \end{pmatrix} \tag{4.11}$$

Considering imaginary and real parts, this is equivalent to three equations. If one chooses 0*@°@*,60*@°@*,120*@°@* for the basis functions, then:

$$k_j(\alpha) = \frac{1}{3}(1 + 2\cos(2(\alpha - \alpha_j)) \tag{4.12}$$

Figure 4.8 shows a filter structure that corresponds to this example:

The design of the basis filters can be conveniently done in the Fourier domain. First a 2D angulary symmetric filter can be designed from a 1D prototype. Then a set of angular variations is imposed. For instance, four oriented filters can be obtained using the symmetric filter multiplied by $\cos^3(\varphi - \alpha_j)$ (φ is the azimuthal angle). The impulse response of the four filters are obtained by inverse Fourier transform.

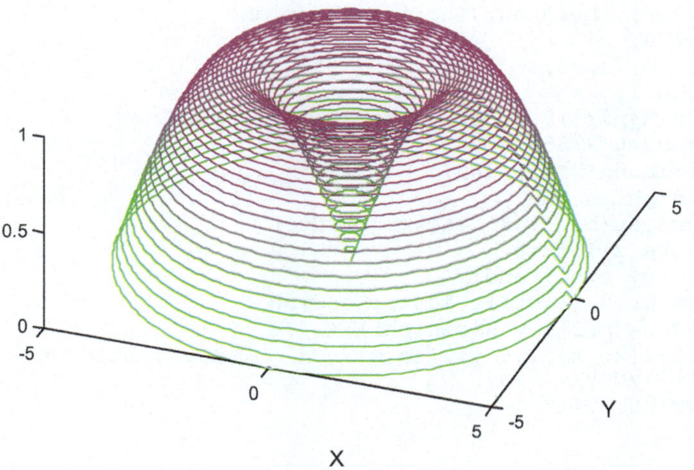

Fig. 4.9 A simple symmetrical filter

For example, let us use the simple symmetrical filter depicted in Fig. 4.9, which is based on a rotated sine.

Program 4.5 A simple symmetrical filter

```
%A simple symmetrical filter
% using polar coordinates
figure(1)
xo=0; yo=0; zso=0;
for R=0:0.1:5,
  zs=sin(R*pi/5);
  for nang=0:2:358,
    phi=(nang*2*pi)/360;
    cg=0;
    if zs>0, cg=1; end;
    clr=[zs,(1-zs),zs]; %color coding
    x=R*cos(phi); y=R*sin(phi);
    plot3([xo x],[yo y],[zso zs],'Color',clr); hold on;
    xo=x; yo=y; zso=zs;
   view(20,50);
  end;
end;
title('Symmetrical filter')
xlabel('X'); ylabel('Y');
```

Program 4.6 One of the four oriented functions

```
% One of the four oriented functions
% based on polar coordinates
figure(1)
```

```
[X,Y]=meshgrid(-5:0.1:5, -5:0.1:5);
Z=0*(X+Y);
mesh(X,Y,Z); hold on; %horizontal plane
alpha=pi/4;
xo=0; yo=0; zo=0;
for R=0:0.1:5,
  zs=sin(R*pi/5); %the symmetrical filter
  for nang=0:2:358,
    phi=(nang*2*pi)/360;
    beta=phi-alpha;
    z=zs*cos(beta)^3; %asym. product
    cz=abs(z); cr=0;
    if z>0, cr=1; end;
    clr=[cr,0,(1-cz)]; %color coding
    x=R*cos(phi); y=R*sin(phi);
    plot3([xo x],[yo y],[zo z],'Color',clr); hold on;
    xo=x; yo=y; zo=z;
    view(20,30);
  end;
end;
title('F1 (45?'))
xlabel('X'); ylabel('Y');
```

Figure 4.10 shows one of these four oriented functions

Several applications are suggested in [75]. First, steerable filters can be applied for analysis of local orientation, by measuring the so-called "oriented energy". Another application takes advantage from steerability to adapt the filter to the imaging aspects of interest, like in medicine for the study of veins. Also, steerable filters can be applied for contour detection. And, steerable filters can be used in 3D.

Interestingly, the part of [75] devoted to contour detection establishes a comparison with the Canny's edge detection method. It is noted that a filter optimized for use with an edge will give incorrect results with other features, like for instance lines. The opposite also occurs, when optimized for lines the filter will give incorrect results with edges. The [75] comments converge in favor of local energy measures for the study of different types of features related to contours.

4.3.1.2 Further Developments

Steerable *wedge* filters were introduced by [170]. This article remarks that the original steerable filters are almost always either symmetric or anti-symmetric, and this causes ambiguity in certain analysis situations, like for instance in case of junctions.

The next two figures reproduces results presented in [170]. Both depict the squared responses of a bank of conventional steerable filters to certain image inputs. These squared responses can be regarded as "*orientation maps*".

Figure 4.11 shows examples of orientation maps corresponding to a line and a cross. These maps are correct.

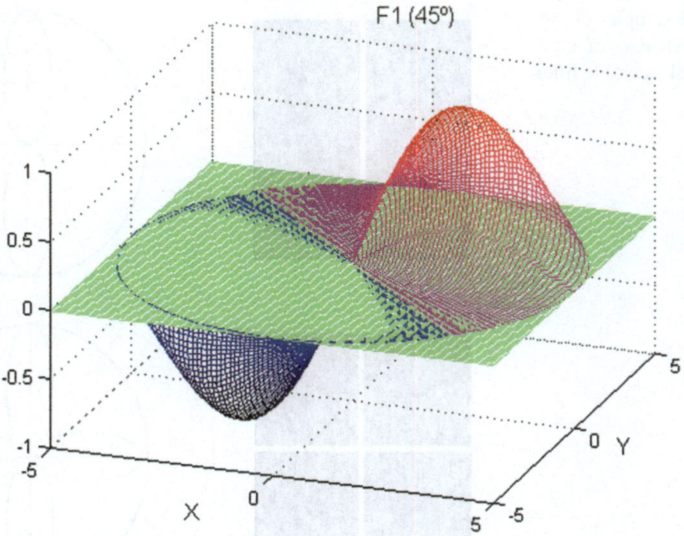

Fig. 4.10 One of the four oriented functions

Now, let us show in Fig. 4.12 some incorrect responses. Case (a) is half a line; case (b) is a corner; case (c) is a junction. In all these cases the correct oriantation maps should have been asymmetric.

The design of the steerable wedge filters is based on the use of 2 N filters, so that the angular portion of each filter is:

$$h_e(\theta) = \sum_{n=1}^{N} w_n \cos(n\theta), \quad h_o(\theta) = \sum_{n=1}^{N} w_n \sin(n\theta) \tag{4.13}$$

There must be a 90° phase shift between the angular portions of the two filters (Hilbert transform).

Refer to [170] for details about the design of the interpolation functions and the radial function.

Figure 4.13 reproduces some of the results shown in [170] using a set of 18 steerable wedge filters.

In the case of junctions, the local analysis requires high orientational resolution. A lot of basis filters may be necessary. In [219] an approximate orientation steerability was proposed, using angular Gaussians. The number of basis filters is considerably decreased. The filters are designed in the spatial domain. With this approach two signatures are provided characterizing the closeness to lines or to edges. The article contains an interesting review of steerable filters, including the point of view of *deformable kernels* [142].

In a more abstract and general level [189] made an important contribution placing the creation of steerable functions in the context of Lie groups. The article shows

Fig. 4.11 Examples of adequate responses of a conventional steerable filter bank

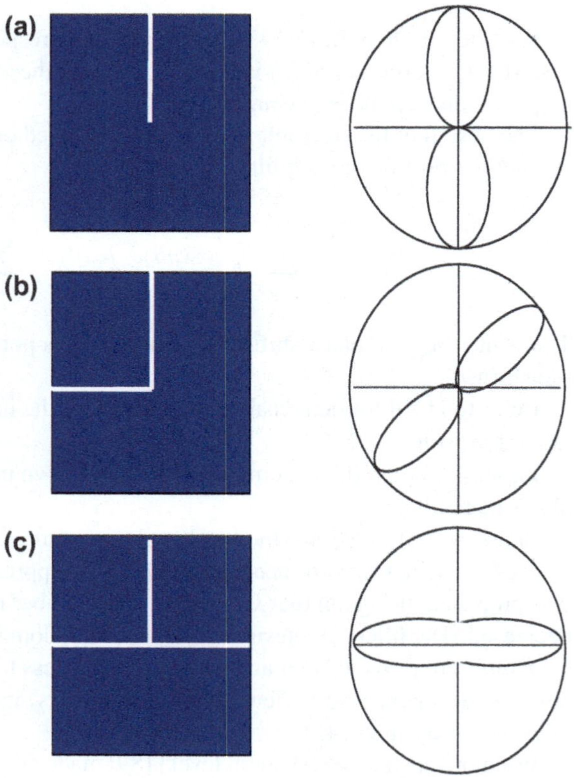

Fig. 4.12 Examples of incorrect responses

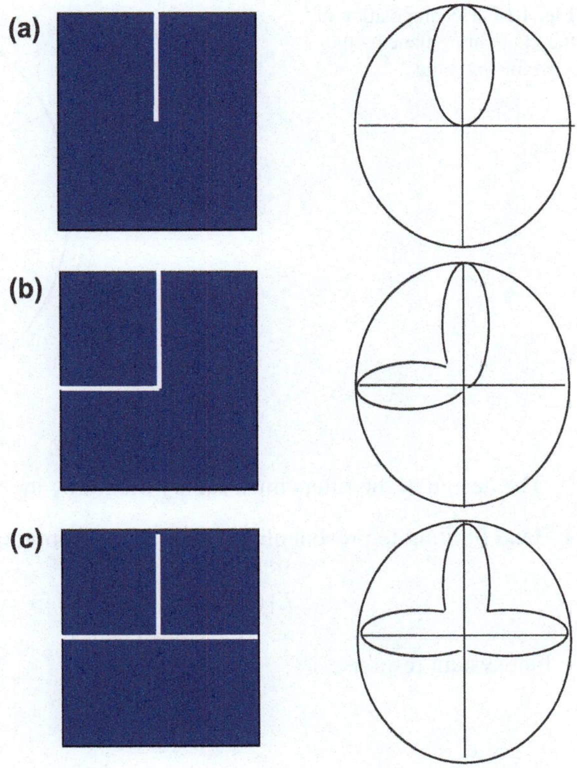

Fig. 4.13 Examples of better responses by using steerable wedge filters

examples of generation of steering polynomials. Essentially, the proposed method for computing basis functions is based on the Taylor expansion.

An interesting combination of Canny's method and steerable filters was proposed in [92]. In particular, one of the applications discussed in the paper was the detection of corners.

4.3.2 *Steerable Pyramid*

A first mention of steerable pyramids was done in [75]; later on, more details were given in [171]. The steerable pyramid combines a recursive multi-scale decomposition into sub-bands, and an orientation decomposition using steerable filters. The result of this combination can be represented on the 2D Fourier domain as depicted in Fig. 4.14.

Figure 4.15 sketches the structure of the steerable pyramid in terms of filters, taking as example a decomposition into K orientations. The pyramid is recursively constructed by inserting a copy of the dashed part in place of the solid circle.

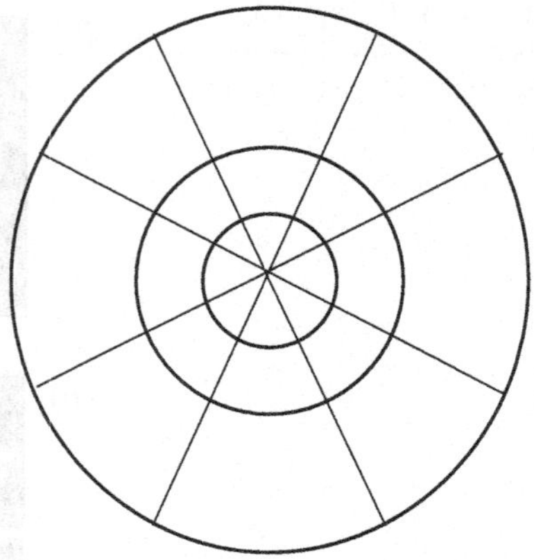

Fig. 4.14 Decomposition of the 2D Fourier plane by the steerable pyramid

The design of the filters must satisfy the following conditions:

- Band limiting to prevent aliasing due to subsampling:

$$L_1(\boldsymbol{\omega}) = 0, \quad for\, |\boldsymbol{\omega}| > \pi/2 \tag{4.14}$$

- Flat system response:

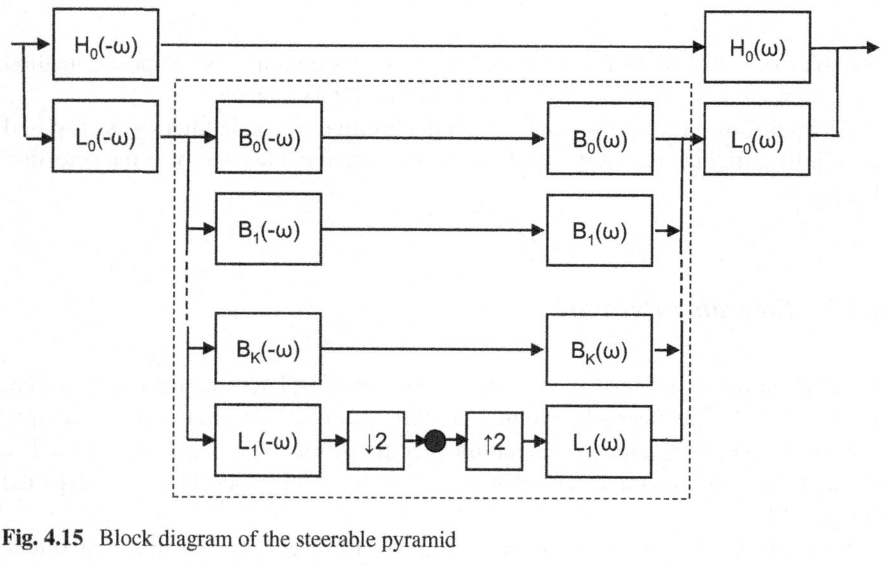

Fig. 4.15 Block diagram of the steerable pyramid

$$|H_0(\omega)|^2 + |L_0(\omega)|^2 \left[|L_1(\omega)|^2 + \sum_{j=1}^{K} |B_j(\omega)|^2 \right] = 1 \tag{4.15}$$

- Recursion:

$$|L_1(\omega/2)|^2 = |L_1(\omega/2)|^2 \left[|L_1(\omega)|^2 + \sum_{J=1}^{K} |B_j(\omega)|^2 \right] \tag{4.16}$$

typical choice would be: $L_0(\omega) = L_1(\omega/2)$.

In addition, the steerability condition can be expressed as:

$$B_j(\omega) = B(\omega)\,(-j\,\cos(\theta - \theta j))^K \tag{4.17}$$

where $\theta_j = \pi \cdot j/(K+1), \; j = 0, 1, \ldots, K.$
and:

$$B(\omega) = \sqrt{\sum_{j=1}^{K} |B_j(\omega)|^2} \tag{4.18}$$

A simplified design -with second derivatives- of the pyramid has been given in [212], based on raised cosine filters, using the following choices (in polar coordinates):

1. First low-pass filter

$$L_0(\omega) = \begin{cases} 1, \; |\omega| < \omega_1 \\ \sqrt{\frac{1}{2}(1 + \cos(\frac{\omega - \omega_1}{\omega_{max} - \omega_1} \cdot \pi))}\;, & \omega_1 \le |\omega| \le \omega_{max} \\ 0, \; |\omega| > \omega_{max} \end{cases} \tag{4.19}$$

where ω_{max} is the band limit of the image, and ω_1 is the cut-off frequency of the filter.

2. High-pass filter

$$H_0(\omega) = \begin{cases} 0, \; |\omega| < \omega_1 \\ \sqrt{\frac{1}{2}(1 - \cos(\frac{\omega - \omega_1}{\omega_{max} - \omega_1} \cdot \pi))}\;, & \omega_1 \le |\omega| \le \omega_{max} \\ 1, \; |\omega| > \omega_{max} \end{cases} \tag{4.20}$$

3. Second low-pass filter

$$L_1(\omega) = \begin{cases} 1, \; |\omega| < \omega_0 \\ \sqrt{\frac{1}{2}(1 - \cos(\frac{\omega - \omega_1}{\omega_{max} - \omega_1} \cdot \pi))}\;, & \omega_0 \le |\omega| \le \omega_1 \\ 0, \; |\omega| > \omega_1 \end{cases} \tag{4.21}$$

where ω_0 is the cut-off frequency of this filter.

4. Band-pass filters

$$R(\omega) = \begin{cases} 0, \ |\omega| < \omega_0 \\ \sqrt{\frac{1}{2}(1 - \cos(\frac{\omega - \omega_1}{\omega_{\max} - \omega_1} \cdot \pi))}\,, & \omega_0 \leq |\omega| \leq \omega_1 \\ 0, \ \omega_1 < |\omega| < \omega_{\max} \end{cases} \tag{4.22}$$

For second derivatives the minimum number of basis functions is 3, and so the band-pass filters are:

$$B_j(\omega, \theta) = R(\omega)\, \cos^2(\theta - \frac{2\pi j}{3}), \ j = 0, 1, 2. \tag{4.23}$$

In order to prepare for an example of steerable pyramid, a set of MATLAB functions has been developed on the basis of the filters just described. By using these functions, the code for the pyramid can be considerably simplified.

Function 4.7 stp_HI

```
function [hi,FHI]=stp_HI(FIMG,N,W1,WMAX)
% High-pass image filtering in 2D Fourier domain
% FIMG-the 2D FFT of the image (NxN size)
FHI=zeros(N,N); %space for filtered image
ny=0;
for wy=-pi:2*pi/(N-1):pi,
  ny=ny+1; nx=0;
  for wx=-pi:2*pi/(N-1):pi,
    nx=nx+1;
    w=sqrt(wx^2+wy^2);
    if w<W1, H=0;
    else
      if w>WMAX, H=1;
      else
        aux=cos(pi*(w-W1)/(WMAX-W1));
        H=sqrt(0.5*(1-aux));
      end;
    end;
    FHI(ny,nx)=FIMG(ny,nx)*H; %filtering in Fourier domain
  end;
end;
hi=ifft2(FHI); %inverse Fourier transform of filtered image
```

Function 4.8 stp_LI

```
function [li,FLI]=stp_LI(FIMG,N,W1,WMAX)
% Low-pass image filtering in 2D Fourier domain
% FIMG-the 2D FFT of the image (NxN size)
FLI=zeros(N,N); %space for filtered image
ny=0;
for wy=-pi:2*pi/(N-1):pi,
  ny=ny+1; nx=0;
  for wx=-pi:2*pi/(N-1):pi,
    nx=nx+1;
    w=sqrt(wx^2+wy^2);
    if w<W1, L=1;
    else
      if w>WMAX, L=0;
      else
        aux=cos(pi*(w-W1)/(WMAX-W1));
        L=sqrt(0.5*(1+aux));
      end;
    end;
    FLI(ny,nx)=FIMG(ny,nx)*L; %filtering in Fourier domain
  end;
end;
li=ifft2(FLI); %inverse Fourier transform of filtered image
```

Function 4.9 stp_BI

```
function [bi,FBI]=stp_BI(FIMG,N,W0,W1,ALPHA)
% Band-pass oriented image filtering in 2D Fourier domain
% FIMG-the 2D FFT of the image (NxN size)
FBI=zeros(N,N); %space for filtered image
ny=0; cs=1;
for wy=-pi:2*pi/(N-1):pi,
  ny=ny+1; nx=0;
  for wx=-pi:2*pi/(N-1):pi,
    nx=nx+1;
    w=sqrt(wx^2+wy^2);
    if w<W0, R=0;
    else
      if w>W1, R=0;
      else
        aux=cos(pi*(w-W0)/(W1-W0));
        theta=atan2(wy,wx);
        cs=cos(theta-ALPHA)^2;
        R=sqrt(0.5*(1-aux));
      end;
    end;
    FBI(ny,nx)=FIMG(ny,nx)*R*cs; %filtering in Fourier domain
  end;
end;
bi=ifft2(FBI); %inverse Fourier transform of filtered image
```

The Program 4.10 provides a steering pyramid implementation example. It uses the functions already introduced. The execution of the program may require a few

Fig. 4.16 Example of image analysis with a steerable pyramid: the first decomposition into high and low frequency

minutes. Three oriented basis functions have been considered, corresponding to 0°, 120° and 240°.

The photograph of a typical Holland mill has been chosen, for evident reasons of orientation features.

Figure 4.16 shows the outputs of the first decomposition of the image into high-frequency and low-frequency components.

Figure 4.17 shows the outputs corresponding to oriented filtering according with 0°, 120° and 240° orientations. Although the image features are not precisely aligned with these angles, it can be anyway noticed the distinct effect of each oriented filter. The image L1, that will be downsampled by half in both x and y coordinates for the next pyramid level, is shown at the bottom of Fig. 4.17.

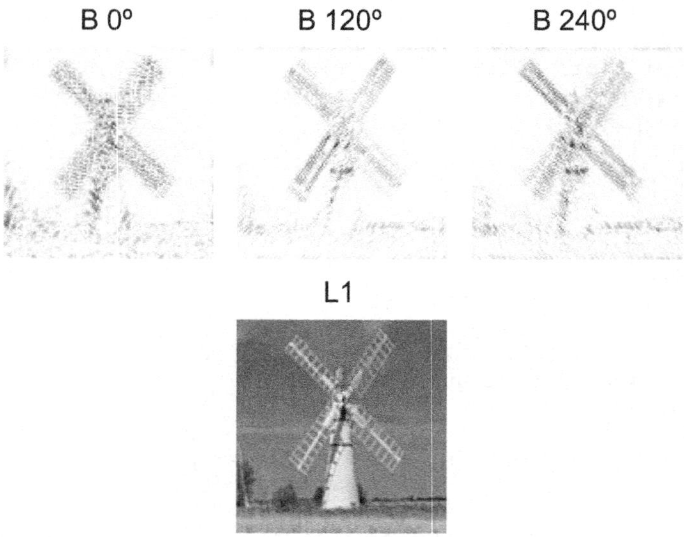

Fig. 4.17 Example of image analysis with a steerable pyramid: the second decomposition into 3 oriented bands and low-pass

Program 4.10 Steerable pyramid

```
% Steerable pyramid
%read the 256x256 image file into a matrix:
ufg=imread('holland1.jpg');
fg=double(ufg); %convert to float
N=256;
Ffg=fftshift(fft2(fg)); %2D Fourier transform of the image
%First decomposition------------------------------------------
w1=pi/2; wmax=pi;
[h0i,H0F]=stp_HI(Ffg,N,w1,wmax); %high-pass
[l0i,L0F]=stp_LI(Ffg,N,w1,wmax); %low-pass
M=256;
figure(1)
subplot(1,2,1)
ahi=abs(h0i); %only real values
nhi=(M/max(max(ahi)))*ahi; %normalize image range
uhi=uint8(M-nhi); %convert to unsigned 8-bit
imshow(uhi); %display the filtered image
title('H0');
subplot(1,2,2)
ali=abs(l0i); %only real values
nli=(M/max(max(ali)))*ali; %normalize image range
uli=uint8(nli); %convert to unsigned 8-bit
imshow(uli); %display the filtered image
title('L0');
%Second decomposition------------------------------------------
w0=pi/4; w1=pi/2;
[b10i,b10F]=stp_BI(L0F,N,w0,w1,0); %band-pass (0°)
[b11i,b11F]=stp_BI(L0F,N,w0,w1,2*pi/3); %band-pass (120°)
[b12i,b12F]=stp_BI(L0F,N,w0,w1,4*pi/3); %band-pass (240°)
[l1i,L1F]=stp_LI(L0F,N,w0,w1); %low-pass
L1F=L1F(1:2:end, 1:2:end);%subsampling
figure(2)
subplot(2,3,1)
ahi=abs(b10i); %only real values
nhi=(M/max(max(ahi)))*ahi; %normalize image range
uhi=uint8(M-nhi); %convert to unsigned 8-bit
imshow(uhi); %display the filtered image
title('B 0 deg');
subplot(2,3,2)
ahi=abs(b11i); %only real values
nhi=(M/max(max(ahi)))*ahi; %normalize image range
uhi=uint8(M-nhi); %convert to unsigned 8-bit
imshow(uhi); %display the filtered image
title('B 120 deg');
subplot(2,3,3)
ahi=abs(b12i); %only real values
nhi=(M/max(max(ahi)))*ahi; %normalize image range
uhi=uint8(M-nhi); %convert to unsigned 8-bit
imshow(uhi); %display the filtered image
title('B 240 deg');
subplot(2,3,5)
ahi=abs(l1i); %only real values
nhi=(M/max(max(ahi)))*ahi; %normalize image range
uhi=uint8(nhi); %convert to unsigned 8-bit
imshow(uhi); %display the filtered image
title('L1');
```

A extended treatement of steerable systems is found in [169] or [23]. Some recent research on the topics of this section is [200] on orientation spaces, and [196] about an unifying approach.

Available software can be obtained from MATLAB Central, like for instance "Steerable Gaussian Filters", or a "Simplified Steerable Pyramid". MATLAB Toolboxes with steerable filters and pyramids are "matlabPyrTools", and the Piotr's MATLAB Toolbox. The web page of Ken Castleman provides software and documentation. The Biomedical Imaging Group from Laussane (EPFL) offers animated demos in Java.

4.4 Application of Wavelets to Images

Images are represented with matrices. The 1-D wavelet transform can be applied to images in several ways. In the so-called *standard decomposition*, the transform is applied to all the columns, for all scales, and then to all the rows of the result. This is represented in Fig. 4.18.

Another way, which is called *nonstandard decomposition*, is to apply the transform to all rows, for the first scale, and then to apply the transform to the columns of the result. Figure 4.19 shows what is happening expressed as an image with four quadrants; labels L, for low pass, and H, for high pass, have been added.

The nonstandard process continues by decomposing the LL quadrant into four, by application of the transform first to rows and then to columns. And so on, scale after scale. This is represented in Fig. 4.20.

Fig. 4.18 Concept of the standard image decomposition using wavelets

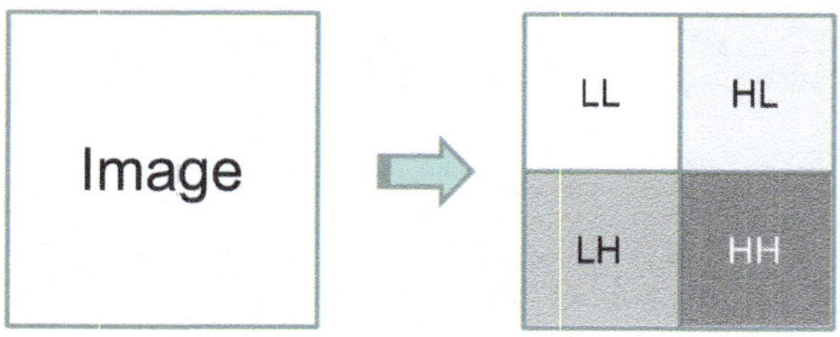

Fig. 4.19 The first step of the non standard decomposition using wavelets

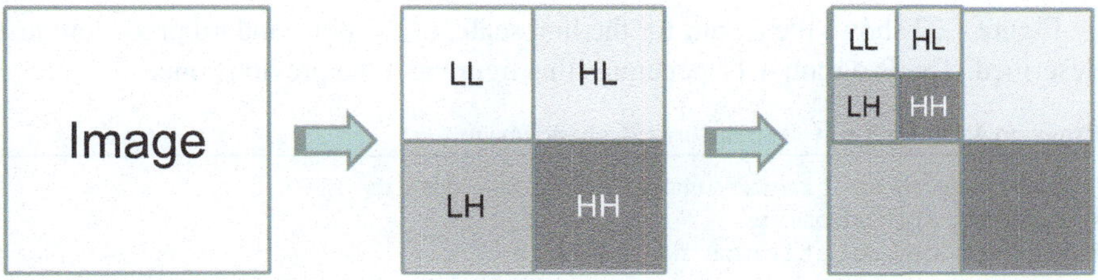

Fig. 4.20 The second step of the non standard decomposition using wavelets

Usually the preferred method is the nonstandard decomposition. Actually, this section uses this method. For the sake of simplicity, the Haar transform is chosen for the examples to be given; the reader could try other wavelets in order to compare the results.

4.4.1 Application to a Test Image

The nonstandard method obtains results with certain peculiarities. A good way for highlighting this, is to build a simple test figure and see the results of the wavelet decomposition.

Figure 4.21 shows a synthetic image to be used for our first basic example of image processing using wavelets. It is black & white image, represented by a 256×256 matrix with zeros (black) and ones (white).

Fig. 4.21 A synthetic image to be transformed

Figure 4.22 shows the result, for the first scale, of the nonstandard procedure just described. The Program 4.11 generates this figure and the previous one.

Program 4.11 Haar wavelet transform of a basic image

```
% Haar wavelet transform of a basic image
% square and rhombus
% direct calculations: no filters
%
%Original image
fg=ones(256,256); %white plane
%the rhombus
for n=1:32,
  fgx=64+(-n:n); fgy=96+n;
  fg(fgx,fgy)=0; %one triangle
  fg(fgx,256-fgy)=0; %the other triangle
end
%the square
fg(128:196,96:160)=0;
%
figure(1)
imshow(fg); %plots the binary image
title('Haar wavelet transform example');
ylabel('original');
%
%Step1
%1 scale wavelet transform of rows
c=1/sqrt(2);
lfg1=zeros(256,128); %space for image
hfg1=zeros(256,128); %""
for nn=1:256,
  auxL=fg(nn,1:2:256)+fg(nn,2:2:256);
  auxH=fg(nn,1:2:256)-fg(nn,2:2:256);
  lfg1(nn,1:128)=c*auxL; %image L
  hfg1(nn,1:128)=c*auxH; %image H
end;
%Step 2
%1 scale wavelet transform of columns of previous step
llfg1=zeros(128,128); %space for image
hlfg1=zeros(128,128); %""
hhfg1=zeros(128,128); %space for image
lhfg1=zeros(128,128); %""
%columns of L
for nn=1:128,
  auxL=lfg1(1:2:256,nn)+lfg1(2:2:256,nn);
  auxH=lfg1(1:2:256,nn)-lfg1(2:2:256,nn);
  llfg1(1:128,nn)=c*auxL; %image LL
  hlfg1(1:128,nn)=c*auxH; %image HL
end;
%columns of H
for nn=1:128,
  auxL=hfg1(1:2:256,nn)+hfg1(2:2:256,nn);
  auxH=hfg1(1:2:256,nn)-hfg1(2:2:256,nn);
  lhfg1(1:128,nn)=c*auxL; %image LH
  hhfg1(1:128,nn)=c*auxH; %image HH
```

```
end;
%normalization to bw
A=zeros(128,128);
Nllfg1=A; Nlhfg1=A; Nhlfg1=A; Nhhfg1=A;
for nn=1:128,
  for mm=1:128,
    if abs(llfg1(nn,mm))>0, Nllfg1(nn,mm)=1; end;
    if abs(lhfg1(nn,mm))>0, Nlhfg1(nn,mm)=1; end;
    if abs(hlfg1(nn,mm))>0, Nhlfg1(nn,mm)=1; end;
    if abs(hhfg1(nn,mm))>0, Nhhfg1(nn,mm)=1; end;
  end;
end;
%Figures
figure(2)
subplot(2,2,1)
imshow(Nllfg1); %LL image
title('LL');
subplot(2,2,2)
imshow(Nlhfg1); %LH image
title('HL');
subplot(2,2,3)
imshow(Nhlfg1); %HL image
title('LH');
subplot(2,2,4)
imshow(Nhhfg1); %HH image
title('HH');
```

What it is immediately noticeable in Fig. 4.22 is that horizontal details are emphasized in the quadrant HL, that vertical details came to LH, and diagonals were stressed in HH. This is the typical result of the nonstandard decomposition.

To close the analysis synthesis loop, it remains to recover the image from the wavelet decomposition. This is done with the Program 4.12. Figure 4.23 shows the successful result of image recovering.

Program 4.12 Haar wavelet transform of a basic image

```
% Haar wavelet transform of a basic image
% recovery of image
% square and rhombus
% direct calculations: no filters
%
%----------------------------------------
% First the transformed images
%
%Original image
fg=ones(256,256); %white plane
%the rhombus
for n=1:32,
  fgx=64+(-n:n); fgy=96+n;
  fg(fgx,fgy)=0; %one triangle
  fg(fgx,256-fgy)=0; %the other triangle
end
%the square
fg(128:196,96:160)=0;
```

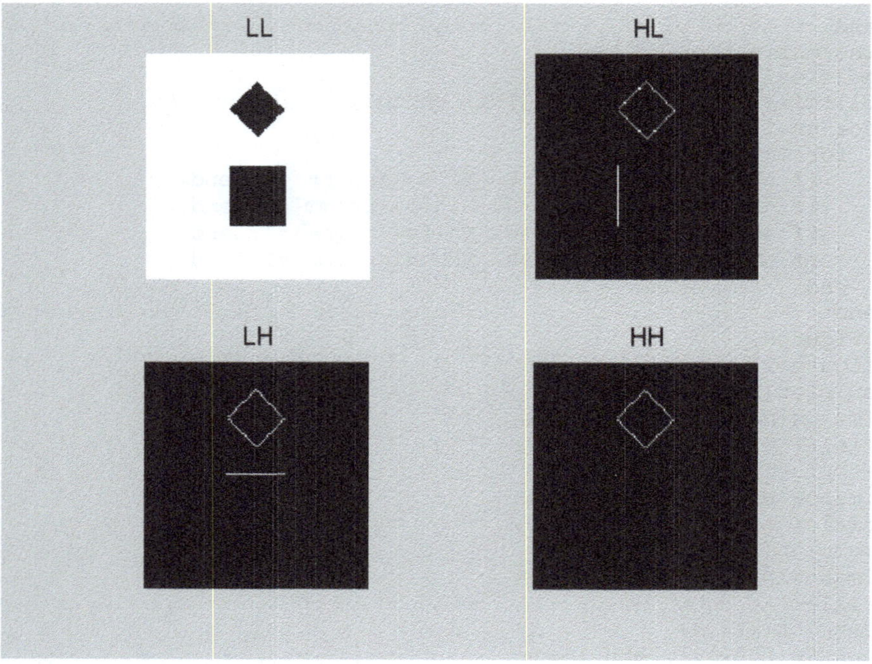

Fig. 4.22 The four images generated by a Haar transform

```
%
%Step1
%1 scale wavelet transform of rows
c=1/sqrt(2);
lfg1=zeros(256,128); %space for image
hfg1=zeros(256,128); %""
for nn=1:256,
  auxL=fg(nn,1:2:256)+fg(nn,2:2:256);
  auxH=fg(nn,1:2:256)-fg(nn,2:2:256);
  lfg1(nn,1:128)=c*auxL; %image L
  hfg1(nn,1:128)=c*auxH; %image H
end;
%Step 2
%1 scale wavelet transform of columns of previous step
llfg1=zeros(128,128); %space for image
hlfg1=zeros(128,128); %""
hhfg1=zeros(128,128); %space for image
lhfg1=zeros(128,128); %""
%columns of L
for nn=1:128,
  auxL=lfg1(1:2:256,nn)+lfg1(2:2:256,nn);
  auxH=lfg1(1:2:256,nn)-lfg1(2:2:256,nn);
  llfg1(1:128,nn)=c*auxL; %image LL
  hlfg1(1:128,nn)=c*auxH; %image HL
end;
%columns of H
```

Fig. 4.23 Image recovered from its Haar transform

```
for nn=1:128,
  auxL=hfg1(1:2:256,nn)+hfg1(2:2:256,nn);
  auxH=hfg1(1:2:256,nn)-hfg1(2:2:256,nn);
  lhfg1(1:128,nn)=c*auxL; %image LH
  hhfg1(1:128,nn)=c*auxH; %image HH
end;
%----------------------------------------
% Second the recovery of original image
% from LL,LH, HL and HH
%
rfg=zeros(256,256); %space for image
%
%recovery of L
for nn=1:128,
  auxL(1:2:256)=llfg1(1:128,nn)+hlfg1(1:128,nn);
  auxL(2:2:256)=llfg1(1:128,nn)-hlfg1(1:128,nn);
  lfg1(1:256,nn)=c*auxL; %image L
end;
%recovery of H
for nn=1:128,
  auxH(1:2:256)=lhfg1(1:128,nn)+hhfg1(1:128,nn);
  auxH(2:2:256)=lhfg1(1:128,nn)-hhfg1(1:128,nn);
  hfg1(1:256,nn)=c*auxH; %image H
end;
%recovery of original
for nn=1:256,
  auxL(1:2:256)=lfg1(nn,1:128)+hfg1(nn,1:128);
  auxL(2:2:256)=lfg1(nn,1:128)-hfg1(nn,1:128);
  rfg(nn,1:256)=c*auxL'; %image H
end;
```

Fig. 4.24 The four images generated by a Haar transform that uses filters

```
figure(1)
imshow(rfg);
title('recovered image');
```

As it has been shown in the chapter on wavelets, the wavelet transform could be implemented in several ways, like for instance using filters. Next program, Program 4.13, presents the analysis of the test figure using Haar filters. Actually the filtering has been done using the *conv2()* MATLAB function. The result, as depicted in Fig. 4.24 is the same.

Program 4.13 Haar wavelet transform of square and rhombus

```
% Haar wavelet transform of a basic image
% square and rhombus
% Using filters
%
%Haar filter
c=1/sqrt(2);
h0=[c c]; %low-pass filter
h1=[-c c]; %high-pass filter
%
fg=ones(256,256); %white plane
%the rhombus
for n=1:32,
  fgx=64+(-n:n); fgy=96+n;
  fg(fgx,fgy)=0; %one triangle
  fg(fgx,256-fgy)=0; %the other triangle
end
%the square
```

```
fg(128:196,96:160)=0;
%
%Step1
%1 scale wavelet transform of rows
lfg1=conv2(fg,h0); %low-pass filtering of rows
lfg1=lfg1(:,2:2:end); %downsampling to get L
hfg1=conv2(fg,h1); %high-pass filtering of rows
hfg1=hfg1(:,2:2:end); %downsampling to get H
%Step 2
%1 scale wavelet transform of columns of previous step
llfg1=conv2(lfg1,h0'); %low-pass filtering of L columns
llfg1=llfg1(2:2:end,:); %downsampling
hlfg1=conv2(lfg1,h1'); %high-pass filtering of L columns
hlfg1=hlfg1(2:2:end,:); %downsampling
lhfg1=conv2(hfg1,h0'); %low-pass filtering of H columns
lhfg1=lhfg1(2:2:end,:); %downsampling
hhfg1=conv2(hfg1,h1'); %high-pass filtering of H columns
hhfg1=hhfg1(2:2:end,:); %downsampling
%normalization to bw
A=zeros(128,128);
Nllfg1=A; Nlhfg1=A; Nhlfg1=A; Nhhfg1=A;
for nn=1:128,
  for mm=1:128,
    if abs(llfg1(nn,mm))>0, Nllfg1(nn,mm)=1; end;
    if abs(lhfg1(nn,mm))>0, Nlhfg1(nn,mm)=1; end;
    if abs(hlfg1(nn,mm))>0, Nhlfg1(nn,mm)=1; end;
    if abs(hhfg1(nn,mm))>0, Nhhfg1(nn,mm)=1; end;
  end;
end;
tfg1=zeros(256,256); %space for compound image
tfg1(1:128,1:128)=Nllfg1; %LL image
tfg1(1:128,129:256)=Nlhfg1; %LH image
tfg1(129:256,1:128)=Nhlfg1; %HL image
tfg1(129:256,129:256)=Nhhfg1; %HH image
%Figure
figure(1)
imshow(tfg1);
title('Haar image transform');
```

A program with the recovery of the image using filters is included in Appendix A.

4.4.2 Application to a Photograph

Indeed, the nonstandard method can be applied to photographs. Figure 4.25 shows a photograph that is chosen for a wavelet analysis exercise.

The nonstandard method can be easily implemented at several scales. For example, Fig. 4.26 shows the result of carrying the analysis to 2 scales. The figure has been generated with the Program 4.14 It is interesting to study how horizontal, vertical and diagonal details project to the figure quadrants.

Program 4.14 Haar 2-Level wavelet transform of a picture

```
% Haar 2-Level wavelet transform of an image
% Young image
% Using filters
%
%Haar filter
c=1/sqrt(2);
h0=[c c]; %low-pass filter
h1=[-c c]; %high-pass filter
%
% The image
ufg=imread('Chica.tif'); %read the image file into a matrix
fg=double(ufg); %convert to float
fg=fg-mean(mean(fg)); %zero mean
[Nr,Nc]=size(fg);
tfg=zeros(Nr,Nc); %space for wavelet plane
xfg=fg;
for nL=1:2, %levels
Nv=floor(Nr/2^nL); Nh=floor(Nc/2^nL);
lfg=zeros(2*Nv,Nh); hfg=zeros(2*Nv,Nh);
llfg=zeros(Nv,Nh); hlfg=zeros(Nv,Nh);
lhfg=zeros(Nv,Nh); hhfg=zeros(Nv,Nh);
%
%Step1
%wavelet transform of rows
aux=conv2(xfg,h0); %low-pass filtering of rows
lfg=0.5*aux(:,2:2:end); %downsampling to get L
aux=conv2(xfg,h1); %high-pass filtering of rows
hfg=0.5*aux(:,2:2:end); %downsampling to get H
%Step 2
%wavelet transform of columns of previous step
aux=conv2(lfg,h0'); %low-pass filtering of L columns
llfg=aux(2:2:end,:); %downsampling
aux=conv2(lfg,h1'); %high-pass filtering of L columns
hlfg=aux(2:2:end,:); %downsampling
aux=conv2(hfg,h0'); %low-pass filtering of H columns
lhfg=aux(2:2:end,:); %downsampling
aux=conv2(hfg,h1'); %high-pass filtering of H columns
hhfg=aux(2:2:end,:); %downsampling
%save on wavelet plane
V1=1:Nv; V2=Nv+1:2*Nv; H1=1:Nh; H2=Nh+1:2*Nh; %ranges
tfg(V1,H1)=llfg; tfg(V1,H2)=lhfg;
tfg(V2,H1)=hlfg; tfg(V2,H2)=hhfg;
xfg=llfg; %prepare next level
end;
%Figure
figure(1)
imshow(tfg,[-20 20]);
title('2-level Haar image transform');
```

In many cases, the image contained in the LL quadrant is still useful, with enough details for certain applications. This suggests basic ideas for image compression.

Fig. 4.25 A photograph to be transformed

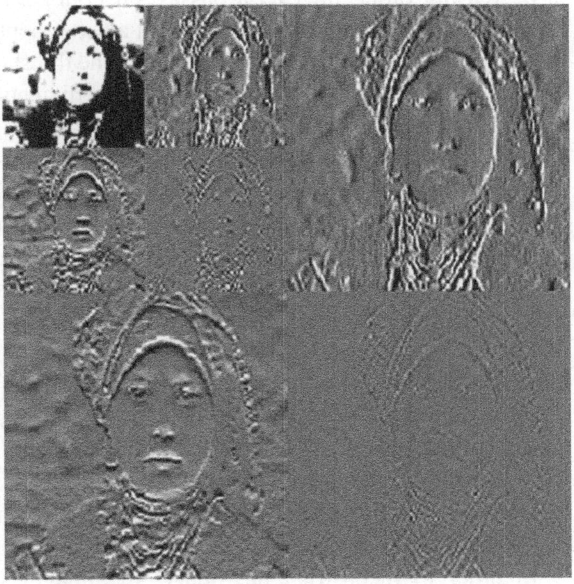

Fig. 4.26 The 8 sub-images generated by a 2-level Haar transform

4.4.3 Some Wavelet-Based Algorithms for Image Coding and Compression

In general, natural images have a low pass spectrum. When decomposing an image into four quadrants using wavelets, the "energy density" is larger in the LL quadrant: the wavelet coefficients are larger. These coefficients are more important.

A lot of practical research has been done for obtaining good image compression, which means high compression rate while keeping image quality. Some of the methods that have been proposed use wavelet analysis and then encoding of the obtained coefficients. As far as some information loss could be tolerated, the smaller wavelet coefficients could be neglected, thus getting more compression.

4.4.3.1 The EZW algorithm

The *Embedded Zerotree Wavelet* encoder was introduced in 1993 [166]. An intuitive explanation of EZW is [198], which will guide this subsection. After decomposing the image at several scales using wavelets, the wavelet coefficients are encoded in decreasing order, in several passes. Of course, this can be done with a direct encoding of the coefficients into bits, but a better approach is to use a threshold and just signal if the coefficients are larger than the threshold or not. The first pass uses a large threshold, identifies and encodes the coefficients above the threshold, and then removes them from the image. The next pass lowers the threshold and repeats the process. This is continued until all coefficients have been encoded.

If a established sequence of thresholds is used, it would be not necessary to transmit them to the decoder. For example, the thresholds could follow a sequence of powers of two: this is called bitplane coding. EZW uses this coding.

For better encoding efficiency, EZW uses zerotrees. It is based on a parent-child structure that can be observed in the wavelet analysis of images. The left hand size of Fig. 4.27 visualizes with arrows from parent to children this structure, the right hand side formalizes the structure as a quadtree (each parent node has four children).

In an approximated way, each parent coefficient can be computed as weighted average of the four children. Then, it is highly probable that if a parent is insignificant, the four children and the rest of descendants also are. This case is a zerotree and can be signaled with just one symbol (T). There are images with large grey zones that can be encoded as zerotrees.

In order to reconstruct the image, the image is scanned in a specific way. Figure 4.28 shows a typical order, called Morton scan.

Suppose a very simple example with only 4×4 wavelet coefficients. The lowest resolution level is a set of four wavelet coefficients. Denote as LLx this set. Considering the previous decomposition level, there are twelve more coefficients, corresponding to the sets LHx, HLx, and HHx. Now, suppose that one has compared the

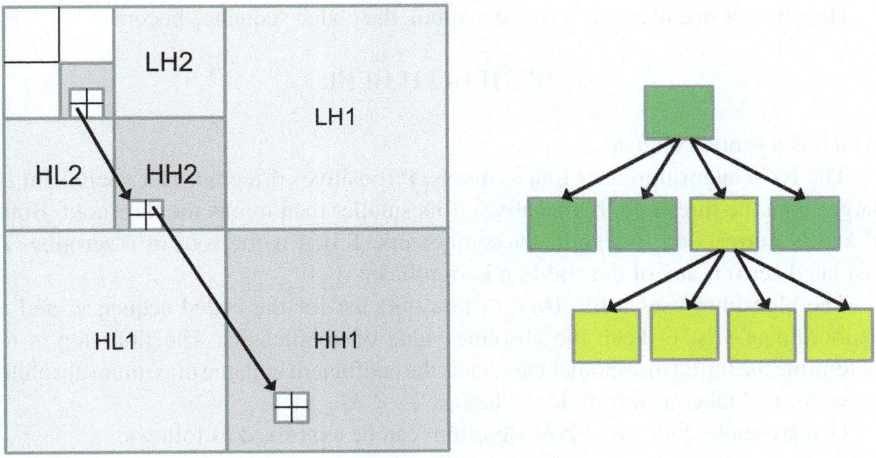

Fig. 4.27 Parent and children: quadtree

Fig. 4.28 Scanning order (Morton scan)

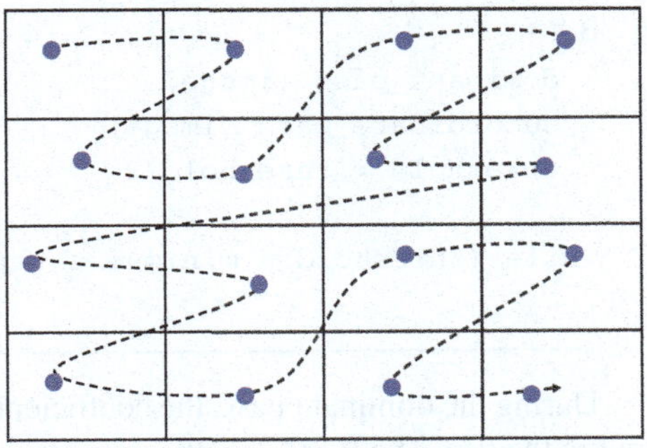

Fig. 4.29 Evaluation of wavelet coefficients, with respect to threshold

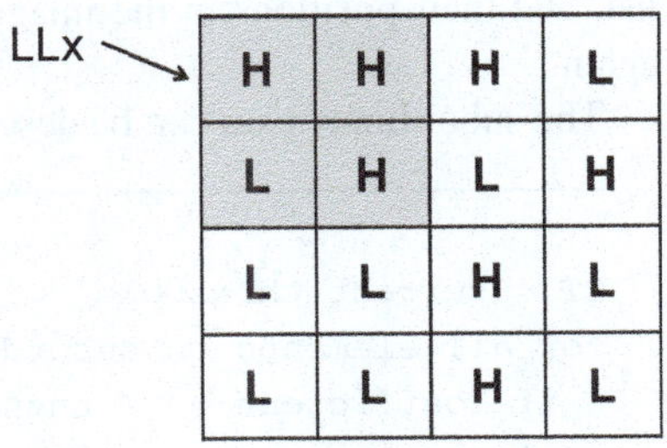

coefficients with a threshold, marking with H if the absolute value of the coefficient is higher than the threshold, or L if not. Figure 4.29 depicts one possible result.

Following the Morton scan order, the encoding result can be expressed with the sequence below:

HHLH HLLH LLLL HLHL

However, if one uses the zerotree symbol, the coded sequence becomes:

HHTH HLLH HLHL

which is a shorter sequence.

The EZW algorithm uses four symbols. P (positive) if the wavelet coefficient is larger than the threshold. N (negative) if its smaller than minus the threshold. Both P and N correspond to significant coefficients. T if it is the root of a zerotree. Z (isolated zero) if any of the children is significant.

The algorithm uses a fifo (first-in first-out) list for the coded sequence, and a "subordinate" list to keep the absolute value of coefficients. The first step is to determine the initial threshold. One selects the coefficient with the maximum absolute value M, and takes as threshold the largest $2^j < M$.

Using pseudo-code, the EZW algorithm can be expressed as follows:

```
threshold = initial_value
do {
  dominant_pass (image)
  subordinate_pass (image)
  threshold = threshold/2
}
while (threshold > minimum threshold)
```

During the dominant pass, the coefficients are encoded with the four symbols and put in the fifo. The P or N coefficients are included (without sign) in the subordinate list, and their positions in the image are filled with zeroes to prevent being coded again.

The subordinate pass can be described with the following pseudo-code:

```
CT = current threshold
for all elements on subordinate list do{
  if coefficient > CT then
   {output a 1
    coefficient= coefficient - CT
   }
  else output a 0
}
```

See [198] for a detailed numerical example. An application case is described in [188]. There are several MATLAB implementations available on Internet, from

Rice University, from MATLAB central, and from Leuven University (kuleuven). A tutorial on image compression, covering EZW is provided in [197].

4.4.3.2 The SPIHT algorithm

As suggested by [206], before describing the SPIHT algorithm, it is convenient to comment the STW (*Spatial-orientation Tree Wavelet*) algorithm. STW was introduced in 1993 [156]; it also considers zerotrees, but uses a state-transition model for encoding.

Take any of the wavelet coefficients $w(m)$, where m is an index according with the established scan order. Denote as $D(m)$ the set of quadtree descendants of $w(m)$. A significance function $S(m)$ is defined as max $|w(n)|$, $w(n) \in D(m)$ (if $D(m)$ is empty, $S(m) = \infty$).

Suppose that the current threshold is T. Four states are defined, two for insignificant values, I_R and I_V, and two for significant values, S_R and S_V:

$$m \in I_R \; iff \; |w(m)| < T, \; S(m) < T \tag{4.24}$$

$$m \in I_V \; iff \; |w(m)| < T, \; S(m) \geq T \tag{4.25}$$

$$m \in S_R \; iff \; |w(m)| \geq T, \; S(m) < T \tag{4.26}$$

$$m \in S_V \; iff \; |w(m)| \geq T, \; S(m) \geq T \tag{4.27}$$

The idea is to encode the transitions between states, as one proceeds from one threshold to the next. Figure 4.30 shows the state transition diagram.

Transitions take place when the threshold is decreased. Once a location m has arrived to S_V, it remains there. A simple binary coding can represent the transitions.

The STW algorithm uses a dominant list and a refinement list. The algorithm starts by choosing a threshold T_0 such all $|w(m)| < T_0$ and at least one $|w(m)| \geq T_0/2$. In the beginning the refinement list is empty. The initial contents of the dominant list is the coefficients at the lowest level resolution: LLx, LHx, HLx, and HHx.

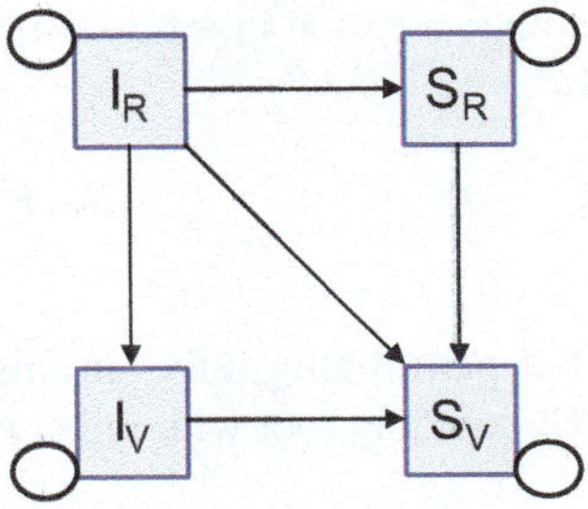

Fig. 4.30 STW state transition diagram

Here is a description of the STW algorithm using pseudo-code:

```
- Step 1: threshold update: T_(k) = T_(k-1)/2
- Step 2: dominant pass:
    Do:
      Get next m in dominant list
      Save old state So. Find new state Sn
      Output code for sate transition So-->Sn
      If Sn != So then do{
        If So != S_R and Sn !=I_V, then
          Append index m to refinement list
          Output sign of w(m) and set w_Q(m)=T_(k)
        If So !=I_V and Sn != S_R, then
          Append child indices of m to dominant list
        If Sn = S_V then
          Remove index m from dominant list}
    Loop until end of dominant list
- Step 3: refinement pass
    For each w(m) do{
      If |w(m)| inside [w_Q(m), w_Q(m)+ T_(k)], then
        Output bit 0
      Else if |w(m)| inside [w_Q(m)+ T_(k), w_Q(m)+ 2 T_(k),
       then
        Output bit 1
        Replace value of w_Q(m) by w_Q(m)+ T_(k)
      }
- Step 4: back to step 1
```

The *Set Partitioning in Hierarchical Trees* (SPIHT) algorithm was introduced in 1996 [157]. It is an important algorithm, widely used.

The algorithm employs three lists: the list of insignificant sets (LIS), the list of insignificant pixels (LIP), and the list of significant pixels (LSP). Given a set F of image location indexes, a significance function is defined so that:

$$S_T(F) = \begin{cases} 1, \; if \; \max\limits_{n \in F} |w(n)| \geq T \\ 0, \; if \; \max\limits_{n \in F} |w(n)| < T \end{cases} \tag{4.28}$$

Set partitioning refers in this case to a distinction concerning quadtree members. Choose any root with index m. Three sets are defined as follows:

- $D(m)$: descendant indices
- $C(m)$: child indices
- $G(m)$: grandchildren (descendants which are not children)

Notice that $G(m) = D(m)\text{-}C(m)$.

An index m is said to be significant or insignificant if the corresponding $w(m)$ is significant or insignificant with respect to the current threshold. It is in the LIS set where partitioning is recorded; in this set, the index m denotes either $D(m)$ or $G(m)$; in the first case the index m is said to be of type D, and in the second case of type G.

The algorithm starts by choosing a threshold T_0 such all $|w(m)| < T_0$ and at least one $|w(m)| \geq T_0/2$. The initial contents of SIP is LLx, LSP is empty, LIS includes all the indices in LLx that have descendants (assigning type D to them all).

In the sorting pass, the LIP is always visited first. For each coefficient in LIP, a significance test against the current threshold is done. If a coefficient is insignificant, a 0 is sent. If a coefficient is significant, a 1 is sent, and then a sign bit (0, + ; 1, −) is sent; and then the coefficient is moved to LSP.

The second action of the sorting pass focuses on LIS. A significance test is applied to each set in LIS. If a set is insignificant, a 0 is sent. If a set is significant, a 1 is sent and then the set is partitioned into $C(m)$ and $G(m)$. Each of the members of $C(m)$ are processed, if it is insignificant, a 0 is sent and the coefficient is moved to LIP; if it is significant, a 1 and then a sign bit are sent, and the coefficient is moved to LSP. If $G(m)$ is significant, then it is partitioned into four sets of descendants, and then each is processed in the same way (significance test and partitioning).

Here is a description of the SPIHT algorithm in pseudo-code. Notation is simplified by writing *Sk* in lieu of S_{T_k}.

```
- Step 1: threshold update: T_(k) = T_(k-)/2
- Step 2: sorting pass:
    For each m in LIP do:
      Output Sk(m)
      If Sk(m) = 1, then
        Move m to end of LSP
        Output sign of w(m)
        set w_Q(m)= T_(k)
    Continue until end of LIP

    For each m in LIS do:
      Output Sk(D(m))
      If Sk(D(m)) = 1, then
        For each n into C(m)
          Output Sk(n)
          If Sk(n) = 1, then
            Append n to LSP
```

```
            Output sign of w(n)
            set w_Q(m)= T_(k)
          Else If Sk(n) = 0, then
            Append n to LIP
        If G(m) != 0, then
          Move m to end of LIS as type G
        Else
          Remove m from LIS
      Else if m is of type G, then
        Output Sk(G(m))
        If Sk(G(m)) = 1, then
          Append C(m) to LIS, all type D indices
          Remove m from LIS
    Continue until end of LIS
- Step 3: refinement pass
    For each w(m) do
      If |w(m)| into [w_Q(m),w_Q(m)+ T_(k)], then
        Output bit 0
      Else if |w(m)| into [w_Q(m)+ T_(k) w_Q(m)+ 2 T_(k)],
       then
         Output bit 1
         Replace value of w_Q(m) by w_Q(m)+ T_(k)
- Step 4: back to step 1
```

Usually the algorithm implementations use a pair of position coordinates (i, j) instead of a number m.

Let us put a simple example [161]. The set of wavelet coefficients is the following (Fig. 4.31):

4.4.3.3 First Pass

The value of the threshold is set to $2^4 = 16$. The lists are initialized as follows:

LIP: (0,0) $\rightarrow$ 26; (0, 1) $\rightarrow$ 6; (1, 0) $\rightarrow$ -7; (1, 1) $\rightarrow$ 7
LIS: (0,1)D; (1,0)D; (1,1)D
LSP: {}

Fig. 4.31 Example of wavelet coefficients

26	6	13	10
-7	7	6	4
4	-4	4	-3
2	-2	-2	0

Examine LIP: the coefficient at (0,0) is significant, greater than 16, thus output a 1 and then a 0 to indicate the coefficient is positive. Move the coordinate (0,0) to LSP. The next three coefficients are insignificant, thus output three zeroes, and leave the coefficients in LIP.

Examine LIS: the descendants of (0,1) are (13, 10, 6, 4), none is significant, thus output a 0. The descendants of (0,1) and the descendants of (1,1) are all insignificant, thus output two zeroes.

In the refinement: do nothing.

Therefore, in this pass 8 bits have been transmitted:

10000000

4.4.3.4 Second Pass

After the first pass, the lists are now:

LIP: (0,1) → 6; (1,0) → −7; (1,1) → 7
LIS: (0,1)D; (1,0)D; (1,1)D
LSP: (0,0)→26

The threshold is set to $2^3 = 8$, (n = 3).

Examine LIP: all coefficients are insignificant, output three zeroes.

Examine LIS: two of the descendants of (0,1) are significant; thus $D(0, 1)$ is significant, output a 1. The first child is significant positive, thus output a 1 followed by a 0. The same happens with the second child, output 1 then 0. Move the coordinates of these two children to LSP. The other two children are insignificant, output a 0 for each, move them to LIP.

Since $G(0,1) = \{\}$, remove (0,1)D from LIS.

The other two elements of LIS, with coordinates (1,0) and (1,1), have no significant descendants; thus output two zeroes.

In the refinement, the content of LSP from the previous pass is examined. It has only one element, with value +26. This value is 11**1**0101 in binary (the first 1 is the sign bit), the nth MSB bit (in this case, the 3rd bit) is 1 (it has been highlighted in bold); thus, output a 1.

Therefore, in the second pass 13 bits have been transmitted:

0001101000001

4.4.3.5 Third Pass

After the second pass, the lists are now:

LIP: (0,1) → 6; (1,0) → −7; (1,1) → 7; (1,2) → 6; (1,3) → 4
LIS: (0,1)D; (1,1)D
LSP: (0,0) → 26; (0,2) → 13; (0,3) → 10

The threshold is set to $2^2 = 4$, (n = 2).

Examine LIP: all coefficients are significant, output 1011101010. Move all coefficients to LSP.

Examine LIS. $D(0,1)(4, -4, 2, -2)$ is significant, output a 1. Due to the child 4, output a 1 then 0; due to the child -4, output 1 then 1. Move the coordinates of these children to LSP. Due to 2 and -2, output two zeroes, and move these children to LIP.

Likewise, $D(1,1)(4, -3, -2, 0)$ is significant, output a 1. Due to the child 4, output 1 then 0. Due to -3, 2, 0, output three zeroes, and move the three children to LIP.

The refinement of 13 and 10, gives a 1 then a 0.

Therefore, in the third pass 25 bits have been transmitted:

1011101010110110011000010

After the third pass, the lists are now:

LIP: (3,0) $\rightarrow$ 2; (3,1) $\rightarrow$ -2; (2,3) $\rightarrow$ -3; (3,2) $\rightarrow$ -2; (3,3) $\rightarrow$ 0
LIS: {}
LSP: (0,0) $\rightarrow$ 26; (0,2) $\rightarrow$ 13; (0,3) $\rightarrow$ 10; (0,1) $\rightarrow$ 6; (1,0) $\rightarrow$ -7;
(1,1) $\rightarrow$ 7;
(1,2) $\rightarrow$ 6; (1,3) $\rightarrow$ 4; (2,0) $\rightarrow$ 4; (2,1) $\rightarrow$ -4; (2,2) $\rightarrow$ 4

One of the main features of the SPIHT algorithm is that the ordering data are not explicitely transmitted. This is not necessary, because the encoding process can be reversed by the decoder on the basis of the transmitted data. This is called embedded coding.

According with the technical literature, the results of SPIHT are really satisfactory. In the presence of data corruption, possibly caused by noise, the algorithm can have problems.

Several improvements of the algorithm have been proposed, like the simplified version of [52], or the modifications in [222]. Wavelet packets can be combined with SPIHT, like in [178]. Some extensions for 3D, color, video signals have been proposed [38, 139].

There is a SPIHT web page with documentation and software, also in MATLAB. A lot of information is provided in the web page of W.A. Pearlman. The MATLAB Central for file interchange offers several SPIHT implementations (the Wavelet Toolbox includes also this algorithm).

Along the years after SPIHT was introduced, other data encoding and compression algorithms have been proposed. See [74, 182] for state of the art surveys.

Some advanced topics and software can be found in the web page of J.E. Fowler.

4.5 New Wavelets for Images

Sparse signal representations, in the form of vectors or matrices, only have some non-zero coefficients, the rest of coefficients being zero.

When a 1-D signal is represented by wavelet analysis, a vector of coefficients is obtained. In general, it is desired to capture the information with a sparse representation: a few significant coefficients. This is convenient for compression and for transmission purposes.

It has been noticed [179], that in 2-D wavelet analysis, the edges of images are repeated at scale after scale, spoiling sparseness. While wavelet analysis works well on point singularities (1-D signal discontinuities, 2-D corners), they fail on higher dimensional singularities.

The problem with wavelets is their limited direction selectivity and iso-tropic scaling. It would be better to have wavelets with different directions, and not isotropic but elongated.

With this in mind, the research is actively working on sparse representations of images. Several wavelet variants have been proposed, departing from different points of view. This section intends to introduce the topic, with special attention to some well-received alternatives.

4.5.1 Perspective

Since so many wavelet variants have been created, it is opportune to start this section with a brief perspective of the topic. More details, with many bibliographic references, can be found in [124, 153].

Three main centres of activity could be identified, around Donoho and Candès; Vetterli and Do; Mallat, Le Pennec and Peyré.

In 1998, Donoho introduced *wedgelets*. It uses squares bisected by straight lines, at different scales. Regular edge discontinuities can be optimally approximated, A sub-optimal approximation technique was provided. The research extended the concept, proposing *wedgeprints*, *platelets*, and *surflets*. Wedgelets will be briefly described in the next subsection.

In 1999, Donoho and Candés introduced *ridgelets*. It is an anisotropic geometric wavelet transform. A ridgelet is a translated and dilated wavelet in one direction, and fixed in the perpendicular direction. It provides an optimal representation of straight line singularities. To analyze local singularities (lines, curves) a reasonable way is to decompose the image into parts, and then apply ridgelet transform to the subimages. This is the idea of *curvelets*, which were introduced by the same two authors in 2000.

First curvelets were not easy to implement. A *second generation* of curvelets were introduced in 2003. They are based on a frequency plane tiling. Two approaches were proposed for digital implementation. A next subsection on curvelets will give details on it.

It was noticed that curvelets lead to significant representation redundancy. In order to alleviate this problem, Vetterli and Do introduced *contourlets*, in 2002. They are based on a discrete directional filter bank structure. The transform has a simple discrete formulation, easy to implement. Contourlets are, like the curvelets, elongated and oscillatory along their width. There is a *second generation* of contourlets, proposed in 2006. A multi-dimensional version is the *surfacelet*.

Both curvelets and contourlets are non-adaptive. In 2005, Le Pennec and Mallat introduced the *bandelets*, which are signal-adaptive transforms. Soon after, a *second generation* of bandelets was proposed by Peyré and Mallat. The bandelets could be visualized in the space domain as a flow of arrows parallel to edges.

A fourth centre of activity can be added to those just mentioned, under the name *shearlets*. The shearlets were introduced in 2005–2006 [64, 84, 103] by several authors: Labate, Kutyniok, Easley, Guo, Lim, etc. It is based on higher dimension wavelets, that can be used for a frequency-plane decomposition using wedge-shaped tiles.

There are other proposed variants that will be described in the last subsection.

4.5.2 Wedgelets

The wedgelet approximation is a geometry based image representation [61]. It uses a quadtree decomposition of the image, according with dyadic squares (Fig. 4.32).

Denote as δ the side length of the smallest square considered. Mark square sides with spacing δ, like it is depicted in Fig. 4.33.

Now, it is possible to connect pairs of marked points in the same square with line segments. Figure 4.34 shows some examples:

These line segments were denoted as '*edgelets*' in first Donoho papers, and then it changed to '*beamlets*'. Armed with these line segments, it is possible to obtain piecewise approximations of curves, and so it can be applied to the mathematical representation of image contours.

A '*wedgelet*' (w) is a function on a dyadic square S that is piecewise constant on either side of an edgelet through S. The function is defined by four parameters,

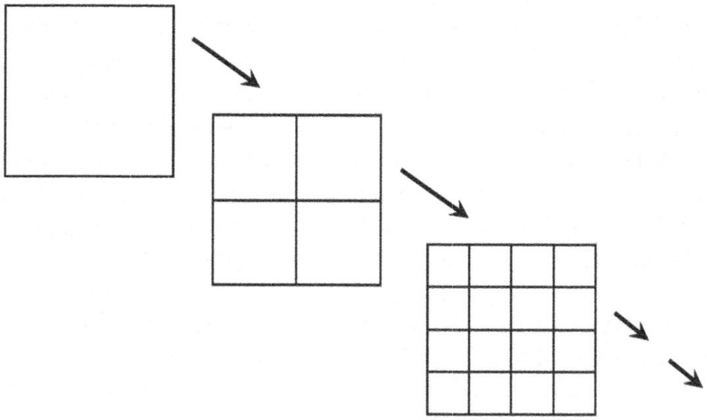

Fig. 4.32 Quadtree decomposition into dyadic squares

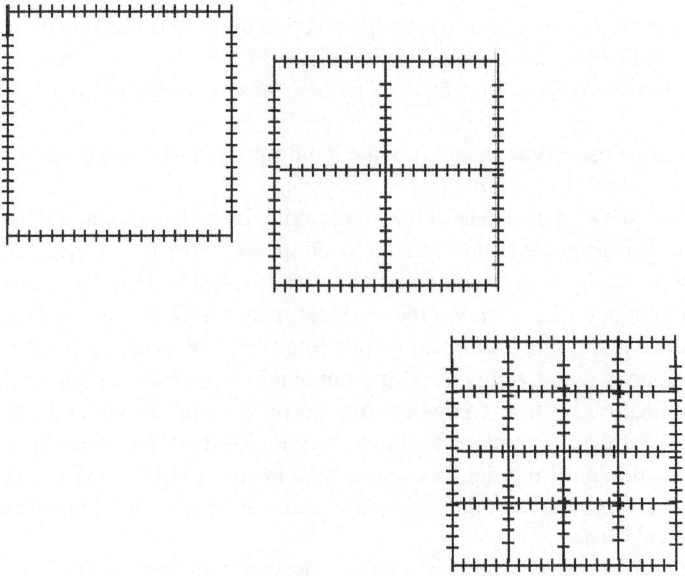

Fig. 4.33 Marks on the dyadic squares

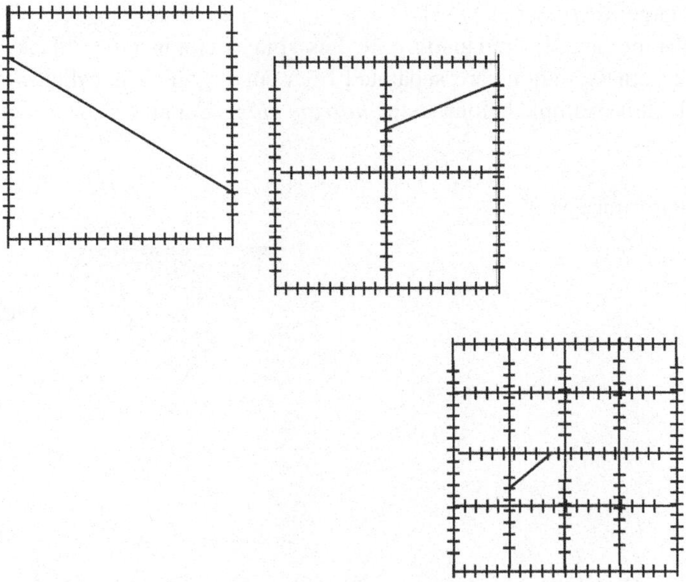

Fig. 4.34 Examples of line segments from mark to mark in the same square

the points (v_1, v_2) where the edgelet intersects with S, and the values that w takes above, c_a, and below, c_b, the edgelet. See Fig. 4.35.

A function that is constant over all S is a *degenerate* wedgelet (there is no edgelet in S).

Suppose that there is an image A in the intial square, and that the quadtree decomposition is applied (Fig. 4.36).

Take for instance a fragment of the image, contained in a particular dyadic square. The fragment is projected onto a wedgelet as shown in the Fig. 4.37.

For the wedgelet projection, an orientation (v_1, v_2) is found such as the error between the image fragment and the wedgelet is minimized. The constants c_a, and c_b are found by averaging in the corresponding two regions of the fragment.

It may happen that dividing a certain square into four, does not improve too much the approximation by four wedgelets instead of just one. In general, the approximation of the image by wedgelets should be judiciously done, combining economy of representation and sufficient accuracy. That means a kind of mini-max problem, minimize the maximum among two penalties: too expensive, too imprecise. See [56, 152] about this issue.

Later on this book will address image representation approaches based on '*dictionaries*', which are collections of pieces that can be combined for the expression of images. Frequently wedgelets are considered as a dictionary, sometimes included as part of larger dictionaries [153].

Consider the case pictured in Fig. 4.38. It is a binary (for instance, black and white) image. The binary regions are separated by a curve, which is called a '*horizon*'. Images like this example belong to the *horizon class* of images.

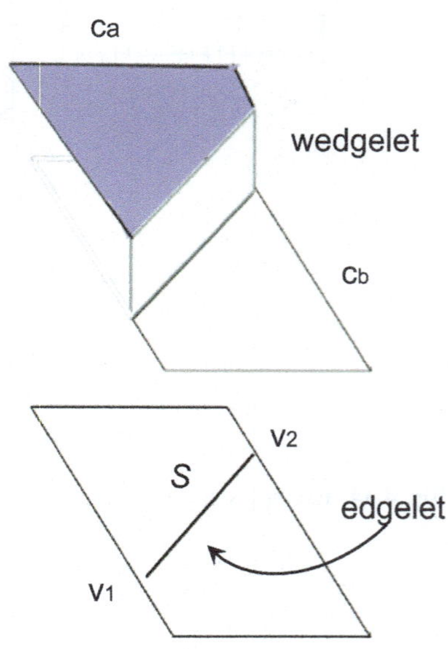

Fig. 4.35 Representation of a wedgelet

Fig. 4.36 Image and quadtree decomposition

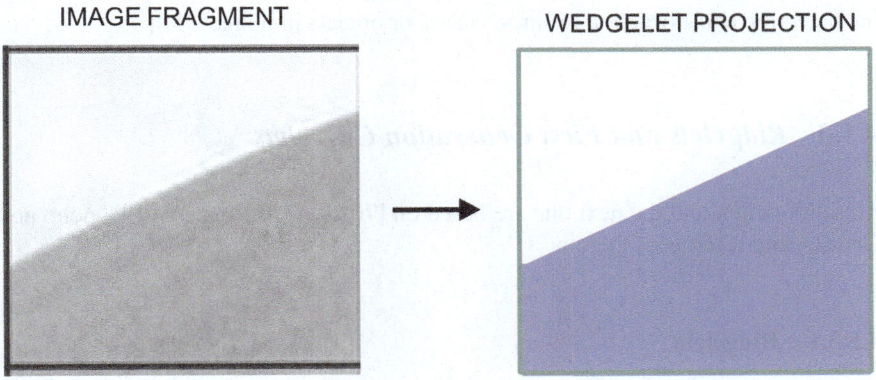

Fig. 4.37 Projection of an image fragment onto a wedgelet

Horizon images are well suited for representation in terms of wedgelets. Moreover, it is no longer necessary to use $w\,(v_1,\ v_2,\ c_a,\ c_b)$; a simpler version $w\,(v_1,\ v_2)$ can be used instead.

A MATLAB Toolbox, called BeamLab, is available via Internet. It includes implementations of beamlet and wedgelet analysis. There is also a web site related to this toolbox, which includes documentation and scientific papers.

There is a version of wedgelets adapted for the use of the same lifting scheme employed for wavelets [76].

Some of the extensions of this topic are '*platelets*, [208], '*surflets*' [27], and '*arclets*' [146]. Second order wedgelets, with curves instead of lines for the edgelets, have been proposed [110].

Fig. 4.38 An example of horizon image

A numer of applications of wedgelets have been published, like for instance their use for appearance models of human faces [43], for image denoising [46], and for the detection and extraction of lines, curves or objects in images [58]

4.5.3 *Ridgelets and First Generation Curvelets*

This subsection and the next one are based on [70, 124], the papers of Donoho and Candès, and references therein.

4.5.3.1 Ridgelets

Since the introduction of ridgelets [19], a significant number of papers have considered ridgelet applications, improvements, and related topics.

In order to obtain a ridgelet, a smooth univariate function ψ is chosen. The function satisfies the following admissibility condition:

$$\int \frac{|\Psi(\omega)|^2}{|\omega|^2}\, d\omega < \infty \tag{4.29}$$

It is assumed that ψ is normalized, so the above expression becomes equal to 1.

The bivariate ridgelet is defined by:

$$\psi_{a,b,\theta}(\mathbf{x}) = a^{-1/2}\, \psi((x\,\cos\theta + y\sin\theta - b)/a) \tag{4.30}$$

where $\mathbf{x}$ are 2-D coordinates, $a > 0$ and real, b is a real, and θ corresponds to a rotation $\theta \in [0, 2\pi])$.

This function is a constant along lines $x \cos\theta + y \sin\theta = const.$ and it is a wavelet ψ in the orthogonal direction (scale and shift parameters of the wavelet are a and b respectively). Therefore, the ridgelet can be rotated, and the wavelet part can be dilated and scaled.

The coefficients of the ridgelet transform of a function $f()$ are given by:

$$R_f(a, b, \theta) = \int \psi_{a,b,\theta}(\mathbf{x})\, f(\mathbf{x})\, d\mathbf{x} \tag{4.31}$$

The exact reconstruction can be achieved with:

$$f(\mathbf{x}) = \int_0^{2\pi} \int_{-\infty}^{\infty} \int_0^{\infty} R_f(a, b, \theta)\, \psi_{a,b,\theta}(\mathbf{x})\, \frac{da}{a^3}\, db\, \frac{d\theta}{4\pi} \tag{4.32}$$

The idea behind the ridgelet definition is related with the Radon transform. Since the Radon transform of $f()$ is:

$$\Re(\theta, r) = \int f(\mathbf{x})\, \delta(x \cos\theta + y \sin\theta - r)\, dx\, dy \tag{4.33}$$

where δ is the Dirac distribution.

Then:

$$R_f(a, b, \theta) = \int \Re(\theta, r)\, a^{-1/2} \psi((r - b)/a))\, dr \tag{4.34}$$

Therefore the ridgelet transform can be defined as the application of a 1-D wavelet to the Radon transform. Actually, the purpose of the ridgelet transform is to translate singularities along lines into point singularities where wavelets are proficient.

Figure 4.39 is a representation of a ridgelet along a certain inclined direction on the image domain.

Several discrete ridgelet transforms have been proposed. The basic procedure is to obtain a discrete Radon transform (DRT), and then apply a 1-D wavelet.

Based on the projection slice theorem, it is possible to obtain the DRT as follows: compute first the 2-D Fourier transform of the image, then convert from Cartesian to polar grid by interpolation (gridding). Finally, an inverse 1-D Fourier transform is applied to the rays. This was the first approach of Donoho and collaborators (see [179]).

Once the DRT is obtained, a 1-D wavelet is applied to obtain the ridgelet coefficients. The complete chain of calculations is summarized [8] as a block diagram in Fig. 4.40.

This approach result in non-orthogonal ridgelets. In other words, they are redundant, so the transform has a larger size than the original image.

Fig. 4.39 Representation of a ridgelet

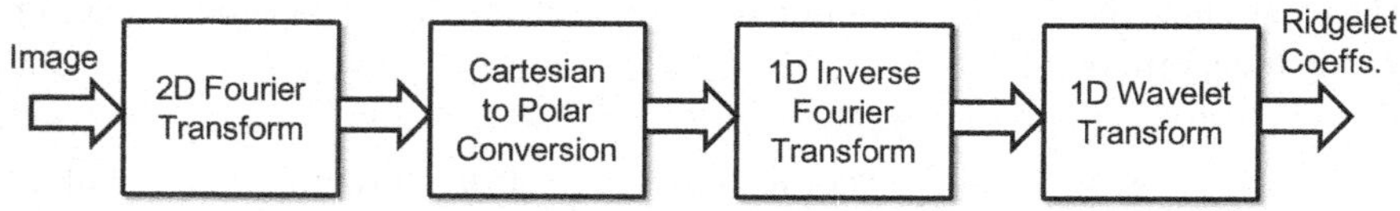

Fig. 4.40 Transformation steps

In an effort to obtain orthogonal ridgelets, Do and Vetterli [54] proposed a calculation of the DRT on the image space, called finite Radon transform (FRAT). A more general definition of lines was proposed, to obtain an optimal ordering of FRAT coefficients. Although FRAT is redundant, the 1-D wavelet transform was applied in such a way that the result, called finite ridgelet transform (FRIT), was orthonormal. The problem with this approach is that it is not geometrically faithful.

Looking at pictures in detail, at pixel level, the difficulty of recognizing a line becomes clear, and this important for computing the DRT. Actually this is one of the main points discussed in [54]. It was proposed in [22] to use '*discrete analytical lines*' in the 2-D Fourier domain, these lines have an arithmetical thickness that can be adapted for specific applications. Based on this approach, a discrete analytical ridgelet transform (DART) was developed. The DART is redundant, but less than the pseudo-polar ridgelet.

Once the DRT is obtained, it remains a second step: the 1-D wavelet transform. It was realized that this transform should be compact in the Fourier domain. The research has used the Meyer wavelet, the Symlet, or a re-normalized B3-spline (in this case, the wavelet coefficients were directly computed in the Fourier space).

After the introduction of wedgelets, the research of Donoho, Candés and collaborators, has proposed a number of interesting extensions and improvements. In [57] *orthonormal wedgelets* were introduced. The article [179] describes in detail the obtention of an *approximate digital ridgelet transform*, and introduces *local ridgelets,* which can be organized as ridgelet pyramids. According with the summary of [70], local ridgelets use a decomposition of the image into overlapping blocks, and thus the redundancy factor becomes 4.

The contribution [9] has exerted a large influence on the directional analysis of images based on the Radon transform, in the form of '*Fast Slant Stack*'. Many papers cite this reference. In particular [60] defines the digital ridgelet transform as the 1-D Meyer wavelet transform of the Slant Stack Radon transform. The polar grid was replaced by a pseudo-polar grid. This article, [60], is particularly interesting, since it includes a retrospective analysis of his research, and includes a final discussion about a new form of obtaining the ridgelet transform, by zero-padding the image to twice its length, then shearing and projecting, and then applying the 1-D wavelet. The reverse transform is also described. A related development of ideas is [73], which proposes a digital implementation of ridgelet packets.

Ridgelets are efficient to represent lines that go through the entire image. However, real images usually have smaller edges (lines or curves). This is why a second step was taken, and a first generation of curvelets was proposed.

There are some available ridgelet implementations in MATLAB, like the Ridgelet and Curvelet first generation Toolbox, the BeamLab, and the FRIT Toolbox (that corresponds to [54]).

4.5.3.2 Curvelets (First Generation)

Many papers have been published about curvelets, in their various forms. One of these, the [21] article, includes many interesting insights and pertinent references. When introducing curvelets [21] starts by considering multiscale ridgelets, using ortho-ridgelets of "length" 1 and "width" as fine as desired; the multiscale ridgelets renormalize and transport the ortho-ridgelets, obtaining a set of all orientations, lengths and finer widths. The result is a too much large dictionary. For the construction of curvelets, the dictionary was restricted to multiscale ridgelets with $width \approx length^2$. To obtain such multiscale ridgelets, a specific decomposition of the image into sub-bands was devised.

For the sub-band decomposition, a sequence of filters Φ_0 (low-pass) and $\Psi_{2s} = 2^{4s}\Psi(2^{2s})$, $s = 0, 1, 2, \ldots$ (band-pass) is built.

The curvelet transform of the image f follows three steps:

- Decomposition into sub-bands
- Smooth partitioning of each sub-band into squares
- Ridgelet analysis of each square

For the smooth partitioning a smooth window is designed, so the side of each square is 2^{-s}. The sub-band s contains ridges of width 2^{-2s} that live inside these squares.

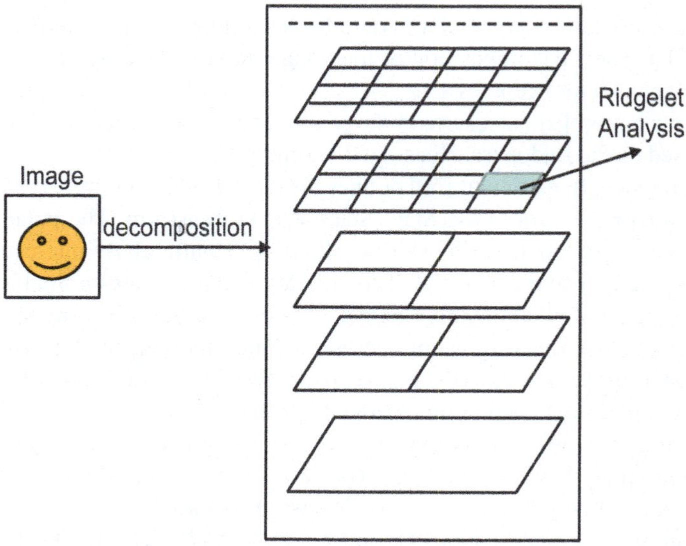

Fig. 4.41 Procedure of the first generation curvelet transform

In this way, a scaling law is respected. This law is: $width \approx length^2$. It can be regarded as a (parabolic) aspect ratio.

Figure 4.41 depicts the procedure for the curvelet analysis.

As it has been described in the chapter on wavelets, the wavelet analysis uses a decomposition into dyadic sub-bands $[2^s, 2^{s+1}]$; this could be taken as the standard decomposition. However, in the case of curvelets, the image is decomposed into $[2^s, 2^{s+2}]$ sub-bands, which is a non-standard way [70, 179].

After smooth partitioning, the resulting squares are renormalized to unit scale [179].

Implementation details of a first toolbox version, called Curvelet 256, are given in [59]. The subband decomposition was actually done using 2-D wavelet, and merging the two dyadic sub-bands $[2^s, 2^{s+1}]$ and $[2^{s+1}, 2^{s+2}]$; according with the comments of [59] this kind of decomposition injects directional artifacts. The complete curvelet analysis of a 256×256 image generates about 1 M coefficients, which means a redundancy factor around 16.

The inversion of the curvelet transform includes four steps:

- Ridgelet synthesis
- Renormalization
- Smooth integration of squares
- Sub-band recomposition

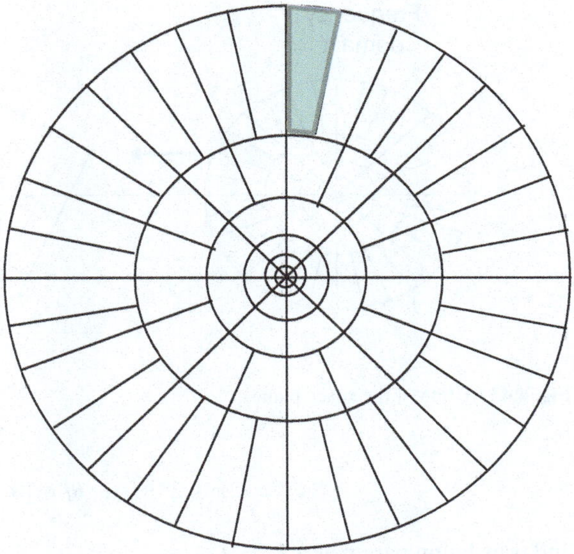

Fig. 4.42 Tiling of the frequency plane

4.5.4 *Curvelets (Second Generation)*

Suppose you have applied the 2-D Fourier transform to your image. Recall from Sect. 4.4 that low frequencies go to the centre of the frequency domain plane, and high frequencies take a radial distance from it.

4.5.4.1 Continuous-Time Curvelets

The second generation of curvelets considers a tiling of the frequency plane. The continuous-time curvelet is based on a polar tiling [6, 120], as shown in Fig. 4.42.

The tiles can be obtained by intersections of radial windows W, and angular windows V. Then, a tile U (a polar 'wedge'), like the tile highlighted in the figure, could be expressed as:

$$U_j\ (r,\ \theta)\ =\ 2^{(-3\,j/4)}\ W(2^{-j}r)\ V(\frac{2^{\lfloor j/2 \rfloor}\theta}{2\pi}) \tag{4.35}$$

where $\lfloor j/2 \rfloor\ =\ floor(j/2)$ and j is the scale.

The mother curvelet is defined in the Fourier domain as:

$$\Psi_j(\omega)\ =\ U_j(\omega) \tag{4.36}$$

Let us consider an equispaced set of rotation angles θ_l $(l = 0,\ 1,\ 2,\ \ldots)$:

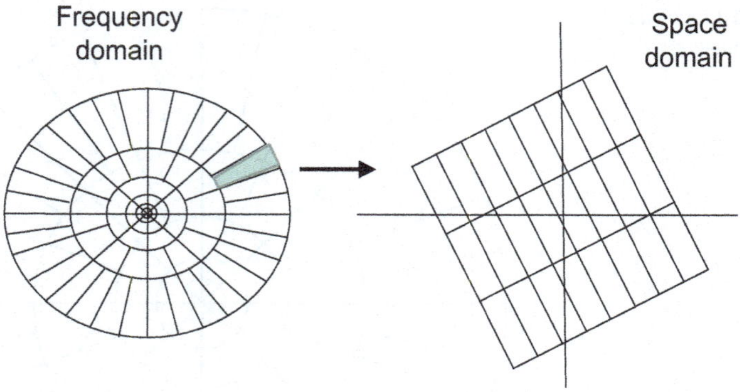

Fig. 4.43 Tiling of the space plane

$$\theta_l = 2\pi \cdot 2^{-\lfloor j/2 \rfloor}, \quad \theta_l \in [0,\ 2\pi] \tag{4.37}$$

and translation parameters $k = (k_x,\ k_y)$.

Then, in the space domain, the curvelets are:

$$\psi_{j,l,k}(\mathbf{x}) = \psi_j(R_\theta\,(\mathbf{x} - \mathbf{x}_k^{(j,l)}\,)) \tag{4.38}$$

where R_θ is the rotation by θ radians.

Curvelets are almost orthogonal, they are tight frames with a moderate redundancy. Curvelets are local both in the space domain and in the frequency plane. They are multi-scale and multi-directional. They are anisotropic, looking as ellipsoids, oscillatory along their short direction and smooth along the other direction. At fine scale, they look as needles.

Choose any of the frequency domain tiles, with a certain scale and angle. The corresponding tiling of the space domain is shown in Fig. 4.43. The dimensions of the rectangles are determined by the parabolic aspect ratio.

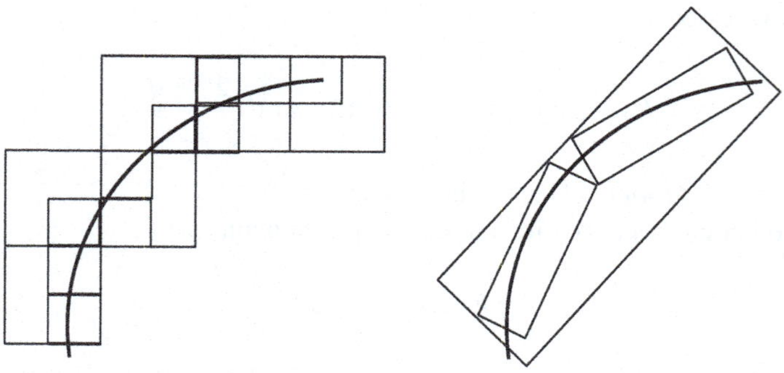

Fig. 4.44 Representation of a curve: (*left*) with wavelets, (*right*) with curvelets

Fig. 4.45 Curve and curvelets

The main idea of curvelets is depicted in Fig. 4.44. By using 2-D wavelets (right) the representation of a curve takes many coefficients. If you use curvelets, instead, a few coefficients would be enough.

Figure 4.45 visualizes three particular situations, corresponding to a curve being represented by curvelets. On the left, one curvelet remains out from the curve, the corresponding transform coefficient, *dj*, will be zero. On the centre, curvelet and curve intersects; the coefficient *dj* would have a small non-zero value. On the right, curvelet and curve have large overlapping, the coefficient *dj* would have a large value.

Some specific choices of V and W windows have been proposed in the literature. For example [120] describes the use of the following Meyer windows:

$$V(\theta) = \begin{cases} 1 & |\theta| \leq 1/3 \\ \cos\left(\frac{\pi}{2}\, v(3\,|\theta| - 1)\right) & 1/3 \leq |\theta| \leq 2/3 \\ 0 & else \end{cases} \tag{4.39}$$

$$W(r) = \begin{cases} \cos\left(\frac{\pi}{2}\, v(5 - 6r)\right) & 2/3 \leq r \leq 5/6 \\ 1 & 5/6 \leq r \leq 4/3 \\ \cos\left(\frac{\pi}{2}\, v(3r - 4)\right) & 4/3 \leq r \leq 5/3 \\ 0 & else \end{cases} \tag{4.40}$$

where $v(x)$ is a smooth function with $v(x) + v(1-x) = 1$, and $v(x) = 0,\ x \leq 0$, and $v(x) = 1,\ x \geq 1$.

For instance, a candidate for $v(x)$ is the following:

$$v(x) = \begin{cases} 0 & x \leq 0 \\ \frac{s(x-1)}{s(x-1)+s(x)} & 0 < x < 1 \\ 1 & x \geq 1 \end{cases} \tag{4.41}$$

with:

$$s(x) = \exp - \left(\frac{1}{(1+x)^2} + \frac{1}{(1-x)^2} \right) \tag{4.42}$$

Figure 4.46 shows a 3D view of the basic curvelet obtained by the product of V and W. A top view is shown in Fig. 4.47. This corresponds to the curvelet support on the frequency domain, which is very approximately the desired tile. Both figures have

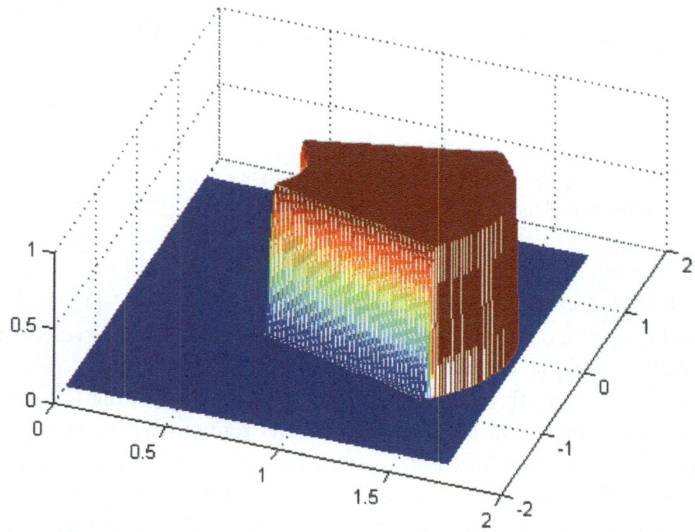

Fig. 4.46 3D view of the basic curvelet example

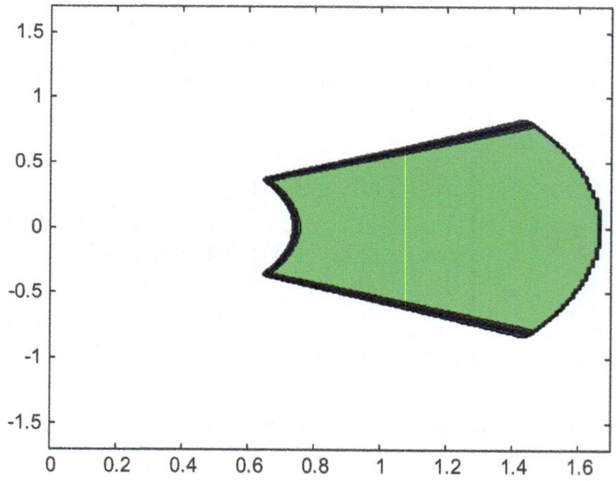

Fig. 4.47 Top view of the basic curvelet

been obtained with the Program A.10, which has been included in the Appendix A. An interesting exercise for the reader could be to represent the windows V and W.

4.5.4.2 Digital Curvelet Transform

Under the name 'Fast Discrete Curvelet Tansform' [20] introduced a digital implementation of the second generation curvelets. Actually, it contains two different

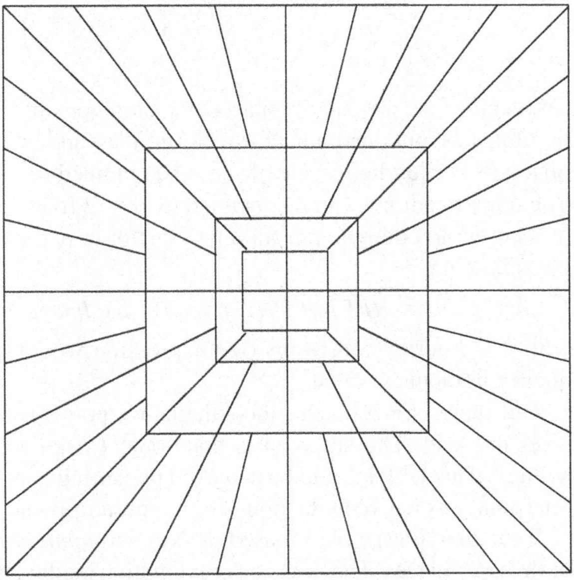

Fig. 4.48 Pseudo-polar tiling of the frequency plane

digital implementations, using nonequispaced FFT or a new technique based on data wrapping. In both cases, the polar tiling of the frequency domain was replaced by a pseudo-polar tiling, which is depicted in Fig. 4.48.

Instead of polar wedge tiles, one has to use trapezoids. In these trapezoids, the frequencies are such that:

$$\left\{(\omega_1, \omega_2) : 2^j \leq \omega_1 \leq 2^{j+1}, -2^{-j/2} \cdot \frac{2}{3} \leq \omega_2/\omega_1 \leq 2^{-j/2} \cdot \frac{2}{3}\right\} \tag{4.43}$$

The trapezoidal tile can be built as follows:

$$U_j\;(\boldsymbol{\omega}) = 2^{(-3j/4)}\, W(2^{-j}\omega_1)\, V(\frac{2^{\lfloor j/2 \rfloor}\omega_2}{\omega_1}) \tag{4.44}$$

where $V()$ can be the same as in the continuous-time curvelet, and:

$$W_j(\boldsymbol{\omega}) = \sqrt{\Phi_{j+1}^2(\boldsymbol{\omega}) - \Phi_j^2(\boldsymbol{\omega})} \tag{4.45}$$

where:

$$\Phi_j(\omega_1, \omega_2) = \phi(\omega_1)\, \phi(\omega_2) \tag{4.46}$$

with $\phi(\omega_1)$, $\phi(\omega_2)$ low-pass one dimensional windows.

The polar tiling of Fig. 4.42 is generated by tile rotations and translations. The pseudo-polar tiling, instead, is generated by tile translations and *shearings*. The shear matrix is:

$$S_\theta = \begin{pmatrix} 1 & 0 \\ -\tan\theta & 1 \end{pmatrix} \tag{4.47}$$

The angles θ are not equally spaced, but the slopes are. Each coronae of the pseudopolar tiling has four quadrants: East, North, West and South. Take for instance the North quadrant, the left half of the tiles can be obtained from symmetry with the right half. The other quadrants can be obtained by $\pm\pi/2$ rotations.

In order to compute the curvelet transform, one would like to use expressions as the following:

IFFT (FFT(Curvelet) FFT(Image))

with 2-D Fourier transforms (the expression would be even simpler by using frequency domain curvelets).

But there is a problem, the calculation points (the samples) are not placed on a regular grid. The situation is that some *Unequispaced FFT* (UFFT) should be applied. This UFFT could be achieved by getting a regular grid taking samples from interpolations between the nodes of the pseudo-polar grid.

The other alternative is based on *data wrapping*, so ellipses fit in parallelograms broken into pieces. The resulting computation method is relatively fast. The technique can be illustrated as follows. Figure 4.49 shows the result of extracting data from a tile (one applies the pertinent frequency domain curvelet as a mask over the 2-D FFT of the image).

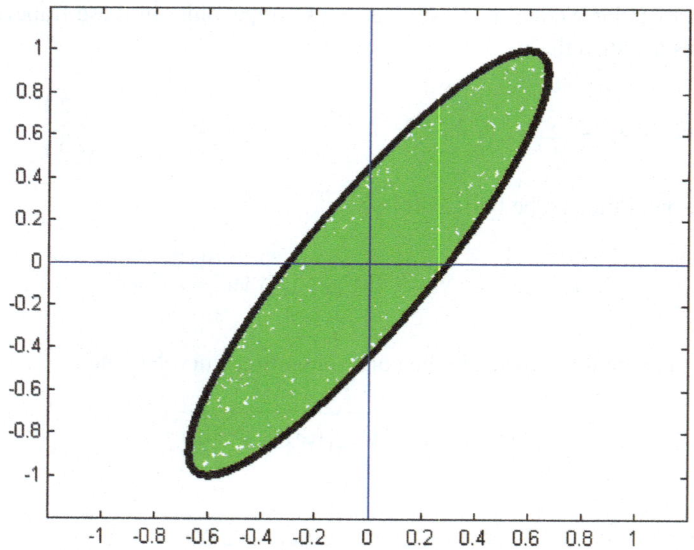

Fig. 4.49 A data tile

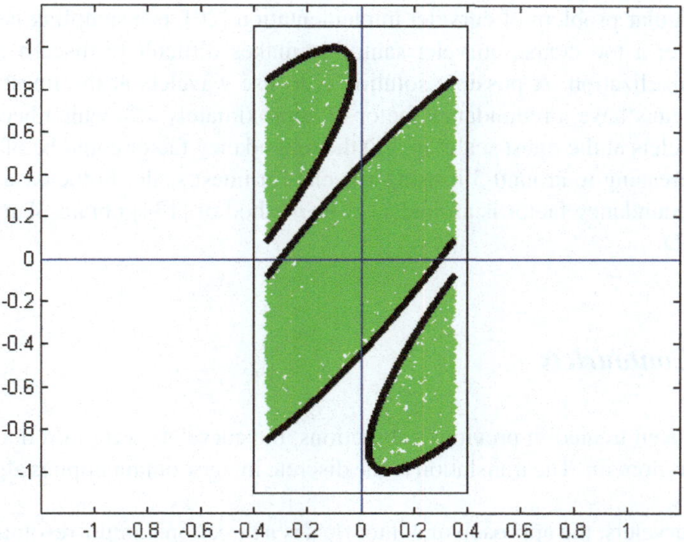

Fig. 4.50 Wrapped data

Now, the data tile X is translated to a rectangle, with the correct aspect ratio, and is wrapped using periodizitation; suppose the sides of the rectangle are H and V, the wrapped data are obtained by simple index reordering:

$$WX(n_1 \bmod H, n_2 \bmod V) = X(n_1, n_2) \tag{4.48}$$

Figure 4.50 shows the result of data wrapping.

Finally, to obtain the curvelet coefficients, an inverse 2-D Fourier transform is applied to the wrapped data. The process of data extraction, wrapping and inverse transform is repeated for all the tiles.

There is a MATLAB toolbox, called CurveLab, for the fast discrete curvelet transform. This toolbox includes UFFT implementation and data wrapping implementation. In addition to the extended technical descriptions given in [20], the reader can find in [107] many details, also geometrical, which are important for the coding. Succint treatements of the second generation curvelet transform can be found in [2, 70, 120].

Another implementation of curvelets is offered in the MATLAB toolbox called Toolox (*sic.*).

Some aspects of the second generation curvelets, as described in [20] and implemented in CurveLab, deserve ameliorations. A number of improvements have been proposed. For instance [210], with a 3-D low-redundancy fast curvelet transform, and [134] with a uniform discrete curvelet transform. Both proposals offer MATLAB implementations through Internet.

A particular problem of curvelet implementation [20], is a sampling issue at the finest scale; a too coarse curvelet sampling makes difficult to discern directions because pixelization. A possible solution is to use wavelets at the finest scale. In 2-D, curvelets have a redundancy factor of approximately 7.2, which becomes 2.8 using wavelets at the finest scale. In 3-D the redundancy factor could be of the order of 40, decreasing to around 5 if using wavelets at finest scale. In the case of [210], the 3-D redundancy factor is around 11. The method of [134] obtains 4 in 2-D, and 54/7 in 3-D.

4.5.5 *Contourlets*

As it has been treated in previous subsections, the curvelets were introduced in the continuous domain. The translation to the discrete image domain implies algorithmic difficulties.

Like curvelets, the approach of contourlets is a directional multi-resolution transform. But the difference is that contourlets are directly constructed in the discrete image domain. Contourlets are not a discretized version of curvelets.

4.5.5.1 Multi-directional Filters

Contourlets are based on a new contribution to 2-D multi-directional filters. This topic has attracted a lot of interest. A most cited milestone of this area is [10], which constructed a 2-D directional filter bank that can be used for a tiling of the frequency plane as represented in Fig. 4.51.

Fig. 4.51 Frequency plane tiling using multi-directional filters

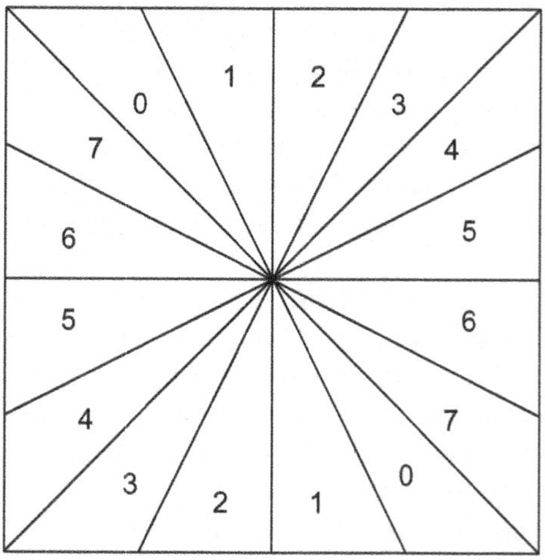

The filter bank introduced in [10] is implemented using image modulation and quincunx diamond filters. In order to get the desired frequency partition, a complicated binary tree structure is proposed.

In his doctoral Thesis [53] introduced a new filter bank construction, with a simpler tree expansion and avoiding modulation. The design is based on two pieces: a two-channel quincunx filter bank with fan filters, depicted in Fig. 4.52, and a −45° shearing operator.

Let us describe in more detail the new construction. Recall that quincunx sampling causes 45° rotation. Due to Noble identities, if you have a quincunx sampling followed by a fan filter, and interchange them, the fan filter is transformed into an equivalent quadrant filter; this is represented in Fig. 4.53.

Consider now the structure depicted in Fig. 4.54. The sampling ratios are such that $Q_0 \cdot Q_1 = 2\,I$, which is downsampling by two in each dimension. The figure shows two decomposition levels, the first one with two filters, and the second one with four filters. The second level work as quadrant filters.

With the configuration shown in Fig. 4.54, a decomposition into four directions is achieved. Figure 4.55 shows the result in terms of frequency plane tiling.

In order to obtain a decomposition into eight directions (Fig. 4.51), a third level is added to the structure. The idea is to apply shearing before fan filters. The complete filter bank is obtained by adding four filter banks as shown in Fig. 4.56.

See [135] for an improved implementation of the eight directions filter bank, where a phase correction of the filters is included.

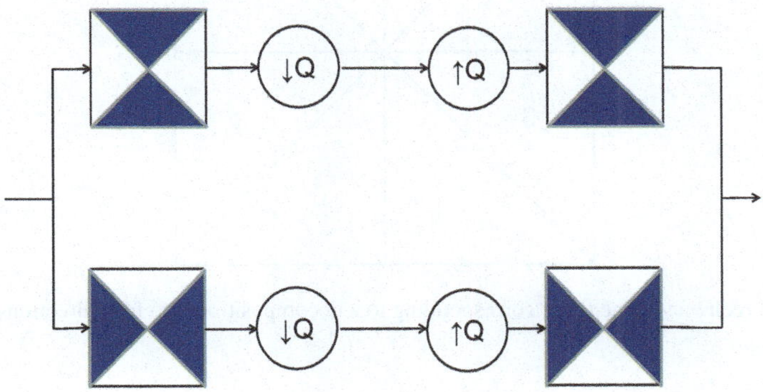

Fig. 4.52 Filter bank with fan filters

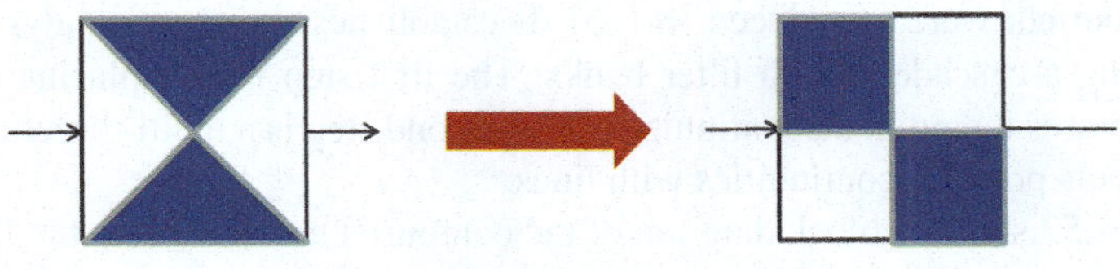

Fig. 4.53 From fan filter to quadrant filter (after interchange with quincux sampling)

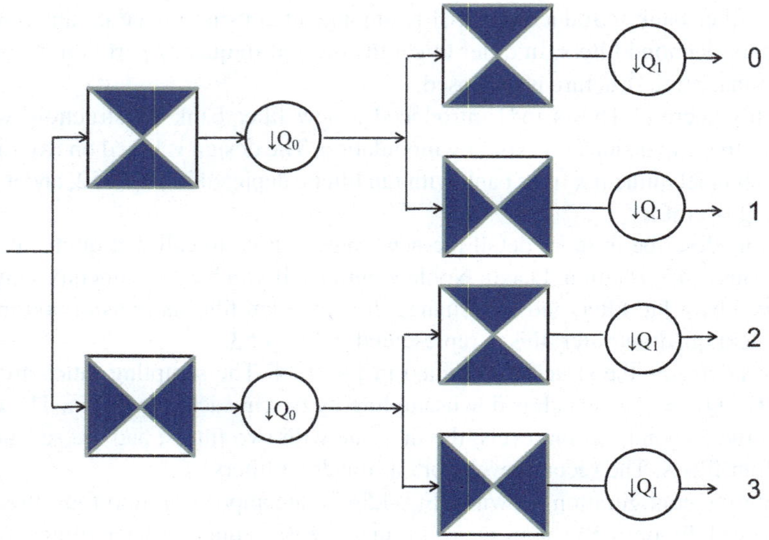

Fig. 4.54 The first two levels of the decomposition

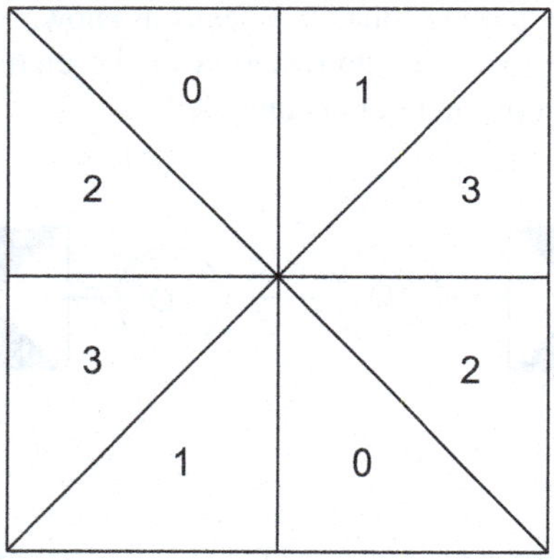

Fig. 4.55 Frequency plane tiling corresponding to a decomposition into four directions

4.5.5.2 Contourlets

The contourlets were introduced in [55]. It establishes an image analysis in two steps, using a cascade of two filter banks. The first step is a Laplacian pyramid, which captures the point discontinuities. The second step is a multi-directional filter that connects point discontinuities with lines.

Figure 4.57 shows a block diagram of the contourlet analysis, with the Laplacian pyramid filters on the left, and the application of multi-directional filters on the right.

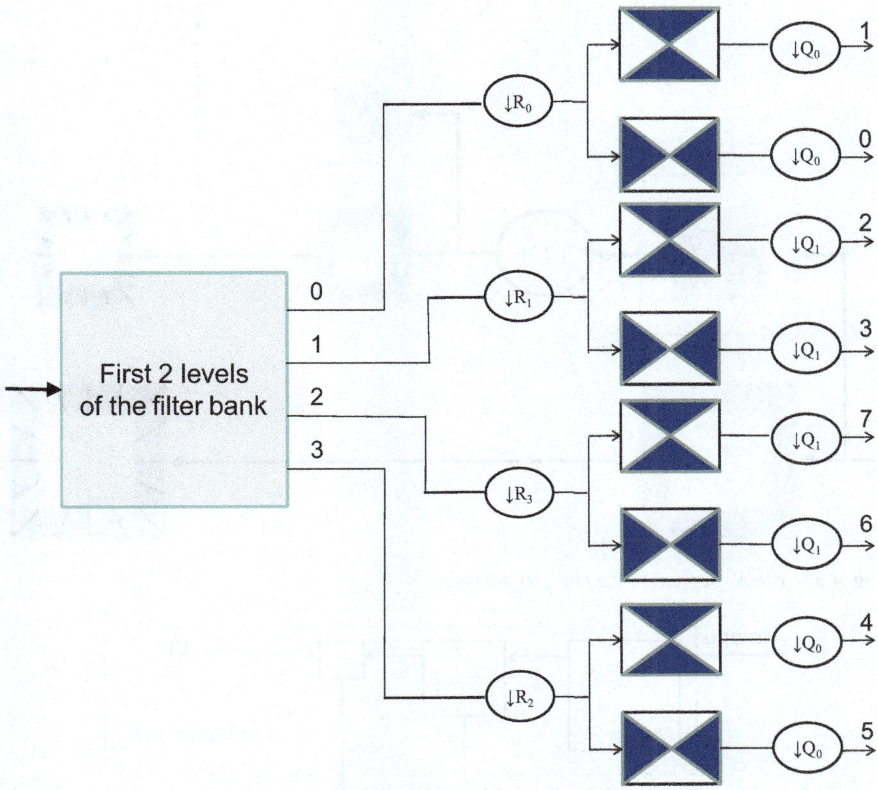

Fig. 4.56 Filter bank for the eight directions

Other authors prefer an equivalent graphical description of the curvelet transform, as shown in Fig. 4.58. It highlights the presence of mostly horizontal components and mostly vertical components.

With proper design of the filters, the contourlet transform provides a tight frame [55]. The discrete contourlet transform has a redundancy factor that is less than 4/3.

Contourlets have elongated supports at various scales, directions and aspect ratios. They can efficiently approximate smooth contours.

According with [53] it is important to stress that the contourlet expansions are defined on rectangular grids (the image space). Contourlets are defined via iterated filter banks. It opens many possibilities for further refinements on spatial resolution; for example as contourlet packets.

One of the contents of Minh N. Do web page is the Contourlet Toolbox in MATLAB.

In the already mentioned [135], some improvements are also introduced with respect the Laplacian pyramid, in order to alleviate aliasing problems of the contourlet.

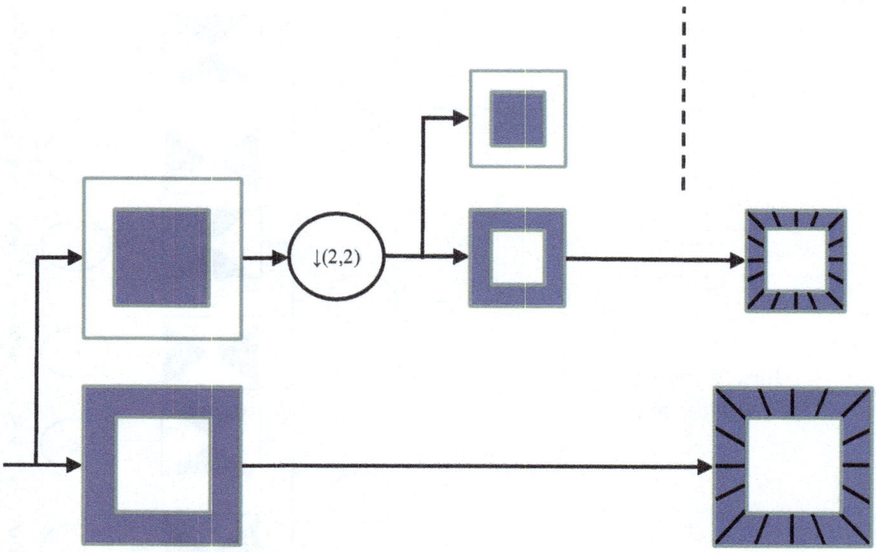

Fig. 4.57 Block diagram of the curvelet analysis

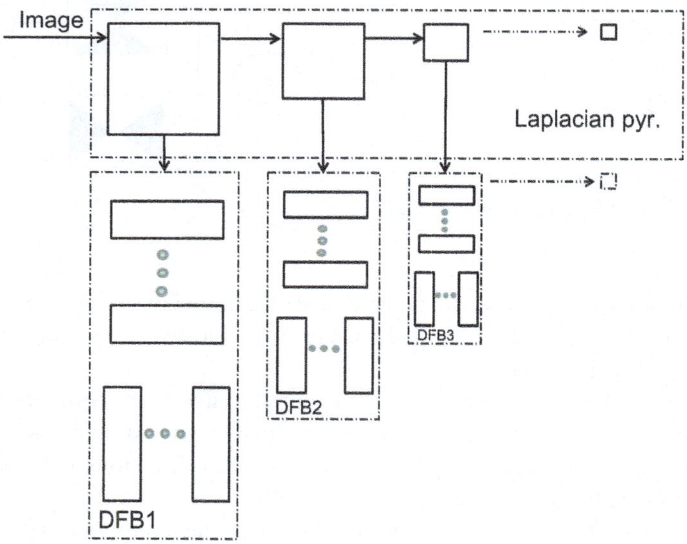

Fig. 4.58 Curvelet analysis structure

Instead of using two separated steps [115] proposed a combined iterated nonseparable filter bank for both steps. In this way, the so called CRISP contourlets obtain nonredundant image representation.

By considering directional vanishing moments [40] introduced a filter bank design that obtains better visual quality.

A further contribution of Yue Lu is 3-D directional filter banks and surfacelets [116]. Both the web pages of Minh N. Do and of Yue Lu, link to a MATLAB Toolbox with the name SurfBox. The surfacelets have a 3-D redundancy factor of 4.02.

An alternative to the already described frequency plane partition was introduced in [136], obtaining a simpler directional filter bank called uniform DFB.

Most papers on multiscale transforms cite the significant contribution of [172]. It recommends the use of *translation-invariant* wavelet transforms. Experience has shown that, in particular, this is important for image denoising.

In consequence, the research has proposed several ways to obtain translation - invariant contourlets. In the case of [41], the idea was to use nonsubsampled filter banks; and in the case of [69], an algorithmic solution combined with a modified filter bank is considered. Both papers are complemented with new MATLAB Toolboxes, which can be found in the Minh N. Do, and R. Eslami, web pages.

4.5.6 Bandelets

While curvelets and contourlets are fixed representation, bandelets serve for an *adaptive* representation. The main idea is to detect edges and use a kind of oriented wavelet for each edge. Detailed description of bandelets is given in [143–145]. Since bandelets use many nested iterations, it is convenient to create MATLAB functions in order to simplify the programs. We closely follow the implementation details offered by [144].

The bandelet analysis takes two steps.

(1) In the first step, a standard 2-D wavelet analysis is applied to the image. The result is a representation on the wavelet plane where edges are emphasized. For example, let us choose the image shown in Fig. 4.59.

Fig. 4.59 Image

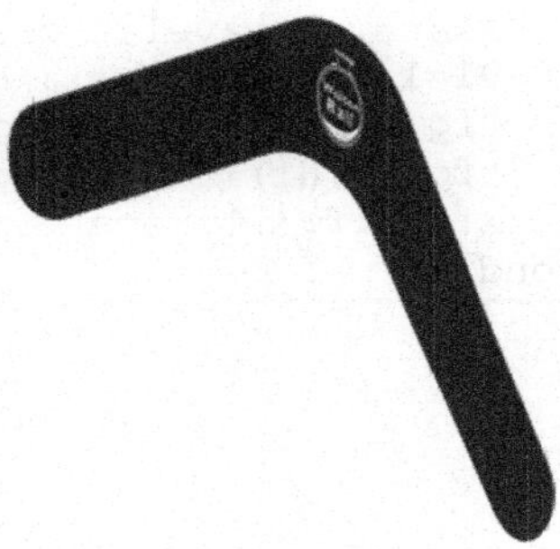

A MATLAB function was defined that loads the boomerang image, applies the 2D Haar transform, and returns the wavelet image (named *tfg*).

Function 4.15 B_2D_Haar

```
function tfg=B_2D_Haar()
% Haar 2-Level wavelet transform of an image
% Boomerang image
% Using filters
% Haar filter:
c=1/sqrt(2);
h0=[c c]; %low-pass filter
h1=[-c c]; %high-pass filter
% The image:
ufg=imread('Bmg1.tif'); %read the image file into a matrix
fg=double(ufg); %convert to float
fg=fg-mean(mean(fg)); %zero mean
[Nr,Nc]=size(fg);
tfg=zeros(Nr,Nc); %space for wavelet plane
xfg=fg;
for nL=1:2, %levels
  Nv=floor(Nr/2^nL); Nh=floor(Nc/2^nL);
  lfg=zeros(2*Nv,Nh); hfg=zeros(2*Nv,Nh);
  llfg=zeros(Nv,Nh); hlfg=zeros(Nv,Nh);
  lhfg=zeros(Nv,Nh); hhfg=zeros(Nv,Nh);
  %
  %Step1
  %wavelet transform of rows
  aux=conv2(xfg,h0); %low-pass filtering of rows
  lfg=0.5*aux(:,2:2:end); %downsampling to get L
  aux=conv2(xfg,h1); %high-pass filtering of rows
  hfg=0.5*aux(:,2:2:end); %downsampling to get H
  %Step 2
  %wavelet transform of columns of previous step
  aux=conv2(lfg,h0'); %low-pass filtering of L columns
  llfg=aux(2:2:end,:); %downsampling
  aux=conv2(lfg,h1'); %high-pass filtering of L columns
  hlfg=aux(2:2:end,:); %downsampling
  aux=conv2(hfg,h0'); %low-pass filtering of H columns
  lhfg=aux(2:2:end,:); %downsampling
  aux=conv2(hfg,h1'); %high-pass filtering of H columns
  hhfg=aux(2:2:end,:); %downsampling
  %save on wavelet plane
  V1=1:Nv; V2=Nv+1:2*Nv; H1=1:Nh; H2=Nh+1:2*Nh; %ranges
  tfg(V1,H1)=llfg; tfg(V1,H2)=lhfg;
  tfg(V2,H1)=hlfg; tfg(V2,H2)=hhfg;
  xfg=llfg; %prepare next level
end;
```

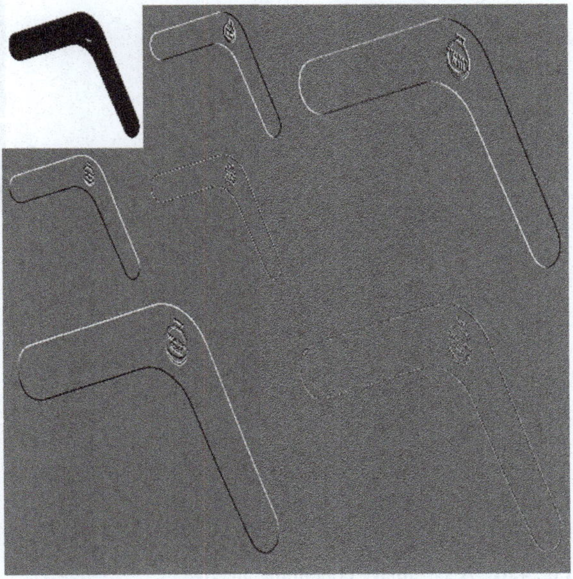

Fig. 4.60 2-level Haar wavelet transform of the image

Figure 4.60 shows the 2-level Haar wavelet transform of the image. The figure has been generated with the Program 4.16, which calls the function previously listed.

Program 4.16 Haar 2-Level wavelet transform of a boomerang picture

```
% Haar 2-Level wavelet transform of an image
% Boomerang image
% Using filters
tfg=B_2D_Haar; %function call
%display
figure(1)
imshow(tfg,[-20 20]);
title('2-level Haar image transform');
```

Our next task is to study sub-images of the wavelet plane. These are images $2^j \times 2^j$ constituting a regular grid. Select any of them; for instance the sub-image represented in Fig. 4.61. If there is an edge in the sub-image, the next task is to determine the direction of the edge (one may suppose a straight-line approximation).

An edge direction searching procedure can be applied. Take for example a certain direction, draw a perpendicular *PP*'. See Fig. 4.62. A reordering of wavelet data is now done as follows. The sub-image represents a set of values $f(\mathbf{x}_i)$, where $\mathbf{x}_i$ are the points belonging to the sub-image. Project all points $\mathbf{x}_i$ on *PP*'. The projections form an ordered set along *PP*'. Then, it is possible to draw an ordered set of values $f(i)$, corresponding to the order of point projections along *PP*'. This is a 1-D signal, which will be denoted as F; Fig. 4.63 shows how F may look like. This figure has been obtained with the Program 4.17.

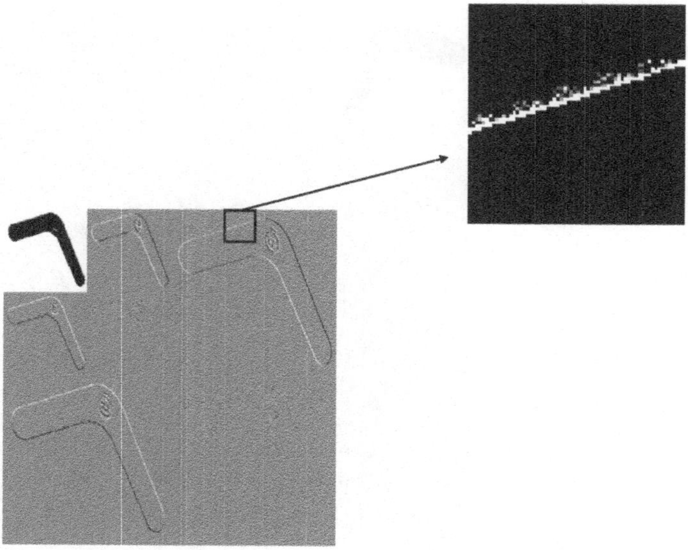

Fig. 4.61 A sub-image containing an edge

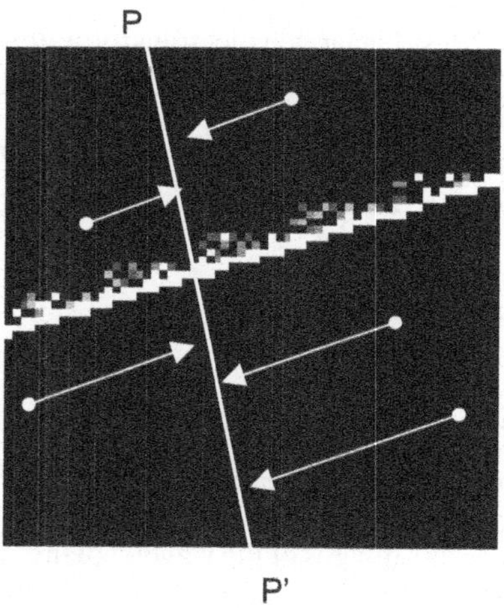

Fig. 4.62 Projections on PP'

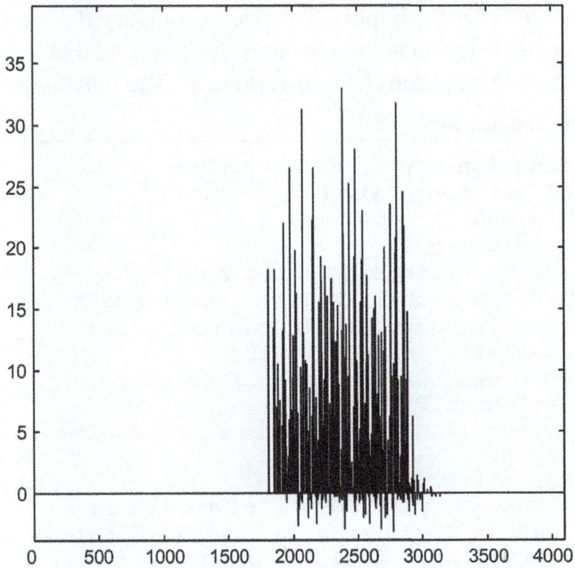

Fig. 4.63 The 1-D signal obtained from projection on PP'

Program 4.17 Project square on PP'

```
% Project square on PP'
tfg=B_2D_Haar; %function call
%the square
rbeg=1; Nr=size(tfg,1);
cbeg=5*Nr/8;
cw=(Nr/8)-1;
sq=tfg(rbeg:rbeg+cw,cbeg:cbeg+cw);
%projection
alpha=(120*pi)/180; %rads
np=cw+1;
[YY,XX]=meshgrid(1:np,1:np);
p=(-sin(alpha)*XX(:))+ (cos(alpha)*YY(:));
[aux,ix]=sort(p); %ascending order
F=sq(ix); %ordered values
figure(1)
plot(F,'k');
axis([0 length(F) 1.2*min(F) 1.2*max(F)]);
title('Example of 1-D function F');
```

Next, one applies 1-D wavelet analysis to F, yielding a set of coefficients B (bandelet coefficients), which are depicted in Fig. 4.64. It can be shown that if the direction guess is wrong, this implies large coefficients. The best edge direction fit corresponds to the smallest coefficients. This criterion paves the way for a suitable direction searching procedure.

In order to simplify the code to compute B sets, a MATLAB function has been created. The function obtains F for a specified projection angle, and then it applies the 1-D Haar wavelet transform to F. The function is listed below:

Function 4.18 B_set

```
function wty=B_set(sq,alpha)
%B set for a given direction
% alpha in rads
np=size(sq,1);
[YY,XX]=meshgrid(1:np,1:np);
p=(-sin(alpha)*XX(:))+ (cos(alpha)*YY(:));
[aux,ix]=sort(p); %ascending order
F=sq(ix); %ordered values
%1-D Haar wavelet transform of F
Ns=length(F);
K=floor(log2(Ns)); %number of scales
wty=F;
for n=1:K,
  aux1= wty(1:2:Ns-1) + wty(2:2:Ns);
  aux2= wty(1:2:Ns-1) - wty(2:2:Ns);
  wty(1:Ns)=[aux1,aux2]/sqrt(2);
  Ns=Ns/2;
end;
```

Figure 4.64 has been obtained with the Program 4.19, which uses the previous function. It shows the B sets for 130° projection angle. This angle has been specified in one of the sentences of the program, and can be edited for any other desired angle. Recall that the B sets are obtained with a 1D wavelet transform of F; in this case we used the Haar transform.

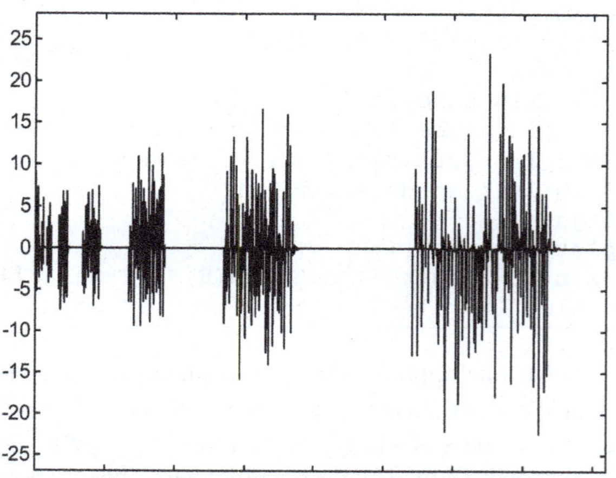

Fig. 4.64 Example of B set, obtained with the Haar transform of the 1-D projection signal

Program 4.19 B set for a given direction

```
% B set for a given direction
tfg=B_2D_Haar; %function call
th=5; %threshold
%the square
rbeg=1; Nr=size(tfg,1);
cbeg=5*Nr/8;
cw=(Nr/8)-1;
sq=tfg(rbeg:rbeg+cw,cbeg:cbeg+cw);
%angle
alpha=(130*pi)/180; %rads
wty=B_Bset(sq,alpha); %function call
%display
figure(1)
plot(wty,'k');
axis([0 length(wty) 1.2*min(wty) 1.2*max(wty)]);
title('Example of B set');
%evaluation
Rs=wty.*(abs(wty)<th); %residual
Sa=sum(abs(wty(:)>th)); %value above threshold
Lg=sum(Rs(:).^2)+(Sa*(th^2)); %Lagrangian
```

Notice that the last sentences of Program 4.19 make an evaluation of the projection result. A Lagrangian is computed. The best direction corresponds to the minimum Lagrangian value.

In order to explore B sets for a set of different direction angles, a MATLAB function has been created. It is listed below. The 180° are divided into *NA* parts. For each angle, the Lagrangian is computed.

Function 4.20 B_Bestdir

```
function SLG=B_Bestdir(sq,th,NA)
%Find best direction, exploring on NA angles
SLG=zeros(1,NA); %space for Lagrangians
for nx=1:NA,
%projection
alpha=(nx*pi)/NA; %rads
wty=B_Bset(sq,alpha); %function call
%evaluation
Rs=wty.*(abs(wty)<th); %residual
Sa=sum(abs(wty(:)>th)); %value above threshold
Lg=sum(Rs(:).^2)+(Sa*(th^2)); %Lagrangian
SLG(nx)=Lg;
end;
```

Using the previous function, it is easy to do the following experiment: divide the 180° into 60 parts. Obtain the Lagrangian for each projection angle. Plot the values of the Lagrangians and compare to see whether there is a minimum.

Figure 4.65 shows the results in function of the direction angle. Obviously there is a minimum (at 108°), indicating what is the best direction. The figure has been generated with the Program 4.21, which calls the previous function.

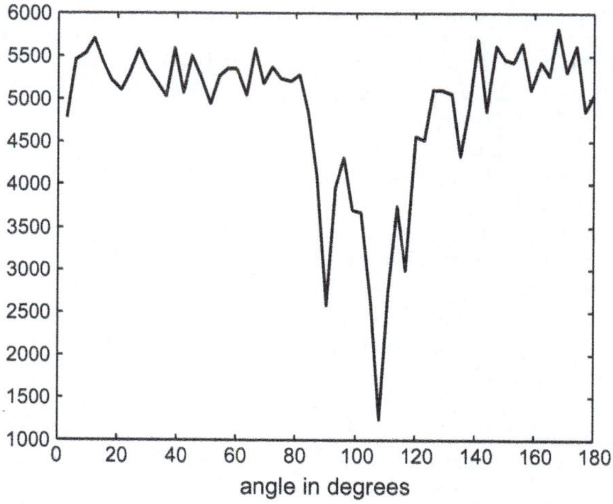

Fig. 4.65 The Lagrangian is obtained for several directions; the minimum shows the best direction

Program 4.21 Explore a set of directions, display Lagrangians

```
% Explore a set of directions, display Lagrangians
tfg=B_2D_Haar; %function call
th=5; %threshold
NA=60; %number of angles to test
%the square
rbeg=1; Nr=size(tfg,1);
cbeg=5*Nr/8;
cw=(Nr/8)-1;
sq=tfg(rbeg:rbeg+cw,cbeg:cbeg+cw);
SLG=B_Bestdir(sq,th,NA); %function call
[minL,iL]=min(SLG);
%display
figure(1)
xa=(1:NA).*(180/NA);
plot(xa(:),SLG(:),'k');
title('Lagrangian vs. angle');
xlabel('angle in degrees');
%print best angle
angle=(iL*180)/NA
```

(2) In the second step the objective is to reduce the size of the image representation. Large sub-image squares with no edges, do not deserve to dig into smaller subdivisions. Therefore, there is an exploration to determine what sub-images should be selected.

A frequently used method for such work is quad-tree exploration. It is also used in robotics; for example in garbage harvesting scenarios, where a mobile robot grabs empty cans. A large square A is divided into four equal squares $B1 \ldots B4$. If any of

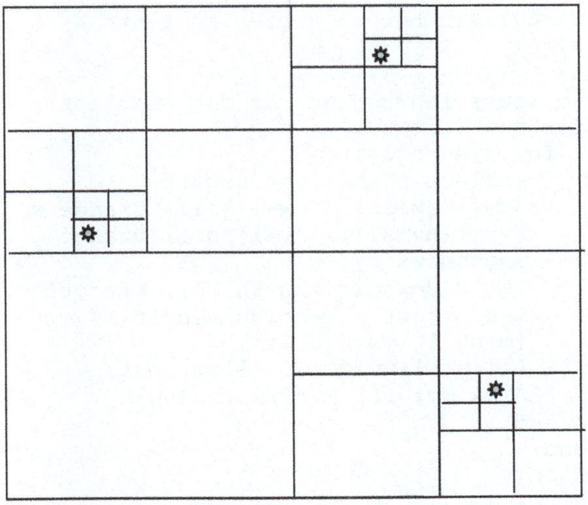

Fig. 4.66 Example of quadtree

these has something of interest, it is divided into four squares $C1 \ldots C4$; and so on. Fig. 4.66 shows an example of quadtree.

This method is applied to the analyzed wavelet plane, considering bandelet coefficients in each sub-image (for all scales). Actually, contrary to the usual way, is applied in a bottom-up manner. First, a complete study of all smallest sub-images is done, obtaining all corresponding Lagrangians. Second, the quad tree is built. The target is to save image representation bits.

Program 4.22 makes the study of all smallest sub-images. The result is saved in the matrix A, which contains all best projection angles, and in the matrix L, which contains the Lagrangians. Figure 4.67 shows the contents of these matrices in pseudo-colours. The shape of the boomerang can be recognized.

Notice that Program 4.24 only explores the LH part for the first subdivision of the wavelet plane into LL, LH, HL, HH parts. This is done here for illustration purposes. In reality, the bandelet analysis must analyze all subdivisions of the wavelet plane, at all the scales considered.

Program 4.22 Smallest sub-images analysis

```
% Smallest sub-images analysis
% for the larger LH
tfg=B_2D_Haar; %function call
th=5; %threshold
sq=tfg(257:512,1:256); %LH square
n=size(sq);
Kmax=8; %256 = 2^8
%smallest square
Kmin=4; %exponent
sl=2^Kmin; %square side
nsl=n(1)/sl; %number of squares in a row or column
Q=zeros(n)+Kmin; A=zeros(n); L=zeros(n/sl);
[YY,XX]=meshgrid(1:sl,1:sl);
```

```
NA=60; %number of angles to test
disp('analysis of smallest squares');
%-----------------------------------------------
%compute lagrangians of all smallest squares
for nx=0:nsl-1,
  for ny=0:nsl-1,
    %select a smallest square
    wx=1+(nx*sl):((nx+1)*sl); %range x
    wy=1+(ny*sl):((ny+1)*sl); %range y
    ssq=sq(wx,wy);
    SLG=B_Bestdir(ssq,th,NA); %function call
    %save best direction result
    [minL,iL]=min(SLG);
    L(1+nx,1+ny)=minL; %lagrangian
    A(wx,wy)=iL; %index to angle
  end;
end;
disp('--Lagrangians of smallest squares');
disp('--already computed');
%display
figure (1)
subplot(1,2,1)
imagesc(A);
title('best angles')
subplot(1,2,2)
imagesc(L)
title('Lagrangians')
```

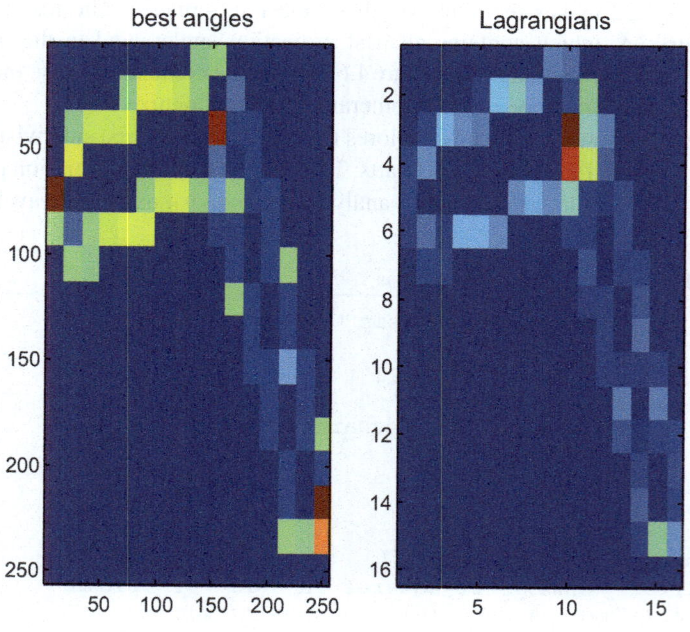

Fig. 4.67 Result of the LH smallest sub-images analysis

After the analysis of smallest sub-images, a quadtree is built. A simple bottom up procedure is used. According with the quadtree structure, sub-images are joined as groups of 4. If a group has better Lagrangian (less than the sum of the Lagrangians of the four group members), then it is accepted.

A MATLAB function is defined that implements the two steps: analysis of all smallest sub-images, and quadtree build-up. It is listed below. The information of the result is contained in a matrix Q that describes the quadtree, and a matrix A of best angles.

Function 4.23 B_calc_quadt

```
function [Q,A]=B_calc_quadt(sq,th,Kmin,Kmax)
% quadtree calculation
n=size(sq);
sl=2^Kmin; %smallest square side
nsl=n(1)/sl; %number of squares in a row or column
Q=zeros(n)+Kmin; A=zeros(n); L=zeros(n/sl);
[YY,XX]=meshgrid(1:sl,1:sl);
NA=60; %number of angles to test
%-------------------------------------------
%compute lagrangians of all smallest squares
for nx=0:nsl-1,
  for ny=0:nsl-1,
    %select a smallest square
    wx=1+(nx*sl):((nx+1)*sl); %range x
    wy=1+(ny*sl):((ny+1)*sl); %range y
    ssq=sq(wx,wy);
    SLG=B_Bestdir(ssq,th,NA); %function call
    %save best direction result
    [minL,iL]=min(SLG);
    L(1+nx,1+ny)=minL; %lagrangian
    A(wx,wy)=iL;
  end;
end;
%-------------------------------------------
% merging 4 squares into 1 when appropriate
% bottom-up iteration
gamma=0.15;
for j=Kmin+1:Kmax, %from small to bigger squares
  sl=2^j;
  nsl=n(1)/sl; %number of squares in a row or column
  [YY,XX]=meshgrid(1:sl,1:sl);
  Laux=zeros(n/sl); %for new lagrangians
  for nx=0:nsl-1,
    for ny=0:nsl-1,
      %select a square
      wx=1+(nx*sl):((nx+1)*sl); %range x
      wy=1+(ny*sl):((ny+1)*sl); %range y
      ssq=sq(wx,wy);
      %sum the lagrangians of 4 smaller squares, and gamma
      sx=1+2*nx; sy=1+2*ny;
      Lsum=L(sx,sy)+L(1+sx,sy)+L(sx,1+sy)+L(1+sx,1+sy)+gamma;
      SLG=B_Bestdir(ssq,th,NA); %function call
```

```
      %best direction result
      [minL,iL]=min(SLG);
      %merging if appropriate
      if minL<Lsum,
        Laux(1+nx,1+ny)=minL;
        Q(wx,wy)=j;
        A(wx,wy)=iL;
      else
        Laux(1+nx,1+ny)=Lsum;
      end;
    end;
  end;
  L=Laux;
end;
```

Equipped with the previous function, a complete work on the largest LH part of the wavelet plane has been done, Program 4.24. Figure 4.68 shows the result concerning the *A* matrix. It tells you that some grouping has been achieved, and some predominant directions are present in important regions of the image.

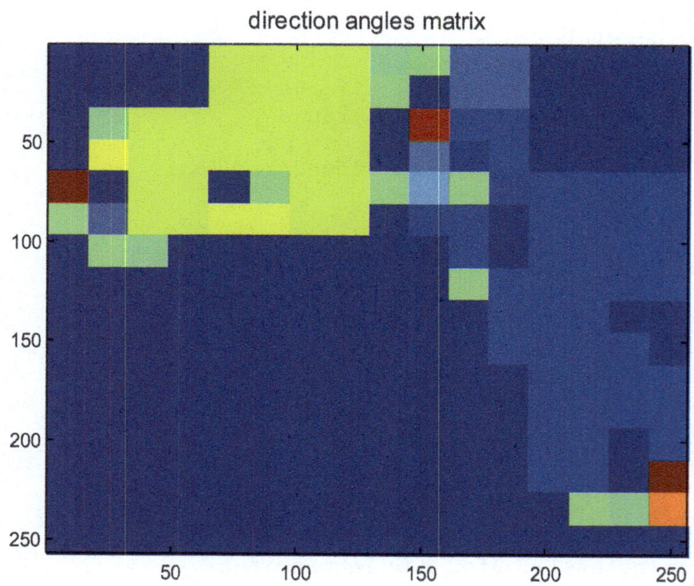

Fig. 4.68 The A matrix after complete work on largest LH part

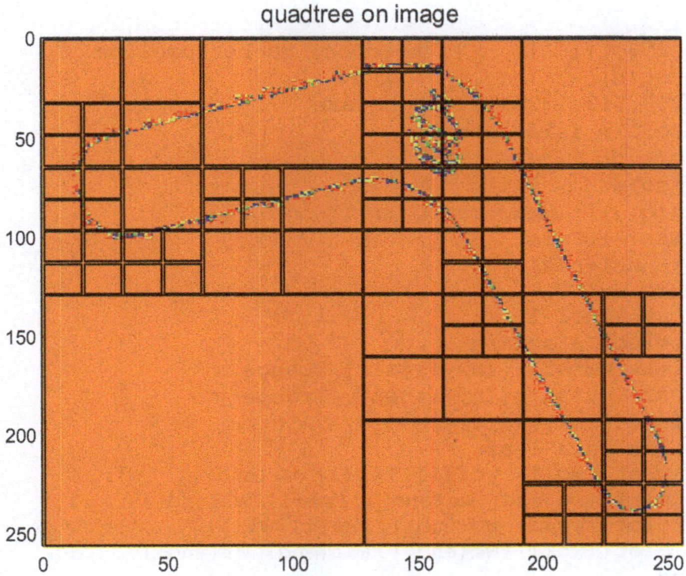

Fig. 4.69 A view of the obtained quadtree

Program 4.24 Quadtree analysis for larger LH

```
% Quadtree analysis for larger LH
% several levels
tfg=B_2D_Haar; %function call
th=5; %threshold
sq=tfg(257:512,1:256); %LH square
Kmax=8; %256 = 2^8
%smallest square
Kmin=4; %exponent
disp('please wait');
[Q,A]=B_calc_quadt(sq,th,Kmin,Kmax); %function call
%display
figure(1)
imagesc(A);
title('direction angles matrix');
```

A motivating activity now is to interpret the contents of the matrix Q, which is a condensed description of a quadtree.

Program 4.25 uses some detailed work to interpret Q, build the quadtree, and draw the quadtree over the image on LH. Figure 4.69 shows the result, using the 'prism' option for the colours, so the image profile shows interesting details.

Program 4.25 Draw the quadtree

```
% Draw the quadtree
% use after the qadtree analysis
% analysis and plot of Q (the quadtree desciption)
```

```
figure(1)
plot([1 256],[1 1],'g'); hold on; %a bottom line
h=gca; set(h,'YDir','reverse');
sq=tfg(257:512,1:256); %LH square
n=size(sq,1); %side length
colormap('prism');
imagesc(sq);
Kmin=min(Q(:)); Kmax=max(Q(:));
for j=Kmax:-1:Kmin,
  sl=2^j; nsl=n/sl;
  for nx=0:nsl-1,
    for ny=0:nsl-1,
      %select a square
      wl=1+(nx*sl):((nx+1)*sl); %range x
      wc=1+(ny*sl):((ny+1)*sl); %range y
      if (Q(wl(1),wc(1)))==j,
        %it is a leaf
        plot([wc(1) wc(1)],[wl(1) wl(end)],'k');
        plot([wc(end) wc(end)],[wl(1) wl(end)],'k');
        plot([wc(1) wc(end)],[wl(1) wl(1)],'k');
        plot([wc(1) wc(end)],[wl(end) wl(end)],'k');
      end
    end
end
end
axis([0 256 0 256]); title('quadtree on image');
```

To increase the efficiency, once the LH (for instance) quadtree is determined, it is used on HL and HH at the same scale.

Once the quadtrees are established for all the subdivisions of the wavelet plane, the process continues by analyzing the sub-images of the quadtrees, at all scales, to obtain best angles.

The final result is a description of the image in terms of quadtrees coefficients the best angles, and B sets.

4.5.7 Shearlets

Wavelets are functions of two variables, scale and translation, that can be used for a convenient tiling of the time-frequency plane. Shearlets are functions of three variables that can be used for a pseudo-polar tiling of the 2-D frequency plane. The third variable controls the orientation of the shearlets.

There are at least two main research groups working on shearlets, and this motivates diverse flavours and slight differences in the mathematical formulations. The subsection is based on [64, 102, 218].

4.5.7.1 Continuous Shearlets

The continuous shearlets are a version of higher dimension wavelets. Let us introduce them in mathematical terms.

An *affine family* generated by the function $\psi \in L^2(\Re)$ is a collection of functions of the form:

$$\{\, \psi_{a,t}(x) = a^{-1/2}\, \psi(a^{-1}\,(x - t)) : a > 0,\ t \in \Re\} \tag{4.49}$$

If all functions $f \in L^2(\Re)$ can be recovered via the reproducing formula:

$$f = \int\int < f, \psi_{a,t} > \ \psi_{a,t}\, dt\, \frac{da}{a^2} \tag{4.50}$$

then ψ is a continuous wavelet.

In these equations, a is scale, and t translation. Note the slight change of notation with respect to scale: it was denoted as s in other chapters, now is a (the s will be used for other purposes).

The mathematical framework can be extended to 2-D by considering the affine system:

$$\{\, \psi_{M,t}(x) = |\det M|^{-1/2}\, \psi(M^{-1}\,(x - t)) : M \in G,\ t \in \Re^2\} \tag{4.51}$$

where $\psi \in L^2(\Re^2)$ and G is a subset of the group of 2×2 invertible matrices. As before, if any function $f \in L^2(\Re^2)$ can be recovered via the corresponding reproducing formula, then ψ is a continuous 2-D wavelet.

Extensions to more dimensions can be formulated in the same way.

2-D continuous shearlets are a non-isotropic version of the 2-D continuous wavelet. The matrix M is chosen as follows:

$$M_{as} = \begin{pmatrix} a & \sqrt{a}\, s \\ 0 & \sqrt{a} \end{pmatrix} \tag{4.52}$$

This matrix can be factored as $B \cdot A$, with:

$$A = \begin{pmatrix} a & 0 \\ 0 & \sqrt{a} \end{pmatrix};\quad B = \begin{pmatrix} 1 & s \\ 0 & 1 \end{pmatrix} \tag{4.53}$$

B is the shear matrix, and A is the anisotropic dilation matrix; s is a shearing parameter.

Therefore 2-D continuus shearlets constitute an affine system with functions of three variables:

$$\{\, \psi_{a,s,t}(x) = |\det M_{as}|^{-1/2}\, \psi(M_{as}^{-1}\,(x - t)) : t \in \Re^2\} \tag{4.54}$$

The generating function ψ is a well localized function. A particular form is chosen, with the following expression in the 2-D Fourier domain:

$$\Psi(\omega) = \Psi_1(\omega_1) \cdot \Psi_1(\frac{\omega_2}{\omega_1}) \tag{4.55}$$

Both Ψ_1 and Ψ_2 are smooth functions. The support of Ψ_1 is $[-2,\ -1/2] \cup [1/2,\ 2]$, and the support of Ψ_2 is $[-1,\ 1]$ (some authors refer to this second function as a "bump" function [100, 128]).

Then, the shearlets have the following expression in the 2-D frequency domain:

$$\Psi_{a,s,t}(\omega) = a^{3/4}\, e^{-2\pi\omega\, t}\, \Psi_1(a\,\omega_1)\, \Psi_2(a^{-1/2}(\frac{\omega_2}{\omega_1} - s)) \tag{4.56}$$

The support of the shearlets are:

$$\{\omega_1;\ \omega_2\} = \left\{[-\frac{2}{a},\ -\frac{1}{2a}] \cup [\frac{1}{2a},\ \frac{2}{a}];\ [\frac{\omega_2}{\omega_1} - s] \leq \sqrt{a}\right\} \tag{4.57}$$

Therefore, the shearlets have frequency support on a pair of trapezoids, at various scales, symmetric with respect to the origin, and oriented along a line with slope s.

The values of s are restricted to the interval $[-1, 1]$. With this condition, the shearlets will only cover the horizontal cone. To complete the plane covering, a second set of shearlets is added for the vertical cone. The vertical shearlets are obtained by $\pi/2$ rotation of the horizontal shearlets. This approach is called 'cone-adapted shearlets' [101, 218].

Figure 4.70 shows some horizontal and vertical shearlets.

By dilation, shearing and translation of a single generating function ψ, shearlets provide a complete tiling of the pseudo-polar frequency plane. In virtue of the matrix A, shearlets have a parabolic aspect ratio in the frequency domain.

Figure 4.71 displays the decomposition of the pseudo-polar plane into a horizontal and a vertical cone.

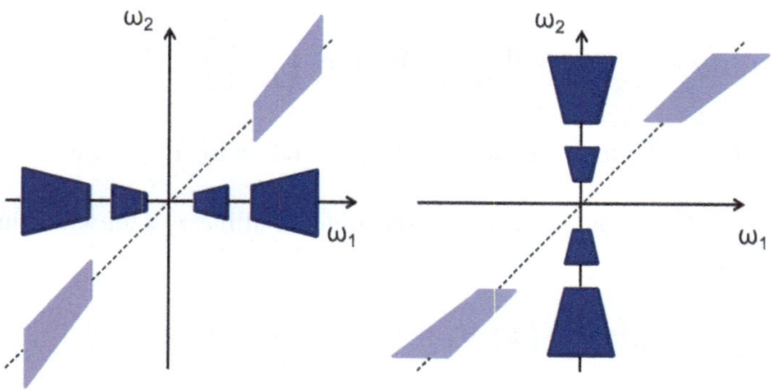

Fig. 4.70 Examples of horizontal and vertical shearlets

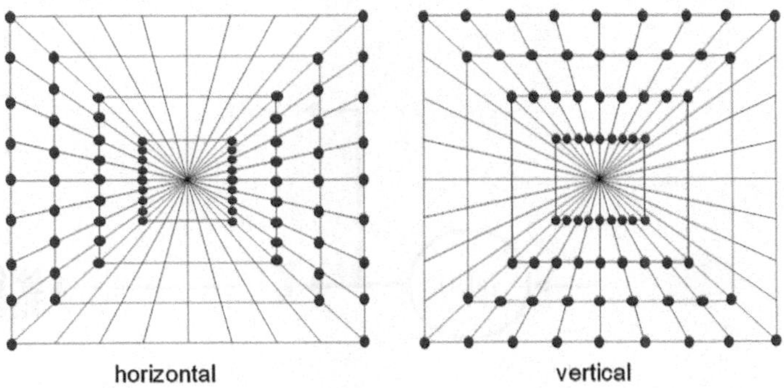

Fig. 4.71 Horizontal and vertical cones

The shearlet transform has the ability to identify not only the location of image singularities, but also their orientation.

One of the distinctive features of shearlets is te use of shearing in place of rotation. In this way, a suitable link between continuous and digital versions is devised, since the shear matrix preserves the integer lattice. In comparison with curvelets, shearlets are normally associated to a fixed translation lattice, while curvelets are not.

4.5.7.2 Digital Implementations

One of the first digital implementations was proposed in [64]. A discretized version of the M_{as} matrix was formulated as $M_{jl} = B^l A^j$, with:

$$A = \begin{pmatrix} 4 & 0 \\ 0 & 2 \end{pmatrix}; \; B = \begin{pmatrix} 1 & 1 \\ 0 & 1 \end{pmatrix} \tag{4.58}$$

A filter bank structure was designed to obtain the shearlet based decomposition. The first level was a Laplacian pyramid, with the low pass filtered image a (1/4,1/4) of the size of the original image. A 2-D Fourier transform is applied to the high pass filtered image, and then corresponding shearlet trapezoidal masks are applied for each angular direction.

Figure 4.72 shows a block diagram of the digital shearlet transform.

Later on, another digital implementation was proposed in [102], claiming that it is a rational design of the digital transform. It is observed that the introduction of a special set of coordinates in the 2-D frequency space, requires a cascade of several operations: 2-D Fourier transform, change of variables, weighting.

The weighting is done with the square root of the Jacobian of the change of variables. It can be regarded as a sampling density compensation.

Actually, in the pseudo-polar plane, the density of samples is larger near the center. For this reason, the conventional pseudo-polar Fourier transform is not an isometry. As said in [9], fortunately it is possible to obtain a near-isometry (and even isometry) by combining oversampling with proper weighting.

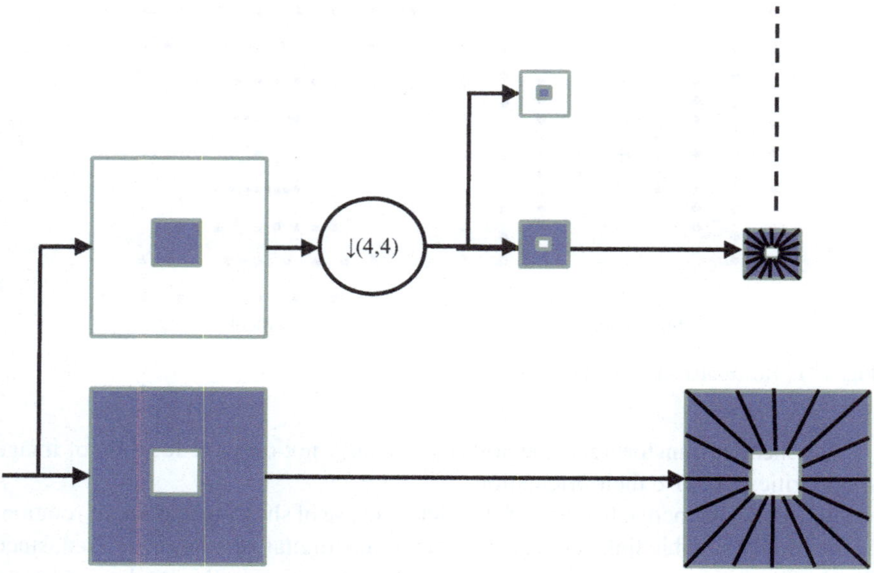

Fig. 4.72 Block diagram of the proposed digital shearlet analysis

An important part of the work described in [102] corresponds to obtaing a good combination of oversampling and weighting. An oversampling ratio of 16 at least is recommended.Three weighting choices are introduced.

The mentioned authors have developed in MATLAB the ShearLab Toolbox, based on their proposed digital implementation.

Fig. 4.73 The frequency plane is decomposed into four regions

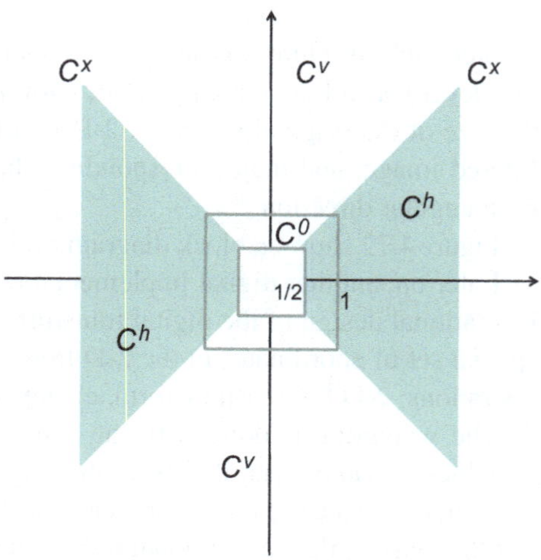

In his tutorial on shearlets [86] included a third implementation, offering also MATLAB code, called FFST: *Fast Finite Shearlet Transform*. The 2-D frequency plane was decomposed into four regions, as depicted in Fig. 4.73.

Regions C^h and C^v are the horizontal and vertical cones. Region $C^0$is a low-frequency part. Regions C^x are the seam lines between the cones. There is overlapping of the regions.

A scaling function is defined for the low frequency region. conventional shearlets are defined for the cones. For the seam lines, two versions are considered, one keeps the form of the conventional shearlet, the other provides a smooth construction.

This implementation has a larger oversampling factor, compared to ShearLab. It gives 61 images of the same size as the original image.

Some modifications and extensions of the shearlet transform have been proposed, including in some cases new implementations. For instance the compactly supported shearlet systems of [109], with implementation; the low redundant version of [77]; the extension to 3-D for video application in [132]; or the hyperbolic shearlets proposed by [66].

Shearlets are at the crossroad of several conceptual topics. They are a particular case of composite wavelets. In addition, because the connection with affine transformations, the theory of groups has also entered into scene. And, it also have to do with crystallographic wavelets.

A review of five years of research on shearlets is offered in [101].

4.5.8 Other Wavelet Variants

Obviously, the research creativity concerning wavelets for image analysis has produced an exceedingly large number of alternatives. Being not possible to describe them all here, nevertheless some of them are briefly introduced below.

The web page of Laurent Duval includes a long list of wavelet variants, together with abstracts and links. Some authors use the term *X-lets* to refer to directional wavelets.

4.5.8.1 Ripplets

Type-I ripplets generalize the scaling law of curvelets by adding two parameters: support c and degree d [215]. The ripplet generating function is, in the frequency domain:

$$P_a\,(r,\omega) = \frac{1}{\sqrt{c}}\, a^{\frac{1+d}{2d}}\; W(a\cdot r)\; V\,(\frac{a^{1/d}}{c\cdot a}\cdot\omega) \tag{4.59}$$

where $W(r)$ is the radial window, with support *[1/2, 2]*, and $V(\omega)$ is the angular window, with support $[-1, 1]$.

The aspect ratio of ripplets is $width \approx length^d$. For $c = 1$ and $d = 2$, ripplets have a parabolic aspect ratio, like curvelets; for $d = 3$, ripplets have cubic scaling, and so forth.

The discretization of the ripplet transform is based on discretization of the parameters; a is sampled at dyadic intervals.

Type-II ripplets are based on a generalization of the Radon transform [214]. Ridgelets are a particular case of type-II ripplets, for $d = 1$. For $d = 2$, the ripplets look like a curved ridgelet.

Some medical applications have been reported, taking advantage of a better approximation to curved singularities [44].

MATLAB code is available from D.O. Wu web page.

4.5.8.2 Directionlets

The directionlets are lattice-based directional wavelets. As seen in another section of the chapter, one uses a lattice generating matrix for the mathematical representation. This representation is not unique: different matrices can represent the same lattice.

In a lattice, there are directions that join sets of nodes. The idea of directionlets is to use skewed wavelets, following these directions. Figure 4.74 shows an example of grid, with a vector basis and two evident suitable directions for the wavelets.

The chosen directions can be represented with a pair of integer numbers, n_1, n_2. The corresponding wavelet will have an anisotropy ratio of $\rho = n_1/n_2$. Therefore,

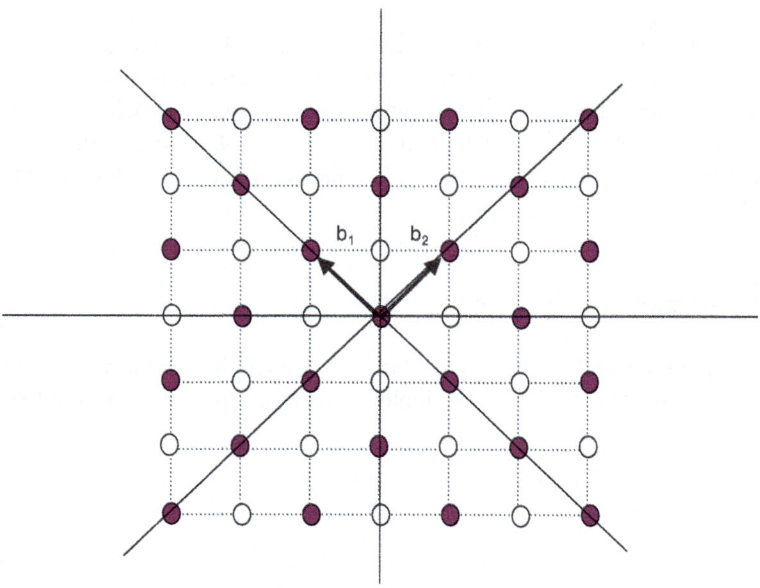

Fig. 4.74 Suitable directions on a image grid

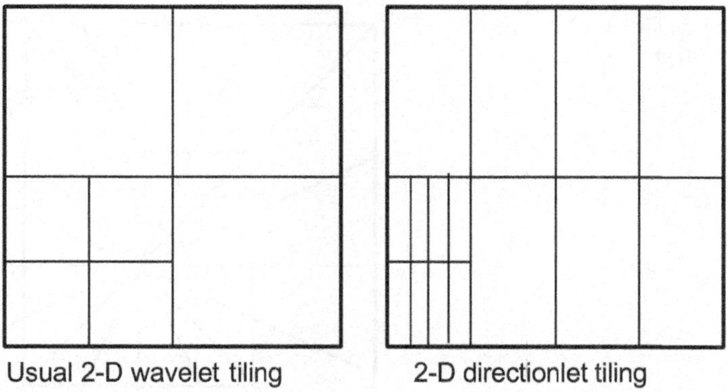

Fig. 4.75 Directionlet tiling is different from the usual 2-D wavelet tiling

the tiling of the 2-D frequency plane will be different from the usual 2-D wavelet transform. Figure 4.75 shows an example, comparing both tilings.

Images are decomposed into segments, and then different transform directions are chosen in each segment.

Reference [202] is a detailed introduction to directionlets. The web page of V. Velisavljevic contains links to several papers. Some applications have already been published [111, 165].

4.5.8.3 Directional Haar wavelet

A kind of directional Haar wavelet has been proposed, being supported on space domain triangles [98]. This wavelet can be considered as a special composite wavelet. A simple dilation matrix, $A = 2I$, and a finite collection of shear matrices are used.

Fig. 4.76 Coarsest level image partition

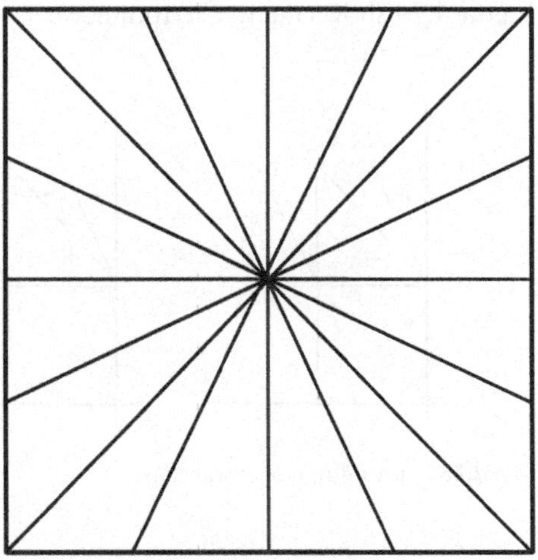

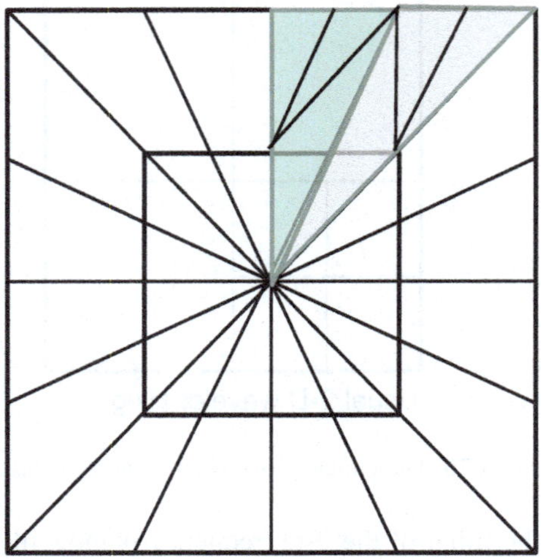

Fig. 4.77 Second level of the decomposition

For the definition of scaling functions, which is the coarsest decomposition level, the image is partitioned as shown in Fig. 4.76.

Once at the second level of the decomposition, part of the situation is as depicted in Fig. 4.77.

The construction of three directional wavelets is represented in Fig. 4.78.

Detailed mathematical expressions are provided in [98], in order to design a filter bank for the transform.

4.5.8.4 Tetrolets

The tetrolet transform is a kind of adaptive 2-D Haar transform. The image is decomposed into 4 × 4 blocks. Then, a tiling of each block is made using tetrominoes. Figure 4.79 shows the five tetrominoes.

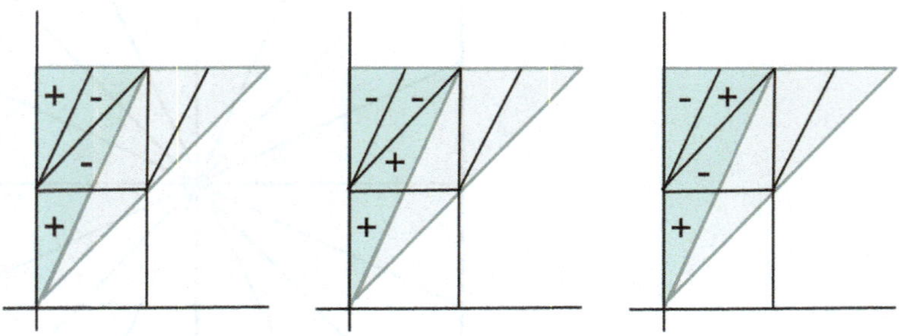

Fig. 4.78 Three directional wavelets

Fig. 4.79 Tetrominoes

Fig. 4.80 A tiling of the 4×4 block using tetrominoes

There are 22 fundamental tilings of the 4×4 blocks using tetrominoes. Figure 4.80 shows for example one of these possible tilings.

Once the image is decomposed into blocks, the tetrolet transform obtains the best tetrolet tiling match for each block.

The development of mathematical expressions for the implementation of the tetrolet transform using a filter bank, is described in [97].

4.6 Complex Wavelets

From the beginning of wavelets some shortcomings were noticed, in part due to their oscillatory nature. In particular, they are not translation invariant and have directionality limitations. These are problems already contemplated in previous pages.

It was also noticed that the Fourier transform is free from these drawbacks. It is translation invariant, and is highly directional. This is because the Fourier transform is based on:

$$e^{j\,\omega t} = \cos(\omega t) + j\,\sin(\omega t) \tag{4.60}$$

The real and imaginary parts of this complex function form a Hilbert transform pair (they are in quadrature). The signal $e^{j\omega t}$ is analytical: its spectrum is completely on the positive frequency axis; there is no symmetrical part on the negative frequency axis.

When there are signal shifts, the amplitude of the Fourier transform remains unaltered, while the shifts are encoded on the transform phase.

In order to adhere to these advantages, a complex wavelet is considered:

$$\psi_c(t) = \psi_r(t) + j\,\psi_i(t) \tag{4.61}$$

where the wavelet $\psi_r$is even, and the wavelet ψ_i is odd. Both wavelets should form a Hilbert transform pair (not all wavelets can be combined for this).

Figure 4.81 shows an example of two wavelets that can be used to approximate a Hilbert pair. Actually, most of the the spectrum of the complex wavelet formed with this pair lies on the right-hand side, as shown in Fig. 4.82.

The program that has been developed to generate Figs. 4.81 and 4.82 has been included in Appendix A. It is similar to one of the programs in the chapter on wavelets (the one that was used to generate LeGall wavelet).

Although the idea seems simple and effective, there are important difficulties for the implementation. A popular and intuitive path for the solution is to use a double tree filter bank, one of the trees for the real wavelet part, and the other for the imaginary wavelet part; this is called the dual-tree complex wavelet transform.

Two main aspects will be considered in this section. One is related to implementation issues. The second is devoted to 2-D application, in which some directionality can be achieved. The section is based on a well-kown article [164], and some additional literature that is cited later on.

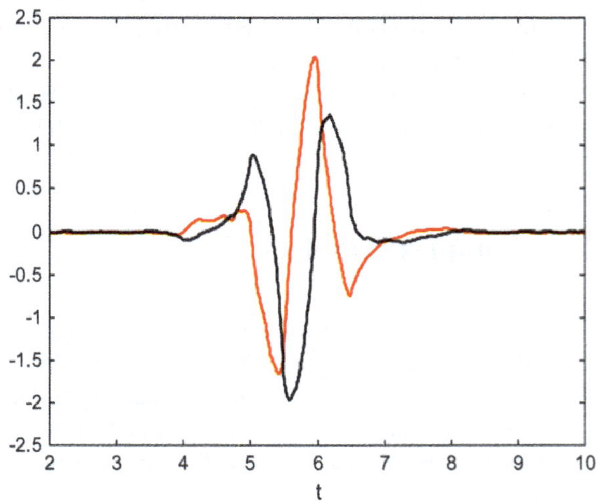

Fig. 4.81 Example of wavelets for Hilbert pair approximation

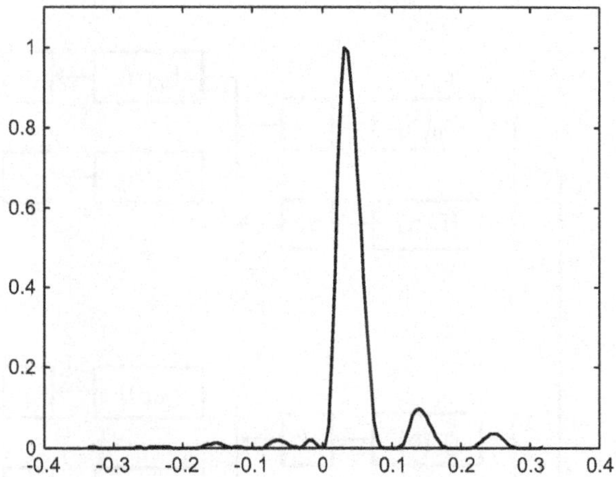

Fig. 4.82 Spectrum of the complex wavelet

4.6.1 Implementation Issues

It happens that analytic wavelets must have infinite support. However, one wants finite support wavelets. Hence, a compromise is needed, trying to get close to be analytic while having suitable finite support.

Basic design alternatives fail to obtain the desired complex wavelet [164]. Fortunately, the dual-tree complex wavelet introduced by Kingsbury in 1998 (see [95] for extended details), provides a successful and simple solution. Two wavelet filter banks obtain the real wavelet and the imaginary wavelet; the filters are jointly designed in order to get an approximately analytic complex wavelet. Perfect reconstruction is achieved by another dual-tree. Figure 4.83 shows a block diagram with the analysis dual tree.

Figure 4.84 shows the block diagram of the synthesis dual tree, which is symmetrical to the analysis. PR is achieved in each of the two trees, and then the results are averaged.

It has been found that, to obtain the approximately analytic complex wavelet, the coefficients of the low-pass filters of both trees, should obey to the following condition:

$$g_0(n) \approx h_0(n - 0.5) \tag{4.62}$$

More rigorously, in the Fourier domain:

$$G_0(e^{j\omega}) = e^{-j\,0.5\,\omega}\, H_0(e^{j\omega}) \tag{4.63}$$

Put in other way:

$$|G_0(e^{j\omega})| = |H_0(e^{j\omega})| \tag{4.64}$$

$$\arg G_0(e^{j\omega}) = \arg H_0(e^{j\omega}) - 0.5\,\omega \tag{4.65}$$

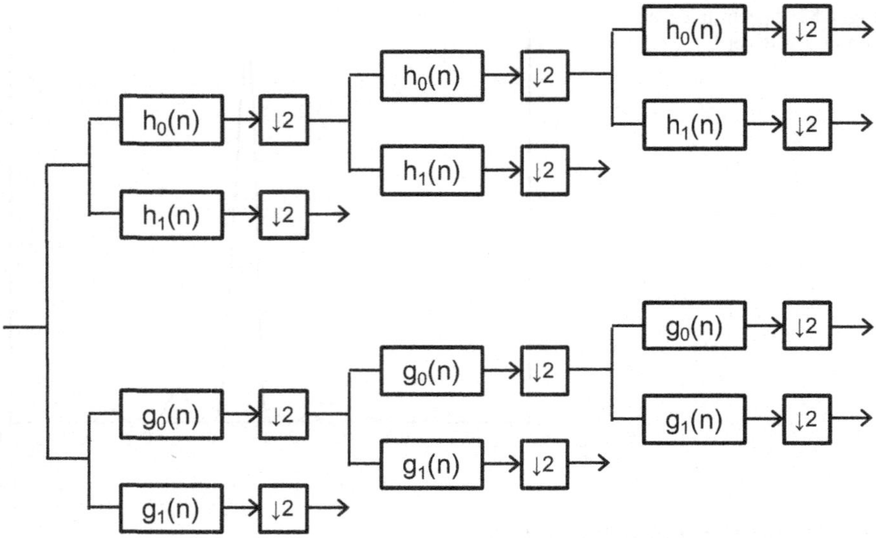

Fig. 4.83 Block diagram of the analysis dual tree

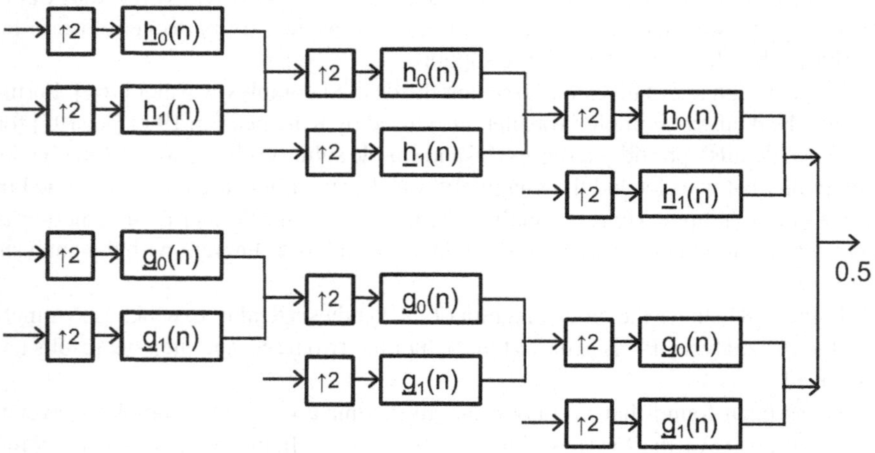

Fig. 4.84 Block diagram of the synthesis dual tree

Normally, it is desired to use short filters. Of course, one could take any wavelet filter $h_0(n)$ and then design a $g_0(n)$ to comply with the established conditions; but in some cases this lead to long $g_0(n)$ filter. In [164], three solutions for the design are offered.

- The first solution is based on *linear-phase biorthogonal* filters:

$$h_0(n) = h_0(N - 1 - n) \tag{4.66}$$

$$g_0(n) = g_0(N - n) \tag{4.67}$$

with N an odd number, $h_0(n)$ symmetric odd-length FIR filter, and $g_0(n)$ symmetric even-length FIR filter. In this way one obtains:

$$|G_0(e^{j\omega})| \neq |H_0(e^{j\omega})| \tag{4.68}$$

$$\arg G_0(e^{j\omega}) = \arg H_0(e^{j\omega}) - 0.5\,\omega \tag{4.69}$$

Thus, the design effort aims at alleviating the amplitude difference.

- The second solution is called *the Q-shift* solution. It starts with:

$$g_0(n) = h_0(N - 1 - n) \tag{4.70}$$

where the length of $h_0(n)$is N, which is an even number. In this case:

$$|G_0(e^{j\omega})| = |H_0(e^{j\omega})| \tag{4.71}$$

$$\arg G_0(e^{j\omega}) \neq \arg H_0(e^{j\omega}) - 0.5\,\omega \tag{4.72}$$

The expression (4.70) implies that:

$$\arg G_0(e^{j\omega}) = -\arg H_0(e^{j\omega}) - (N - 1)\,\omega \tag{4.73}$$

Supposing that the filters were such as $\arg G_0(e^{j\omega}) \approx \arg H_0(e^{j\omega}) - 0.5\,\omega$, then:

$$\arg H_0(e^{j\omega}) \approx -0.5\,(N - 1)\,\omega + 0.25\,\omega \tag{4.74}$$

Therefore, $h_0(n)$ should be approximately symmetric around $n = 0.5(N - 1) - 0.25$; a quarter away from the centre, and so the *Q-shift* name.

With the Q-shift solution, the imaginary part of the complex wavelet is the time-reversed version of the real part:

$$\psi_g(t) = \psi_h(N - 1 - t) \tag{4.75}$$

- The third solution is a *common-factor* solution.

The filters have a common factor:

$$h_0(n) = f(n) * d(n) \tag{4.76}$$

$$g_0(n) = f(n) * d(L - n) \tag{4.77}$$

In the Fourier domain:

$$H_0(z) = F(z)\,D(z) \tag{4.78}$$

$$G_0(z) = F(z)\, z^{-L} D(1/z) \tag{4.79}$$

Then:

$$|G_0(e^{j\omega})| = |H_0(e^{j\omega})| \tag{4.80}$$

$$\arg G_0(e^{j\omega}) \neq \arg H_0(e^{j\omega}) - 0.5\,\omega \tag{4.81}$$

One should alleviate the phase difference. Write:

$$G_0(z) = H_0(z)\, A(z) \tag{4.82}$$

with: $A(z) = \frac{z^{-L} D(1/z)}{D(z)}$

The amplitude of $A(z)$ must be 1, and $D(z)$ must be chosen so that:

$$\arg A(e^{j\omega}) \approx -0.5\,\omega \tag{4.83}$$

For this design step, it is suggested to use Thiran's filters [191].

There is still another implementation issue. The first stage of the dual tree must be different from the other stages. This is related to a interlacing property that arises as consequence of the $g_0(n) \approx h_0(n - 0.5)$ condition. This condition implies that the scaling functions are mutually delayed by a half sample: $\phi_g(n) \approx \phi_h(t - 0.5)$. Therefore, the integer translates of $\phi_g(t)$ lay in the middle of the integer translates of $\phi_h(t)$. The two scaling functions are thus interlaced.

Suppose that one uses different filters for each stage of the dual tree, as depicted in Fig. 4.85 shows the block diagram of the analysis dual tree.

Let us look at the upper chain of low-pass filters. At stage j, one has the equivalent system depicted in Fig. 4.86.

The total low-pass filter, from the filter bank input to the output of stage j, is:

$$H_T^{(j)}(z) = H_0^{(1)}(z) \cdot H_0^{(2)}(z^2) \ldots H_0^{(j)}(z^{2^{j-1}}) \tag{4.84}$$

A similar expression is found for the lower chain of low-pass filters.

At stage 1, the translates by 2 of $g_T^{(1)}(n)$ must be between the translates by 2 of $h_T^{(1)}(n)$. That is:

$$g_T^{(1)}(n) \approx h_T^{(1)}(n - 1) \tag{4.85}$$

At stage 2, the translates by 4 of $g_T^{(2)}(n)$ must be between the translates by 4 of $h_T^{(2)}(n)$. That is:

$$g_T^{(2)}(n) \approx h_T^{(2)}(n - 2) \tag{4.86}$$

At stage 3:

$$g_T^{(3)}(n) \approx h_T^{(3)}(n - 4) \tag{4.87}$$

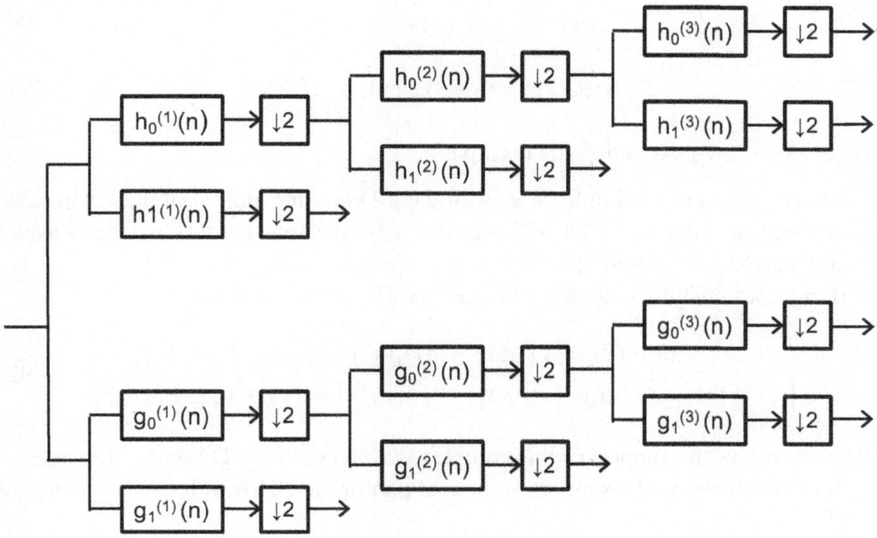

Fig. 4.85 Block diagram of the analysis dual tree

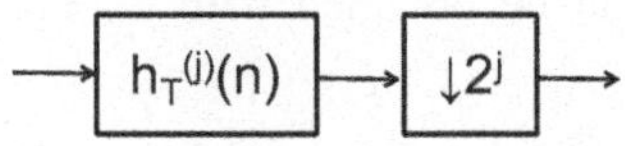

Fig. 4.86 Stage j equivalent system

and so on.

Let us come back to $g_T^{(1)}(n) \approx h_T^{(1)}(n-1)$. Half sample delay should not be used for the first stage. Instead, the recipe is to use the same filters and one sample delay between the upper and the lower trees.

For the other stages, it is always found that for the j stage:

$$g_0^{(j)}(n) \approx h_0^{(j)}(n-0.5) \tag{4.88}$$

The dual tree representation is 2 times redundant (in other words: expansive).

4.6.2 2-D Application

When the nonstandard 2-D wavelet transform was applied to images, as it was described in a previous section of this chapter, LH, HL and HH wavelets were used. These wavelets can be expressed as follows:

$$\psi_1(x, y) = \phi(x)\,\psi(y), \quad (LH) \tag{4.89}$$

$$\psi_2(x, y) = \psi(x)\,\phi(y), \quad (HL) \tag{4.90}$$

$$\psi_3(x, y) = \psi(x)\,\psi(y), \quad (HH) \tag{4.91}$$

where $\phi(\,)$is low-pass and $\psi(\,)$is high-pass.

Figure 4.87 shows a simplistic view of the 2-D Fourier support of these wavelets. Notice that the support of the HH wavelet is on the four corners, so this wavelet cannot discern −45° from 45°.

If one uses analytic complex wavelets, the HH case becomes:

$$\begin{aligned}\psi_3(x, y) &= \psi_c(x)\,\psi_c(y) = [\,\psi_h(x) + j\psi_g(x)] \cdot [\,\psi_h(y) + j\psi_g(y)] \\ &= \{\psi_h(x)\,\psi_h(y) - \psi_g(x)\,\psi_g(y)\} + j\,\{\psi_h(x)\,\psi_g(y) + \psi_g(x)\,\psi_h(y)\}\end{aligned} \tag{4.92}$$

Fig. 4.88 shows the support of this complex wavelet on the 2-D Fourier domain.

Figure 4.89 shows the support of the real part of the HH wavelet. It is oriented at −45°.

The +45° direction can be obtained with $\psi_3(x, y) = \psi_c(x)\,\psi_c(y)^*$ (the star means complex conjugate).

Four more directions can be obtained with $\phi_c(x)\,\psi_c(y)$, $\phi_c(x)\,\psi_c(y)^*$, $\psi_c(x)\,\phi_c(y)$, $\psi_c(x)\,\phi_c(y)^*$.

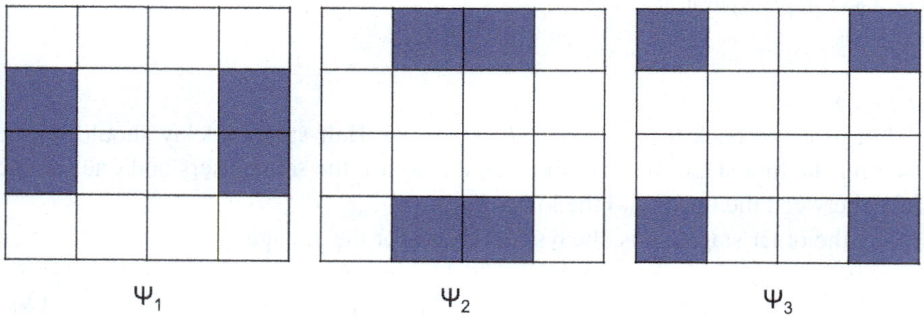

Fig. 4.87 Support of the LH, HL and HH wavelets on the Fourier plane

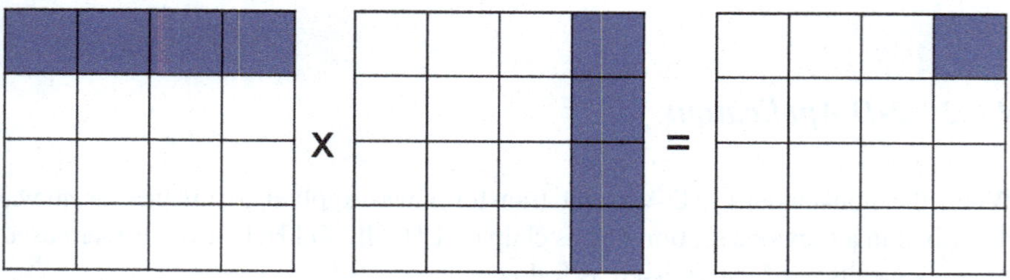

Fig. 4.88 Support of the HH complex wavelet on the Fourier plane

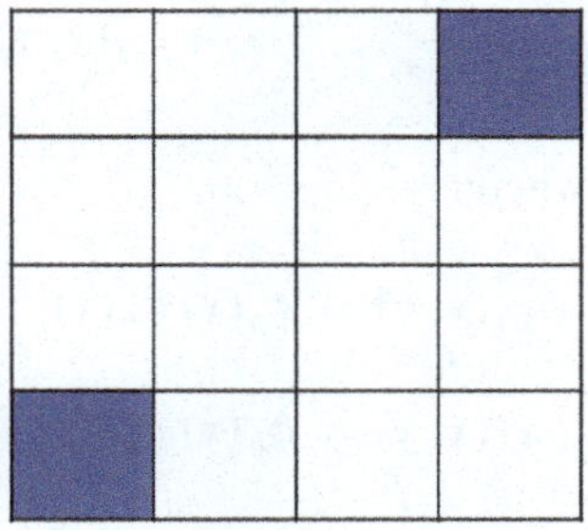

Fig. 4.89 Support of the real part of the complex HH wavelet

In consequence, by taking real parts, one gets six wavelets (for $i = 1 \ldots 3$):

$$\psi_i(x, y) = \frac{1}{\sqrt{2}}(\psi_{1,i}(x, y) - \psi_{2,i}(x, y)) \tag{4.93}$$

$$\psi_{i+3}(x, y) = \frac{1}{\sqrt{2}}(\psi_{1,i}(x, y) + \psi_{2,i}(x, y)) \tag{4.94}$$

where:

$$\psi_{1,1}(x, y) = \phi_h(x)\,\psi_h(y);\ \psi_{1,2}(x, y) = \psi_h(x)\,\phi_h(y);\ \psi_{1,3}(x, y) = \psi_h(x)\,\psi_h(y)$$

$$\psi_{2,1}(x, y) = \phi_g(x)\,\psi_g(y);\ \psi_{2,2}(x, y) = \psi_g(x)\,\phi_g(y);\ \psi_{2,3}(x, y) = \psi_g(x)\,\psi_g(y)$$

Figure 4.90 shows the six directions that can be obtained, using real parts.

A possible implementation can be to use $\{h_0(n), h_1(n)\}$ for one real nonstandard 2-D wavelet transform, and to use $\{g_0(n), g_1(n)\}$ for another. One obtains two HL, two LH and two HH subbands. To obtain an oriented wavelet transform, use the sum and difference of each pair of subbands. This is called *real oriented 2-D dual-tree transform*. This is not a complex transform, and it is not analytic. It is 2 times expansive.

Notice that taking imaginary parts, one can obtain another set of six oriented wavelets (for $i = 1 \ldots 3$):

$$\psi_i(x, y) = \frac{1}{\sqrt{2}}(\psi_{3,i}(x, y) - \psi_{4,i}(x, y)) \tag{4.95}$$

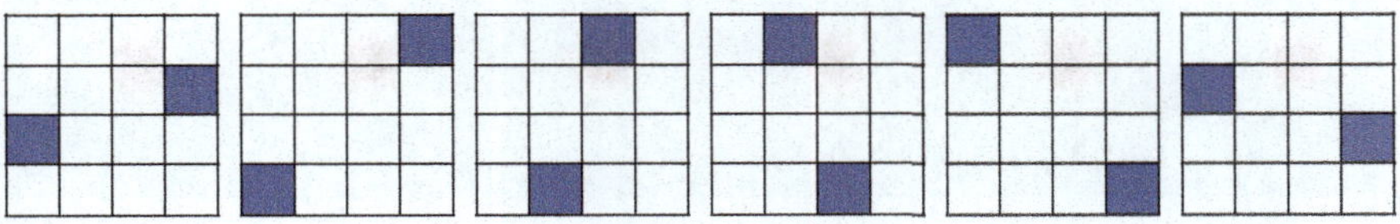

Fig. 4.90 The six wavelet directions that can be obtained

$$\psi_{i+3}(x, y) = \frac{1}{\sqrt{2}}(\psi_{3,i}(x, y) + \psi_{4,i}(x, y)) \quad (4.96)$$

where:

$$\psi_{3,1}(x, y) = \phi_g(x)\,\psi_h(y);\ \psi_{3,2}(x, y) = \psi_g(x)\,\phi_h(y);\ \psi_{3,3}(x, y) = \psi_g(x)\,\psi_h(y)$$

$$\psi_{4,1}(x, y) = \phi_h(x)\,\psi_g(y);\ \psi_{2,2}(x, y) = \psi_h(x)\,\phi_g(y);\ \psi_{2,3}(x, y) = \psi_h(x)\,\psi_g(y)$$

A *complex oriented 2-D dual-tree transform* can be formed by using the wavelets BB for the real part, and the wavelets CC for the imaginary parts. It is a 4 times expansive transform, and it is approximately analytic.

Figure 4.91 shows the six complex wavelets: the first row is the real part, the second row is the imaginary part, and the third row is the magnitude (which has no oscillatory behavior). This figure has been obtained with a long program that has been included in Appendix A.

The reader can find more information on complex wavelets in the Thesis [167], in particular the Chap. 3. Another Thesis of interest is [204], which focuses in image texture analysis.

Non-redundant versions of the complex wavelets have been proposed in [71] and the Thesis [201]. Both references are correlated. The Thesis focuses on seismic analysis applications.

In [164] there are sections telling that the dual-tree complex wavelet has near shift invariance, and near rotation invariance. The ulterior research has insisted in getting closer to these invariances. See, for instance, the Thesis [1], where one finds pages on analytic discrete wavelet transform, degree of shift invariance, and a hyperanalytic wavelet transform.

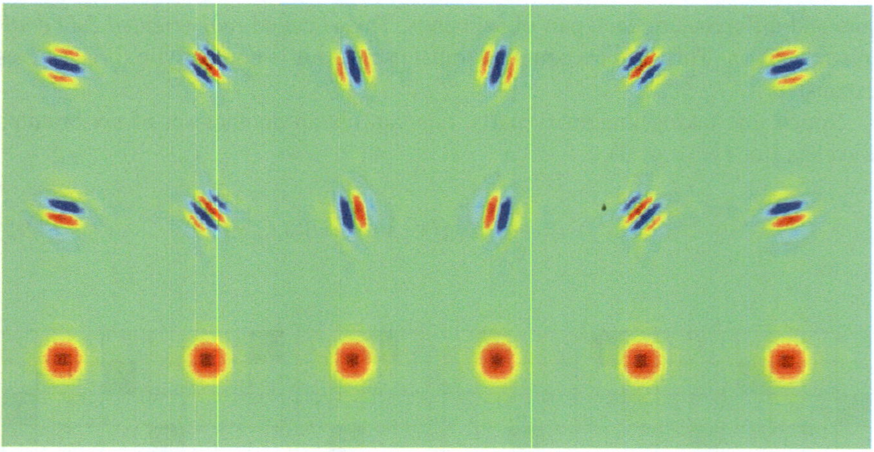

Fig. 4.91 The six complex wavelets

Phaselets were introduced [78] in 2003. Later on, the article [79] was devoted to phaselets and framelets. The phaselet transform includes the dual-tree complex wavelet as a particular case. A certain relationship of wavelet phases in the Fourier domain is established, which implies nearly shift invariance.

MATLAB software for implementation of dual-tree complex wavelet can be found in the web page of Brooklin Poly, the home page of LTFAT, and in MATLAB Central.

4.7 Experiments

Among the typical applications of wavelets, image denoising and image compression are of outmost importance. In this section a number of experiments is proposed, related with these topics.

All the experiments are based on a long program that has been included at the end of the section. For each experiment some editing of the program is done, and it will be concisely described.

The program itself is developed in simple terms. It could have been done in a recursive form, with certain loops, but it has been preferred to follow a linear processing sequence.

4.7.1 2 Level Haar Decomposition of the Image

A picture with clear features, like vertical, diagonal and horizontal lines, has been selected. The first task is to obtain the 2D wavelet transform of the image. For simplicity, a decomposition of only two levels has been considered, using the Haar transform.

The program has been edited as follows: *Kmix = 1* (fragment A) to put noise to zero; only figures (1) and (3) are relevant.

Figure 4.92 shows the selected picture and the results of the wavelet analysis. The HH regions show diagonal features, the LH region emphasizes horizontal traits, and the HL regions the vertical traits.

4.7.2 Fine Noise Is Added. Denoising Is Applied

In this experiment, white noise is added at pixel level, and then hard denoising is applied by establishing a threshold and setting pixels to zero if under the threshold.

The program has been edited so in the fragment (A) the one-line sentence for white noise generation is activated, and *Kmix = 0.8*. You should deactivate the generation of patched noise, by adding *%* to all lines included in the double for loop.

Fig. 4.92 The original picture and its 2-level wavelet analysis

Figure 4.93 shows the noisy image, it corresponds to figure (1) in the program. The picture becomes more obscure and granular due to the noise. The right hand side shows its 2-level wavelet transform (figure (3) in the program): the noise can be clearly observed in the first and the second level of the image decomposition, specially in the HH regions.

The denoising is activated by setting thresholds in the fragment (C).

Figure 4.94 shows the denoised picture (figure (2) in the program, and its 2-level wavelet transform (figure (6)). Due to the thresholds, the HH regions only have a few noise pixels, and the LH and HL regions still show horizontal and vertical picture features mixed with some noise pixels.

Fig. 4.93 Noisy image and its 2-level wavelet transform

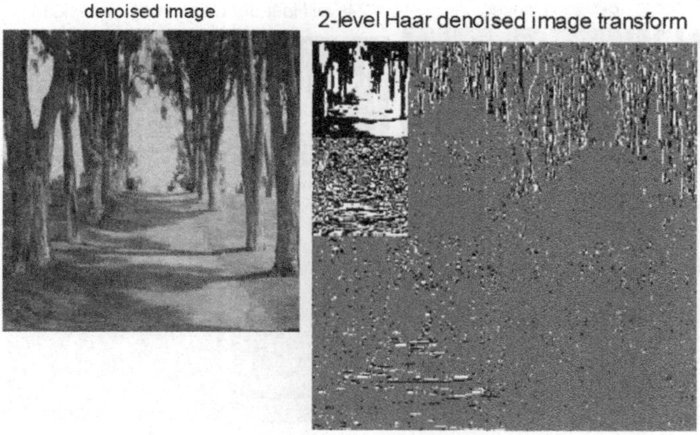

Fig. 4.94 Denoised image and its 2-level wavelet transform

4.7.3 Patched Noise Is Added. Denoising Is Applied

In this experiment the noise is made with 2 by 2 square patches with the four pixels having the same value. The values are generated at random.

This kind of noise is generated by activating, in the fragment (A) of the program, the lines included in the double for loop, and deactivating with a % the one-line sentence for white noise generation.

Figure 4.95 shows the noisy image (figure (1)) and its 2-level wavelet transform (figure (3)). Notice that the noise came to the HH, LH and HL regions of the second level. No noise came to the first level. This is due to the size of the patches. If 4×4 patches were used, a decomposition into 3 levels would be required to observe this noise.

Fig. 4.95 Image with patched noise and its 2-level wavelet transform

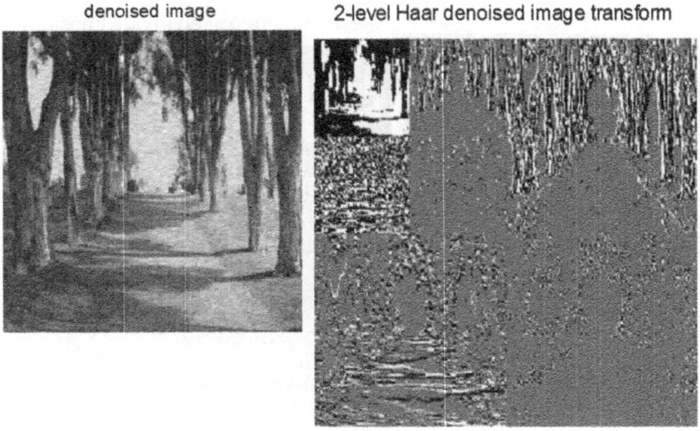

Fig. 4.96 Denoised image and its 2-level wavelet transform

It only makes sense to use hard denoising at the second decomposition level, which is governed by the second threshold in fragment (C) of the program.

Figure 4.96 shows the denoised image (figure (2)) and its 2-level wavelet transform (figure (6)).

4.7.4 Display of LL Regions, No Noise

In this experiment the noise is deactivated, *Kmix = 1* (fragment (A)), and the three figures included in fragment (B) are activated.

The purpose of this experiment is to show the LL region at level 1, and the LL region at level 2. Figure 4.97 shows the original picture compared with the other LL images represented with the same size.

Fig. 4.97 The original picture and the LL1 and LL2 images

There are two main points to emphasize. The first is that for many purposes, LL1 provides similar impression with only 25 % of information bits. And in applications with icons, LL2 could be sufficient with even less information bits (about 6 % of the original). The second point is that pixelization increases from original to LL1, and from LL1 to LL2.; this could be used for photographic effects.

Program 4.26 Image denoising experiment

```
% Image denoising experiment
%Read a 256x256 image
ufg=imread('trees1.jpg'); %read the image file into a matrix
fg=double(ufg); %convert to float
x=fg-mean(mean(fg)); %zero mean
[Nr,Nc]=size(fg);
tfg=zeros(Nr,Nc); %space for wavelet plane
%(A) ==========================================
v=zeros(256,256);
%or add fine noise
%v=50*randn(256,256);
%or add patched noise
pa=ones(2,2);
for ni=1:128,
  for nj=1:128,
    mi=1+(ni-1)*2; mj=1+(nj-1)*2;
    v(mi:mi+1,mj:mj+1)=50*randn(1,1)*pa;
  end;
end;
%mix image and noise
Kmix=0.8; %to be edited
x=Kmix*x; v=(1-Kmix)*v;
fg=x+v;
%==========================================
%show the noisy image
figure(1)
imshow(fg,[min(min(fg)),max(max(fg))]);
title('noisy image');
%Wavelet analysis----------------------------------------
% Haar 2-Level wavelet image transform
c=1/sqrt(2);
xfg=fg;
%Level 1--------------------------
%Step1
N=256; M=128;
%1 scale wavelet transform of rows
lfg1=zeros(N,M); hfg1=zeros(N,M);
for nn=1:N,
  auxL=fg(nn,1:2:N)+fg(nn,2:2:N);
  auxH=fg(nn,1:2:N)-fg(nn,2:2:N);
  lfg1(nn,1:M)=c*auxL; hfg1(nn,1:M)=c*auxH;
end;
%Step 2
%1 scale wavelet transform of columns of previous step
llfg1=zeros(M,M); hlfg1=zeros(M,M);
hhfg1=zeros(M,M); lhfg1=zeros(M,M);
```

```
%columns of L
for nn=1:M,
  auxL=lfg1(1:2:N,nn)+lfg1(2:2:N,nn);
  auxH=lfg1(1:2:N,nn)-lfg1(2:2:N,nn);
  llfg1(1:M,nn)=c*auxL; %image LL
  hlfg1(1:M,nn)=c*auxH; %image HL
end;
%columns of H
for nn=1:M,
  auxL=hfg1(1:2:N,nn)+hfg1(2:2:N,nn);
  auxH=hfg1(1:2:N,nn)-hfg1(2:2:N,nn);
  lhfg1(1:M,nn)=c*auxL; %image LH
  hhfg1(1:M,nn)=c*auxH; %image HH
end;
%save on wavelet plane
tfg(1:M,M+1:N)=lhfg1;
tfg(M+1:N,1:M)=hlfg1; tfg(M+1:N,M+1:N)=hhfg1;
%Level 2---------------------------
fg=llfg1;
%Step1
N=128; M=64;
%1 scale wavelet transform of rows
lfg2=zeros(N,M); hfg2=zeros(N,M);
for nn=1:N,
  auxL=fg(nn,1:2:N)+fg(nn,2:2:N);
  auxH=fg(nn,1:2:N)-fg(nn,2:2:N);
  lfg2(nn,1:M)=c*auxL; hfg2(nn,1:M)=c*auxH;
end;
%Step 2
%1 scale wavelet transform of columns of previous step
llfg2=zeros(M,M); hlfg2=zeros(M,M);
hhfg2=zeros(M,M); lhfg2=zeros(M,M);
%columns of L
for nn=1:M,
  auxL=lfg2(1:2:N,nn)+lfg2(2:2:N,nn);
  auxH=lfg2(1:2:N,nn)-lfg2(2:2:N,nn);
  llfg2(1:M,nn)=c*auxL; %image LL
  hlfg2(1:M,nn)=c*auxH; %image HL
end;
%columns of H
for nn=1:M,
  auxL=hfg2(1:2:N,nn)+hfg2(2:2:N,nn);
  auxH=hfg2(1:2:N,nn)-hfg2(2:2:N,nn);
  lhfg2(1:M,nn)=c*auxL; %image LH
  hhfg2(1:M,nn)=c*auxH; %image HH
end;
%save on wavelet plane
tfg(1:M,1:M)=llfg2; tfg(1:M,M+1:N)=lhfg2;
tfg(M+1:N,1:M)=hlfg2; tfg(M+1:N,M+1:N)=hhfg2;
%(B) ====================================
% display of analysis result--------
%total wavelet plane
figure(3)
imshow(tfg,[-20 20]);
title('2-level Haar image transform');
```

```
%figure(4)
%imshow(llfg1,[min(min(llfg1)), max(max(llfg1))]);
%title('LL1, Haar image transform');
%figure(5)
%imshow(llfg2,[min(min(llfg2)), max(max(llfg2))]);
%title('LL2, Haar image transform');
%============================================
%(C) ========================================
% Hard denoising
th=30; %threshold (edit)
for ni=1:128,
  for nj=1:128,
    if abs(hhfg1(ni,nj))<th, hhfg1(ni,nj)=0; end;
    if abs(hlfg1(ni,nj))<th, hlfg1(ni,nj)=0; end;
    if abs(lhfg1(ni,nj))<th, lhfg1(ni,nj)=0; end;
  end;
end;
th=60; %threshold (edit)
for ni=1:64,
for nj=1:64,
  if abs(hhfg2(ni,nj))<th, hhfg2(ni,nj)=0; end;
  if abs(lhfg2(ni,nj))<th, lhfg2(ni,nj)=0; end;
  if abs(lhfg2(ni,nj))<th, lhfg2(ni,nj)=0; end;
end;
end;
%display of thresholding result---------
dtfg=zeros(256,256);
%save on denoised wavelet plane
N=256; M=128;
dtfg(1:M,1:M)=llfg1; dtfg(1:M,M+1:N)=lhfg1;
dtfg(M+1:N,1:M)=hlfg1; dtfg(M+1:N,M+1:N)=hhfg1;
N=128; M=64;
dtfg(1:M,1:M)=llfg2; dtfg(1:M,M+1:N)=lhfg2;
dtfg(M+1:N,1:M)=hlfg2; dtfg(M+1:N,M+1:N)=hhfg2;
figure(6)
imshow(dtfg,[-20 20]);
title('2-level Haar denoised image transform');
%===========================================
%Wavelet synthesis---------------------------------------
rfg=zeros(256,256); %space for image
%recovery from level 2--------------------
%
N=128; M=64;
%recovery of L
for nn=1:M,
  auxL(1:2:N)=llfg2(1:M,nn)+hlfg2(1:M,nn);
  auxL(2:2:N)=llfg2(1:M,nn)-hlfg2(1:M,nn);
  lfg2(1:N,nn)=c*auxL; %image L
end;
%recovery of H
for nn=1:M,
  auxH(1:2:N)=lhfg2(1:M,nn)+hhfg2(1:M,nn);
  auxH(2:2:N)=lhfg2(1:M,nn)-hhfg2(1:M,nn);
```

```
  hfg2(1:N,nn)=c*auxH; %image H
end;
%recovery of original
for nn=1:N,
  auxL(1:2:N)=lfg2(nn,1:M)+hfg2(nn,1:M);
  auxL(2:2:N)=lfg2(nn,1:M)-hfg2(nn,1:M);
  rfg(nn,1:N)=c*auxL'; %image H
end;
llfg1=rfg;
%recovery from level 1--------------------
%
N=256; M=128;
%recovery of L
for nn=1:M,
  auxL(1:2:N)=llfg1(1:M,nn)+hlfg1(1:M,nn);
  auxL(2:2:N)=llfg1(1:M,nn)-hlfg1(1:M,nn);
  lfg1(1:N,nn)=c*auxL; %image L
end;
%recovery of H
for nn=1:M,
  auxH(1:2:N)=lhfg1(1:M,nn)+hhfg1(1:M,nn);
  auxH(2:2:N)=lhfg1(1:M,nn)-hhfg1(1:M,nn);
  hfg1(1:N,nn)=c*auxH; %image H
end;
%recovery of original
for nn=1:N,
  auxL(1:2:N)=lfg1(nn,1:M)+hfg1(nn,1:M);
  auxL(2:2:N)=lfg1(nn,1:M)-hfg1(nn,1:M);
  rfg(nn,1:N)=c*auxL'; %image H
end;
% display------------------
figure(2)
imshow(rfg,[min(min(rfg)),max(max(rfg))]);
title('denoised image');
```

4.8 Applications

The purpose of this section is to give an idea of the type of applications that wavelet variants have found in 2D processing. Pointers to sources of more detailed information will be included.

Typical image processing applications of wavelets are denoising and compression. Actually, denoising has become a sort of benchmark for comparison of image processing methods. Compression has an evident consumer interest.

In addition, many applications of wavelet variants for 2D signals and images are related to visual diagnosis in several scientific, engineering and medical fields.

4.8.1 Denoising

The application of wavelets for image denoising was introduced in the 90's. Since then a lot of research activity has been devoted to this application. An interesting review of image denoising using wavelets is included in the Thesis [36]. The denoising methods are divided into three categories: thresholding, shrinkage, and other estimation approaches. Using this division, a number of denoising methods can be classified as follows:

- Thresholding
  - Universal thresholding
  - Stein's unbiased risk estimation (SURE)
  - Cross validation
  - BayesShrink
- Shrinkage
  - MMSE
  - Bivariate shrinkage using level dependency
  - Neighbour dependency
  - Adaptive Bayesian wavelet shrinkage (ABWS)
  - Markov Random Field
  - Hidden Markov Tree
- Other
  - Gaussian scale mixture

Relevant references for all these methods is provided by [36]. This reading can be complemented with [150]. Another, more general and abstract, source of information is the doctoral Thesis [1]. The denoising goal is to obtain good estimates of the original picture pixels. There will be some estimation error. The average of this error on all pixels is called '*risk*'. For the minimization of this risk one could use a Bayesian approach, or some optimization approach.

In the seminal work of Donoho, two thresholding rules were considered: hard and soft. Hard-thresholding puts to zero all wavelet coefficients smaller than the threshold and leaves unchanged the rest of coefficients; soft-thresholding shrinks this rest of coefficients by the threshold value. If one uses the same threshold value for all subbands, this is global thresholding. If one uses a different threshold value for each subband, this is subband thresholding.

The four methods included in the thresholding category differ in the way the value of the threshold is computed. The universal threshold is $\sigma\sqrt{2\log N}$, where σ is the noise variance and N is the size of the image. The '*SureShrink*' method is based on SURE, and is a subband thresholding. The '*BayesShrik*' method is based on knowledge of image prior probability. Cross-validation is another subband thresholding method, based on *leaving-out one* technique.

In the shrinkage methods the noisy coefficients are decreased by a shrinkage factor. The linear minimum mean-square error (MMSE) method is equivalent to the Wiener filter, which will be studied in another chapter of this book. The ideas of EZW can be applied to assume interscale relationships: this is the basis of the method of bivariate shrinkage using level dependency. Similarly, large wavelet coefficients would probably have large neighbour coefficients, and this is taken into account by the neighbour dependency method. In [37] both neighbour and level dependency is considered. Markov random fields can be used for image prior models. Also, hidden Markov models can be used as models of wavelet coefficients statistics, being useful to retain statistical inter-dependencies.

An interesting discussion of denoising fundamentals is [186]. The chapter of [68] offers a good exposition of the initial denoising methods, and includes a comparison of some wavelet types –Daubechies, symlets, etc.—when used for denoising. A number of papers compare different thresholding and shrinkage methods, like for instance [154].

A popular approach for building image prior models, in a Bayesian context, is based on Gaussian mixtures, that can be used in the wavelet domain as in the important article of [147] and more recently in [85].

Apart from global or subband thresholding, *adaptive thresholding* has been also proposed by the research, like for instance in [28, 29]. A recent contribution that is adaptive and edge-preserving is [42]. In general an important aspect is to consider the image statistics, more or less locally; the article [149] includes an interesting overview of this aspect. Other approaches take into account image features for a selective denoising intervention [10].

In addition to the issues related to denoising decisions and actions, there is another question: which wavelet is better [118], or what variant or alternative to choose. Almost all proposed image decomposition methods have been applied for denoising. A fair comparison is difficult, since there are several types of images, faces, nature, buildings, etc., and there are also different qualitative judgement criteria (of course a quantitative criterion can be the noise/signal ratio).

Let us introduce some papers related to the methods included in this chapter.

- The use of Laplacian pyramid for denoising is reported in [133], and in [88] for medical application.
- On the basis of a Laplacian probability distribution prior [148] proposes a denoising method using steerable pyramid. Also, a kind of steerable pyramid, using complex wavelets, is proposed by [17] with application to image denoising.
- The use of complex ridgelets for denoising is introduced in [34], it compares favourably with the basic methods while preserving sharp edges.
- In the already cited paper [179] curvelets are applied for image denoising, attaining a good perceptual quality. An application of curvelets for the enhancement of SAR urban images obtained from satellites is described in [162]. Poisson noise removal using fast curvelets is described in [137]. The article [190] is a deep work on context adaptation based on statistical analysis of curvelet coefficients.

- Image denoising with contourlets is treated in [69, 125]. An application of contourlets for speckle reducing of medical ultrasound images is described in [90].
- In [65] shearlets are successfully applied for image denoising. A 3D version of shearlets is applied in [104] for video denoising.
- The Thesis [173] includes a review of denoising methods, and then focuses on the use of tetrolets for denoising as main contribution.
- Many combinations of methods have been proposed, like for instance the use of wavelet-curvelet in [183] (this paper includes an ample comparison table), or a two-level adaptive approach using Gaussian scale mixtures [83].
- The use of complex wavelets is investigated in the Thesis [1]. A survey of dual-tree complex wavelets for denoising is offered in [159].

Nowadays denoising is a wide topic that includes wavelets and many other methods [129]. It is recommended to see the [18] review, although it contains aspects that will be considered later in this book. For a intuitive perspective the reader is invited to check the delicious article of [106]. A good representation of some contemporary views is the Thesis [31], which includes a chapter on fundamental limits of image denoising. A practical professional prospect is given in [47].

Concerning extensive works on denoising with wavelets, the Thesis [117] contemplates many interesting facets, including fast video denoising, arbitrary transform-domain processing, and denoising of fluorescence microscopy data. The article of [35] focuses again, after a distillation of so many studies, on thresholding, whith interesting results. For a profound philosophic-mathematical treatement see [216].

4.8.2 Compression

Imagine the case of a horizontal line made with 300 black pixels. To transmit this line one could send 300 bytes, or just 'HK300' (horizontal black line 300 pixels long). This is just a simple example telling that image compression is related to coding. In addition it is well known that humans can be 'cheated'; for instance movies use a number of still frames per second to give an animation impression. It has been shown in the previous experiments with 2D wavelet analysis that high-frequency components (on the wavelet plane) can be discarded, and still one sees a fairly good picture, perhaps because one unconciously 'fills the holes' or just do not pay enough attention. Some of the contemporary approaches for image compression do use human vision system (HSV) models. Lossy compression methods are based on almost unnoticeable information discarding.

Like in the case of denoising, there are many image compression methods. There are entire books on the topic [158, 161], and several extensive reviews are available [67, 123]. The panoply of compression methods includes today vector quantization (VQ), fractal compression, wavelets, and others. A comparison of methods is provided by [33]. In relation to fractal compression, see [160]. By the way, 'panoply' is also the name of a NASA software for Earth maps display.

Concerning basic aspects of coding, from the times of telegraph with the Morse code, or even before, it has been noticed that savings can be obtained by using less bits for commonly seen values; this is called entropy coding. An example of it is Huffman coding. Also, if there is significant correlation between successive samples, predictive coding can be used, so only the residual between actual and predicted values is encoded.

Because personal computers, in all contemporary forms, are so widely used, people is already familiar with image coding formats: vectorial or bitmap, GIF, PNG, TIFF, etc. For a comparison of these formats, see [207] or reports included in the ECE 533 course projects (12.9.2. Internet Link lists).

What people perhaps do not know, is that in many cases wavelets are there, because an algorithm like EZW, SPIHT or similar is being used. References already mentioned about these algorithms [74, 156, 157, 166, 182, 188], and JPEG2000 [175, 197], are opportune in this context. The reader is invited to continue with other algorithms like WDR and ADWR [206], SPECK [91], and EBCOT [187].

From a more conceptual point of view, a recommended reference is the tutorial [203] on wavelet approximation and compression. A very interesting article, cited many times, on interactions between harmonic analysis and data compression is [62]. Another important, very cited, mathematical article is [26], which puts wavelet approximation in the context of variational optimization.

For the practitioner, interesting papers are [174] with a survey of image compression patents (digital cameras, color, printers, etc.) [82] with a compression performance comparison of different wavelets, or [176] that includes details on compression implementation using MATLAB:

Some of the wavelet variants papers already cited include image compression applications. Let us add to them the article of [30] with a new oriented wavelet transform.

The HSV model is considered for the compression of grayscale and of color images in the Thesis [14].

Image compression is important in medicine, and there is an extensive literature on this aspect, like chapters in books [45, 130] and papers [192]. The articles [163, 181, 211] offer specific reviews, while [127] extensively shows current trends.

And it is really satisfactory to find that wavelet image compression is being applied in Mars exploration [93].

A source of MATLAB code is the Image Compression Tools for MATLAB, available on Internet.

4.8.3 Image Registration

Image registration is the process of finding correspondence between points in two images of a scene [80]. This is an important process in medicine, where images of the same target taken with different methods, like X-ray and echography, should be aligned. Another illustrative example is to combine several photographs as a mosaic to get a larger view (recall some NASA maps). The origins of image registration are related to aerial surveillance.

In more abstract terms, in the case of alignment one has two images A and B, and one knows that:

$$B = g(f(A) \tag{4.97}$$

where $g()$ is a 1D intensity transformation, and $f()$ is a 2D spatial transformation. The problem is to find the two functions $g()$ and $f()$, such that B and the transformed A match. It is an inverse problem.

Some of the spatial transformations that can be considered are, rigid, affine, projective, or perspective transformations.

A basic hint for registration is to find corresponding points or features in two images, before trying to establish full image cross-correlation (perhaps by extensive searching).

Due to the importance of image registration there are many related papers. A general survey with many references is [223], while the more recent survey [50] includes modern aspects. In the medical context [72] provides an introduction to the topic; facets of the theory and practice is given by [39]; an important survey is [122], and another survey about deformable image registration is [177].

There is a Medical Registration Toolbox (MIRT) in MATLAB. Also, the MATLAB Computer Vision System Toolbox contains functions for image registration. One of the research centres involved in medical image registration is the Department of Informatik of the Friedrich-Alexander-Universität.

Wavelets are useful for localization of features in order to establish correspondences [138]. Moreover, the orientation analysis capabilities of the tools introduced in this chapter are being realized, as reported on the use of curvelets for image registration in [155], aiming at medical application, and [2] with a more general view. The steerable pyramid is used in [205] for remote (satellite) sensing applications.

Consider the following case: two consecutive photographs of a moving target are taken. It is, for instance, a car in the middle of some traffic. Image registration can be applied here for several purposes; one of them being motion estimation of that car. This topic, motion estimation, is also interesting in astrophysics, or for the study of fluids, etc. The book [184] includes, in Chap. 7, a good description of nuclear aspects of the topic. The motion estimation of the continuum body, like human organs, ocean flows, clouds, etc. is considered in the survey [112]. A series of block-matching algorithms are described, with practical purposes, in [140]. An important application of motion estimation is video compression, see the Thesis [108].

The article [213] provides a technique for image registration using a wavelet-based motion model. A comparison of wavelets for motion estimation of video sequences is given by [220]. The use of wavelets and motion estimation for video coding is considered in [94, 113].

A good advanced treatise of wavelets and optical flow is the Thesis [16]. The application of wavelets for fluid motion estimation is described in [48, 49]. In the [119] article, curvelet-based snakes are used for the detection and tracking of geophysical fluids, with an example of sea surface temperature data in a certain zone.

4.8.4 Seismic Signals

The origin of wavelets is directly linked to the analysis of seismic signals, so it is not strange that many papers on wavelets and seismography applications have been published. Of course, it is not the purpose of this subsection to review them all, but just to cite some illustrative contributions.

A word of caution is needed at the beginning. It happens that the term ‘seismic wavelet’ exist, and it is not related to the mathematical wavelets studied in this book. See [168] for a quick tutorial, or the Thesis [51] for more details.

Now, let us focus on mathematical wavelets (the wavelet transform).

A seismic earthquake signal is composed by several different seismic waves, called phases, that can be employed to characterize the type of the signal. Examples of frequently cited phases are P-phase and S-phase; in terms of arrival time to the seismic detector, P-phase arrives before than the S-phase. An important reference of wavelet application for phase identification is [7]. The more recent article [221] shows relevant advances on this area.

Interesting applications of wavelet analysis are [114] with application to early tsunami warning [24, 25] for the geometry determination of reservoirs [3] for signal classification related to volcanic origin. Links to other interesting contributions are included in the web page of the Wavelet Seismic Inversion Lab.

There are many applications of curvelets. Actually, a good starting point to see what is happening here is provided by the web pages of Felix Hermann, Martijn V. Hoop, Huub Douma. Most cited papers are [63, 89].

Other contributions that use curvelets are [32] about seismic migration and demigration (the seismic image captured in a station is a seismic migrated image subject to a particular propagation velocity); [217] about interpolation of nonuniformly sampled records; [99] and [185] about denoising.

4.8.5 Other Applications

A number of industrial applications of wavelets is reviewed in [193]. A large collection of wavelet applications is contained as chapters in the books [10–13].

The web page of Jean-Luc Stark is a good entry point to access software and papers related to the application of wavelets to astronomy. Two new 3D curvelet decompositions are applied in [209] for astronomical filament detection and for data restoration. An illustrative article on astronomical image representation with curvelets is [180]. In the book [199] on wavelets in physics there are chapters devoted to astrophysical applications, study of turbulences, solid state physics, medicine and physiology, etc.

Biomedical applications were reviewed in [195], which is a frequently cited article; the book [4] comprises a series of chapters with different wavelet applications in medicine and biology. A more recent review centered on biomedical images is [105], which covers magnetic resonance imaging, 3D ultrasound, tomography, mammography, tumor detection, etc. In March 2003 there was a IEEE T. Medical Imaging special issue on wavelets.

Some illustrative examples of medical applications of wavelets might be, [131] with the analysis of heart sounds, or [126] with the analysis of blood flow. Tomography applications are described in [151], X-ray, and [194], emission. An interesting case of lung imaging is [96]. The Thesis [81] deals with the denoising of cardiac PET data. Wavelets, curvelets and ridgelets are applied for medical image segmentation in [5, 6].

Vibration and fatigue is a major concern in several engineering sectors, like in aviation, civil structures, machinery, etc. An special issue of the Intl. J. of Wavelets, Multiresolution and Information Processing was devoted, in July 2009, to the application of wavelets to condition-based maintenance. A large review of machine monitoring using wavelets is [141]. An adaptive wavelet transform method is applied by [15] to identify cracks in gears.

Textile applications of wavelets are increasing. The topic of fabric defect detection is reviewed in [121].

4.9 Resources

4.9.1 MATLAB

Usually the toolboxes include tables of coefficients for the different wavelet families.

4.9.1.1 Toolboxes

- Laplacian pyramid Toolbox:
  http://www.ifp.illinois.edu/~minhdo/software/
- Piotr's MATLAB Toolbox:
  http://vision.ucsd.edu/~pdollar/toolbox/doc/
- matlabPyrTools:
  http://www.cns.nyu.edu/lcv/software.php
- Ridgelet and Curvelet first generation Toolbox:
  http://www.mathworks.com/matlabcentral/fileexchange/31559-ridgelet-and-curvelet-first-generation-toolbox

- BeamLab 200:
  http://www-stat.stanford.edu/~beamlab/
- FRIT Toolbox:
  http://www.ifp.illinois.edu/~minhdo/software/
- Toolox:
  http://www.ceremade.dauphine.fr/~peyre/matlab/wavelets/content.html
- CurveLab:
  contact through
  http://www.curvelet.org
- Uniform discrete curvelet transform demo:
  https://sites.google.com/site/nttruong/
- 3-D fast low-redundancy curvelet:
  http://jstarck.free.fr/cur3d.html
- Contourlet Toolbox:
  http://www.ifp.illinois.edu/~minhdo/software/
- Sharp contourlet Toolbox:
  http://lu.seas.harvard.edu
- Nonsubsampled contourlet Toolbox:
  http://www.ifp.illinois.edu/~minhdo/software/
- Surfbox Toolbox:
  http://www.ifp.illinois.edu/~minhdo/software/
- PDTDFB (shiftable complex directional pyramid) Toolbox:

  http://www.mathworks.com/matlabcentral/fileexchange/22398-pdtdfb-toolbox
- Translation-invariant contourlet:
  www.ece.rochester.edu/~eslami/
- ShearLab:
  www.shearlab.org
- FFST: Fast Finite Shearlet Transform:
  http://www.mathematik.uni-kl.de/imagepro/members/haeuser/ffst/
- Local Shearlet Toolbox:
  http://www.math.uh.edu/~dlabate/software.html

4.9.1.2 Matlab Code

- EZW (Rice University):
  www.owlnet.rice.edu/~elec539/Projects99/BACH/proj1/code.html
- EZW (Leuven University):

http://people.cs.kuleuven.be/~adhemar.bultheel/WWW/WAVE/ezw/matlab.html

- EZW (MATLAB Central):
  http://www.mathworks.com/matlabcentral/fileexchange/39651-ezwembedded-zerotree-wavelet
- SPIHT (Rensselaer Poly. Inst.):
  www.cipr.rpi.edu/research/SPIHT/spiht3.html#mat-spiht
- SPIHT (MATLAB Central):
  http://www.mathworks.com/matlabcentral/fileexchange/4808-spiht
- another SPIHT (MATLAB Central):
  http://www.mathworks.com/matlabcentral/fileexchange/22552-spiht-algoritm-based-image-compression
- Gabriel Peyre:
  http://www.mathworks.es/matlabcentral/fileexchange/authors/14044
- Pascal Getreuer:
  http://www.getreuer.info/
  (see also):
  http://www.mathworks.com/matlabcentral/fileexchange/authors/14582
- Demetrio Labate:
  http://www.math.uh.edu/~dlabate/software.html
  (includes video enhancement using 3D shearlets)
- D.O. Wu:
  www.wu.ece.ufl.edu
- Brooklin Poly:
  http://eeweb.poly.edu/iselesni/WaveletSoftware/
- LTFAT home page:
  http://www.ltfat.sourceforge.net/
- Image Compression Tools for MATLAB:
  http://www.ux.uis.no/~karlsk/ICTools/ictools.html

4.9.2 Internet

The page of Michael Elad has links to several toolboxes, including PPFT. The same happens with the pages of Dave Donoho or Emmanuel Candès (beamlets, curvelets, etc.), and of Minh Do (contourlets, Laplacian pyramid, etc.).

See the page of Xiaoming Huo for software and papers related to beamlets.

4.9.2.1 Web Sites

- K. Castleman:
  http://home.earthlink.net/~castleman
- Biomedical Imaging Group:
  http://bigwww.epfl.ch/demo/steerable/
- W.A. Pearlman:
  http://www.cipr.rpi.edu/~pearlman/
- SPIHT (Rensselaer Poly. Inst.):
  www.cipr.rpi.edu/research/SPIHT/
- J.E. Fowler:
  https://www.ece.msstate.edu/~fowler/
- Beamlab:
  www-stat.stanford.edu/~beamlab/
- Xiaoming Huo:
  www2.isye.gatech.edu/~xiaoming/
- Michael Elad:
  www.cs.technion.ac.il/~elad/index.html
- Dave Donoho:
  http://www-stat.stanford.edu/~donoho
- Emmanuel Candès:
  www-stat.stanford.edu/~candes/
- Jean-Luc Starck:
  http://jstarck.free.fr/jstarck/Home.html
- Minh N. Do:
  www.ifp.illinois.edu/~minhdo
- Yue M. Lu:
  http://lu.seas.harvard.edu
- Ramin Eslami:
  http://www.ece.rochester.edu/~eslami/
- Shearlets:
  http://www.shearlet.org
- Fast Multipole Methods:
  http://www.fastmultipole.org
- D.O. Wu:
  http://www.wu.ece.ufl.edu
- Dep. Informatik, Friedich-Alexander-Universität:
  http://www5.cs.fau.de/research/areas/medical-image-registration/

- Hermann, F.:
  http://www.slim.eos.ubc.ca
- De Hoop, M.V.:
  http://www.math.purdue.edu/~mdehoop/site
- Douma, H. :
  http://www.cwp.mines.edu/~huub/
- The Wavelet Seismic Inversion Lab.:
  http://www.cwp.mines.edu/~zmeng/waveletlab/waveletlab.html

4.9.2.2 Link Lists

- Curvelet.org:
  http://www.curvelet.org
- Shearlet.org:
  http://www.shearlet.org
- Laurent Duval web page:
  http://www.laurent-duval.eu/
- ECE533 Digital Image Processing Projects:
  http://homepages.cae.wisc.edu/~ece533/project/f06/index.html

References

1. I. Adam, *Complex Wavelet Transform: Application to Denoising*. Ph.D. thesis, Politehnica University of Timisoara and Universitè de Rennes (2010)
2. M. Alam, T. Howlander, and M. Rahman. Entropy-based image registration method using the curvelet transform. *Signal, Image and Video Processing* (2012)
3. P. Alasonati, J. Wassermann, M. Ohrnberger, Signal classification by wavelet-based hidden Markov models: Application to seismic signals of volcanic origin. in *Statistics in Volcanology*, pp. 1–27 (COSIV, 2006) Chapter 13
4. A. Aldroubi and M. Unser (eds.), *Wavelets in Medicine and Biology* (CRC Press, 1996)
5. S. AlZubi, *3D Multiresolution Statistical Approaches for Accelerated Medical Image and Volume Segmentation*. Ph.D. thesis, Brunel University, London (2011)
6. S. AlZubi, N. Islam, M. Abbod, Multiresolution analysis using wavelet, ridgelet, and curvelet transforms for medical image segmentation. Int. J. Biomedical Imaging, **2011**, 1–18, 2011. ID: 136034
7. K.S. Anant, F.U. Dowla, Wavelet transform methods for phase identification in three-component seismograms. Bull. Seismol. Soc. Am. **87**, 1598–1612 (1997)
8. S. Arivazhagan, K. Gowri, L. Ganesan, Rotation and scale-invariant texture classification uisng log-polar and ridgelet transform. J. Pattern Recognit. Res. **1**, 131–139 (2010)
9. A. Averbuch, R.R. Coifman, D.L. Donoho, M. Israeli, J. Walden, *Fast Slant Stack: A Notion of Radon Transform for Data in a Cartesian Grid Which is Rapidly Computible, Algebraically Exact, Geometrically Faithful and Invertible* (Dept. Statistics, Stanford University, USA, Technical report, 2001)

10. R.H. Bamberger, M.J.T. Smith, A filter bank for the directional decomposition of images: Theory and design. IEEE T. Signal Processing, 40, 4, 882–893 (1992)
11. D. Balenau (ed.), *Discrete Wavelet Transforms—Theory and Applications* (InTech, 2011)
12. D. Balenau (ed.), *Advances in Wavelet Theory and Their Applications in Engineering, Physics and Technology* (InTech, 2012)
13. D. Balenau (ed.), *Wavelet Transforms and Their Recent Applications in Biology and Geoscience* (InTech, 2012)
14. A.P. Beegan, Wavelet-based image compression using human visual system models. Master's thesis, Virginia Tech. (2001)
15. A. Belsak, J. Flasker, Adaptive wavelet method to identify cracks in gears. EURASIP J. Adv. Sign. Process. 1–8 (2010). ID. 879875
16. C. Bernard, *Wavelets and Ill Posed Problems: Optic Flow and Scattered Data Interpolation*. Ph.D. thesis, MINES Paris Tech. (1998)
17. A.A. Bharath, J. Ng, A steerable complex wavelet construction and its application to image denoising. IEEE T. Image Process. **14**(7), 948–959 (2005)
18. A. Buades, B. Coll, J.M. Morel, A review of image denoising algorithms with a new one. SIAM J. Multiscale Model. Simul. **4**(2), 490–530 (2005)
19. E. Candès. *Ridgelets: Theory and Applications*. Ph.D. thesis, Stanford University (1998)
20. E. Candès, L. Demanet, D. Donoho, L. Ying, Fast discrete curvelet transforms. Multiscale Model. Simul. SIAM **5**, 861–899 (2006)
21. E.J. Candès, F. Guo, New multiscale transforms, minimum total variation synthesis: Applications to edge-preserving image reconstruction. J. Sign. Process. Image Video Coding Deyond Standards **8**(11), 1519–1543 (2002)
22. P. Carre, E. Andres, Discrete analytical ridgelet transform. Sign. Process. **84**(11), 2165–2173 (2004)
23. K. Castleman, *Digital Image Processing* (Pearson, 1995)
24. M. Castro de Matos, O. Davogustto, C. Cabarcas, K. Marfurt, Improving reservoir geometry by integrating continuous wavelet transform seismic attributes, in *Proceedings of the SEG Las Vegas Annual Meeting*, pp. 1–5 (2012)
25. M. Castro de Matos, P.L.M. Manassi, Osorio, P.R. Schroeder Johan, Unsupervised seismic facies analysis using wavelet transform and self-organizing maps. Geophysics **72**(1), 9–21 (2007)
26. A. Chambolle, R.A. DeVore, N.Y. Lee, B.J. Lucier, Nonlinear wavelet image processing: Variational problems, compression, and noise removal through wavelet shrinkage. IEEE T. Image Process. **7**(3), 319–335 (1998)
27. V. Chandrasekaran, *Surflets: A Sparse Representation for Multidimensional Functions Containing Smooth Discontinuities* (In Proc. Intl. Symp, Information Theory, 2004)
28. S.G. Chang, B. Yu, M. Vetterli, Adaptive wavelet thresholding for image denoising and compression. IEEE Trans. Image Process. **9**(9), 1532–1546 (2000)
29. S.G. Chang, B. Yu, M. Vetterli, Spatially adaptive wavelet thresholding with context modeling for image denoising. IEEE Trans. Image Process. **9**(9), 1522–1531 (2000)
30. V. Chappelier, C. Guillemot, Oriented wavelet transform for image compression and denoising. IEEE T. Image Process. **15**(10), 2892–2903 (2006)
31. P. Chatterjee, *Patch-based Image Denoising and Its Performance Limits*. Ph.D. thesis, University of California at Santa Cruz (2011)
32. H. Chauris and T. Nguyen. Seismic demigration/migration in the curvelet domain. *Geophysics*, 73(2):35–46, 2008.
33. C.C. Chen, On the selection of image compression algorithms. Proc. IEEE Int. Conf. Pattern Recognit. **2**, 1500–1504 (1998)
34. G.Y. Chen, B. Kégl, Image denoising with complex ridgelets. Pattern Recognit. **40**, 578–585 (2007)
35. C. Chesneau, J. Fadili, J.L. Starck, Stein block thresholding for image denoising. Appl. Comput. Harmonic Anal. **28**, 67–88 (2010)

36. D. Cho, Image denoising using wavelet transforms. Master's thesis, Concordia University, Canada (2004)
37. D. Cho, T.D. Bui, G. Chen, Image denoising based on wavelet shrinkage using neighbor and level dependency. Int. J. Wavelets, Multiresolut. Inf. Process. **7**(3), 299–311 (2009)
38. E. Christophe, W.A. Pearlman, Three-dimensional SPIHT coding of volume images with random access and resolution scalability. EURASIP J. Image Video Process. **2008**(id:248905), 1–13 (2008)
39. W.R. Crum, T. Hartkens, D.L. Hill, Non-rigid image registration: Theory and practice. Br. J. Radiol. **77**, 140–153 (2004)
40. A.L. Cunha, M.N. Do, On two-channel filter banks with directional vanishing moments. IEEE Trans. Image Process. **16**(5), 1207–1219 (2007)
41. A.L. Cunha, J. Zhou, M.N. Do, The nonsubsampled contourlet transform: Theory, design, and applications. IEEE Trans. Image Process. **15**(10), 3089–3101 (2006)
42. R.D. da Silva, R. Minetto, W.R. Schwartz, Adaptive edge-preserving image denoising using wavelet transforms. Pattern Anal. Appl. 1–14 (2012)
43. S. Darkner, R. Larsen, M.B. Stegmann, B. Ersboll, *Wedgelet enhanced appearance model*, in Proceedings of the Computer Vision and Pattern Recognition, Workshop (2004), pp. 177–180
44. S. Das, M. Chowdhury, M.K. Kundu, Medical image fusion based on ripplet transform type i. Prog. Electromagn. Res. B **30**, 355–370 (2011)
45. C. Delgorge-Rosenberg, C. Rosenberger, Evaluation of medical image compression, in Fotiadis Exarchos, Papadopoulos, editor, *Handbook of Research on Advanced Techniques in Diagnostic Imaging and Biomedical Applications* (IGI Global, 2009)
46. L. Demaret, F. Friedrich, H. Führ, T. Szygowski, Multiscale wedgelet denoising algorithms. Proc. SPIE **5914**(XI-12) (2005)
47. M. DeNies, *Survey of Image Denoising Algorithms and Noise Estimation* (2012) Denies Video Software: http://www.deniesvideo.com/whitepapers.htm
48. P. Derian, P. Heas, C. Herzet, E. Memin, Wavelet-based fluid motion estimation, in Proceedings of the 3rd International Conference Scale Space and Variational Methods in Computer Vision, pp. 737–748 (2011)
49. P. Derian, P. Heas, E. Memin, Wavelets to reconstruct turbulence multifractals from experimental image sequences, in Proceedings of the 7th International Symposium on Turbulence and Shear Flow Phenomena (TSFP), pp. 1–6 (2011)
50. M. Deshmukh, U. Bhosle, A survey of image registration. Intl. J. Image Process. **5**(3), 245–269 (2011)
51. A.K. Dey, An analysis of seismic wavelet estimation. Master's thesis, University of Calgary (1999)
52. J.R. Ding, J.F. Yang, A simplified SPIHT algorithm. J. Chin. Inst. Eng. **31**(4), 715–719 (2008)
53. M.N. Do, *Directional Multiresolution Image Representation*. Ph.D. thesis, Department of Communication Systems, Swiss Federal Institute of Technology Lausanne (2001)
54. M.N. Do, M. Vetterli, The finite ridgelet transform for image representation. IEEE Trans. Image Process. **12**(1), 16–28 (2003)
55. M.N. Do, M. Vetterli, The contourlet transform: An efficient directional multiresolution image representation. IEEE Trans. Image Process. **14**(12), 2091–2106 (2005)
56. D.L. Donoho, Wedgelets: Nearly-minimax estimation of edges. Ann. Stat. **27**, 859–897 (1999)
57. D. L. Donoho. Orthonormal ridgelets and linear singularities. *SIAM J. Math. Anal.*, 31(5):1062–1099, 2000.
58. D.L. Donoho, Applications of beamlets to detection and extraction of lines, curves and objects in very noisy images, in *Proceedings IEEE-EURASIP Biennial Intl. Wkshp. Nonlinear Signal and Image Processing 2001* (2001)
59. D. L. Donoho and M. R. Duncan. Digital curvelet transform: Strategy, implementation and experiments. In *Proc. SPIE*, volume 4056, pages 12–29, 2000.
60. D.L. Donoho, A.G. Flesia, Digital ridgelet transform based on true ridge functions. Stud. Comput. Math. **10**, 1–30 (2003)

61. D.L. Donoho, X. Huo, Beamlets and multiscale image analysis, in *Multiscale and Multiresolution Methods, Lecture Notes in Computational Science and Engineering*, vol. 20 (LNCSE Springer 2001)
62. D.L. Donoho, M. Vetterli, R.A. DeVore, I. Daubechies, Data compression and harmonic analysis. IEEE T. Inf. Theory **44**(6), 2435–2476 (1998)
63. H. Douma, M.V. deHoop, Leading-order seismic imaging using curvelets. Geophysiscs, 72(6), 231–248 (2007)
64. G. Easley, D. Labate, W. Lim, Optimally sparse image representations using shearlets, in *40th Asilomar Conference on Signals, Systems and Computers*, pp. 974–978 (2006). Monterey
65. G.R. Easley, D. Labate, F. Colonna, Shearlet-based total variation diffusion for denoising. IEEE T. Image Process. **18**(2), 260–268 (2009)
66. G.R. Easley, D. Labate, V.M. Patel, Hyperbolic shearlets, in *Proceedings of the IEEE International Conference Image Processing* (2012)
67. O. Egger, P. Fleury, T. Ebrahimi, M. Kunt, High-performance compression of visual information –a tutorial review- part I: Still pictures. Proc. IEEE **87**(6), 976–1011 (1999)
68. B. Ergen, Signal and image denoising using wavelet transform, in *Advances in Wavelet Theory and Their Applications in Engineering, Physics and Technology*, de.by D. Baleanu (InTech Europe, 2012. Chap. 21)
69. R. Eslami, H. Radha, Translation-invariant contourlet transform and its application to image denoising. IEEE Trans. Image Process. **15**(11), 3362–3374 (2006)
70. M.J. Fadili, J.L. Starck, Curvelets and ridgelets, in *Encyclopedia of Complexity and Systems Science*, vol. 3 (Springer, 2007), pp. 1718–1738
71. F.C.A. Fernandes, R.L.C. van Spaendonck, C.S. Burrus, A new framework for complex wavelet transforms. IEEE Trans. Sign. Process. **51**(7), 1825–1837 (2003)
72. B. Fisher, J. Modersitzki, Ill posed medicine—an introduction to image registration. Inverse Prob. **24**, 1–19 (2008)
73. A.G. Flesia, H. Hel-Or, A. Averbuch, E.J. Candès, R.R. Coifman, D.L. Donoho, Digital implementation of ridgelet packets, in *Beyond Wavelets*, ed. by G. Welland (Academic Press, 2003)
74. J.E. Fowler, Embedded wavelet-based image compression: State of the art. Inf. Technol. **45**(5), 256–262 (2003)
75. W.T. Freeman, E.H. Adelson, The design and use of steerable filters. IEEE T. Pattern Anal. Mach. Intell. **13**(9), 891–906 (1991)
76. R.L.G. Claypoole, R.G. Baraniuk, A multiresolution wedgelet transform for image processing, in *Wavelet Applications in Signal and Image Processing VIII* ed. by M.A. Unser, A. Aldroubi, A.F. Laine, vol. 4119 (Proc. SPIE, 2000), pp. 253–262
77. B. Goossens, J. Aelterman, H. Luong, A. Pizurica, W. Philips, Efficient design of a low redundant discrete shearlet transform, in *Proceedings of the IEEE International Workshop Local and Non-local Approximation in Image Processing*, pp. 112–124 (2009)
78. R.A. Gopinath, The phaselet transform-an integral redundancy nearly shift-invariant wavelet transform. IEEE T. Sign. Process. **51**(7), 1792–1805 (2003)
79. R.A. Gopinath, Phaselets of framelets. IEEE T. Sign. Process. **53**(5), 1794–1806 (2005)
80. A. A. Goshtasby. *Image Registration: Principles, Tools and Methods*. Springer, 2012.
81. G.C. Green, Wavelet-based denoising of cardiac PET data. Master's thesis, Carleton University, Canada (2005)
82. S. Grgic, M. Grgic, B. Zovko-Cthlar, Performance analysis of image compression using wavelets. IEEE T. Ind. Electron. **48**(3), 682–695 (2001)
83. J. A. Guerrero-Colon and J. Portilla. Two-level adaptive denoising using Gaussian scale mixtures in overcomplete oriented pyramids. In *Proc. IEEE ICIP*, volume 1, pages 105–108, 2005.
84. K. Guo, G. Kutyniok, D. Labate, Sparse multidimensional representations using anisotropic dilation and shear operators, in *Proceedings of the International Conference on the Interactions between Wavelets and Splines*, pp. 189–201 (2005)

85. D.K. Hammond, E.P. Simoncelly, Image denoising with an orientation-adaptive Gaussian scale mixture model, in *Proceedings of the IEEE International Conference Image Processing*, pp 1433–1436 (2006)
86. S. Häuser, *Fast Finite Shearlet Transform: A Tutorial* (University of Kaiserslautern, 2011). arXiv:1202.1773
87. D.J. Heeger, *Notes on Steerable Filters*. New York University, Notes on Motion Estimation, www.cns.nyu.edu, Psych 267/CS 348D/EE365, 1998. http://citeseerx.ist.psu.edu/viewdoc/download?doi=10.1.1.88.9897&rep=rep1&type=pdf
88. M. Hensel, T. Pralow, R.R. Grigat, Real-time denoising of medical X-ray image sequences: Three entirely different approaches, in *Proceedings of the ICIAR*, pp. 479–490 (2006)
89. F.J. Hermann, G. Hennenfent, Non-parametric seismic data recovery with curvelet frames. Geophys. J. **173**, 233–248 (2008)
90. P.S. Hiremath, P.T. Akkasaligar, S. Badiger, Speckle reducing contourlet transform for medical ultrasound images. World Acad. Sci Eng. Technol. **56**, 1217–1224 (2011)
91. A. Islam, W.A. Pearlman, An embedded and efficient low-complexity hierarchical image coder. Proc. SPIE **3653** (1999)
92. M. Jacob, M. Unser, Design of steerable filters for feature detection using Canny-like criteria. IEEE Trans. Pattern Anal. Mach. Intell. **26**(6), 1007–1019 (2004)
93. A. Kiely, M. Klimesh, The ICER progressive wavelet image compressor. Technical report, JPL NASA, 2003. IPN Progress Report1-46
94. H.S. Kim, H.W. Park, Wavelet-based moving-picture coding using shift-invariant motion estimation in wavelet domain. Sign. Process. Image Commun. **16**, 669–679 (2001)
95. N.G. Kingsbury, Complex wavelets for shift invariant analysis and filtering of signals. J. Appl. Comput. Harmonic Anal. **10**(3), 234–253 (2001)
96. P. Korfiatis, S. Skiadopoulos, P. Sakellaropoulos, C. Kalogeropoulou, L. Costaridou, Combining 2D wavelet edge highlighting and 3D thresholding for lung segmentation in thin-slice CT. Br. J. Radiol. **80**, 996–1005 (2007)
97. J. Krommweh, Tetrolet transform: A new adaptive Haar wavelet algorithm for sparse image representation. J. Vis. Commun. Image Represent. **21**, 364–374 (2010)
98. J. Krommweh, G. Plonka, Directioal haar wavelet frames on triangles. Appl. Comput. Harmonic Anal. **27**, 215–234 (2009)
99. V. Kumar, J. Oueity, R. M. Clowes, and F. Hermann. Enhancing crustal reflection data through curvelet denoising. *Tectonophysics*, 508:106–116, 2011.
100. G. Kutyniok, Clustered sparsity and separation of cartoon and texture. SIAM J. Imaging Sci. **6**(2), 848–874 (2013)
101. G. Kutyniok, D. Labate, Shearlets: The first five years. Technical report, Oberwolfachn. 44/. (2010)
102. G. Kutyniok, M. Sharam, X. Zhuang, Shearlab: A rational design of a digital parabolic scaling algorithm. SIAM J. Imaging Sci. **5**(4), 1291–1332 (2012)
103. D. Labate, W. Lim, G. Kutyniok, G. Weiss, Sparse multidimensional representations using shearlets. Proc. SPIE Wavelets XI, 254–262 (2005)
104. D. Labate, P. Negi, 3d discrete shearlet transform and video denoising. Proc. SPIE **8138** (2011)
105. A.F. Laine, Wavelets in temporal and spatial processing of biomedical images. Ann. Rev. Biomed. Eng. **2**, 511–550 (2000)
106. M. Lebrun, M. Colom, A. Buades, J.M. Morel, Secrets of image denoising cuisine. Acta Numer. **21**, 475–576 (2012)
107. Q. Li, J. Shen, *Report on Implementing Fast Discrete Curvelet Transform* (Florida State University, Department Scientific Computing, 2007). http://people.sc.fsu.edu/~ql05/report_files/Report_FDCT_Wrapping.pdf
108. Z. Li, *New Methods for Motion Estimation with Applications to Low Complexity Video Compression*. Ph.D. thesis, Purdue University (2005)
109. W.Q. Lim, The discrete shearlet transform: A new directional transform and compactly supported shearlet frames. IEEE Trans. Image Process. **19**, 1166–1180 (2010)

110. A. Lisowska. Second order wedgelets in image coding. In *Proc. IEEE EUROCON 2007*, pages 237–244, 2007.
111. J. Liu, G. Liu, Y. Wang, W. He, A watermarking algorithm based on direction of image specific edge, in *Proceedings of the IEEE 3rd International Congress on Image and Signal Processing*, pp 1146–1150 (2010)
112. W. Liu, E. Ribeiro, A survey on image-based continuum-body motion estimation. Image Vision Comput. **29**, 509–523 (2011)
113. Y. Liu, K.N. Ngan, Fast multiresolution motion estimation algorithms for wavelet-based scalable video coding. Sign. Process. Image Commun. **22**, 448–465 (2007)
114. O.G. Lockwood, H. Kanamori, Wavelet analysis of the seismograms of the 2004 Sumatra-Andaman earthquake and its application to tsunami early warning. Geochem. Geophys. Geosyst. **7**(9), 1–10 (2006)
115. Y. Lu, M.N. Do, CRISP-contourlets: A critically-sampled directional multiresultion image representation, in *Proceedings of the of SPIE Conference on Wavelet Applications in Signal and Image Processing X* (San Diego 2003), pp. 655–665
116. Y. Lu, M.N. Do, Multidimensional directional filter banks and surfacelets. IEEE Trans. Image Process. **16**(4), 918–931 (2007)
117. F. Luisier, *The SURE-LET Approach to Image Denoising*. Ph.D. thesis, Ecole Polytechnique Fédrale de Lausanne (2010)
118. F. Luisier, T. Blu, B. Forster, M. Unser, Which wavelet bases are the best for image denoising? in *Proceedings of the SPIE Conference Mathematical Imaging*, vol. 5914, pp. 59140E–1 to 59140E–12 (2005)
119. J. Ma, A. Antoniadis, F.X. Le Dimet, Curvelet-based snake for multiscale detection and tracking of geophysical fluids. IEEE T. Geosci. Remote Sens. **44**(12), 3626–3638 (2006)
120. J. Ma, G. Plonka, The curvelet transform. IEEE Sign. Process. Mgz **27**(2), 118–133 (2010)
121. P.M. Mahajan, S.R. Kolhe, P.M. Patil, A review of automatic fabric defect detection techniques. Adv. Comput. Res. **1**(2), 18–29 (2009)
122. J.B.A. Maintz, M.A. Viergever, A survey of medical image registration. Med. Image Anal. **2**(1), 1–36 (1998)
123. A. Mammeri, B. Hadjou, A. Khoumsi, A survey of image compression algoritms for visual sensor networks. ISRN Sens. Netwo. ID **760320**, 1–19 (2012)
124. O. Marques, *Practical Image and Video Processing Using MATLAB* (J. Wiley, 2011)
125. B. Matalon, M. Elad, M. Zibulevsky, Improved denoising of images using modeling of a redundant contourlet transform. Proc. SPIE **5914**, 617–628 (2005)
126. P. May. Wavelet analysis of blood flow singularities by using ultrasound data. Technical report, Stanford University, 2002. Center for Turbulence Research Annual Research Briefs 2002.
127. G. Menegaz, Trends in medical image compression. Curr. Med. Imaging Rev. **2**(2), 1–20 (2006)
128. F.G. Meyer, R.R. Coifman, Brushlets: a tool for directional image analysis and image compression. Appl. Comput. Harmonic Anal. **4**(2), 147–187 (1997)
129. M. C. Motwani, M. C. Gadiya, R. C. Motwani, and F. C. Jr. Harris. Survey of image denoising techniques. In *Proc. GSPx*, 2004.
130. A. Nait-Ali, C. Cavaro-Menard, *Compression of Biomedical Images and Signals* (John Wiley, 2008)
131. H. Nazeran, Wavelet-based segmentation and feature extraction of heart sounds for intelligent PDA-based phonocardiography. Methods Inf. Med. **46**, 1–7 (2007)
132. P.S. Negi, D. Labate, 3D discrete shearlet transform and video processing. IEEE Trans. Image Process. **21**(6), 2944–2954 (2012)
133. H.T. Nguyen, N. Linh-Trung, The Laplacian pyramid with rational scaling factors and application on image denoising, in *Proceedings of the International Confernce Information Science, Signal Processing and their Applications* (2010), pp. 468–471
134. T.T. Nguyen, H. Chauris, Uniform discrete curvelet transform. IEEE Trans. Signal Process. **58**(7), 3618–3634 (2010)

135. T. T. Nguyen, Y. Liu, H. Chauris, S. Oraintara, Implementational aspects of the contourlet filter bank and application in image coding. EURASIP J. Adv. Signal Process., **2008**(ID 373487), 1–18 (2008)
136. T.T. Nguyen, S. Oraintara, A directional decomposition: Theory, design, and implementation, in *Proceedings of the IEEE Intrnational Symposium Circuits and Systems (ISCAS)*, vol. 3, pp. 281–284 (2004)
137. S. Palakkai, K.M.M. Prabhu, Poisson image denoising using fast discrete curvelet transform and wave atom. Signal Process. **92**(9), 2002–2017 (2012)
138. C. Paulson, E. Soundararajan, D. Wu, Wavelet-based image registration. Proc. SPIE **7704** (2010)
139. W.A. Pearlman, B.J. Kim, Z. Xiong, Embedded video subband coding with 3D SPIHT, in *Wavelet Image and Video Compression* ed. by P.N. Topiwala (Springer, 2002), pp. 397–432
140. H. Peinsipp, *Implementation of a Java Applet for Demonstration of Block-matching Motion-estimation Algorithms* (Technical report, Mannheim University, Dep. Informatik, 2003)
141. Z.K. Peng, F.L. Chu, Application of the wavelet transform in machine condition monitoring and fault diagnostics: A review with bibliography. Mech. Syst. Signal Process. **18**(2), 199–221 (2004)
142. P. Perona, Steerable-scalable kernels for edge detection and junction analysis. Image Vision Comput. **10**(10), 663–672 (1992)
143. G. Peyré, S. Mallat, Discrete bandelets with geometric orthogonal filters, in *Proceedings of the IEEE International Conference Image Processing*, vol. 1 (ICIP I- 2005), pp. 65–68
144. G. Peyré, S. Mallat, *A Matlab Tour of Second Generation Bandelets* (2005). www.cmap.polytechnique.fr/~peyre/BandeletsTutorial.pdf
145. G. Peyré, S. Mallat, Surface compression with geometric bandelets. Proc. ACM SIGGRAPH'05 601–608 (2005)
146. P. Pongpiyapaiboon, Development of efficient algorithm for geometrical representation based on arclet decomposition. Master's thesis, Technische Universität Munich (2005)
147. J. Portilla, V. Strela, J. Wainwright, E.P. Simoncelli, Image denoising using scale mixtures of Gaussians in the wavelet domain. IEEE T. Image Process. **12**(11), 1338–1351 (2003)
148. H. Rabbani, Image denoising in steerable pyramid domain based on a local Laplace prior. Pattern Recogn. **42**, 2181–2193 (2009)
149. S.M.M. Rahman, K. Hasan, Md. Wavelet-domain iterative center weighted median filter for image denoising. Signal Process. **83**, 1001–1012 (2003)
150. U. Rajashekar, E.P. Simoncelli, Multiscale denoising of photographic images, in *The Essential Guide to Image Processing* ed. by A. Bovik (Chapter 11. Academic Press, 2009)
151. M. Rantala, S. Vänskä, S. Järvenpää, M. Kalke, M. Lassas, J. Moberg, S. Siltanen, Wavelet-based reconstruction for limited-angle X-ray tomography. IEEE T. Med. Imaging **25**(2), 210–217 (2006)
152. J.K. Romberg, M. Wakin, R. Baraniuk, Multiscale wedgelet image analysis: Fast decomposition and modeling. Proc. IEEE Int. Conf. Image Process. **3**, 585–588 (2002)
153. R. Rubinstein, A.M. Bruckstein, M. Elad, Dictionaries for sparse representation modeling. Proc. IEEE **98**(6), 1045–1057 (2010)
154. S.D. Ruikar, D.D. Doye, Wavelet based image denoising technique. Int. J. Adv. Comput. Sci. Appl. **2**(3), 49–53 (2011)
155. M.N. Safran, M. Freiman, M. Werman, L. Joskowicz, Curvelet-based sampling for accurate and efficient multimodal image registration. Proc. SPIE **7259** (2009)
156. A. Said, W.A. Pearlman, Image compression using the spatial-orientation tree, in *Proceedings of the IEEE International Symposium Circuits and Systems*, pp. 279–282 (1993)
157. A. Said, W. A. Pearlman, A new, fast, and efficient image codec based on set partitioning in hierarchical trees. IEEE Trans. Circ. Syst. Video Technol. **6**(3), 243–250 (1996)
158. D. Salomon, *Handbook of Data Compression* (Springer, 2009)
159. R.K. Sarawale, S.R. Chougule, Survey of image denoising methods using dual-tree complex DWT and double-density complex DWT. Intl. J. Advanced Research in Computer. Eng. Technol. **1**(10), 121–126 (2012)

160. D. Saupe, R. Hamzaoui, A review of the fractal image compression literature. Comput. Graph. **28**(4), 268–276 (1994)
161. K. Sayood, *Introduction to Data Compression* (Morgan Kaufmann, 2012)
162. A. Schmitt, B. Wessel, A. Roth, Curvelet approach for SAR image denoising, structure enhancement, and change detection. Int. Arch. Photogrammetry **38**(part 3/W4), 151–156 (2009)
163. E. Seeram. Irreversible compression in digital radiology. a literature review. *Radiography*, 12(1):45–59, 2006.
164. I.W. Selesnick, R.G. Baraniuk, N.G. Kingsbury, The dual-tree complex wavelet transform. *IEEE Signal Processing Magazine*, pp. 123–151 (2005)
165. R. Sethunadh, T. Thomas, Image denoising using SURE-based adaptive thresholding in directionlet domain. Signal Image Process. (SIPIJ) **3**(6), 61–73 (2012)
166. J.M. Shapiro, Embedded image coding using zerotrees of wavelet coefficients. IEEE Trans. Signal Process. **41**(12), 3445–3462 (1993)
167. P.D. Shukla, Complex wavelet transforms and their applications. Master's thesis, University of Strathclyde (2003)
168. R. Simn and R. White. Phase, polarity and the interpreter's wavelet. *First Break*, 20:277–281, 2002.
169. E.P. Simoncelli, E.H. Adelson, *Subband Transforms* (Kluwer Academic Publishers, 1990)
170. E.P. Simoncelli, H. Farid, Steerable wedge filters for local orientation analysis. IEEE Trans. Image Process. **5**(9), 1377–1382 (1996)
171. E.P. Simoncelli, W.T. Freeman, The steerable pyramid: A flexible architecture for multi-scale derivative computation. Proc. IEEE Int. Conf. Image Process. **3**, 444–447 (1995)
172. E.P. Simoncelli, W.T. Freeman, E.H. Adelson, D.J. Heeger, Shiftable multiscale transforms. IEEE Trans. Inf. Theory **38**(2), 587–607 (1992)
173. M.K. Singh, Denoising of natural images using the wavelet transform. Master's thesis, San José State University (2010)
174. V. Singh, Recent patents o image compression—a survey. Recent Pat. Sign. Process. **2**, 47–62 (2010)
175. A.N. Skodras, C.A. Christopoulos, T. Ebrahimi, JPEG2000: The upcoming still image compression standard. Pattern Recogn. Lett. **22**(12), 1337–1345 (2001)
176. M.S. Song, Wavelet image compression. Contemp. Math. 1–33 (2006)
177. A. Sotiras, C. Davatazikos, N. Paragios, *Deformable Medical Image Registration; A Survey* (INRIA Research, Technical report, 2012)
178. N. Sprljan, S. Grgic, and M. Grgic. Modified SPIHT algorithm for wavelet packet image coding. *Real-Time Imaging*, 11(5–6):378–388, 2005.
179. J.L. Starck, E. Candès, D. Donoho, The curvelet transform for image denoising. IEEE Trans. Image Process. **11**(6), 670–684 (2002)
180. J.L. Starck, D. Donoho, E. Candès, Astronomical image representation by the curvelet transform. Astron. and Astrophys. **398**, 785–800 (2003)
181. M.G. Strintzis, A review of compression methods for medical images in PACS. Int. J. Med. Inf. **52**, 159–165 (1998)
182. R. Sudhakar, Ms. R. Karthiga, S. Jayaraman, Image compression using coding and wavelet coefficients—a survey. ICGST-GVIP J. **5**(6), 25–38 (2005)
183. P.D. Swami, A. Jain, Segmentation based combined wavelet-curvelet approach for image denoising. Int. J. Inf. Eng. **2**(1), 32–37 (2012)
184. R. Szeliski, *Computer Vision: Algorithms and Applications* (Springer, 2010)
185. G. Tang, J. Ma, Application of total-variation-based curvelet shrinkage for three-dimensional seismic data denoising. IEEE Geosci. Remote Sens. Lett. **8**(1), 103–107 (2011)
186. C. Taswell, The what, how, and why of wavelet shrinkage denoising. Comput. Sci. Eng. **2**(3), 12–19 (2000)
187. D. Taubman, High performance scalable image compression with EBCOT. IEEE T. Image Process. **9**(7), 1158–1170 (2000)

188. N. Tekbiyik, H.S. Tozkoparan, Embedded zerotree wavelet compression. Technical report, Eastern Maditerranean University, 2005. B.S. Project
189. P.C. Teo, Y. Hel-Or, Lie generators for computing steerable functions. Pattern Recognit. Lett. **19**, 7–17 (1998)
190. L. Tessens, A. Pizurica, A. Alecu, A. Munteanu, W. Philips, Context adaptive image denoising through modeling of curvelet domain coefficients. J. Electron. Imaging **17**(3), 033021–1 to 033021–17 (2008)
191. J. P. Thiran. Recursive digital filters with maximally flat group delay. *IEEE T. Circuit Theory*, 18(6):659–664, 1971.
192. A.S. Tolba, Wavelet packet compression of medical images. Digital Sign. Process. **12**(4), 441–470 (2002)
193. F. Truchetet, O. Laligant, A review on industrial applications of wavelet and multiresolution based signal-image processing. J. Electron. Imaging **17**(3), 1–11 (2008)
194. F.E. Turkheimer, M. Brett, D. Visvikis, V.J. Cunningham, Multiresolution analysis of emission tomography images in the wavelet domain. J. Cereb. Blood Flow Metab. **19**, 189–208 (1999)
195. M. Unser, A. Aldroubi, A review of wavelets in biomedical applications. Proc. IEEE **84**(4), 626–638 (1996)
196. M. Unser, N. Chenouard, A unifying parametric framework for 2D steerable wavelet transforms. SIAM J. Imaging Sci. **6**(1), 102–135 (2013)
197. B.E. Usevitch, A tutorial on modern lossy wavelet image compression: Foundations of JPEG 2000. *IEEE Signal Processing Mgz.*, pp. 22–35 (2001)
198. C. Valens, *Embedded Zerotree Wavelet Encoding*. http://www.mindless.com, 1999. http://140.129.20.249/~jmchen/wavelets/Tutorials/c.valens/ezwe.pdf
199. J.C. van den Berg, *Wavelets in Physics* (Cambridge University Press, 2004)
200. M. van Ginkel, *Image Analysis Using Orientation Space Based on Steerable Filters*. PhD thesis, TU Delft (2002)
201. R.L.C. van Spaendonck, *Seismic Applications of Complex Wavelet Transforms*. Ph.D. thesis, TU Delft (2003)
202. V. Velisavljevic, B. Beferull-Lozano, M. Vetterli, P.L. Dragotti, Directionlets: Anisotropc multi-directional representation with separable filtering. IEEE Trans. Image Process. **15**(7), 1916–1933 (2006)
203. M. Vetterli, Wavelets, approximation and compression. *IEEE Signal Processing Mgz.*, pp 59–73 (2001)
204. A.P.N. Vo, *Complex Directional Wavelet Transforms: Representation, Statistical Modeling and Applications*. Ph.D. thesis, The University of Texas at Arlington (2008)
205. M. Wahed, GhS El-tawel, A.G. El-karim, Automatic image registration technique of remote sensing images. Int. J. Adv. Comput. Sci. Appl. **4**(2), 177–187 (2013)
206. J.S. Walker, Wavelet-based image compression, in *Transforms and Data Compression Handbook*, ed. by Yip Rao (CRC Press, 2000)
207. R.H. Wiggins III, H.R. Davidson, C. Harnsberger, J.R. Lauman, P.A. Goede, Image file formats: Past, present, and future. Radio Graph. **21**(3), 789–798 (2001)
208. R.M. Willett, R.D. Nowak, Platelets: a multiscale approach for recovering edges and surfaces in photon-limited medical imaging. IEEE Trans. Med. Imaging **22**(3), 332–350 (2003)
209. A. Woiselle, J.L. Starck, J. Fadili, 3D curvelet transforms and astronomical data restoration. Appl. Comput. Harmonic Anal. **28**(2), 171–188 (2010)
210. A. Woiselle, J.L. Starck, J. Fadili, 3-D denoising and inpainting with the low-redundancy fast curvelet transform. J. Math. Imaging Vision **39**, 121–139 (2011)
211. S.T.C. Wong, L. Zaremba, D. Gooden, H.K. Huang, Radiologic image compression—a review. Proc. IEEE **83**(2), 194–219 (1995)
212. Q. Wu, M.A. Schulze, K.R. Castleman, Steerable pyramid filters for selective image enhancement applications, in *Proceedings of the IEEE International Symposium Circuits and Systems*, vol. 5, pp. 325–328 (1998)
213. Y.T. Wu, T. Kanade, C.C. Li, J. Cohn, Image registration using wavelet-based motion model. Int. J. Comput. Vision **38**(2), 129–152 (2000)

214. J. Xu, D. Wu, Ripplet transform type II transform for feature extraction. IET Image Process. **6**(4), 374–385 (2012)
215. J. Xu, L. Wu, Yang, D. Wu, Ripplet: A new transform for image processing. J. Vis. Commun. Image Represent. **21**, 627–639 (2010)
216. Y. Hel-Or, D. Shaked, A discriminative approach for wavelet denoising. IEEE Trans. Image Process. **17**(4), 443–457 (2008)
217. P. Yang, J. Gao, W. Chen, Curvelet-based POCS interpolation of nonuniformly sampled seismic records. J. Appl. Geophys. **79**, 90–99 (2012)
218. S. Yi, D. Labate, G.R. Easley, H. Krim, A shearlet approach to edge analysis and detection. IEEE T. Image Process. **18**(5), 929–941 (2009)
219. W. Yu, K. Daniilidis, G. Sommer, Approximate orientation steerability based on angular Gaussian. IEEE Trans. Image Process. **18**(2), 193–205 (2001)
220. J. Zan, M.O. Ahmad, M.N.S. Swamy, Comparison of wavelets for multiresolution motion estimation. IEEE T. Circ. Syst. Video Technol. **16**(3), 439–446 (2006)
221. H. Zhang, C. Thurber, C. Rowe, Automatic P-wave arrival detection and picking with multiscale wavelet analysis for single-component recordings. Bull. Seismol. Soc. Am. **93**(5), 1904–1912 (2003)
222. W. Zhang. Several kinds of modified SPIHT codec, in *Discrete Wavelet Transforms- Algorithms and Applications*, ed. by H. Olkkonsen (Intech, 2011)
223. B. Zitova, J. Flusser, Image registration methods: A survey. Image Vis. Comput. **21**, 977–1000 (2003)

Part II
Data-Based Actions: Adaptive Filtering, Modelling, Analysis, and Classification

Chapter 5
Adaptive Filters and Observers

5.1 Introduction

In signal processing, as well as in other fields, it is always advisable to take advantage of all the 'a priori' knowledge available about the problem in hand. It is also convenient to be able to express this knowledge with mathematical models, and provide the means for these models to be used for problem solving.

This chapter begins with the *Wiener* filter, which is based on linear estimation. The filter uses models of the noise and the signal. A main application of the filter is to deal with signals contaminated with additive noise.

As an introduction to the issues to be treated, let us refer to a simple example. It is asked to measure the length of a table. Several measurements are made, and the arithmetic mean is taken as result. This is an example of estimation of a constant L:

$$y(n) = L + v(n) \tag{5.1}$$

where $y(n)$ are the measurements, and $v(n)$ are disturbances (errors).

The least-square estimation of L is the mean:

$$\hat{L}(N) = \frac{1}{N} \sum_{K=1}^{N} y(k) \tag{5.2}$$

To compute this estimation, a set of N measurements is used. This is "batch-mode" estimation.

The expression (5.2) can be reformulated as:

$$\hat{L}(N) = \hat{L}(N-1) + \frac{1}{N}(y(N) - \hat{L}(N-1)) \tag{5.3}$$

This is a recursive method ("recursive estimation"). Notice that the current estimate is obtained from the previous estimate plus a correction. As it will be confirmed in this

© Springer Science+Business Media Singapore 2017

J.M. Giron-Sierra, *Digital Signal Processing with Matlab Examples, Volume 2*,
Signals and Communication Technology, DOI 10.1007/978-981-10-2537-2_5

and the following chapters, the same format—previous value plus correction—will appear with different applications.

Batch procedures are applied when you have already the data, like for instance a photograph to be processed. Recursive procedures are applied for on line signals, like for instance a telephone conversation being enhanced in real time.

As summer comes, the table would become longer because of dilatation. If you use winter data mixed with summer data, something wrong is obtained. In case of using recursive estimation along the year, it would be opportune to use a "forgetting factor" to concede more importance to recent data and not to old data. This is an example of adaptation.

Notice that the estimation needs enough data to converge inside a desired confidence bound. In the case of the table length, it would be good to let the estimation converge before table length changes become relevant (for instance, estimate day after day). This is a general issue: let the estimation converge faster than significant variable changes. In certain circumstances, this is a serious problem: think for example about estimation of airplane altitude while the plane is climbing.

An important part of this chapter is devoted to images and to Bayesian processing. These are topics that attract a lot of research activity. As it will be mentioned later on, the initial problem with the Hubble Space Telescope motivated a lot of interest about image deblurring. Nowadays, image processing is a key for many important applications.

This chapter also deals with state observation. It happens that you can obtain input and output data from a system, and it is required to obtain information about the internal states of the system. During our research about fermentation processes, the case was that data could be obtained, in real time, about control inputs such temperature, pressure, etc., and about outputs such pH, CO_2, etc. The problem was to estimate the concentration of certain substances along time. It was possible to obtain data about these concentrations, but using chemical analysis, taking some hours delay. Instead, on the basis of a state variable model, it was possible to get an estimate of these concentrations in real time, which is important to have good control on the process.

The material covered in this chapter has important practical applications. At the same time, it provides a basis for the developments to be presented in the next chapters of the trilogy.

5.2 The Wiener Filter

Frequently the signals are corrupted by additive noise, so the measurement obtained, $y(n)$, is related to the original signal, $x(n)$, as follows:

$$y(n) = x(n) + v(n) \tag{5.4}$$

where $v(n)$ is the noise.

The problem is to estimate $x(n)$ from $y(n)$, as best as possible.

All variables, $x(n)$, $y(n)$ and $v(n)$ are zero-mean.

In the following, discrete time auto-correlations and cross-correlations (for instance, $R_{xy}(m)$) will appear. Similarly to the continuous case, the z-transform of the discrete cross-correlation is the discrete cross power spectrum density (for instance, $S_{xy}(m)$).

5.2.1 *Problem Statement. Transfer Function*

Denote $\hat{x}(n)$ the estimation of $x(n)$. Define the error of the estimation as:

$$e(n) = x(n) - \hat{x}(n) \tag{5.5}$$

A criterion to measure the quality of the estimation is the mean-square error (MSE):

$$\Sigma = E(e^2) \tag{5.6}$$

($E()$, expected value)

We want to minimize Σ. This is called minimum mean-square error (MMSE) estimation.

Let us choose *linear estimation*, so that:

$$\hat{x}(n) = \sum_{k=-\infty}^{\infty} h(k)\, y(n-k) \tag{5.7}$$

Clearly, the estimator can be seen as a digital filter, with impulse response $h(k)$.

The MSE criterion is:

$$\Sigma = E\left\{[\sum_{k=-\infty}^{\infty} h(k)\, y(n-k)) - x(n)]^2\right\} \tag{5.8}$$

The minimum value can be obtained by setting:

$$\frac{\partial \Sigma}{\partial h(m)} = 0 \tag{5.9}$$

for all values of m.

Then:

$$\frac{\partial \Sigma}{\partial h(m)} = E\left\{2[\sum_{k=-\infty}^{\infty} h(k)\, y(n-k)) - x(n)]\, y(n-m)\right\} = 0 \quad (5.10)$$

Therefore:

$$E(e(n)\, y(n-m)) = 0\,, \;\; \forall m \quad (5.11)$$

which is the **orthogonality principle**. For the optimal filter, the error is orthogonal to the measurements.

Notice that the left hand side of (5.11) is the cross-correlation of $e(n)$ and $y(n)$:

$$\begin{aligned} R_{ey}(m) &= E(e(n)\, y(n-m)) = E((x(n) - \hat{x}(n))\, y(n-m)) = \\ &= R_{xy}(m) - R_{\hat{x}y}(m) = 0 \end{aligned} \quad (5.12)$$

Hence:

$$R_{xy}(m) = R_{\hat{x}y}(m)\,, \;\; \forall m \quad (5.13)$$

Taking into account (5.7):

$$R_{\hat{x}y}(m) = \sum_{k=-\infty}^{\infty} h(k)\, R_{yy}(m-k) = R_{xy}(m) \quad (5.14)$$

Using the z-transform:

$$H(z)\, S_{yy}(z) = S_{xy}(z) \quad (5.15)$$

Therefore, the transfer function of the optimal filter (the estimator) is:

$$H(z) = \frac{S_{xy}(z)}{S_{yy}(z)} \quad (5.16)$$

(*Wiener filter*)

Now suppose that signal and noise are uncorrelated, that is:

$$R_{xv}(n, m) = E(x(n)\, v(m)) = 0 \quad (5.17)$$

Then:

$$\begin{aligned} R_{yy}(n, m) &= E(y(n)\, y(m)) = E[(x(n) + v(n))(x(m) + v(m))] = \\ &= E(x(n) + x(m)) + E\,(v(n)\, v(m)) = R_{xx}(n, m) + R_{vv}(n, m) \end{aligned} \quad (5.18)$$

And:

$$R_{xy}(n, m) = E(x(n)\, y(m)) = E\,(x(n)\,(x(m) + v(m)) = R_{xx}(n, m) \quad (5.19)$$

In this case, the Wiener filter is:

$$H(z) = \frac{S_{xx}(z)}{S_{xx}(z) + S_{vv}(z)} \quad (5.20)$$

In continuous time domain, the expression of the filter is similar:

$$H(s) = \frac{S_{xx}(s)}{S_{xx}(s) + S_{vv}(s)} \quad (5.21)$$

Let us highlight that we use knowledge about the power spectra of the original signal and the noise.

5.2.1.1 Example of Sine + Noise Signal

A first example is the case of a sine plus noise signal. It is treated with the Program 5.1, which plots two figures. Figure 5.1 shows the signal to be filtered; the additive noise is Gaussian.

Figure 5.2 shows the successful result obtained with the frequency-domain Wiener filter.

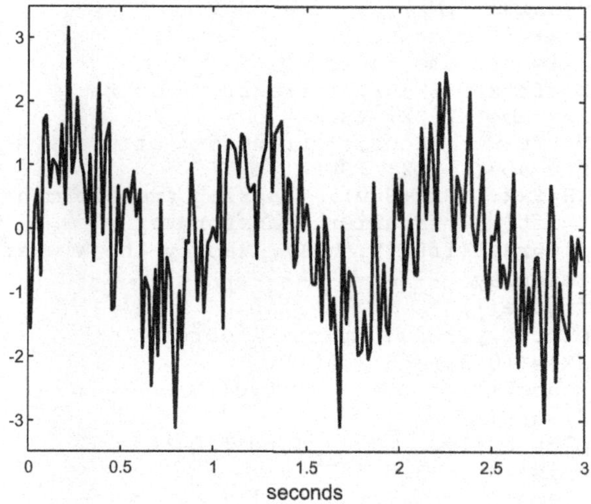

Fig. 5.1 Sine + noise signal

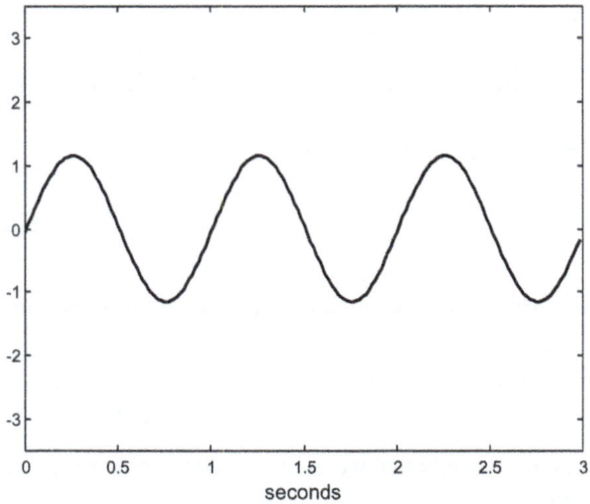

Fig. 5.2 The filtered signal

Program 5.1 Frequency domain Wiener filtering, sine signal + noise

```
%Frequency domain Wiener filtering
% sine signal + noise
fx=1; %signal frequency in Hz
wx=2*pi*fx; %signal frequency in rad/s
fs=60; %sampling frequency in Hz
tiv=1/fs; %time interval between samples;
t=0:tiv:(3-tiv); %time intervals set
x=sin(wx*t); %signal data set
Nx=length(x);
v=randn(1,Nx); %normal noise
y=x+v; %sine+noise
%Wiener computations
X=fft(x); %Fourier transform of x
Sxx=abs(X).^2; %Sxx
V=fft(v); %Fourier transform of v
Svv=abs(V).^2; %Svv
WH=Sxx./(Sxx+Svv); %Fourier transform of the Wiener filter
Y=fft(y); %Fourier transform of y
fly=real(ifft(Y.*WH)); %apply the Wiener filter
%display---------------------------
figure(1)
plot(t,y,'k'); %plots figure
axis([0 3 -3.5 3.5]);
xlabel('seconds'); title('sine+noise signal');
figure(2)
plot(t,fly,'k'); %plots figure
axis([0 3 -3.5 3.5]);
xlabel('seconds'); title('filtered signal');
```

5.2.1.2 Example of Noisy Image

Another interesting example is the case of a photograph including additive noise. Program 5.2 deals with this case. Figure 5.3 shows the noisy original image.

Figure 5.4 shows the good result obtained with the Wiener filter.

Fig. 5.3 Noisy image

Fig. 5.4 The filtered image

Program 5.2 Frequency domain Wiener filtering, image + noise

```
% Frequency domain Wiener filtering
% Image + noise
ux=imread('antena1.tif');
[Ny,Nx]=size(ux);
%signal and noise between 0 and 1
x=im2double(ux);
vn=abs(0.4*randn(Ny,Nx)); v=mod(vn,1);
X=fftshift(fft2(x)); %Fourier transform of x
%image+noise
Kmix=0.7; %to be edited
x=Kmix*x; v=(1-Kmix)*v;
y=x+v;
%Wiener computations
Sxx=abs(X).^2; %Sxx
V=fftshift(fft2(v)); %Fourier transform of v
Svv=abs(V).^2; %Svv
WH=Sxx./(Sxx+Svv); %Fourier transform of the Wiener filter
Y=fftshift(fft2(y)); %Fourier transform of y
fly=abs(ifft2(Y.*WH)); %apply the Wiener filter
%signals between 0 and 1
miy=min(min(y)); y=y-miy; %y>=0
may=max(max(y)); y=y/may; %y<=1
mify=min(min(fly)); fly=fly-mify; %fly>=0
mafy=max(max(fly)); fly=fly/mafy; %fly<=1
Uy=im2uint8(y); %convert to uint8 for display
Ufly=im2uint8(fly); %convert to uint8 for display
%display---------------------------
figure(1)
imshow(Uy); %plots figure
title('image+noise');
figure(2)
imshow(Ufly); %plots figure
title('filtered image');
```

Later on, in this chapter, there is a section devoted to images with emphasis on deblurring.

5.2.2 *Versions of the Filter*

Let us look again to the linear estimation:

$$\hat{x}(n) = \sum_{k=-\infty}^{\infty} h(k)\, y(n-k) \tag{5.22}$$

This expression corresponds to a non-causal filter. If you want to design a causal filter, a general expression could be:

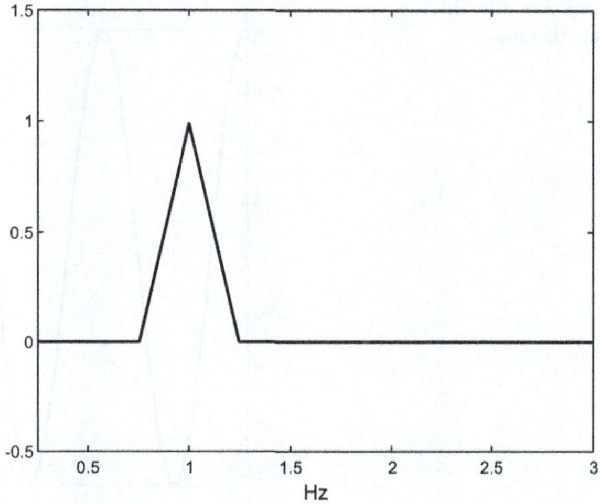

Fig. 5.5 Frequency response of the filter

$$\hat{x}(n) = \sum_{k=A}^{B} h(k)\, y(n-k) \tag{5.23}$$

where A is a finite value, and B is a positive value.

Supposing you want to design a FIR filter, it is pertinent to consider truncation and time-shifting. In order to enter into details, consider again the example of the sine + noise signal (Fig. 5.1) and the Wiener filter given by Eq. (5.20). The Program 5.3 focuses on the frequency response of the Wiener filter for this example, and the corresponding FIR filter coefficients.

Figure 5.5 shows the frequency response. It is a sharp band pass filter, tuned to the sine part of the signal.

Figure 5.6 shows the corresponding 240 FIR filter coefficients (deduced from 240 signal samples). These coefficients have been denoted as wh in the program. As it was expected, they conform to a sinusoidal pattern. A vertical dotted line has been added to mark the centre of symmetry of the FIR coefficients.

Finally, Fig. 5.7 shows the output of the FIR filter. The transient of the filter is clearly noticeable.

Program 5.3 Wiener filtering: sine signal + noise

```
% Wiener filtering
% sine signal + noise
% Details: bandwidth and filter coeffs.
fx=1; %signal frequency in Hz
wx=2*pi*fx; %signal frequency in rad/s
fs=60; %sampling frequency in Hz
tiv=1/fs; %time interval between samples;
t=0:tiv:(4-tiv); %time intervals set
```

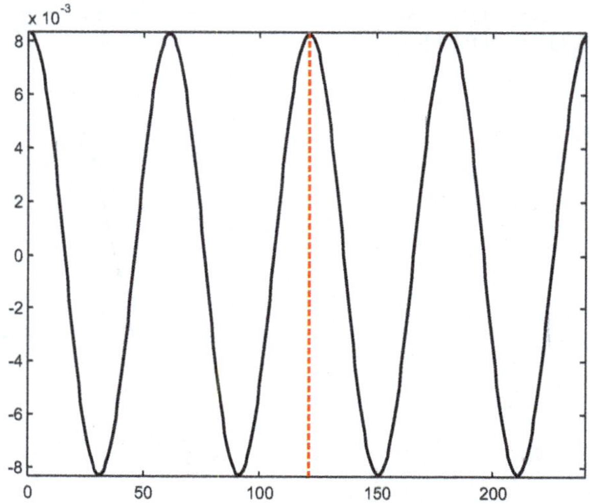

Fig. 5.6 FIR filter coefficients

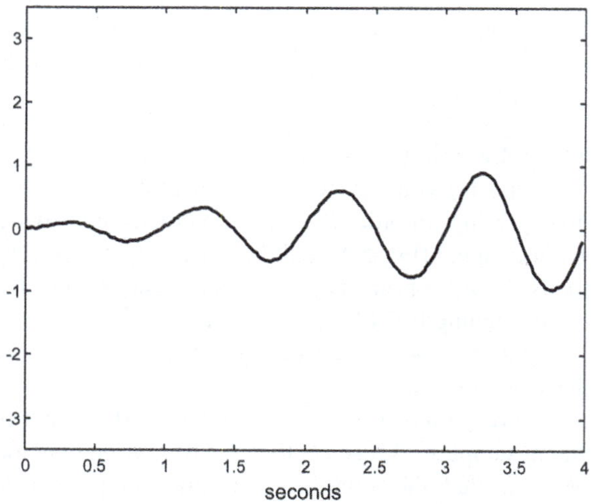

Fig. 5.7 Response of the filter

```
x=sin(wx*t); %signal data set
Nx=length(x);
v=randn(1,Nx); %normal noise
y=x+v; %sine+noise
%Wiener computations
X=fft(x); %Fourier transform of x
Sxx=abs(X).^2; %Sxx
V=fft(v); %Fourier transform of v
Svv=abs(V).^2; %Svv
WH=Sxx./(Sxx+Svv); %Fourier transform of the Wiener filter
wh=real(ifft(WH)); %filter impulse response
fly=filter(wh,1,y); %apply the Wiener filter
%display---------------------------
```

```
figure(1)
fiv=60/Nx;
fq=0:fiv:3;
plot(fq,WH(1:length(fq)),'k'); %plots figure
axis([fiv 3 -0.5 1.5]);
xlabel('Hz'); title('Frequency response of the filter');
figure(2)
plot(wh,'k'); hold on; %plots figure
limh=2/Nx;
plot([1+Nx/2 1+Nx/2],[-limh limh],'r--');
axis([0 Nx -limh limh]);
title('Filter coefficients');
figure(3)
plot(t,fly,'k'); %plots figure
axis([0 4 -3.5 3.5]);
xlabel('seconds'); title('filtered signal');
```

Usually, a shorter (truncated) version of the FIR filter is preferred. Time shifting is also welcome. This causes changes of the limits $A - B$ in the expression (5.23).

It is possible to take advantage of symmetry in order to save computational effort. For the FIR filter, depart from the following expression:

$$\hat{x}(n) = \sum_{k=0}^{N} h(k)\, y(n-k) \tag{5.24}$$

To minimize the MSE set the differential of the error to zero, and apply the orthogonality principle. Then:

$$\sum_{k=0}^{N} h(k)\, R_{yy}(m-k) = R_{xy}(m)\ ,\ \ m = 0,\, 1,\, 2,\, \ldots N \tag{5.25}$$

In matrix form:

$$\begin{pmatrix} R_{yy}(0)\ R_{yy}(1)\ \ldots\ R_{yy}(-N) \\ R_{yy}(1)\ R_{yy}(0)\ \ldots\ R_{yy}(-N+1) \\ R_{yy}(2)\ R_{yy}(1)\ \ldots\ R_{yy}(-N+2) \\ \ldots \\ R_{yy}(N)\ R_{yy}(N-1)\ \ldots\ R_{yy}(0) \end{pmatrix} \cdot \begin{pmatrix} h(0) \\ h(1) \\ . \\ . \\ h(N) \end{pmatrix} = \begin{pmatrix} R_{xy}(0) \\ R_{xy}(1) \\ . \\ . \\ R_{xy}(N) \end{pmatrix} \tag{5.26}$$

which can be expressed as:

$$R_y \cdot \mathbf{h} = \mathbf{r}_{xy} \ \rightarrow\ \mathbf{h} = R_y^{-1}\, \mathbf{r}_{xy} \tag{5.27}$$

(the "normal equations")

The auto-correlation matrix is of Toeplitz type.

The MMSE is:

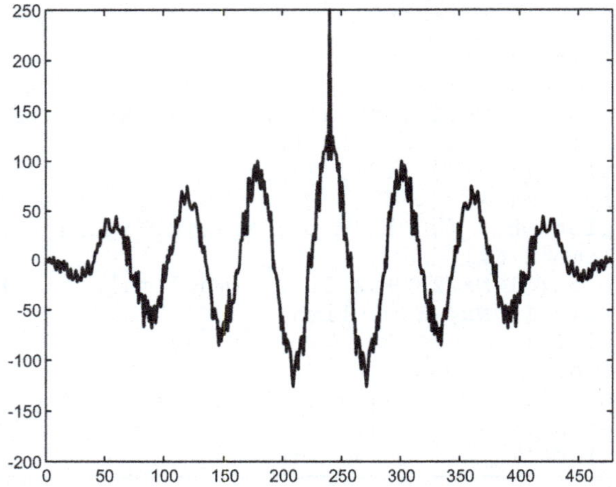

Fig. 5.8 Auto-correlation of y

$$E(e^2) = E([\hat{x} - x]^2) = E(x^2) - E(\hat{x}x) = E(x^2) - \mathbf{h}^T \mathbf{r}_{xy} \tag{5.28}$$

(the last term has been obtained considering the orthogonality principle)

From the point of view of theory, the Eqs. (5.25)–(5.28) can be extended to deal with the following case:

$$\sum_{k=0}^{\infty} h(k)\, R_{yy}(m-k) = R_{xy}(m)\ ,\ \ m = 0,\ 1,\ 2,\ \ldots\infty \tag{5.29}$$

(*Wiener-Hopf equation*)

Up to now, the Wiener filter has been obtained in a frequency domain context, as in Eq. (5.20). Now, Eq. (5.27) can be used for a time domain approach. In order to gain insight into this alternative, consider one more time the example of sine plus noise (Fig. 5.1).

Program 5.4 shows with the Figs. 5.8 and 5.9 the auto-correlation of the signal y and the cross-correlation of x and y. Notice the mirror symmetry.

Program 5.4 Cross-correlations: sine signal + noise

```
% Cross-correlations
% sine signal + noise
fx=1; %signal frequency in Hz
wx=2*pi*fx; %signal frequency in rad/s
fs=60; %sampling frequency in Hz
tiv=1/fs; %time interval between samples;
t=0:tiv:(4-tiv); %time intervals set
x=sin(wx*t); %signal data set
Nx=length(x);
v=randn(1,Nx); %normal noise
y=x+v; %sine+noise
```

```
%Cross-correlations
syy=xcorr(y); %symmetrical auto-correlation sequence
sxy=xcorr(x,y); %symmetrical cross correlation
%display----------------------------
figure(1)
plot(syy,'k'); %auto-correlation
axis([0 479 -200 250]);
title('auto-correlation of y');
figure(2)
plot(sxy,'k'); %cross-correlation
title('x-y cross-correlation');
axis([0 479 -150 150]);
```

Continuing with the example, the Program 5.5 computes the FIR filter coefficients according with (5.27). In this program the number of FIR filter coefficients to be computed has been limited to 50. Actually, it could be larger, up to the number of signal samples, but the result is not much improved.

Some aspects of the Program 5.5 deserve a comment. First, notice the use of the *toeplitz()* MATLAB function. And note also how the program selects the upper half of the auto-correlation of y and the cross-correlation of x and y, to save computing effort. Then, a symmetrical FIR filter is obtained by flipping and concatenation.

Figure 5.10 shows the upper half values of the FIR filter coefficients, as obtained from computation.

Figure 5.11 shows the values of the symmetrical FIR filter coefficients, obtained by flipping and concatenation.

Figure 5.12 shows the filtered signal.

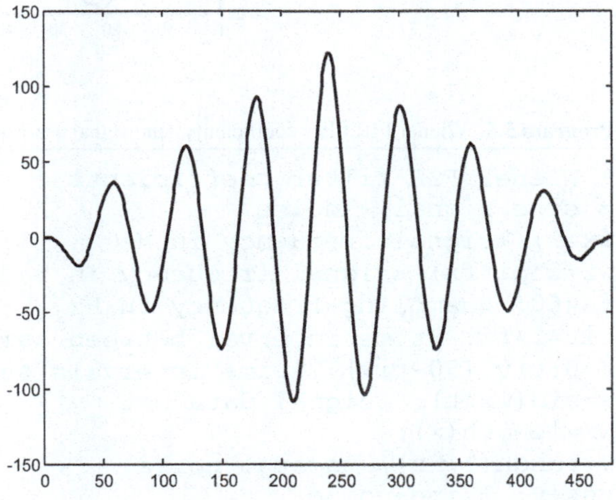

Fig. 5.9 Cross-correlation of x and y

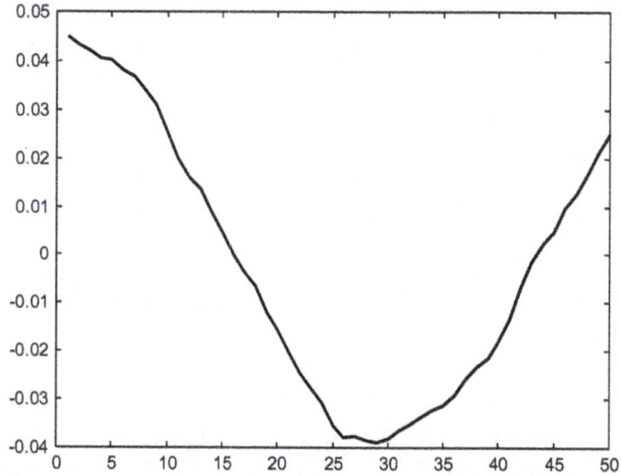

Fig. 5.10 Upper half of FIR filter coefficients

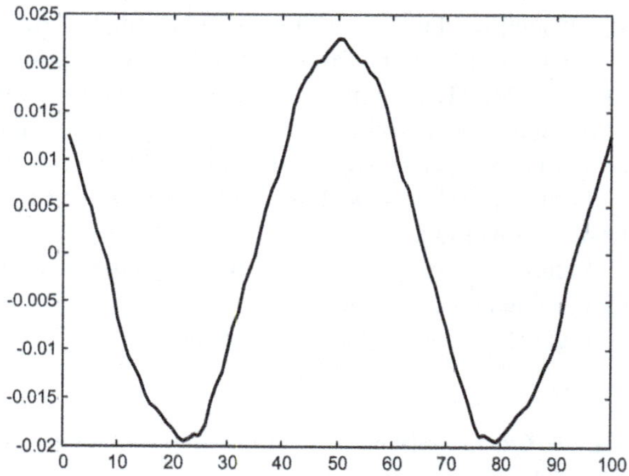

Fig. 5.11 Symmetrical FIR filter coefficients

Program 5.5 Wiener FIR filter coefficients, sine signal + noise

```
% Wiener FIR filter coefficients
% sine signal + noise
fx=1; %signal frequency in Hz
wx=2*pi*fx; %signal frequency in rad/s
fs=60; %sampling frequency in Hz
tiv=1/fs; %time interval between samples;
t=0:tiv:(50-tiv); %time intervals set
x=sin(wx*t); %signal data set
Nx=length(x);
v=randn(1,Nx); %normal noise
y=x+v; %sine+noise
Nh=50; %number of FIR coeffs
%Wiener computations
syy=xcorr(y); %symmetrical auto-correlation sequence
```

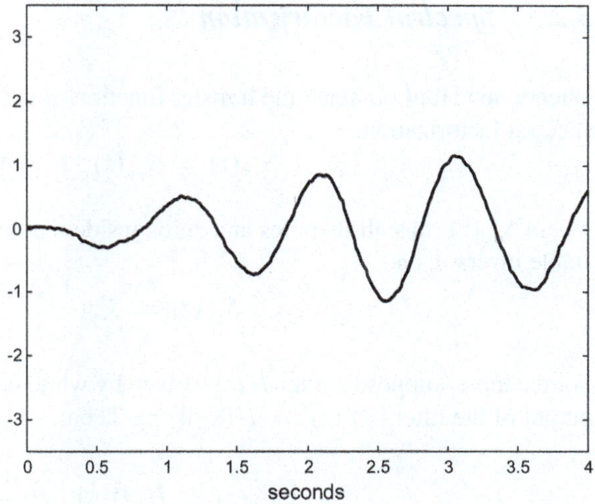

Fig. 5.12 The filtered signal

```
Ry=toeplitz(syy(Nx:Nx+Nh-1)); %Ry matrix
sxy=xcorr(x,y); %symmetrical cross correlation
rxy=sxy(Nx:Nx+Nh-1); %rxy vector
wh=Ry\rxy'; %Wiener FIR coeffs.
%append for FIR symmetry
rwh=wh';
lwh=fliplr(rwh);
sh=0.5*[lwh rwh]; %symmetrical FIR
fly=filter(sh,1,y); %apply the Wiener filter
%display-----------------------------
figure(1)
plot(wh,'k'); %computed FIR coefficients (right-side)
title('FIR coefficients (right-side)');
figure(2)
plot(sh,'k'); %Symmetrical FIR coefficients
title('Symmetrical FIR coefficients');
figure(3)
plot(t,fly,'k'); %filtered signal
axis([0 4 -3.5 3.5]);
xlabel('seconds'); title('filtered signal');
```

5.2.3 Spectral Factorization

Wiener and Hopf obtained the transfer function of the optimal causal filter by using spectral factorization:

$$S_{yy}(z) = S_{yy}^{+}(z) \cdot S_{yy}^{-}(z) \tag{5.30}$$

where $S_{yy}^{+}(z)$ has all its poles and zeros inside the unit circle (it is stable and with stable inverse), and:

$$S_{yy}^{-}(z) = S_{yy}^{+}(\frac{1}{z})^{T} \tag{5.31}$$

For example, suppose a filter $H(z)$ driven by white noise w, with $S_w(z) = 1$. The output of the filter is $Y(z) = H(z)\, W(z)$. Then:

$$S_{yy}(z) = H(z) \cdot 1 \cdot H(\frac{1}{z})$$

If for instance:

$$H(z) = \frac{z-7}{z-0.3}$$

Then:

$$S_{yy}(z) = \frac{(z-7)\,(\frac{1}{z}-7)}{(z-0.3)\,(\frac{1}{z}-0.3)}$$

And:

$$S_{yy}^{+}(z) = \frac{(\frac{1}{z}-7)}{(z-0.3)}\,, \quad S_{yy}^{-}(z) = \frac{(z-7)}{(\frac{1}{z}-0.3)}$$

Define the operator $\{\}_{+}$, which takes the causal part. For example, consider a time function $b(n)$ with (two-sided) z-transform $B(z)$, then:

$$\{B(z)\}_{+} = \sum_{k=0}^{\infty} b(k)\, z^{-k} \tag{5.32}$$

This is a one-sided version of the z-transform that takes the causal part, that is: the past.

Now, recall the Wiener-Hopf equation (5.29). It can be formulated as follows:

$$R_{xy}(m) - \sum_{k=0}^{\infty} h(k)\, R_{yy}(m-k) = f(m)\,, \quad m = 0,\, 1,\, 2,\, \ldots\infty \tag{5.33}$$

where $f(m)$ is a strictly anti-causal function ($f(m) = 0,\ \forall m \geq 0$).

Define $h(k) = 0$ for $k < 0$. Then:

$$S_{xy}(z) - H(z)\, S_{yy}(z) = F(z) \tag{5.34}$$

Using the factorization:

$$\frac{S_{xy}(z)}{S_{yy}^{-}(z)} - H(z)\, S_{yy}^{+}(z) = \frac{F(z)}{S_{yy}^{-}(z)} \tag{5.35}$$

Now, apply the $\{\}_{+}$ operator to both sides:

$$\left\{ \frac{S_{xy}(z)}{S_{yy}^{-}(z)} \right\}_{+} - H(z)\, S_{yy}^{+}(z) = 0 \tag{5.36}$$

And then, the final result is:

$$H(z) = \frac{1}{S_{yy}^{+}(z)} \left\{ \frac{S_{xy}(z)}{S_{yy}^{-}(z)} \right\}_{+} \tag{5.37}$$

Example:

Let be a signal x such that:

$$x(n+1) = 0.4\, x(n) + w(n)$$

where w is white noise with $\sigma_w^2 = 0.64$.

This signal can be modelled as a transfer function $A(z)$ driven by w:

$$A(z) = \frac{1}{z - 0.4}$$

Then:

$$S_{xx}(z) = \sigma_w^2\, A(z)\, A(z^{-1}) = \frac{0.64}{(z-0.4)\,(z^{-1}-0.4)} = \frac{0.64}{(1-0.4z)(1-0.4z^{-1})}$$

The situation is that one obtains a signal y that is related to x as follows:

$$y(n) = x(n) + v(n)$$

where v is white noise with $\sigma_v^2 = 1$.

Then:

$$S_{xy}(z) = S_{xx}(z) + S_{xv}(z) = S_{xx}(z)$$

$$S_{yy}(z) = S_{xx}(z) + S_{vv}(z) = S_{xx}(z) + 1$$

The spectral factorization of S_{yy} is:

$$S_{yy}(z) = \frac{0.64}{(1-0.4z)(1-0.4z^{-1})} + 1 = \frac{0.64 + (1-0.4z)(1-0.4z^{-1})}{(1-0.4z)(1-0.4z^{-1})} =$$

$$= \frac{0.0938(1-4.266z)(1-4.266z^{-1})}{(1-0.4z)(1-0.4z^{-1})} =$$

$$0.0938 \cdot \frac{1-4.266\,z}{1-0.4z^{-1}} \cdot \frac{1-4.266\,z^{-1}}{1-0.4z} = 0.0938 \cdot S^{+}_{yy}(z) \cdot S^{-}_{yy}(z)$$

Now:

$$\left\{\frac{S_{xy}(z)}{S^{-}_{yy}(z)}\right\}_{+} = \left\{\frac{\frac{0.64}{(1-0.4z)(1-0.4z^{-1})}}{\frac{(1-4.266z^{-1})}{(1-0.4z)}}\right\}_{+} = \left\{\frac{0.64}{(1-0.4z^{-1})(1-4.266z^{-1})}\right\}_{+}$$

The causal part can be obtained by partial fraction expansion:

$$\left\{\frac{S_{xy}(z)}{S^{-}_{yy}(z)}\right\}_{+} = \left\{\frac{-0.066}{(1-0.4z^{-1})} + \frac{0.706}{(1-4.266z^{-1})}\right\}_{+} = \frac{-0.066}{(1-0.4z^{-1})}$$

Therefore, the Wiener filter is:

$$H(z) = \frac{1}{S^{+}_{yy}(z)}\left\{\frac{S_{xy}(z)}{S^{-}_{yy}(z)}\right\}_{+} = \frac{(1-0.4z^{-1})}{0.0938\,(1-4.266z)} \cdot \frac{-0.066}{(1-0.4\,z^{-1})} =$$
$$= \frac{-0.704}{(1-4.266z)} = \frac{0.704\,z^{-1}}{4.266(1-\frac{z^{-1}}{4.266})} = \frac{0.165\,z^{-1}}{(1-0.234\,z^{-1})}$$

Which means:

$$\hat{x}(n) = 0.234\,\hat{x}(n-1) + 0.165\,\hat{y}(n-1)$$

5.2.4 The Error Surface

This section is considering an optimization problem: to find a filter that minimizes the estimation error. The solutions already given have been found via analysis; however it is also convenient to delineate a way for search-based optimization. For this reason, let us study the estimation error, which can be written as follows:

$$\Sigma = E(e^2) = E\,(x^2) + \mathbf{h}^T R_y\,\mathbf{h} - 2 \cdot \mathbf{h}^T\,\mathbf{r}_{xy} \tag{5.38}$$

(the minimum value is attained when $R_y\,\mathbf{h} = \mathbf{r}_{xy}$)

The right hand side is a quadratic function. The expression can be further shortened as follows:

$$\Sigma = \Sigma_{\min} + \tilde{h}^T\,R_y\,\tilde{h} \tag{5.39}$$

with:

$$\tilde{h} = h - h_{opt}; \quad (h_{opt} = R_y^{-1}\,\mathbf{r}_{xy}) \tag{5.40}$$

A tridimensional example could be the following:

$$\Sigma = 14 + (h(0)\ h(1))\begin{pmatrix} 3.8 & 0.9 \\ 0.9 & 3.8 \end{pmatrix}\begin{pmatrix} h(0) \\ h(1) \end{pmatrix} - 2(h(0)\ h(1))\begin{pmatrix} 1.7 \\ 0.6 \end{pmatrix}$$

Figure 5.13 shows the error surface. Notice that there is a minimum value. This figure has been obtained with the Program 5.6.

Figure 5.14, also obtained with the Program 5.6, shows a contour plot of the error surface,

Fig. 5.13 Error surface

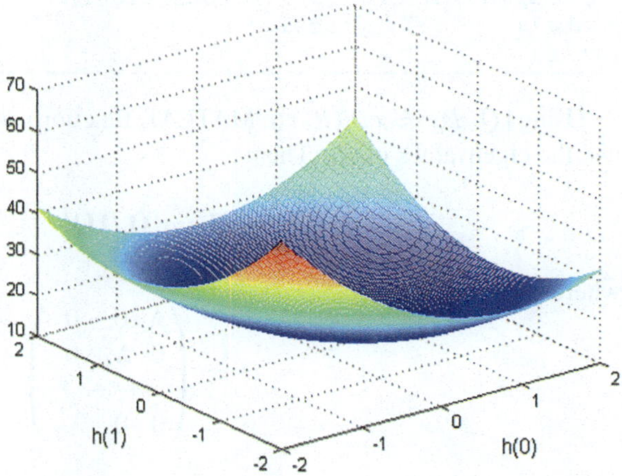

Fig. 5.14 Contour plot of the error

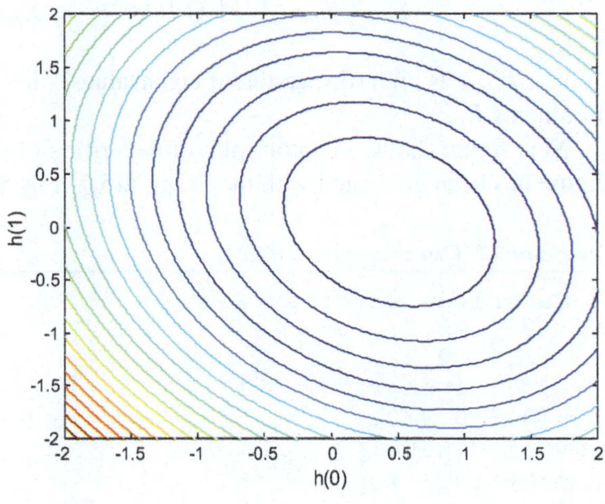

Program 5.6 Error surface

```
% Error surface
h0=-2:0.02:2;
h1=-2:0.02:2;
[H0,H1]=meshgrid(h0,h1);
Ry=[3.8 0.9;0.9 3.8];
rxy=[1.7;0.6];
aux1=(Ry(1,1)*H0.^2)+(Ry(1,2)*H0.*H1)+(Ry(2,1)*H1.*H0)...
+(Ry(2,2)*H1.^2);
aux2=(H0*rxy(1))+(H1*rxy(2));
ers=14+aux1-2*aux2;
%display---------------------------
figure(1)
mesh(H0,H1,ers); %plots figure
xlabel('h(0)'); ylabel('h(1)');title('error surface');
figure(2)
contour(H0,H1,ers,20); %plots figure
xlabel('h(0)'); ylabel('h(1)');
title('error surface (contour)');
```

Using $[Q, D] = eig(R_y)$ (a MATLAB function) you can obtain the eigenvectors and the eigenvalues of R_y. Then:

$$R_y = \mathbf{Q}\, D\, \mathbf{Q}^T \tag{5.41}$$

where:

$$D = \begin{pmatrix} \lambda_1 & 0 & 0 \\ 0 & \lambda_2 & 0 \\ - & - & - \\ 0 & 0 & \lambda_N \end{pmatrix} \tag{5.42}$$

Therefore:

$$\Sigma = \Sigma_{\min} + \tilde{h}^T\, \mathbf{Q}\, D\, \mathbf{Q}^T\, \tilde{h} = \Sigma_{\min} + \boldsymbol{v}^T\, D\, \boldsymbol{v} \tag{5.43}$$

If $\lambda_{max} / \lambda_{min}$ is high (dispersion of eigenvalues), this is an indication of bad conditioning of R_y.

Next figure shows a contour plot of the error surface according with (5.43). The figure has been generated with the Program 5.7 (Fig. 5.15).

Program 5.7 Canonical error surface

```
% Canonical error surfae
v0=-2:0.02:2;
v1=-2:0.02:2;
[V0,V1]=meshgrid(v0,v1);
Ry=[3.8 0.9;0.9 3.8];
rxy=[1.7;0.6];
hop=Ry\rxy; %optimal h
Emin=14-(hop'*rxy); %minimum error
```

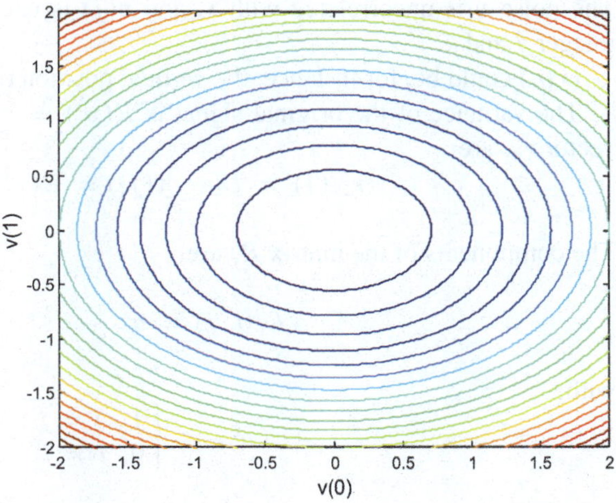

Fig. 5.15 Contour plot of the canonical error

```
[Q,D]=eig(Ry); %eigen
aux1=(D(1,1)*V0.^2)+(D(2,2)*V1.^2);
ers=Emin+aux1;
%display--------------------------
figure(1)
contour(V0,V1,ers,20); %plots figure
```

In the case of R_y a $2 \, x \, 2$ matrix, it can be expressed as:

$$R_y = \begin{pmatrix} \sigma^2 & r\sigma^2 \\ r\sigma^2 & \sigma^2 \end{pmatrix} \tag{5.44}$$

Then:

$$D = \begin{pmatrix} (1+r)\sigma^2 & 0 \\ 0 & (1-r)\sigma^2 \end{pmatrix} \tag{5.45}$$

And:

$$\frac{\lambda_{max}}{\lambda_{min}} = \frac{1+r}{1-r} \tag{5.46}$$

5.2.5 A Simple Example of Batch Mode and Recursive Mode

It is illustrative to see in detail the following simple case. Suppose that in:

$$y(n) = x(n) + v(n) \tag{5.47}$$

The noise v is uncorrelated with x, and $v(j)$ is uncorrelated with $v(k)$ for every integer j and k.

($x(n)$ could be, for instance, the vertical position of a buoy subject to waves).

The variance of the original signal is $E(x^2) = \sigma_x^2$, and the components of the vector $\mathbf{r}_{xy}$ are:

$$r_{xy}(j) = E(x \cdot y(j)) = E(x^2) = \sigma_x^2 \tag{5.48}$$

The components of the matrix R_y are:

$$R_y(i, j) = \sigma_x^2 + \sigma_v^2 \cdot \delta_{ij} \tag{5.49}$$

with:

$$\delta_{ij} = \begin{cases} 1, & i = j \\ 0, & i \neq j \end{cases}$$

Therefore:

$$\begin{pmatrix} \sigma_x^2 + \sigma_v^2 & \sigma_x^2 & \sigma_x^2 & \ldots & \sigma_x^2 \\ \sigma_x^2 & \sigma_x^2 + \sigma_v^2 & \sigma_x^2 & \ldots & \sigma_x^2 \\ - & - & - & & \\ \sigma_x^2 & \sigma_x^2 & \sigma_x^2 & \ldots & \sigma_x^2 + \sigma_v^2 \end{pmatrix} \begin{pmatrix} h(0) \\ h(1) \\ \cdot \\ \cdot \\ h(N) \end{pmatrix} = \begin{pmatrix} \sigma_x^2 \\ \sigma_x^2 \\ \cdot \\ \cdot \\ \sigma_x^2 \end{pmatrix} \tag{5.50}$$

The solution is: $h(0) = h(1) = h(2) = \ldots = h(N)$
with:

$$h(i) = \frac{\sigma_x^2}{L\sigma_x^2 + \sigma_v^2} = \frac{1}{L + \rho} \tag{5.51}$$

where:

$$L = N + 1; \quad \rho = \frac{\sigma_v^2}{\sigma_x^2} \tag{5.52}$$

(ρ is the noise to signal ratio)

Therefore:

$$\hat{x} = \frac{1}{L + \rho} \sum_{k=0}^{N} y(k) \tag{5.53}$$

And:

$$\Sigma = E(e^2) = \frac{\sigma_v^2}{L + \rho} \tag{5.54}$$

If $\rho << 1$ then $\hat{x}$ is the simple mean.

The example just studied is a case of batch mode processing. Now, let us proceed with a recursive mode.

Let us estimate the signal as observation data are coming. The following estimation could be used:

$$\hat{x}(n) = \sum_{k=0}^{n} h(n, k)\, y(n-k) \tag{5.55}$$

Both the coefficients h and the estimate $\hat{x}$ will depend on n.

Suppose two consecutive times k and $k+1$. Write:

$$h(i, k+1) = \frac{1}{L+\rho+1}; \quad h(i, k) = \frac{1}{L+\rho} \tag{5.56}$$

$$\Sigma(k+1) = \frac{\sigma_v^2}{L+\rho+1}; \quad \Sigma(k) = \frac{\sigma_v^2}{L+\rho} \tag{5.57}$$

Then:

$$\frac{\Sigma(k+1)}{\Sigma(k)} = \frac{h(i, k+1)}{h(i, k)} = \frac{L+\rho}{L+\rho+1} \tag{5.58}$$

$$\hat{x}(k+1) = \frac{L+\rho}{L+\rho+1}\,\hat{x}(k) + \frac{1}{L+\rho+1}\,y(k+1) \tag{5.59}$$

Now, substitute:

$$f(k+1) = \frac{L+\rho}{L+\rho+1}; \quad g(k+1) = \frac{1}{L+\rho+1}$$

Then:

$$\hat{x}(k+1) = f(k+1)\,\hat{x}(k) + g(k+1)\,y(k+1) \tag{5.60}$$

But: $f(k+1) = 1 - g(k+1)$

Therefore:

$$\hat{x}(k+1) = \hat{x}(k) + g(k+1)\,(y(k+1) - \hat{x}(k)) \tag{5.61}$$

This recursive estimation has the same form as in (5.3): the second term on the right hand of (5.61) depends on the difference between measurement and estimation.

5.3 Recursive Estimation of Filter Coefficients

There are applications where it is possible to get typical records of the original signal x and the corresponding contaminated signal y. For instance, the case of a telephone link where you inject a test signal at the input, and you record the output at the other end of the link. With these data, one can obtain the proper Wiener filter. This could be done in several ways, in batch mode or in recursive mode.

In this section several recursive methods will be introduced. These methods are frequently used, and pave the way for adaptive versions of the Wiener filter (it will be introduced in the next section).

Some of the recursive methods involve matrix inversion. It may happen that the inversion shows numerical difficulties. For this reason, the theory tried to circumvent possible problems by using the matrix inversion lemma, which will be included in this section. When using this lemma, the expressions take a peculiar form that will be met in this chapter and in the following chapters of the trilogy.

The recursive expressions will use a vector of present and past values of y. The notation for this vector is:

$$\mathbf{y}(m) = [y(m), y(m-1), \ldots, y(m - Nh)]^T \tag{5.62}$$

where Nh is the length of the FIR filter to be computed.

Here is a simple way to write a recursive computation of R_y and $\mathbf{r}_{xy}$:

$$R_y(n) = \sum_{k=1}^{n} \mathbf{y}(k)\, \mathbf{y}^T(k) = R_y(n-1) + \mathbf{y}(n)\, \mathbf{y}^T(n) \tag{5.63}$$

$$\mathbf{r}_{xy}(n) = \mathbf{r}_{xy}(n-1) + x(n)\, \mathbf{y}(n) \tag{5.64}$$

The first section introduces an analytical recursive method, denoted *'Recursive Least Squares'* (RLS). The second subsection focuses on search-based methods.

5.3.1 The RLS Method

The right expression in (5.27) was:

$$\mathbf{h} = R_y^{-1}\, \mathbf{r}_{xy} \tag{5.65}$$

The idea of RLS is to use a recursive computation of R_y^{-1}, instead of matrix inversion. Then one needs an expression like:

$$R_y^{-1}(n) = R_y^{-1}(n-1) + update(n) \tag{5.66}$$

Introduce the **matrix inversion lemma**:

Let A and B be two positive-definite $N \times N$ matrices related by:

$$A = B^{-1} + C\, D^{-1}\, C^T \tag{5.67}$$

where D is a positive definite $M \times M$ matrix, and C is a $N \times M$ matrix.

The matrix inversion lemma states that:

$$A^{-1} = B - BC\,(D + C^T BC)^{-1}\, C^T B \tag{5.68}$$

(Note: there are other alternative formulations in the literature)

Application to our case:

Let:

$$\begin{array}{c} R_y(n) = A; \;\; R_y^{-1}(n-1) = B \\ \mathbf{y}(n) = C; \;\; D = I \end{array} \tag{5.69}$$

(according to (5.63) the chosen matrices fulfil (5.67))

Then:

$$R_y^{-1}(n) = R_y^{-1}(n-1) - \frac{R_y^{-1}(n-1)\mathbf{y}(n)\,\mathbf{y}^T(n) R_y^{-1}(n-1)}{1 + \mathbf{y}^T(n) R_y^{-1}(n-1)\mathbf{y}(n)} \tag{5.70}$$

Denote:

$$\begin{array}{l} L(n) = R_y^{-1}(n) \\ \mathbf{K}(n) = \frac{R_y^{-1}(n-1)\mathbf{y}(n)}{1+\mathbf{y}^T(n)R_y^{-1}(n-1)\mathbf{y}(n)} \end{array} \tag{5.71}$$

Then:

$$L(n) = L\,(n-1) - \mathbf{K}(n)\,\mathbf{y}^T(n)\,L(n-1) \tag{5.72}$$

Notice that (5.71):

$$\mathbf{K}(n) = [L\,(n-1) - \mathbf{K}(n)\,\mathbf{y}^T(n)\,L(n-1)]\,\mathbf{y}(n) = L(n)\mathbf{y}(n) \tag{5.73}$$

Now, let us use the recursion for the filter coefficients:

$$\begin{array}{l} \mathbf{h}(n) = R_y^{-1}(n)\,\mathbf{r}_{xy}(n) = L(n)\,\mathbf{r}_{xy}(n) = L(n)\,[\mathbf{r}_{xy}(n-1) + x(n)\,\mathbf{y}(n)] = \\ = [(L(n-1) - \mathbf{K}(n)\,\mathbf{y}^T(n)\,L(n-1))\mathbf{r}_{xy}(n-1)\,] + \mathbf{K}(n)\,x(n) \end{array} \tag{5.74}$$

But: $\mathbf{h}(n-1) = L(n-1)\,\mathbf{r}_{xy}(n-1)$

Therefore:

$$\begin{array}{l} \mathbf{h}(n) = \mathbf{h}(n-1) + \mathbf{K}(n)\,[x(n) - \mathbf{y}^T(n)\,\mathbf{h}(n-1)] = \\ = \mathbf{h}(n-1) + \mathbf{K}(n)\,e(n) \end{array} \tag{5.75}$$

Let us summarize with an algorithm (the order of the filter is limited to Nh).

Algorithm:

- Initial values: $L(0) = I; \;\; \mathbf{h}^T(0) = (1, 1, 1\ldots)/Nh$
- For n = 1, 2, ...

- Compute $\mathbf{K}(n)$ according with (5.71)
- $e(n) = (x(n) - \mathbf{y}^T(n)\,\mathbf{h}(n-1)\,)$
- $\mathbf{h}(n) = \mathbf{h}(n-1) + \mathbf{K}(n)\,e(n)$
- $L(n) = L(n-1) - \mathbf{K}(n)\,\mathbf{y}^T(n)L(n-1)$.

The Program 5.8 applies the RLS method to the same example of sine plus additive noise (Fig. 5.1). The program generates three figures. Figure 5.16 shows the evolution of the error: it rapidly converges to a bounded value. Notice that the estimation error must be of random nature, corresponding to unpredictable noise.

Figure 5.17 shows the FIR filter coefficients (again limited to 50 coefficients).

Figure 5.18 shows the filtered signal.

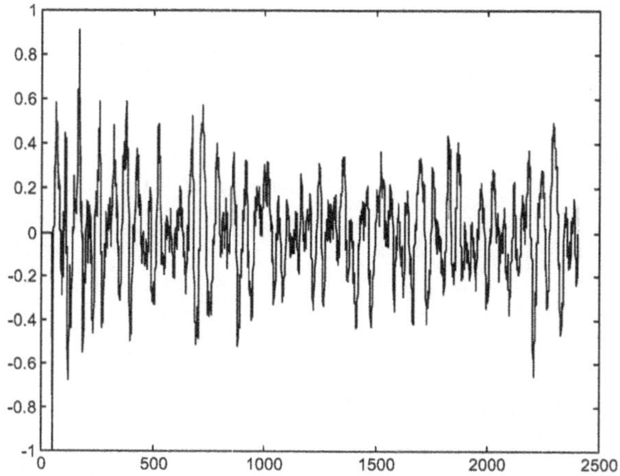

Fig. 5.16 Error evolution

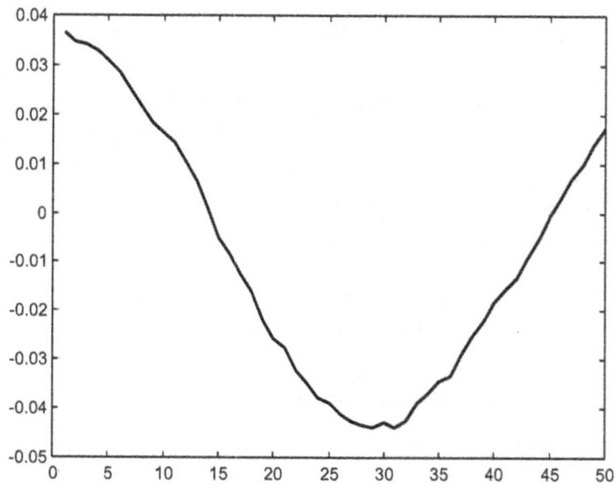

Fig. 5.17 Upper half of FIR filter coefficients

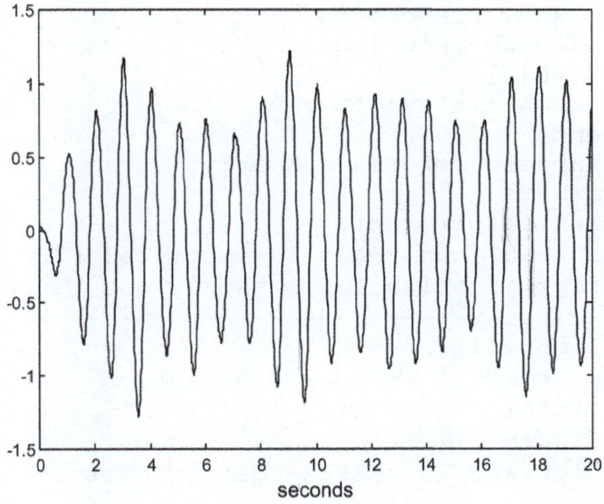

Fig. 5.18 The filtered signal

Program 5.8 RLS, sine signal + noise

```
% RLS
% sine signal + noise
fx=1; %signal frequency in Hz
wx=2*pi*fx; %signal frequency in rad/s
fs=60; %sampling frequency in Hz
tiv=1/fs; %time interval between samples;
t=0:tiv:(40-tiv); %time intervals set
x=sin(wx*t); %signal data set
Nx=length(x);
v=randn(1,Nx); %normal noise
y=x+v; %sine+noise
%FIR coeff. computation
Nh=50; %number of FIR coefficients
wh=ones(Nh,1)/Nh; %FIR coefficients initialization
yy=zeros(Nh,1); %y vector
er=zeros(1,Nx); %for error recording
fo=zeros(1,Nx); %filter output
L=eye(Nh);
K=zeros(Nh,1);
for nn=Nh:Nx-1,
  %actualization of yy
  yy=y(nn:-1:(nn-Nh+1))'; %take a segment and reverse
  %K(n)
  K=(L*yy)/(1+(yy'*L*yy));
  %filter output
  fo(nn)=wh'*yy;
  %error
  er(nn)=x(nn)-fo(nn);
```

```
   %new FIR coeffs
   wh=wh+(K*er(nn));
   %L(n)
   L=L-((K*yy')*L);
end;
%append for FIR symmetry
rwh=wh';
lwh=fliplr(rwh);
sh=0.5*[lwh rwh]; %symmetrical FIR
fly=filter(sh,1,y); %apply the Wiener filter
%display---------------------------
figure(1)
plot(er,'k'); %error evolution
title('error evolution');
figure(2)
plot(wh,'k'); %FIR coefficients
title('FIR coefficients (right-side)');
figure(3)
np=60*20;
plot(t(1:np),fly(1:np),'k'); %filtered signal
xlabel('seconds'); title('filtered signal');
```

5.3.2 Search-Based Methods

Let us go back to the error surface (5.38). The expression was:

$$\Sigma = E(e^2) = E\,(x^2) + \mathbf{h}^T R_y\,\mathbf{h} - 2 \cdot \mathbf{h}^T\,\mathbf{r}_{xy} \tag{5.76}$$

The problem is to find the minimum of this surface (the coefficients of the filter that minimize Σ).

Most search based methods use the gradient of Σ with respect to $\mathbf{h}$.

The scientific literature proposes several deterministic or heuristic alternatives for the search. In this section, two representative deterministic methods have been selected.

5.3.2.1 Steepest Descent

The steepest descent method uses the gradient in a direct way. Here is the recursive formulation:

$$\mathbf{h}(n) = \mathbf{h}(n-1) + \mu\,\left(-\frac{\partial \Sigma(n-1)}{\delta \mathbf{h}(n-1)}\right) \tag{5.77}$$

where μ is the adaptation step size.

The gradient is:

$$\frac{\partial \Sigma(n-1)}{\delta \mathbf{h}(n-1)} = 2\, R_y(n-1)\, \mathbf{h}(n-1) \,-\, 2\, \mathbf{r}_{xy}(n-1) \tag{5.78}$$

Therefore:

$$\mathbf{h}(n) = \mathbf{h}(n-1) + \mu\, [\mathbf{r}_{xy}(n-1) \,-\, R_y(n-1)\, \mathbf{h}(n-1)] \tag{5.79}$$

(μ absorbs the factor 2)

Using the notation introduced for the error surface (in Sect. 5.2.4):

$$\tilde{h}(n) = [I \,-\, \mu\, R_y(n-1)]\, \tilde{h}(n-1) = [I \,-\, \mu\, \mathbf{Q}\, D\, \mathbf{Q}^T]\, \tilde{h}(n-1) \tag{5.80}$$

Then:

$$\tilde{v}(n) = [I \,-\, \mu\, D]\, \tilde{v}(n-1) \tag{5.81}$$

And individual equations can be easily extracted:

$$v_k(n) = [I \,-\, \mu\, \lambda_k]\, v_k(n-1) \tag{5.82}$$

The iterative search converges if: $-1 < \; [I \,-\, \mu\, \lambda_k] \; < 1$

5.3.2.2 LMS

While the steepest descent method uses averaged error (Σ), the *'Least Mean-Square'* (LMS) method uses the instantaneous squared error:

$$\mathbf{h}(n) = \mathbf{h}(n-1) + \mu\, \left(-\frac{\partial e_i^2(n-1)}{\partial \mathbf{h}(n-1)}\right) \tag{5.83}$$

where the instantaneous error is:

$$e_i = (x - \mathbf{h}^T\, \mathbf{y}) \tag{5.84}$$

The LMS method was introduced by *Widrow and Hoff* in 1960, and it is since then widely used. It is simple to compute. The gradient of the squared instantaneous error is:

$$\begin{aligned} &\tfrac{\partial e_i^2(n-1)}{\partial \mathbf{h}(n-1)} = \tfrac{\partial}{\partial \mathbf{h}(n-1)} [x(n-1) - \mathbf{h}^T(n-1)\, \mathbf{y}(n-1)\,]^2 = \\ &= -2\, \mathbf{y}(n-1)\, [x(n-1) - \mathbf{h}^T(n-1)\, \mathbf{y}(n-1)\,] = \\ &= -2\, \mathbf{y}(n-1)\, e_i(n-1) \end{aligned} \tag{5.85}$$

Direct substitution in (5.83) gives:

$$\mathbf{h}(n) = \mathbf{h}(n-1) + \mu\,(\mathbf{y}(n-1)\,e_i(n-1)) \tag{5.86}$$

(the factor 2 is absorbed by μ)

Continuing with the same example, the Program 5.9 applies LMS to the sine plus noise signal. Figure 5.19 shows the evolution of the error, with an initial damped oscillation followed by the bounded random error. The damping of the oscillation depends on the value assigned to μ.

Figure 5.20 shows the FIR filter coefficients. As before, the number of coefficients is limited to 50. The reader could extend this number and check that the plot of the coefficients is a sinusoid.

Figure 5.21 shows the filtered signal.

Fig. 5.19 Error evolution

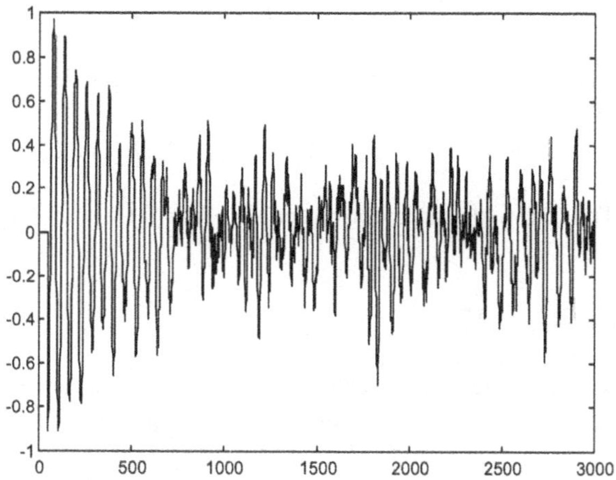

Fig. 5.20 Upper half of FIR filter coefficients

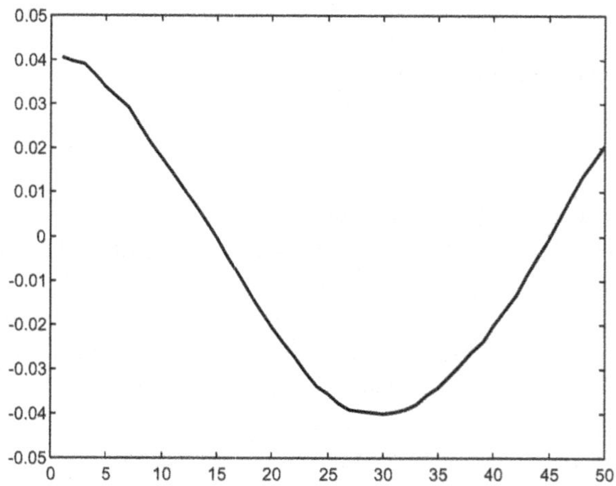

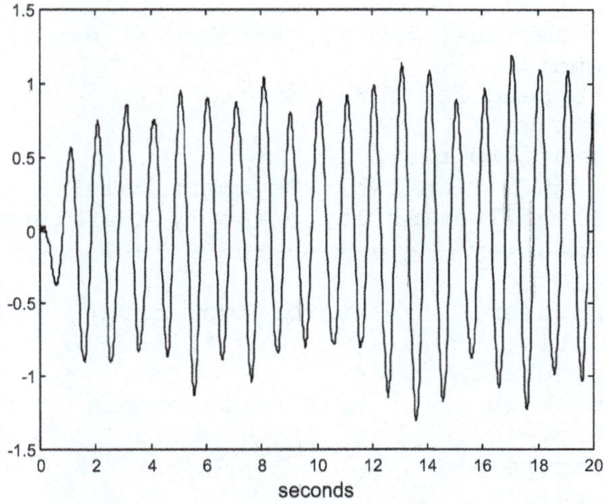

Fig. 5.21 The filtered signal

Program 5.9 LMS, sine signal + noise

```
% LMS
% sine signal + noise
fx=1; %signal frequency in Hz
wx=2*pi*fx; %signal frequency in rad/s
fs=60; %sampling frequency in Hz
tiv=1/fs; %time interval between samples;
t=0:tiv:(50-tiv); %time intervals set
x=sin(wx*t); %signal data set
Nx=length(x);
v=randn(1,Nx); %normal noise
y=x+v; %sine+noise
%FIR coeff. computation
Nh=50; %number of FIR coefficients
wh=zeros(Nh,1); %FIR coefficients initialization
yy=zeros(Nh,1); %y vector
er=zeros(1,Nx); %for error recording
hwh=zeros(1,Nx); %for coefficient recording
fo=zeros(1,Nx); %filter output
hwh(1)=wh(1); %initial value
mu=1/(100*Nh); %learning rate
for nn=Nh:Nx-1,
  %actualization of yy
  yy=y(nn:-1:(nn-Nh+1))'; %take a segment and reverse
  %filter output
  fo(nn)=wh'*yy;
  %error
  er(nn)=x(nn)-fo(nn);
  %new FIR coeffs
  wh=wh+((mu*er(nn))*yy);
```

```
    hwh(nn+1)=wh(1); %record of new wh(1)
end;
%append for FIR symmetry
rwh=wh';
lwh=fliplr(rwh);
sh=0.5*[lwh rwh]; %symmetrical FIR
fly=filter(sh,1,y); %apply the Wiener filter
%display--------------------------------
figure(1)
plot(er,'k'); %error evolution
title('error evolution');
figure(2)
plot(wh,'k'); %FIR coefficients
title('FIR coefficients (right-side)');
figure(3)
np=60*20;
plot(t(1:np),fly(1:np),'k'); %filtered signal
xlabel('seconds'); title('filtered signal');
figure(4)
plot(hwh,'k'); %evolution of wh(1)
title('evolution of filter coefficient h(1)');
```

The above program (Program 5.9) generates also the Fig. 5.22, which shows the evolution of the filter first coefficient along the adaptation transient. This is an illustrative plot that is useful for discussion of the LMS convergence.

Notice that the mean of the coefficient follows an exponential curve that converges to a final constant. At the same time, there are fluctuations of the coefficient value around the mean. Similar adaptation curves can be observed for the rest of the filter coefficients.

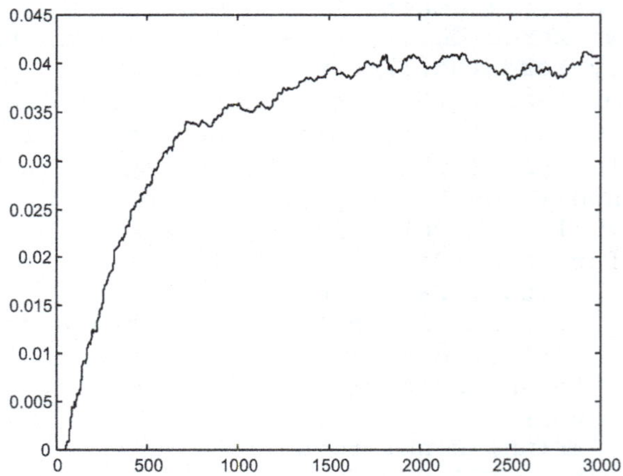

Fig. 5.22 Evolution of the filter first coefficient h(1)

The reader is invited to play with the values of μ, using the Program 5.9. It will be realized that a decrease of μ causes a slower adaptation transient and less fluctuation of the coefficients.

From the user point of view, a key aspect of the LMS filter performance is the error. The mean square error (MSE) obtained by the Wiener filter could be taken as a reference minimum value $\varepsilon_{\min}$.

After the adaptation transient the LMS filter obtains a steady-state MSE, also known as the mean asymptotic square error (MASE). Denote this error as ε_{LMS}.

There is a steady-state excess MSE given by:

$$\varepsilon_{exc} = \varepsilon_{LMS} - \varepsilon_{\min} \tag{5.87}$$

Also, there is a 'misadjustment', M, defined as:

$$M = \frac{\varepsilon_{exc}}{\varepsilon_{\min}} \tag{5.88}$$

The behaviour of the LMS filter depends on the auto-correlation matrix R_y of the input signal. In particular, the misadjustment is given by:

$$M = \frac{\frac{1}{2}\,\mu\, tr(R_y)}{(1 - \frac{1}{2}\,\mu\, tr(R_y))} \tag{5.89}$$

where $tr(R_y)$ is the trace of the matrix R_y (the trace is the sum of the diagonal elements of the matrix). The use of $tr(R_y)$ has a practical advantage since it can be computed based on input signal power, easy to estimate or to measure (more details in the Sect. 5.8).

If μ is sufficiently small, the misadjustment value could be approximated with:

$$M \approx \frac{1}{2}\,\mu\; tr(R_y) \tag{5.90}$$

The trace of a square matrix is invariant to a change of basis, therefore:

$$tr(R_y) = \sum_{i=1}^{N} \lambda_i \tag{5.91}$$

where λ_i are the eigenvalues of R_y.

Now, let us study the fluctuations of the LMS filter coefficients. Denote:

$$\eta(n) = \mathbf{h}(n) - \mathbf{h}_W \tag{5.92}$$

where $\mathbf{h}_W$ are the coefficients of the Wiener filter (the optimal filter).

Then:

$$E(\eta(n)) = (I - \mu R_y)\, E(\eta(n-1)) = (I - \mu R_y)^n\, E(\eta(0)) \tag{5.93}$$

The mean of the fluctuations converge to zero if:

$$\lim_{n \to \infty} (1 - \mu \lambda_i)^n = 0, \ \ \forall i \tag{5.94}$$

The adaptation involves a combination of transients; one for each eigenvalue. The transient corresponding to the lowest eigenvalue is the slowest, and therefore dominant, transient. This is a main aspect of the LMS adaptation time. For small values of μ, the time constants associated to the eigenvalues could be approximated with:

$$\tau_i \approx \frac{1}{\mu \lambda_i} \tag{5.95}$$

Denote:

$$C(n) = E(\eta(n)\, \eta(n)^T) \tag{5.96}$$

where the matrix $C(n)$ serves as measure of the fluctuations size.

Then, for small values of μ:

$$C(n+1) \approx C(n) - \mu\,(R_y\, C(n) + C(n) R_y) + \mu^2\, \sigma_v^2\, R_y \tag{5.97}$$

A sufficient condition for convergence of $C(n)$ as n tends to infinity is that:

$$0 < \mu < \frac{1}{tr(R_y)} \tag{5.98}$$

The LMS specialists recommend more stringent upper bounds, like for instance:

$$0 < \mu < \frac{2}{3\, tr(R_y)} \tag{5.99}$$

in order to ensure the stability of the LMS filter.

5.3.2.3 NLMS

The convergence speed of the LMS filter depends on the input signal power. To make it independent, signal normalization is introduced. The expression of the normalized LMS (NLMS) filter is the following:

$$\mathbf{h}(n) = \mathbf{h}(n-1) + \frac{\mu}{|\mathbf{y}(n-1)|^2}\, (\mathbf{y}(n-1)\, e_i(n-1)) \tag{5.100}$$

NLMS is the common choice for non-stationary situations. The MATLAB programs in the next section, provide examples of using NLMS.

To avoid division problems when the denominator is close to zero, a constant ε is included as follows:

$$\mathbf{h}(n) = \mathbf{h}(n-1) + \frac{\mu}{\varepsilon + |\mathbf{y}(n-1)|^2} \left(\mathbf{y}(n-1)\, e_i(n-1)\right) \tag{5.101}$$

5.4 Adaptive Filters

An important observation is that the filter coefficients are computed from characteristics of the input signal. The computation scheme can be designed to translate possible changes of the input signal characteristics into adequate changes of the filter coefficients: the filter adapts to the signal.

This adaptative feature can be exploited for cases with unknown aspects so it would be difficult to pre-compute the filter, but still it would be possible to let the filter find its way. This concept will become clear with the examples given in this section.

A synthesis of the filtering methodology explained in Sect. 5.3. could be expressed with the diagram shown in Fig. 5.23. The filter coefficients are automatically modified along time to minimize the error.

Next figure (Fig. 5.24) shows a simpler, modified version of the diagram. The notation has been slightly changed according with the standard literature on adaptive filters. The signal $d(n)$ is the *desired signal.*

This section considers four basic adaptive filter structures. Although the examples will be FIR filters, it is also possible to use IIR and even non-linear filters.

A classical reference book for adaptive signal processing is [47]. Newer books on this matter are [15, 16, 41], and [36] with MATLAB examples. The article [19]

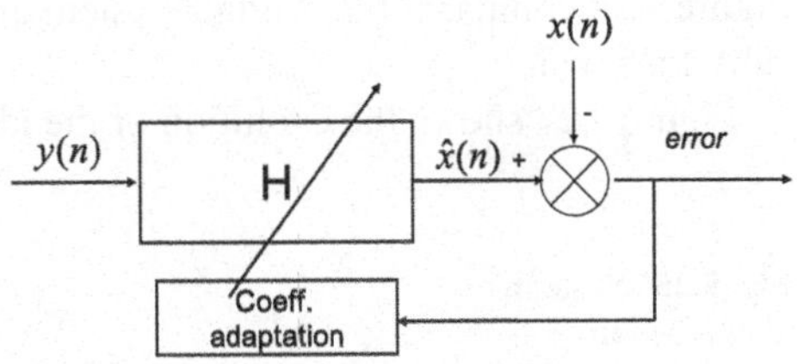

Fig. 5.23 Block diagram of a recursive filter

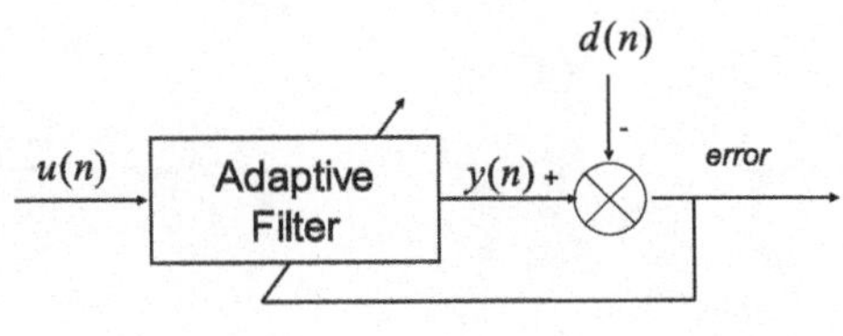

Fig. 5.24 Interpretation as adaptive filter

presents a comprehensive review of three decades of linear filtering theory. Analog adaptive filters are reviewed in [6]. The historical importance of adaptive filters is emphasized in [10].

5.4.1 System Identification

In many practical applications it is convenient to obtain a filter with input/output behaviour similar to the behaviour of an unknown system. For instance an auditorium: $u(n)$ could be caused by a loudspeaker and $d(n)$ could be the sound heard at a certain place in the room: the unknown system is the sound transmission from the loudspeaker to the place.

Other examples to be mentioned are related to transmission channels. It may happen that the adaptive filter is located at the end of the channel, so $y(n)$ is available, but not $u(n)$. A remedy for that is to use when adequate a predefined $u(n)$ segment that the adaptive filter knows.

Telephones have some problems. One of them is coupling. Suppose a conversation between telephone 1, you, and telephone 2. You say a word, the telephone 2 repeats aloud this word, and this is picked by mike 2, and then the word comes back to you. This may represent an echo or, in a worst case, an unpleasant audio oscillation. What it is usually done is to put an adaptive filter that cancels the echo. Refer to diagram in Fig. 5.25; the adaptive filter obtains from $u(n)$, the telephone 2 ear, a signal $y(n)$ that cancels out $d(n)$, which is your word as heard by mike 2.

Echo cancellation is an important application, like for instance for hands-free mobile phones.

Notice that the same delay should happen from $u(n)$ to $d(n)$, and from $u(n)$ to $y(n)$, in order to get perfect zero error. The designer has to specify an adequate order of the filter according with the expected delay in the case at hand.

As a simple example, a FIR filter has been chosen to be the unknown system to be identified. Program 5.10 successfully applies NLMS for system identification. Figure 5.26 compares the original system and the identified system: they are practically identical.

Figure 5.27 shows the evolution of the identification error.

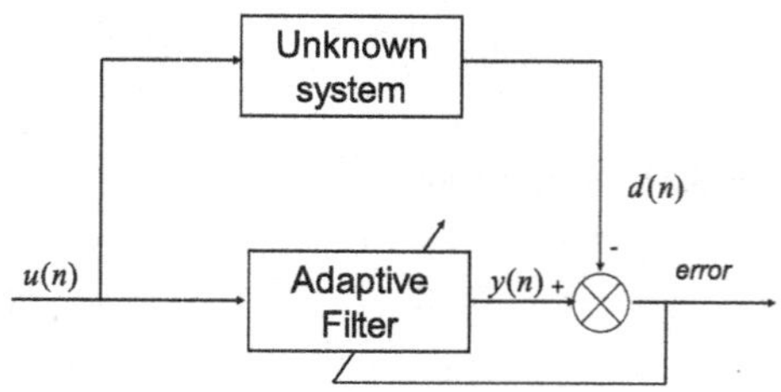

Fig. 5.25 System identification

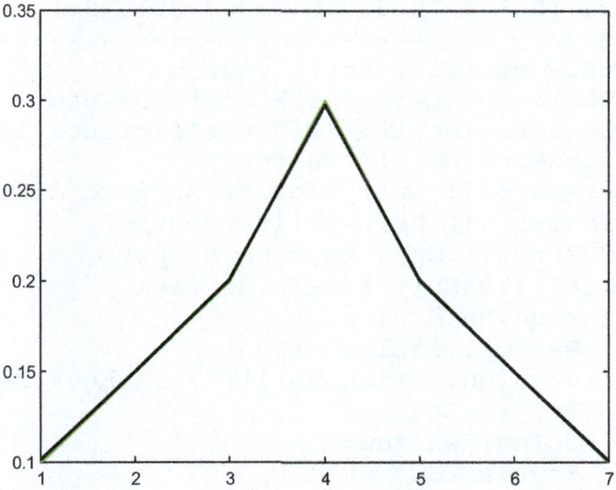

Fig. 5.26 Original and identified FIR coefficients

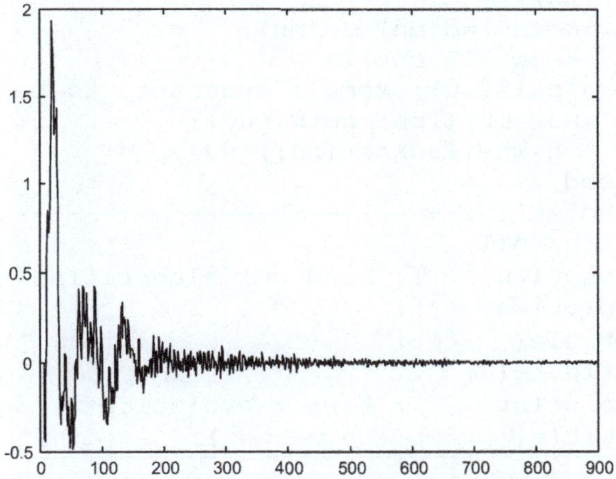

Fig. 5.27 Error evolution

Program 5.10 System Identifications using NLMS, sine signal + noise

```
% System Identifications using NLMS
% sine signal + noise
fx=1; %signal frequency in Hz
wx=2*pi*fx; %signal frequency in rad/s
fs=60; %sampling frequency in Hz
tiv=1/fs; %time interval between samples;
t=0:tiv:(15-tiv); %time intervals set
x=sin(wx*t); %signal data set
Nx=length(x);
v=randn(1,Nx); %normal noise
u=x+v; %sine+noise
%The Unknown System--------------------------
```

```
wu=[0.1 0.15 0.2 0.3 0.2 0.15 0.1]';
%----------------------------------------------
%FIR coeff. identification
Nh=7; %number of FIR coefficients
wh=zeros(Nh,1); %FIR coefficients initialization
uu=zeros(Nh,1); %u vector
er=zeros(1,Nx); %for error recording
fo=zeros(1,Nx); %filter output
d=zeros(1,Nx); %system output
mu=1/(10*Nh); %learning rate
for nn=Nh:Nx-1,
  %actualization of uu
  uu=u(nn:-1:(nn-Nh+1))'; %take a segment and reverse
  %filter output
  fo(nn)=wh'*uu;
  %system output
  d(nn)=wu'*uu;
  %error
  er(nn)=d(nn)-fo(nn);
  %new FIR coeffs
  ipsi=0.01; %small constant, to be edited
  aux=mu/(ipsi+norm(uu));
  wh=wh+((aux*er(nn))*uu);
end;
%display---------------------------
figure(1)
plot(wu,'g'); hold on; %Identified FIR coefficients
plot(wh,'k');
title('Identified FIR coefficients');
figure(2)
plot(er,'k'); %error evolution
title('error evolution');
```

The reader is invited to change the order of the adaptive filter, to see how it identifies the unknown system. Also, it is interesting to experiment with the input signal, which should contain enough excitation of the system in order to get it identified.

5.4.2 *Inverse System Identification*

The diagram in Fig. 5.28 represents a basic configuration for inverse system identification. The adaptive filter is asked to be the inverse of the unknown system, so the chain of the unknown system and the inverse filter would be equivalent to a delay. Half of the delay is caused by the unknown system, and the other half by the filter.

Actually, the ideal communication channel should have unit gain in all frequencies. The idea of the adaptive filter is to compensate for the frequency response of a channel (the unknown system), so approaching the ideal performance. An important application of this concept is equalization [31, 37].

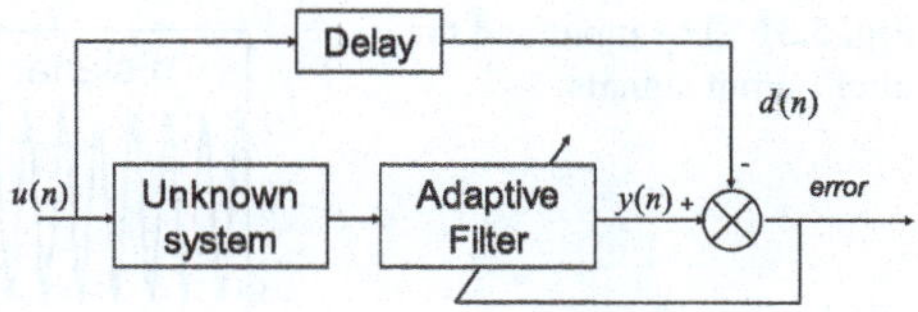

Fig. 5.28 Inverse system identification

The same FIR filter as before has been chosen as unknown system. Program 5.11 applies the concept expressed in diagram in Fig. 5.28. The input signal has less noise than in the previous example, in order to avoid convergence problems. Figure 5.29 shows the evolution of the error.

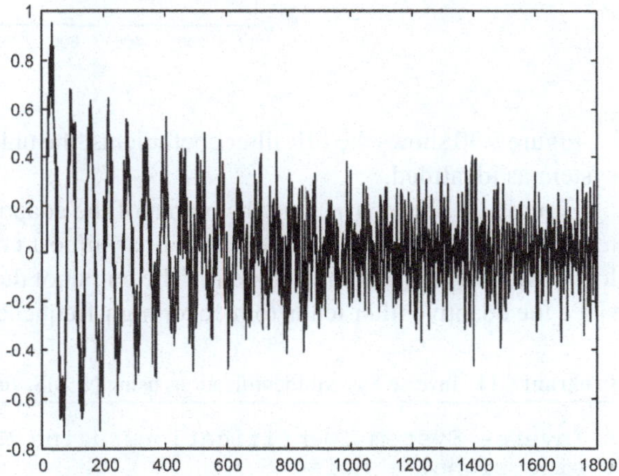

Fig. 5.29 Error evolution

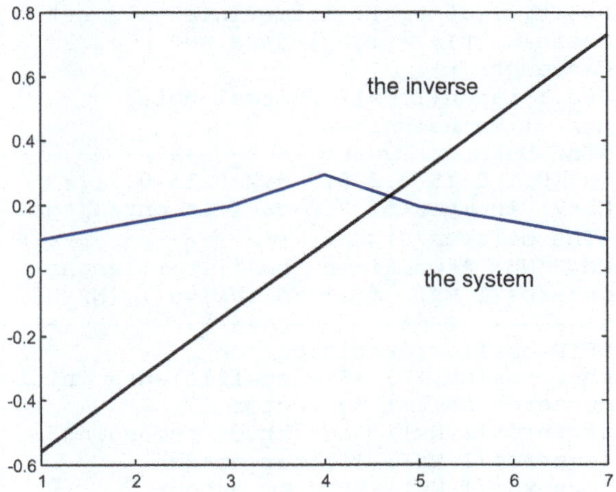

Fig. 5.30 The system and the identified FIR coefficients

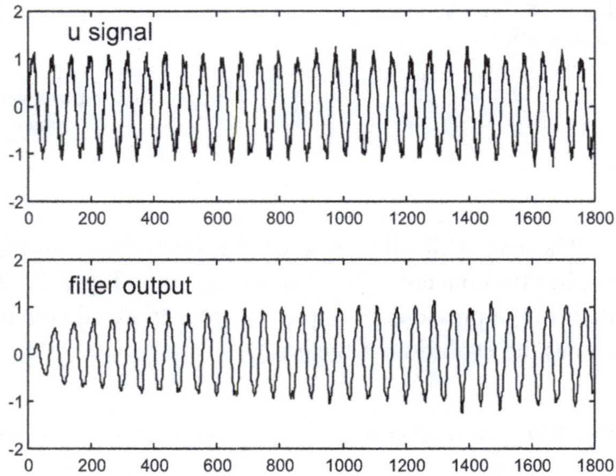

Fig. 5.31 The input and the filter output signals

Figure 5.30 shows the FIR filter coefficients, the unknown system, and the inverse system as identified.

Figure 5.31 shows $u(n)$ and the output of the adaptive filter. The reader is invited to plot the input of the adaptive filter: it has almost no noise, since it comes from a low-pass filter (the unknown system). The output of the filter is more similar to $u(n)$, since the adaptive filter tends to recover high-frequencies.

Program 5.11 Inverse System Identifications using NLMS, sine signal + noise

```
% Inverse System Identifications using NLMS
% sine signal + noise
fx=1; %signal frequency in Hz
wx=2*pi*fx; %signal frequency in rad/s
fs=60; %sampling frequency in Hz
tiv=1/fs; %time interval between samples;
t=0:tiv:(30-tiv); %time intervals set
x=sin(wx*t); %signal data set
Nx=length(x);
v=0.1*randn(1,Nx); %normal noise
u=x+v; %sine+noise
%The Unknown System-------------------------
wu=[0.1 0.15 0.2 0.3 0.2 0.15 0.1]';
Nh=7; %number of FIR coefficients (equal to system order)
%The delayed signal-------------------------
Nd=2*Nh; %two times the filter lenght
d=zeros(1,Nx); d((1+Nd):Nx)=u(1:(Nx-Nd));
%-------------------------------------------
%FIR coeff. identification
wh=zeros(Nh,1); %FIR coefficients initialization
uu=zeros(Nh,1); %u vector
er=zeros(1,Nx); %for error recording
fo=zeros(1,Nx); %filter output
so=zeros(1,Nx); %system output
```

Fig. 5.32 Noise cancellation

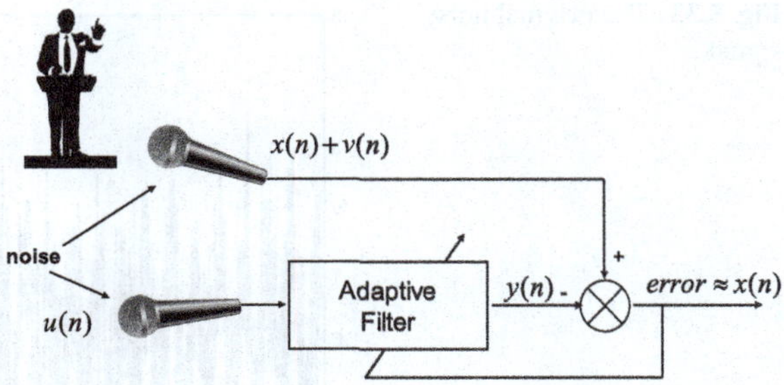


```
mu=1/(10*Nh); %learning rate
for nn=Nh:Nx-1,
  %actualization of uu
  uu=u(nn:-1:(nn-Nh+1))'; %take a segment and reverse
  %system output
  so(nn)=wu'*uu;
  %filter output
  ss=so(nn:-1:(nn-Nh+1))';
  fo(nn)=wh'*ss;
  %error
  er(nn)=d(nn)-fo(nn);
  %new FIR coeffs
  ipsi=0.01; %small constant, to be edited
  aux=mu/(ipsi+norm(ss));
  wh=wh+((aux*er(nn))*ss);
end;
%display---------------------------
figure(1)
plot(er,'k'); %error evolution
title('error evolution');
figure(2)
plot(wu,'b'); hold on; %the system
plot(wh,'k'); %the identified inverse
title('Identified FIR coefficients');
figure(3)
subplot(2,1,1)
plot(u,'k'); title('u signal');
subplot(2,1,2)
plot(fo,'k'); title('filter output');
```

5.4.3 Noise Cancellation

Suppose a speech taking place with a background noise coming from some place apart. It is possible with a second microphone to capture this noise.

The diagram in Fig. 5.32 depicts the situation. An adaptive filter has been included. Of course, if $u(n)$ and $v(n)$ were equal, no filter would be needed, it would be just

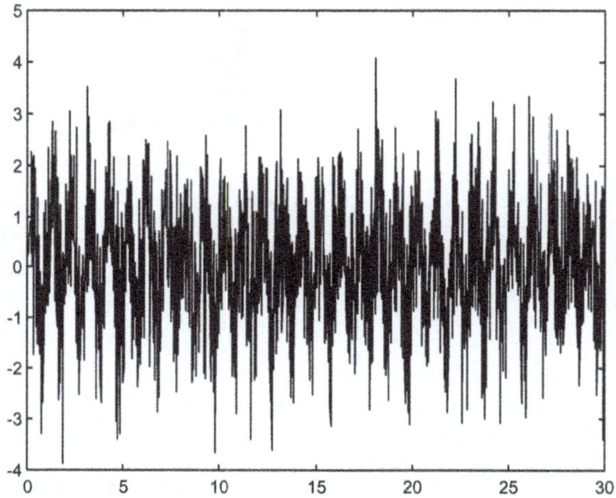

Fig. 5.33 The original noisy signal

a matter of subtracting signals from both microphones. But the sound takes time to travel. In the depicted case, the noise arrives before to $u(n)$ than to $v(n)$. The mission of the adaptive filter is to add a delay and to compensate for air transmission effects. The ideal behaviour would be that $y(n)$ equals $v(n)$, and so the speech is recovered as the filtering error.

Phase correction or locking is a common practice in telecommunications. It is a matter of delay adaptation with some other improvements, and it falls in the domain of adaptive filters.

Program 5.12 considers a simple case with 12 sampling periods as delay between $u(n)$ and $v(n)$. Figure 5.33 shows the noisy signal to be filtered.

Figure 5.34 shows the filtered signal. NLMS has been applied; the initial filter adaptation is clearly noticeable.

Figure 5.35 shows the right half of the adaptive FIR filter coefficients, Most action of the filter is to have unit gain and an adequate delay

Finally, Fig. 5.36 shows the noise estimated by the filter. This noise must cancel out the noise included in the speech.

Program 5.12 Noise cancellation using LMS, sine signal + noise

```
% Noise cancellation using LMS
% sine signal + noise
fx=1; %signal frequency in Hz
wx=2*pi*fx; %signal frequency in rad/s
fs=60; %sampling frequency in Hz
tiv=1/fs; %time interval between samples;
t=0:tiv:(30-tiv); %time intervals set
x=sin(wx*t); %signal data set
Nx=length(x);
u=randn(1,Nx); %reference (normal) noise
Nd=12; %delay between noises
```

```
v=zeros(1,Nx); v(1,(1+Nd):Nx)=u(1,1:(Nx-Nd)); %delayed noise
z=x+v; %sine+noise
%FIR coeff. computation
Nh=Nd; %number of FIR coefficients
wh=zeros(Nh,1); %FIR coefficients initialization
vv=zeros(Nh,1); %v vector
er=zeros(1,Nx); %for error recording
fo=zeros(1,Nx); %filter output
mu=1/(10*Nh); %learning rate
for nn=Nh:Nx-1,
  %actualization of vv
  vv=v(nn:-1:(nn-Nh+1))'; %take a segment and reverse
  %filter output
  fo(nn)=wh'*vv;
  %error
  er(nn)=z(nn)-fo(nn);
  %new FIR coeffs
  epsi=0.01; %small constant, to be edited
  aux=mu/(epsi+norm(vv));
  wh=wh+((aux*er(nn))*vv);
end;
%append for FIR symmetry
rwh=wh';
lwh=fliplr(rwh);
sh=0.5*[lwh rwh]; %symmetrical FIR
fly=filter(sh,1,v); %apply the Wiener filter
%display--------------------------
figure(1)
plot(t,z,'k'); %noisy signal
title('noisy signal');
figure(2)
plot(t,er,'k'); %filtered signal (the error)
title('filtered signal');
```

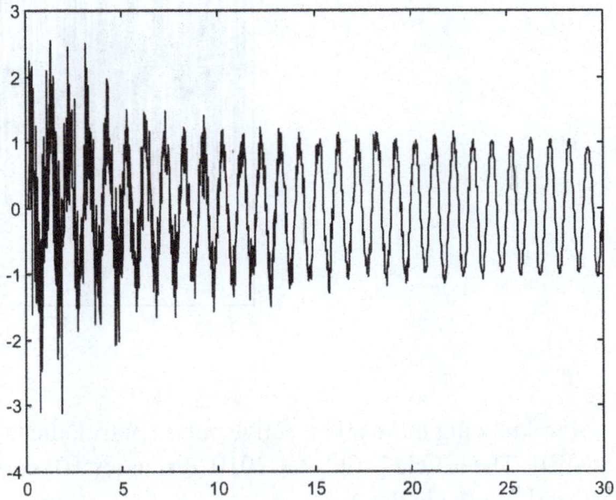

Fig. 5.34 The filtered signal

```
figure(3)
plot(wh,'k'); %FIR coefficients
title('FIR coefficients (right-side)');
figure(4)
plot(t,fly,'k'); %predicted noise
xlabel('seconds'); title('predicted noise');
```

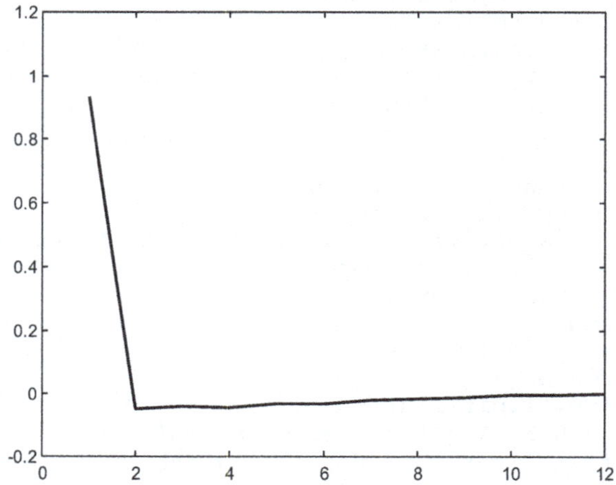

Fig. 5.35 FIR filter coefficients

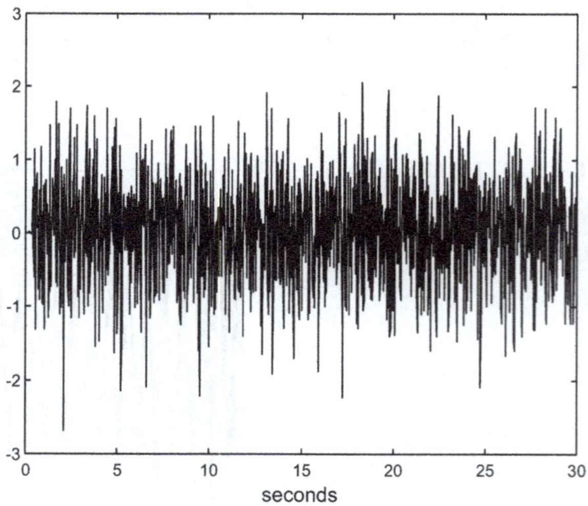

Fig. 5.36 The estimated noise (filter output).

For knowing more about active noise control, there are tutorial reviews from [23], and [9]. The article [46], year 2010, discusses 50 years of this topic, the state of the art, and future challenges.

5.4.4 Linear Prediction

Another important application is linear prediction, [50]. Figure 5.37 shows a diagram with the concept. If the adaptive filter is able to compensate for the delay, then it can predict $u(n)$ from *u(n-delay)*.

In certain speech processing and transmission applications, data rates can be much reduced if only the error and the filter coefficients are transmitted (vocoder principle).

Program 5.13 deals with an example of 7 sampling periods delay. Figure 5.38 shows the evolution of the error.

Figure 5.39 shows the FIR filter coefficients.

Finally, Fig. 5.40 shows the $u(n)$ signal and the filter output, which should be a delayed version of $u(n)$.

Fig. 5.37 Prediction

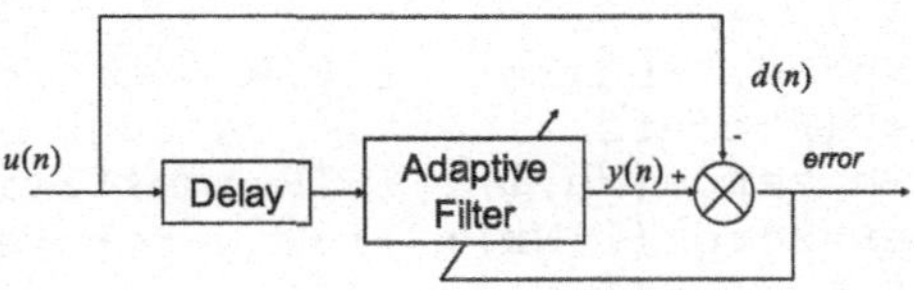

Fig. 5.38 Error evolution

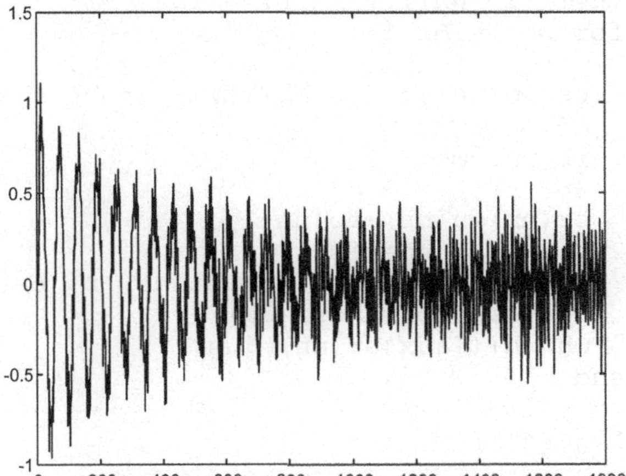

Program 5.13 Linear Prediction using NLMS, sine signal + noise

```
% Linear Prediction using NLMS
% sine signal + noise
fx=1; %signal frequency in Hz
wx=2*pi*fx; %signal frequency in rad/s
fs=60; %sampling frequency in Hz
tiv=1/fs; %time interval between samples;
t=0:tiv:(30-tiv); %time intervals set
x=sin(wx*t); %signal data set
Nx=length(x);
v=0.1*randn(1,Nx); %normal noise
u=x+v; %sine+noise
%Delayed signal-------------------------
Nd=7; %delay length
ud=zeros(1,Nx); ud((1+Nd):Nx)=u(1:(Nx-Nd));
%-----------------------------------------
Nh=Nd; %number of FIR coefficients
%FIR coeff. identification
wh=zeros(Nh,1); %FIR coefficients initialization
er=zeros(1,Nx); %for error recording
fo=zeros(1,Nx); %filter output
mu=1/(10*Nh); %learning rate
for nn=Nh:Nx-1,
  %actualization of ud
  us=ud(nn:-1:(nn-Nh+1))'; %take a segment and reverse
  %filter output
  fo(nn)=wh'*us;
  %error
  er(nn)=u(nn)-fo(nn);
  %new FIR coeffs
  ipsi=0.01; %small constant, to be edited
  aux=mu/(ipsi+norm(us));
  wh=wh+((aux*er(nn))*us);
end;
%display---------------------------
figure(1)
plot(er,'k'); %error evolution
title('error evolution');
figure(2)
plot(wh,'k'); %the FIR coefficients
title('Identified FIR coefficients');
figure(3)
subplot(2,1,1)
plot(u,'k'); title('u signal');
subplot(2,1,2)
plot(fo,'k'); title('filter output');
```

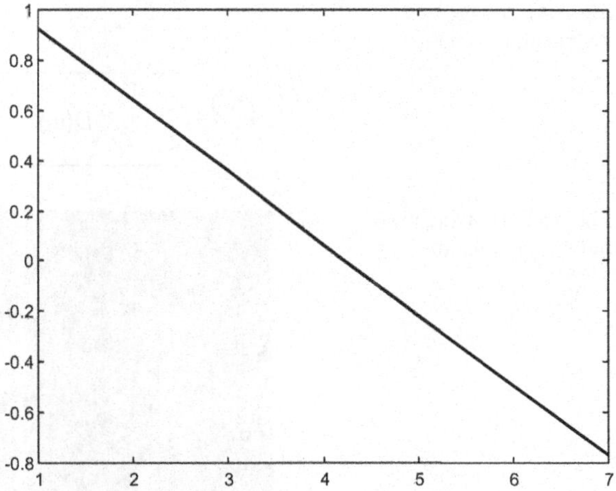

Fig. 5.39 Identified FIR coefficients

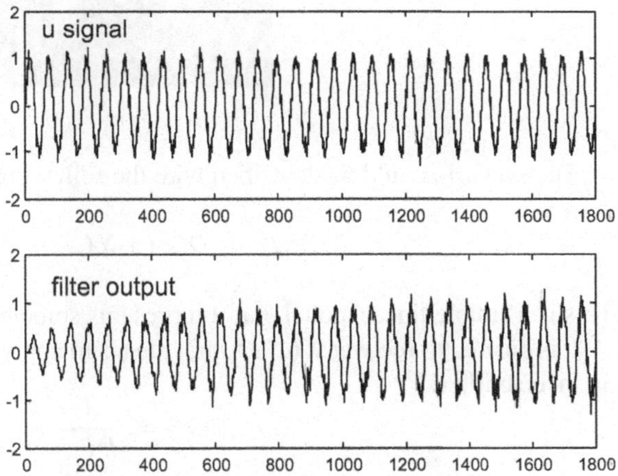

Fig. 5.40 The input and the filter output signals

5.5 Image Deblurring

A very interesting application of the Wiener filter is image deblurring. The issue of image enhancement is nowadays a large and active research field, plenty of subtle details. This section could serve as a basic introduction to the topic.

Consider the scenario depicted in Fig. 5.41, which is taken in the frequency domain. An image $Y(\omega)$ is obtained from an original scene $X(\omega)$ (another image). The block $D(\omega)$ represents a 'degradation filter', which involves camera deficiencies and, perhaps, motion effects (both object and camera motion). There is also some additive noise $V(\omega)$.

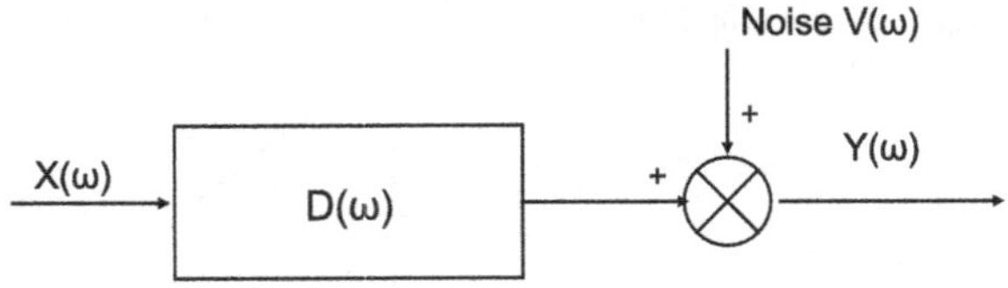

Fig. 5.41 Image degradation scenario

Fig. 5.42 Blurring mask, frequency domain

The scenario could be described with the following expression:

$$Y(\omega) = D(\omega) \cdot X(\omega) + V\omega) \quad (5.102)$$

This is a simple linear model, but it represents some basic facts.

It is desired to recover $X(\omega)$. A first idea is to apply **inverse filtering**. That is, use an inverse filter:

$$L(\omega) = \frac{1}{D(\omega)} \quad (5.103)$$

Then:

$$L(\omega)\, Y(\omega) = X(\omega) + \frac{V\omega)}{D(\omega)} \quad (5.104)$$

This approach has a serious problem. Usually $D(\omega)$ is a low-pass filter, so the inverse is a high-pass filter which tends to amplify the noise.

Let us try an example. The Program 5.14 obtains a blurred and noisy image and then applies the inverse filter. The blur is obtained with a blurring mask, shown in Fig. 5.42, in the frequency domain. To avoid inversion problems, all components of the mask are non-zero.

Fig. 5.43 Blurred and noisy image

Fig. 5.44 Result of inverse filtering

The blurred and noisy image (perhaps too much, but it is for the shake of an imposing example) is shown in Fig. 5.43.

Finally, Fig. 5.44 shows the result of inverse filtering, the noise becomes predominant, although it could be taken as an intriguing photographic effect.

Program 5.14 Inverse filter, blurred image with noise

```
% Inverse filter
% Blurred image with noise
clear all;
ux=imread('antena1.tif');
[Ny,Nx]=size(ux);
%signal and noise between 0 and 1
x=im2double(ux);
vn=abs(0.2*randn(Ny,Nx)); v=mod(vn,1);
X=fftshift(fft2(x)); %Fourier transform of x
%frequency domain blurring filter (circular mask)
dx=(0:(Nx-1))-round(Nx/2); dy=(0:(Ny-1))-round(Ny/2);
%a grid with the same size as image, with center (0,0):
```

```
[gx,gy]=meshgrid(dx,dy);
R=sqrt((gx.^2)+(gy.^2));
BF=0.2+(R<15); %circle (low-pass filter)
bx=abs(ifft2(X.*BF)); %blurred image
Kmix=0.75; %to be edited
bx=Kmix*bx; v=(1-Kmix)*v;
y=bx+v; %noise + blurred image
%Inverse filtering
Y=fftshift(fft2(y)); %Fourier transform of y
fly=abs(ifft2(Y./BF)); %apply the inverse filter
%signals between 0 and 1
miy=min(min(y)); y=y-miy; %y>=0
may=max(max(y)); y=y/may; %y<=1
mify=min(min(fly)); fly=fly-mify; %fly>=0
mafy=max(max(fly)); fly=fly/mafy; %fly<=1
Uy=im2uint8(y); %convert to uint8 for display
Ufly=im2uint8(fly); %convert to uint8 for display
%display---------------------------
figure(1)
imshow(BF); %plots figure
title('blurring mask');
figure(2)
imshow(Uy); %plots figure
title('blurred image with noise');
figure(3)
imshow(Ufly); %plots figure
title('deblurred with inverse filter');
```

The Wiener filter is applied to this case as follows:

$$H(\omega) = \frac{D(\omega)^* \cdot S_{xx}(\omega)}{|D(\omega)|^2 \cdot S_{xx}(\omega) + S_{vv}(\omega)} = \frac{1}{D(\omega)} \left[\frac{|D(\omega)|^2}{|D(\omega)|^2 + \frac{S_{vv}(\omega)}{S_{xx}(\omega)}} \right] \tag{5.105}$$

The right-hand side of the equation includes the inverse filter multiplied by another term with depends on the noise/signal rate. With no noise, the Wiener filter becomes an inverse filter.

The Program 5.15 applies the Wiener filter to the same blurred and noisy image as before. Figure 5.45 shows the good result obtained.

Fig. 5.45 Result of Wiener filter

Program 5.15 Frequency domain Wiener filtering, blurred image with noise

```
% Frequency domain Wiener filtering
% Blurred image with noise
clear all;
ux=imread('antena1.tif');
[Ny,Nx]=size(ux);
%signal and noise between 0 and 1
x=im2double(ux);
vn=abs(0.2*randn(Ny,Nx)); v=mod(vn,1);
X=fftshift(fft2(x)); %Fourier transform of x
%frequency domain blurring filter (circular mask)
dx=(0:(Nx-1))-round(Nx/2); dy=(0:(Ny-1))-round(Ny/2);
%a grid with the same size as image, with center (0,0):
[gx,gy]=meshgrid(dx,dy);
R=sqrt((gx.^2)+(gy.^2));
BF=0.2+(R<15); %circle (low-pass filter)
bx=abs(ifft2(X.*BF)); %blurred image
Kmix=0.75; %to be edited
bx=Kmix*bx; v=(1-Kmix)*v;
y=bx+v; %noise + blurred image
%Wiener computations
Sxx=abs(X).^2; %Sxx
V=fftshift(fft2(v)); %Fourier transform of v
Svv=abs(V).^2; %Svv
%Wiener filter
Wnum=conj(BF).*Sxx;
Wden1=(abs(BF).^2).*Sxx;
WH=Wnum./(Wden1+Svv); %Fourier transform of the Wiener filter
Y=fftshift(fft2(y)); %Fourier transform of y
fly=abs(ifft2(Y.*WH)); %apply the Wiener filter
%signal between 0 and 1
mify=min(min(fly)); fly=fly-mify; %fly>=0
mafy=max(max(fly)); fly=fly/mafy; %fly<=1
Ufly=im2uint8(fly); %convert to uint8 for display
%display------------------------------
figure(1)
imshow(Ufly); %plots figure
title('deblurred with Wiener filter');
```

It may be difficult to determine the $S_{vv}(\omega)$ term. A common practice is just to use a constant, to be empirically determined.

5.5.1 Motion blur

The experience of getting a moved photograph because the object has moved and/or because camera shake is very common. A simple representation of this effect could be done in terms of the degradation filter.

Suppose the object has moved from (x, y) to $(x - \alpha,\ y - \beta)$ in T seconds. Then, the degradation filter is:

$$D(u, v) = T\,\frac{\sin \pi(\alpha u + \beta v)}{\pi(\alpha u + \beta v)} \cdot \exp(-j\pi(\alpha u + \beta v)) \tag{5.106}$$

where two filter coordinates (u, v) have been considered.

In our case, both are supposed identical to ω

Program 5.16 offers an example of motion blur and how to filter out this effect with the Wiener filter. Figure 5.46 shows the image with noise and the motion blur, and Fig. 5.48 shows the result of the deblurring filter. The reader is invited to perform experiments here.

Notice in Fig. 5.46 that you cannot read the plate because of the blurring. The next figure (Fig. 5.47) shows the frequency domain blur mask that has been applied

Fig. 5.46 Image with noise and motion blur

Fig. 5.47 The motion blur mask

to simulate the motion blur. The effect of the filter is to duplicate and shift several times the same image.

Figure 5.48 shows the result of the Wiener filter. The image has been deblurred and now you can read the car's plate.

Program 5.16 Frequency domain Wiener filtering, image with motion blurring

```
% Frequency domain Wiener filtering
% Image with motion blurring
clear all;
ux=imread('car1.tif');
[Ny,Nx]=size(ux);
%signal and noise between 0 and 1
x=im2double(ux);
vn=abs(0.1*randn(Ny,Nx)); v=mod(vn,1);
X=fftshift(fft2(x)); %Fourier transform of x
%motion description
T=1; %time in seconds
a=9; %vertical shift in pixels
b=5; %horizontal shift in pixels
%frequency domain motion blur filter (circular mask)
for nu=1:Ny,
  for nv=1:Nx,
    aux1=(pi*(a*nu+b*nv))/(Nx+Ny);
    BF(nu,nv)=(T*sin(aux1)*exp(-j*aux1))/aux1;
  end;
end;
BF=2*pi*BF/T; %normalization
bx=abs(ifft2(X.*BF)); %blurred image
Kmix=0.85; %to be edited
bx=Kmix*bx; v=(1-Kmix)*v;
y=bx+v; %noise + blurred image
%Wiener computations
```

```
Sxx=abs(X).^2; %Sxx
V=fftshift(fft2(v)); %Fourier transform of v
Svv=abs(V).^2; %Svv
%Wiener filter
Wnum=conj(BF).*Sxx;
Wden1=(abs(BF).^2).*Sxx;
WH=Wnum./(Wden1+Svv); %Fourier transform of the Wiener filter
Y=fftshift(fft2(y)); %Fourier transform of y
fly=abs(ifft2(Y.*WH)); %apply the Wiener filter
%signals between 0 and 1
miy=min(min(y)); y=y-miy; %y>=0
may=max(max(y)); y=y/may; %y<=1
mify=min(min(fly)); fly=fly-mify; %fly>=0
mafy=max(max(fly)); fly=fly/mafy; %fly<=1
Uy=im2uint8(y); %convert to uint8 for display
Ufly=im2uint8(fly); %convert to uint8 for display
%display----------------------------
figure(1)
imshow(abs(BF)); %plots figure
title('motion blur mask');
figure(2)
imshow(Uy); %plots figure
title('Image with motion blur');
figure(3)
imshow(Ufly); %plots figure
title('deblurred image');
```

Fig. 5.48 Result of the Wiener filter

An interesting article, recommended to the reader, is [11] on removing camera shake from a single photograph. A complementary problem is treated in [28], which corresponds to photographs of moving objects.

5.6 More Adaptive Filters and Some Mathematical Aspects

Adaptive filters are so suggestive and flexible that many extensions and new applications have been proposed.

With the desire of offering a more complete panorama of adaptive filters, this section introduces illustrative examples of the many adaptive filtering algorithms that the reader may find in both scientific and industrial ambients.

Some authors have proposed unifying views, based on the mathematics of linear equations and matrix inversion. Since this could be useful to gain insight into the topic, a subsection is included for mathematical aspects and then another subsection for the unifying perspective.

5.6.1 LMS Variants

Recall the LMS expression:

$$\mathbf{h}(n) = \mathbf{h}(n-1) + \mu\,(\mathbf{y}(n-1)\,e_i(n-1)) \tag{5.107}$$

The standard LMS uses a constant μ. Depending on the application purpose, it could be beneficial to have a variable μ. Some changes of the vectors included in the expression could be suitable. Some authors have studied the use of other errors. Also the filter length could be variable.

Let us introduce some examples of the proposed LMS variants.

5.6.1.1 Leaky LMS

The filter adaptation expression is modified as follows:

$$\mathbf{h}(n) = (1-\mu\,\gamma)\,\mathbf{h}(n-1) + \mu\,(\mathbf{y}(n-1)\,e_i(n-1)) \tag{5.108}$$

The factor $(1 - \mu\gamma)$ is called *'leakage factor'*, where $0 \leq \gamma < 1/N$ (the filter has N coefficients).

It can be shown that:

$$\lim_{n\to\infty} E(\mathbf{h}(n)) = (R_y + \gamma\,I)^{-1}\,\mathbf{r}_{xy} \tag{5.109}$$

It is the same effect as if white noise with variance γ was added to the input. In the case of nonpersistent excitation (sometimes the input signal is zero) the standard LMS has problems, but not the leaky LMS.

Recall that the coefficients of the Wiener filter are:

$$\mathbf{h}_W(n) = R_y^{-1}\, \mathbf{r}_{xy} \tag{5.110}$$

Compared to the eigenvalues of LMS, which are the same as the Wiener filter, the eigenvalues of leaky LMS are $(\lambda_1 + \gamma, \ \ldots, \lambda_N + \gamma)$. Thus, the slow transient due to λ_{min} could be accelerated.

The word *'leaky'* indicates that the energy leaks out, so the filter coefficients are allowed to decay.

Since in the limit the leaky LMS filter coefficients are not equal to the Wiener coefficients, the leaky algorithm is biased. In [20] it was proposed to use a variable leaky algorithm: γ changes along the filter action, so it is decreased when the error is larger than the LMS error, and vice versa.

5.6.1.2 Sign-Based Algorithms

The implementation of the adaptive filter is simplified by taking only signs, using the *sgn* function:

$$sgn(x) = \begin{cases} 1, & x > 0 \\ 0, & x = 0 \\ -1, & x < 0 \end{cases} \tag{5.111}$$

There are three main LMS variants based on sign:

- The sign-error algorithm (SE-LMS):

$$\mathbf{h}(n) = \mathbf{h}(n-1) + \mu\, sgn(e_i(n-1)) \cdot \mathbf{y}(n-1) \tag{5.112}$$

- The sign-data algorithm (SD-LMS):

$$\mathbf{h}(n) = \mathbf{h}(n-1) + \mu\, sgn(\mathbf{y}(n-1)) \cdot e_i(n-1) \tag{5.113}$$

- The sign-sign algorithm (SS-LMS):

$$\mathbf{h}(n) = \mathbf{h}(n-1) + \mu\, sgn(\mathbf{y}(n-1)) \cdot sgn(e_i(n-1)) \tag{5.114}$$

It has been shown that SE-LMS is more robust than LMS for impulsive noise environment. However, instability problems have been demonstrated for SD-LMS and SS-LMS. The sign-based algorithms converge more slowly than LMS and have bias. The SD-LMS has the advantage that convergence is independent of input signal amplitude.

5.6.1.3 Variable Step-Size

The misadjustment and the coefficient fluctuations of the LMS filter can be decreased by taking a small value of μ. Unfortunately, decreasing μ also decreases the adaptation speed. Soon after LMS was introduced, it was proposed to apply some '*gear-shifting*': a large value of μ at start-up to obtain fast convergence rate, and then switching to a smaller value of μ to improve the steady-state response.

The general form of variable step-size algorithms is the following:

$$\mathbf{h}(n+1) = \mathbf{h}(n) + \mu(n)\,(\mathbf{y}(n)\,e_i(n)) \tag{5.115}$$

(notice the slight changes we made with respect to indexes).

To complete the algorithm an expression of the evolution of $\mu(n)$ should be given.

Two branches could be distinguished: algorithms based on LMS, and algorithms based on NLMS.

- *Algorithms based on LMS:*
  - The variable step size algorithm (VSS-LMS) of [24]:
    The step size is updated as follows:

    $$\mu(n+1) = \alpha\,\mu(n) + \delta\, e_i^2(n) \tag{5.116}$$

    where $0 < \alpha < 1$
  - The MVSS-LMS algorithm of [1]:
    The step size is updated as follows:

    $$\mu(n+1) = \alpha\,\mu(n) + \delta\, p^2(n) \tag{5.117}$$

    where $p(n)$ is a local estimate of the auto-correlation of the error:

    $$p(n) = \beta\, p(n-1) + (1-\beta)\, e_i(n)\, e_i(n-1) \tag{5.118}$$

- *Algorithms based on NLMS:*
  - The VS-NLMS algorithm of [43]:
    The filter expression is:

    $$\mathbf{h}(n+1) = \mathbf{h}(n) + \mu(n)\frac{\mathbf{y}(n)}{\varepsilon + |\mathbf{y}(n)|^2}\, e_i(n) \tag{5.119}$$

    The step size is updated as follows:

    $$\mu(n) = \mu_{\max}\frac{|\mathbf{p}(n)|^2}{\delta + |\mathbf{p}(n)|^2} \tag{5.120}$$

where $\mathbf{p}(n)$ is given by:

$$\mathbf{p}(n) = \beta\,\mathbf{p}(n-1) + (1-\beta)\frac{\mathbf{y}(n)}{\varepsilon + |\mathbf{y}(n)|^2}\,\mathbf{e}(n) \tag{5.121}$$

and:

$$\mathbf{e}(n) = [e_i(n),\, e_i(n-1),\, \ldots,\, e_i(n-N+1), \tag{5.122}$$

- The NPVSS algorithm of [5]:
  The filter expression is:

$$\mathbf{h}(n+1) = \mathbf{h}(n) + \mu(n)\,\mathbf{y}(n)e_i(n) \tag{5.123}$$

The power of the error is estimated with:

$$\hat{\sigma}_e^2(n) = \beta\,\hat{\sigma}_e^2(n-1) + (1-\beta)\,e_i^2(n) \tag{5.124}$$

The step size is updated as follows; if $\hat{\sigma}_e^2 \geq \sigma_v$:

$$\mu(n) = \frac{1}{\varepsilon + |\mathbf{y}(n)|^2}\left(1 - \frac{\sigma_v}{\delta + \hat{\sigma}_e^2(n)}\right) \tag{5.125}$$

elsewhere: $\mu(n) = 0$.

- The GNGD algorithm of [32]:
  The filter expression is:

$$\mathbf{h}(n+1) = \mathbf{h}(n) + \frac{\mu}{\varepsilon(n) + |\mathbf{y}(n)|^2}\,\mathbf{y}(n)e_i(n) \tag{5.126}$$

where:

$$\varepsilon(n) = \varepsilon(n-1) - \rho\,\mu\frac{e_i(n)\,e_i(n-1)\;\mathbf{y}^T(n)\,\mathbf{y}(n-1)}{\varepsilon(n-1) + |\mathbf{y}(n-1)|^2} \tag{5.127}$$

- The RR-NLMS algorithm of [8]:
  It is a modified version of GNGD, where:

$$\varepsilon'(n) = \varepsilon(n-1) - \rho\,sgn\,(e_i(n)\,e_i(n-1)\;\mathbf{y}^T(n)\,\mathbf{y}(n-1)) \tag{5.128}$$

and:

$$\varepsilon(n) = \begin{cases} \varepsilon'(n)\,, & \varepsilon'(n) \geq \varepsilon_{\min} \\ \varepsilon_{\min}\,, & \varepsilon'(n) < \varepsilon_{\min} \end{cases} \tag{5.129}$$

- The GSER algorithm of [27]:

The filter expression is:

$$\mathbf{h}(n+1) = \mathbf{h}(n) + \frac{\hat{\sigma}_e^2(n)\,\mu}{\varepsilon + \hat{\sigma}_e^2(n)|\mathbf{y}(n)|^2}\,\mathbf{y}(n)e_i(n) \tag{5.130}$$

where:

$$\hat{\sigma}_e^2(n) = \beta\,\hat{\sigma}_e^2(n-1) + (1-\beta)\,e_i^2(n) \tag{5.131}$$

The performance of variable step-size NLMS algorithms, is compared in [26].

5.6.2 *Other Adaptive Filters*

5.6.2.1 Affine Projection Algorithm (APA)

The affine projection filters use multiple, delayed input signal vectors. An excitation matrix is built as follows:

$$Y(n) = [\mathbf{y}(n),\, \mathbf{y}(n-1),\, \ldots, \mathbf{y}(n-N+1)] \tag{5.132}$$

The filter expression is the following:

$$\mathbf{h}(n+1) = \mathbf{h}(n) + \mu\,\frac{Y(n)}{Y^T(n)\,Y(n)}\,\mathbf{e}(n) \tag{5.133}$$

where:

$$\mathbf{e}(n) = [e(n),\, e(n-1),\, \ldots,\, e(n-N+1)]^T \tag{5.134}$$

The convergence of APA is better than NLMS for colored input signals.

5.6.3 *Mathematical Aspects*

This subsection considers in particular certain aspects of mathematical topics that concerns adaptive filtering.

Estimation is in general related to systems of linear equations. Depending on the estimation framework, there could exist more equations that variables, equal, or less. The matrix with the equation coefficients could be square, or not. If the matrix is square, the solution of the equations could be found with matrix inversion. There are pseudo-inverses of non-square matrices.

Some applications of signal processing get into difficulties with equations and matrix inversion. Iterative or search based methods provide ways to overcome these difficulties. The solutions sought could be interpreted as optimization of certain objective functions, like for instance minimizing a defined error.

5.6.3.1 Systems of Linear Equations

Let us consider a system of linear equations:

$$A\,x = b \tag{5.135}$$

(where A is a matrix, x and b are vectors)

There are several numerical procedures to solve the system. For instance, by LU decomposition, when possible (not every matrix has a LU decomposition). The idea is to decompose A into the product of a right-triangular matrix U and a left-triangular matrix L:

$$A = LU \tag{5.136}$$

The system is solved in two steps.

$$L\,y = b \tag{5.137}$$

$$U\,x = y \tag{5.138}$$

The main advantage of the decomposition is that the two steps are easy to solve, and with few operations.

Among the methods for obtaining L and U, the most used are Doolittle, Crout and Cholesky decompositions. For symmetric matrices Cholesky is preferred.

In addition to direct methods there are iterative methods. They are based in fixed point iteration:

$$x = Mx \tag{5.139}$$

The procedure is to start with an initial guess x_0 and then to iterate $x_{k+1} = M\,x_k$. If M is a contraction ($\|M\| < 1$), the iterations converge to the solution.

Coming back to the system of linear equations, let us apply the Richardson's iteration. The system of equations can be written as follows:

$$A\,x = I\,x + (A - I)\,x = b \tag{5.140}$$

Thus:

$$I\,x = (I - A)\,x + b \tag{5.141}$$

So here the iteration matrix M is $(I - A)$.

Suppose that $\|I - A)\| > 1$, so the iteration does not converge. In this case, it is still possible to apply **preconditioning** of the problem. One chooses another matrix L, and:

$$L\,A\,x = I\,x + (L\,A - I)\,x = L\,b \tag{5.142}$$

Then:

$$I\,x = (I - L\,A)\,x + L\,b \tag{5.143}$$

The matrix L is chosen to be simple, for instance diagonal, and such that $\|I - LA)\| < 1$.

Other iterative methods use *splitting* of A:

$$A = A_1 + A_2 \tag{5.144}$$

Thus, the iteration is:

$$x_{k+1} = A_1^{-1}(b - A_2 x_k) \tag{5.145}$$

(A_1^{-1} should be easy to compute)

The Jacobi method chooses a diagonal $A_1^{-1} = D$ and $A_2 = L + U$.

The Gauss-Seidel method chooses $A_1 = D + L$ and $A_2 = U$.

Apart from the methods already described, there are other, denoted as Krylov methods, based on different types of searching. Some of these methods will be described below as optimization procedures.

5.6.3.2 Singular Value Decomposition (SVD)

Let us focus for a moment on square-symmetric matrices, for instance the matrix M. A well-known decomposition can be made based on eigenvalues and eigenvectors.

$$M\,X = X\,\Lambda \tag{5.146}$$

where X is a matrix of eigenvectors and Λ is the diagonal matrix of eigenvalues.

Consider a real $m \times n$ matrix A. Both $A^T A$ and $A\,A^T$ are square-symmetric. They can be decomposed as follows:

$$A^T A\,V = V\,D \tag{5.147}$$

$$A\,A^T\,U = U\,D' \tag{5.148}$$

The eigenvectors of $A^T A$, contained in the matrix V, are called the 'right' singular vectors. The eigenvectors of $A\,A^T$, contained in the matrix U, are called the 'left' singular vectors.

The same decompositions can be expressed as:

$$A^T A = V\,D\,V^T \tag{5.149}$$

$$A\,A^T = U\,D'\,U^T \tag{5.150}$$

where D is a $n \times n$ diagonal matrix with the eigenvalues of $A^T A$, and V is a $n \times n$ orthogonal matrix with the eigenvectors of $A^T A$ as columns. Likewise, D' is a $m \times m$ diagonal matrix with the eigenvalues of $A\,A^T$, and U is a $m \times m$ orthogonal matrix with the eigenvectors of $A\,A^T$ as columns.

Any real $m \times n$ matrix A can be decomposed as:

$$A = U\,S\,V^T \tag{5.151}$$

(*singular value decomposition*)

where:

- U is $m \times m$ and orthogonal. Its columns are eigenvectors of $A\,A^T$.
- V is $n \times n$ and orthogonal. Its columns are eigenvectors of $A^T A$.
- S is $m \times n$ diagonal:

$$S = diag(\sigma_1,\ \sigma_2,\ \ldots,\ \sigma_r,\ 0,\ \ldots,\ 0) \tag{5.152}$$

The entries of S are called *singular values.*

Notice that:

$$A^T A = V\,S^T\,U^T U\,S\,V^T = V\,diag(\sigma_1^2,\ \sigma_2^2,\ \ldots,\ \sigma_r^2,\ 0,\ \ldots,\ 0)\,V^T \tag{5.153}$$

Example (1):

Let:

$$A = \begin{pmatrix} 1 & 1.5 \\ 2 & 2 \end{pmatrix} \tag{5.154}$$

Figure 5.49 shows that the image of a circle, applying $A\,x$, is an ellipse. The figure has been generated with the Program 5.17. This program uses the *svd()* MATLAB function to compute the SVD decomposition of A. It also uses the *cond()* MATLAB

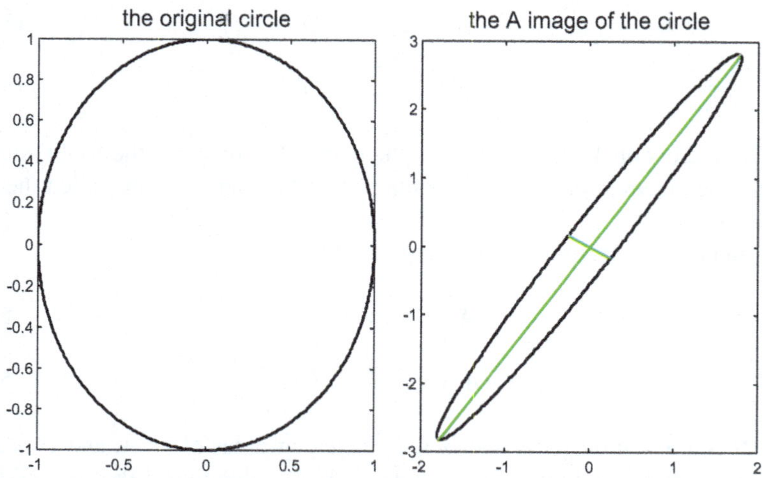

Fig. 5.49 The *ellipse* on the *right* is the A image of the *circle* on the *left*. The *lines* inside the ellipse correspond to the singular values

function to compute the condition number of the matrix A; details about this number will be given later on. Using the information given by the singular vectors, the program draws to perpendicular lines inside the ellipse. The length of these lines is given by the singular values. The matrix A has two singular values: one is large, corresponding to the long axis of the ellipse, and the other is small, corresponding to the short axis. The relation between singular values corresponds to the ellipse eccentricity.

The Program 5.17 is intended as a tool for the reader to explore changes in the matrix A. Some lines of the program are not terminated with the semicolon in order to print values of interest; the reader could do the same with other lines.

Program 5.17 SVD axes example

```
% SVD axes example
A=[1 1.5; 2 2]; %the A matrix
%a circle
R=1; %radius
fi=0:0.01:(2*pi);%angle
x1=R*cos(fi); x2=R*sin(fi);
x=[x1;x2];
%the image of the circle, using A
px=A*x;
%the SVD decomposition
[u,s,v]=svd(A);
%print diagonal of s (the singular values)
diag(s)
%print condition(A)
cond(A)
%display
figure(1)
subplot(1,2,1)
plot(x(1,:),x(2,:),'k');
title('the original circle');
subplot(1,2,2)
plot(px(1,:),px(2,:),'k'); hold on;
L=s(1,1); %the first singular value
plot([-L*u(1,1) L*u(1,1)],[-L*u(2,1) L*u(2,1)],'g');
L=s(2,2); %the second singular value
plot([-L*u(1,2) L*u(1,2)],[-L*u(2,2) L*u(2,2)],'g');
title('the A image of the circle');
```

Using the SVD decomposition, one can write:

$$A v_j = \sigma_j \, u_j, \; j = 1, 2, \ldots, N \tag{5.155}$$

where v_j are the columns of V, u_j are the columns of U, and σ_j are the singular values.

The *null space* of a matrix A is the set of vectors x such that $A\,x\ = 0$. The *range* of A is the set of b such that $A\,x\ = b$ has a solution. If $\sigma_j = 0$ then v_j is in the null space of A, if $\sigma_j \neq 0$ then v_j is in the range of A.

The matrix A can be expanded as follows:

$$A = \sum_{j=1}^{N} \sigma_j u_j\, v_j^T \tag{5.156}$$

Sometimes there is a combination of large and small (even zero) singular values. The matrix A could be approximated taking only the large singular values in the expansion.

Example (2):

Let:

$$A = \begin{pmatrix} 1 & 3 & 1 \\ 2 & 5 & 2 \\ 1 & 3 & 1 \end{pmatrix} \tag{5.157}$$

Using the Program 5.18 one computes the SVD decomposition. The singular values are:

$$\boldsymbol{\sigma} = (7.4113,\ 0.2699,\ 0) \tag{5.158}$$

The matrix A has rank = 2. Since the first singular values is notably larger than the second, it would be possible to approximate the matrix A by the first term of the expansion:

$$A_1 = \begin{pmatrix} 1.0925 & 2.9311 & 1.0925 \\ 1.8933 & 5.0796 & 1.8933 \\ 1.0925 & 2.9311 & 1.0925 \end{pmatrix} \tag{5.159}$$

Program 5.18 Some mathematics, example

```
% Some mathematics (1)
%3x3 example
A=[1 3 1;
2 5 2;
1 3 1];
[u,s,v]=svd(A);
A1=s(1,1)*u(:,1)*v(:,1)';
A2=s(2,2)*u(:,2)*v(:,2)';
A3=s(3,3)*u(:,3)*v(:,3)';
Ap=A1+A2+A3;
M=A*A';
N=A'*A;
[U,D1]=eig(M);
S1=sqrt(diag(D1));
[V,D2]=eig(N);
```

```
S2=sqrt(diag(D2));
%check values of the matrices
s
A1
```

If A is square non-singular (all singular values are nonzero), the inverse of the matrix could be obtained via the SVD decomposition:

$$A^{-1} = V\, S^{-1}\, U^T \tag{5.160}$$

with:

$$S^{-1} = diag(\frac{1}{\sigma_1}, \frac{1}{\sigma_2}, ..., \frac{1}{\sigma_N}) \tag{5.161}$$

This is a good way, in numerical terms, for obtaining the inverse of A.

When A is singular, it is still possible to use a **pseudo-inverse**, like the following:

$$A^+ = V\, S^+\, U^T \tag{5.162}$$

with:

$$S^+ = diag(\frac{1}{\sigma_1}, \frac{1}{\sigma_2}, ..., \frac{1}{\sigma_r}, 0, ..., 0) \tag{5.163}$$

The zeros correspond to zero or very small (under a certain threshold) singular values.

One of the properties of the pseudo-inverse is that: $A\, A^+ A = A$.

The pseudo-inverse can be used in case of non-square A.

The Program 5.19 finds the inverse of a non-singular matrix, and then the pseudo-inverse of a singular matrix.

Program 5.19 Some mathematics: inverse and pseudoinverse

```
% Some mathematics (2)
% inverse and pseudoinverse
%---------------------------------
%2x2 example, nonsingular matrix
A=[1 2;
3 4];
[u,s,v]=svd(A);
si=diag(1./diag(s)); %the inverse of s
Ainv=v*(si)*u'; %inversion using SVD
%check of inversion
A*Ainv
%---------------------------------
%2x2 example, singular matrix
B=[1 2;
1 2];
[U,S,V]=svd(B);
%pseudo-inverse of S
```

```
SI=zeros(2,2);
for n=1:2,
  if S(n,n)>0.001, %small threshold
    SI(n,n)=1/S(n,n);
  else
    SI(n,n)=0;
  end;
end;
Binv=V*SI*U'; %pseudo-inversion using SVD
%check of pseudo-inversion
B*Binv*B
```

Let us look again at the linear system of equations:

$$A\,x\,=\,b \tag{5.164}$$

Using the mentioned pseudo-inverse property:

$$A\,A^{+}Ax\,=\,b \tag{5.165}$$

Thus:

$$A\,A^{+}\,b\,=\,b \tag{5.166}$$

Therefore, $x\,=\,A^{+}\,b$ is a solution of the problem. All the solutions are given by:

$$x\,=\,A^{+}b\,+\,\mathrm{N}(A) \tag{5.167}$$

where $N(A)$ is the null space of A.

Another expression of the pseudo-inverse is:

$$if\; m \geq n,\;\; A^{+}\,=(A^{T}A)^{-1}\,A^{T} \tag{5.168}$$

$$if\; m \leq n,\;\; A^{+}\,=\,A^{T}(A^{T}A)^{-1} \tag{5.169}$$

To see it:

$$A^{T}A\,=\,V\,S^{2}\,V^{T}\,;\quad (A^{T}A)^{-1}\,=\,V\,S^{-2}\,V^{T} \tag{5.170}$$

$$(A^{T}A)^{-1}\,A^{T}\,=\,V\,S^{-2}\,V^{T}V\,S\,U^{T}\,=\,V\,S^{-1}\,U^{T} \tag{5.171}$$

Moreover:

$$A\,x\,=b\;\rightarrow\;A^{T}A\,x\,=\,A^{T}\,b\rightarrow\;x\,=\,(A^{T}\,A)^{-1}\,A^{T}b \tag{5.172}$$

The SVD decomposition will be revisited in a next chapter, in relation with principal component analysis.

5.6.3.3 Ill-Posed Problems

Depending on the matrix A numerical difficulties may appear when trying to solve the system of equations:

$$A\,x = b_0 \tag{5.173}$$

Suppose there is an error in the data, so the equations change as follows:

$$A\,x = b_0 + \delta\,b \tag{5.174}$$

Then:

$$x = A^{-1}\,b + A^{-1}\delta\,b = x_0 + \delta\,x \tag{5.175}$$

After some algebra, it is found that:

$$\frac{\|\delta\,x\|}{\|x_0\|} = \frac{\|\delta\,b\|}{\|b_0\|} \cdot \frac{\sigma_{\max}}{\sigma_{\min}} \tag{5.176}$$

The *condition number* of a matrix A is defined as:

$$cond\,(A) = \frac{\sigma_{\max}}{\sigma_{\min}} \tag{5.177}$$

If the condition number is large, it means that errors are amplified: a bad situation for solving the system of equations (or for computing the inverse of the matrix).

It is the experience of many signal processing applications, like echo cancelling, image processing, electro medicine, etc., that the auto-correlation matrix frequently has a large condition number.

According to Hadamard, a well-posed problem has a unique solution, and this solution depends continuously on the problem data. This last condition is not fulfilled if small variations of data cause large variations of the solution (solution instability).

A classical way for solution stabilization is **regularization**. Two popular regularization techniques are the Tikhonov method and truncated SVD.

The difficulties with matrix inversion $A^{-1} = V\,S^{-1}\,U^T$ arise when some entries of:

$$S^{-1} = diag(\frac{1}{\sigma_1}, \frac{1}{\sigma_2}, \ldots, \frac{1}{\sigma_N}) \tag{5.178}$$

become too large because some singular values are close to zero. The idea of regularization is to multiply the singular values by a function $w()$ such that:

$$w(\sigma)\,\sigma^{-1} \to 0 \;\; as \;\; \sigma \to 0 \tag{5.179}$$

The Tikhonov regularization uses:

$$w(\sigma) = \frac{\sigma^2}{\sigma^2 + \alpha} \tag{5.180}$$

The truncated SVD (TSVD) employs a threshold (see Program 5.19):

$$\begin{aligned} w(\sigma) &= 1, \ for \ \sigma > \alpha \\ w(\sigma) &= 0, \ for \ |\sigma| \leq \alpha \end{aligned} \tag{5.181}$$

It can be shown that the Tikhonov regularization is equivalent to use:

$$\hat{x} = (A^T A + \alpha I)^{-1} A^T b \tag{5.182}$$

as solution of the linear system of equations.

Indeed there is a subsequent issue: to choose a suitable α. A popular method is based on the curve:

$$L \Leftrightarrow \log(|\hat{x}|) \ \ vs. \ \log(|A\hat{x} - b|) \tag{5.183}$$

This curve is called the *L-curve*, because usually is shaped like an L letter. The corner of the L corresponds to the good value of α. It has evident connection with the Pareto front.

In the simple case of the inversion of a certain scalar variable $y \rightarrow 1/y$, the regularization could be: $y \rightarrow 1/(y + \delta)$ (where δ is a scalar value). This is the case of the frequency domain Wiener filter, which can be seen as a regularized inverse filter. Likewise, many variations of the LMS include regularization, like for instance the NLMS.

5.6.3.4 Least-Square Problem. Normal Equations

Actually, the solution of $A\,x = b$ could be approximated via optimization. The *least square problem* is to find the solutions minimizing:

$$\| A\,x - b \|^2 \tag{5.184}$$

Gauss and Lebesgue discovered that these solutions are given by:

$$A^T A\,x = A^T b \tag{5.185}$$

which are called *'normal equations'*.

The Program 16.6.4 shows how to solve the normal equations using Cholesky factorization. Notice the use of the *chol()* MATLAB function.

Program 5.20 Some mathematics: example of Cholesky factorization

```
% Some mathematics (3)
% Example of Cholesky factorization
%normal equation: (A'*A)*x = A'*b
%it will become: M*x=r
A=[5.2 7.7 2.6;
3.4 1.1 0.7;
4.2 6.3 5.8;];
b= [3 2 1]';
M=A'*A;
r=A'*b;
%Cholesky factorization
C=chol(M);
%problem solution (value of x)
x= C\(C'\r);
%print values
x
C
```

The condition number of the left-hand side of the normal equations is:

$$cond(A^T A) = (cond(A))^2 \tag{5.186}$$

Therefore, numerical difficulties may happen, also when using the Cholesky factorization. For this case, it is better to use the QR decomposition. The columns of the matrix Q are orthonormal. The matrix R is upper triangular. A theorem establishes that the solution set of the least squares problem is identical to the solution set of:

$$R\,x = Q^T b \tag{5.187}$$

The Program 5.21 shows how to solve the normal equations using QR decomposition by means of the *qr()* MATLAB function.

Program 5.21 Some mathematics: example of QR factorization

```
% Some mathematics (4)
% Example of QR factorization
%normal equation: (A'*A)*x = A'*b
A=[5.2 7.7 2.6;
3.4 1.1 0.7;
4.2 6.3 5.8;];
b= [3 2 1]';
%QR factorization
[Q,R]=qr(A,0);
V=Q'*b; %right-hand side of equation
%problem solution (value of x)
x= R\V;
```

```
%print values
x
Q
R
```

5.6.3.5 Search-Based Optimization

In general, optimization is an important topic. The research has found a lot of methods to solve a variety of optimization problems. Concerning adaptive filters, iterative and search-based optimization methods are of particular interest.

The target of optimization may be to maximize, or to minimize, a certain objective function. A good specification of the objective function is a key aspect for having success.

If you have practiced some trekking, through mountains and valleys, you already know many peculiar issues of search-based optimization, like for instance that a peak may hide another, higher peak. Search may fall into traps: local maxima or minima.

In general the iterative search chooses at each step k, a direction $\mathbf{p}_k$ and a step length α_k:

$$\mathbf{x}_{k+1} = \mathbf{x}_k + \alpha_k \, \mathbf{p}_k \tag{5.188}$$

The steepest-descent method has been already described, in relation to recursive estimation of filter coefficients. It uses the gradient:

$$\mathbf{p}_k = -\nabla \, e(\mathbf{x}_k) \tag{5.189}$$

Figure 5.50 shows an example of how the steepest descent advances towards the minimum in an error surface. This figure, and the next one, has been obtained with

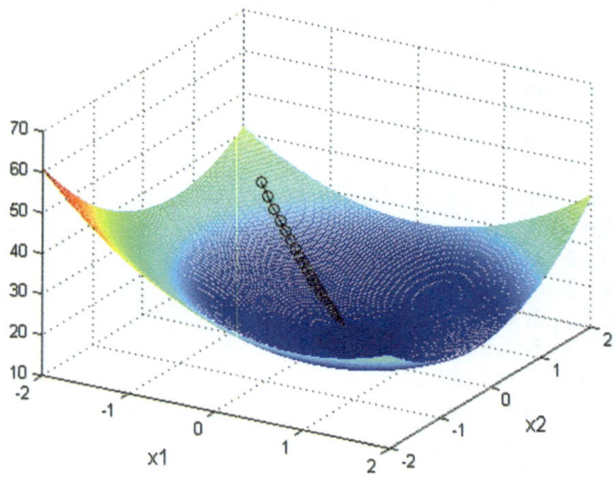

Fig. 5.50 Steepest descent on the error surface

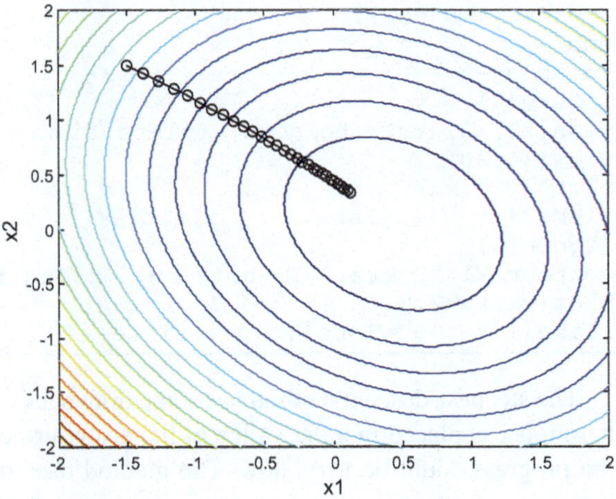

Fig. 5.51 Steepest descent on error contours

the Program 5.22. The error surface is the same as in Fig. 5.13, but from another point of view. See Sect. 5.3.2 for details about the gradient.

Figure 5.51 depicts a contour plot in 2D of the steepest descent motion towards the optimum.

Program 5.22 Steepest-descent steps

```
% Steepest-descent steps
x1=-2:0.02:2;
x2=-2:0.02:2;
[X1,X2]=meshgrid(x1,x2);
A=[3.8 0.9;0.9 3.8];
b=[1.7;0.6];
aux1=(A(1,1)*X1.^2)+(A(1,2)*X1.*X2)+(A(2,1)*X2.*X1)+...
(A(2,2)*X2.^2);
aux2=(X1*b(1))+(X2*b(2));
ers=14+aux1-2*aux2;
Np=30; %number of steps
%space for the record of steps
vx1=zeros(1,Np);
vx2=zeros(1,Np);
ver=zeros(1,Np);
%doing the steps
px=[-1.5;1.5]; %initial point
mu=0.02; %step size
for nn=1:Np,
  %storing:
  vx1(nn)=px(1);
  vx2(nn)=px(2);
  aux1=px'*A*px; aux2=2*px'*b;
  ver(nn)=14+aux1-2*aux2;
  %next step
  aux=b-(A*px);
```

```
  px=px+(mu*aux);
end;
%display---------------------------
figure(1)
mesh(X1,X2,ers); hold on; %plots figure
view(30,30);
plot3(vx1,vx2,ver,'-ok');
xlabel('x1'); ylabel('x2');title('error surface');
figure(2)
contour(X1,X2,ers,20); hold on; %plots figure
plot(vx1,vx2,'-ok');
xlabel('x1'); ylabel('x2');title('steps on the error plot');
```

The steepest-descent method has some drawbacks. For instance, if the solution at iteration k is placed in a flat valley of the error surface, and far from the minimum, the progress could be very slow. The method uses fixed step size, but it could be better to use adaptive step size. Also, at iteration k it may happen that the gradient does not point to the minimum.

A number of alternatives have been proposed to obtain a better direction $\mathbf{p}_k$. For instance the *conjugate gradient* method:

$$\mathbf{p}_k = -\nabla e(\mathbf{x}_k) + \beta_k \, \mathbf{p}_{k-1} \tag{5.190}$$

where the parameter β_k is intended for making $\mathbf{p}_k$ and $\mathbf{p}_{k-1}$ approximately conjugate.

Now, consider the following Taylor series:

$$\nabla e(\mathbf{x}+\mathbf{p}) = \nabla e(\mathbf{x}) + \nabla^2 e(\mathbf{x})\, \mathbf{p} + \ldots \tag{5.191}$$

Set $\nabla e(\mathbf{x} + \mathbf{p}) = 0$, then obtain direction:

$$\mathbf{p}_k = -(\nabla^2 e(\mathbf{x}))^{-1} \cdot \nabla e(\mathbf{x}) \tag{5.192}$$

this is the *Newton's method*. It uses the information that the Hessian gives about the curvature of $e()$. The Hessian is the $\nabla^2(e(\mathbf{x}))$ term.

It may happen that the direction given by the Hessian is not a descent. There are modifications of the Newton's method to avoid this problem: for instance, regularized versions of the method.

Quasi-Newton methods approximate the Hessian by a certain matrix G_k. For instance, the *secant method* uses:

$$G_k(\mathbf{x}_{k+1} - \mathbf{x}_k) = \nabla e_{k+1} - \nabla e_k \tag{5.193}$$

If the function to optimize (the error in our case) is a sum of squared terms:

$$e(\mathbf{x}) = \sum r_i^2(\mathbf{x}) \tag{5.194}$$

like for instance: $e(\mathbf{x}) = \sum (b_i - Ax_i)^2$

Then, one computes the Jacobian matrix J, which has the following entries:

$$J_{ij}(\mathbf{x}) = \left.\frac{\partial r_i}{\partial x_j}\right|_{\mathbf{x}} \tag{5.195}$$

Now, the following approximation is used:

$$\nabla\, e(\mathbf{x}+\mathbf{p}) \approx \nabla\, e(\mathbf{x}) + \nabla^2\, e(\mathbf{x})\,\mathbf{p} \;\approx\; J^T\mathbf{r} + J^T\, J\,\mathbf{p} \tag{5.196}$$

And so, the normal equations are obtained:

$$(J^T J)\,\mathbf{p} = -J^T\,\mathbf{r} \tag{5.197}$$

This is the *Gauss-Newton method.*

Levenberg proposed a kind of regularization, as follows:

$$(J^T J + \lambda\, I)\,\mathbf{p} = -J^T\,\mathbf{r} \tag{5.198}$$

Marquardt introduced a scaling of the regularization:

$$(J^T J + \lambda\, diag(J^T\, J)\,\mathbf{p} = -J^T\,\mathbf{r} \tag{5.199}$$

This is the popular *Levenberg-Marquardt method.*

Finally, at least some mention should be done to derivative-free algorithms, like heuristic search, direct search, line search, etc. In particular, heuristic search has motivated a lot of research, with proposals such as genetic algorithms, simulated annealing, tabu search, etc. These methods have important merits for complicated problems, perhaps of multi-objective nature and subject to nonlinearities, constraints, etc.

An Appendix on optimization has been included in the third book, so the mathematical topics just briefly outlined, are treated there in more detail.

5.6.4 Unifying Perspective

In a series of papers, Husøy and Abadi have introduced a unifying approach encompassing most adaptive filter algorithms, [17]. The papers give detailed developments. Here, in this subsection, a short summary is given.

Start from the Wiener filter:

$$R_y \cdot \mathbf{h} = \mathbf{r}_{xy} \tag{5.200}$$

This system of equations could be iteratively solved. Let us apply Richardson's method:

$$\mathbf{h}(k+1) = h(k) + \mu(\mathbf{r}_{xy} - R_y h(k)) \quad (5.201)$$

Again, it would be interesting to study the fluctuations of the filter coefficients, according with:

$$E(\eta(n+1)) = (I - \mu R_y)\, E(\eta(n)) \quad (5.202)$$

In order to accelerate the convergence, a preconditioning could be applied:

$$C R_y \cdot \mathbf{h} = C \mathbf{r}_{xy} \quad (5.203)$$

So,

$$E(\eta(n+1)) = (I - \mu C R_y)\, E(\eta(n)) \quad (5.204)$$

The idea of the preconditioning could be to avoid large differences between eigenvalues, so it would be advisable to set C as an approximate inverse of R_y.

Usually adaptive filters use real-time estimations of R_y and $\mathbf{r}_{xy}$. For example, an estimation of R_y could be $\mathbf{y} \cdot \mathbf{y}^T$, and an estimation of $\mathbf{r}_{xy}$ could be $x \cdot \mathbf{y}$:

$$\mathbf{h}(k+1) = h(k) + \mu \mathbf{y}(x - \mathbf{y}^T h(k)) \quad (5.205)$$

which is the LMS filter.

More degrees of freedom for adaptive algorithms could be obtained by using other ways of estimation. For instance, using a signal matrix, like (recall the affine projection algorithm, APA):

$$Y(n) = [\mathbf{y}(n), \mathbf{y}(n-1), \ldots, \mathbf{y}(n-N+1)] \quad (5.206)$$

and a vector of inputs:

$$\mathbf{x}(n) = [x(n), x(n-1), \ldots, x(n-N+1)]^T \quad (5.207)$$

To extend the generalization, filtering could be added to the estimation, and so:

$$R_y(n) = Y(n)\, F\, F^T\, Y(n) \quad (5.208)$$

and:

$$\mathbf{r}_{xy}(n) = Y(n)\, F\, F^T x(n) \quad (5.209)$$

To simplify the notation, let us write:

$$Y_F(n) = X(n)\, F; \quad x_F(n) = F^T x(n); \quad e_F(n) = F^T e(n) \quad (5.210)$$

A general expression is obtained using preconditioning and generalized estimation:

$$\begin{aligned} \mathbf{h}(k+1) &= \mathbf{h}(k) + \mu\, C(n)(\mathbf{r}_{xy}(n) - R_y(n)\mathbf{h}(k)) = \\ &= \mathbf{h}(k) + \mu\, C(n)\, Y_F(n)\, e_F(n) \end{aligned} \tag{5.211}$$

Now, several preconditioning strategies could be applied. The most simple is to use a constant $C(n)$. An example of this strategy is LMS. Another strategy is to use a regularized inverse of R_y:

$$C(n) = (\varepsilon I + R_y(n))^{-1} = (\varepsilon I + Y_F(n) Y_F^T(n))^{-1} \tag{5.212}$$

It is shown that using different estimation alternatives, this preconditioning corresponds to NLMS, or a regularized APA, or some other adaptive filters.

It is also possible to estimate R_y in a way that its invertibility is assured. Then, the preconditioning could be $C(n) = R_y^{-1}(n)$. An example of such alternative is RLS.

5.7 Bayesian Estimation: Application to Images

The point of view of Bayes is becoming more and more important in the fields of data and signal processing, especially when there are uncertainties, perturbations, noise.

This is the case, for example, of imaging systems, which combine optics and electronics. Important applications of these systems are found in microscopy, astronomy, medicine, etc. Good image quality, with adequate characteristics, is desired. It has been demonstrated that digital processing could improve the performances of imaging systems. In certain cases, this improvement belongs to the category of inverse problems [38].

In this section, the particular case of degraded images is chosen. Indeed, image restoring by software means is good news for consumer applications, and for professional purposes. At the same time, from academic perspective, this is an illustrative problem concerning Bayesian methodology.

An image degradation model could be used. For instance, the model employed in Sect. 5.6:

$$Y(\omega) = D(\omega) \cdot X(\omega) + V\omega) \tag{5.213}$$

We want to obtain a good estimate of X using Y and the model.

In Bayesian terms one could write:

$$P(X|Y) = \frac{P(Y|X)\, P(X)}{P(Y)} \tag{5.214}$$

where $P(X)$ is the 'prior' and $P(Y|X)$ corresponds to image degradation. Usually $P(Y)$ is considered a normalization factor.

The term $P(Y|X)$ is the **likelihood** of X given Y.

The term $P(X|Y)$ is the posterior probability of X given Y. Therefore:

$$posterior \propto prior \times likelihood \tag{5.215}$$

In terms of probability density functions (PDFs):

$$f_X(X|Y) \propto f_X(X) \times L(X|Y) \tag{5.216}$$

where $L(X|Y)$ is the likelihood of X given Y.

The original image X can be estimated as the maximum of $P(X|Y)$, this is the maximum 'a posteriori' (MAP) estimation.

If the prior is flat, then the MAP estimation is the same as the maximum likelihood (ML) estimation.

Note that if the prior is not flat, it may take the MAP away from the ML.

It may happen that at the beginning of an application, the model is not known. It would be convenient to determine a model, establishing its parameters $\boldsymbol{\theta}$.

This task of model identification could also be described in Bayesian terms:

$$f_\theta(\boldsymbol{\theta}|Y) = \frac{f_Y(Y|\boldsymbol{\theta})\, f_\theta(\boldsymbol{\theta})}{f_Y(Y)} \tag{5.217}$$

There are applications where both model parameters and input variables must be estimated on-line.

Soon, this section will focus on an iterative algorithm that is used for image restoration. But before, it is interesting to sketch a connection with the structure of the Wiener filter.

Let us come back to:

$$y = x + v \tag{5.218}$$

Suppose that x and v are zero-mean Gaussian random variables. Specifically:

$$f_x\,(x) = \frac{1}{\sigma_x\sqrt{2\,\pi}}\,\exp(-x^2/2\,\sigma_x^2) \tag{5.219}$$

$$f_v\,(v) = \frac{1}{\sigma_v\sqrt{2\,\pi}}\,\exp(-v^2/2\,\sigma_v^2) \tag{5.220}$$

Suppose one observation of y is obtained. In Bayesian terms, one has the following:

- Prior:
  $$f_x\,(x) = \frac{1}{\sigma_x\sqrt{2\,\pi}}\,\exp(-x^2/2\,\sigma_x^2)$$
- Likelihood:
  $$L\,(x|y) = \frac{1}{\sigma_v\sqrt{2\,\pi}}\,\exp(-y-x)^2/2\,\sigma_v^2)$$
- Posterior:
  $$f_x\,(x|y) \propto \exp[(-x^2/2\sigma_x^2) - (-y-x)^2/2\,\sigma_v^2)]$$

If you write the posterior as:

$$f_x(x|y) \propto \exp(-(x-\mu)^2/2\sigma^2) \tag{5.221}$$

Developing and comparing the exponents of the posteriors, it can be deduced that:

$$\sigma = \frac{\sigma_x^2 \sigma_v^2}{\sigma_x^2 + \sigma_v^2} \tag{5.222}$$

$$\mu = \frac{\sigma_x^2}{\sigma_x^2 + \sigma_v^2} \cdot y \tag{5.223}$$

The value of μ is the Bayesian estimate of x. This estimate is similar in structure to the Wiener filter.

The expression of the variance can be written as follows:

$$\sigma = \frac{1}{\frac{1}{\sigma_x^2} + \frac{1}{\sigma_v^2}} \tag{5.224}$$

Notice that this is similar to the resistance of two resistors in parallel. Clearly:

$$\sigma < \min(\sigma_x, \sigma_v) \tag{5.225}$$

(the MAP estimate is more trustful than the priors).

In case the mean of x is nonzero, the corresponding Bayesian estimate is:

$$\mu = \frac{\sigma_v^2 \mu_x + \sigma_x^2 y}{\sigma_x^2 + \sigma_v^2} = [\frac{\mu_x}{\sigma_x^2} + \frac{y}{\sigma_v^2}]\ \sigma \tag{5.226}$$

This expression can be interpreted as a weighted compromise between the expected prior mean and the observed value y. We will be back to this kind of expression in the third book.

5.7.1 Introduction to Image Restoration

There are companies that take your old photographs, perhaps broken, or decoloured, or out of focus, etc., and obtain for you improved copies. This is a kind of activity that belongs to image restoration in a broad sense.

For this section, our attention will be restricted, however, to image degradation that can be represented in mathematical terms, with emphasis on Fourier.

The initial images from the Hubble Space Telescope—a satellite that was launched in 1990—were disappointing, blurred. It was realized that the primary mirror of the telescope had been polished with some spherical aberration. The flaw was tiny, about 1/50 of a paper sheet thickness, but it was spoiling a very expensive mission. The scientific community began an intensive effort on computerized image restoration. It was successful, and it created a lot of research momentum in favour of new, powerful restoration technologies.

People involved in optical systems frequently use *'point-spread functions'* (PSF). In the case of a telescope, the PSF describes the two-dimensional distribution of light in the telescope focal plane, the light coming from astronomical point sources.

More in general, the PSF is the impulse response of the imaging system (for instance a camera with a lens).

The *'optical transfer function'* (OTF) is two-dimensional; amplitude gain and phase versus spatial frequency (cycles pre picture width). For instance, 680 lines of a BW screen can display up to 680 alternate black and white lines. The *'modulation transfer function'* (MTF) is the magnitude of the OTF (the phase component is excluded).

The Image Processing Toolbox of MATLAB provides the functions *otf2psf()* and *psf2otf()* to convert from OTF to PSF or vice-versa.

The PSF can model many image degradation phenomena. Some of these models are now briefly introduced.

5.7.2 Uniform Out-of-Focus Blur

The image of any point source is a circle, denoted as *'circle of confusion'* (COC), with a certain radius R. A simple PSF model would be:

$$d(x, y) = \begin{cases} \frac{1}{\pi R^2}, \; if \sqrt{x^2 + y^2} < R \\ 0, \; elsewhere \end{cases} \tag{5.227}$$

5.7.3 Atmospheric Turbulence Blur

Several models have been proposed for atmospheric blur. A frequently used PSF model is Gaussian:

$$d(x, y) = C \exp(-\frac{x^2 + y^2}{2\sigma^2}) \tag{5.228}$$

where σ determines the spread of the blur.

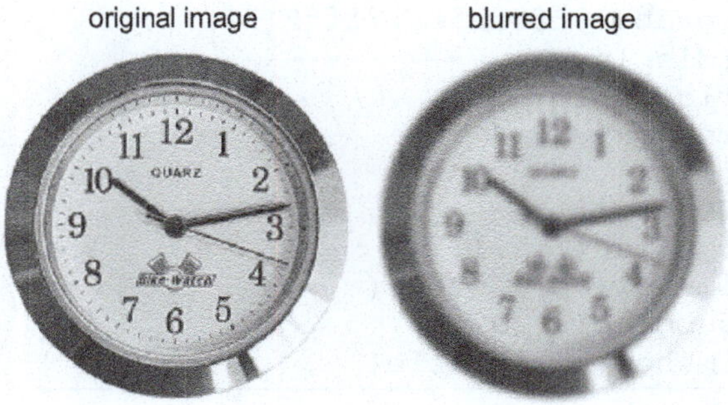

Fig. 5.52 Original and blurred pictures

5.7.4 Linear Motion Blur

The length of motion L is velocity multiplied by exposure time. The velocity as seen by the camera. The PSF is:

$$d(x, y) = \begin{cases} \frac{1}{L}, \; if \sqrt{x^2 + y^2} \le L/2 \\ 0, \; elsewhere \end{cases} \tag{5.229}$$

The variables x and y are related: $y = x \cdot \tan \alpha$; where α is the angle of the linear motion.

The Program 5.23 loads a picture, applies an atmospheric turbulence blur, and displays the result. Figure 5.52 shows the original picture and the blurred picture. The reader is invited to modify the PSF, or try the out-of-focus, and see the results.

Program 5.23 PSF example (atmospheric turbulence)

```
% PSF example (atmospheric turbulence)
%load the image
ip=imread('clock1.jpg');
ip=ip(:,:,1); %select one plane
op=im2double(ip); %convert to float
%create the PSF
[oL,oC]=size(op);
L=oL-1; C=oC-1; %number-1 of rows and columns
sigma=3;
[x,y]=meshgrid(-C/2:C/2, -L/2:L/2);
argm=-((x.^2)+(y.^2))/(2*sigma*sigma);
d=exp(argm);
md=sum(d(:));
d=d/md; %the normalized PSF
%image blurring caused by PSF
OP=fft2(op);
D=fft2(d);
BP=D.*OP; %using Fourier for convolution
```

```
bp=abs(ifftshift(ifft2(BP)));
%display-------------------
figure(1)
mesh(x,y,d);
title('Gaussian PSF');
figure(2)
subplot(1,2,1)
imshow(op); title('original image')
subplot(1,2,2)
imshow(bp); title('blurred image')
```

Next figure shows a 3D view of the PSF used for atmospheric turbulence blur (Fig. 5.53).

5.7.5 The Lucy-Richardson Algorithm (RLA)

RLA has been independently proposed by Richardson (1972) [39] and Lucy (1974) [29] for astronomy image applications. It was derived using Bayesian arguments. Later on, the same algorithm has been derived by other authors for medical applications (emission tomography). Now it is used for many applications involving telescopes, microscopes, cameras, and medical instruments.

Let us write again the image restoration problem in Bayesian terms:

$$f_X(X|Y) \propto f_X(X) \times L(X|Y) \tag{5.230}$$

RLA considers photon counting in the CCD sensor pixels, so the noise is (mostly) of Poisson type. Astronomical objects are supposed equi-probable, so the prior is flat (this is called an uninformative prior). Therefore the MAP estimation is a maximum

Fig. 5.53 Atmospheric turbulence Gaussian PSF

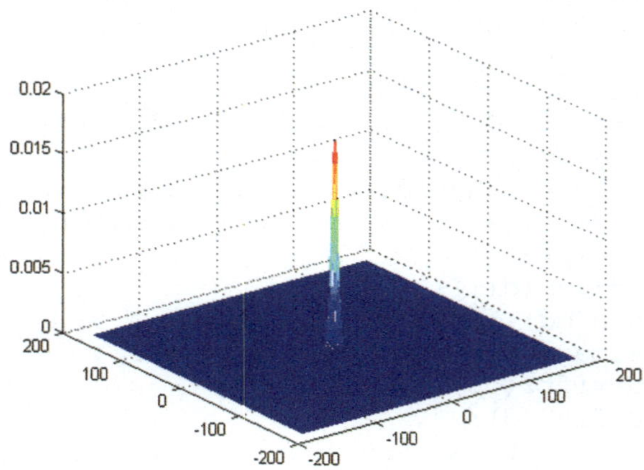

likelihood (ML) estimation. The RLA algorithm is obtained by building the likelihood of a set of Poisson processes, and finding the maximum by equalling the differential to zero. The solution is obtained by fixed point iteration.

Supposing a two-dimensional array of pixels, with indexes i, j, the image model would be:

$$y(i, j) = d(i, j) * x(i, j) + n(i, j) \tag{5.231}$$

where * denotes convolution, $d(i, j)$ is the PSF, $n(i, j)$ is noise; $x(i, j)$ is the original image, and $y(i, j)$ is the observed image.

The Poisson likelihood would be:

$$P(y|x) = \prod_{i,j} (\frac{1}{y(i, j)!} ((d * x)(i, j))^{y(i,j)} \exp(-(d * x)(i, j)) \tag{5.232}$$

Taking logarithms:

$$\ln P(y|x) = const. - \sum_{i,j} ((d * x)(i, j)) + \sum_{i,j} y(i, j) \ln((d * x)(i, j)) \tag{5.233}$$

By taking the negative logarithm of the likelihood, one obtains the Kullback-Leibler divergence:

$$D_{KL}(x) = -\ln P(x|y),$$

the ML estimate is given by:

$$\hat{x}_{ML} = \arg\min (D_{KL}(x)) \tag{5.234}$$

Let us obtain this argument:

$$\frac{\partial \ln (P(y|x)(i, j))}{\partial x(i, j)} = 0 \tag{5.235}$$

From the Karush-Kuhn-Tucker (KKT) conditions, which are necessary and sufficient for the previous equation, one deduces that:

$$\left[\frac{y}{d * x} * d^A\right](i, j) = 1 \tag{5.236}$$

where d^A is the adjoint: a spatially reversed version of d that is $d\,(-i,-j)$.

This equation can be solved by the following iteration:

$$x^{(n+1)}(i, j) = \left[\frac{y}{d * x^{(n)}} * d^A\right](i, j) \cdot x^{(n)}(i, j) \tag{5.237}$$

This is the Lucy-Richardson algorithm. Using text, it can be written as follows:

$$image^{(n+1)} = image^{(n)} \cdot \left[\frac{original\,image}{image^{(n)} * PSF} * PSF^{A} \right] = image^{(n)} \cdot CF \tag{5.238}$$

Usually, the convolutions are computed using Fourier transforms. The factor CF can be seen as a *correction factor*.

Program 5.24 gives an example of how to implement the Lucy-Richardson algorithm. The program uses a blurred image of Saturn, as the images taken by the Hubble Telescope (before refurbishing), and obtains a deblurred image. Figure 5.54 shows the original image, and Fig. 5.55 the restored image. A PSF corresponding to atmospheric turbulence was chosen (the same PSF of the previous example).

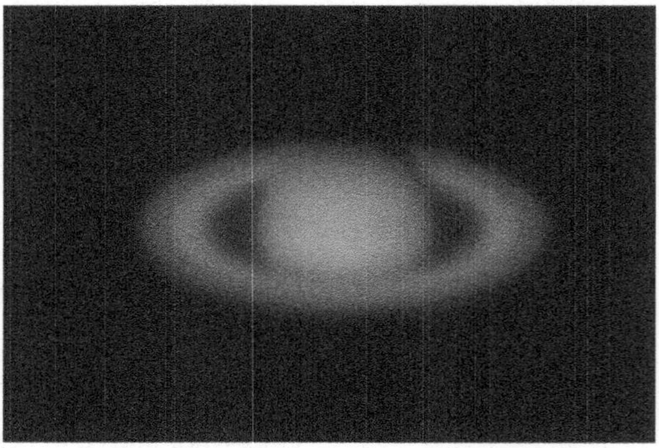

Fig. 5.54 Original blurred picture

Fig. 5.55 Restored image

Program 5.24 Lucy-Richardson example

```
% Lucy-Richardson example
%load the image
ip=imread('saturn.tif');
op=im2double(ip); %convert to float
%create the PSF
[oL,oC]=size(op);
L=oL-1; C=oC-1; %number-1 of rows and columns
sigma=2;
[x,y]=meshgrid(-C/2:C/2, -L/2:L/2);
argm=-((x.^2)+(y.^2))/(2*sigma*sigma);
d=exp(argm);
md=sum(d(:));
d=d/md; %the normalized PSF
%the degradation OTF
D=fft2(d);
%preparation for LR
mp=medfilt2(op); %median filtering
ep=mp; %initial estimated image
%the LR iterations
for nn=1:5,
  %denominator (convolution using Fourier)
  EP=fft2(ep);
  BEP=D.*EP;
  bep=abs(ifftshift(ifft2(BEP)));
  %num/den
  r=ep./(0.00001+bep); %simple regularization
  %correction vector (convolution using Fourier)
  R=fft2(r);
  CV=D.*R;
  cv=abs(ifftshift(ifft2(CV)));
  %next estimate
  ep=cv.*ep;
end;
%display----------------
figure(1)
imshow(op); title('original image')
figure(2)
imshow(ep); title('restored image');
```

The Bayesian image restoration in astronomy is reviewed in [34].

Several authors use a different notation for the LRA. It is convenient to include their expressions, in order to give a more complete view of the topic. They use a one-dimensional set of pixels, so the image model is:

$$\mathbf{y} = H\,\mathbf{x} + \mathbf{n} \tag{5.239}$$

where H is the PSF matrix.

The Poisson likelihood is:

$$P(\mathbf{y}|\mathbf{x}) = \prod_i \frac{1}{y_i!} \left((\sum_j h_{ij}\, x_j)^{y_i} \exp(-\sum_j h_{ij}\, x_j) \right) \tag{5.240}$$

The optimization condition is:

$$H^T \left[\frac{\mathbf{y}}{H\,\mathbf{x}} \right] = 1 \tag{5.241}$$

And the LRA is:

$$\mathbf{x}^{(n+1)} = (H^T \left[\frac{\mathbf{y}}{H\,\mathbf{x}} \right])\, \mathbf{x}^{(n)} \tag{5.242}$$

5.7.6 *Other Aspects of the Topic*

Because images are important for many applications, the research on image restoring has been intensively working and there are many significant contributions on a variety of aspects. Some of them have been selected for this subsection.

5.7.6.1 More Iterative Algorithms

There are iterative algorithms that resemble the LMS. For instance, the Jansson-Van Cittert method:

$$x^{(n+1)} = x^{(n)} + \alpha\,(y - d * x^{(n)}) \tag{5.243}$$

where α is usually 1.

In Fourier space:

$$X^{(n+1)} = X^{(n)} + \alpha\,(Y - D \cdot X^{(n)}) \tag{5.244}$$

The Landweber method, which is frequently used, is another example of this kind of algorithms:

$$\mathbf{x}^{(n+1)} = \mathbf{x}^{(n)} + \gamma\, H^t\, (\mathbf{y} - H\,\mathbf{x}^{(n)}) \tag{5.245}$$

In order to ensure stability:

$$0 < \gamma < \frac{2}{\sigma_{\max}} \tag{5.246}$$

where $\sigma_{\max}$ is the largest singular value of H. Faster convergence is obtained with larger values of λ; the optimal value is:

$$\gamma_{opt} = \frac{2}{\sigma_{\max} + \sigma_{\min}} \tag{5.247}$$

Some adaptive step-size schemes have been proposed for faster convergence.

In general, the image restoration problem is a matter of estimation that, if noise is absent, takes the form:

$$H\,\mathbf{x} = \mathbf{y} \tag{5.248}$$

This is the kind of problem that has been considered in the Sect. 5.6.3 on mathematical aspects. Several image restoration methods are based on the algorithms introduced in that subsection. In practical application it is convenient to study the residual: that is, the difference between $\mathbf{y}$ and $H\,\mathbf{x}$.

In cases with small image degradation, logarithms could be approximated with a second order polynomial. Then, the KL divergence could be written as follows:

$$D_{KL}(x) = -\ln\, P(x|y) \approx \frac{1}{2} \sum_i \frac{1}{y_i} |H\,x_i - y_i|^2, \tag{5.249}$$

Therefore you have here a least-square problem, and any of the iterative methods described in the Sect. 5.6.3. could be applied. In general this approach is called '*weighted least-square method*'. It is used in medical imaging.

Image restoration can also be cast as an optimization problem. In this context, it is usual to consider constraints, and perhaps the inclusion of several targets (multi-objective optimization). An important example, pertaining to image improvement, is a set of methods based on *total variation* (TV). Absolute values of image gradients are added up, obtaining a certain value D. The criterion to minimize would be:

$$J = \|\mathbf{y} - H\,\mathbf{x}\|^2 + \lambda\, D \tag{5.250}$$

The parameter λ (a Lagrange multiplier) is adjusted to penalize gradients caused by noise, while preserving edges. This method can be applied for one-dimensional signals, and it is especially suitable for piece-wise signals. It can be interpreted as an instance of Tikhonov regularization.

Apart from the mentioned methods, let us note that the MAP could be directly obtained by solving:

$$\frac{\partial\, \ln\, P(X|Y)}{\partial X} = 0 \tag{5.251}$$

For such a problem, steepest descent, conjugate gradient, or other iterative optimization algorithms could be used. Along the iterations, it is important to take into account that pixels cannot have negative values.

Of course, heuristic optimization methods could be applied for image restoring.

5.7.6.2 Blind Image Restoration

In many real-life applications the PSF is not known a priori. Some actions should be taken for PSF determination. They could be off-line procedures, but is is also possible to combine on-line PSF determination with image restoration.

One of the proposals for this on-line combined estimation of PSF and image is based upon LRA. It is an alternation. During the first k iterations, the LRA is used to estimate the image, then, during another k iterations, the LRA is used to estimate the PSF; then, back to image estimation, and so on. The alternation uses two versions of the LRA:

$$x^{(n+1)}(i,j) = \left[\frac{y}{d * x^{(n)}} * d^A\right](i,j) \cdot x^{(n)}(i,j) \tag{5.252}$$

$$d^{(n+1)}(i,j) = \left[\frac{y}{d^{(n)} * x} * d^{(n)A}\right](i,j) \cdot d^{(n)}(i,j) \tag{5.253}$$

This approach is coincident with the so-called *'expectation maximization'* (EM) algorithm, which will be described in a next chapter.

Sometimes it is possible to have some initial idea of the PSF expression, so the estimation task would be simplified. This is denoted as *myotic*, or semi-blind, approach. For instance, if you know that a Gaussian PSF is adequate for the application at hand, then the PSF estimation problem is simplified to just a few parameters.

5.7.6.3 Gradients

Images could have some peculiar characteristics, and that may be helpful for restoration purposes. Many photographs are related to people, nature, buildings, crafts, etc. A lot of information, in these photographs, is given by edges. As it has been shown in the chapter on image processing, edges are related to high frequencies. Important hints about information contents could be obtained by studying the gradients.

For illustration purposes, an example of image gradients is presented in the next three figures. Figure 5.56 shows an image to analyze. This a small size image, good for fast computations.

Figure 5.57 shows a 3D view of the image gradients.

Sometimes detailed studies of gradients are in order. Figure 5.58 zooms around the statue nose. Using the *quiver()* MATLAB function, the local gradients are represented as arrows.

Fig. 5.56 An image to be studied

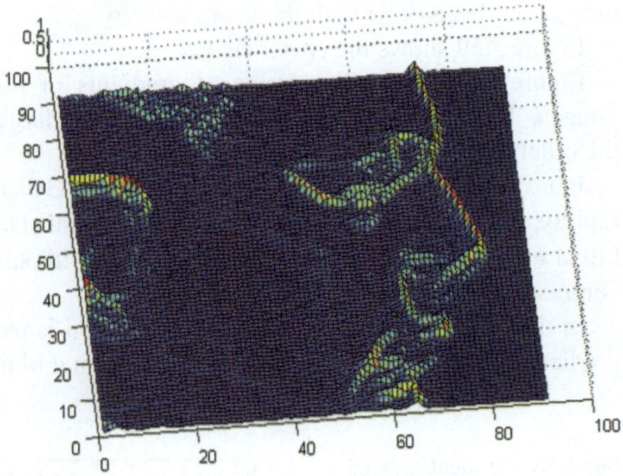

Fig. 5.57 3D view of the image gradients

Program 5.25 Image gradients example

```
% Image gradients example
%load the image
ip=imread('Head.jpg');
op=im2double(ip); %convert to float
figure(1)
imshow(ip);
figure(2)
N=90; H=42; V=40;
lp=op(V:V+N,H:H+N); %select part
[L,C]=size(lp);
[x,y]=meshgrid(1:1:L,C:-1:1);
```

```
[vx,vy]=gradient(lp);
mv=abs(vx+(j*vy));
surf(x,y,mv)
view(-5.5,86);
title('3D view of the gradients');
figure(3)
N=34; H=94; V=62;
lp=op(V:V+N,H:H+N); %select part
[L,C]=size(lp);
[x,y]=meshgrid(1:1:L,C:-1:1);
[vx,vy]=gradient(lp);
quiver(x,y,vx,vy,'k');
axis([1 N 1 N]);
title('Zoom on nose gradients');
```

Actually, the research is using histograms of gradients (HG). For instance, the HG of astronomy images and the HG of natural images have significant differences. Let us compare two images that have similar size. The computation of the HG for both images has been done with the Program 5.26.

Figure 5.59 shows the two images.

Figure 5.60 shows histograms of gradients of the natural and the astronomy images. The HG of the astronomy image has a large peak, well over the peak of the other HG.

It has been found in many examples that the HG has a sharp peak and then falls off rapidly, with long tails at larger gradients. Specialists say that many images contain broad and somewhat homogeneous surfaces, with small gradients, and a few high contrast edges that yield large gradients.

In the case of motion blur, terrestrial motion is usually horizontal. The vertical gradients could be used to estimate the horizontal blur.

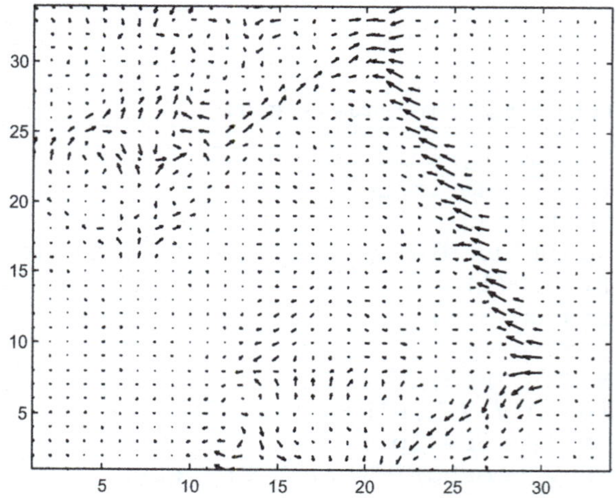

Fig. 5.58 Gradients around the nose

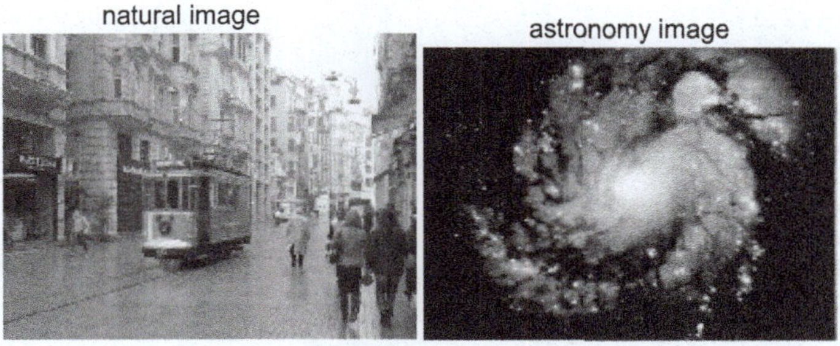

Fig. 5.59 Natural and astronomy images

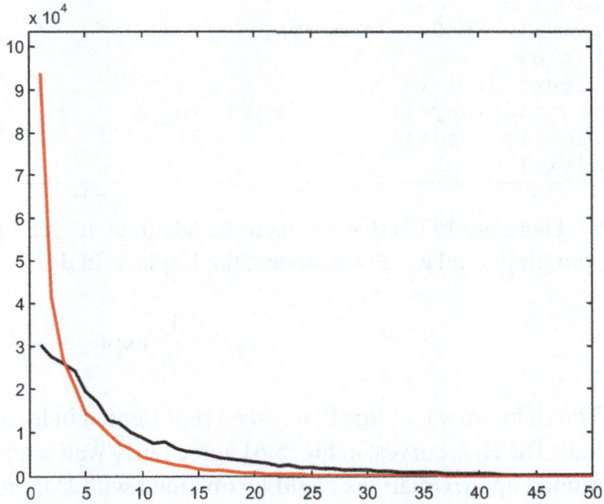

Fig. 5.60 Histograms of gradients, natural image in *black*, astronomy image in *red*

It is possible to produce high-resolution images from low-resolution ones, by using the gradients as predictors to fill vacant pixels. This is called *super-resolution*.

Gradients could also be used for film coloring, and for denoising purposes.

Let us continue with the example of the natural image and the astronomy image. Both HGs have so-called 'heavy-tails'. These tails can be studied in more detail by using logarithms. Figure 5.61 shows the logarithms of the HGs.

Program 5.26 Image HOG example

```
% Image HOG example
%load the images
ip=imread('Street1.jpg');
ip=ip(:,:,1); %select one plane
np=im2double(ip); %convert to float
ip=imread('Nebula1.jpg');
ip=ip(:,:,1); %select one plane
ap=im2double(ip); %convert to float
figure(1)
```

```
subplot(1,2,1)
imshow(np); title('natural image');
subplot(1,2,2)
imshow(ap); title('astronomy image');
figure(2)
%natural image
[vx,vy]=gradient(np); %gradients
mv=abs(vx+(j*vy)); %gradient amplitude
[hny]=hist(mv(:),100); %histogram
%astronomy image
[vx,vy]=gradient(ap); %gradients
mv=abs(vx+(j*vy)); %gradient amplitude
[hay]=hist(mv(:),100); %histogram
plot(hny,'k'); hold on;
plot(hay,'r');
axis([0 50 0 1.1*max(hay)]);
title('HoG of natural (black) and astronomy (red) images');
figure(3)
plot(log(hny+1),'k'); hold on;
plot(log(hay+1),'r')
title('log HoG of natural (black) and astronomy (red) images');
```

There are PDFs that seems to be adequate to get a probability distribution corresponding to a HG. For instance, the Laplace PDF:

$$f_x(X) = \frac{1}{2b}\exp(-\frac{|x-\mu|}{b}) \tag{5.254}$$

The right branch of this PDF, when represented in logarithms, is a descending straigt line. The HGs curves, in Fig. 5.61 agree fairly well with Laplacian straight lines. More refined approximations could be obtained with Lorentzian PDF, or Huber PDF, etc. Most of these PDFs are particular cases of the Gibb's distribution:

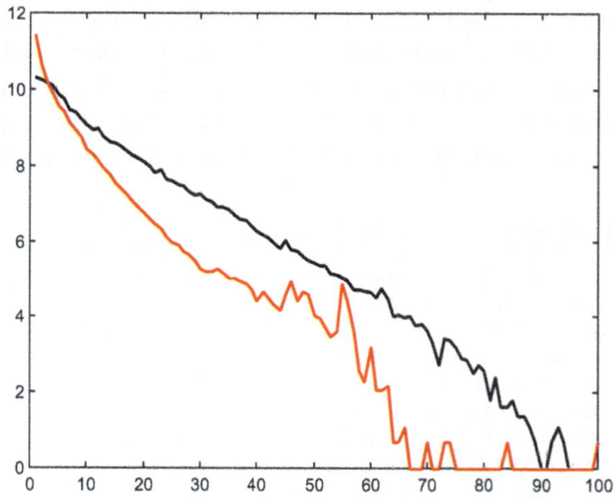

Fig. 5.61 Logarithms of the HGs; natural image in *black*, astronomy image in *red*

$$f_x(X) = \frac{1}{K}\,\exp(-\alpha\, U(x)) \qquad (5.255)$$

where $U(x)$ is a potential function.

Part of the research has proposed to fit the HG with a mix of Gaussians. Non-Gaussian can be approximated reasonably well by a finite sum of Gaussian densities.

Departing from an estimation of the gradient PDF, it is possible to integrate and obtain an estimation of the image PDF. This could be useful for image priors and posteriors, and for the determination of PSF. Also, the gradient PDF gives a measure of image sharpness that can be used for re-focusing.

Laplacian distributions will appear again in a next chapter, when analyzing principal and independent components of signals.

Let us note that some literature focuses on *histograms of oriented gradients* (HOG), with more or less emphasis on the orientation.

5.7.6.4 Priors

The Lucy-Richardson method supposes a flat prior. If this is not the case, an alternative to explore is to suppose a Gaussian prior. Let us come back to:

$$f_X(X|Y) \;\propto\; f_X(X) \;\times\; L(X|Y) \qquad (5.256)$$

Taking negative logarithms:

$$-\ln[f_X(X|Y)] \;\propto\; -\ln[f_X(X)] \;-\; \ln[L(X|Y)] \qquad (5.257)$$

Assume that there is additive Gaussian measurement noise. In this case:

$$-\ln[L(X|Y)] \;=\; \frac{1}{2\,\sigma_v^2}\,|\mathbf{y} \;-\; H\,\mathbf{x}|^2 \qquad (5.258)$$

This is a chi-squared distribution.

If the prior is Gaussian, then:

$$-\ln[f_X(X)] \;=\; \frac{1}{2\,\sigma_x^2}\,|\mathbf{x} \;-\; \mu_x|^2 \qquad (5.259)$$

Therefore, for Gaussian noise and prior, the MAP can be obtained by solving the following set of linear equations (the result of differentials equal to zero):

$$-\;\frac{1}{\sigma_v^2}\,H^T\,(\mathbf{y} \;-\; H\,\mathbf{x}) + \;\frac{1}{\sigma_x^2}\,(\mathbf{x} \;-\; \mu_x) = 0 \qquad (5.260)$$

Specialists remark that the effect of the prior is to drive the MAP solution towards $\boldsymbol{\mu}_x$ and away from the ML solution. This is in consonance with the example introduced at the beginning of this section (related with the Wiener filter).

In more general terms, with a simpler notation ($-\ln[f_x()]$ is replaced by $\phi()$), the MAP equation could be written as follows:

$$\phi_{MAP}(\mathbf{x}) = \phi_{ML}(\mathbf{x}) + \mu\,\phi_{PRIOR}(\mathbf{x}) \tag{5.261}$$

where μ is called the *hyper-parameter*. Again, one finds an optimization problem. The ϕ_{PRIOR} term could be interpreted as a regularization term. With this perspective, there is an additional issue: to specify a suitable hiper-parameter value. This could be done, for instance, using *L-curves*.

Let us focus on priors proposed in the literature.

The *maximum entropy method* (MEM) proposes to use:

$$f_x(X) = \exp(-\alpha\, H(X)) \tag{5.262}$$

Or, equivalently:

$$\phi_{PRIOR}(\mathbf{x}) = \alpha\, H(\mathbf{x}) \tag{5.263}$$

where $H()$ is the entropy. Obviously, the last two expressions correspond to Gibb's distribution.

Some of the commonly used entropy functions are:

$$H(X) = -\sum_{i,j} \ln\,[x(i,j)] \tag{5.264}$$

$$H(X) = -\sum_{i,j} x(i,j)\cdot \ln\,[x(i,j)] \tag{5.265}$$

$$H(X) = \sum_{i,j} x(i,j) - M(i,j) - x(i,j)\cdot \ln\,[x(i,j)/M(i,j)] \tag{5.266}$$

where M is a model: usually a flat image.

As it has been said before, other proposed priors are based on Laplace, Lorentzian, or Huber, etc., distributions.

Another type of priors are obtained considering contiguous pixels, much in the way gradients are computed. For instance the so-called simultaneous auto-regressive model (SAR):

$$f_x(X) = \exp(-\frac{\alpha}{2}\,\Gamma) \tag{5.267}$$

where:

$$\Gamma = \sum_{i,j} x(i,j) - \frac{1}{4}[x(i,j+1) + x(i,j-1) + x(i+1,j) + x(i-1,j)] \tag{5.268}$$

The conditional autoregressive model (CAR) uses another expression for Γ:

$$\Gamma = \sum_{i,j} [x(i,j) - x(i,j+1)]^2 + [x(i,j) - x(i+1,j)]^2 \tag{5.269}$$

Other priors are being proposed, according with the application needs.

5.7.6.5 Improvements and Extensions

Some of the deblurring examples given in this chapter show noticeable *ringing*. This is due to the use of the DFT and the presence of edges, like the picture borders. The ISP toolbox of MATLAB includes the function *edgetaper()*, which is used to preprocess the image before deblurring. It removes high-frequency drop-off at the image boundaries.

It has been realized that the Lucy-Richardson algorithm converges quite slowly. Some acceleration methods have been proposed. For instance, a line search, or any suitable extrapolation procedure, could be applied in each LRA iteration, to determine the optimal step size for that iteration (this is a sort of adaptation). Another idea is to use the following expression:

$$\mathbf{x}^{(n+1)} = (H^T \left[\frac{\mathbf{y}}{H\,\mathbf{x}}\right]^q)\,\mathbf{x}^{(n)} \tag{5.270}$$

Where the exponent q takes values between 1 and 3. More than 3 causes instability. In continuation with this idea, adaptive schemes have been devised to start with $q = 3$, and gradually decrease this value towards 1 according with the convergence.

Some modifications of the LRA have been introduced to take into account 'a priori' knowledge about image points, like for example distant stars in astronomy. In this case, the real image is divided into two parts $\mathbf{x}^p + \mathbf{x}^m$, where $\mathbf{x}^p$ contain the points. The LRA deals in parallel with the two parts. Ringing caused by $\mathbf{x}^p$ is minimized. Entropic priors are used for $\mathbf{x}^m$. This algorithm is called PLUCY. Extensions of PLUCY are CPLUCY and GIRA, [44].

Another algorithm used in astronomy is ISRA ('image space reconstruction algorithm'). It is similar to LRA, but using Gaussian prior.

Coming back to motion deblurring, there is an interesting proposal [45] for including projective motion paths in Richardson-Lucy algorithms, with good results.

5.8 Observers

State variable models could be used to give information on the system states based on inputs and outputs. This is the mission of so-called *'observers'*. They are useful when some or all the internal states of a system are difficult to measure in real time. Some industrial applications use the term *'software sensor'* for such systems.

This section focuses on a fundamental contribution in this area, which is the Luenberger observer, [30]. It belongs to a linear systems framework. Nowadays there are also several proposals of observers for non linear systems.

5.8.1 *The Luenberger Observer*

The situation considered is represented by the diagram in Fig. 5.62. There is a linear system, with a certain number of state variables and some inputs and outputs. The observer is another linear system which reads the inputs and the outputs of the system. The observer gives an estimate of the system states.

The state variable model of the system is:

$$\begin{aligned} \bar{x}(n+1) &= A\bar{x}(n) + B\bar{u}(n) \\ \bar{y}(n) &= C\,\bar{x}(n) \end{aligned} \tag{5.271}$$

Luenberger proposed the following observer:

$$\bar{x}_e(n+1) = A\,\bar{x}_e(n) + B\,\bar{u}(n) + K\,(\bar{y}(n) - C\,\bar{x}_e(n)) \tag{5.272}$$

where $\mathbf{x}_e$ is the observer's estimation of the system state.

The observer uses the same matrices of the system model. The last term to the right—the term with K—is intended for estimation improvement action.

It may happen that the initial conditions of the system were unknown. For this and other reasons, the observer's estimation may differ from the system state. An

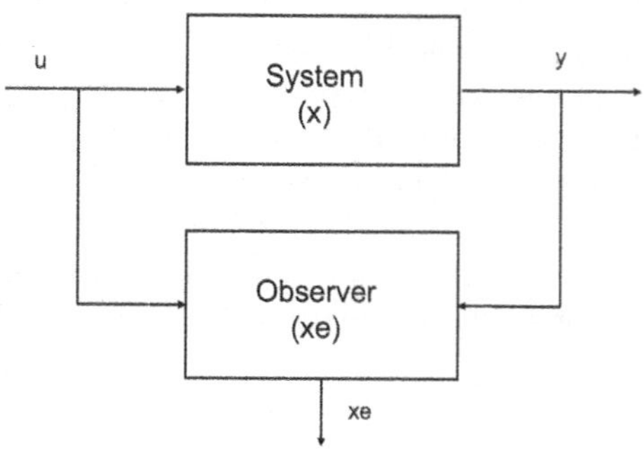

Fig. 5.62 Observer concept

estimation error can be defined as follows:

$$\mathbf{e}(n) = \mathbf{y}(n) - C\,\mathbf{x}_e \tag{5.273}$$

Clearly, the last term to the right in the observer's equation tries to improve the next estimation in function of the current error. The designer has to choose a value for K. With some algebra it is easy to show that:

$$\mathbf{e}(n+1) = (A - K\,C)\,\mathbf{e}(n) \tag{5.274}$$

According with this equation, the error dynamics is governed by K: in other words, the error will converge to zero quickly or slowly depending on K. A prudential decision should be taken in function of the confidence about the model used and the input-output measurements.

An intuitive example is a two-tank system, as depicted in Fig. 5.63 (and in the first book). Both tanks communicate through a pipe with resistance R_1. The input is liquid that falls into tank1, the output is liquid that leaves tank2 through a pipe with resistance R_2.

Next equations are a basic model of the system:

$$\begin{aligned} A_1 \frac{dh_1}{dt} &= \frac{1}{R_1}(h_2 - h_1) + u(t) \\ A_2 \frac{dh_2}{dt} &= -\frac{1}{R_1}(h_2 - h_1) - \frac{1}{R_2} h_2 \end{aligned} \tag{5.275}$$

where $A_1 = 1$, $A_2 = 1$, $R_1 = 0.5$, $R_2 = 0.4$.

From these equations, we can write the state space model using the following matrices:

$$A = \begin{pmatrix} -\frac{1}{R_1 A_1} & \frac{1}{R_1 A_1} \\ \frac{1}{R_1 A_2} & \frac{1}{A_2}(\frac{1}{R_1} + \frac{1}{R_2}) \end{pmatrix} \quad B = \begin{pmatrix} \frac{1}{A_1} \\ 0 \end{pmatrix}$$

$$C = (0\ 1) \tag{5.276}$$

The states are $x_1 = h_1(t)$ and $x_2 = h_2(t)$.

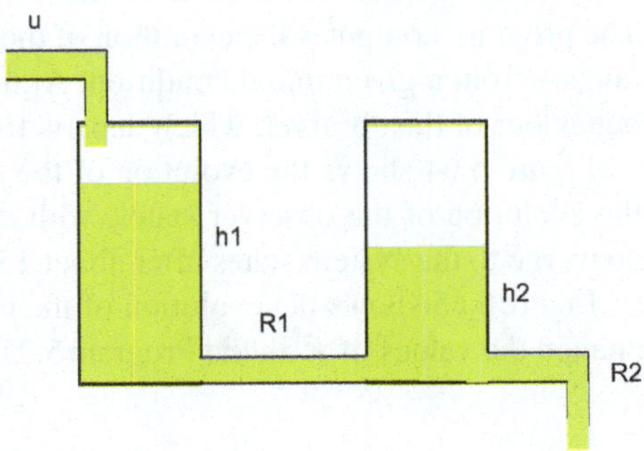

Fig. 5.63 A two-tank system example

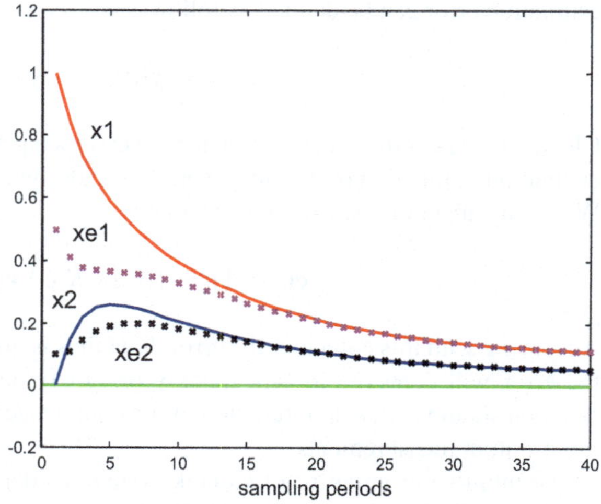

Fig. 5.64 System and observer state evolution

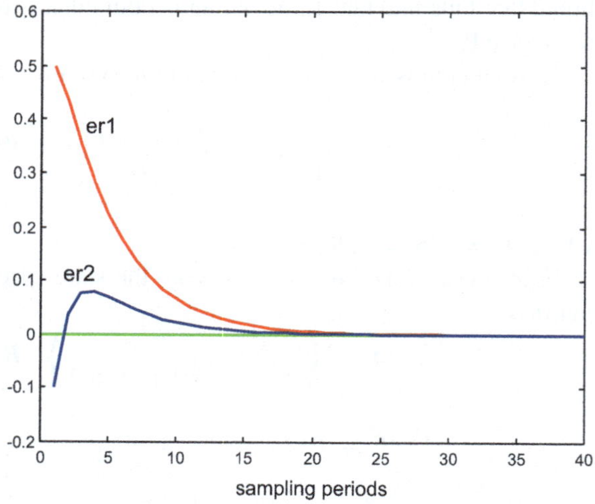

Fig. 5.65 Observation error evolution

Now an observer is attached to the system. Program 5.27 includes the observer. The program computes the evolution of the two-tank system—the height of the two tanks—from a given initial condition. At the same time, the program computes the behaviour of the observer, which departs from another initial condition.

Figure 5.64 shows the evolution of the system states, in continuous curves, and the evolution of the observer states, with dotted curves. The states of the observer converge to the system states after about 15 samples.

Figure 5.65 shows the evolution of the error along time. The reader is invited to change the values of K in the Program 5.27 to study changes in the error dynamics.

Program 5.27 observer example

```
%observer example
%state space system model (2 tank system):
A1=1; A2=1; R1=0.5; R2=0.4;
A=[-1/(R1*A1) 1/(R1*A1); 1/(R1*A2) -(1/A2)*((1/R1)+(1/R2))];
B=[1/A1; 0]; C=[0 1]; D=0;
Ts=0.1; %sampling period
csys=ss(A,B,C,D); %setting the continuous time model
dsys=c2d(csys,Ts,'zoh'); %getting the discrete-time model
[a,b,c,d]=ssdata(dsys); %retrieves discrete-time model matrices
% system simulation preparation
Nf=40; %simulation horizon
x1=zeros(1,Nf); % for x1(n) record
x2=zeros(1,Nf); % for x2(n) record
y=zeros(1,Nf); % for y(n) record
x=[1;0]; % state vector with initial tank levels
u=0.1; %constant input
% observer simulation preparation
K=[0.3; 0.3]; %observer constants
er=zeros(2,Nf); % for error record
xe1=zeros(1,Nf); % for xe1(n) record
xe2=zeros(1,Nf); % for xe2(n) record
xe=[0.5; 0.1]; % observer state vector with initial values
%behaviour of the system and the observer after initial state
% with constant input u
for nn=1:Nf,
  x1(nn)=x(1); x2(nn)=x(2); %recording the system state
  y(nn)=c*x; %recording the system output
  xe1(nn)=xe(1); xe2(nn)=xe(2); %recording the observer state
  er(:,nn)=x-xe; %recording the error
  xn=(a*x)+(b*u); %next system state
  x=xn; %system state actualization
  xen=(a*xe)+(b*u)+(K*(y(nn)-(c*xe))); %next observer state
  xe=xen; %observer state actualization
end;
% display of states evolution
figure(1)
plot([0 Nf],[0 0],'g'); hold on; %horizontal axis
plot([0 0],[-0.2 1.2],'k'); %vertical axis
plot(x1,'r'); %plots x1
plot(x2,'b'); %plots x2
plot(xe1,'mx'); %plots xe1
plot(xe2,'kx'); %plots xe2
xlabel('sampling periods');
title('system and observer states');
% display of error evolution
figure(2)
plot([0 Nf],[0 0],'g'); hold on; %horizontal axis
plot([0 0],[-0.2 0.6],'k'); %vertical axis
plot(er(1,:),'r'); %plots x1 error
plot(er(2,:),'b'); %plots x2 error
xlabel('sampling periods');
title('observation error');
```

5.8.2 Noises

Most times, at system level, there will be process and measurement noises. It is convenient to have an idea of the behavior of the observer in such circumstances.

A program, included in the Appendix A, has been developed to study this scenario. Next figures have been obtained with this program.

Figure 5.66 shows the evolution of the system and the observer states, starting from the same initial conditions as in the previous subsection.

Figure 5.67 shows the evolution of the observation error.

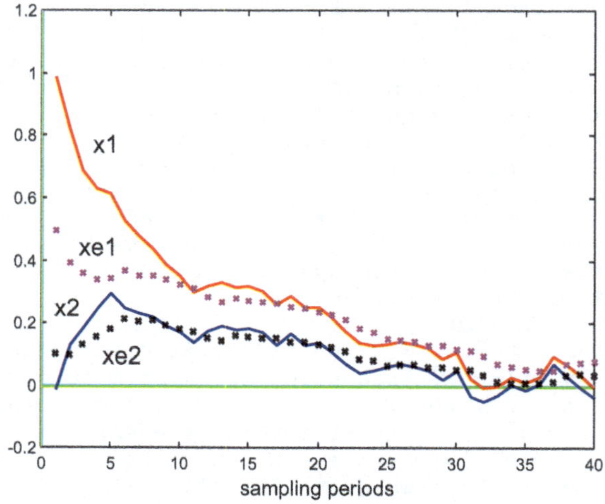

Fig. 5.66 System and observer state evolution in presence of noise

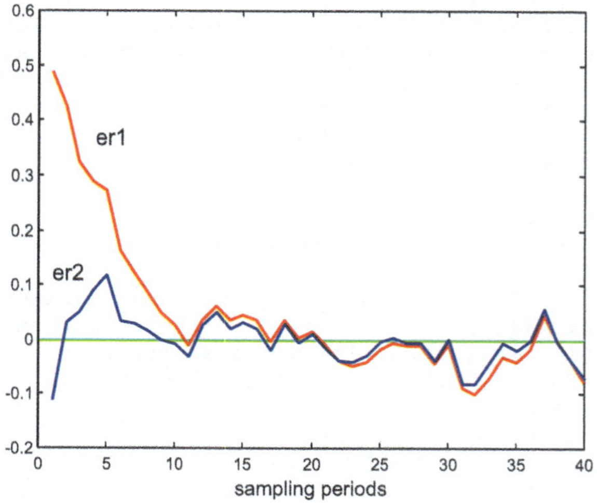

Fig. 5.67 Observation error evolution

5.9 Experiments

5.9.1 Eigenvalues of Signals

It is important to be aware that the matrix R_y depends on the filter input signal. In case of stationary signals, the matrix is constant along time: this is the ideal situation considered for the Wiener filter. However, in practice most signals are non-stationary, but it may be possible to consider reasonably stationary signal segments and use a block by block filtering approach.

The following examples are intended to show how the eigenvalues of R_y evolve according with the signal characteristics.

5.9.1.1 Quasi-white Noise

It is difficult to generate a "perfect" white noise. For such a noise, the auto-correlation sequence of the signal should be 0,0…0,1,0,…0,0. Only a one surrounded by zeros. In consequence, the matrix R_y should be diagonal, with ones in the diagonal. Clearly, all eigenvalues should be one.

The Program 5.28 selects a Gaussian noise as generated by MATLAB, and generates 200 consecutive blocks of 50 samples of this signal. Then, the program computes the matrix R_y for each block, and its eigenvalues. Figure 5.68 depicts the 200 sets of eigenvalues along time. Should the noise be "perfect-white", all eigenvalues would be one. But, it is clear, this is not the case.

It is suggested for the reader to visualize the auto-correlation sequence to see how close is to the ideal. By the way, notice that in the program the auto-correlation sequence has been normalized, dividing by N_d.

The program also prints a series of interesting values, which are computed for the last block of the signal. First, the mean of the N_d eigenvalues, which is also given by $R_y(1, 1)$. Second, the "energy in the filter", which can be obtained with the trace of R_y, or adding up the eigenvalues, or computing the energy of the N_d signal samples inside the filter. Finally, the program prints the condition number of R_y.

Program 5.28 Record of eigenvalues of noise

```
%Record of Eigenvalues of Noise
Nf=200; %number of data-frames
Nd=50; %number of data per frame
%signal
y=randn(Nf*Nd,1); %data
%for recording
veig=zeros(Nd,Nf);
%by frames
for nn=0:(Nf-1),
  aux1=1+(nn*Nd); aux2=(nn+1)*Nd;
  vy=y(aux1:aux2);
```

```
  syy=xcorr(vy)/Nd; %symmetrical auto-correlation sequence
  Ry=toeplitz(syy(Nd:(2*Nd)-1)); % auto-correlation matrix
  E=eig(Ry); %eigenvalues
  veig(:,nn+1)=E;
end
%display
figure(1)
nn=1:Nf;
plot(nn,veig(:,nn),'.k');
title('eigenvalues along data blocks');
xlabel('number of data block');
%print
disp('mean eigenvalue')
mean(E)
Ry(1,1)
disp('energy in the filter')
sum(E)
sum(vy.^2)
trace(Ry)
disp('cond(Ry)');
cond(Ry)
```

5.9.1.2 Music

The next case is a short piece of instrumental music. It is a Bag Pipe. The reader is invited to hear this signal.

The Program 5.29 obtains the eigenvalues of the consecutive blocks of this signal (each block has 100 samples). This program only differs from the previous one by the first sentences, devoted to have a signal to be analyzed. Figure 5.69 shows the

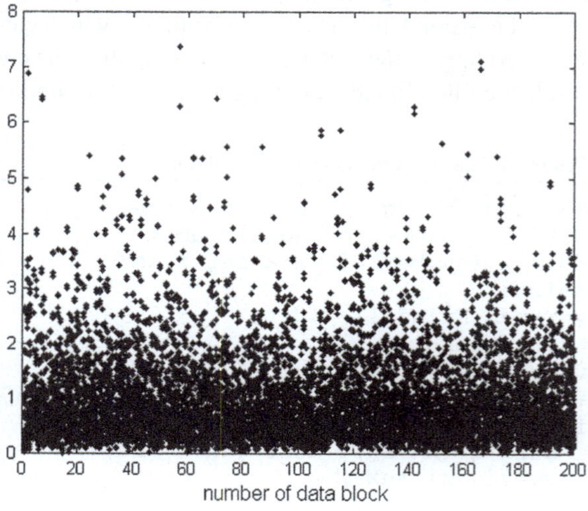

Fig. 5.68 Eigenvalues along data blocks. "White" noise

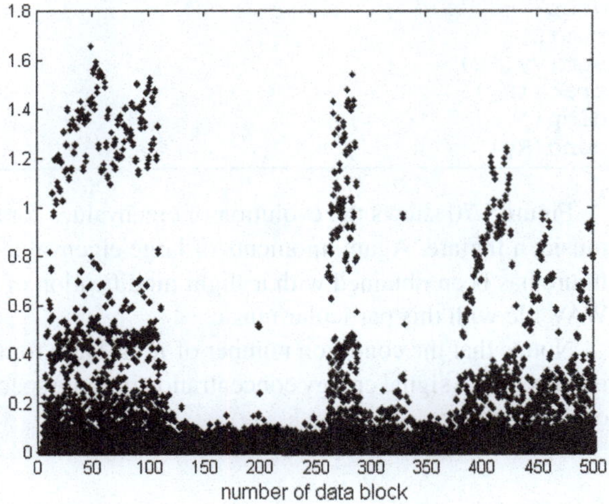

Fig. 5.69 Eigenvalues along data blocks Bag pipes

results. Corresponding to strokes in the music, there are moments with high spread of eigenvalues.

Program 5.29 Record of eigenvalues of music

```
% Record of Eigenvalues of Music
%signal
[y1,fs1]=wavread('BagPipes.wav'); %read wav file
Nf=500; %number of data frames
Nd=100; %number of data per frame
y=y1(1:(Nf*Nd)); %signal segment
%for recording
veig=zeros(Nd,Nf);
%by frames
for nn=0:(Nf-1),
  aux1=1+(nn*Nd); aux2=(nn+1)*Nd;
  vy=y(aux1:aux2);
  syy=xcorr(vy)/Nd; %symmetrical auto-correlation sequence
  Ry=toeplitz(syy(Nd:(2*Nd)-1)); % auto-correlation matrix
  E=eig(Ry); %eigenvalues
  veig(:,nn+1)=E;
end
%display
figure(1)
nn=1:Nf;
plot(nn,veig(:,nn),'.k');
title('eigenvalues along data blocks');
xlabel('number of data block');
%print
disp('last data frame info:');
disp('mean eigenvalue')
mean(E)
Ry(1,1)
```

```
disp('energy in the filter')
sum(E)
sum(vy.^2)
trace(Ry)
disp('cond(Ry)');
cond(Ry)
```

Figure 5.70 shows the evolution of eigenvalues for another piece of instrumental music: a fanfare. Again, moments of large eigenvalue spread can be observed. The figure has been obtained with a slight modification of the Program 5.29, to load the WAV file with this particular music.

Notice that the condition number of R_y is large for the two music examples. This is related with signal energy concentration in some specific frequencies (high tonality of the signal).

5.9.2 Water

Finally, the sound of water is considered. By a simple change of the Program 5.29 a WAV file with water sound is loaded. Figure 5.71 shows the evolution of eigenvalues along time. It seems a good noise.

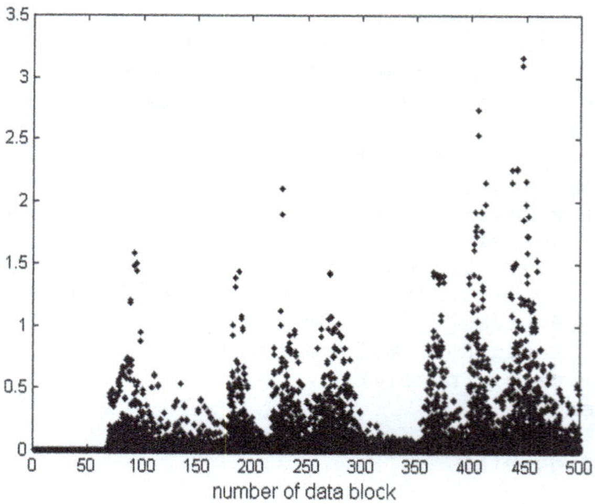

Fig. 5.70 Eigenvalues along data blocks. Fanfare

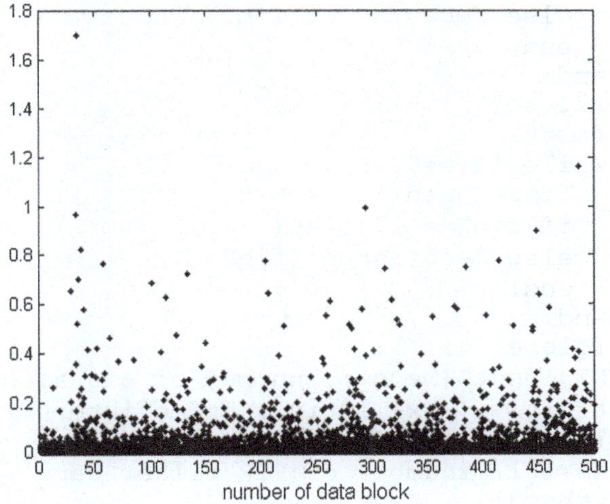

Fig. 5.71 Eigenvalues along data blocks. Water

5.9.3 *Fetal Heart Rate Monitoring*

Electrocardiograms (ECG) are important in medical contexts. The ECG signals could suffer from interferences, outliers, bad contacts, etc. Indeed good filtering is desired, and many efforts have been invested on it.

The example we are to consider is somewhat different from cleaning the ECG. It is the case of a two hearts being mixed in the ECG. We wish to separate two ECGs. An electrode on the thorax gets the mother's ECG. A second electrode on the abdomen gets the mixed ECGs.

In this case we use the simple method. A kind of model of mother's ECG transmission from thorax to abdomen is obtained. The model is inverted to obtain from the mixed ECG an estimation of the corresponding mother's ECG. Then, by simple subtraction one gets an estimation of the fetal ECG.

Next figure shows the mother's ECG and the mixed ECG (Fig. 5.72).

The MATLAB code for obtaining the fetal ECG estimate is simple, as it can be seen in the Program 5.30. The transmission model is denoted as G. Figure 5.73 shows the result: the fetal ECG, which pace is faster than of the mother's ECG.

Program 5.30 ECG example

```
% ECG example
fs=1000; %samplig frequency in Hz
tiv=1/fs; %time interval between samples
%read ECG files
fer=0;
while fer==0,
   fid2=fopen('ecg_mother.txt','r');
   if fid2==-1, disp('read error')
```

```
    else Mecg=fscanf(fid2,'%f \r\n'); fer=1;
    end;
end;
fclose('all');
fer=0;
while fer==0,
    fid2=fopen('ecg_both.txt','r');
    if fid2==-1, disp('read error')
    else Becg=fscanf(fid2,'%f \r\n'); fer=1;
    end;
end;
fclose('all');
Nx=length(Mecg); %number of signal samples
t=0:tiv:((Nx-1)*tiv); %time intervals set
%prepare for filtering
Nh=121; %number of FIR filter taps
y=Becg;
x=Mecg;
G=convmtx(x,Nh); %model of Mecg(T) --> Mecg(A)
G=G((Nh-1)/2 + [1:Nx],:); %for zero delay
xe=G\y; %estimated Mecg using Becg
f=y-(G*xe); %fetal ECG
%display----------
figure(1)
subplot(2,1,1)
plot(t,x,'k'); title('Thorax ECG');
subplot(2,1,2)
plot(t,y,'k'); title('Abdominal ECG');
xlabel('s');
figure(2)
plot(t,f,'k'); title ('fetal ECG');
xlabel('s');
```

Notice that this experiment involves matrix inversion. The results could be spoiled, in real applications, by noise.

5.10 Some Motivating Applications

Most of the sections in this chapter immediately suggest possible applications, like echo suppression, modem equalization, noise cancelling, image improvement, etc. However, we would like to mention some examples of application that suppose a grain of ingenuity, and which connect with modern activities.

A method for noise reduction for dual-microphone mobile phones is described in [18]; the method is based on the Wiener filter and a model of the noise.

The use of equalizers in acoustic modems, for air and underwater applications, is included in the Master Thesis [21].

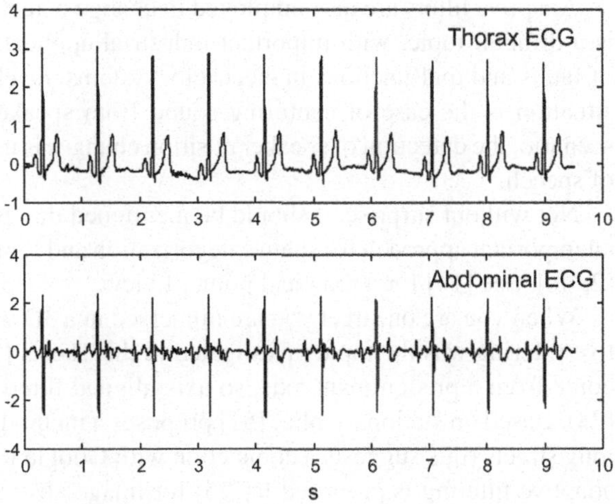

Fig. 5.72 Mother's ECG, and mixed ECG

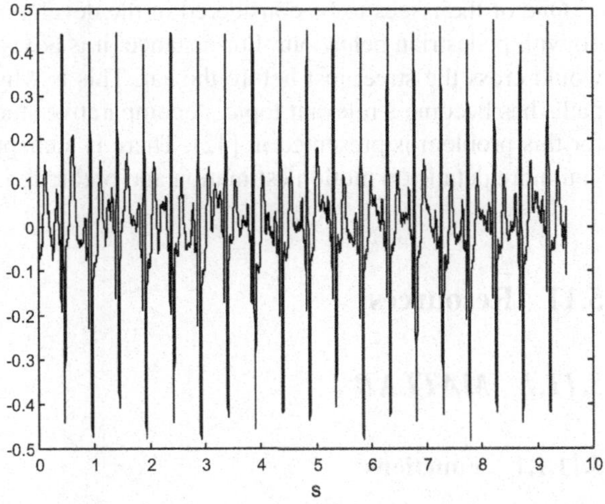

Fig. 5.73 The fetal ECG

Combinations of wavelets and Wiener filtering have been proposed for medical applications, like [35] for electrocardiographic (ECG) signal denoising, or [49] for tomographic image reconstruction.

Speech communication is a traditional application field, with many associated research publications. Some new proposals are the use of multi-channel Wiener filters, like in the Thesis [3], or the development of Mel-scale Wiener filters for comfortable communication inside an automobile [22]. See the article [7] for new insights on speech noise reduction.

Adaptive filters are also employed in change detection applications, [4, 12]. This is a difficult topic, with important industrial applications (detection and diagnosis of faults and malfunctions in electricity systems, machinery, etc.). A quite different situation is the case of capturing sound from speakers that move in a room or a scenario, the detection of speaker position changes is used in [48] for dereverberation of speech.

Not without surprise, it should be mentioned that Bayesian filtering is becoming an important approach for spam categorization and for the elimination of junk e-mail, [2, 40]. See [14] for a practical point of view.

When you are on street you are immersed in a 3D world. Multidimensional adaptive filtering was the topic investigated in the Thesis [13]. Scene illumination could come from a predominant axis, so axis-aligned filtering would become opportune, [33]. Based on stationary blur, [51] proposes a method for estimation of the depth of long streets (this suggests a connection with Google maps street views). A structure adaptive filtering is proposed in [25] for image abstraction, which produces a kind of posterizing.

One of the issues to be considered in the development of driver-less cars has to do with pedestrian behaviour. For instance, it is advisable to predict if a pedestrian would cross the street just before the car. This is why the prediction of pedestrian paths has become a relevant topic. A comparative study of using Bayesian filtering for this problem is presented in [42]. There is a chapter in this book that will give you more details on motion estimation and prediction.

5.11 Resources

5.11.1 MATLAB

5.11.1.1 Functions

The Image Processing Toolbox offers suitable functions for important aspects of the chapter:

fspecial()
wiener2()—2D adaptive noise removal filtering
deconvwr()—Deblur image using Wiener filter
deconvblind()—Deblur image using blind deconvolution
deconvreg()—Deblur image using regularized filter
deconvlucy()—Deblur image using Lucy-Richardson method
psf2otf(), otf2psf()—Convert between PSF and OTF
edgetaper()—Taper discontinuities along image edges

5.11.1.2 Toolboxes

- Regularization Tools:
  http://www.imm.dtu.dk/~pcha/Regutools/
- Image Reconstruction Toolbox (Including NUFFT):
  http://web.eecs.umich.edu/~fessler/code/mri.htm
- RestoreTools:
  http://www.mathcs.emory.edu/~nagy/RestoreTools/
- Biosig: biomedical signal processing (EEG, ECG, etc.):
  http://biosig.sourceforge.net/
- Change detection and adaptive filtering toolbox:
  http://users.isy.liu.se/en/rt/fredrik/sigmoid/

5.11.1.3 Matlab Code

- Deblurring Images:
  http://www2.compute.dtu.dk/~pcha/HNO/
- Removing camera shake from a single image:
  http://cs.nyu.edu/~fergus/research/deblur.html
- Matlab code for "Non-uniform Deblurring for Shaken Images":
  http://www.di.ens.fr/willow/research/deblurring/deblur_nonuniform_v0.1.html
- A General Framework for Regularized, Similarity-based Image Restoration (Univ. Santa Cruz):
  https://users.soe.ucsc.edu/~aminkh/KernelRestoration.html
- Deblurring Text Images via L0-Regularized Intensity and Gradient Prior:
  https://eng.ucmerced.edu/people/zhu/CVPR14_text.html
- Computer Vision Softwares:
  http://202.118.75.4/LiMing/CV_software.html

5.11.2 Internet

5.11.2.1 Web Sites

- Examples of deconvtv-image deblurring and denoising:
  http://people.seas.harvard.edu/~schan/deconvtv_folder/deconvtv_image.html
- Patch-based Locally Optimal Wiener Filtering for
  Image Denoising (PLOW) (Univ. Santa Cruz):
  https://users.soe.ucsc.edu/~priyam/PLOW/
- Speech coding:
  http://markus-hauenstein.de/sigpro/codec/codec.shtml

- Efficient Deblurring for Shaken and Partially Saturated Images: http://www.di.ens.fr/willow/research/saturation/
- Changyin Zhou: http://www1.cs.columbia.edu/~changyin/
- Jiaya JIA: http://www.cse.cuhk.edu.hk/leojia/index.html

5.11.2.2 Link Lists

- Inverse problems: http://web.iitd.ac.in/~vvksrini/links.html
- Deconvolution software: http://www.deconvolve.net/DNLinks.html

References

1. T. Aboulnasr, K. Mayyas, A robust variable step-size LMS-type algorithm: analysis and simulations. IEEE T. Sig. Proc. **45**(3), 631–639 (1997)
2. I. Androutsopoulos, J. Koutsias, K.V. Chandrinos, G. Paliouras, C.D. Spyropoulos, *An evaluation of naive Bayesian anti-spam filtering*. Software and Knowledge Engineering Laboratory, National Centre for Scientific Research (Demokritos), Athens, Greece, 2000. arXiv preprint arXiv:cs/0006013
3. M. Awais, Multichannel Wiener filtering for speech enhancement in modulation domain. Master's thesis, Blekinge Institute of Technology, Sweden, 2010
4. M. Basseville, I.V. Nikiforov, *Detection of Abrupt Changes: Theory and Application*, vol. 104 (Prentice Hall, Englewood Cliffs, 1993)
5. J. Benesty, H. Rey, L.R. Vega, S. Tressens, A nonparametric VSS NLMS algorithm. IEEE Signal Process. Lett. **13**(10), 581–584 (2006)
6. A. Carusone, D.A. Johns, Analogue adaptive filters: past and present. IEE Proc. Circuits Devices Syst. **147**(1), 82–90, (2000)
7. J. Chen, J. Benesty, Y. Huang, S. Doclo, New insights into the noise reduction Wiener filter. IEEE Trans. Audio Speech Lang. Process. **14**(4), 1218–1234 (2006)
8. Y.S. Choi, H.C. Shin, W.J. Song, Robust regularization for normalized LMS algorithms. IEEE T. Circuits Syst. II **53**(8), 627–631 (2006)
9. S.J. Elliot, P.A. Nelson, Active noise control. IEEE Signal Process. Mgz. **10**(4), 12–35 (1993)
10. D. Erdogmus, J.C. Principe, From linear adaptive filtering to nonlinear information processing. IEEE Signal Process. Mgz., 14–33 (2006)
11. R. Fergus, B. Singh, A. Hertzmann, S. Roweis, W.T. Freeman, Removing camera shake from a single photograph. ACM. Trans. Graphics **25**(3) (2006)
12. F. Gustafsson, F. Gustafsson, *Adaptive Filtering and Change Detection*. (Wiley, 2000)
13. L. Haglund, *Adaptive Multidimensional Filtering*. PhD thesis, Linköpings University, (1991)
14. R. Haskins, Bayesian spam-filtering techniques. LOGIN: The USENIX Mag. **28**(3), 1–7 (2003)
15. S. Haykin, *Adaptive Filter Theory*. (Prentice Hall, 2002)
16. S. Haykin, B. Widrow, *Least-Mean-Square Adaptive Filters*. (Wiley, 2003)
17. J.H. Husøy, M.S.E. Abadi, Unified approach to adaptive filters and their performance. IET Sig. Process. **2**(2), 97–109 (2008)
18. M. Jeub, C. Herglotz, C. Nelke, C. Beaugeant, P. Vary, in *Noise Reduction for Dual-Microphone Mobile Phones Exploiting Power Level Differences*. Proceedings of IEEE International Conference on Acoustics, Speech and Signal Processing (ICASSP), (2012), pp. 1693–1696
19. T. Kailath, A view of three decades of linear filtering theory. IEEE T. Inf. Theory **20**(146) (1974)

20. M. Kamenetsky, B. Widrow, in *A Variable Leaky LMS Adaptive Algorithm*, in Proceedings of 38th IEEE Asilomar Conference on Signals, Systems and Computers, vol. 1 (2004), pp. 125–128
21. E. Karlsson, Software acoustic modem. Master's thesis, Linköpings University (2013)
22. H.S. Kim, Y.M. Cho, H.J. Kim, Speech enhancement via mel-scale Wiener filtering with a frequency-wise voice activity detector. J. Mech. Sci. Technol. **21**(5), 708–722 (2007)
23. S.M. Kuo, D.R. Morgan, Active noise control: a tutorial review. Proc. IEEE **87**(6), 943–973 (1999)
24. R.H. Kwong, E.W. Johnston, A variable step size LMS algorithm. IEEE T. Signal Process. **40**(7), 1633–1642 (1992)
25. J.E. Kyprianidis, J. Döllner, *Image Abstraction By Structure Adaptive Filtering*. (Theory and Practice of, Computer Graphics, 2008), pp. 51–58
26. J. Lee, J-W Chen, H-C. Huang, Performance comparison of variable step-size NLMS algorithms, in *Proceedings of the World Congress on Engineering and Computer Science*, vol. 1, (2009)
27. J. Lee, H.C. Huang, Y.N. Yang, The generalized square-errorregularized LMS algorithm, in *Proceedings of WCECS* (2008), pp. 157–160
28. A. Levin, P. Sand, T.S. Cho, F. Durand, W.T. Freeman, Motion-invariant photography. ACM. Trans. Graphics **27**(3) (2008)
29. L.B. Lucy, An iterative method for the rectification of observed distributions. Astronomical J. **79**, 745–754 (1974)
30. D.G. Luenberger, *Introduction to Dynamic Systems: Theory, Models, and Applications*. (Sons, 1979)
31. G. Malik, A.S. Sappal, Adaptive equalization algorithms: An overview. Int. J. Adv. Comput. Sci. Appl. **2**(3), 62–67 (2011)
32. D.P. Mandic, A generalized normalized gradient descent algorithm. IEEE Signal Process. Lett. **11**(2), 115–118 (2004)
33. S.U. Mehta, B. Wang, R. Ramamoorthi, F. Durand, Axis-aligned filtering for interactive physically-based diffuse indirect lighting. ACM Trans. Graphics (TOG) **32**(4), 1–11 (2013)
34. R. Molina, J. Núñez, F. Cortijo, J. Mateos, Image restoration in astronomy: A Bayesian review. IEEE Signal Process. Mgz. (2001), pp. 12–29
35. N. Nikolaev, Z. Nikolov, A. Gotchev, K. Egiazarian, *Wavelet Domain Wiener Filtering for ECG Denoising Using Improved Signal Estimate*, in Proceedings of IEEE International Conference on Acoustics, Speech, and Signal Processing, ICASSP'00, vol. 6, (2000), pp. 3578–3581
36. A.D. Poularikas, Z.M. Ramadan, *Adaptive Filtering Primer with MATLAB* (Taylor & Francis, CRC, 2006)
37. S.U. Qureshi, Adaptive equalization. Proc. IEEE **73**(9), 1349–1387 (1985)
38. A. Ribes, F. Schmitt, Linear inverse problems in imaging. IEEE Signal Process. Mgz. **84** (2008)
39. W.H. Richardson, Bayesian-based iterative method of image restoration. J. Optical Soc. Am. **62**, 55–59 (1972)
40. M. Sahami, S. Dumais, D. Heckerman, E. Horvitz, *A Bayesian Approach to Filtering Junk e-mail*. Stanford University, (1998). http://ai.stanford.edu/users/sahami/papers-dir/spam.pdf
41. A.H. Sayed, *Fundamentals of Adaptive Filtering*. (Wiley, 2003)
42. N. Schneider, D.M. Gavrila, Pedestrian path prediction with recursive Bayesian filters: A comparative study, in *Pattern Recognition*. (Springer Berlin Heidelberg, 2013), pp. 174–183
43. H.C. Shin, A.H. Sayed, W.J. Song, Variable step-size NLMS and affine projection algorithms. IEEE Signal Process. Lett. **11**(2), 132–135 (2004)
44. J.L. Starck, E. Pantin, Deconvolution in astronomy: a review. Publ. Astron. Soc. Pac. **114**, 1051–1069 (2002)
45. Y-W Tai, P. Tan, L. Gao, M.S. Brown, Richardson-Lucy deblurring from scenes under projective motion path. IEEE T. Pattern Anal. Mach. Intell. **33**(8), 1603–1618 (2011)
46. T. Van Waterschoot, M. Moonen, Fifty years of acoustic feedback control: State of the art and future challenges. Proc. IEEE **99**(2), 288–327 (2010)
47. B. Widrow, S. Steam, *Adaptive Signal Processing*. (Prentice Hall, 1985)

48. T. Yoshioka, H. Tachibana, T. Nakatani, M. Miyoshi, in *Adaptive Dereverberation of Speech Signals with Speaker-Position Change Detection.* Proceedings of IEEE International Conference on Acoustics, Speech and Signal Processing, ICASSP 2009, (2009), pp. 3733–3736
49. R. Zdunek, Z. He, A. Cichocki, in *Tomographic Image Reconstruction from Limited-View Projections with Wiener Filtered FOCUSS Algorithm*, Proceedings of 5th IEEE International Symposium on Biomedical Imaging: From Nano to Macro, ISBI (2008), pp. 768–771
50. J.R. Zeidler, Performance analysis of LMS adaptive prediction filters. Proc. IEEE **78**(12), 1781–1806 (1990)
51. J.Y. Zheng, M. Shi, Scanning depth of route panorama based on stationary blur. Int. J. Comput. Vis. **78**(2–3), 169–186 (2008)

Chapter 6
Experimental Modelling

6.1 Introduction

Mathematical models are interesting and useful for several purposes. In the domain considered in this book, models can refer to signals, to data, or to systems. During scientific and professional treatment of certain problems, an important task is to obtain mathematical models from experiments, and this is the matter contemplated in this chapter.

To put an example, imagine you were establishing the law $F = m\,a$. To this end, you design a series of experiments, applying forces and measuring accelerations. Once data were obtained, a linear data fitting would be applied, and the input/output model $F = m\,a$ would be obtained. Some precision issues would be expected.

In the case of linear dynamical systems, it would be adequate to use transfer functions or state space for the models. Time-series models could be also appropriate.

As it was made clear by the state space approach, there are observability and controllability aspects to be taken into account. If we apply stimuli, the stimuli must be appropriate (rich enough) to excite all system behaviours.

The name given to the activity for getting a model is '*system identification*'. It includes experiment design, including stimuli design, and processing of results.

While the comments above focused on systems, there are also signals and data sets that can be modelled. For instance, it would be important to establish that a signal consists of a deterministic signal buried in noise. Also, it would be interesting for forecasting purposes to detect a trend or a pattern. In a way, models provide a kind of data compression.

Since this is an important theme with many practical implications and uses, it is nowadays quite extensive, with a lot of specific aspects that could be object of study. Evidently, this large dimension cannot be afforded in the limits of this chapter, but it would be pertinent to present some fundamentals together with a series of examples in MATLAB.

© Springer Science+Business Media Singapore 2017

J.M. Giron-Sierra, *Digital Signal Processing with Matlab Examples, Volume 2*,
Signals and Communication Technology, DOI 10.1007/978-981-10-2537-2_6

Along the chapter relevant literature will be referenced. An important book is [23], which can be complemented with [1, 19, 31, 34]. The status of the scientific efforts on system identification is reviewed in [24]. An illustrative field of application, acoustic systems, is specifically treated in [17].

There are several Toolboxes that could be used for experimental modelling. Some of them are mentioned by the end of the chapter. As you will notice, instead of using these Toolboxes we preferred to include a series of MATLAB programs for the aspects considered. In order to moderate the size of this chapter, some of these programs have been displaced to Appendix A.

6.2 Data Fitting

Many experimental studies have to deal with data fitting, especially if a certain law or mathematical relationship is tried to be established. Therefore, it is opportune to pay attention to this topic. For instance, data fitting is extensively used in data mining or to extract signals from noise.

There are MATLAB built-in functions that can be used for data fitting. In addition there are specific toolboxes for such purposes. Let us focus on the MATLAB built-in functions.

In general the data fitting is an optimization problem. A certain curve has to be mathematically generated to fit the data, minimizing a fitting error. This error can be defined in several ways, according to different fitting approaches, like least squares or maximum likelihood.

With some extent, data fitting is an art. There are hypothesis to test, related for instance to the mathematical function to be tried for the fitting. Also, it is convenient to pre-process the data, since non-sense data perhaps should be eliminated. Most times it is a good idea to plot the data in different ways, for example using logarithms or trigonometric functions.

It is interesting to consider that we can have the case of $y = f(x)$, where x and y are data, or in particular $y = f(t)$, so y is a signal.

The function *polyfit()* finds the coefficients of a polynomial of degree n, which you specify when calling the function, so the polynomial fits the data in a least squares sense. The Program 6.1 offers a simple example where the data are generated with a 2nd degree polynomial and some noise, and then *polyfit()* is invoked to fit the data with a polynomial (a 2nd degree was specified). Figure 6.1 shows a satisfactory result, confirmed with the numerical data provided also by the Program 6.1.

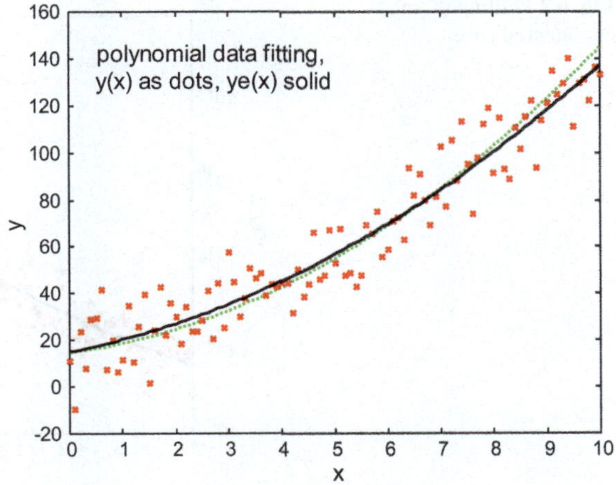

Fig. 6.1 Polynomial fitting example

Program 6.1 Polynomial fitting example

```
% polynomial fitting example
x=0:0.1:10; %independent variable data set
y=x.^2+3*x+15; %example of y=f(x)
N=length(x); %number of data points
di=randn(1,N); %random disturbance
ym=y+(10*di); %adding some disturbance to data
[P,S]=polyfit(x,ym,2); %polinomial fitting of disturbed data
ye=polyval(P,x); %estimated y values
%display of results
figure(1)
plot(x,y,':g'); hold on; %plots original data
plot(x,ym,'xr'); %plots disturbed data
plot(x,ye,'k'); %plots the fitting
title('polynomial data fitting, y(x) as dots, ye(x) solid');
xlabel('x'); ylabel('y');
cor=corrcoef(ym,ye); %fit quality checking
cor
P
S
```

Suppose the experimental data you get are of exponential nature. Still it is possible to use *polyfit()*. It is just a matter of taking logarithms of the data, and then try the polynomial fit. The Program 6.2 gives an example based on the previous case. Figure 6.2 shows the results, using logarithms in the vertical axis. Indeed the same idea of conversion can be applied to other types of functions, using their inverses.

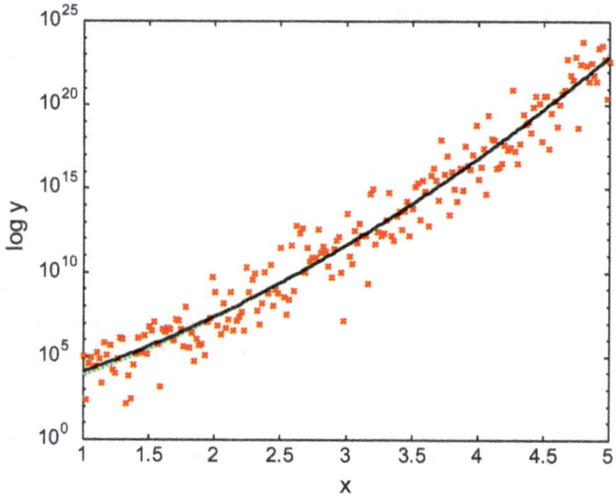

Fig. 6.2 Fitting of an exponential case

Program 6.2 Fitting exponential data

```
% fitting exponential data
x=1:0.02:5; %independent variable data set
y=exp(x.^2+5*x+3); %example of y=exp(f(x))
N=length(x); %number of data points
di=randn(1,N); %random disturbance
ym=y.*exp(3*di); %adding some disturbance to data
lym=log(ym); %get polynomial version
[P,S]=polyfit(x,lym,2); %polinomial fitting
lye=polyval(P,x); %estimated f(x) values
ye=exp(lye); %estimated data
%display of results
figure(1)
semilogy(x,y,':g'); hold on; %plots original data
semilogy(x,ym,'xr'); %plots disturbed data
semilogy(x,ye,'k'); %plots the fitting
title('polynomial fitting of exponential data,
y(x) as dots, ye(x) solid');
xlabel('x'); ylabel('log y');
cor=corrcoef(ym,ye); %fit quality checking
cor
P
S
```

For more general cases, the problem can be solved using optimization functions. For instance, the *fminbnd()* or *fminsearch()* MATLAB built-in functions. Program 6.3 deals with an example where the data are sinusoidal. The function *fminbnd()* is invoked, and the results obtained, as shown in Fig. 6.3, are quite satisfactory.

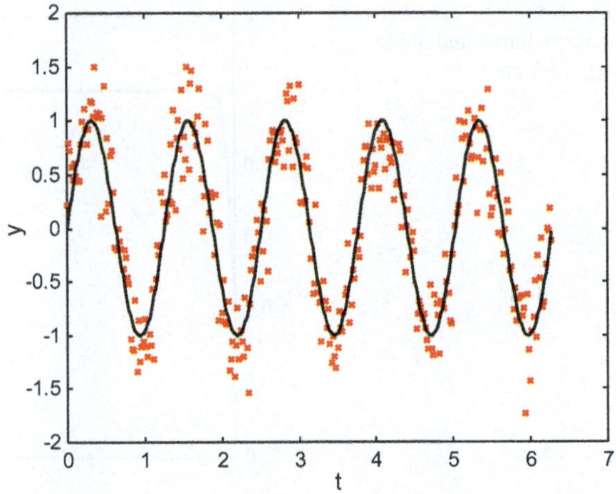

Fig. 6.3 Example of specific function fitting

Program 6.3 Fitting a sinusoid

```
% fitting a sinusoid
t=0:0.02:(2*pi); %independent variable data set
w=5; %parameter
y=sin(w*t); %example of y=sin(wt)
N=length(t); %number of data points
di=randn(1,N); %random disturbance
ym=y+(0.3*di); %adding some disturbance to data
%function to be minimised by x:
ferror=inline('sum(abs(ym-sin(t*x)))');
%find x for minimum error:
[wex ferrorx]=fminbnd(ferror,1,7,[],t,ym);
ye=sin(wex*t); %estimated data
%display of results
figure(1)
plot(t,y,':g'); hold on; %plots original data
plot(t,ym,'xr'); %plots disturbed data
plot(t,ye,'k'); %plots the fitting
title('fitting of sinusoidal data, y(t) as dots, ye(t) solid');
xlabel('t'); ylabel('y');
cor=corrcoef(ym,ye); %fit quality checking
wex
ferrorx
cor
```

6.3 Coherence. Delays

Let us prepare for experimental modelling, in which it is important to take care of the nature of signals providing the desired information. Two issues will be considered next: coherence and delays.

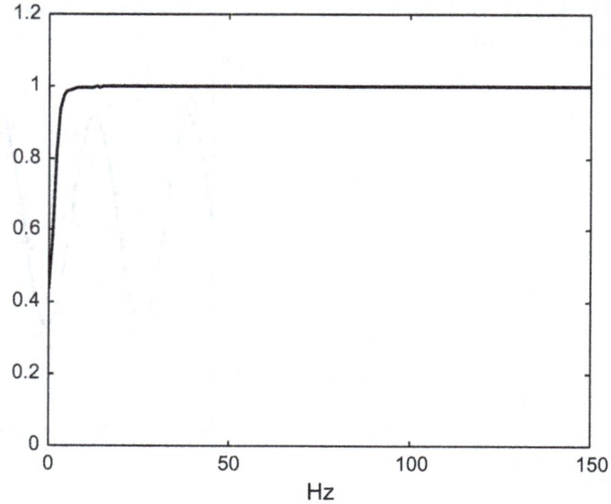

Fig. 6.4 Coherence between system output and noise input for case 2

6.3.1 Coherence Between Two Signals

Suppose you want to measure with a microphone the noise from a loudspeaker which is far from you; perhaps in large hall, or in a classical train station. It may happen that the microphone was located in a place where interferences or obstacles cause a null; so the noise captured by the microphone does not correspond to the loudspeaker. This case can be detected by *coherence* between the loudspeaker signal and the microphone signal.

Coherence is important in radar applications. It is also important in biomedical measurement of several signals, like the electrocardiogram, electroencephalogram, myograms, etc. when the research tries to establish relationships with same excitations.

As simple example of coherence study, the Program 6.4 computes the coherence between noise input and system output for the case 2. The Program uses the MATLAB Signal Processing Toolbox function *cohere()*. Figure 6.4 shows the result.

Program 6.4 Coherence between system output and noise input

```
%Coherence between system output and noise input
%DTrF case 2
% continuous time transfer function:
num2Ns=[10 0]; den2Ns=[1 3 10]; %G2(s)=10s/(s^2+3s+10);
fs=300; %sampling frequency in Hz
%discrete transfer function (from the continuous case)
[num2Nz,den2Nz]= impinvar(num2Ns,den2Ns,fs); %G2(z)
%response of G2 to noise
tiv=1/fs; %time interval between samples;
t=0:tiv:(60-tiv); %time intervals set (60 seconds)
N=length(t); %number of data points
u=randn(N,1); %random input signal data set
```

```
y=filter(num2Nz,den2Nz,u); %G2(z) response to noise
[chr fchr]=cohere(u,y,fs); %coherence computation
%display coherence
figure(1)
plot((fchr*fs/2),chr,'k');
xlabel('Hz');
axis([0 150 0 1.2]);
title('coherence of input and output signals: case 2')
```

6.3.2 Delays

A practical problem that may appear when getting experimental data is now studied with some detail. This is the case of delays.

Suppose you are measuring a signal, perhaps again with a microphone. It may happen that also echoes—delayed versions of the signal—would enter into the measurements.

Another type of scenarios corresponds to the determination of system models, getting information of the system responses to certain inputs. Sometimes you can measure inputs and outputs. Sometimes it is only possible to measure outputs. These circumstances depend on experimental conditions: data can be obtained from laboratory experiments or from real world.

In the case of measuring inputs and outputs, delays may exert a strong influence on frequency response up to the point of masking the dynamic behaviour of the system. Some of the examples in this section are addressed to this aspect.

What inputs can be applied in laboratory, or used in real world, depend on varied limitations. Typical inputs are impulses, sinusoids, or noise. Imagine getting data of the dynamic behaviour of an airplane. The pilot could try an impulse, but it really sounds dangerous. He may also try to impose a sinusoidal input, but it looks not very comfortable. What an on-board data acquisition system can get easily are responses when there are turbulences.

Let us focus on delays between input and output. An example of pure delay is a conveyor taking apples from an input to an output, using say 10 s in the travel. Another example is a microphone catching the sound of a loudspeaker from 30 m distance.

A linear system with delay τ has the following frequency response:

$$D(j\omega) = e^{-j\omega\tau}\, G(j\omega) \tag{6.1}$$

According with Eq. (6.1) the delay contributes to the frequency response with an exponential term. In a Bode diagram, the magnitudes of $D(j\omega)$ and $G(j\omega)$ are the same, since the magnitude of the exponential is 1. However the phases of $D(j\omega)$ and $G(j\omega)$ are different, the exponential term adds $-(j\omega\tau)$ phase to the $G(j\omega)$ phase.

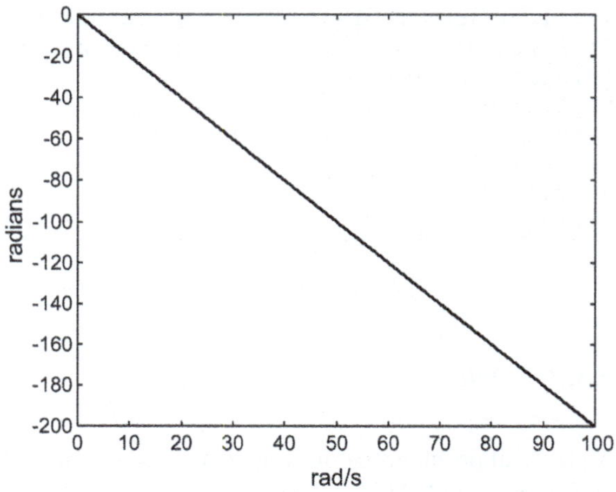

Fig. 6.5 Phase caused by 2 s delay

Figure 6.5 (Program 6.5) shows the added phase due to a delay of 2 s. It is quite significant. Notice that scales are linear.

Program 6.5 Phase of pure time delay

```
% phase of pure time delay
Td=2; %time delay in seconds
w=0.1:0.1:100; %frequency data set in rad/s
phased=-Td*w; %phase induced by time delay
plot(w,phased,'k'); %plots phased vs. w
title('phase caused by a delay of 2 seconds');
xlabel('rad/s'); ylabel('radians');
%taking last part of the phase line:
dtangent=(phased(1000)-phased(100))/(w(1000)-w(100));
eTd=-dtangent; %Td estimate
eTd %result display
```

Let us prepare an experiment. A noise signal $u(t)$ is generated, and then a second signal $y(t)$ is obtained by delaying 2 s $u(t)$. It is like if $y(t)$ was an echo of $u(t)$. The Program 6.6 creates the two signals. Figure 6.6 depicts the result.

Program 6.6 Noise and pure time delay

```
% noise and pure time delay
Td=2; %time delay in seconds
%input signal
fs=20; %sampling frequency in Hz
tiv=1/fs; %time interval between samples;
tu=0:tiv:(60-tiv); %time intervals set (60 seconds)
Nu=length(tu); %number of data points
u=randn(Nu,1); %random input signal data set
%output signal
NTd=Td*fs; %number of samples along Td
```

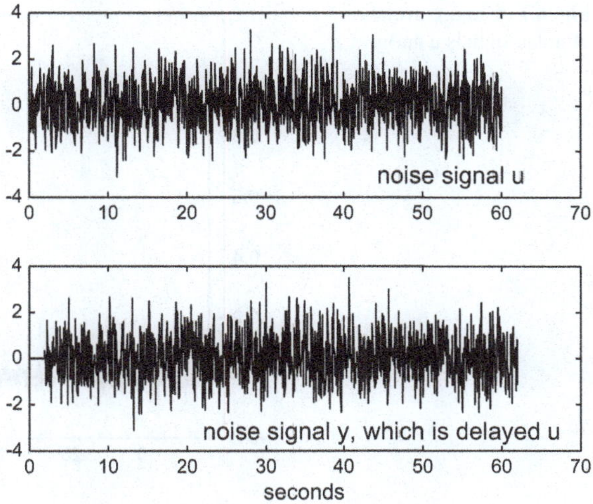

Fig. 6.6 Noise signals u and y; y is delayed u

```
Ny=Nu+NTd;
y=zeros(1,Ny);
y((NTd+1):Ny)=u(1:Nu); %y is u delayed Td seconds
ty=0:tiv:(60+Td-tiv); %time intervals set (60+Td seconds)
%plot input and output
subplot(2,1,1)
plot(tu,u,'k'); %plots input
axis([0 70 -4 4]);
title('noise signal u')
subplot(2,1,2)
plot(ty,y,'k'); %plots output (has delay)
axis([0 70 -4 4]);
title('noise signal y, which is delayed u');
xlabel('seconds');
```

Cross-correlation helps us to detect that $u(t)$ and $y(t)$ are correlated, and also that $y(t)$ is a 2 s echo of $u(t)$. Program 6.7 uses *xcorr()* for this purpose, obtaining the result shown in Fig. 6.7. Notice the spike at 2 s.

Program 6.7 Cross correlation in noise and pure time delay

```
% Cross correlation in noise and pure time delay
Td=2; %time delay in seconds
%input signal
fs=20; %sampling frequency in Hz
tiv=1/fs; %time interval between samples;
tu=0:tiv:(60-tiv); %time intervals set (60 seconds)
Nu=length(tu); %number of data points
u=randn(Nu,1); %random input signal data set
%output signal
NTd=Td*fs; %number of samples along Td
Ny=Nu+NTd;
```

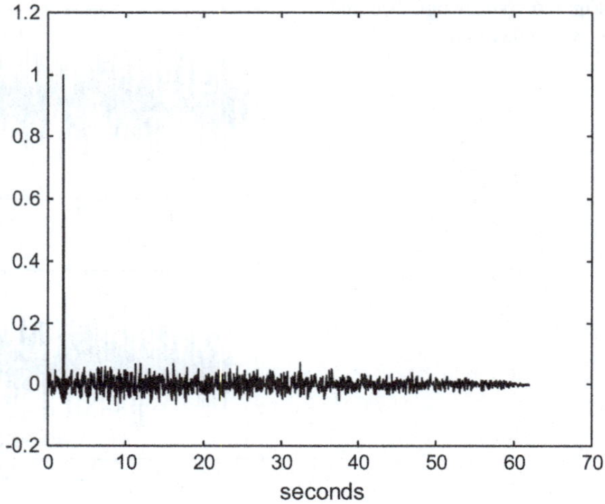

Fig. 6.7 Cross-correlation of noise signals u and y

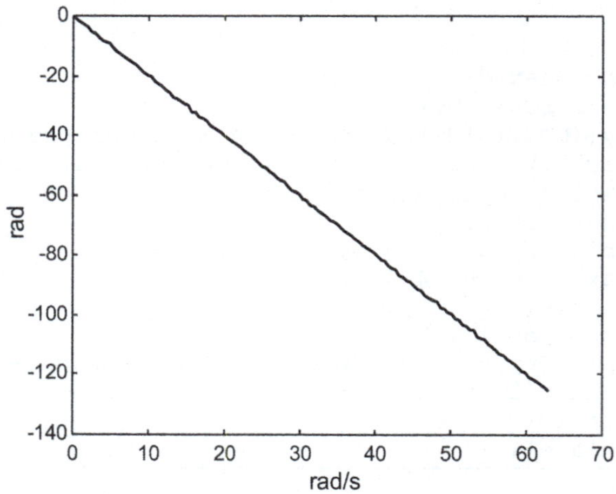

Fig. 6.8 Estimation of phase caused by delay between u and y

```
y=zeros(1,Ny);
y((NTd+1):Ny)=u(1:Nu); %y is u delayed Td seconds
ty=0:tiv:(60+Td-tiv); %time intervals set (60+Td seconds)
ac=xcorr(u,y); %cross-correlation of u and y
ac=ac/Ny; %normalization
Nac=length(ac); mNac=ceil(0.5*Nac); %to plot half ac
plot(ty,ac(mNac:Nac),'k');
title('cross-correlation'); xlabel('seconds');
[V K]=max(ac);
eNTd=K-mNac;
eTd=eNTd/fs
```

Now, let us use the *tfe()* function to estimate the frequency response of the 'system' having $u(t)$ as input and $y(t)$ as output. The 'system' is in this case a pure delay. The Program 6.8 makes this work, obtaining the Fig. 6.8 and numerical information of estimated delay. The delay is estimated computing the tangent of the phase curve (actually a straight line). Compare with Fig. 6.5. Notice that *tfe()* deals with noise signals, and that it obtains the frequency response in terms of data—magnitude and phase-sets.

Program 6.8 Frequency response estimation in noise and pure time delay

```
% Frequency response estimation in noise and pure time delay
Td=2; %time delay in seconds
%input signal
fs=20; %sampling frequency in Hz
tiv=1/fs; %time interval between samples;
tu=0:tiv:(60-tiv); %time intervals set (60 seconds)
Nu=length(tu); %number of data points
u=randn(Nu,1); %random input signal data set
%output signal
NTd=Td*fs; %number of samples along Td
Ny=Nu+NTd;
y=zeros(1,Ny);
y((NTd+1):Ny)=u(1:Nu); %y is u delayed Td seconds
nfft=256; %length of FFT
window=hanning(256); %window function
%transfer function estimate with stable data:
[GE,FE]=tfe(u(61:Nu),y(61:Nu),nfft,fs,window);
phased=unwrap(angle(GE));
w=FE*2*pi; %frequency in rad/s
%plots phase of estimated frequency response of the system:
plot(w,phased,'k');
title('estimation of the phase caused by the delay')
xlabel('rad/s'),; ylabel('rad');
%taking last part of the phase line:
dtangent=(phased(100)-phased(10))/(w(100)-w(10));
eTd=-dtangent; %Td estimate
eTd %result display
```

Now suppose that experimental frequency response data are obtained in a certain scenario. The data are magnitude and phase data sets corresponding to a set of frequencies. It happens that the data corresponds to a pure delay, but we are not aware of this.

Take as example of frequency response experimental data, the results obtained by *tfe()* in the previous Program 6.8.

Let us proceed to estimate a transfer function model from the experimental data, using *invfreqs()*. The Program 6.9 obtains the result, and shows several aspects in three figures.

Notice that we needed a high degree in transfer function numerator and denominator polynomials, to obtain a good fitting.

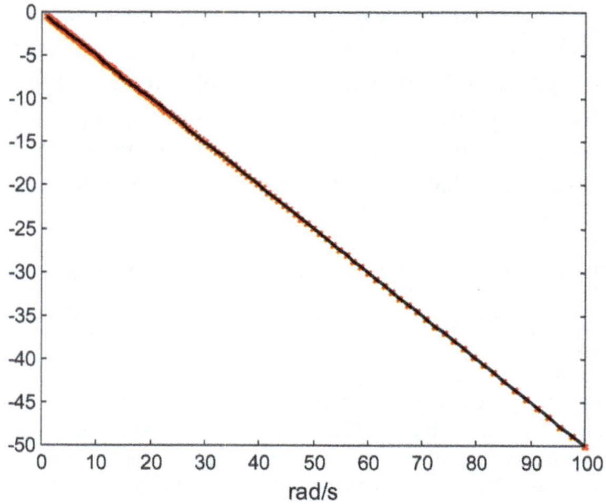

Fig. 6.9 Approximation of delay frequency response

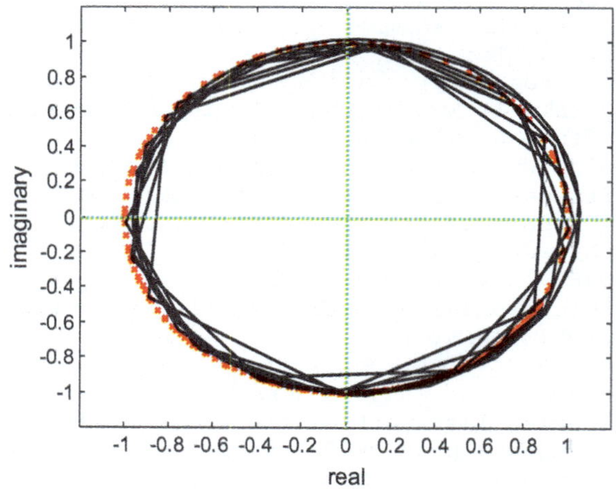

Fig. 6.10 Approximation of delay frequency response

Figure 6.9 shows in linear scales the phase data and the phase corresponding to the estimated transfer function. The agreement is satisfactory.

Figure 6.10 presents another view of the experimental data and the frequency response of the estimated transfer function. The complex plane is used. A pure delay means in this plane a circling curve (the dotted circle in the figure). The frequency response (the straight lines in the figure) approximates well the data, since they are in the corners.

Figure 6.11 shows a pole-zero map of the transfer function. Poles (left-hand side of the complex plane) and zeros (right-hand side) mirror approximately each other. This is characteristic of delay transfer function approximations, like in the case of Padé approximations. In other words, if you see such type of pole-zero maps, you can suppose there is a delay in the system you are modelling.

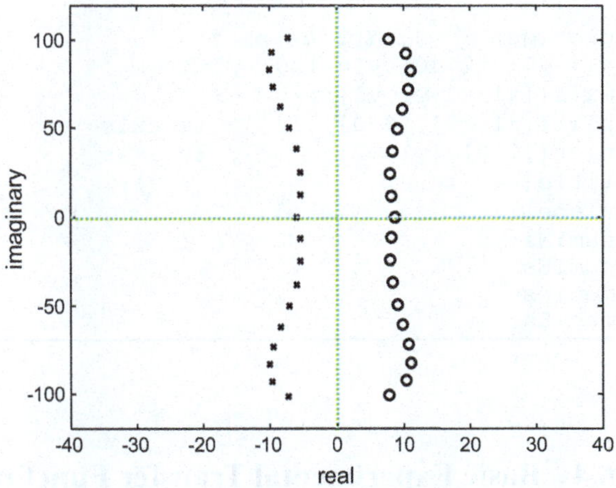

Fig. 6.11 Pole-zero map of TrF

Program 6.9 Approximation of delay by transfer function

```
% Approximation of delay by transfer function
Td=0.5; %pure delay in seconds
wr=logspace(0,2,200); %frequency values for response (rad/s)
Hdly=exp(-j*Td*wr); %frequency response corresponding to Td
%using invfreqs to obtain a TrF approximation
na=19; %denominator degree
nb=19; %numerator degree
[num,den]=invfreqs(Hdly,wr,nb,na); %TrF computation
Htrf=freqs(num,den,wr); %TrF frequency response
%compare phase of frequency responses cooresponding
% to the delay and to TrF
figure(1)
plot(wr,unwrap(angle(Hdly)),'xr'); hold on;
plot(wr,unwrap(angle(Htrf)),'k');
x1=-1.2; x2=1.2; y1=-1.2; y2=1.2;
title('phase of frequency responses
corresponding to delay and TrF');
xlabel('rad/s');
%compare frequency responses cooresponding
% to the delay and to TrF
figure(2)
plot(Hdly,'xr'); hold on;
plot(Htrf,'k');
x1=-1.2; x2=1.2; y1=-1.2; y2=1.2;
axis([x1 x2 y1 y2]);
plot([x1 x2],[0 0],':g'); %x axis
plot([0 0],[y1 y2],':g'); %y axis
title('frequency responses corresponding to delay and TrF');
xlabel('real'); ylabel('imaginary');
%pole-zero map of TrF
figure(3)
gz=roots(num); gp=roots(den);
plot(gz,'ok'); hold on; %zeros plot
```

```
plot(gp,'xk'); %poles plot
x1=-40; x2=40; y1=-120; y2=120;
axis([x1 x2 y1 y2]);
plot([x1 x2],[0 0],':g'); %x axis
plot([0 0],[y1 y2],':g'); %y axis
title('zeros and poles of TrF');
xlabel('real'); ylabel('imaginary');
num1Ns
num1Es
den1Ns
den1Es
```

6.4 Basic Experimental Transfer Function Modelling

Consider that we want to determine the model of a system in terms of transfer function. Inputs can be applied and measured. Corresponding responses can be measured at the output. Recall that only linear systems have transfer functions, and that this is a 'black-box' approach: apply an input excitation, measure the response and establish an output/input mathematical relation. Transfer functions describe the dynamical behaviour of a system.

This section will review three different scenarios of experimental modelling with transfer functions, according with three types of inputs: impulse, sinusoidal, or noise.

6.4.1 Two Simple Transfer Function Examples

Two transfer functions (TrF) will be used as examples, to simulate the system to be modelled. One is a first order TrF:

$$G1(s) = \frac{10}{s + 10} \tag{6.2}$$

The other is a second order TrF, which corresponds to a damped oscillator.

$$G2(s) = \frac{10\,s}{s^2 + 3\,s + 10} \tag{6.3}$$

Depending on the scenario, the continuous time $G(s)$ or the discrete time $G(z)$ will be used. The latter is obtained from $G(s)$ with the function *c2d*() with the option '*zoh*':

Obtaining a transfer function is a good prize; with the transfer function we can predict the response of the system to any input.

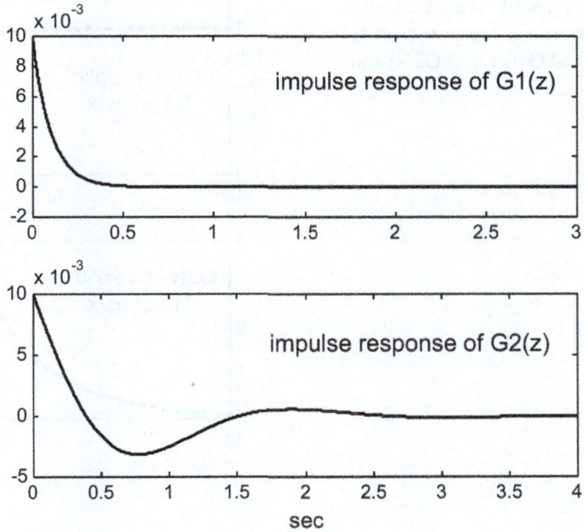

Fig. 6.12 Impulse responses of G1(z) and G2(z)

The modelling approach can be applied to signals. The idea is to suppose that the signal is generated by the transfer function in response to a known input. This concepts can be further exploited for signal compression purposes.

In general it is important that the input excitation be rich. It may happen that the system has unknown resonance frequencies, and these frequencies must be excited to be detected. This may be case, for example, of a bridge that potentially may be resonant to certain earthquake motions. Excitation richness refers to having a sufficiently broad bandwidth in the range of interest. A narrow impulse does have a large bandwidth, as shown by its Fourier transform. White noise has also a large bandwidth, theoretically infinite.

Figure 6.12 shows the impulse response of $G1(z)$ and $G2(z)$. This figure has been obtained with the Program 6.10.

Program 6.10 Transfer functions for study: impulse response

```
% Transfer functions for study: impulse response
% continuous time transfer functions:
num1Ns=10; den1Ns=[1 10]; %G1(s)=10/(s+10)
num2Ns=[10 0]; den2Ns=[1 3 10]; %G2(s)=10s/(s^2+3s+10);
fs=1000; %sampling frequency in Hz
Ts=1/fs; %sampling period in seconds
%discrete transfer functions (from the continuous cases)
[num1Nz,den1Nz]= impinvar(num1Ns,den1Ns,fs); %G1(z)
[num2Nz,den2Nz]= impinvar(num2Ns,den2Ns,fs); %G2(z)
%impulse responses
t=0:Ts:(4-Ts); %sampling times data set (4 seconds)
Ns=length(t); %number of samples
subplot(2,1,1)
[h1Nz,t1]=impz(num1Nz,den1Nz,Ns); %G1(z) impulse response
plot(t,h1Nz'b'); hold on;
```

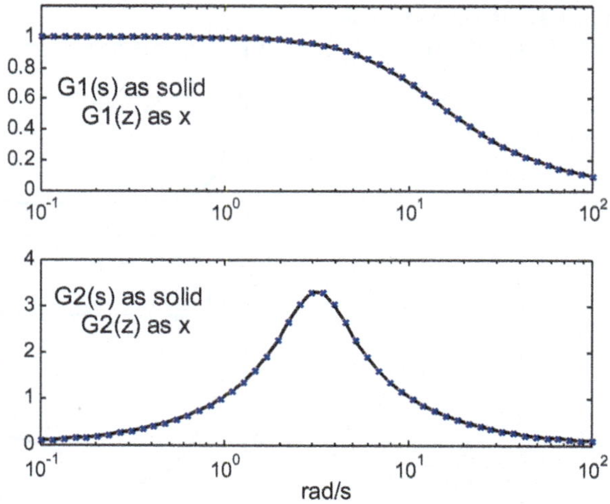

Fig. 6.13 Amplitude of frequency responses of G1(s) and G1(z), and G2(s) and G2(z)

```
title('impulse response of G1(z)');
subplot(2,1,2)
[h2Nz,t2]=impz(num2Nz,den2Nz,Ns); %G2(z) impulse response
plot(t,h2Nz'b'); hold on;
title('impulse response of G2(z)');
xlabel('sec')
```

To complete the view of the reference systems, Fig. 6.13 shows the amplitude frequency response of $G1(s)$ and $(G1(z)$ on the same plot, on top, and the same with $G2(s)$ and $G2(z)$ at the bottom. This figure has been generated with the Program 6.11.

Program 6.11 Transfer functions for study: frequency response

```
% Transfer functions for study: frequency response
% continuous time transfer functions:
num1Ns=10; den1Ns=[1 10]; %G1(s)=10/(s+10)
num2Ns=[10 0]; den2Ns=[1 3 10]; %G2(s)=10s/(s^2+3s+10);
fs=1000; %sampling frequency in Hz
Ts=1/fs; %sampling period in seconds
%discrete transfer functions (from the continuous cases)
[num1Nz,den1Nz]= impinvar(num1Ns,den1Ns,fs); %G1(z)
[num2Nz,den2Nz]= impinvar(num2Ns,den2Ns,fs); %G2(z)
%frequency responses
wr=logspace(-1,2); %frequency values for response (rad/s)
subplot(2,1,1)
H1Ns=freqs(num1Ns,den1Ns,wr); %G1(s) frequency response
semilogx(wr,abs(H1Ns),'k'); hold on;
%G1(z) frequency response:
H1Nz=freqz(num1Nz,den1Nz,wr/(2*pi),fs);
semilogx(wr,abs(H1Nz),'xb');
title('G1(s) as solid & G1(z) as x');
axis([0.1 100 0 1.2]);
subplot(2,1,2)
```

Fig. 6.14 Impulse response experiment

```
H2Ns=freqs(num2Ns,den2Ns,wr); %G2(s) frequency response
semilogx(wr,abs(H2Ns),'k'); hold on;
%G2(z) frequency response:
H2Nz=freqz(num2Nz,den2Nz,wr/(2*pi),fs);
semilogx(wr,abs(H2Nz),'xb');
title('G2(s) as solid & G2(z) as x');
xlabel('rad/s')
axis([0.1 100 0 4]);
```

6.4.2 Obtaining a Transfer Function Model from Impulse Response

Let us do the following experiment: obtaining the impulse responses of $G1(z)$ and $G2(z)$, and checking that these transfer functions can be determined from these responses.

Figure 6.14 depicts the experiment. An impulse is applied to the system, and the response (a series of impulses) is recorded for further analysis.

Among the functions offered by the MATLAB Signal Toolbox, let us select *prony()* to deal with the first case, $G1(z)$.

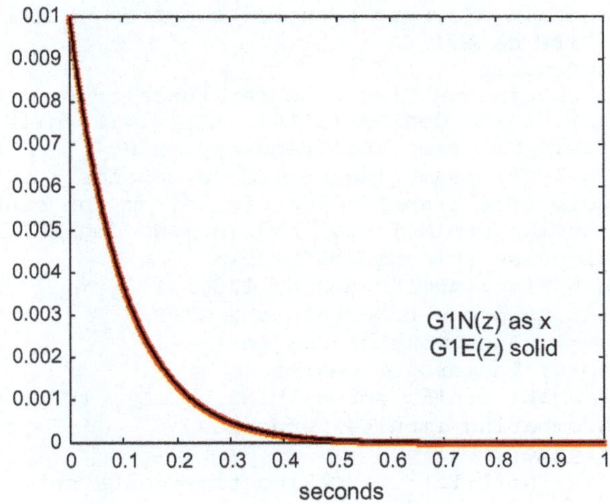

Fig. 6.15 Comparison of impulse responses of original and estimated G1(z)

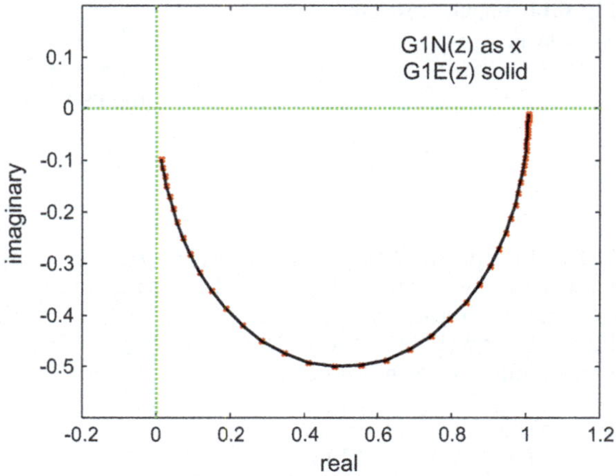

Fig. 6.16 Comparison of frequency responses of original and estimated G1(z)

The Program 6.12 gives us two figures with the results of using *prony()*. Figure 6.15 compares the impulse response of $G1(z)$ and the impulse response of the estimated transfer function $G1(z)$.

Figure 6.16 compares the frequency response of $G1(z)$ and the frequency response of the estimated transfer function $G1(z)$. The results are compared on the complex plane, to make visible magnitude and phase.

The Program 6.12 gives also numerical results about the numerator an denominator of $G1(z)$ and $G1(z)$.

Both the figures and numerical results are quite satisfactory.

Notice in the Program 6.12 that the user has to guess the order of the numerator and denominator polynomials. This may be not easy in real modelling situations, which also usually have noise and perhaps not very good signal sampling.

Program 6.12 Obtain discrete transfer function (DTrF) from impulse response

```
%Obtain discrete transfer function (DTrF) from impulse response
%DTrF case 1
%no noise
% continuous time transfer function:
num1Ns=10; den1Ns=[1 10]; %G1(s)=10/(s+10)
fs=1000; %sampling frequency in Hz
Ts=1/fs; %sampling period in seconds
%discrete transfer function (from the continuous case)
[num1Nz,den1Nz]= impinvar(num1Ns,den1Ns,fs); %G1(z)
%impulse response of G1(z)
h1Nz=impz(num1Nz,den1Nz,128,fs);
%using prony to obtain the DTrF
na=1; %denominator degree
nb=0; %numerator degree
[num1Ez,den1Ez]=prony(h1Nz,nb,na); %DTrF computation
%comparing impulse responses
figure(1)
t=0:Ts:(1-Ts); %sampling times data set (1 second)
```

```
Ns=length(t);
h1Nz=impz(num1Nz,den1Nz,Ns); %impulse response of G1(z)
h1Ez=impz(num1Ez,den1Ez,Ns); %impulse response of \^G1(z)
plot(t,h1Nz,'xr',t,h1Ez,'k'); %plots both impulse responses
title('impulse response, G1N(z)as x & G1E(z) solid');
xlabel('seconds');
%comparing frequency responses
figure(2)
wr=logspace(-1,2); %frequency values for response (rad/s)
%G1(z) frequency response
H1Nz=freqz(num1Nz,den1Nz,wr/(2*pi),fs);
plot(H1Nz,'xr'); hold on;
%^G1(z) frequency response
H1Ez=freqz(num1Ez,den1Ez,wr/(2*pi),fs);
plot(H1Ez,'k');
x1=-0.2; x2=1.2; y1=-0.6; y2=0.2;
axis([x1 x2 y1 y2]);
plot([x1 x2],[0 0],':g'); %x axis
plot([0 0],[y1 y2],':g'); %y axis
title('complex frequency response, G1N(z)as x & G1E(z) solid');
xlabel('real'); ylabel('imaginary');
num1Nz
num1Ez
den1Nz
den1Ez
```

Now, let us use the MATLAB Signal Toolbox *stmcb()* to estimate a model $G2(z)$ from the impulse response of $G2(z)$.

Like in the previous case, the Program A.18 gives us two figures with the results. Figure 6.17 compares the impulse response of $G2(z)$ and the impulse response of the estimated transfer function $G2(z)$.

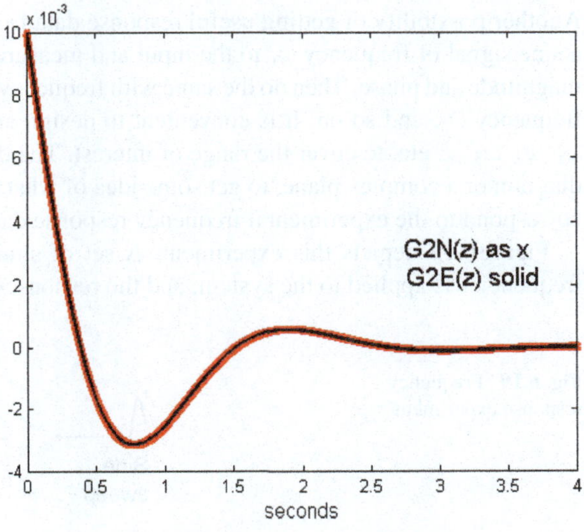

Fig. 6.17 Comparison of impulse responses of original and estimated G2(z)

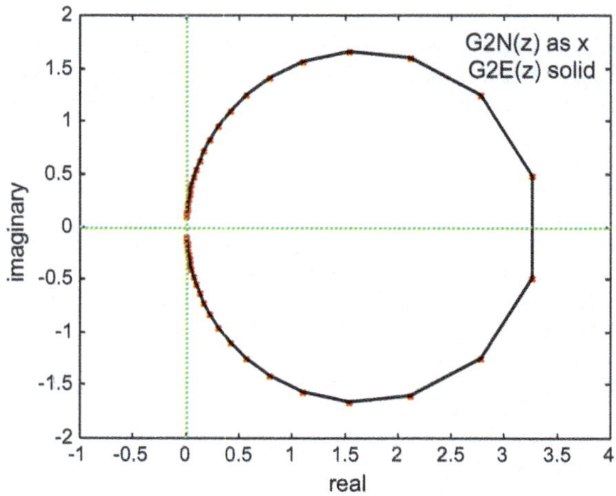

Fig. 6.18 Comparison of frequency responses of original and estimated G2(z)

And Fig. 6.18 compares in the complex plane the frequency response of $G2(z)$ and the frequency response of the estimated transfer function $G2(z)$. There is a circling path that corresponds to the resonance peak.

The Program A.18 is similar to the previous program, and also gives numerical results about the numerator an denominator of $G2(z)$ and $G2(z)$. Like before, both the figures and numerical results are satisfactory.

The Program A.18 has been included in Appendix A.

6.4.3 Obtaining a Transfer Function Model from Sine Sweep

Another possibility of getting useful response data from a linear system is to apply a sine signal of frequency ω_1 to the input and measure the output/input relations of magnitude and phase. Then do the same with frequency ω_2, and then repeat this using frequency ω_3, and so on. It is convenient to design an adequate set of frequencies $\omega_1, \omega_2, \omega_3 \ldots$ etc. to cover the range of interest. The data set can be plot on a Bode diagram or a complex plane, to get some idea of what type of transfer function may correspond to the experimental frequency response data.

Figure 6.19 depicts this experiment. A set of sinusoidal signals with different frequencies is applied to the system, and the responses are analyzed.

Fig. 6.19 Frequency response experiment

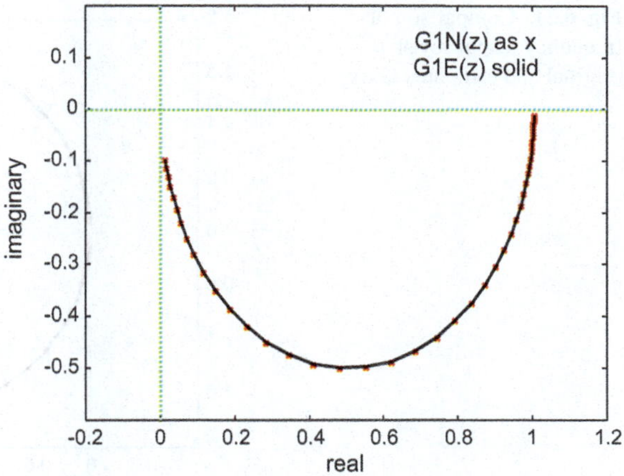

Fig. 6.20 Comparison of frequency responses of original and estimated G1(z)

Let us suppose that experimental frequency response data have been obtained from the two plants selected as examples in this section.

The first plant will be considered as discrete, $G1(z)$, and the second plant as continuous time, $G2(s)$.

Program 6.13 applies *invfreqz()* to estimate a model from the frequency response data of $G1(z)$. Again, the user has to guess the order of $G1(z)$ numerator and denominator polynomials. Figure 6.20 compares on the complex plane the frequency response of the estimated transfer function $G1(z)$, and the experimental frequency response data. The program gives also numerical results about numerator and denominator of $G1(z)$. Good modelling results are obtained.

Program 6.13 Obtain discrete transfer function (DTrF) from discrete frequency response

```
%Obtain discrete transfer function (DTrF)
% from discrete frequency response
%DTrF case 1
%no noise
% continuous time transfer function:
num1Ns=10; den1Ns=[1 10]; %G1(s)=10/(s+10)
fs=1000; %sampling frequency in Hz
Ts=1/fs; %sampling period in seconds
%discrete transfer function (from the continuous case)
[num1Nz,den1Nz]= impinvar(num1Ns,den1Ns,fs); %G1(z)
wr=logspace(-1,2); %frequency values for response (rad/s)
%discrete frequency response of G1(z)
%G1(z) frequency response:
H1Nz=freqz(num1Nz,den1Nz,wr/(2*pi),fs);
%using invfreqz to obtain the DTrF
na=1; %denominator degree
nb=0; %numerator degree
W=wr/fs; %normalized frequency values 0..pi rad/s
[num1Ez,den1Ez]=invfreqz(H1Nz,W,nb,na); %DTrF computation
%^G1(z) frequency response:
H1Ez=freqz(num1Ez,den1Ez,wr/(2*pi),fs);
```

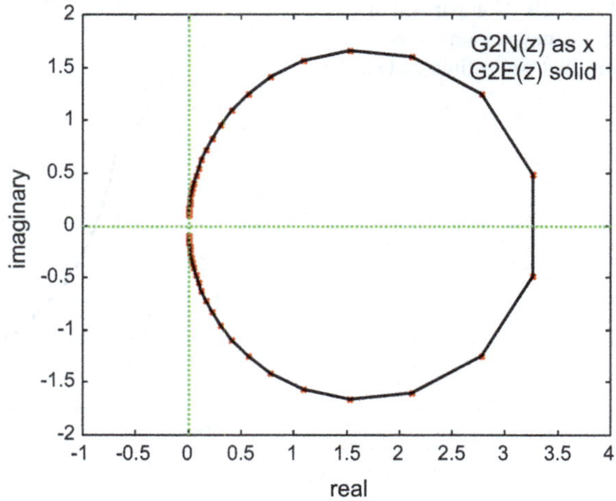

Fig. 6.21 Comparison of frequency responses of original and estimated G2(z)

```
%comparing frequency responses
plot(H1Nz,'xr'); hold on;
plot(H1Ez,'k');
x1=-0.2; x2=1.2; y1=-0.6; y2=0.2;
axis([x1 x2 y1 y2]);
plot([x1 x2],[0 0],':g'); %x axis
plot([0 0],[y1 y2],':g'); %y axis
title('complex frequency response, G1N(z)as x & G1E(z) solid');
xlabel('real'); ylabel('imaginary');
num1Nz
num1Ez
den1Nz
den1Ez
```

Program A.19 applies *invfreqs()* to estimate a model from the frequency response data of $G2(s)$. Like before, the user has to guess the order of $G2(s)$ numerator and denominator polynomials.

Figure 6.21 compares on the complex plane the frequency response of the estimated transfer function $G2(s)$, and the experimental frequency response data. Again good modelling results are obtained.

The Program A.19 has been included in Appendix A.

Fig. 6.22 Noise input and response of G1(z)

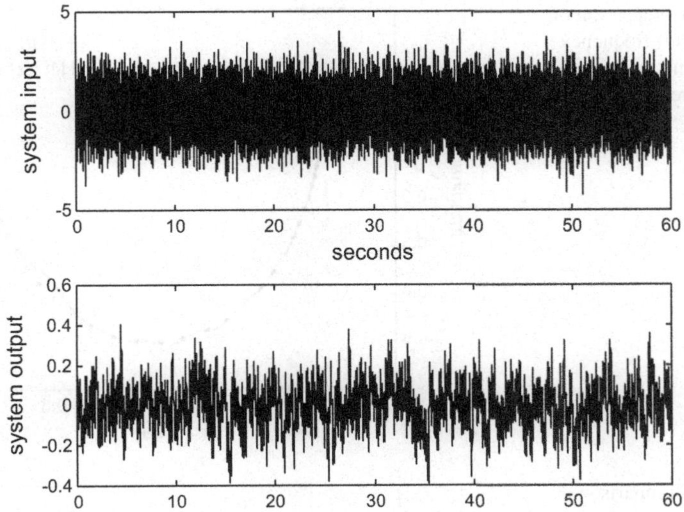

Fig. 6.23 Noise response experiment

6.4.4 Obtaining a Transfer Function Model from Response to Noise

A third alternative is that the system is subject to noise at the input, so the system response is a sort of filtered noise. Both input and output noises are recorded, and with these data we have to obtain the system transfer function.

Figure 6.23 depicts the experiment. Noise is applied to the input, and the response is recorded for analysis.

We shall proceed in two steps. First the MATLAB Signal Toolbox *tfe()* function is used to determine a set of estimated magnitude and phase frequency response data. Then, as in Sect. 6.4.2, *invfreqz()* is used to determine a transfer function.

The first case, corresponding to $G1(z)$, is treated by the Program 6.14. The program produces the next three figures. The first figure, Fig. 6.22, shows the system input and output; the input is noise generated with *randn()*, and the output filtered noise. Both input and output are fed into *tfe()*.

Figure 6.24 shows on the complex plane a solid curve with the results obtained by *tfe()*. It is a data set of estimated magnitudes and phases of the system frequency response. The figure includes points of the frequency response of $G1(z)$for comparison purposes.

Notice that the estimated frequency response is not completely smooth, although it approximates fairly well the frequency response of $G1(z)$.

Now, *invfreqz()* is fed with the estimated frequency response data obtained by *tfe()*. Figure 6.25 compares the frequency response of the transfer function estimated by *tfe()*, $G1(z)$, and the frequency response of $G1(z)$. The agreement is quite satisfactory.

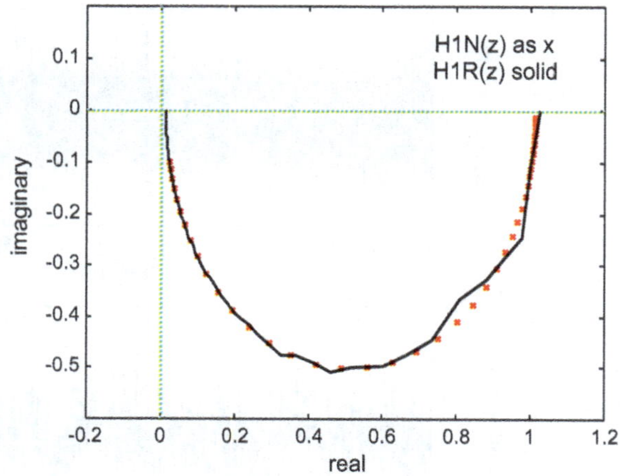

Fig. 6.24 Comparison of original G1(z) frequency response, and estimated frequency response

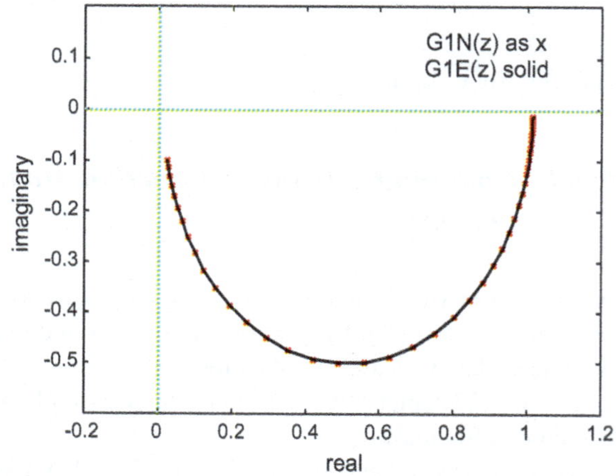

Fig. 6.25 Comparison of frequency responses of original and estimated G1(z)

Program 6.14 Obtain discrete transfer function (DTrF) from discrete frequency response to noise

```
%Obtain discrete transfer function (DTrF)
% from discrete frequency response to noise
%DTrF case 1
% continuous time transfer function:
num1Ns=10; den1Ns=[1 10]; %G1(s)=10/(s+10);
fs=400; %sampling frequency in Hz
%discrete transfer function (from the continuous case)
[num1Nz,den1Nz]= impinvar(num1Ns,den1Ns,fs); %G1(z)
%discrete frequency response of G1(z)
wr=logspace(-1,2); %frequency values for response (rad/s)
%G1(z) frequency response:
H1Nz=freqz(num1Nz,den1Nz,wr/(2*pi),fs);
%response of G1 to noise
tiv=1/fs; %time interval between samples;
t=0:tiv:(60-tiv); %time intervals set (60 seconds)
```

```
N=length(t); %number of data points
u=randn(N,1); %random input signal data set
y=filter(num1Nz,den1Nz,u); %G1(z) response to noise
%display of input and output signals
figure(1)
subplot(2,1,1)
plot(t,u,'k'); %input u plot
xlabel('seconds'); ylabel('system input');
title('input and output signals: case 1')
subplot(2,1,2)
plot(t,y,'k'); %output y plot
ylabel('system output');
%-------------------------------------------------
%frequency response estimate,using tfe
nfft=2048; %length of FFT
window=hanning(nfft); %window function
%frequency response estimate:
[H1Rz,F1Rz]=tfe(u(1000:N),y(1000:N),nfft,fs,window);
%comparing original and estimated frequency responses
figure(2)
plot(H1Nz,'xr'); hold on;
plot(H1Rz,'k');
x1=-0.2; x2=1.2; y1=-0.6; y2=0.2;
axis([x1 x2 y1 y2]);
plot([x1 x2],[0 0],':g'); %x axis
plot([0 0],[y1 y2],':g'); %y axis
title('complex frequency response,
H1N(z)as x & H1R(z) solid');
xlabel('real'); ylabel('imaginary');
%-------------------------------------------------
%using invfreqz to obtain the DTrF
na=1; %denominator degree
nb=0; %numerator degree
W=F1Rz*2*pi/fs; %normalized frequency 0..pi
[num1Ez,den1Ez]=invfreqz(H1Rz,W,nb,na); %DTrF computation
%^G1(z) frequency response:
H1Ez=freqz(num1Ez,den1Ez,wr/(2*pi),fs);
%comparing G1(z) and ^G1(z) frequency responses
figure(3)
plot(H1Nz,'xr'); hold on;
plot(H1Ez,'k');
axis([x1 x2 y1 y2]);
plot([x1 x2],[0 0],':g'); %x axis
plot([0 0],[y1 y2],':g'); %y axis
title('complex frequency response,
G1N(z)as x & G1E(z) solid');
xlabel('real'); ylabel('imaginary');
num1Nz
num1Ez
den1Nz
den1Ez
```

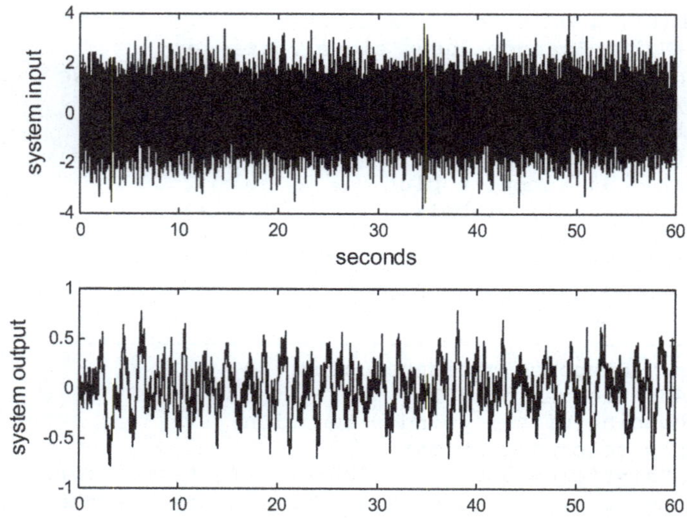

Fig. 6.26 Noise input and response of G2(z)

Fig. 6.27 Comparison of original G2(z) frequency response, and estimated frequency response

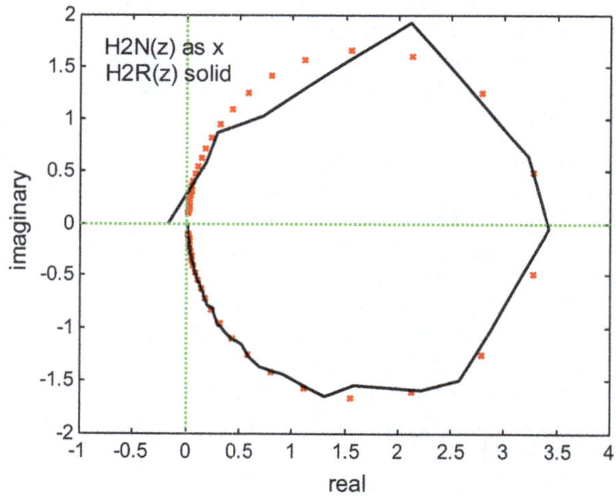

The second case, corresponding to $G2(z)$, is treated by the Program A.20, following the same steps as before, producing also three figures. Figure 6.26 shows the system input and output.

Figure 6.27 shows the data set obtained by *tfe()* with the estimated frequency response. There are noticeable divergences with respect to the frequency response of $G2(z)$, however these divergences occur mainly in the right-hand side of the 'circle', and this zone corresponds to the resonance peak; what happens is that the damping here is slightly underestimated.

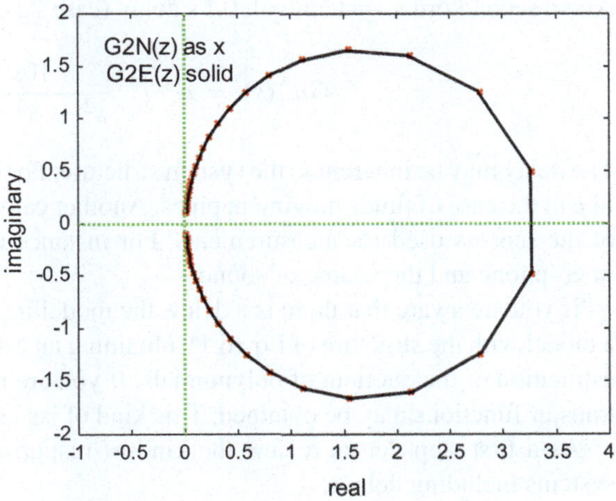

Fig. 6.28 Comparison of frequency responses of original and estimated G2(z)

Figure 6.28 shows the final result obtained by *invfreqz()*, comparing the frequency response of the estimated transfer function, and the frequency response of $G2(z)$. The agreement is really satisfactory.

The Program A.20 has been included in Appendix A.

6.5 The Case of Transfer Functions with Delay

There are systems that combine dynamic behaviour and delay. Figure 6.29 depicts a block diagram corresponding to this case, which is not unusual in real life.

The purpose of this section is to illustrate what you can expect at the hour of experimental modelling when there is delay.

6.5.1 Two Simple Examples

As examples, let us continue with a modification of the previous cases. For instance a first order system with 0.5 s delay (let us denote it as case 1d):

$$Gd1(s) = e^{-j0.5\,s}\,\frac{10}{s+10} \tag{6.4}$$

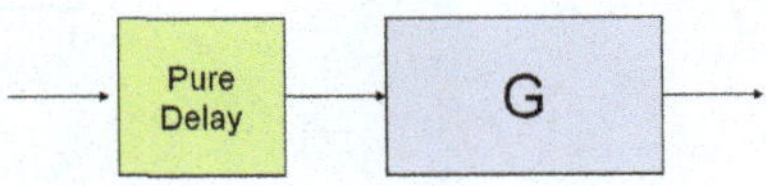

Fig. 6.29 System with pure delay

And a second order system with 0.5 s delay (case 2d):

$$Gd2(s) = e^{-j0.5\,s} \frac{10s}{s^2 + 3s + 10} \tag{6.5}$$

The delay may be inherent to the system structure. For instance transportation delays, like in the case of fluids moving in pipes. Another cause of delay may be the location of the sensors used for measurements. For instance the delay due to distance of a microphone and the source of sound.

If you are aware that there is a delay, the modelling effort can be driven to obtain a model with the structure of Eq. (6.1), obtaining an estimate of the delay and then an estimation of the quotient of polynomials. If you are not aware of the delay, strange transfer functions may be obtained. This kind of issues is the subject of this section.

As a first step, let us review the kind of responses that can be expected from systems including delays.

6.5.2 Responses of Case 1d

Figure 6.30, which has been generated with the Program 6.15, depicts the impulse response of case 1d. It is just a time-shifted version of the impulse response of case 1, as shown on top in Fig. 6.12. Of course, the direct measurement of the time-shift in the impulse response would be clearly useful to obtain a model of the exponential part in expressions like (6.4).

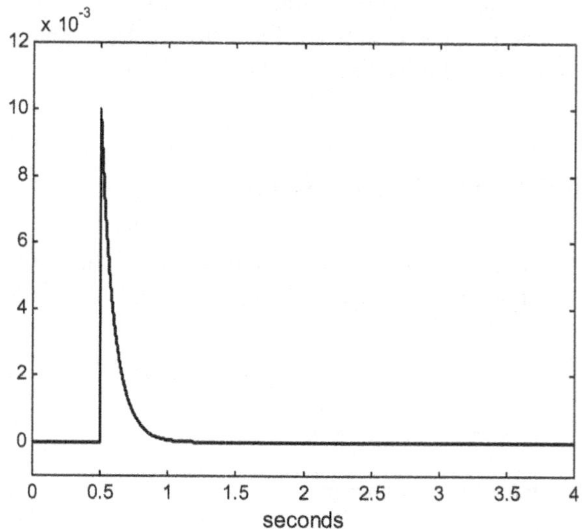

Fig. 6.30 Impulse response of case 1d

Program 6.15 Transfer function + delay, for study: impulse response

```
% Transfer function + delay, for study: impulse response
%case 1d
Td=0.5; %pure delay in seconds
% continuous time transfer function:
num1Ns=10; den1Ns=[1 10]; %G1(s)=10/(s+10)
fs=1000; %sampling frequency in Hz
Ts=1/fs; %sampling period in seconds
Nd=ceil(Td/Ts); %number of samples corresponding to the delay
%discrete transfer function (from the continuous cases)
[num1Nz,den1Nz]= impinvar(num1Ns,den1Ns,fs); %G1(z)
num1Nz=[zeros(1,Nd) num1Nz]; %adding delay to G1(z)
%impulse response
t=0:Ts:(4-Ts); %sampling times data set (4 seconds)
Ns=length(t); %number of samples
[h1Nz,t1]=impz(num1Nz,den1Nz,Ns); %G1(z) impulse response
plot(t,h1Nz,'b'); hold on;
title('impulse response of G1(z)');
xlabel('seconds')
axis([0 4 0 0.012]);
```

The most interesting aspect of Fig. 6.31, which shows the frequency response of case 1d, is the steep descent of the phase curve. To highlight this, the figure shows at the same time the frequency response of case 1. Notice that the magnitude curves are the same, but the phase curves diverge quite a lot. This is an indication of having delay.

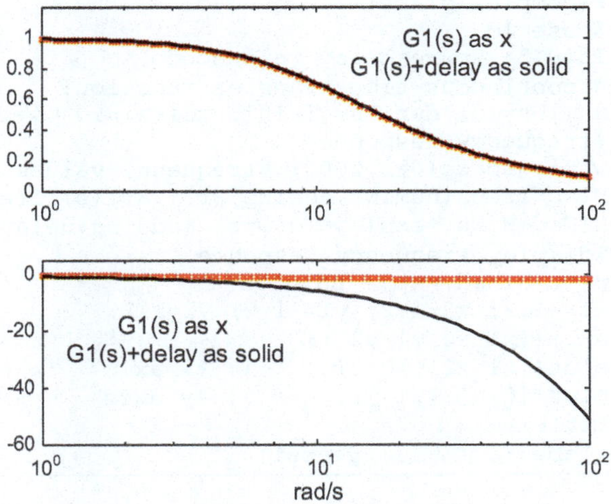

Fig. 6.31 Comparison of frequency responses of case 1 and case 1d

Program 6.16 Transfer function + delay, for study: frequency response (Bode diagram)

```
% Transfer function + delay, for study: frequency response
%(Bode diagram)
%case 1d
Td=0.5; %pure delay in seconds
% continuous time transfer function:
num1Ns=10; den1Ns=[1 10]; %G1(s)=10/(s+10)
%frequency response
wr=logspace(0,2,80); %frequency values for response (rad/s)
H1Ns=freqs(num1Ns,den1Ns,wr); %G1(s) frequency response
H1Nsd=H1Ns.*exp(-j*Td*wr); %adding delay to G1(s)
subplot(2,1,1)
semilogx(wr,abs(H1Ns),'xr'); hold on;
semilogx(wr,abs(H1Nsd),'k'); hold on;
title('G1(s) as x {&} G1(s)+delay as solid');
axis([1 100 0 1.2]);
subplot(2,1,2)
semilogx(wr,angle(H1Ns),'xr'); hold on;
semilogx(wr,unwrap(angle(H1Nsd)),'k'); hold on;
xlabel('rad/s')
axis([1 100 -60 5]);
```

The frequency response of case 1d is also shown on the complex plane by the Fig. 6.32 (Program 6.17). Here the delay causes a spiral, which again is an indication of delay.

Program 6.17 Transfer function + delay, for study: frequency response (complex plane)

```
% Transfer function + delay, for study: frequency response
%(complex plane)
%case 1d
Td=0.5; %pure delay in seconds
% continuous time transfer function:
num1Ns=10; den1Ns=[1 10]; %G1(s)=10/(s+10)
%frequency response
wr=logspace(0,2,200); %frequency values for response (rad/s)
H1Ns=freqs(num1Ns,den1Ns,wr); %G1(s) frequency response
H1Nsd=H1Ns.*exp(-j*Td*wr); %adding delay to G1(s)
%display frequency response
plot(H1Nsd,'k'); hold on;
x1=-1.2; x2=1.2; y1=-1.2; y2=1.2;
axis([x1 x2 y1 y2]);
plot([x1 x2],[0 0],':g'); %x axis
plot([0 0],[y1 y2],':g'); %y axis
title('G1(s)+delay complex frequency response');
xlabel('real'); ylabel('imaginary');
```

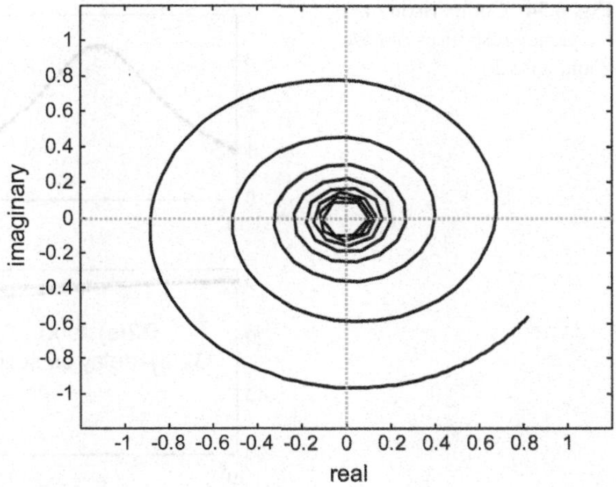

Fig. 6.32 Frequency response of case 1d in the complex plane

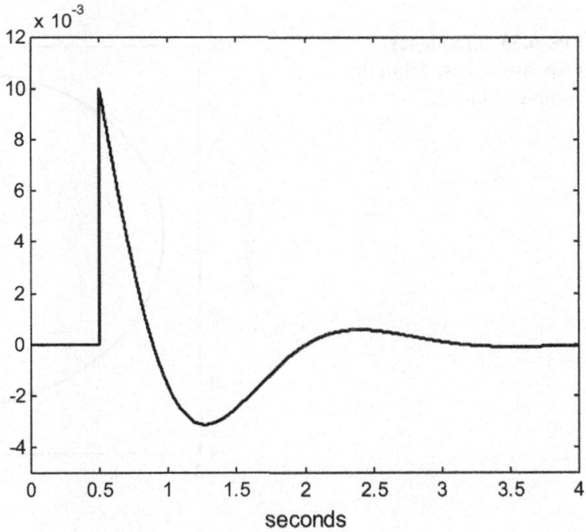

Fig. 6.33 Impulse response of case 2d

6.5.3 Responses of Case 2d

The responses of case 2d follow similar patterns as in case 1d. For instance, the Fig. 6.33 (Program A.21) depicts the impulse response of case 2d, which is a time-shifted version of the impulse response of case 2 (bottom of Fig. 6.12).

The three Programs A.21, A.22, and A.23, corresponding to this subsection have been included in Appendix A.

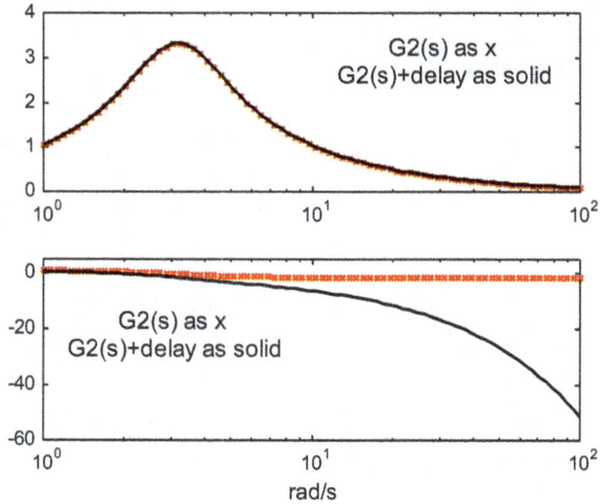

Fig. 6.34 Comparison of frequency responses of case 2 and case 2d

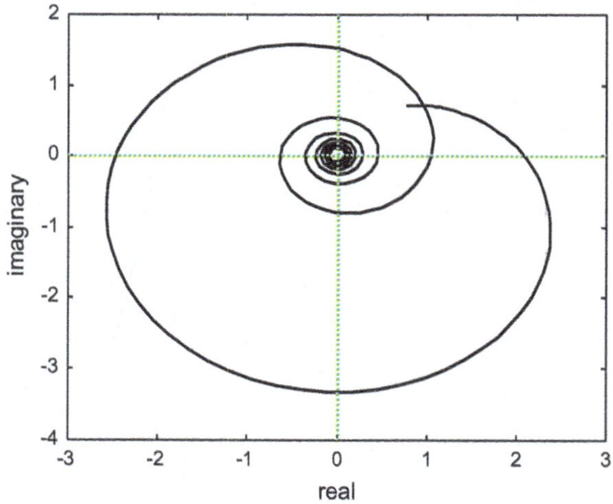

Fig. 6.35 Frequency response of case 2d in the complex plane

The frequency response of case 2d is compared in Fig. 6.34 with the frequency response of case 2. Magnitude curves are equal, but phase curves diverge, since the phase of case 2d plunges down due to the delay. Figure 6.34 has been generated with the Program A.22.

The frequency response of case 2d is also plotted on the complex plane, Fig. 6.35 (Program A.23). The dynamic behaviour of the system is visible at the beginning of the curve (low frequencies). Then, as frequency increases, the curve becomes a deep spiral because of the delay.

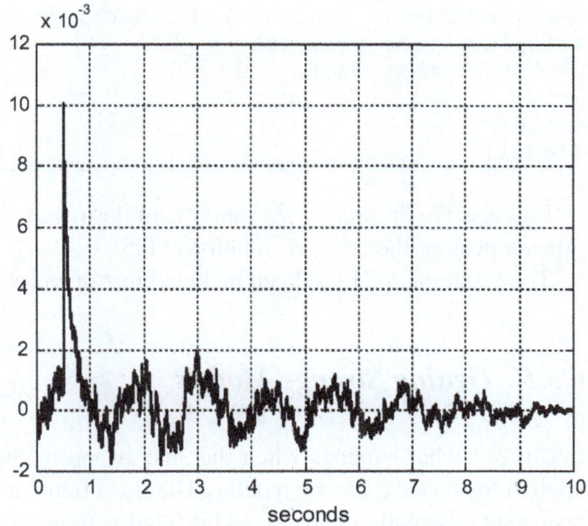

Fig. 6.36 Detecting the delay in case 1d

6.5.4 Detecting the Delay

The response to noise can be useful to detect delays, by means of cross-correlation between the input and the output. Let us show the results in cases 1d and 2d.

Figure 6.36, depicts the cross-correlation between input and output for the case 1d. The figure has been generated with the Program 6.18, which uses the function *xcorr()* to compute the cross-correlation. There is a neat peak at 0.5 s corresponding to the system delay.

Program 6.18 Transfer function + delay, for study: noise response

```
% Transfer function + delay, for study: noise response
%case 1d
Td=0.5; %pure delay in seconds
% continuous time transfer function:
num1Ns=10; den1Ns=[1 10]; %G1(s)=10/(s+10)
fs=1000; %sampling frequency in Hz
Ts=1/fs; %sampling period in seconds
Nd=ceil(Td/Ts); %number of samples corresponding to the delay
%discrete transfer function (from the continuous cases)
[num1Nz,den1Nz]= impinvar(num1Ns,den1Ns,fs); %G1(z)
num1Nz=[zeros(1,Nd) num1Nz]; %adding delay to G1(z)
%response of G1 to noise
tiv=1/fs; %time interval between samples;
t=0:tiv:(10-tiv); %time intervals set (10 seconds)
N=length(t); %number of data points
u=randn(N,1); %random input signal data set
y=filter(num1Nz,den1Nz,u); %G1(z) response to noise
ac=xcorr(u,y); %cross-correlation of u and y
ac=ac/N; %normalization
```

```
Nac=length(ac); mNac=ceil(0.5*Nac); %to plot half ac
%display cross-correlation
plot(t,ac(mNac:Nac),'k');
grid;
title('G1(z)+delay noise response cross-correlation');
xlabel('seconds');
```

Figure 6.37 (Program A.24) shows similar results for the case 2d: there is a significant peak of the cross-correlation at 0.5 s.

The Program A.24 has been included in Appendix A.

6.5.5 Getting Strange Models

Let us see what happens when the sine sweep method described in Sect. 6.4.3 is applied to a system having a delay. That is, a frequency response of the system has been experimentally obtained, and is fitted with *invfreqs()* or *invfreqz()*. The models obtained (transfer functions) are not wrong, but they look strange because many poles and real-positive zeros are needed to model the delay, so even it can mask the dynamic behaviour of the system.

The Program 6.19 applies *invfreqs()* to obtain a model of case 1d from the frequency response of this case. A lot of poles and zeros are needed for a good fitting. The program generates two figures. The first of these, Fig. 6.38, compares the frequency response of case 1d, and the frequency response of the obtained model. There are small divergences.

Figure 6.39 depicts the pole-zero map of the obtained model. There is a large set of pole-zero pairs that clearly corresponds to a model of the delay, similar to Fig. 6.11.

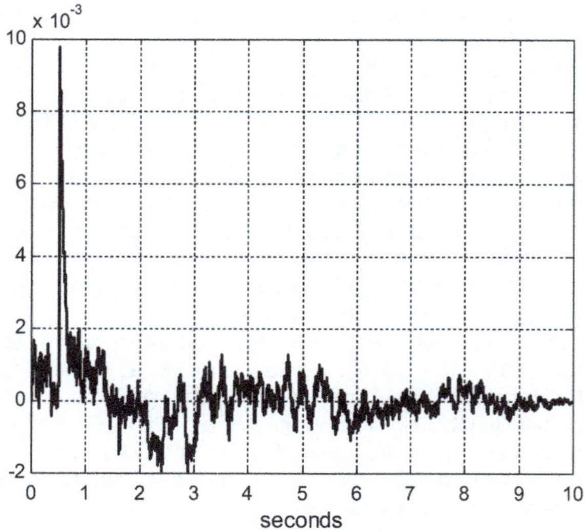

Fig. 6.37 Detecting the delay in case 2d

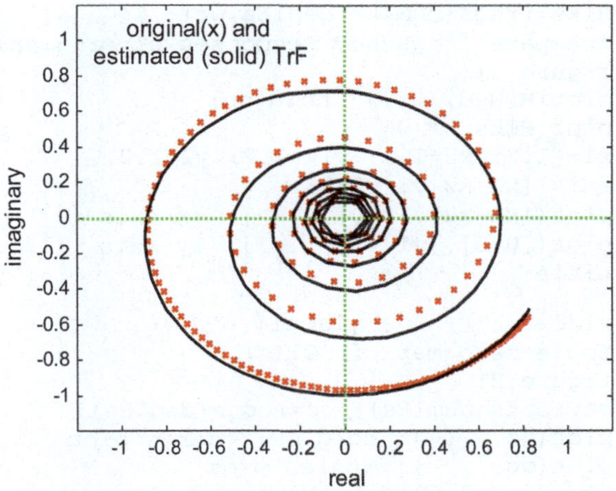

Fig. 6.38 Comparison of frequency responses of original and estimated case 1d

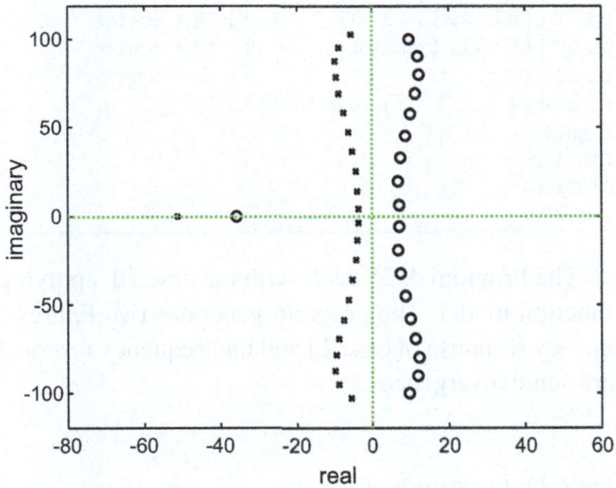

Fig. 6.39 Pole-zero map of estimated TrF for case 1d

Program 6.19 Strange model from frequency response of transfer function with delay

```
% Strange model from frequency response of
% transfer function with delay
%case 1d
Td=0.5; %pure delay in seconds
% continuous time transfer function:
num1Ns=10; den1Ns=[1 10]; %G1(s)=10/(s+10)
%frequency response
wr=logspace(0,2,200); %frequency values for response (rad/s)
H1Ns=freqs(num1Ns,den1Ns,wr); %G1(s) frequency response
H1Nsd=H1Ns.*exp(-j*Td*wr); %adding delay to G1(s)
%using invfreqs to obtain the TrF
na=24; %denominator degree
nb=20; %numerator degree
[num1Es,den1Es]=invfreqs(H1Nsd,wr,nb,na); %TrF computation
```

```
H1Es=freqs(num1Es,den1Es,wr); %^G1(s) frequency response
%compare frequency responses of original and estimated TrF
figure(1)
plot(H1Nsd,'xr'); hold on;
plot(H1Es,'k');
x1=-1.2; x2=1.2; y1=-1.2; y2=1.2;
axis([x1 x2 y1 y2]);
plot([x1 x2],[0 0],':g'); %x axis
plot([0 0],[y1 y2],':g'); %y axis
title('frequency responses of original(x)
and estimated (solid) TrF');
xlabel('real'); ylabel('imaginary');
%pole-zero map of ^G1(s)
figure(2)
gz=roots(num1Es); gp=roots(den1Es);
plot(gz,'ok'); hold on; %zeros plot
plot(gp,'xk'); %poles plot
x1=-80; x2=60; y1=-120; y2=120;
axis([x1 x2 y1 y2]);
plot([x1 x2],[0 0],':g'); %x axis
plot([0 0],[y1 y2],':g'); %y axis
title('zeros and poles of estimated TrF');
xlabel('real'); ylabel('imaginary');
num1Ns
num1Es
den1Ns
den1Es
```

The Program A.25 deals with the case 2d, applying *invfreqs()* to obtain a transfer function model. The program generates two figures. Figure 6.40 compares the frequency response of case 2d and the frequency response of the obtained model. There are some divergences.

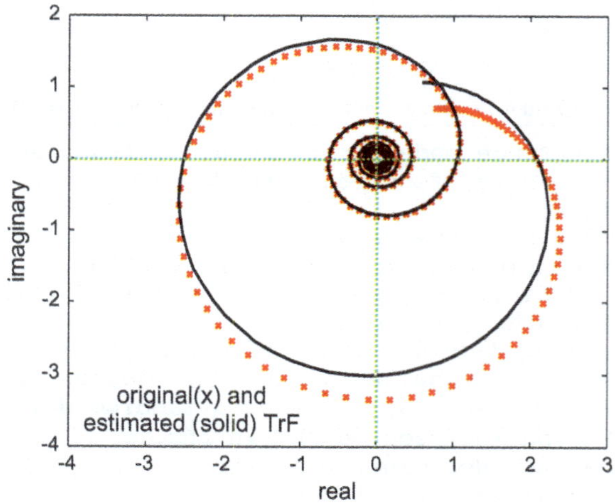

Fig. 6.40 Comparison of frequency responses of original and estimated case 2d

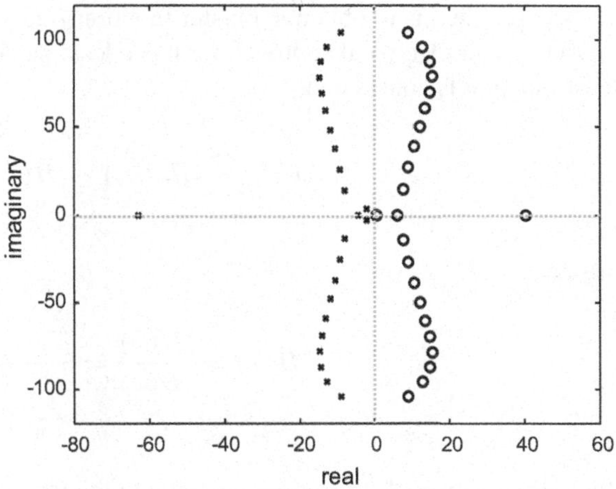

Fig. 6.41 Pole-zero map of estimated TrF for case 2d

Figure 6.41 shows the pole-zero map of the obtained model. As in case 1d, there is a large set of pole-zero pairs that reproduces the pattern in Fig. 6.11.

The Program A.25 has been included in Appendix A.

A survey of time-delay estimation methods can be found in [6].

6.6 Methods for Frequency-Domain Modelling

In this section a more general treatment of frequency-domain modelling, in terms of transfer functions, will be introduced.

Frequency-domain modelling methods enjoy important advantages. There are convenient graphical representations of frequency responses that help to visualize the quality of a transfer-function model with respect to the experimental data. No initial state of the system is required. Noise could be adequately managed. In general, the acquisition of experimental data is not complicated.

The author of the present book has some experience with ship motion modelling using scale ships in a towing tank facility: regular waves of different frequencies were generated, and the motion responses of the ship were measured. In order to obtain frequency response data, a series of 24 experiments—each for a different wave frequency-was done. Each experiment took 1 h, since one has to wait for the water to become flat before a new wave train was generated.

Suppose a single-input, single–output (SISO) linear system. By some experimental means, applying suitable inputs to the system, one has obtained a set of frequency response data (amplitude and phase) corresponding to a set of frequencies $\Omega = \{\omega_1, \omega_2, \ldots, \omega_n\}$. Denote the measured frequency responses as $H_m(\Omega)$.

The problem is to obtain a transfer function $H(\omega)$ that could be used as a model of the system. A typical approach for it is a least-square fitting, where the following cost function is considered:

$$J = \sum_{k=1}^{n} |H_m(\omega_k) \; - \; H(\omega_k)|^2 \tag{6.6}$$

Since:

$$H(\omega) \; = \; \frac{N(\omega)}{D(\omega)} \; = \; \frac{\sum_{j=0}^{M} \alpha_j \, s^j}{\sum_{i=0}^{N} \beta_i \, s^i} \tag{6.7}$$

the minimization of J is a non-quadratic in the parameters problem.

Many alternatives for this nonlinear least-squares minimization have been proposed. An important review of the solutions already suggested before year 1994 has been presented by R. Pintelon and colleagues in [29]. More recently, Pintelon and Schoukens have published a book, [30], on frequency-domain system identification.

6.6.1 The Levi's Approximation

As early as 1959, Levi introduced a change in the problem formulation, such that the minimization becomes linear [22]. By multiplying with the denominator of the transfer function, a new cost function was defined:

$$J_L = \sum_{k=1}^{n} |H_m(\omega_k) D(\omega_k) \; - \; N(\omega_k)|^2 \tag{6.8}$$

As reported in the literature, this scheme may lead to poor numerical conditioning, especially when the frequency span and/or the model order become large. Also, it tends to emphasize high frequency errors and to neglect to some extent low frequency errors.

Program 6.20 provides an example of Levi's approximation. Figure 6.42 shows the results. The frequency response data to be fitted are represented with X marks. The result of the approximation is shown with a continuous curve. A plant with a second order polynomial in the numerator and the product of second order polynomials in the denominator has been chosen for the test. Notice that a function has been embedded in the code (modern versions of MATLAB support the '@...' format).

Program 6.20 Example of Levi's approximation

```
% Example of Levi's approximation
%(frequency domain modelling)
% a simple test case ----------------
num=[1 0.2 1];
den1=[1 0.1 0.1];
den2=[1 0.5 7];
den=conv(den1,den2);
Ho=tf(num,den);
w=logspace(-1,1,100);
%frequency-domain response
Hm=freqresp(Ho,w);
Hm=squeeze(Hm)'; % "measured" data to be fitted
%error function
erf=@(x)sum(abs((Hm.*((j*w).^4+x(1)*(j*w).^3+...
x(2)*(j*w).^2+x(3)*(j*w)+x(4)))...
-(x(5)*(j*w).^2+x(6)*(j*w)+x(7))).^2);
[x,fiterr]=fminsearch(erf,ones(7,1));
esnum=[x(5) -x(6) x(7)];
esden=[1 -x(1) x(2) -x(3) x(4)];
Hes=tf(esnum,esden);
%figure(1)
bode(Ho,'rx'); hold on;
bode(Hes,'b');
grid
title('Bode diagram: (x) data, (continuous) approximation');
```

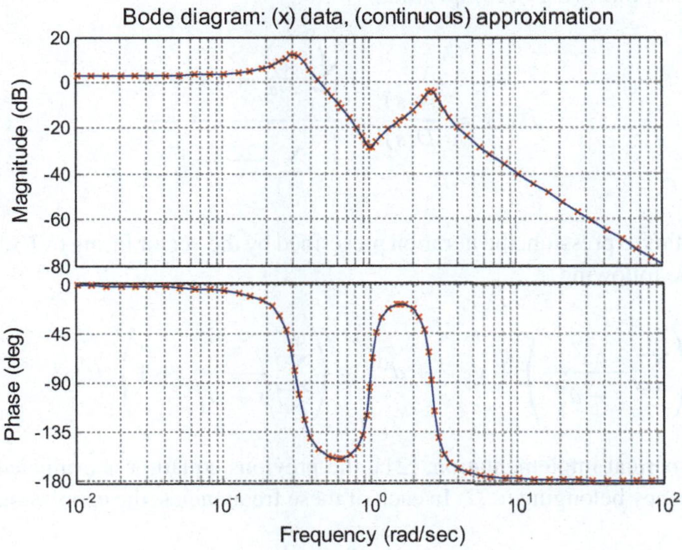

Fig. 6.42 Frequency domain modelling via Levi's approximation

6.6.2 The SK Iterative Weighted Approach

In 1963, Sanathanan and Koerner introduced an iterative weighted approach, which pays better attention to low frequency errors, [32]. Many authors refer to this method as the SK algorithm. Each iteration has to minimize the following:

$$J_{SK}^{(i)} = \sum_{k=1}^{n} \frac{|H_m(\omega_k) D^{(i)}(\omega_k) - N^{(i)}(\omega_k)|^2}{|D^{(i-1)}(\omega_k)|^2} \tag{6.9}$$

Usually, the starting values of the parameters are taken from the Levi's approximation.

Numerous modifications of this procedure have been proposed. Most of them use a weighting factor, as follows:

$$J_W^{(i)} = \sum_{k=1}^{n} |W^{(i-1)}(\omega_k)|^2 \cdot |H_m(\omega_k) D^{(i)}(\omega_k) - N^{(i)}(\omega_k)|^2 \tag{6.10}$$

The SK method is applied for control-relevant multivariable system identification in [20], which is related to the CR-IDENT Toolbox.

6.6.3 The Vector Fitting (VF) Approach

Consider the following decomposition:

$$H(s) = \frac{N(s)}{D(s)} = \frac{\sum_{j=1}^{N} \frac{c_j}{s-a_j} + s\,e + d}{\sum_{j=1}^{N} \frac{\gamma_j}{s-a_j} + 1} \tag{6.11}$$

Based on this expression, the iteration prescribed by the vector fitting (VF) algorithm, [15], is the following

$$\left(\sum_{j=1}^{N} \frac{c_j^{(i)}}{s - a_j^{(i)}} \right) + s\,e^{(i)} + d^{(i)} \approx \left(\sum_{j=1}^{N} \frac{\gamma_j^{(i)}}{s - a_j^{(i)}} + 1 \right) H(s) \tag{6.12}$$

In the numerical implementation, [21], the previous equation is evaluated for each *k-th* frequency belonging to Ω. In each of these frequencies, the expression is equivalent to:

$$\mathbf{A}_k \, \mathbf{x}^T = b_k \tag{6.13}$$

with:

$$b_k = H(s_k) \tag{6.14}$$

$$\mathbf{A}_k = \{\frac{1}{s_k - a_1^{(i)}}, \ldots, \frac{1}{s_k - a_N^{(i)}}, \; s, \; 1, \frac{-H(s_k)}{s_k - a_1^{(i)}}, \ldots, \frac{-H(s_k)}{s_k - a_N^{(i)}} \} \tag{6.15}$$

$$\mathbf{x} = \{c_1^{(i)}, \ldots, c_N^{(i)}, e^{(i)}, d^{(i)}, \gamma_1^{(i)}, \ldots, \gamma_N^{(i)}\} \tag{6.16}$$

The equations can be organized as follows:

$$[\mathbf{A}_1^T \; \mathbf{A}_2^T \ldots \mathbf{A}_n^T]^T \; \mathbf{x}^T = [b_1 \; b_2 \ldots b_n]^T \tag{6.17}$$

which is an over-determined system of equations, which can be solved through normal equations or a QR decomposition.

The VF algorithm starts with an initial choice of the poles. For instance, complex conjugate pairs with small real parts and the imaginary parts evenly spaced on the adequate frequency range.

In each iteration, the algorithm takes a residue identification step and a pole relocation step. The first step computes the coefficients (the residues) by solving Eq. (6.13). The second step selects as new poles the zeros of:

$$\sigma(s) = \left(\sum_{j=1}^{N} \frac{\gamma_j^{(i)}}{s - a_j^{(i)}} + 1 \right) \tag{6.18}$$

A simple way for obtaining these zeros is to compute the eigenvalues of the following matrix:

$$\Psi = \begin{bmatrix} a_1^{(i)} & \cdots & & & \\ & a_2^{(i)} & & & \\ & & a_3^{(i)} & & \\ & & & \ddots & \\ & & & & a_N^{(i)} \end{bmatrix} - \begin{bmatrix} 1 \\ 1 \\ \vdots \\ 1 \end{bmatrix} \begin{bmatrix} \gamma_1^{(i)} \\ \gamma_2^{(i)} \\ \vdots \\ \gamma_N^{(i)} \end{bmatrix} \tag{6.19}$$

In case a pole was unstable, it can be flipped against the imaginary axis to get a stable pole.

See the web page on Vector Fitting (address given in the resources section at the end of the chapter) for more details of improvements and implementations. Based on one of these implementations a program has been developed (Program A.26) for the fitting of a test example. This program has been included in Appendix A.

Figure 6.43 shows the good result of model fitting for the chosen example.

The VF algorithm has got wide acceptance due to its efficiency and good data fitting performances. It has been shown in [16] that the VF iteration can be regarded as a reformulated SK iteration.

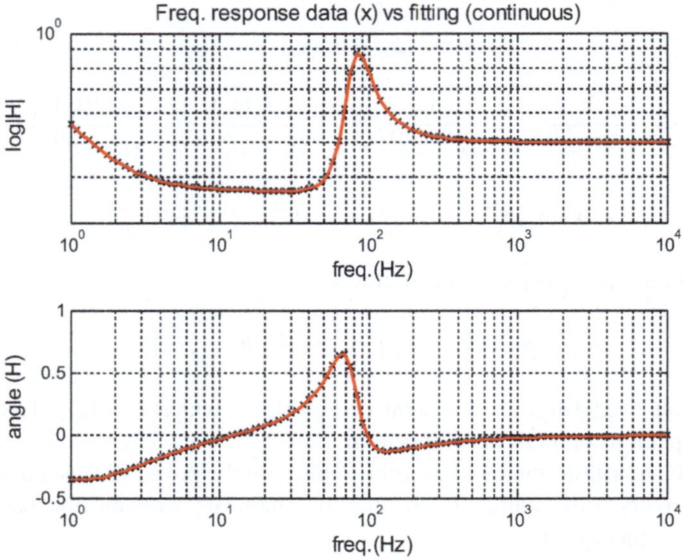

Fig. 6.43 Frequency domain modelling using vector fitting

According with reported experiences, the convergence of the VF algorithm may deteriorate in the presence of noise. Some modifications of the algorithm have been proposed, like for instance [12], to alleviate this problem.

The research is working on improvements of iterative methods, [11, 35], trying to cope with difficult modelling scenarios, like in the case of vibration modes of mechanical structures, or cases related to transmission lines and transformers [13, 14].

There are other frequency domain system identification approaches and extensions, like for instance frequency domain state-space system identification [9], or multiple-input multiple output systems [3, 8]. An extensive study of frequency domain identification for dynamical structures with oscillation modes is [7].

6.7 Methods for Time-Series Modelling

The scenario to be considered in this section is the following: there is a single-input single-output system, and it is possible to record the responses of the system to a series of input data. A mathematical model of the system is proposed, with a certain structure. The problem is to estimate the parameters of the model, using input and output data.

As an estimation problem, it can be treated as an optimization problem. An estimation error is defined; and a criterion to optimize is stated in terms of this error (typically, using squares). For this reason, the methods to be introduced in this section can be seen as a continuation of the matter concerning the Wiener filter.

Parameter estimation is a nuclear part of a more general topic, already mentioned in the introduction to this chapter: model (or system) identification. When you face a system for the first time, and you want to obtain a suitable model of its dynamics, you should design a series of experiments, providing input signals rich enough to excite all the dynamics hidden in the system. For example, a new airplane: you want to know how it responds to aileron motions; motions that could be soft or brisk. Then, perhaps based on first principles (physics) you may have an idea of the suitable model structure. And now, using experimental data and the model structure, it comes to estimate the parameters. Depending on the results, you may need to modify the model structure and try again the parameter estimation. All these activities constitute system identification.

Most theory on parameter estimation has been developed using ARMAX models. The results are easily translated to other types of models.

The general expression of the ARMAX model is:

$$A(q^{-1})\, y(t) = B(q^{-1})\, u(t) + C(q^{-1})\, e(t) \tag{6.20}$$

where $e(t)$ is white noise.

The present response of the system is:

$$\begin{aligned} y(t) &= -a_1\, y(t-1) - a_2\, y(t-2) - \ldots - a_n\, y(t-n) + \\ &+ b_0 u(t) + b_1\, u(t-1) + b_2\, u(t-2) + \ldots + b_m\, u(t-m) + \\ &+ c_0 e(t) + c_1\, e(t-1) + c_2\, e(t-2) + \ldots + c_l\, e(t-l) \end{aligned} \tag{6.21}$$

Denote:

$$\boldsymbol{\theta} = \begin{pmatrix} a_1 \\ a_2 \\ . \\ a_n \\ b_0 \\ b_1 \\ . \\ b_m \\ c_0 \\ c_1 \\ . \\ c_l \end{pmatrix} \; ; \quad \mathbf{d}(t) = \begin{pmatrix} -y(t-1) \\ -y(t-2) \\ . \\ -y(t-n) \\ u(t) \\ u(t-1) \\ . \\ u(t-m) \\ e(t) \\ e(t-1) \\ . \\ e(t-l) \end{pmatrix} \tag{6.22}$$

Then, the ARMAX model can be written as follows:

$$y(t) = \mathbf{d}^T(t) \cdot \boldsymbol{\theta} \tag{6.23}$$

The parameter estimation problem is to estimate $\boldsymbol{\theta}$ using input $u()$ and output $y()$ data.

It is opportune to establish an estimation error. For instance, the following:

$$er(t, \hat{\theta}) = y - \mathbf{d}^T(t) \cdot \hat{\theta} \tag{6.24}$$

where $\hat{\theta}$ are the estimated parameters.

Other error definitions could be considered. The error introduced in (6.24) is called "*equation error*".

A fundamental reference for system identification is [23]. Most system identification Toolboxes, like for instance IDTOOL [10], are based on that book.

6.7.1 Basic Identification Methods

Depending on the application, it would be possible to apply batch mode estimation, or recursive estimation. In the case of batch mode, there are two steps: first to gather sufficient data samples, and then to compute the estimated parameters. In the case of recursive estimation, the estimation of parameters is made in line with the data stream.

1. **Batch mode parameter estimation**

   Along time, a set of N measurements of input and output data have been obtained. Successive equations can be written as follows:

$$\begin{aligned} y(1) &= \mathbf{d}^T(1) \cdot \hat{\theta} + er(1, \hat{\theta}) \\ y(2) &= \mathbf{d}^T(2) \cdot \hat{\theta} + er(2, \hat{\theta}) \\ &\dots \\ y(N) &= \mathbf{d}^T(N) \cdot \hat{\theta} + er(N, \hat{\theta}) \end{aligned} \tag{6.25}$$

In a more compact form:

$$\mathbf{y} = D \cdot \hat{\theta} + \mathbf{er}(\hat{\theta}) \tag{6.26}$$

In order to minimize the sum of the squared errors, the following *"normal equations"* should be solved:

$$D^T D \cdot \hat{\theta} = D^T \cdot \mathbf{y} \tag{6.27}$$

   Denote: $P = [D^T D]^{-1}$
   Therefore:

$$\hat{\theta} = P \cdot D^T \cdot \mathbf{y} \tag{6.28}$$

1. **Recursive mode parameter estimation**

   As before, let us minimize the sum of the squared errors. By subtraction of successive batch mode estimations, the following recursive expression is obtained:

$$\hat{\theta}(k+1) = \hat{\theta}(k) + \chi(k+1)\,[y(k+1) - \mathbf{d}^T(k+1)\cdot\hat{\theta}(k)] \quad (6.29)$$

where:

$$\chi(k+1) = P(k+1)\cdot\mathbf{d}(k+1) = \frac{P(k)\,\mathbf{d}(k+1)}{1 + \mathbf{d}^T(k+1)\,P(k)\,\mathbf{d}(k+1)} \quad (6.30)$$

From above:

$$P(k+1) = P(k) - \chi(k+1)\,\mathbf{d}^T(k+1)\,P(k) \quad (6.31)$$

The typical recursive estimation algorithm starts from:

$$\hat{\theta}(0) = \mathbf{0};\; P(0) = \alpha\, I\,,\;\; \alpha >> 0 \quad (6.32)$$

and then it uses experimental data to compute $\chi(k+1)$ and $P\,(k+1)$, then $\hat{\theta}(k+1)$, and so on (for k = *0, 1, 2...*).

An example of recursive parameter identification is included next. Program 6.21 implements the algorithm just described, with successful results. Figure 6.44 shows the evolution of the error along the iterations of the identification method, which rapidly converges. The reader is invited to try other plants and see the identification results.

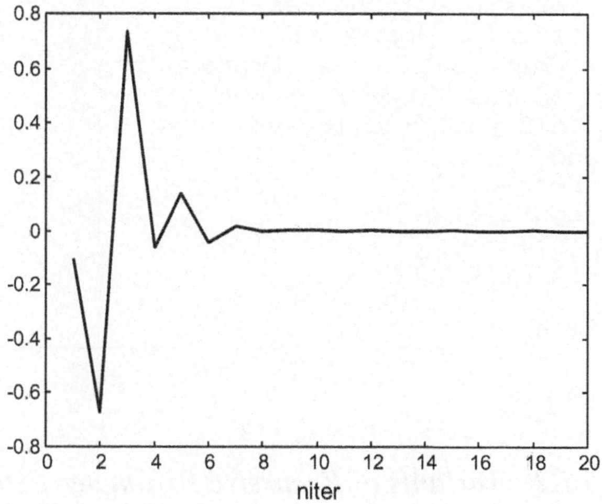

Fig. 6.44 Evolution of error along identification iterations

Program 6.21 Example of ARMA parameter estimation

```
% Example of ARMA parameter estimation
% (time-series ARMAX modelling)
% a simple test case -----------------
% [a1 a2 b1 b2 c1]
thetao=[0.9 0.7 0.5 0.3 0.2]';
N=length(thetao);
% iterative estimation of parameters
M=20;
erec=zeros(1,M); %for error recording
thetaes=zeros(N,1); % estimated parameters
d=zeros(N,1); % vector of inputs/outputs
P=diag(1000*ones(N,1));
for nn=1:M,
  up=randn(1,1); % input at present
  ep=0.5*randn(1,1); % noise at present
  % actualization of d
  d(4)=d(3); d(3)=up;
  d(5)=ep;
  yp=d'*thetao; % plant output at present
  % parameter estimation steps:
  aux=1+(d'*P*d);
  kappa=(P*d)/aux;
  P=P-(kappa*d'*P);
  err=yp-(d'*thetaes);
  erec(nn)=err;
  thetaes=thetaes+(kappa*err);
  %actualization of d
  d(2)=d(1); d(1)=-yp;
end
thetaes
figure(1)
plot(erec,'k');
title('Error evolution along parameter
identification iterations');
xlabel('niter');
```

6.7.2 *Variants of Recursive Parameter Estimation*

Suppose you have a large unmanned airplane. It happens that along the flight the fuel consumption makes the dynamic behavior of the airplane change. You want a good control of the airplane at all times, so it is convenient to use current information of the airplane parameters (that change along time). Then, it seems opportune to apply recursive parameter estimation in real-time. This is the basic idea of adaptive control, which requires much care for successful results. One of the problems of the recursive algorithm already introduced is that it may tend to 'sleep', in the sense that it may not adapt well to the parameter changing. Many alternatives have been proposed to respond to this drawback. A radical one is to add a supervision level that detects the 'sleep' and then applies a reset to the identification algorithm. Another way, that could be used in combination with the supervision, is to use forgetting factors.

The idea of forgetting factors is to give more importance to the most recent information, and forget old-obsolete estimations. For example, in the case of *exponential forgetting*, the update of P is made as follows:

$$P(k+1) = \frac{1}{\lambda}\left\{P(k) - \frac{P(k)\mathbf{d}(k+1)\,\mathbf{d}^T(k+1)P(k)}{\lambda + \mathbf{d}^T(k+1)P(k)\,\mathbf{d}(k+1)}\right\} \tag{6.33}$$

where $0 < \lambda < 1$ is a forgetting factor.

In the case of *variable exponential forgetting*, the value of λ changes according with:

$$\lambda(k+1) = \lambda_0\,\lambda(k) + 1 - \lambda_0 \tag{6.34}$$

where $\lambda(0) = \lambda_0 \in (0.95\ 0.99)$.

Another alternative is to use a *fixed directional forgetting*, which uses the following update:

$$P(k+1) = \left\{P(k) - \frac{P(k)\mathbf{d}(k+1)\,\mathbf{d}^T(k+1)P(k)}{\varepsilon^{-1} + \mathbf{d}^T(k+1)P(k)\,\mathbf{d}(k+1)}\right\} \tag{6.35}$$

with:

$$\varepsilon(k) = \lambda - \frac{1-\lambda}{\mathbf{d}^T(k+1)P(k)\mathbf{d}(k)} \tag{6.36}$$

and $0 < \lambda < 1$.

In order to avoid problems when the noise is not so ideally zero-mean Gaussian, an *instrumental variable* $\mathbf{z}$ can be used. This variable is introduced in the updating of P as follows:

$$\chi(k+1) = \frac{P(k)\,\mathbf{d}(k+1)}{1 + \mathbf{d}^T(k+1)\,P(k)\,\mathbf{z}(k+1)} \tag{6.37}$$

A typical choice of the instrumental variable is:

$$\mathbf{z}(k) = [u(k-1)\ u(k-2)\ \ldots u(k-na-nb)]^T \tag{6.38}$$

Another choice, which is model dependent, could be:

$$\mathbf{z}(k) = [y_u(k-1)\ y_u(k-2)\ \ldots y_u(k-na)\ \ u(k-1)\ \ldots u(k-nb)]^T \tag{6.39}$$

with:

$$\begin{aligned} y_u(k) = \hat{b}_1(k)\, u(k-1) + \ldots + \hat{b}_{nb}(k) u(k-nb) - \\ -\hat{a}_1(k) y_u(k-1) - \ldots - \hat{a}_{na}(k) y_u(k-na) \end{aligned} \quad (6.40)$$

where one uses current parameter estimates $\hat{b}_1(k) \ \ldots \hat{a}_{na}(k)$.

See [26] for other variants of the recursive parameter estimation algorithm.

6.8 Experiments

The two experiments included in this section illustrate important aspects of experimental modeling. The first one is an example of having just a data series, so there is no input but one could consider that a white noise input generates the data.

The second experiment put the focus on an important issue: can we arbitrarily choose the order of the model?

6.8.1 AR Model Identification of Canadian Lynx Data

Time-series modeling is important in contexts where you have data that change along time, and that perhaps obey to a certain internal law (or a mix of laws) in a certain degree. A typical case would be weather forecasting, in which you can expect seasonal characteristics. Likewise, sales of swimming apparel would show good moments around summer, and feeble numbers during winter. Electricity consuming along days, weeks, and year would have certain predictability. And so one could continue with insect populations, sunspots, milk production, stock forecasting, monthly airline passengers, rivers flow, fruit prices, etc.

Most of the cases suggested above would not be responses to an input. However, AR modeling could be done, supposing a white noise input. A deterministic fitting would not be expected, since we place ourselves in a statistical scenario, but the general behavior would be captured.

This experiment is just an AR model estimation for a well-known data series: the Canadian Lynx population from 1821 to 1934. Figure 6.45 displays this series of data.

In order to estimate the AR model, we choose the *aryule( )* function, which belongs to the Signal Processing Toolbox. This function also estimates the variance of the input noise. See Program 6.22, which obtains the Figs. 6.45 and 6.46.

In order to generate responses from the AR model, we use the *filter( )* function.

Figure 6.46 compares the Lynx data with a response of the AR model. Although not exact fitting was expected, both curves have evident similarities.

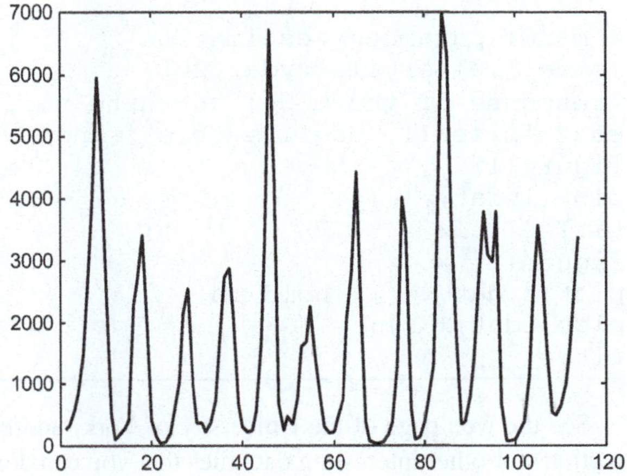

Fig. 6.45 Canadian Lynx population data

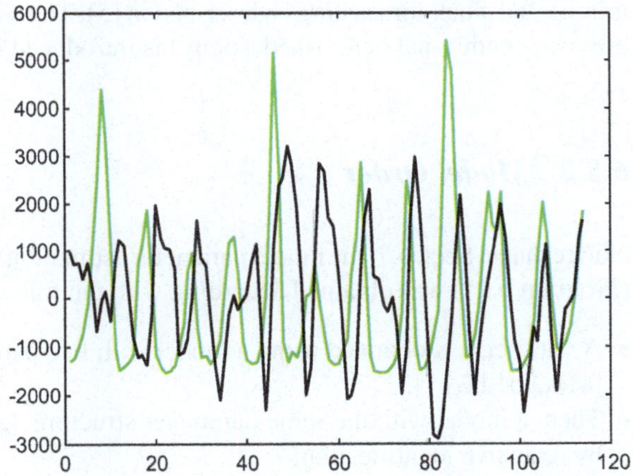

Fig. 6.46 Comparison of detrended data (*green*) and predicted data (*black*)

Program 6.22 Example of AR model estimation

```
% Example of AR model estimation
% Canadian Lynx data
% load Lynx population data
fer=0;
while fer==0,
  fid2=fopen('lynx.txt','r');
  if fid2==-1, disp('read error')
  else lydat=fscanf(fid2,'%f \r\n'); fer=1;
  end;
end;
fclose('all');
N=length(lydat);
MD=mean(lydat);
zydat=lydat-MD; %zero-mean data
```

```
% model parameter estimation
[model1,e]=aryule(zydat,20);
%response of the model to white noise
edat=filter(1,model1,sqrt(e)*randn(N,1));
figure(1)
plot(lydat,'k');
title('Canadian Lynx data');
figure(2)
plot(zydat,'g'); hold on
plot(edat,'k');
title('AR model response')
```

See the web page of the University of York (address included in the Resources section) for other interesting data files that you can download.

For more details on time-series modeling, see [2, 27]. An application for data mining and stock forecasting is described in [5]. The use of GARCH models (autoregressive conditional heteroskedasticity) is introduced in [28].

6.8.2 Model Order

Notice that in Sect. 6.7. the model parameter estimation that was presented in Fig. 6.44 (Program 6.21) was obtained according with the following scenario:

- A data set was obtained using a model with five parameters: [a1 a2 b1 b2 c1].
- Then, a model with the same parameter structure, [a1 a2 b1 b2 c1], was obtained by recursive identification.

In real life, you may not have a clear idea of what is the suitable parameter structure for the model to be estimated.

We propose to the reader to try several different models for the same data input. To this end, a new identification case is stated, and again the recursive identification method is applied. A reference result is obtained, taking the same parameter structure for data generation and for model estimation. This was implemented with the Program 6.23, which is a modified version of Program 6.21 prepared for changing easily the parameter structure of the estimated model.

Notice that now, in the Program 6.23, we are generating a pseudo random binary series (PRBS) of data values. Figure 6.47 shows an instance of these series. PRBS are frequently used for identification purposes (see [18, 33] for more information on PRBS data generation).

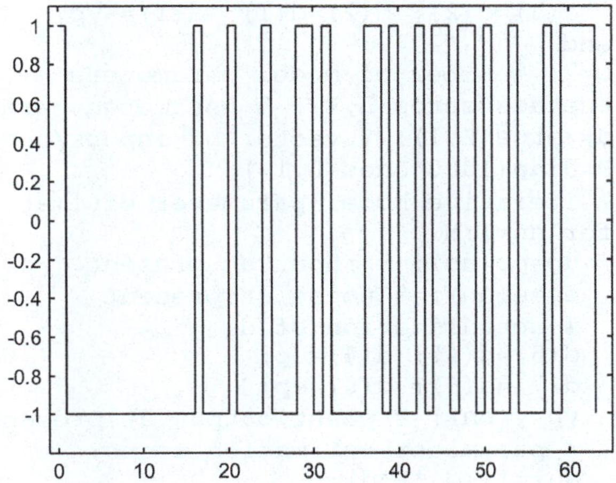

Fig. 6.47 PBRS signal

Program 6.23 Example of ARMA parameter estimation

```
% Example of ARMA parameter estimation
% using PRBS as input
% a test case ------------------
% [a1 a2 a3 b1 b2 c1 c2]
thetao=[0.9 0.7 0.4 0.3 0.8 0.5 0.1]';
N=length(thetao);
d=zeros(N,1);
M=64;
% vector with a series of PRBS values
pr=zeros(M,1);
x=[zeros(1,15) 1];
for k=1:M,
  pr(k)=2*x(16)-1;
  q=x(16)+x(15)+x(13)+x(4);
  x(2:16)=x(1:15);
  x(1)=mod(q,2);
end;
% vector with plant output in response to pr
y=zeros(M,1);
e=zeros(M,1); %noises
for nn=1:M,
  up=pr(nn); % input at present
  ep=0.5*randn(1,1); % noise at present
  e(nn)=ep; %save noise
  % actualization of d
  d(5)=d(4); d(4)=up;
  d(7)=d(6); d(6)=ep;
  yp=d'*thetao; % plant output at present
  y(nn)=yp; %save plant output
  %actualization of d
```

```
  d(3)=d(2); d(2)=d(1); d(1)=-yp;
end
L=7; %number of model parameters
thetaes=zeros(L,1); % estimated parameters
d=zeros(L,1); % vector of inputs/outputs
P=diag(1000*ones(L,1));
% iterative model parameter estimation ----------------
for nn=1:M,
  up=pr(nn); % input at present
  ep=e(nn); % noise at present
  % actualization of d
  d(5)=d(4); d(4)=up;
  d(7)=d(6); d(6)=ep;
  yp=y(nn); % plant output at present
  % parameter estimation steps:
  aux=1+(d'*P*d);
  kappa=(P*d)/aux;
  P=P-(kappa*d'*P);
  err=yp-(d'*thetaes);
  thetaes=thetaes+(kappa*err);
  %actualization of d
  d(3)=d(2); d(2)=d(1); d(1)=-yp;
end
thetaes
figure(1)
xx=0;
for nn=1:M-1,
plot([xx xx+1], [pr(nn) pr(nn)],'k'); hold on;
plot([xx+1 xx+1], [pr(nn) pr(nn+1)],'k');
xx=xx+1;
end;
axis([-1 M+1 -1.2 1.1]);
title('PRBS input');
```

By simple changes in the sentences devoted to the actualization of d in the iterative parameter estimation loop, the Program 6.23 can be used to explore model estimation with different parameter structures. Here is a table with some parameter estimation results:

Struct.	a1..an	b1..bm	c1..cm
A	0.4020, 0.5038	0.3696	0.3904
B	0.6386, 0.5027	0.6780, 0.2943	0.5195
C	0.4389, 0.5092, 0.0873	0.3534	0.4463
D	0.8859, 0.6882, 0.382	0.298, 0.8004	0.5206
E	0.8241, 0.6514, 0.366	0.3025, 0.7797, −0.048	0.4854
F	0.8999, 0.7, 0.4	0.3, 0.8	0.5, 0.1
G	0.8861, 0.6876, 0.3902, −0.056	0.3001, 0.7958, −0.0112	0.5002, 0.0928
H	0.8802, 0.6838, 0.3871, −0.0072	0.2999, 0.7940, −0.0155, 0.0011	0.5001, 0.09

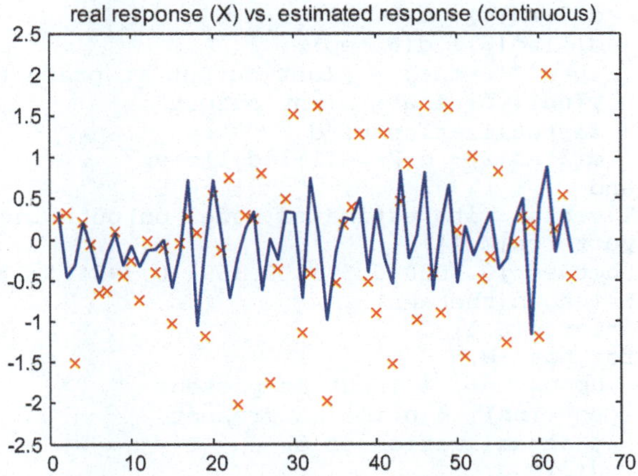

Fig. 6.48 Original plant response (X) versus response of structure A estimated plant (*continuous*)

A second program (Program 6.24) has been included for evaluation of model quality. It computes the responses of the original plant and the modeled plant for the same inputs. Then, it computes the sum of squared errors between data points with same abscissa. Finally, it displays both responses for visual comparison.

Figure 6.48, generated with the Program 6.24, compares responses of the original plant and the estimated plant corresponding to structure A.

Program 6.24 Example of ARMA parameter estimation: comparison

```
% Example of ARMA parameter estimation
% comparison of original and estimated model
% in terms of squared error between output vectors
% the test case -----------------
% [a1 a2 a3 b1 b2 c1 c2]
thetao=[0.9 0.7 0.4 0.3 0.8 0.5 0.1]';
N=length(thetao);
M=64;
% vector with a series of PRBS values
pr=zeros(M,1);
x=[zeros(1,15) 1];
for k=1:M,
  pr(k)=2*x(16)-1;
  q=x(16)+x(15)+x(13)+x(4);
  x(2:16)=x(1:15);
  x(1)=mod(q,2);
end;
% vector with plant output in response to pr
y=zeros(M,1);
e=zeros(M,1); %noises
d=zeros(N,1);
for nn=1:M,
  up=pr(nn); % input at present
  ep=0.5*randn(1,1); % noise at present
  e(nn)=ep; %save noise
  % actualization of d
```

```
  d(5)=d(4); d(4)=up;
  d(7)=d(6); d(6)=ep;
  yp=d'*thetao; % plant output at present
  y(nn)=yp; %save plant output
  %actualization of d
  d(3)=d(2); d(2)=d(1); d(1)=-yp;
end
% vector with estimated model outputs in response to pr
ye=zeros(M,1);
thetaes=[0.4020 0.5038 0.3696 0.3904]'; % estimated parameters
L=length(thetaes);
d=zeros(L,1);
for nn=1:M,
  up=pr(nn); % input at present
  ep=e(nn); % noise at present
  % actualization of d
  d(3)=up;
  d(4)=ep;
  yp=d'*thetaes; % plant output at present
  ye(nn)=yp; %save estimated plant output
  %actualization of d
  d(2)=d(1); d(1)=-yp;
end
Aerr=0;
for nn=1:M,
  Aerr=Aerr+((y(nn)-ye(nn))^2);
end
Aerr
figure(1)
plot(y,'rx'); hold on;
plot(ye,'b');
title('real response (X) vs. estimated response (continuous)');
```

Note that parts of the code in Program 6.24 could have been done using the function *filter( )*.

Next table shows the errors obtained for each of the structures detailed in the previous table.

Structure	Error
A	46.6514
B	46.31
C	48.6
D	0.376
E	0.2148
F	0
G	7 e -5
H	8 e-5

Figure 6.49 compares the original data and the predicted data for the structure F estimated model.

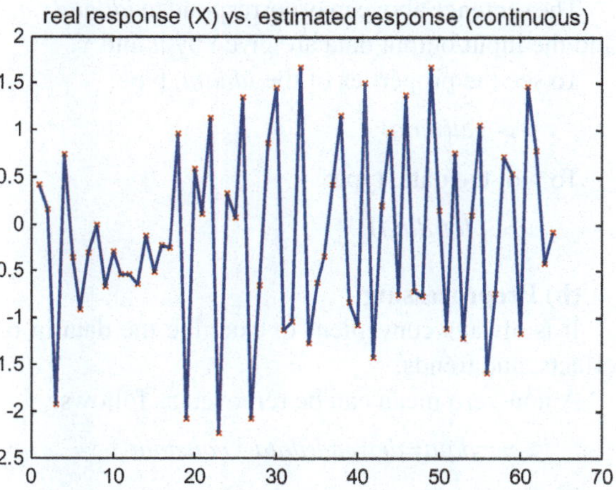

Fig. 6.49 Original plant response (X) versus response of structure F estimated plant (*continuous*)

An important remark is that a higher order model is not necessarily a better model. At the same time, lower order models could be insufficient.

6.9 Introduction to the MATLAB System Identification Toolbox

As shown in the section on Resources, there exists a number of identification Toolboxes. They belong to different fields of interest, like for instance control systems, econometrics, forecasting, etc. In this section we focus on the System Identification Toolbox provided by Mathworks, which can be regarded as a general purpose Toolbox, although it mostly comes from a control systems context.

This is a quick introduction to the Toolbox. It gives you a set of functions that you can use in command mode, or through a GUI.

6.9.1 Identification Steps Using the Toolbox Functions

In most system identification cases a step by step procedure is followed, as we review next.

(a) Data input

Here is an example of loading data from a file and then constitute an *iddata* object:

\>> *load myplant.dat; data=iddata(myplant(:,2), myplant(:,1), 1)*

The sentence above may correspond to *iddata(y,u,Ts)*, where *Ts* is sampling period and the input/output data are given by u and y.

To see the properties of the *iddata*, type:

>> *get(data)*

To plot the data, type:

>> *plot(data)*

(b) Preprocessing

It is always convenient to examine the data in order to detect outliers, aliasing effects, and trends.

A non-zero mean can be removed as follows:

>> *zdata=detrend(data, 'constant')*

Linear trends are removed with:

>> *zdata=detrend(data)*

It would be also opportune to see if there was a good excitation in the frequency range of interest, and if data are coherent. The following sentences give suitable information for that:

>> *yy=zdata.OutputData; uu=zdata.InputData;*
>> *S=spectrum(uu,yy)*

Data filtering could be applied as follows:

>> *fdata=idfilt(zdata, N, wc)*

With the sentence above, an *N-order* Butterworth filter with cut-off frequency *wn* is applied.

Of course, it is also possible to use the *filter( )* and *filtfilt( )* functions of the Signal Processing Toolbox.

(c) Model estimation

Possible delays can be estimated with:

>> *nk=delayest(zdata)*

An ARMAX model estimation can be obtained with the following sentence:

>> *m1= armax(zdata, [na, nb, nc, nk]);*

which corresponds to a model:

$$A(q^{-1})\; y(t) \;=\; B(q^{-1})\, q^{-nk} u(t) + C(q^{-1})\, e(t) \tag{6.41}$$

Other types of models can be estimated, by using, instead of *armax( )*, other functions. Here is a table of the available functions for model estimation:

ar( )	AR models	Least squares method
arx( )	ARX model	Least squares method
armax( )	ARMAX model	Least squares method
ivar( )	AR model	Instrumental variable method
ivx( )	ARX model	Instrumental variable method
oe( )	Output-error model	
bj( )	Box-Jenkins model	
pem( )	General model	

The output-error model has the following expression:

$$y(t) = \frac{B(q^{-1})}{F(q^{-1})}\, u(t) + e(t) \tag{6.42}$$

And the Box-Jenkins model has the following expression:

$$y(t) = \frac{B(q^{-1})}{F(q^{-1})}\, u(t) + \frac{C(q^{-1})}{D(q^{-1})} e(t) \tag{6.43}$$

The results of the model estimation can be examined with:

>> *present(m1)*

There are some functions that help for selecting the orders of model polynomials. For example, the following sentence,

>> *V = arxstruc(zdata, vdata, struc(1:10,1:10,4));*

computes the loss function for ARX models of orders of *na* and *nb* from 10 and fixed delay of 4. The results of this computation can be compared using:

>> *selstruc(V)*

Residual analysis of the model can be done with:

>> *e =resid(m1,zdata)*

6.9.2 Using the GUI

To start up the GUI, type:

>> *ident*

The following window appears (Fig. 6.50).

The first step would be to import data. We take the example, and another small window pops up (Fig. 6.51).

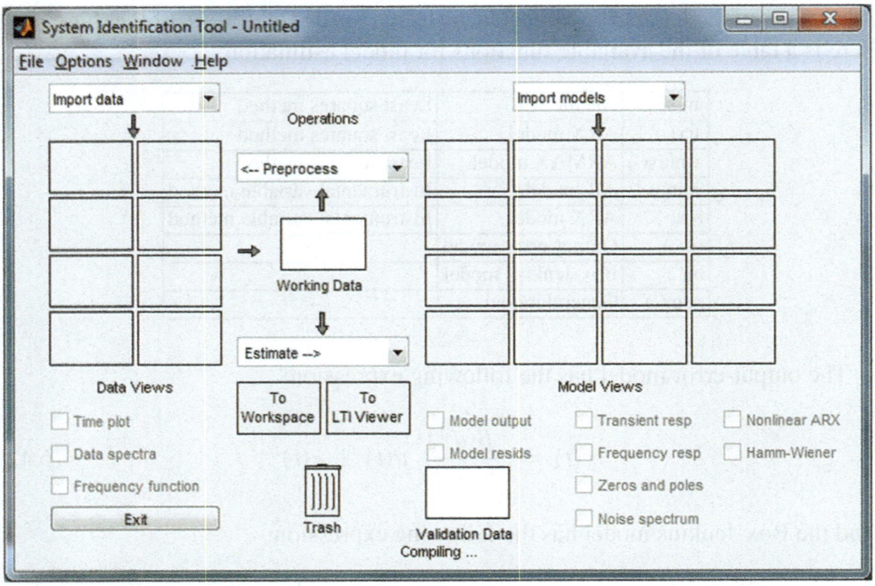

Fig. 6.50 Initial GUI

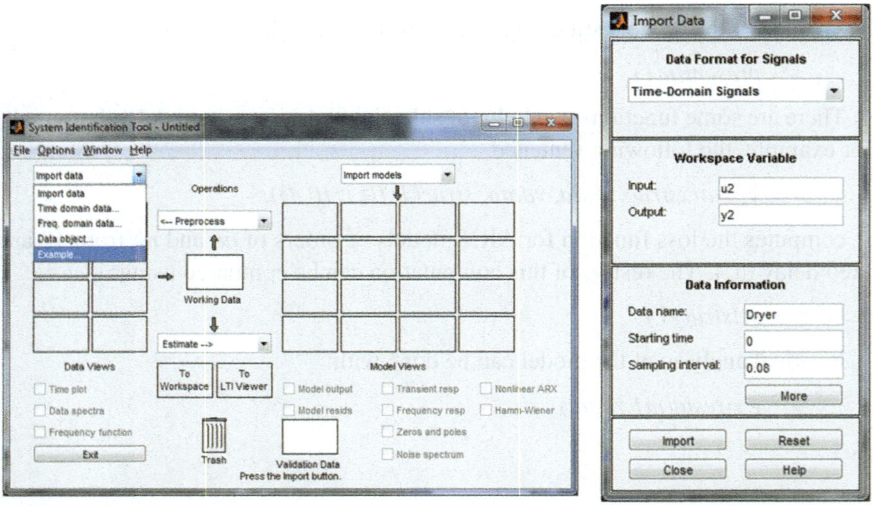

Fig. 6.51 Importing data

We import the data, which includes original data and validation data. The main window is updated as follows (Fig. 6.52).

We can now preprocess the data (Fig. 6.53).

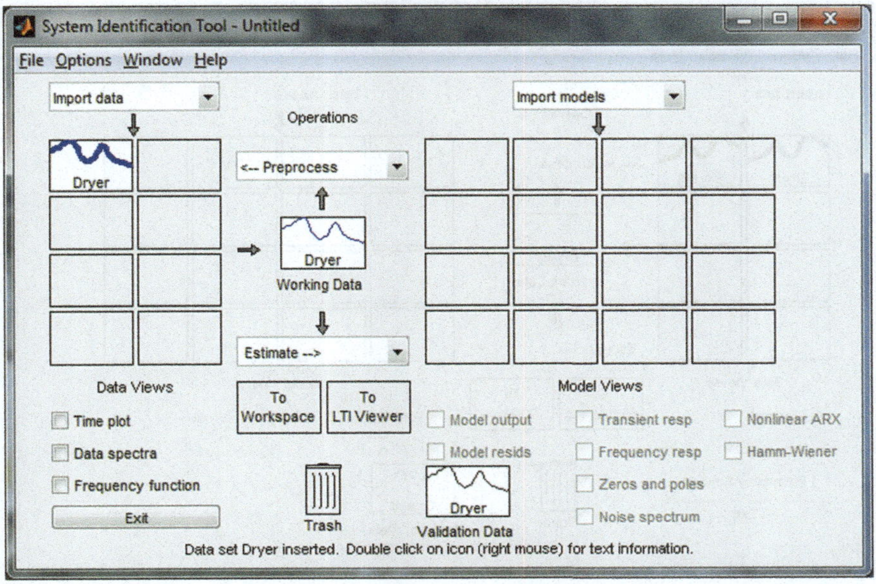

Fig. 6.52 Data were imported

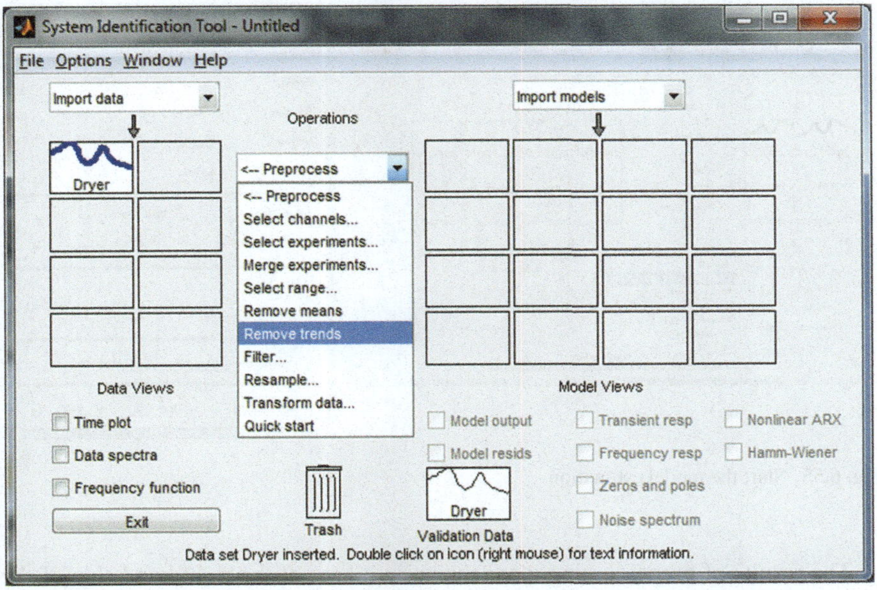

Fig. 6.53 Preprocess the data

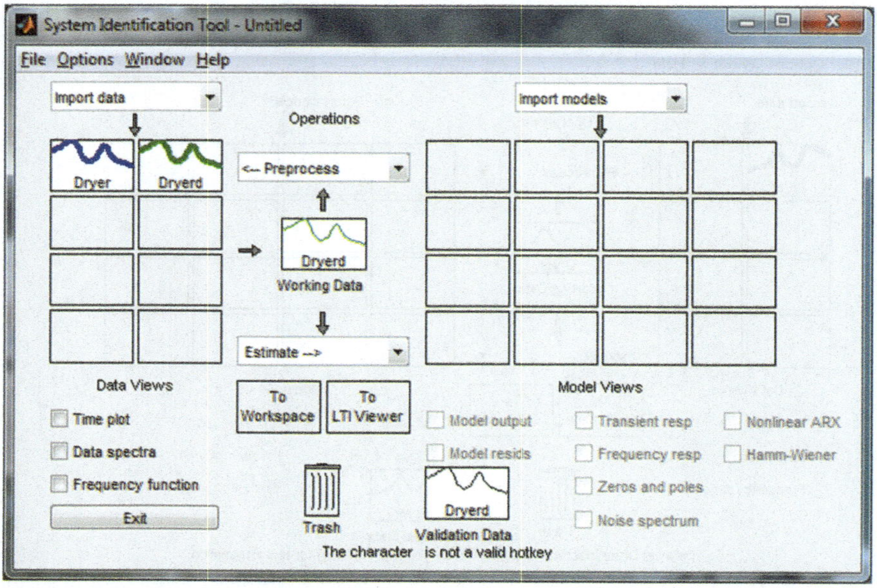

Fig. 6.54 Drag preprocessed data

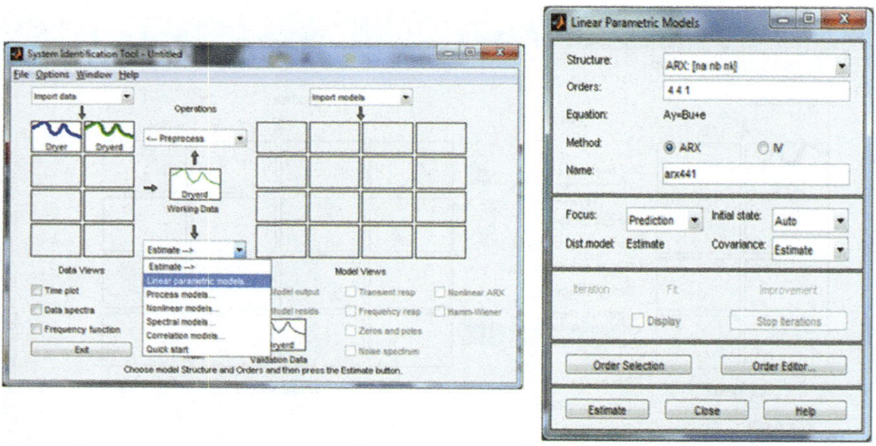

Fig. 6.55 Start the model estimation

The results of preprocessing were shown in a box, at the right hand side of the original data. Now, *you must drag the preprocessed data to the Working Data and Validation Data boxes.* Therefore, you obtain the following window (Fig. 6.54).

Then, we go to the *Estimate* menu and select *Linear parametric models.* A new specific window appears (Fig. 6.55).

We click on *Estimation*, there is some computation work, and then a model appears on a box (right hand side of main window) (Fig. 6.56).

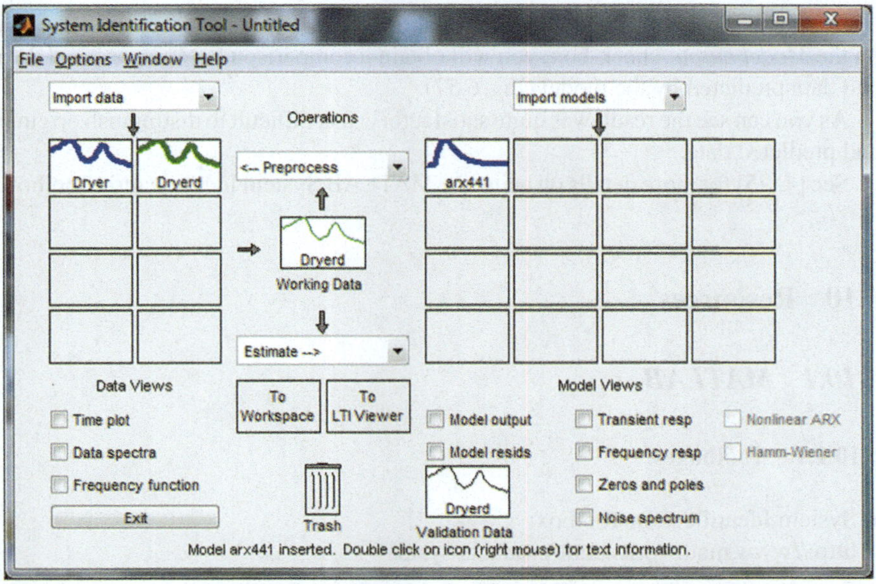

Fig. 6.56 The model was estimated

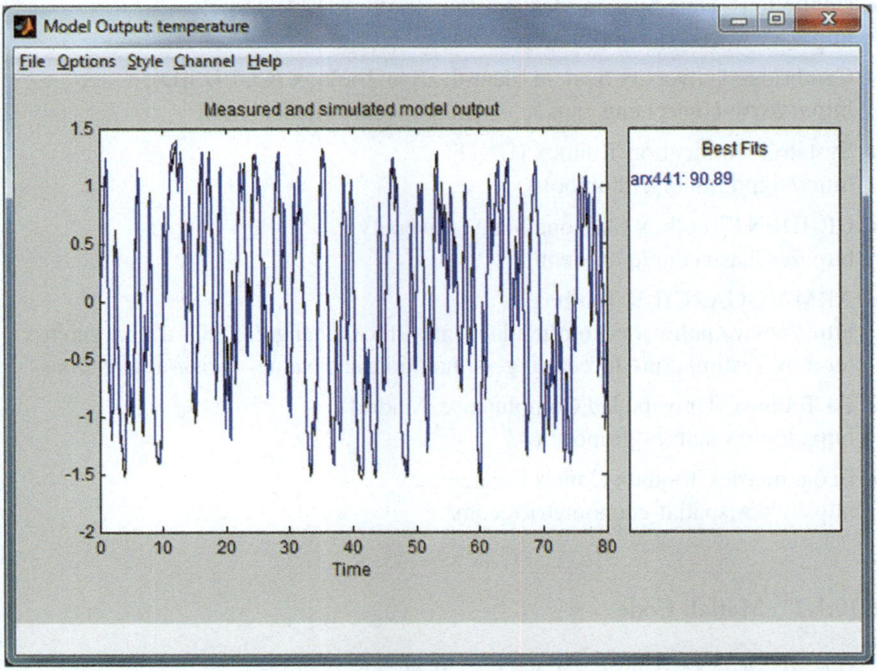

Fig. 6.57 Comparison of original and predicted data

Now, you can check the quality of the estimated model. For example, if you click on the *Model output* check-box, you will obtain a comparison between original data and data predicted by the model (Fig. 6.57).

As you can see the result was quite satisfactory: it is difficult to distinguish original and predicted data.

See [4, 25] for more details on using the MATLAB System Identification Toolbox.

6.10 Resources

6.10.1 MATLAB

6.10.1.1 Toolboxes

- System Identification Toolbox:
  http://www.mathworks.com/products/sysid/
- Frequency Domain System Identification Toolbox (FDIDENT):
  http://www.mathworks.com/connections/product_detail/product_35570.html
- Continuous-Time System Identification Toolbox (CONTSID):
  http://www.iris.cran.uhp-nancy.fr/contsid/
- Cambridge University System Identification Toolbox (CUEDSID):
  http://www-control.eng.cam.ac.uk/jmm/cuedsid/cuedsid.html
- System Identification Toolbox (UNIT):
  http://sigpromu.org/idtoolbox/
- CR-IDENT Toolbox (Arizona State University):
  http://csel.asu.edu/?q=crident
- ARMAX-GARCH-K Toolbox:
  http://www.mathworks.com/matlabcentral/fileexchange/32882-armax-garch-k-toolbox--estimation--forecasting--simulation-and-value-at-risk-applications-
- E4 Toolbox (Universidad Complutense, Madrid):
  https://www.ucm.es/grupoee/e4/
- Econometrics Toolbox (James LeSage):
  http://www.spatial-econometrics.com/

6.10.1.2 Matlab Code

- Vibrationdata Matlab Signal Analysis Package:
  https://vibrationdata.wordpress.com/2013/05/29/vibrationdata-matlab-signal-analysis-package/

- Vibratools:
  see: www.sandv.com/downloads/0302ahli.pdf
- Time-series modelling:
  http://www.math.kth.se/matstat/gru/5b1545/homeworks_2004.html
- Andrew Patton's Matlab code page:
  http://public.econ.duke.edu/~ap172/code.html

6.10.2 Internet

6.10.2.1 Web Sites

- The Vector Fitting Web Site:
  http://www.sintef.no/Projectweb/VECTFIT/
- Time series data files, University of York:
  http://www.york.ac.uk/depts./maths/data/ts/
- Vibration data (beam vibration):
  http://www.vibrationdata.com/beams.html
- Stock prediction software iPredict:
  http://www.ipredict.it/

6.10.2.2 Link Lists

- Software Tools (IFAC links):
  http://tc.ifac-control.org/1/1/links/software-tools

References

1. R.G.K.M. Aarts, *System Identification and Parameter Estimation*. Lecture Notes (University of Twente, 2012). https://www.utwente.nl/ctw/wa/web_dev/old/lectures/113170/notes/notes.pdf
2. R. Adhikari, R.K. Agrawal, *An Introductory Study on Time Series Modeling and Forecasting* (Computer and Systems Sciences Jawaharlal Nehru University New Delhi, India, 2013). arXiv preprint arXiv:1302.6613
3. J.C. Agüero, J.I. Yuz, G.C. Goodwin, Frequency domain identification of MIMO state space models using the EM algorithm, in *Proceedings European Control Conference ECC* (2007)
4. L. Andersson, U. Jönsson, K.H. Johansson, J. Bengtsson, *A Manual for System Identification* (Lund Institute of Technology, 1998). http://www.control.lth.se/media/Education/EngineeringProgram/FRT041/2011/manuallab.pdf
5. J.M. Azevedo, R. Almeida, P. Almeida, Using data mining with time series data in short-term stocks prediction: a literature review. Int. J. Intell. Sci. **2**(4), 176–181 (2012)
6. S. Björklund, A survey and comparison of time-delay estimation methods in linear systems. Ph.D. thesis, Linköping University, 2003

7. B. Cauberghe, Applied frequency-domain system identification in the field of experimental and operational modal analysis. Ph.D. thesis, Vrije University, Brussels., 2004
8. B. Chen, A.P. Petropulu, Frequency domain blind MIMO system identification based on second and higher order statistics. IEEE Trans. Sign. Process. **49**(8), 1677–1688 (2001)
9. C.W. Chen, J.N. Juang, G. Lee, Frequency domain state-space system identification. J. Vibr. Acoust. **116**(4), 523–528 (1994)
10. L. Cirka, M. Fikar, A toolbox for recursive identification of dynamical systems. AT & P J. PLUS2 44–47 (2007)
11. D. Deschrijver, B. Gustavsen, T. Dhaene, Advancements in iterative methods for rational approximation in the frequency domain. IEEE Trans. Power Delivery **22**(3), 1633–1642 (2007)
12. F. Ferranti, Y. Rolain, L. Knockaert, T. Dhaene, Variance weighted vector fitting for noisy frequency responses. IEEE Microwave Wirel. Compon. Lett. **20**(4), 187–189 (2010)
13. B. Gustavsen, C. Heitz, Modal vector fitting: a tool for generating rational models of high accuracy with arbitrary terminal conditions. IEEE Trans. Adv. Packag. **31**(4), 664–672 (2008)
14. B. Gustavsen, C. Heitz, Fast realization of the modal vector fitting method for rational modeling with accurate representation of small eigenvalues. IEEE Trans. Power Delivery **24**(3), 1396–1405 (2009)
15. B. Gustavsen, A. Semlyen, Rational approximation of frequency domain responses by vector fitting. IEEE Trans. Power Delivery **14**(3), 1052–1061 (1999)
16. W. Hendrickx, T. Dhaene, A discussion of rational approximation of frequency domain responses by vector fitting. IEEE Trans. Power Syst. **21**(1), 441–443 (2006)
17. Y. Huang, J. Benesty, J. Chen, Identification of acoustic MIMO systems: challenges and opportunities. Signal Process. **86**, 1278–1295 (2006)
18. K.H. Johansson, *Some Exercises in Systems Identification* (Dept. Automatic Control, Lund University, 2007). http://www.control.lth.se/media/Education/EngineeringProgram/FRT041/2012/exercises.pdf
19. K.J. Keesman, *System Identification* (Springer-Verlag, 2011)
20. H. Lee, D.E. Rivera, Control-relevant curvefitting for plant-friendly multivariable system identification, in *Proceedings IEEE American Control Conference*, pp. 1431–1436 (2005)
21. C.U. Lei, Y. Wang, Q. Chen, N. Wong, A decade of vector fitting development: applications on signal/power integrity, in *Proceedings AIP Conference 2010*, vol. 1285, pp. 435–449 (American Institute of Physics, 2010)
22. E.C. Levi, Complex curve fitting. IRE Trans. Autom. Control **4**(1), 37–44 (1959)
23. L. Ljung, *System Identification: Theory for the User* (Springer-Verlag, 1999)
24. L. Ljung, State of the art in linear system identification: time and frequency domain methods. Proc. IEEE Am. Control Conf. **1**, 650–660 (2004)
25. J. McLellan, *Using Matlab and the System Identification Toolbox to Estimate Time Series Models* (Queens University, 2004). http://www.chemeng.queensu.ca/courses/CHEE436/matlab/documents/using_Matlab.pdf
26. P. Navrátil, V. Bobál, Recursive identification algorithms library, in *Proceedings 17th International Conference on Process Control*, pp. 516–523 (2009)
27. L. Nilsson, O. Seleznjev, *Statistical modeling, simulation and time series* (Umea University, 2009). http://www.tp.umu.se/ModSim/files/StatCompleteLectureNotes.pdf
28. R. Perrelli, *Introduction to ARCH and GARCH Models* (University of Illinois, 2001). http://www.econ.uiuc.edu/~econ472/ARCH.pdf
29. R. Pintelon, P. Guillaume, Y. Rolain, J. Schoukens, H. Van Hamme, Parametric identification of transfer functions in the frequency domain-a survey. IEEE Trans. Autom. Control **39**(11), 2245–2260 (1994)
30. R. Pintelon, J. Schoukens, *System Identification: A Frequency Domain Approach* (Wiley, 2012)
31. G.P. Rao, H. Unbehauen, Identification of continuous-time systems. IEE Proc. Control Theory Appl. **153**(2), 185–220 (2006)
32. C.K. Sanathanan, J. Koerner, Transfer function synthesis as a ratio of two complex polynomials. IEEE Trans. Autom. Control **8**(1), 56–58 (1963)

33. J. Songsiri, *Input Signals* (Lecture Presentation, EE531, Chualongkorn University, 2011). http://jitkomut.lecturer.eng.chula.ac.th/ee531/signal.pdf
34. M. Verhaegen, *Filtering and System Identification. A Least-Squares Approach* (Cambridge University Press, 2007)
35. R. Voorhoeve, T. Oomen, R. van Herpen, M. Steinbuch, On numerically reliable frequency-domain system identification: new connections and a comparison of methods. Proc. IFAC World Congress **19**, 0018–1002 (2014)

Chapter 7
Data Analysis and Classification

7.1 Introduction

Classification is a fundamental activity in many scientific disciplines, and in a large variety of profesional applications. In many circumstances, classification is not an easy task. Analysis tools are needed in order to detect distinctive characteristics, and to compare according with suitable measures.

The aim of this chapter is to introduce a repertory of important methods. A complete treatment is not attempted, since it would require an extensive book by itself. Anyway, this chapter—that includes many lines of MATLAB code—has already a large size.

Some aspects of the chapter have evident connections with pattern recognition techniques. In general it is quite difficult to outperform humans in pattern recognition tasks. However, the help that could be obtained from computers is welcome in cases involving large data sets (like in data mining). Precisely, the first sections of the chapter are devoted to component analysis methods, like PCA or ICA, which enable us to reduce the dimensionality of data analysis problems, putting focus on most significant aspects.

An interesting application example that has been considered in the first part of the chapter, is blind source separation.

Once arrived at the fifth section, attention is paid to data clusters and discrimination. A popular example, with two data clusters corresponding to two types of IRIS flowers, is taken as reference for the introduction of Linear Discriminant Analysis (LDA) and also Support Vector Machines (SVM). The section continues with the K-means method for clustering, and the K-nearest-neighbor method for labeling new data. The section ends with a technique that expands enormously the power of the methods already introduced: the idea is to use kernels, so the discriminative actions in 2D problems were not limited to the use of lines, but can extend to the use of curves (in 3D one would deal with planes or curved surfaces; and in general with hyperplanes or hypersurfaces).

© Springer Science+Business Media Singapore 2017

J.M. Giron-Sierra, *Digital Signal Processing with Matlab Examples, Volume 2*,
Signals and Communication Technology, DOI 10.1007/978-981-10-2537-2_7

The next sections open the view to probabilistic contexts, including the Bayesian methodology. Important methods, like the Expectation-Maximization (EM) algorithm, Bayesian linear regression, Kriging, Gaussian processes, neurons, etc., have been considered.

The chapter ends with some interesting experiments. One corresponds to an important problem: face detection. The other is an example of K-means application for picture color reduction.

7.2 A Basic Idea of Component Analysis

The next two sections will consider two linear analysis techniques that try to discover components in the data. The techniques are denoted as Principal Component Analysis (PCA), and Independent Component Analysis, (ICA). Both are linear, and are expressed with vector and matrix algebra. When using ICA it usually happens that PCA is applied in a first step, for data dimensionality reduction.

These sections will present interesting application examples, like image compression, and blind source separation of mixed sounds or images. As an intuitive introduction to the topic, let us consider an example. One has two records of sound samples, two melodies. It is possible to represent one data set against the other. This is called a '*scatterplot*'. Figure 7.1 shows a scatterplot, for the two melodies read by the Program 7.1. The program also gives us the opportunity to hear the melodies.

In addition, the Program 7.1 also displays the Fig. 7.2. This figure shows three histograms. The top and middle plots correspond to each of the melodies. An inclined axis has been drawn on the scatterplot, Fig. 7.1. The scatterplot data are projected on this axis. The histogram at the bottom of Fig. 7.2 corresponds to this projection.

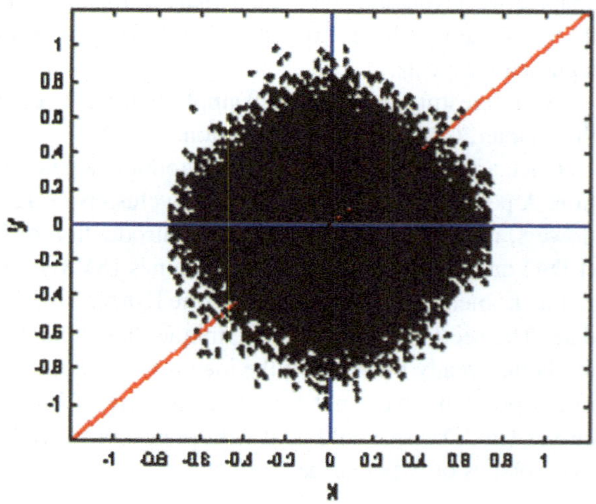

Fig. 7.1 Scatterplot of 2 uncorrelated signals

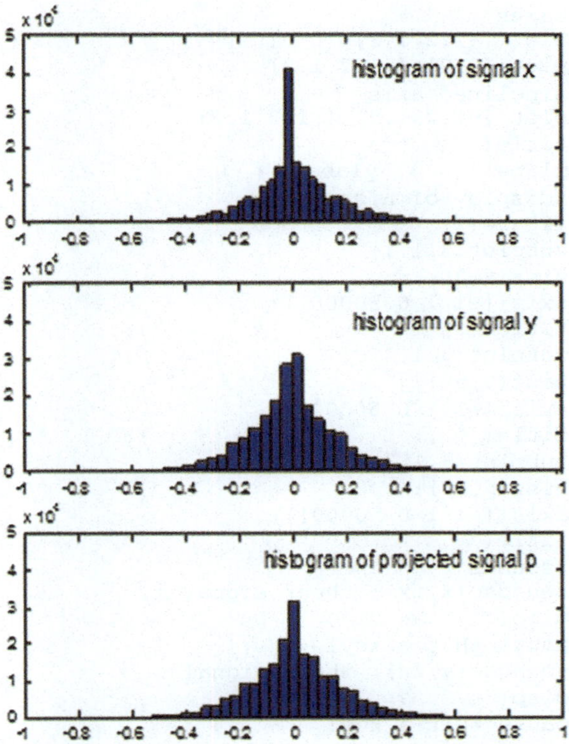

Fig. 7.2 Histograms of signals a and b, and projection p

It looks more Gaussian than the other two histograms. A consequence of the central limit theorem is that a mix of signals is more Gaussian than each separated signal.

Program 7.1 Scatterplot of two audio signals

```
%Scatterplot of two audio signals
%and Histograms of 3 projections: vertical, horizontal and 45°
%you can hear the three projections
%read two audio signals, they have same size
[x,fx]=wavread('wind1.wav'); %read wav file
[y,fy]=wavread('b1.wav'); %read wav file
N=length(x);
alpha=pi/4; %45deg angle in radians
p=zeros(1,N);
%projection of scatterplot on the 45 inclined axis
for i=1:N,
p(i)=(x(i)*cos(alpha)) + (y(i)*sin(alpha));
end;
%display of scatterplot
figure(1)
plot(x,y,'k.'); hold on;
axis([-1.2 1.2 -1.2 1.2]);
```

```
%axes
plot([-1.2 1.2],[0 0],'b');
plot([0 0],[-1.2 1.2],'b');
%inclined axis
plot([-1.2 1.2],[-1.2 1.2],'r');
title('scatterplot of signals x and y');
xlabel('x'); ylabel('y');
%display of histograms
figure(2)
subplot(3,1,1)
hist(x,50);
axis([-1 1 0 50000]);
title('histogram of signal x');
subplot(3,1,2)
hist(y,50);
axis([-1 1 0 50000]);
title('histogram of signal y');
subplot(3,1,3)
hist(p,50);
axis([-1 1 0 50000]);
title('histogram of projected signal p');
%sounds
soundsc(x,fx); %hear signal x
disp('signal x');
pause %Hit a key!
soundsc(y,fx); %hear signal y
disp('signal y');
pause %Hit a key!
soundsc(p,fx); %hear the projected signal
disp('projected signal p');
```

Now let us do an exercise of exploratory analysis [117]. Like before, let us take the scatterplot data and project that on an inclined axis. And let us compute a certain numerical characteristic, for instance the fourth moment of the projected data (denote it as $M4$). This is done for a certain axis inclination angle β. Let us rotate this inclined axis. Figure 7.3 shows a polar plot. Each point of the plot is $M4$ for the corresponding angle β, with β taking values from 0 to 360°. The figure has been obtained with the Program 7.2.

Program 7.2 Exploring the fourth moment as projection axis is rotated

```
%Exploring the fourth moment as projection axis is rotated
%example of two audio signals
%read two audio signals, they have same size
[x,fx]=wavread('wind1.wav'); %read wav file
[y,fy]=wavread('b1.wav'); %read wav file
N=length(x);
A=60; %number of circle partitions
mnt=zeros(1,A+1);
ag=zeros(1,A+1);
i=1:N; %vectorized iterations
for nn=0:A,
```

```
alpha=(2*pi*nn)/A; %angle of projection axis in radians
p=zeros(1,N);
%projection of scatterplot on the inclined axis
p(i)=(x(i)*cos(alpha)) + (y(i)*sin(alpha));
moment4=mean(p.^4); %fourth moment
mnt(nn+1)=moment4; %save result
ag(nn+1)=alpha;
end;
%display
figure(1)
polar(ag,mnt);
title('fourth moment as projection axis rotates');
```

The example just considered is a case of '*Projection Pursuit*' [103, 139, 237]. The idea is to project on a rotating axis, compute a certain value concerning the projection statistics, and draw a polar plot. The objective is to obtain '*interesting*' projection angles. The observed interesting facts could be important for certain practical purposes. For instance, directions of maximal variance, or directions of more or less Gaussianity. More details in [68, 259].

7.3 Principal Component Analysis (PCA)

The principal component analysis (PCA) is a popular method with many applications, like for instance data dimensional reduction, discovery of data structures, or face recognition [69, 156, 281, 310].

Fig. 7.3 Example of projection pursuit

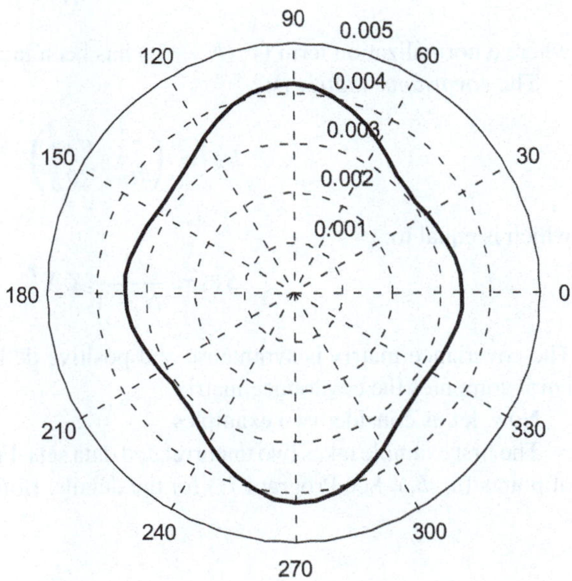

7.3.1 Mathematical Aspects

PCA is related to the covariance matrix, and it is usually computed using the SVD decomposition. Therefore, let us focus briefly on these two topics as an introduction to PCA.

7.3.1.1 The Covariance Matrix. Correlated or Uncorrelated Data

Suppose two sets of measurements, from which it is easy to obtain two zero-mean data sets A and B:

$$\begin{aligned} A &= [a_1,\ a_2,\ ..,\ a_N] \\ B &= [b_1,\ b_2,\ ..,\ b_N] \end{aligned} \tag{7.1}$$

These data sets can be put in matrix form:

$$X = \begin{pmatrix} A \\ B \end{pmatrix} \tag{7.2}$$

(two rows of N data)

The variances of A and B are the following:

$$\sigma_A^2 = < a_i, a_i > \quad , \quad \sigma_B^2 = < b_i, b_i > \tag{7.3}$$

The covariance of A and B is:

$$\sigma_{AB} = < a_i, b_i > = \frac{1}{N-1} A\, B^T \tag{7.4}$$

where a normalization term $(1/(N-1))$ has been included.

The *covariance matrix* is:

$$S_X = \begin{pmatrix} \sigma_A^2 & \sigma_{AB} \\ \sigma_{BA} & \sigma_B^2 \end{pmatrix} \tag{7.5}$$

which is equal to:

$$S_X = \frac{1}{N-1} X\, X^T \tag{7.6}$$

The covariance matrix is symmetric and positive definite. The MATLAB function *cov()* computes the covariance matrix.

Now, let us consider two examples.

The first example takes two uncorrelated data sets. Figure 7.4 shows the scatterplot of points (a_i, b_i). See Program 7.3 for the details. Both data sets have equal PDF.

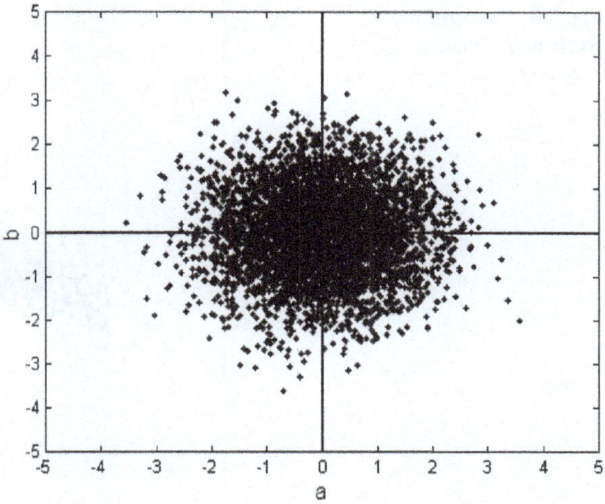

Fig. 7.4 Scatterplot of 2 uncorrelated signals

Program 7.3 Scatterplot example, uncorrelated

```
% Scatterplot example
% 2 uncorrelated variables
N=5000;
a=normrnd(0,1,1,N); a=a-mean(a);
b=normrnd(0,1,1,N); b=b-mean(b);
figure(1)
plot(a,b,'k.'); hold on; %scatterplot
plot([-5 5],[0 0],'k');
plot([0 0],[-5 5],'k');
axis([-5 5 -5 5]);
title('scatterplot: uncorrelated variables');
xlabel('a'); ylabel('b');
X=[a; b]; %data matrix: 2 rows of N values
%print covariance matrix
Sx=cov(X') %covariance matrix
```

When MATLAB executes the last line of the program, it displays the covariance matrix. The expected result is:

0.9796 0.0201
0.0201 1.0075

In this example, the terms out of the main diagonal should be zero, since it is presumed that the two data sets are uncorrelated. However, the real situation is that there was some small correlation.

The second example explicitly includes correlation between both data sets. See Program 7.4 for the details.

Figure 7.5 shows the scatterplot for this case. Compare with the previous scatterplot. While the scatterplot of the uncorrelated data was circular, the scatterplot of

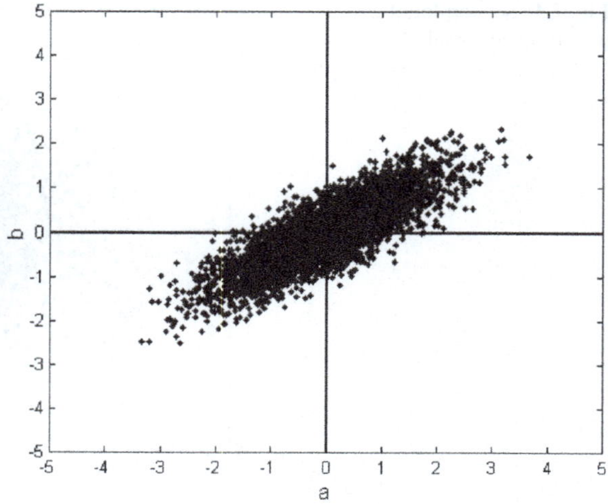

Fig. 7.5 Scatterplot of 2 correlated signals

correlated data becomes elliptical and inclined. Notice that plotting a dataset versus itself would give a straight line with 45° angle.

Program 7.4 Scatterplot example, some correlation

```
% Scatterplot example
% 2 variables with some correlation
N=5000;
a=normrnd(0,1,1,N); a=a-mean(a)
b=(0.4*normrnd(0,1,1,N))+(0.6*a); b=b-mean(b);
figure(1)
plot(a,b,'k.'); hold on; %scatterplot
plot([-5 5],[0 0],'k');
plot([0 0],[-5 5],'k');
axis([-5 5 -5 5]);
title('scatterplot: correlated variables');
xlabel('a'); ylabel('b');
X=[a; b]; %data matrix: 2 rows of N values
%print covariance matrix
Sx=cov(X') %covariance matrix
```

The covariance matrix for this second example is:

1.0102 0.5991
0.5991 0.5156

Clearly, the terms out of the main diagonal take noticeable values: an indication of correlation between the data sets.

7.3.1.2 The Singular Value Decomposition (SVD)

Let A be a $m \times n$ matrix. Then it can be decomposed as follows:

$$A = U\, S\, V^T \qquad (7.7)$$

where U is a $m \times m$ matrix, S is a $m \times n$ matrix, and V is a $n \times n$ matrix. The matrix S is diagonal. U and V are orthonormal.

Denote as λ_i the eigenvalues of $A^T A$. Then, the singular values of A are $\sigma_i = \sqrt{\lambda_i}$. The eigenvectors of $A^T A$ make up the columns of V. The eigenvectors of $A\, A^T$ make up the columns of U. The diagonal of S are the singular values of A. The singular values in S come in descending order along the diagonal.

MATLAB provides the *svd()* function for singular value decomposition. This function has an option for "economy size" decomposition: if $m > n$ only the first columns of U are computed and S is $n \times n$.

7.3.2 Principal Components

The principal components are the eigenvectors of the covariance matrix. The eigenvalues of this matrix are the magnitude of the principal components.

Recall equation (7.6). Let us consider the SVD of $(X^T/sqrt(N-1))$, the eigenvectors of $X\, X^T/(N-1)$ will make up the columns of V. Therefore, in this way one can compute the principal components.

Taking again the example considered in the Program 7.4 and pictured in Fig. 7.5, let us compute the principal components. This is done with the Program 7.5, using the SVD decomposition. Figure 7.6 shows the main principal component: a straight line crossing the elliptical data cloud. The reader is invited to play with the program, looking also at the sizes of *A, U, S, V*, and into the content of *SD*, which is a square diagonal matrix.

Program 7.5 PCA example: correlated variables

```
% PCA example: correlated variables
%data---------------
N=1000; M=2;
D=zeros(M,N);
D(1,:)=normrnd(0,1,1,N);
D(2,:)=(0.6*D(1,:))+(0.4*normrnd(0,1,1,N));
%PCA----------------
me=mean(D,2); %mean of each row
X=D-repmat(me,1,N); %subtract mean in each row
A=X'/sqrt(N-1);
%singular value decomposition
[U,S,V]=svd(A); %V contains principal components
SD=diag(S);
```

```
vr=SD.*SD; %variances
prd=V'*X; %data projection
%display
plot(X(1,:),X(2,:),'k.'); hold on;
plot([-4*V(1,1) 4*V(1,1)],[-4*V(2,1) 4*V(2,1)],'r');
axis([-4 4 -4 4]);
title('PCA main component'); xlabel('x1'); ylabel('x2');
%print the variance
vr
```

Notice that the PCA main component is the axis of maximum variance. It captures significant information, to the point that this may be enough for data interpretation purposes.

Actually the variance of the data projected onto a certain direction $\mathbf{u}$ is given by:

$$var(\mathbf{u}) = \mathbf{u}^T \, S_X \, \mathbf{u} \tag{7.8}$$

And this variance is maximized when:

$$S_X \, \mathbf{u} = \lambda \, \mathbf{u} \tag{7.9}$$

Thus, the direction with maximum variance is an eigenvector (principal component) of the covariance matrix. The variance of the projected data would be:

$$var(\vec{p}) = \mathbf{p}^T \, S_X \, \mathbf{p} = \mathbf{p}^T \, \lambda \, \mathbf{p} = \lambda \tag{7.10}$$

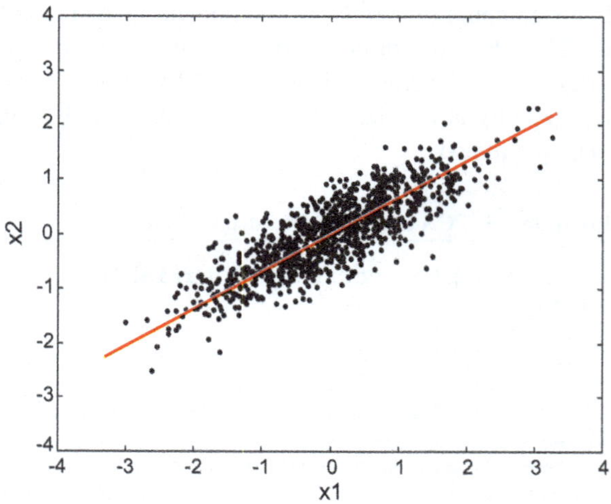

Fig. 7.6 PCA main component

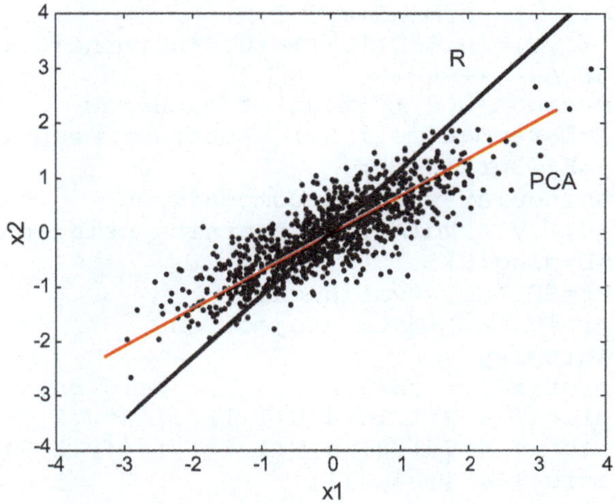

Fig. 7.7 Principal component, and regression line

Another way of computing principal components is by application of the MATLAB *eig()* function to the covariance matrix. This function gives the eigenvectors and the eigenvalues. It is up to the user to sort the eigenvalues and the eigenvectors.

According with specific terminology of PCA practitioners, the product of the eigenvectors and their corresponding singular values gives the '*factor loadings*'.

If one adds the following lines of code to the Program 7.5, it is possible to display also the regression line:

Fragment 7.6 Plot regression line

```
x=X(1,:); y=X(2,:);
%regression
b=regress(x',y');
plot(x,b*x,'k');
```

Figure 7.7 shows the result. Obviously the regression line is not coincident with the PCA main component. This is due to the manner data are projected. In the case of PCA, data are projected perpendicular to the PCA axis. In the case of the regression line, data are projected vertically.

The eigenvectors are mutually orthogonal. Figure 7.8 shows the two principal components for the example being considered. Both principal component lines are perpendicular.

Program 7.7 PCA example: correlated variables

```
% PCA example: correlated variables
% showing the 2 PCA components
%data---------------
N=1000; M=2;
D=zeros(M,N);
```

```
D(1,:)=normrnd(0,1,1,N);
D(2,:)=(0.6*D(1,:))+(0.4*normrnd(0,1,1,N));
%PCA----------------
me=mean(D,2); %mean of each row
X=D-repmat(me,1,N); %subtract mean in each row
A=X'/sqrt(N-1);
%singular value decomposition
[U,S,V]=svd(A); %V contains principal components
SD=diag(S);
vr=SD.*SD; %variances
prd=V'*X; %data projection
%display
plot(X(1,:),X(2,:),'k.'); hold on;
plot([-4*V(1,1) 4*V(1,1)],[-4*V(2,1) 4*V(2,1)],'r');
plot([-4*V(1,2) 4*V(1,2)],[-4*V(2,2) 4*V(2,2)],'r');
axis([-4 4 -4 4]);
title('PCA components'); xlabel('x1'); ylabel('x2');
```

The eigenvectors (principal components) constitute a new basis. Data can be projected into this basis:

$$P = V^T X \tag{7.11}$$

This projection is called '*principal component scores*'. The product of each one and their corresponding singular values gives the '*factor scores*'.

Another view of the mentioned projection into a new basis, is *data rotation* until the variance is maximized along the horizontal axis. This concept is applied, for instance, to image alignment.

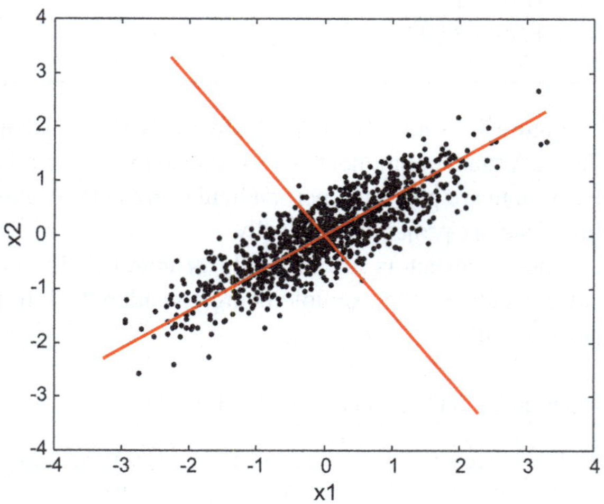

Fig. 7.8 Two PCA components

PCA can be applied to *n-dimensional* problems, with *n* data sets. In many applications, the information given by a subset of the eigenvectors (the 1st eigenvector, the 2nd, etc, till the *mth*) is enough. And so there is a *dimensional reduction*.

7.3.3 Application Examples

7.3.3.1 A 3D Accelerometer

Triaxial accelerometers are used for several interesting applications, like autonomous vehicles (air, land, water) [155, 304]. Suppose we got data from a sinusoidal acceleration in a certain direction in the 3D space. Let us show that PCA could be applied to determine that direction. Program 7.8 is in charge of this example.

The first lines of the program are devoted to generate the three signals. Some noise is added, to simulate real measurement conditions. Of course, we know the acceleration direction. Figure 7.9 shows the acceleration signals.

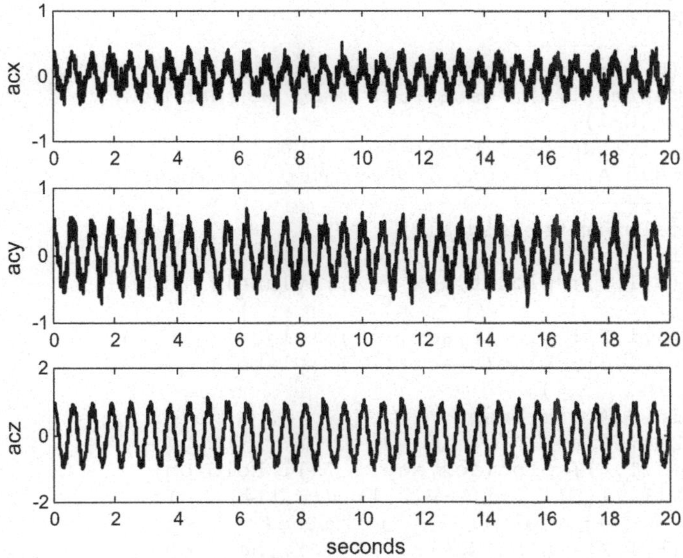

Fig. 7.9 Signals from the 3 axis accelerometer

Program 7.8 PCA of 3D accelerometer

```
%PCA of 3D accelerometer
ha=30; va=60; %horiz. & vert. direction angles (degrees)
%radians
rha=(ha*pi)/180; rva=(va*pi)/180;
%direction projections
diz=sin(rva);
dix=cos(rva)*sin(rha);
diy=cos(rva)*cos(rha);
%acceleration signal, in the direction ha-hv
t=0:0.01:20;
N=length(t);
w=10; %frequency in rad/s
ac=cos(w*t);
%signal projections+noise
acx=(ac*dix)+(0.1*normrnd(0,1,1,N));
acy=(ac*diy)+(0.1*normrnd(0,1,1,N));
acz=(ac*diz)+(0.1*normrnd(0,1,1,N));
%PCA computation
M=3;
D=zeros(M,N);
D(1,:)=acx;
D(2,:)=acy;
D(3,:)=acz;
%PCA---------------
me=mean(D,2); %mean of each row
X=D-repmat(me,1,N); %subtract mean in each row
A=X'/sqrt(N-1);
%singular value decomposition
[U,S,V]=svd(A); %V contains principal components
%display-----------------------
%the accelerometer signals
figure(1)
subplot(3,1,1); plot(t,acx,'k'); ylabel('acx');
title('signals from 3D accelerometer');
subplot(3,1,2); plot(t,acy,'k'); ylabel('acy');
subplot(3,1,3); plot(t,acz,'k'); ylabel('acz');
xlabel('seconds');
%the main PCA component
figure(2)
subplot(3,1,1); plot(acx,acy,'g.'); hold on;
plot([-V(1,1) V(1,1)],[-V(2,1) V(2,1)],'r');
xlabel('acx'); ylabel('acy'); title('main PCA component');
subplot(3,1,2); plot(acx,acz,'g.'); hold on;
plot([-V(1,1) V(1,1)],[-V(3,1) V(3,1)],'r');
xlabel('acx'); ylabel('acz');
subplot(3,1,3); plot(acy,acz,'g.'); hold on;
plot([-V(2,1) V(2,1)],[-V(3,1) V(3,1)],'r');
xlabel('acy'); ylabel('acz');
%print values for comparison
dix/V(1,1)
diy/V(2,1)
diz/V(3,1)
```

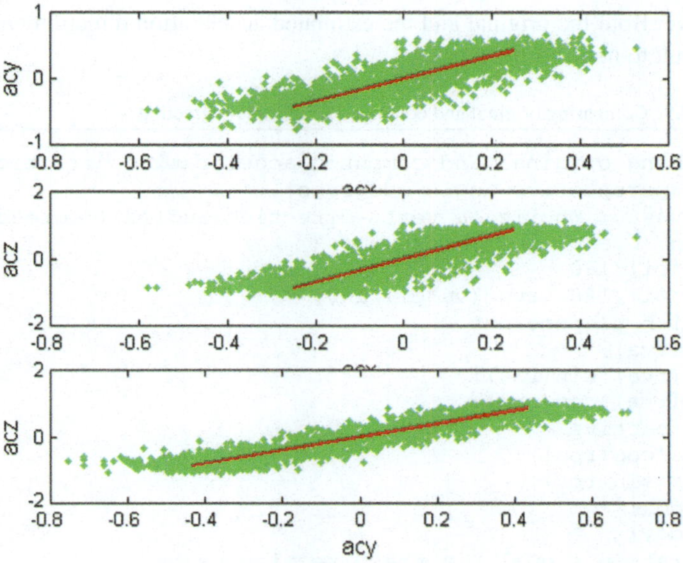

Fig. 7.10 Principal components of the accelerometer signals

The second part of the program computes the principal components. Then, projections of the sampled data on the three coordinate planes are displayed. Figure 7.10 shows the result. The three principal component lines are also displayed.

The program includes three sentences at the end, in order to print the ratios of original and estimated acceleration direction on the three coordinate axes.

With a more detailed work on the problem geometry, it is possible to show in perspective the results of our exercise. This is the task of the Program 7.9, which has many lines in common with the previous program. Figure 7.11 shows the result in 3D

Fig. 7.11 3D reconstruction of the acceleration

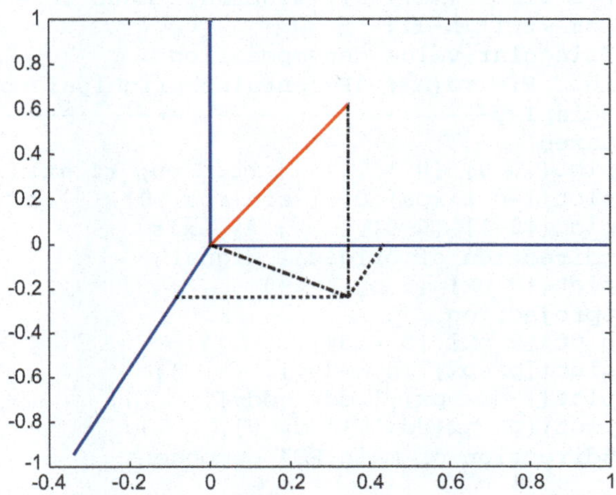

perspective. Both the original and the estimated acceleration direction are painted: it is difficult to notice differences.

Program 7.9 Comparing original and computed acceleration direction

```
% Comparing original and computed acceleration direction
pa=70; %perspective angle (degrees)
ha=30; va=60; %horiz. & vert. direction angles (degrees)
%radians
rpa=(pa*pi)/180;
rha=(ha*pi)/180; rva=(va*pi)/180;
%direction projections
diz=sin(rva);
dix=cos(rva)*sin(rha);
diy=cos(rva)*cos(rha);
%in perspective
ddx=(dix*cos(rpa));
ddy=(dix*sin(rpa));
px=diy-ddx;
py=diz-ddy;
%acceleration signal, in the direction ha-hv
t=0:0.01:20;
N=length(t);
w=10; %frequency in rad/s
ac=cos(w*t);
%signal projections+noise
acx=(ac*dix)+(0.1*normrnd(0,1,1,N));
acy=(ac*diy)+(0.1*normrnd(0,1,1,N));
acz=(ac*diz)+(0.1*normrnd(0,1,1,N));
%PCA computation
M=3;
D=zeros(M,N);
D(1,:)=acx;
D(2,:)=acy;
D(3,:)=acz;
%PCA---------------
me=mean(D,2); %mean of each row
E=D-repmat(me,1,N); %subtract mean in each row
A=E'/sqrt(N-1);
%singular value decomposition
[U,S,P]=svd(A); %P contains principal components
%display-----------------------------------------------
%axes
plot([0 0],[0 1],'b'); hold on; %z axis
plot([-cos(rpa) 0],[-sin(rpa) 0],'b'); %x axis
plot([0 1],[0 0],'b'); %y axis
%direction of original signal
plot([0 px],[0 py],'k');
%projection
plot([0 px],[0 -ddy],'k-.');
plot([px px],[py -ddy],'k-.');
plot([-ddx px],[-ddy -ddy],'k:');
plot([px px+ddx],[-ddy 0],'k:');
%direction of main PCA component
```

```
%direction projections
aix=P(1,1); aiy=P(2,1); aiz=P(3,1);
%in perspective
adx=(aix*cos(rpa));
ady=(aix*sin(rpa));
ax=aiy-adx;
ay=aiz-ady;
plot([0 ax],[0 ay],'r');
title('direction of measured acceleration');
```

7.3.3.2 Images

An important property of the SVD of a matrix A, is that the truncated SVDk is the best rank k approximation to A. More precisely, let A_k be the truncated singular value decomposition:

$$A_k = U_k\, S_k\, V_k^T \tag{7.12}$$

where U_k and V_k mean the first k columns of U and V, and S_k is the first k columns and rows of S.

Then A_k is the matrix that makes $\|A - A_k\|_2$ minimal over all rank k matrices.

This property is fundamental for dimensional reduction, or for data compression.

Let us consider the example of a picture [315]. Program 7.10 computes and displays several truncated SVD of the picture, with matrix ranks $k = 10, 30, 50$ and 100.

Figure 7.12 shows the results. Notice that with relatively few principal components good reconstructions of the picture are obtained.

Program 7.10 Image approximation (compression) with PCA

```
% Image approximation (compression)
keaton=imread('keaton1bw.tif'); %read image file into a matrix
D=im2double(keaton);
[U,S,V]=svd(D); %singular value decomposition
%display of approximated photo
figure(1)
nn=1;
for k=[10 30 50 100];
Dk=U(:,1:k)*S(1:k,1:k)*V(:,1:k)';
subplot(2,2,nn);
nn=nn+1;
imshow(Dk); %display
title(['k=' num2str(k)]);
end;
%scree plot
figure(2)
for n=2:50, %from second eigenvalue
plot(n,S(n,n),'ko'); hold on;
end
title('scree plot (excluding first eigenvalue)');
```

Fig. 7.12 Picture reconstruction using k principal components

```
xlabel('n'); ylabel('eigenvalue');
%print first eigenvalues
S(1,1) %first eigenvalue
```

Fig. 7.13 Screeplot corresponding to the previous picture

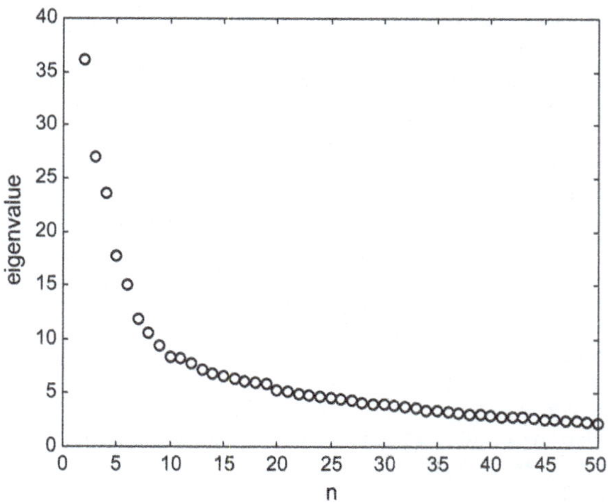

The Program 7.10 also displays the so-called '*screeplot*' corresponding to the picture. Each point in the plot is the value of a eigenvalue, starting with the main principal component, and continuing with the 2nd, the 3rd, etc.

Figure 7.13 depicts the result. The screeplot is useful for estimating the relative importance of successive principal components. It is also of great help for recognizing where to truncate.

7.4 Independent Component Analysis (ICA)

Usually the PCA method does not work well for data with *non-Gaussian* PDFs. In consequence, there has been extensive research on other methods, for the analysis of different types of data or signals. One of these methods is independent component analysis (ICA) [67].

Also, a great interest has been awakened by the '*blind source separation*' (BSS) problem. A particular example of BSS cases is the '*cocktail party problem*'. Several solution procedures have been proposed, including ICA with encouraging results.

This part begins with an intuitive introduction, and then continues with details about independence of components and assumptions of the method. The nutshell of the method is to optimize a criterion. The criterion is a measure of independence. Once a criterion is defined, then several optimization strategies are available.

In accordance with its notable interest, there is abundant literature on ICA, like for instance the books [142, 289], or the papers [140, 141, 144, 221, 236, 246].

7.4.1 Blind Source Separation and the Cocktail Party Problem

Let us consider the following question. Imagine a cocktail party in which the speech of several people is captured by several separated microphones. The problem is to recover the speech of each one.

Choose a simple scenario, with two speakers. A model of what happens with the microphone signals could be stated as follows:

$$\begin{aligned} x_1(t) &= m_{11}\, s_1(t) + m_{12}\, s_2(t) \\ x_2(t) &= m_{21}\, s_1(t) + m_{22}\, s_2(t) \end{aligned} \tag{7.13}$$

where $s_1(t)$ and $s_2(t)$ are the original talk from each speaker, and $x_1(t)$ and $x_2(t)$ are the signals given by the microphones, which capture a mix of the two speeches. The model can be expressed in matrix form:

$$\mathbf{x} = M\,\mathbf{s} \tag{7.14}$$

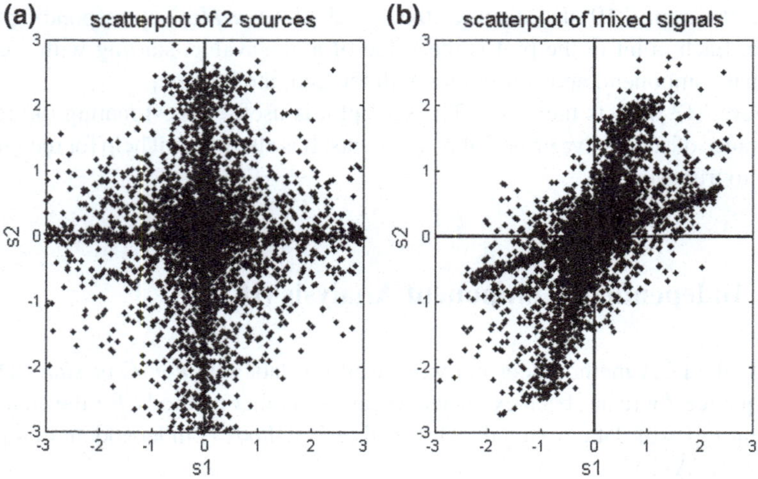

Fig. 7.14 Scatterplot of (*left*) original signals, **b** mixed signals

where M is called the *mixing matrix*.

Now, the problem is to estimate the mixing matrix M and to recover the original speeches $s_1(t)$ and $s_2(t)$.

The case just described is an example of the blind source separation (BSS) problem. In turn, the BSS problem was preceded, time ago, by the '*blind convolution*' problem, which is related to filtering and communications.

In order to illustrate the problem, the Program 7.11 loads two short speeches and plots (Fig. 7.14) a scatterplot of the two speeches (on the left hand side of the figure), and a second scatterplot corresponding to mixed signals (on the right hand side of the figure). Notice that in both scatterplots one could perceive two main axes, which are perpendicular in the case of the sources, and not perpendicular in the case of mixed signals. Another aspect to remark is that speeches are not Gaussian.

Program 7.11 Scatterplot of original sources and mixed signals

```
% Scatterplot of original sources and
% scatterplot of mixed signals
% example of two speeches
%read two sound files
[a,fs1]=wavread('spch1.wav'); %read wav file
[b,fs1]=wavread('spch2.wav'); % " " "
R=2; %reduce data size for clearer picture
a=decimate(a,R);
b=decimate(b,R);
s1=(a-mean(a))'; %zero-mean
s2=(b-mean(b))'; % " " "
vr1=var(s1); s1=s1/sqrt(vr1); %variance=1
vr2=var(s2); s2=s2/sqrt(vr2); %" " "
```

```
s=[s1;s2]; %combine sources
%mix of sources
N=length(s1);
M=[0.7 0.3; 0.2 0.8]; %example of mixing matrix
x=M*s; %mixed signals
%display
figure(1)
subplot(1,2,1);
%scatterplot of sources
plot(s(1,:),s(2,:),'k.'); hold on; %scatterplot
L=3;
%axes
plot([-L L],[0 0],'k');
plot([0 0],[-L L],'k');
axis([-L L -L L]);
title('scatterplot of 2 sources');
xlabel('s1'); ylabel('s2');
subplot(1,2,2);
%scatterplot of mixed signals
plot(x(1,:),x(2,:),'k.'); hold on; %scatterplot
L=3;
%axes
plot([-L L],[0 0],'k');
plot([0 0],[-L L],'k');
axis([-L L -L L]);
title('scatterplot of mixed signals');
xlabel('s1'); ylabel('s2');
```

The model (7.13) can be expressed as follows:

$$\mathbf{s} = M^{-1}\,\mathbf{x} \tag{7.15}$$

If the mixing matrix was invertible and known, it would be easy to recover from the mixed signals the original separated speeches, which are the real independent components.

However, usually the mixing matrix is unknown, and some procedure should be devised to estimate the principal components. For example, as it is now introduced, it is possible to estimate row by row the mixing matrix and so the principal components.

Let us focus on *one* of the independent components, which will be denoted as y:

$$y = \vec{w}^{\,T}\,\mathbf{x} \tag{7.16}$$

Now the problem is to determine $\mathbf{w}$. With a change of variables:

$$\mathbf{q} = M^T\,\vec{w} \tag{7.17}$$

One gets:

$$y = \vec{w}^T \mathbf{x} = \vec{w}^T M\, \mathbf{s} = \vec{q}^T \mathbf{s} \tag{7.18}$$

Key argument: By the central limit theorem it is known that the sum of random variables is more Gaussian than these variables. Therefore, y is more Gaussian than each $s_i(t)$. Indeed, if $y = s_i(t)$, $(i = 1 or 2)$, then it becomes least Gaussian; and in this case only one of the values in $\mathbf{q}$ would be non-zero. The consequence is that maximization of the non-Gaussianity of (7.16) gives us $\mathbf{w}$, and then the independent component.

The procedure can be repeated for each component, selecting the corresponding row of M^{-1}.

As we shall see, a common way for obtaining the independent components involves two steps: whitening and rotation. But before, it is opportune to make a first comparison between PCA and ICA.

7.4.2 PCA and ICA

Let us compare PCA and ICA.

Suppose that $\mathbf{x}$ are the measured data.

- The PCA problem is to find a basis change V such that:

$$\mathbf{x} = V\, \mathbf{p} \tag{7.19}$$

  where $\mathbf{p}$ has uncorrelated components p_i.
- The ICA problem is to find a basis change M such that:

$$\mathbf{x} = M\, \mathbf{s} \tag{7.20}$$

  where $\mathbf{s}$ has independent components s_i.

Both ICA and PCA are linear, as manifested by the matrix-vector algebra being used. The inverse of the M matrix is called the *unmixing matrix*.

The example considered before (Fig. 7.14) is now used for highlighting the differences between PCA and ICA.

Figure 7.15 shows the PCA components of the mix of two speeches.

Compare now the previous figure with Fig. 7.16, which shows the ICA components of the mix of two speeches. Clearly, ICA is better for indicating where data are more concentrated.

Both figures have been obtained with the Program 7.12

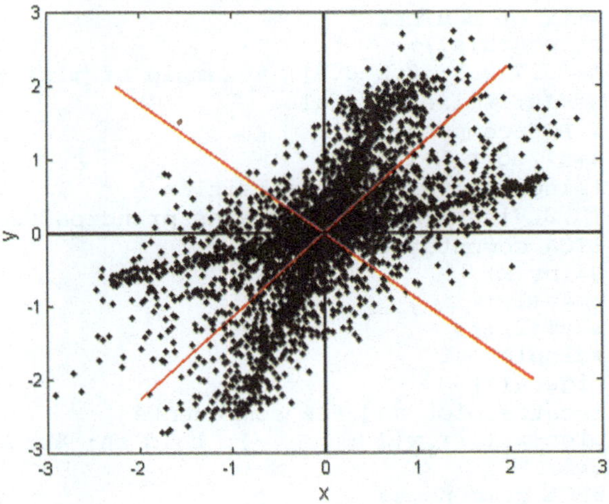

Fig. 7.15 Scatterplot of a mix of two speeches: PCA components

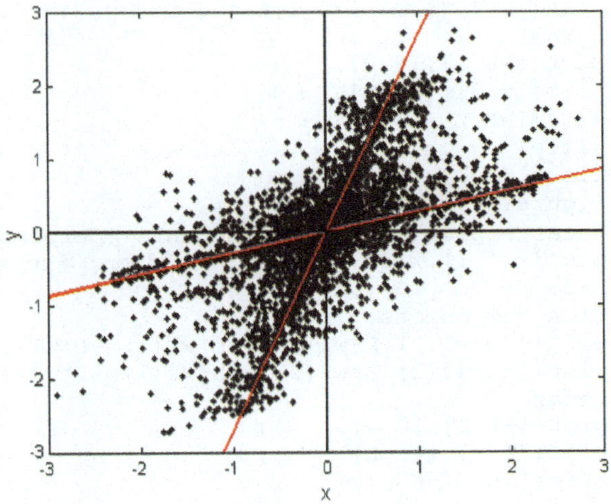

Fig. 7.16 Scatterplot of a mix of two speeches: ICA components

Program 7.12 Comparison of PCA and ICA components

```
% Comparison of PCA and ICA components
% example of two mixed speeches
%read two sound files
[a,fs1]=wavread('spch1.wav'); %read wav file
[b,fs1]=wavread('spch2.wav'); % " " "
R=2; %reduce data size for clearer picture
a=decimate(a,R);
b=decimate(b,R);
s1=(a-mean(a))'; %zero-mean
s2=(b-mean(b))'; % " " "
vr1=var(s1); s1=s1/sqrt(vr1); %variance=1
vr2=var(s2); s2=s2/sqrt(vr2); %" " "
s=[s1;s2]; %combine sources
```

```
%mix of sources
N=length(s1);
M=[0.7 0.3; 0.2 0.8]; %example of mixing matrix
x=M*s; %mixed signals
% PCA computation
A=x'/sqrt(N-1);
%singular value decomposition
[U,S,V]=svd(A); %V contains principal components
%ICA computation
U=inv(M);
w1=U(1,:);
w2=U(2,:);
%display
figure(1)
%scatterplot and PCA components
plot(x(1,:),x(2,:),'k.'); hold on; %scatterplot
L=3;
%PCA components:
plot([-L*V(1,1) L*V(1,1)],[-L*V(2,1) L*V(2,1)],'r');
plot([-L*V(1,2) L*V(1,2)],[-L*V(2,2) L*V(2,2)],'r');
%axes
plot([-L L],[0 0],'k');
plot([0 0],[-L L],'k');
axis([-L L -L L]);
title('scatterplot of 2 mixed speeches: PCA components');
xlabel('x'); ylabel('y');
figure(2)
%scatterplot and ICA components (perpendicular to w)
plot(x(1,:),x(2,:),'k.'); hold on; %scatterplot
L=3;
%ICA components:
plot([-L*w1(2) L*w1(2)],[L*w1(1) -L*w1(1)],'r');
plot([-L*w2(2) L*w2(2)],[L*w2(1) -L*w2(1)],'r');
%axes
plot([-L L],[0 0],'k');
plot([0 0],[-L L],'k');
axis([-L L -L L]);
title('scatterplot of 2 mixed speeches: ICA components');
xlabel('x'); ylabel('y');
```

7.4.3 Whitening

The first step of some ICA algorithms is whitening, or sphering, of the observed data $\mathbf{x}$. Whitening means a linear transformation of $\mathbf{x}$:

$$\mathbf{v} = Q\,\mathbf{x} \tag{7.21}$$

Such that the covariance matrix of $\vec{v}$ is I (unit matrix, diagonal).

The eigen-decomposition of the covariance matrix of **x** can be written as:

$$S_x = U \, \Lambda \, U^T \tag{7.22}$$

A whitening matrix Q can be obtained with:

$$Q = U \, \Lambda^{-1/2} \, U^T \tag{7.23}$$

where the matrix:

$$\Lambda^{-1/2} = diag\,(1/\sqrt{\lambda_1}\,,\; 1/\sqrt{\lambda_2}, \ldots,\; 1/\sqrt{\lambda_n}) \tag{7.24}$$

can be computed easily component by component. Actually, here we may have the opportunity to neglect small eigenvalues, in order to reduce dimensionality.

Now, it is interesting to consider an example. Suppose we have two source signals, which are two uncorrelated uniformly random variables with zero mean and variance equal to one. Notice that it is *not* the case of random variables with Gaussian PDF, but variables with uniform PDF. The two sources are mixed with a mixing matrix, so we have two observed signals.

Figure 7.17 shows the scatterplot of the two sources. It is a square. Figure 7.18 shows the scatterplot of the observed signals (the mix). Both figures have been obtained with the simple Program 7.13. Our problem is to recover the sources with an unmixing matrix.

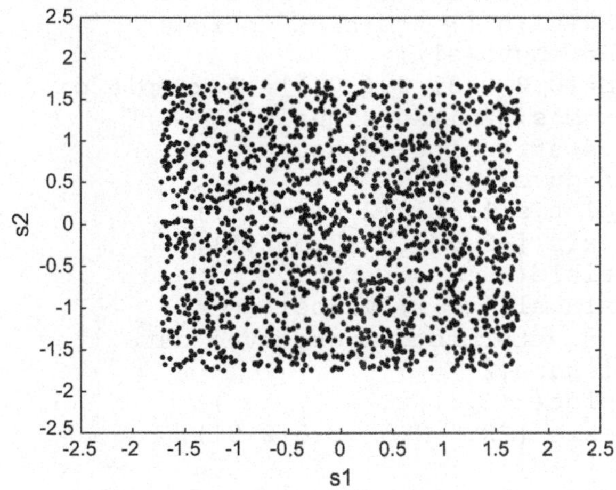

Fig. 7.17 Scatterplot of two uniform random sources

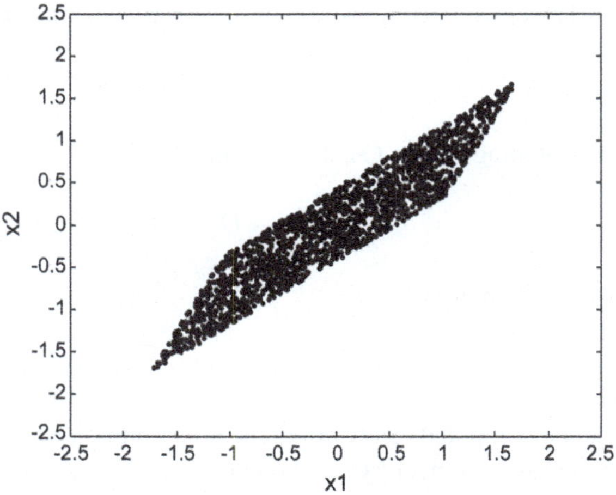

Fig. 7.18 Scatterplot of a mix of the two uniform random sources

Program 7.13 Scatterplots of two sources, and two mixed signals

```
%Scatterplots of two sources, and two mixed signals
% the sources are uniformly random variables
%two uniformly random sources,
N=2000;
s1=rand(1,N); s1=2*(s1-mean(s1)); %zero-mean
s2=rand(1,N); s2=2*(s2-mean(s2)); %" " "
vr1=var(s1); s1=s1/sqrt(vr1); %variance=1
vr2=var(s2); s2=s2/sqrt(vr2); %" " "
s=[s1;s2]; %combine sources
N=length(s1);
M=[0.2 0.8; 0.4 0.6]; %example of mixing matrix
x=M*s; %mixed signals
%scatterplot of sources
figure(1)
plot(s(1,:),s(2,:),'k.');
axis([-2.5 2.5 -2.5 2.5]);
title('scatterplot of two uniformly random sources');
xlabel('s1'); ylabel('s2');
%scatterplot of observed data (the mix)
figure(2)
plot(x(1,:),x(2,:),'k.');
axis([-2.5 2.5 -2.5 2.5]);
title('scatterplot of observed data (the mix)');
xlabel('x1'); ylabel('x2');
%print covariance matrix of sources
Ss=cov(s')
%print covariance matrix of observed signals
Sx=cov(x')
```

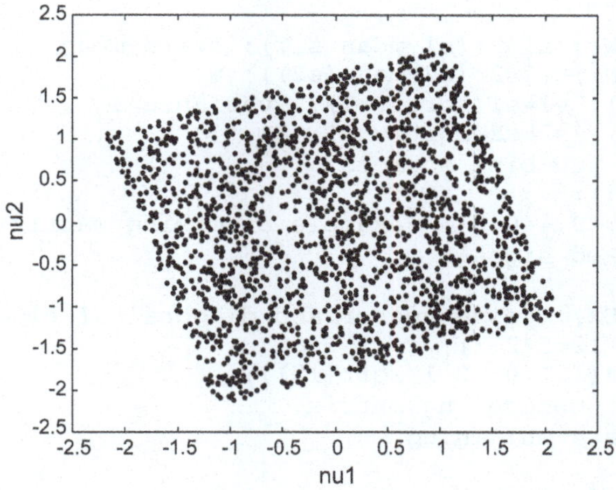

Fig. 7.19 Scatterplot of whitened data

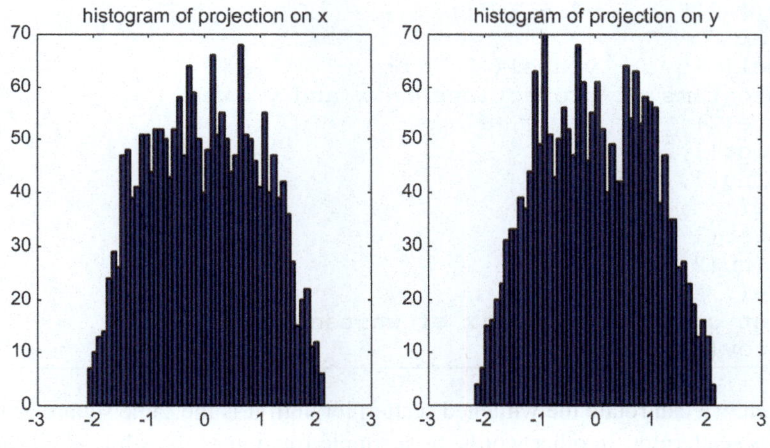

Fig. 7.20 Whitened data: projection on x and y axes

Let us apply whitening to the observed data. The result is shown in Fig. 7.19, which is generated by the Program 7.14. The program also prints the covariance matrix of the whitened data, to confirm that we obtain the identity matrix. Figure 7.20, also generated by the Program 7.14, shows the histograms of the data projected on the coordinate axes. The whitening has the effect that these projections are nearly Gaussians.

Program 7.14 Whitening of scatterplot of two random (uniform) signals

```
%Whitening of scatterplot of two random (uniform) signals
%two uniformly random sources,
N=2000;
```

```
s1=rand(1,N); s1=2*(s1-mean(s1)); %zero-mean
s2=rand(1,N); s2=2*(s2-mean(s2)); %" " "
vr1=var(s1); s1=s1/sqrt(vr1); %variance=1
vr2=var(s2); s2=s2/sqrt(vr2); %" " "
s=[s1;s2]; %combine sources
N=length(s1);
M=[0.2 0.8; 0.4 0.6]; %example of mixing matrix
x=M*s; %mixed signals
Sx=cov(x');
[U,L]=eig(Sx); %eigenvector and diagonal of eigenvalues
l1=L(1,1); l2=L(2,2);
sqL=[1/sqrt(l1) 0; 0 1/sqrt(l2)];
Q=U*sqL*U'; %whitening matrix
nu=Q*x; %data whitening
% display
%scatterplot of whitened data
figure(1)
plot(nu(1,:),nu(2,:),'k.');
axis([-2.5 2.5 -2.5 2.5]);
title('scatterplot of whitened data');
xlabel('nu1'); ylabel('nu2');
%histograms of projections on x and y axes
figure(2)
subplot(1,2,1)
hist(nu(1,:),50);
title('histogram of projection on x');
subplot(1,2,2)
hist(nu(2,:),50);
title('histogram of projection on y');
%print covariance matrix of whitened data
Snu=cov(nu')
```

Then, we can rotate the whitened scatterplot until it is the same square as in the sources scatterplot. In other words, by a simple rotation of the whitened data, it is possible to recover the sources.

The rotation of the scatterplot is done with the Program 7.15. Figure 7.21 shows the result, and Fig. 7.22 shows the histograms of the data projections on the coordinate axes. The initial square (the scatterplot of the two sources) was recovered. The main idea is to rotate the scatterplot trying to get data projections on the coordinate axes as non-gaussian as possible.

In summary, the recovery was done with rotation of whitened data:

$$\mathbf{s} = P \cdot Q \cdot \mathbf{x} \tag{7.25}$$

where P is a rotation matrix.

The Program 7.15 includes a simple rotation, with an angle '*alpha*'. Since in this example we already know the mixing matrix, we computed P as follows:

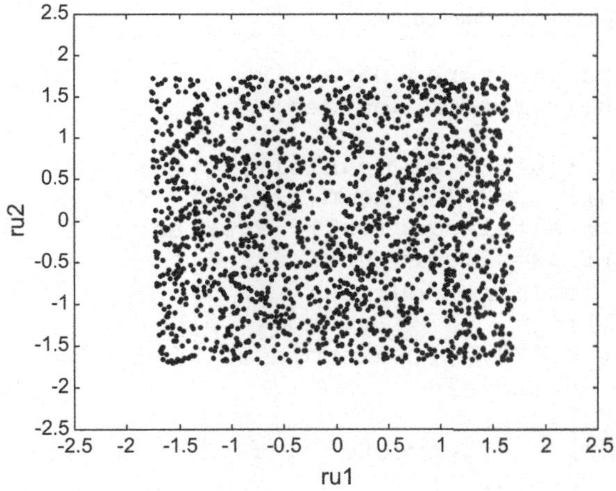

Fig. 7.21 Scatterplot of whitened and rotated data

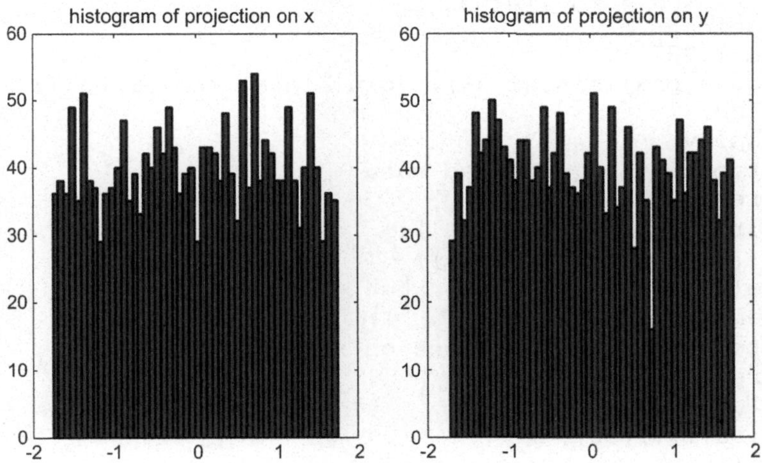

Fig. 7.22 Whitened and rotated data: projection on x and y axes

$$P = M^{-1} \cdot Q^{-1} = \begin{pmatrix} \cos\alpha & \sin\alpha \\ -\sin\alpha & \cos\alpha \end{pmatrix} \tag{7.26}$$

And then it is easy to obtain *alpha*. Of course, in reality you do not know M, and the rotation that maximizes non-Gaussianity of the projections on x and y must be obtained by an algebraic or searching method.

Program 7.15 Rotating the whitened data

```
%Rotating the whitened data
%two uniformly random sources,
N=2000;
s1=rand(1,N); s1=2*(s1-mean(s1)); %zero-mean
s2=rand(1,N); s2=2*(s2-mean(s2)); %" " "
vr1=var(s1); s1=s1/sqrt(vr1); %variance=1
vr2=var(s2); s2=s2/sqrt(vr2); %" " "
s=[s1;s2]; %combine sources
N=length(s1);
M=[0.2 0.8; 0.4 0.6]; %example of mixing matrix
x=M*s; %mixed signals
Sx=cov(x');
[U,L]=eig(Sx); %eigenvector and diagonal of eigenvalues
l1=L(1,1); l2=L(2,2);
sqL=[1/sqrt(l1) 0; 0 1/sqrt(l2)];
Q=U*sqL*U'; %whitening matrix
nu=Q*x; %data whitening
%rotation
alpha=1.887;
P=[cos(alpha) sin(alpha); -sin(alpha) cos(alpha)];
ru=P*nu;
% display
%scatterplot of whitened data
figure(1)
plot(ru(1,:),ru(2,:),'k.');
axis([-2.5 2.5 -2.5 2.5]);
title('scatterplot of rotated whitened data');
xlabel('ru1'); ylabel('ru2');
%histograms of projections on x and y axes
figure(2)
subplot(1,2,1)
hist(ru(1,:),50);
title('histogram of projection on x');
subplot(1,2,2)
hist(ru(2,:),50);
title('histogram of projection on y');
%print covariance matrix of whitened data
Sru=cov(ru')
```

Coming back to mathematics, after whitening, we have:

$$\mathbf{v} = Q\,M\,\mathbf{s} = B\,\mathbf{s} \tag{7.27}$$

Looking at covariance matrices, we see that:

$$S_{\nu} = E(\mathbf{v}\nu^T) = B\,E(\mathbf{s}\,\mathbf{s}^T)\,B^T = B\,S_s B^T = B\,B^T = I \tag{7.28}$$

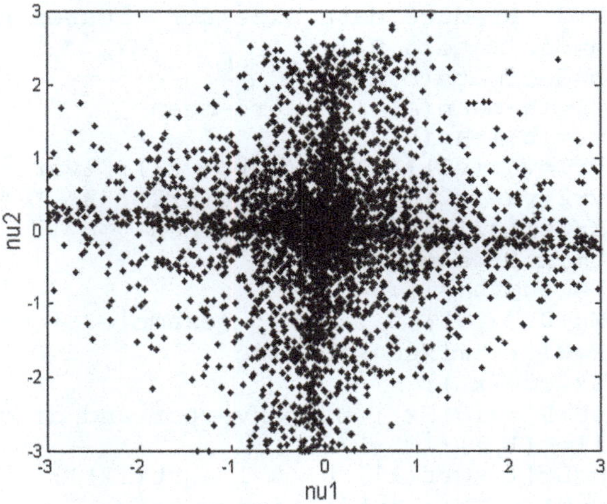

Fig. 7.23 Scatterplot of whitened mix of 2 speeches

(recall that the variances of the sources are one)

Therefore the new mixing matrix B is orthogonal (in two dimensions that means a simple rotation). Since B is orthogonal, we can write:

$$\mathbf{s} = B^T\, \mathbf{v} \tag{7.29}$$

Notice that a rotation of the whitened data does not change the value of the covariance matrix. Therefore, by rotating the whitening matrix we can obtain another whitening matrix: many alternative formulations of the whitening matrix could be given. For instance, some literature uses the following:

$$Q = \Lambda^{-1/2}\, U^T = S^{-1}\, U^T \tag{7.30}$$

where S and U can be obtained with the SVD of $\mathbf{x}$ (recall PCA). Let us check the whitening:

$$S_v = Q\, S_x Q^T = S^{-1}\, U^T S_x\, U\, (S^{-1})^T = S^{-1}\, \Lambda\, (S^{-1})^T = I \tag{7.31}$$

Figure 7.23, which has been generated with the Program 7.16, shows the effect of whitening on the mix of 2 speeches. Recall Figs. 7.15 or 7.16. Now, the cross branches become perpendicular, and the plot is ready for adequate rotation.

Program 7.16 Whitening of the mix of 2 speeches

```
%Whitening of the mix of 2 speeches
%read two sound files
[a,fs1]=wavread('spch1.wav'); %read wav file
[b,fs1]=wavread('spch2.wav'); % " " "
```

```
R=2; %reduce data size for clearer picture
a=decimate(a,R);
b=decimate(b,R);
s1=(a-mean(a))'; %zero-mean
s2=(b-mean(b))'; % " " "
vr1=var(s1); s1=s1/sqrt(vr1); %variance=1
vr2=var(s2); s2=s2/sqrt(vr2); %" " "
s=[s1;s2]; %combine sources
%mix of sources
N=length(s1);
M=[0.7 0.3; 0.2 0.8]; %example of mixing matrix
x=M*s; %mixed signals
Sx=cov(x');
[U,L]=eig(Sx); %eigenvector and diagonal of eigenvalues
l1=L(1,1); l2=L(2,2);
sqL=[1/sqrt(l1) 0; 0 1/sqrt(l2)];
Q=U*sqL*U'; %whitening matrix
nu=Q*x; %data whitening
% display
%scatterplot of whitened data
figure(1)
plot(nu(1,:),nu(2,:),'k.');
axis([-3 3 -3 3]);
title('scatterplot of whitened data');
xlabel('nu1'); ylabel('nu2');
%print covariance matrix of whitened data
Snu=cov(nu')
```

Let us substitute in the previous program the initial fragment devoted to get sound data, for the following fragment:

Fragment 7.17 Source change

```
N=5000;
s1=normrnd(0,1,1,N); s1=2*(s1-mean(s1)); %zero-mean
s2=normrnd(0,1,1,N); s2=2*(s2-mean(s2)); %" " "
vr1=var(s1); s1=s1/sqrt(vr1); %variance=1
vr2=var(s2); s2=s2/sqrt(vr2); %" " "
```

Now, the sources are two Gaussian random data. Recall Fig. 7.4, the scatterplot looks like a circle filled with dots. If we mix the two sources, and then we apply whitening the result is again a circular cloud. This is shown in Fig. 7.24. It is not worth to try now rotation to recover the original scatterplot, there are no hints. This is another way of saying that in case of Gaussian random data, all information is given by second order statistics, there is no more.

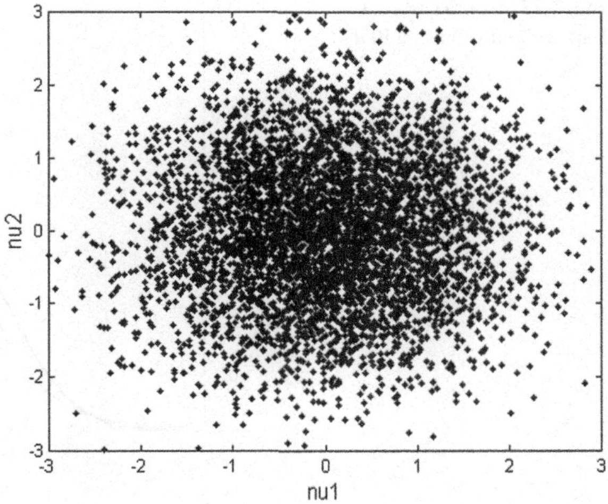

Fig. 7.24 Scatterplot of whitened mix of 2 Gaussians

7.4.4 Determination of Non-Gaussianity

For ICA purposes it is important to determine the non-Gaussianity of data. Some extensions of the concepts already studied in the chapter on statistical aspects are welcome. Let us begin with a closer view of moments.

The *normalized* n-th central moment or standardized moment, is the n-th central moment divided by σ^n.

The third central moment of the signal x is:

$$\mu_3 = E((x-\mu)^3) = \int_{-\infty}^{\infty} (\nu-\mu)^3 \, f_x(\nu-\mu)\, d\nu \tag{7.32}$$

The normalized third central moment is called the '*skewness*'.

Since the skewness takes into account the sign of x, it gives a measure of the asymmetry of the PDF. A PDF skewed to the right has a positive skewness; a PDF skewed to the left has negative skewness.

The fourth central moment of x is:

$$\mu_4 = E((x-\mu)^4) = \int_{-\infty}^{\infty} (\nu-\mu)^4 \, f_x(\nu-\mu)\, d\nu \tag{7.33}$$

The fourth central moment of a Gaussian distribution is $3\,\sigma^4$.

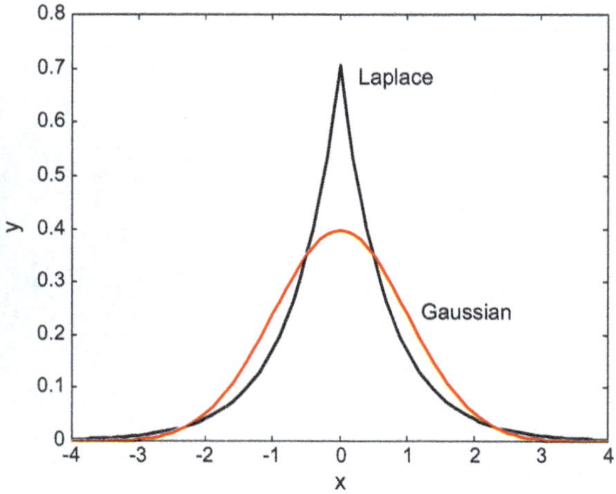

Fig. 7.25 Comparison of Laplace and Gaussian PDF

7.4.4.1 Kurtosis

A Classical Measure of Non-Gaussianity is '*kurtosis*':

$$kurt(x) = E((x-\mu)^4) - 3(E((x-\mu)^2)^2 \quad (7.34)$$

Some authors define kurtosis as the normalized fourth central moment minus 3.

The kurtosis is positive when the PDF of x is super-Gaussian, zero if it is Gaussian, and negative if it is sub-Gaussian. The sub-Gaussian PDFs are flatter than the normal PDF. The super-Gaussian PDFs have a sharper peak than the normal PDF.

Figure 7.25 compares the Laplace and the Gaussian PDFs. The Laplace PDF is super-Gaussian.

Program 7.18 Laplace PDF

```
% Laplace PDF
% compared to Gaussian PDF
x=-4:0.1:4;
yn=normpdf(x,0,1); %normal PDF
k=sqrt(2); %normalization constant
yL=(1/k)*exp(-k*abs(x)); %Laplace PDF
%display
figure(1)
plot(x,yL,'k'); hold on;
plot(x,yn,'r');
title('Laplace PDF compared with Gaussian PDF')
xlabel('x'); ylabel('y');
```

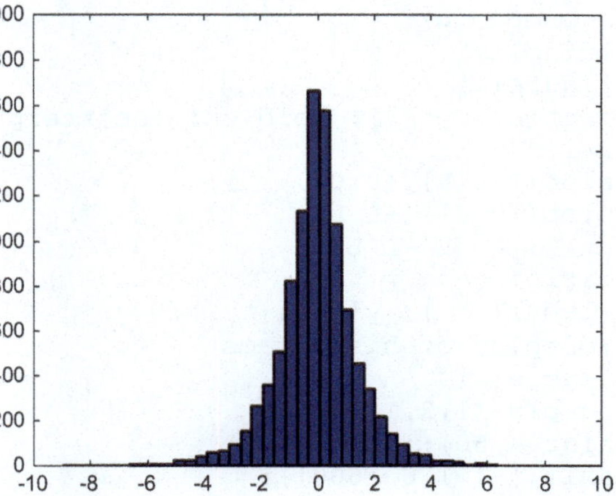

Fig. 7.26 Histogram of Laplace random data

For simulation purposes it is convenient to be able to generate Laplacian random data. Program 7.19 provides a simple generation. Figure 7.26 shows an histogram of the obtained random data.

Program 7.19 Laplace random signal histogram

```
% Laplace random signal
% Histogram
N=10000;
a=rand(1,N); %uniform random data
er=-log(1-a);%exponential random data
lr=er.*sign(rand(1,N)-0.5); %Laplace random data
%display
figure(1)
hist(lr,50);
axis([-10 10 0 2000]);
title('Histogram of Laplace random signal');
```

The kurtosis of speech is usually super-Gaussian, since the silent intervals are abundant, causing the central peak of the x histogram to be higher.

Let us use again the files with the two speeches. Figure 7.27 is again the scatterplot of these signals. Figure 7.28 shows the histograms of the speeches, both look Laplacian.

Program 7.20 Two speeches: scatterplot and histograms

```
% two speeches: scatterplot and histograms
%read two sound files
[a,fs1]=wavread('spch1.wav'); %read wav file
[b,fs1]=wavread('spch2.wav'); % " " "
a=a-mean(a);
```

```
b=b-mean(b);
%display
figure(1)
plot(a,b,'k.'); hold on; %scatterplot
L=1;
plot([-L L],[0 0],'k');
plot([0 0],[-L L],'k');
axis([-L L -L L]);
title('scatterplot: 2 speeches');
xlabel('a'); ylabel('b');
%display of histograms
figure(2)
subplot(1,2,1)
hist(a,50);
axis([-1 1 0 1600]);
title('histogram of signal a');
subplot(1,2,2)
hist(b,50);
axis([-1 1 0 1600]);
title('histogram of signal b');
```

Typically non-Gaussianity is measured by the absolute value of kurtosis. The kurtosis is zero for Gaussian random variables; and is nonzero for most, but not all, non-Gaussian random variables.

The MATLAB Statistics Toolbox provides the functions *skewness()* and *kurtosis()*. If x and y are random variables:

$$kurt(x + y) = kurt(x) + kurt(y) \tag{7.35}$$

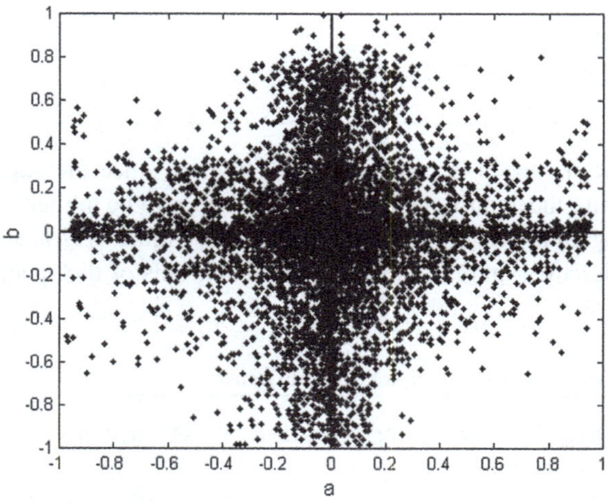

Fig. 7.27 Scatterplot of two speeches

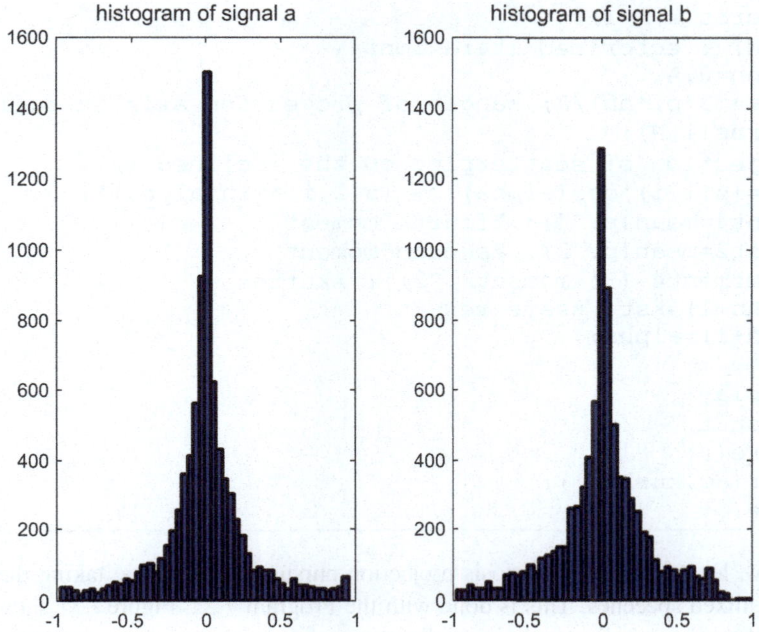

Fig. 7.28 Histograms of the two speeches

$$kurt(\alpha x) = \alpha^4 \, kurt(x) \tag{7.36}$$

It is now interesting to look again at the projection pursuit method, taking kurtosis as the variable to be studied. Figure 7.29 shows this projection pursuit for the case of the two speeches. The figure has been generated with the Program 7.21. It clearly shows the two main directions of data concentration corresponding to each of the speeches.

Program 7.21 Kurtosis projection pursuit example

```
% Kurtosis projection pursuit
% example of two original speeches
%read two sound files
[a,fs1]=wavread('spch1.wav'); %read wav file
[b,fs1]=wavread('spch2.wav'); % " " "
s1=(a-mean(a))'; %zero-mean
s2=(b-mean(b))'; % " " "
vr1=var(s1); s1=s1/sqrt(vr1); %variance=1
vr2=var(s2); s2=s2/sqrt(vr2); %" " "
s=[s1;s2]; %combine sources
N=length(a);
A=60; %number of circle partitions
kur=zeros(1,A+1);
```

```
ag=zeros(1,A+1);
i=1:N; %vectorized iterations
for nn=0:A,
alpha=(2*pi*nn)/A; %angle of projection axis in radians
p=zeros(1,N);
%projection of scatterplot on the inclined axis
p(i)=(s(1,i)*cos(alpha)) + (s(2,i)*sin(alpha));
moment4=mean(p.^4); %fourth moment
moment2=mean(p.^2); %second moment
kst=moment4-(3*(moment2.^2)); %kurtosis
kur(nn+1)=kst; %save result
ag(nn+1)=alpha;
end;
%display
%pursuit
figure(2)
polar(ag,kur,'k');
title('kurtosis as projection axis rotates');
```

Now, let us repeat the kurtosis projection pursuit, but this time taking the case of the mixed speeches. This is done with the Program 7.22. Figure 7.31 shows the result that, again, indicates the main directions of data concentration, corresponding to what can be also observed in the scatterplot (Fig. 7.30).

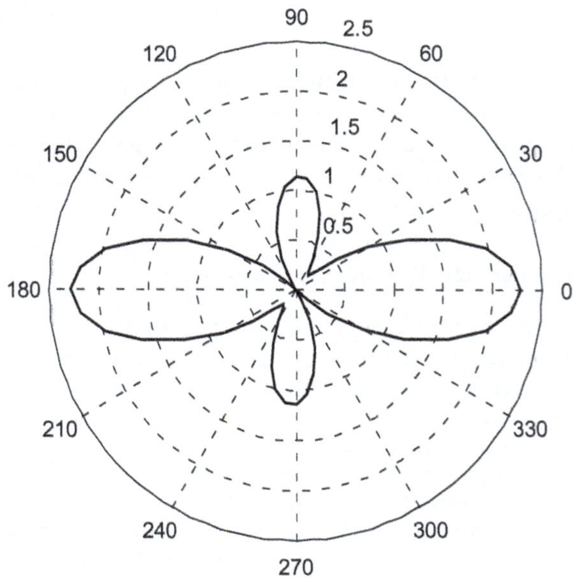

Fig. 7.29 Kurtosis pursuit for the original two speeches

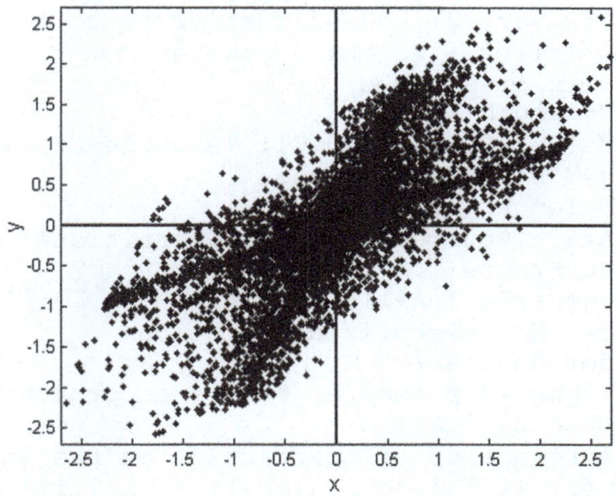

Fig. 7.30 Scatterplot of a mix of two speeches

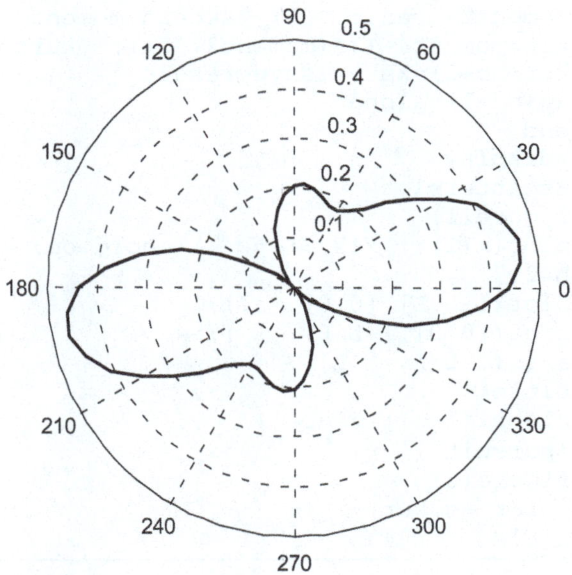

Fig. 7.31 Kurtosis pursuit for the mix of two speeches

Program 7.22 Kurtosis projection pursuit, two mixed speeches

```
% Kurtosis projection pursuit
% example of two mixed speeches
%read two sound files
[a,fs1]=wavread('spch1.wav'); %read wav file
[b,fs1]=wavread('spch2.wav'); % " " "
s1=(a-mean(a))'; %zero-mean
s2=(b-mean(b))'; % " " "
vr1=var(s1); s1=s1/sqrt(vr1); %variance=1
```

```
vr2=var(s2); s2=s2/sqrt(vr2); %" " "
s=[s1;s2]; %combine sources
%mix of sources
N=length(s1);
M=[0.7 0.3; 0.3 0.7]; %example of mixing matrix
x=M*s; %mixed signals
N=length(a);
A=60; %number of circle partitions
kur=zeros(1,A+1);
ag=zeros(1,A+1);
i=1:N; %vectorized iterations
for nn=0:A,
alpha=(2*pi*nn)/A; %angle of projection axis in radians
p=zeros(1,N);
%projection of scatterplot on the inclined axis
p(i)=(x(1,i)*cos(alpha)) + (x(2,i)*sin(alpha));
moment4=mean(p.^4); %fourth moment
moment2=mean(p.^2); %second moment
kst=moment4-(3*(moment2.^2)); %kurtosis
kur(nn+1)=kst; %save result
ag(nn+1)=alpha;
end;
%display
%scatterplot
figure(1)
plot(x(1,:),x(2,:),'k.'); hold on; %scatterplot
L=2.7;
plot([-L L],[0 0],'k');
plot([0 0],[-L L],'k');
axis([-L L -L L]);
title('scatterplot: 2 mixed speeches');
xlabel('x'); ylabel('y');
%pursuit
figure(2)
polar(ag,kur,'k');
title('kurtosis as projection axis rotates');
```

7.4.4.2 Cumulants

There is an elegant way to combine all the moments into a single expression, by using the following moment generation function:

$$\Gamma(\nu) = E(e^{\nu x}) = E(1 + \nu x + ... + \frac{\nu^n x^n}{n!} + ...) = \sum_{k=0}^{\infty} \frac{\nu^k x^k}{k!} \qquad (7.37)$$

(a Taylor expansion of the exponential has been used)

The n-th moment is the n-th derivative of $\Gamma(\nu)$-at the origin.

If we take logarithms we obtain the cumulant generating function:

$$K(\nu) = \log \Gamma(\nu) = \sum_{k=0}^{\infty} \kappa_k \frac{\nu^k}{k\,!} \tag{7.38}$$

The '*cumulants*' κ_i are the coefficients in the Taylor expansion of $K(\nu)$.

All cumulants of order three and higher are zero in the case of the normal (Gaussian) PDF. Therefore nonzero cumulants of order ≥ 3 mean non-Gaussianity.

The first three cumulants equal the first three central moments.

Standarized cumulants:

$$s\kappa_n = \kappa_n \left(\frac{x - \mu}{\sqrt{\mu_2}} \right) \tag{7.39}$$

In the case of the random variable uniformly distributed in [−a, +a] with probability $1/(2a)$:

- Moments:

$$\mu_{2k} = \frac{a^{2k}}{2k + 1} \tag{7.40}$$

- Cumulant of 4th order:

$$\kappa_4 = \frac{a^4}{5} - 3\,\frac{a^4}{9} = -2\,\frac{a^4}{15} \tag{7.41}$$

- Kurtosis:

$$s\kappa_4 = -\frac{6}{5} \tag{7.42}$$

- Odd moments and cumulants are zero

Consider the sum of two independent random variables $s = x + y$. The moment generating function of the sum is:

$$\Gamma_s(\nu) = \Gamma_x(\nu) \cdot \Gamma_y(\nu) \tag{7.43}$$

which is the product.

The cumulant generating function of the sum $s = x + y$ is:

$$K_s(\nu) = K_x(\nu) + K_y(\nu) \tag{7.44}$$

which is the sum. This is convenient for ICA purposes.

Some ICA algorithms use multivariate cumulants. Suppose a data set **u** composed of several series $\{u_1, u_2, \ldots, u_n\}$ of samples; like for example a digitalized electroencephalogram (EEG) obtained with n electrodes. Suppose also that all series are zero mean (if not, just subtract the mean). Here are a series of multivariate cumulants, up to order four:

$$C_i(\mathbf{u}) = E(u_i) = 0 \tag{7.45}$$

$$C_{ij}(\mathbf{u}) = E(u_i u_j) \tag{7.46}$$

$$C_{ijk}(\mathbf{u}) = E(u_i u_j u_k) \tag{7.47}$$

$$\begin{aligned} C_{ijkl}(\mathbf{u}) = E(u_i u_j u_k u_l) - E(u_i u_j)\, E(u_k u_l) - \\ - E(u_i u_k)\, E(u_j u_l) - E(u_i u_l)\, E(u_j u_k) \end{aligned} \tag{7.48}$$

The general expression of cumulant is:

$$C_{ij\ldots z}\,(\mathbf{u}) = \sum (-1)^{p-1}(p-1)! \left[\prod_{\xi=1}^{p} E(\prod_{\alpha \in \nu_l} u_\alpha) \right] \tag{7.49}$$

where the summation extends over all partitions $\{\nu_1, \nu_2, \ldots, \nu_p\}$ of $(i, j, \ldots, z)$.

Cumulants are tensors, which are sets of entries that one can localize using a set of indexes $(i, j, \ldots, z)$. Tensors could be organized according with different algebraic structures, like for instance sets of matrices. A simple and familiar example is a black and white video: each photogram is a matrix, and the video—a tensor—is an ordered set of matrices. In the case of a colour video, each photogram is a tensor (three matrices), and the video is a higher order tensor.

Note that in the case of cumulants, each entry is an expected value (for instance, the average of a set of signal samples).

The diagonal elements of cumulants characterize the distribution of single components. For example, the autocumulants of u_1 are $C_1(\mathbf{u})$, the mean, $C_{11}(\mathbf{u})$, the variance, $C_{111}(\mathbf{u})$, the skewness, and $C_{1111}(\mathbf{u})$, the kurtosis. The off-diagonal elements, with $i\,j\,k\,l \neq i\,i\,i\,i$, are called cross-cumulants; they characterize the statistical dependencies between components [33]. Cumulants are symmetric tensors [67].

Notice that in the bivariate case, the cumulants $C_{ij}(\mathbf{u})$ form the covariance matrix.

7.4.4.3 Negentropy

The more random is a variable, the more unpredictable and unstructured, the larger is its entropy. For continuous variables, the *differential entropy* of a random vector **x** is defined as:

$$H(\mathbf{x}) = -\int_{-\infty}^{\infty} f_{\mathbf{x}}(\nu)\ \log\ f_{\mathbf{x}}(\nu)\ d\nu = -E(\log f_{\mathbf{x}}(\nu)) \qquad (7.50)$$

This definition is also referred to as the Shannon's joint entropy. We see that we enter here in the context of information theory.

Gaussian variables have the largest entropy among all random variables of equal variance. In consequence, entropy can be used to determine non-Gaussianity. However, to obtain a linearly invariant version of entropy, it is more convenient to use '*negentropy*':

$$J(\mathbf{x}) = H(\mathbf{x}_{gauss}) - H(\mathbf{x}) \qquad (7.51)$$

In practice it is difficult to compute the negentropy. Several approximations have been proposed. A classical one is the following:

$$J(x) \approx \frac{1}{12} E((x-\mu)^3)^2 + \frac{1}{48} kurt(x)^2 \qquad (7.52)$$

where both sum terms can be expressed with the third and fourth standardized cumulants.

Other approximations have the general form:

$$J(x) = \sum_{i=1}^{q} k_i [E(G_i(x)) - E(G_i(\xi))]^2 \qquad (7.53)$$

where ξ is a zero-mean Gaussian variable with unit variance; k_i are positive constants, and p an integer. $G_i()$ are nonquadratic functions. For instance, the approximation (7.41) is obtained with $p = 1$ and $G(x) = x^4$. The following functions have been successfully used:

$$G_1(x) = \frac{1}{a} \log \cosh{(a\,x)}\,,\ \ G_2(x) = -\exp(-x^2/2) \qquad (7.54)$$

where $1 \leq a \leq 2$.

In order to compare with Fig. 7.31, another projection pursuit study has been done with respect to the mixed speeches but now using negentropy instead of kurtsosis. Figure 7.32. shows the result, which clearly marks one of the principal components and not so much the other. This figure has been generated with a program that has been included in Appendix A, and which is very similar to Program 7.22, except for the fragment listed after Fig. 7.32 (actually the difference between the two programs is just the lines concerning negentropy)

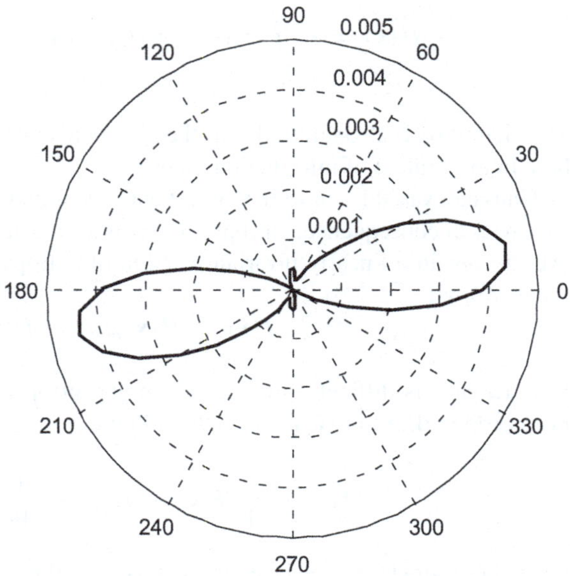

Fig. 7.32 Negentropy pursuit for the mix of two speeches

Fragment 7.23 Negentropy plot

```
moment4=mean(p.^4); %fourth moment
moment2=mean(p.^2); %second moment
moment3=mean(p.^3);
kst=moment4-(3*(moment2.^2)); %kurtosis
ng=((moment3^2)/12)-((kst^2)/48); %negentropy
negt(nn+1)=ng; %save result
ag(nn+1)=alpha;
end;
%display
%pursuit
figure(1)
polar(ag,negt,'k');
title('negentropy as projection axis rotates');
```

7.4.5 Assumptions of the ICA Method. Independence

The problem is that one has several observed signals $x_i(t)$ obtained from a set of sources $s_i(t)$ through a mixing matrix M.

The main assumption in order to apply ICA is that the sources are independent. Also, at most only one of the sources can be Gaussian.

Most of the research on ICA assumes that the number of observed signals is equal to the number of sources (M is square). In the case of more observed signals than sources, it is recommended to apply dimensional reduction.

There are some limitations in ICA. Since any constant multiplying an independent component could be cancelled by dividing the corresponding column of M by the same constant, it is not possible to determine the variance of the independent components: so it is usual to assign unit variance to these components. Also, in principle, ICA does not introduce an order between the independent components.

All the information of zero-mean Gaussian variables is contained in the covariance matrix, which is second-order statistics. This is also the context of PCA: the literature often mentions PCA as a second-order method.

Higher-order statistics (HOS) use information on the distribution of data that is not contained in the covariance matrix. The use of HOS is linked to the presence of non-Gaussian signals.

7.4.5.1 Independence:

The *joint PDF* of a pair of random variables x, y is defined by direct extension of the PDF of a single random variable. Denote it as $f_{xy}(\nu)$. The two variables are *independent* if and only if:

$$f_{xy}(\nu) = f_x(\nu) \cdot f_y(\nu) \tag{7.55}$$

This equation implies that:

$$E(x^p\, y^q) = E(x^p) \cdot E(y^q) \tag{7.56}$$

Let us highlight the difference between independence and uncorrelation. If the variables x, y are uncorrelated then:

$$E(x\, y) = E(x) \cdot E(y) \tag{7.57}$$

Independence is a stronger condition, since (7.29) must be satisfied for all positive integers p and q. Actually, there are examples of uncorrelated variables that are not independent [144].

Notice that correlation is related to the first moment, and independence is related to all moments.

In practical situations it is expected from the nature of sources that they are independent, like in the case of speeches of different people.

7.4.5.2 Mutual Information:

A natural way of checking the independence of x and y is to measure the difference between $f_{xy}(\nu)$ and $f_x(\nu) \cdot f_y(\nu)$. One could use '*Mutual Information*' as a measure of this difference.

The definition of mutual information can be given in terms of differential entropies:

$$MI(x, y) = H(x) - H(x|y) = H(x) + H(y) - H(x, y) \tag{7.58}$$

where $H(x, y)$ is the joint entropy of x and y. The mutual information is a nonnegative, symmetric measure. It is zero if and only if x and y are independent.

A general expression of mutual information is:

$$MI(x_1, x_2, \ldots, x_n) = \sum_{i=1}^{n} H(x_i) - H(\mathbf{x}) \tag{7.59}$$

If x_i are of unit variance and uncorrelated, then:

$$MI(x_1, x_2, \ldots, x_n) = C - \sum_{i=1}^{n} J(x_i) \tag{7.60}$$

where C is a constant. Notice the close relationship between mutual information and negentropy. The minimization of mutual information is thus equivalent to the maximization of non-Gaussianity.

Suppose that: $\mathbf{x} = W\,\mathbf{y}$ (a linear transformation, W is a matrix). An important property of mutual information is that:

$$MI(x_1, x_2, \ldots, x_n) = \sum_{i=1}^{n} H(x_i) - H(\mathbf{y}) - \log|\det W| \tag{7.61}$$

Again, there is a practical problem of computation. Several approximations have been proposed [140]. Some of them are based on Edgeworth expansions; like for example the following:

$$MI(\mathbf{x}) = C + \frac{1}{48}\sum_{i=1}^{n} [4\,\kappa_3(x_i)^2 + \kappa_4(x_i)^2 + 7\,\kappa_4(x_i)^4 - 6\,\kappa_3(x_i)^2\kappa_4(x_i)] \tag{7.62}$$

where x_i are zero-mean and uncorrelated.

7.4.6 Contrast Functions

In practical terms, the ICA problem is to find a matrix M so that the estimated sources (components) are as independent as possible. A function, which measures independence, should be optimized. This function is called *contrast function*.

Of course, the maximum value of a contrast function should be obtained *only* when sources are separated.

Along this section most candidates for contrast functions have been introduced. However, there is one more we wish to include: maximum likelihood estimation.

7.4.6.1 Maximum Likelihood (ML) Estimation:

Suppose the PDFs of the sources are known. Denote as b_i the rows of matrix B (7.29). The idea is to obtain values of b_i that maximizes the probability for the observations (the likelihood):

$$L(B) = \prod_{k=1}^{q} \prod_{i=1}^{n} f_i(b_i^T \cdot \mathbf{v}(t)) \; |\det B| \tag{7.63}$$

In general it is more practical to use the logarithm of the likelihood:

$$\log L(B) = \sum_{k=1}^{q} \sum_{i=1}^{n} \log f_i \, (b_i^T \cdot \mathbf{v}(t)) + q \log |\det B| \tag{7.64}$$

where f_i are the PDFs of the s_i.

Consider expected values:

$$\frac{1}{q} E(\log L(B)) = \sum_{i=1}^{n} E(\log f_i \, (b_i^T \cdot \mathbf{v}(t))) + q \log |\det B| \tag{7.65}$$

Taking into account the definition of differential entropy (7.50), and the expression (7.58), there is a close relationship among the log-likelihood and the negative of mutual information.

7.4.6.2 Relations

Along the section several mutual relations between contrast function candidates have been mentioned. Next table shows a concise view of these links.

	Cumulants	Entropy
Kurtosis	x	
Mutual Info.		x
Max. Likelihood		x

7.4.6.3 Difficulties, Criteria

In general, the estimation of high-order statistics requires more data samples as the order increases. So, in practice, only third or fourth order cumulants are used. The third order cumulants of symmetric PDFs are zero.

The cumulant based estimators may be far from optimal. Because high order cumulants measure primarily the tails of the PDF, and less the middle of the PDF. In addition, higher order cumulants are noticeably sensitive to outliers [140]. These comments apply, for example, to kurtosis.

The statistical properties of the negentropy estimator, using the approximation (7.42), are better than the properties of the estimators based on cumulants.

As said before, contrast functions based on entropy are difficult to estimate, but approximations can be used. Care is needed with (7.60), an approximation to mutual information, since it is valid when the PDF of $\mathbf{x}$ differs not much from the Gaussian PDF.

The problem with maximum likelihood is that it requires to know in advance the PDFs of the sources. An added difficulty is that it may be very sensitive to outliers.

7.4.7 Optimization Algorithms

It has been said [140] that:

ICA method = Contrast function + Optimization algorithm

Many optimization algorithms have been proposed for the ICA context. With some generality, one could say that there are two main alternatives for the optimization of the chosen contrast function:

(a) First a single independent component is obtained. Then the problem is re-sized, and a second independent component is obtained. And so on. In many cases it is not required to obtain all components, but only the most non-Gaussian.
(b) All independent components are obtained at the same time.

The specialists use different names for the two alternatives. Alternative (a) could be called one-unit approach [140], or deflationary method [1, 148, 178]. Alternative (b) could be called multi-unit approach [140], batch mode method [140], symmetric method [1, 178], or parallel approach [148].

Notice that this section has introduced Independent Component Analysis using the example of two mixed speeches. The problem to solve, illustrated with this example, is the recovering of the original signals from the mixed signals, and it is called '*blind source separation*' (*BSS*). Although the example deals with just two signals, the general case may involve many signals. By the way, some authors use the term TITO for two input–two output systems, which is the case of two original signals being mixed to give two outputs.

Given an optimization problem, belonging to alternative (a) or to alternative (b), there are many optimization algorithms that could be applied.

Among the optimization algorithms, there are some of them that can be used—with slight adaptations—in either alternative.

In general, there are iterative (step by step) or direct (one step) optimization algorithms.

An important part of the iterative algorithms are based on gradients. Other algorithms belong to the heuristic search class, like those using genetic algorithms.

Some authors refer to direct algorithms as algebraic methods.

Brief historical perspective

The seminal work on BSS was [134] in 1986, in relation to a neurophysiological application (muscle contraction study). During the 80s most research on this topic was done in France (see the review [159]). In 1994, Comon [67] enounced the concept of independent component analysis and proposed the minimization of adequate cost functions.

In the 90s artificial neural networks captured a lot of research attention. In 1992, Linsker [203] proposed unsupervised learning rules based on information theory. The target of these rules was to maximize the mutual information between inputs and outputs of the neural network. In 1995, Bell and Sejnowski [21] applied stochastic gradients for this maximization, and demonstrated that the procedure—called '*Infomax*'—can solve BSS problems. Soon after, it was shown [48] that this procedure was equivalent to the maximum likelihood estimation approach. Several improvements of Infomax were introduced by Amari [4], using natural gradient, and Cardoso and Laheld [50], using relative gradient.

In 1997, Hyvärinen and Oja [143] introduced the fixed-point or FastICA algorithm, which meant an important thrust for ICA applications and further research.

Another flow of activity yielding algebraic alternatives for ICA was higher-order spectral analysis. An example of this is the JADE algorithm introduced by Cardoso [51].

Now, let us consider with some detail a selection of paradigmatic algorithms, which are the root of many other algorithms that have been proposed as improvements.

In many papers, the notation used for the algorithm descriptions is as follows:

$$\mathbf{y} = W\,\mathbf{x} \tag{7.66}$$

where $\mathbf{x}$ are the mixed signals, W is the unmixing matrix, and $\mathbf{y}$ are the principal components.

7.4.7.1 Infomax

Let us place ourselves in the context of artificial neural networks.

According with the infomax principle, the information flow of the neural network must be maximized, and this is done by maximizing the output differential entropy. Actually, maximum differential entropy implies independent signals [32].

In the case of a neural network with inputs $\mathbf{x}$, and outputs:

$$g_i(\mathbf{w}_i^T \mathbf{x}) \tag{7.67}$$

where $g_i(\cdot)$ are nonlinear functions (usually sigmoids), and $\mathbf{w}_i$ are the weight vectors of the neurons.

The differential entropy to be maximized would be:

$$H(g_1(\mathbf{w}_1^T\mathbf{x}),\ g_2(\mathbf{w}_2^T\mathbf{x}),\ \ldots,\ g_n(\mathbf{w}_n^T\mathbf{x})) = H(\mathbf{y}) \tag{7.68}$$

From the definition of entropy,

$$H(\mathbf{y}) = -\,E(\log(p(\mathbf{y})) \tag{7.69}$$

(using the joint PDF $p(\mathbf{y})$)

After some algebra:

$$H(\mathbf{y}) = H(\mathbf{x}) + \frac{1}{N}\sum_{i=1}^{N}\log(p_{y_i}(y_i)) + \log|W| \tag{7.70}$$

To find the unmixing matrix that maximizes this expression, one has to maximize:

$$h(\mathbf{y}) = \frac{1}{N}\sum_{i=1}^{N}\log(p_i(y_i) + \log|W| \tag{7.71}$$

The gradient associated to this quantity is:

$$\nabla h(\mathbf{y}) = (W^T)^{-1} - \varphi(\mathbf{y})\,\mathbf{x}^T \tag{7.72}$$

where:

$$\varphi(\mathbf{y}) = -\,\frac{\partial p(\mathbf{y})/\partial \mathbf{y}}{p(\mathbf{y})} \tag{7.73}$$

Now, the optimum W is found by iteratively applying:

$$W_{new} = W_{old} + \eta\,\nabla h(\mathbf{y}) \tag{7.74}$$

where η is a small positive number.

The original Infomax algorithm uses:

$$\varphi(\mathbf{y}) = 2\,\tanh(\mathbf{y}) \tag{7.75}$$

and so the algorithm can be expressed as follows:

$$W_{new} = W_{old} + \eta\,[(W^T)^{-1} - 2\,\tanh(W\,\mathbf{x})\,\mathbf{x}^T] \tag{7.76}$$

By taking the approximation (7.75), the Infomax method assumes that the PDFs of the original signals are of the form *sech(y)*, which corresponds to a super-gaussian distribution. Actually, Bell and Sejnowski noted that many real-world signals, including speech, are super-gaussian.

A problem with Infomax it is not able to separate signals with sub-gaussian distributions. Several improvements have been introduced to remediate this difficulty.

Another problem of the algorithm is its slow convergence rate.

Coming back to the nonlinear functions $g_i(\cdot)$, their derivatives should be equal (or good approximations) to the PDFs of the original signals. On this basis, the infomax principle is equivalent to maximum likelihood estimation.

Since Infomax is a gradient based procedure, it may get stuck in a local minimum. In order to avoid this trap, several executions of the algorithm could be run, each departing from different initial values.

An interesting improvement of the algorithm is the use of the "natural" gradient, as follows:

$$W_{new} = W_{old} + \eta\,[\,I - 2\,\tanh(W\,\mathbf{x}) \cdot (W\,\mathbf{x})^T] \tag{7.77}$$

As other algorithms, Infomax requires pre-whitening of data.

7.4.7.2 FastICA

Let us start with a simple task: to extract a single independent component:

$$y = \vec{w}^{\,T}\,\mathbf{x} \tag{7.78}$$

Consider the following approximation to negentropy:

$$J(y) \approx \beta\,[E(G(y)) - E(G(\xi)]^2 \tag{7.79}$$

where $\beta > 0, \;\; \xi\; is\; N(0, 1)$.

Typically the maxima of negentropy are obtained at the maxima of $E(G(y))$, it is opportune to build the following function:

$$F(\mathbf{w}) = E(G(\mathbf{w}^T\mathbf{x}) - \frac{\lambda}{2}(\|\mathbf{w}\|^2 - 1) \tag{7.80}$$

where λ is a Lagrange multiplier.

A Newton-Raphson iteration can be applied for maximization of the function, as follows:

$$\mathbf{w} \leftarrow \mathbf{w} - \left(\frac{\partial^2 F(\mathbf{w})}{\partial \mathbf{w}^2}\right)^{-1} \left(\frac{\partial F(\mathbf{w})}{\partial \mathbf{w}}\right) \tag{7.81}$$

At each step of the iteration we shall divide $\mathbf{w}$ by $\|\mathbf{w}\|$.

The first derivative would be:

$$\left(\frac{\partial F(\mathbf{w})}{\partial \mathbf{w}}\right) = E(\mathbf{x} \cdot g(\mathbf{w}^T \mathbf{x})) - \lambda \mathbf{w} \tag{7.82}$$

where: $g = \partial G / \partial \mathbf{w}$

The stationary values of the function are obtained by equating the derivative to zero. If one multiplies both sides by $\mathbf{w}^T$, then:

$$\lambda = E(\mathbf{w}^T \mathbf{x} \cdot g(\mathbf{w}^T \mathbf{x})) \tag{7.83}$$

The second derivative can be obtained differentiating (7.82):

$$\left(\frac{\partial^2 F(\mathbf{w})}{\partial \mathbf{w}^2}\right) = E(\mathbf{x}\,\mathbf{x}^T \cdot g'(\mathbf{w}^T \mathbf{x})) - \lambda I \approx (E(g'(\mathbf{w}^T \mathbf{x})) - \lambda)\, I \tag{7.84}$$

Using these results, and supposing that the data $\mathbf{x}$ have been sphered, the maximization iteration can be written as follows:

$$\mathbf{w} \leftarrow \mathbf{w} - \frac{E(\mathbf{x}\, g(\mathbf{w}^T \mathbf{x})) - \lambda\, \mathbf{w}}{E(\, g'(\mathbf{w}^T \mathbf{x})) - \lambda} \tag{7.85}$$

Using a simple notation change, the above expression could be written as follows:

$$\mathbf{w}_k = \mathbf{w}_{k-1} - \frac{E_1 - \lambda\, \mathbf{w}_{k-1}}{E_2 - \lambda} \tag{7.86}$$

Then:

$$\mathbf{w}_k\, (\lambda - E_2) = E_1 - \mathbf{w}_{k-1}\, E_2 \tag{7.87}$$

Ignoring the factor $(\lambda - E_2)$, one obtains:

$$\mathbf{w} \leftarrow E(\mathbf{x}\, g(\mathbf{w}^T \mathbf{x})) - \mathbf{w}\, E(g'(\mathbf{w}^T \mathbf{x})) \tag{7.88}$$

In case of using $g(y) = \tanh(y)$ then $g'(y) = (1 - \tanh^2(y))$.

On the basis of the last mathematical expression, the practical FastICA algorithm executes the following steps:

1. Center the data to make the mean zero, and then whiten the result to give $\mathbf{x}$
2. Choose an initial value of $\mathbf{w}$ with unit norm
3. Execute the iteration (7.88), use averages for the expectations
4. Normalize: $\mathbf{w} \leftarrow \mathbf{w}/\,\|\mathbf{w}\|$
5. Go to 3, until convergence

If one wants to obtain more independent components, there are two alternatives:

1. Deflation (one by one):

The single component routine is applied, the new component is orthogonalized, using Gram-Schmidt method, with respect to all the other components already found, and then the new component is normalized.

2. Parallel:

The single component routine is applied in parallel for each independent component to be obtained, and then a symmetric orthogonalization is applied to all components simultaneously.

See details in [148].

The parallel algorithm can be expressed as follows [178]:

$$W^{+} = g(W\mathbf{x}) \cdot \mathbf{x}^{T} - diag[g'(W\mathbf{x})\, 1_N]\, W \tag{7.89}$$

$$W = (W^{+}\, W^{+T})^{-1/2}\, W^{+} \tag{7.90}$$

where 1_N is a N $\times$ 1 vector of 1's.

When kurtosis is used instead of negentropy, the iterative part of the algorithm becomes:

$$\mathbf{w} \leftarrow \mathbf{w} - \frac{1}{3} E(\mathbf{x}\,(\mathbf{w}^{T}\mathbf{x})^{3}) \tag{7.91}$$

$$\mathbf{w} \leftarrow \mathbf{w}\,/\,\|\mathbf{w}\| \tag{7.92}$$

Notice that the algorithm does not use any step size constant. Some authors refer to FastICA as a fixed-point algorithm, but this is subject to discussion.

The FastICA algorithm is frequently used because its simplicity, speed and accuracy. There are several MATLAB toolboxes that include the algorithm. In case of many components, it is recommended to use PCA for data reduction before applying ICA. An interesting performance analysis of FastICA is found in [296]. One of the observations made in this article is that sometimes, not very frequently, the FastICA algorithm gets stuck at saddle points (false double solutions). The article proposes improvements to avoid this problem.

Other interesting practical observations can be found in the Thesis [178]. It says that a problem of the deflation approach is that the estimation of each signal may be biased due to errors in estimation of the previous ones. However, signals that can be well estimated in deflationary way, may be degraded if one uses the parallel approach. The dissertation suggests combining both approaches, taking the parallel approach for a primary estimate, for initialization, and then trying to improve using deflation.

An improvement of the algorithm has been introduced by [328] with the name RobustICA. It uses exact line search, being able to algebraically compute the optimal step size at each iteration.

7.4.7.3 JADE

The JADE algorithm is an important example of tensor-based ICA algorithms. The interest on tensors, and in particular on cumulants, is due to the following property: for a given set $\boldsymbol{u}$ of n sampled signals $\{u_1, u_2, \ldots, u_n\}$, if and only if all signals are statistically independent, the cross-cumulants (the off-diagonal elements) are zero, so the cumulant tensors (of all orders) are diagonal [33, 189].

Therefore, there is a connection between finding independent components and diagonalization, which involves eigenvalues, eigenvectors, and rotations. More specifically, one should focus on '*joint diagonalization*'. Given a set of matrices:

$$(A_1, A_2, \ldots, A_K) \tag{7.93}$$

The problem of joint diagonalization is to find a (unitary) matrix U which makes:

$$U^T A_1 U, \ U^T A_2 U, \ \ldots, \ U^T A_K U \tag{7.94}$$

as diagonal as possible simultaneously.

Although the second-order cumulant tensor is easily diagonalizable by whitening, in general an exact joint diagonalization of higher order cumulants is not possible [67]. Therefore, the research has proposed several approximated joint diagonalizations. One of these is based on the Jacobi optimization method, which is an iterative technique using a sequence of plane rotations. In order to diagonalize a symmetric matrix, Jacobi proposed to minimize the sum of squares of non-diagonal entries.

From the ICA point of view there is another property of cumulants that is important: higher order cumulants of Gaussian variables are zero. Then, higher order cumulants are blind for additive Gaussian noise [189].

In practical terms, the use of higher-order statistics is restricted to cumulants of third and fourth order. In the case of symmetric distributions, third-order cumulants are zero.

The contrast function used by JADE is the following:

$$\Phi_{JADE} = \sum_{i\,j\,k\,l \neq i\,i\,k\,l} (C_{i\,j\,k\,l}(\mathbf{y}))^2 \tag{7.95}$$

where $\mathbf{y}$ are the estimated independent components, and the sum is over all the quadruples $(i\ j\ k\ l)$ of indices with $i \neq j$. This contrast function must be minimized.

The JADE algorithm uses second and fourth order cumulants. The algorithm starts with a whitening step, using the second order cumulant. Then, the Jacobi method is applied, obtaining a rotation that makes the fourth-order cumulants as diagonal as possible.

JADE stands for *'joint approximate diagonalization of eigenmatrices'*. Since the algorithm is computationally expensive, an effort was made to compact the statistical information of cumulants by using eigenmatrices [51] (1993). After some years, in [49] (1999) it was not recommended to use eigenmatrices in case the mixing model has no good accuracy; instead, it was recommended to use '*cumulant matrices*'. Given a matrix M, with entries $m_{i\,j}$ the corresponding cumulant matrix $Q^Y(M)$ has the following entries:

$$[Q^Y(M)]_{i\,j} = \sum_{k\,l} C_{i\,j\,k\,l}(\mathbf{y}) \cdot m_{k\,l} \tag{7.96}$$

Consider now a set of $n \times n$ matrices M_i: M $= [M_1, M_2, \ldots, M_p]$. If M is an orthonormal basis for the linear space of $n \times n$ matrices (therefore $p = n^2$), then the corresponding set of cumulant matrices $Q^Y = [\,Q^Y(M_1),\ Q^Y(M_2),\ \ldots,\ Q^Y(M_p)]$ is a maximal set of cumulant matrices. For any maximal set of cumulant matrices:

$$D_M(V) = \sum_{M_i \in M} Off(V^T\,Q^Z(M_i)\,V) = \Phi_{JADE}(Y) \tag{7.97}$$

where Z are sphered data, V any orthonormal matrix, *Off( )* means off-diagonal, and $Y = V^T\,Z$

Hence, the problem now is to find a matrix V minimizing the contrast function, which corresponds to a joint diagonalization of Q^Z.

The basis of M_i matrices can be built using $e_p \cdot e_q^T$, where e_p is a column vector with a 1 in the pth position and zeros elsewhere, and $1 \leq p, q \leq$ n. The matrix V can be obtained, via Jacobi method, using Givens rotations R_i: $V = R_1\,R_2\,\cdots R_k$

See [265] for a detailed mechanization of the JADE algorithm.

Being not a gradient descent algorithm, JADE does not have the typical problems of using gradients. It also uses very efficient MATLAB functions for the contrast function optimization. It is about 10 times faster than Infomax. Since it is based on kurtosis, JADE works well when kurtosis is adequate. If the signal distributions were skewed, the separating capacity of JADE would degrade. There are tensor-based algorithms that consider third- and fourth-order cumulants, like CuBICA [33]. One of the main problems of JADE is memory requirements for cumulant storage.

In addition to JADE, the article [49] includes a short description of the MaxKurt algorithm, which uses joint diagonalization of matrices with Jacobi method without the need of computing cumulant matrices. [49] also includes a mixed approach (JADE-MaxKurt) called SHIBBS, which uses less cumulant matrices and much less memory. Another interesting article is [189], which refers in mathematical terms to a class of algebraic algorithms: HOEVD, MD, JADE, STOTD.

7.4.7.4 An Algebraic Algorithm

Consider the case of two mixed signals x_1, x_2, which are related to the sources as follows:

$$\begin{bmatrix} x_1 \\ x_2 \end{bmatrix} = \begin{bmatrix} 1 & \alpha \\ \beta & 1 \end{bmatrix} \begin{bmatrix} s_1 \\ s_2 \end{bmatrix} \tag{7.98}$$

The problem is to determine α, β. The algebraic solution is given by the next two equations:

$$\beta = \frac{\alpha\, C_2 - C_3}{\alpha\, C_3 - C_1} \tag{7.99}$$

and,

$$\begin{aligned} &(C_2 C_{10} - C_{11} C_3)\, \alpha^4 + (3\, C_9 C_3 - 3\, C_8 C_2 - C_3 C_{10} + C_1 C_{11})\, \alpha^3 + \\ &+ (3\, C_6 C_2 + 3\, C_8 C_3 - 3\, C_9 C_1 - 3\, C_7 C_3)\, \alpha^2 + \\ &+ (C_5 C_3 + 3\, C_7 C_1 - 3\, C_6 C_3 - C_2 C_4)\, \alpha + (C_3 C_4 - C_1 C_5) = 0 \end{aligned} \tag{7.100}$$

where

$$C_1 = E(x_1^2) - \{E(x_1)\}^2$$

$$C_2 = E(x_2^2) - \{E(x_2)\}^2$$

$$C_3 = E(x_1 x_2) - E(x_1) E(x_2)$$

$$C_4 = E(x_1^4) - E(x_1^3) E(x_1)$$

$$C_5 = E(x_1^3 x_2) - E(x_1^3) E(x_2)$$

$$C_6 = E(x_1^3 x_2) - E(x_1^2\, x_2) E(x_1)$$

$$C_7 = E(x_1^2\, x_2^2) - E(x_1^2\, x_2) E(x_2)$$

$$C_8 = E(x_1^2\, x_2^2) - E(x_1\, x_2^2) E(x_1)$$

$$C_9 = E(x_1\, x_2^3) - E(x_1\, x_2^2) E(x_2)$$

$$C_{10} = E(x_1\, x_2^3) - E(x_1) E(x_2^3)$$

$$C_{11} = E(x_2^4) - E(x_2^3) E(x_2)$$

(15.96)

This is a fast non-iterative algorithm proposed by [323]. It becomes very complex for more than two signals.

7.4.7.5 An Algorithm Based on Eigenvalue Decomposition

This algorithm has been proposed by [250], and has an extremely simple coding in MATLAB: just two lines.

The covariance matrices of the mixed signals and the sources are related as follows:

$$S_x = A\, S_s\, A^H \tag{7.101}$$

where A is the mixing matrix, and H denotes hermitian transpose.

Consider now a matrix Q_s that represents cross-statistics, and such that:

$$Q_x = A\, Q_s\, A^H \tag{7.102}$$

In order to recover the sources, a matrix W must be found, such as $W^H A = I$. With this matrix, the sources are obtained as follows:

$$\mathbf{s} = W^H A\, \mathbf{s} = W^H \mathbf{x} \tag{7.103}$$

Equations (7.101) and (7.102) can be combined to obtain:

$$S_x W = Q_x\, W\, \Lambda \tag{7.104}$$

where it is assumed that: $\Lambda = S_s\, Q_s^{-1}$ is diagonal.

Equation (7.104) is a generalized eigenvalue equation. The entries of Λ correspond to individual source statistics (ratios of diagonal entries of S_s and Q_s). This equation can be used to obtain W, which is the solution sought.

Following the argumentation line of [250] there are three interesting cases: non-stationary sources, non-white sources, and non-Gaussian sources. Corresponding to these cases, there are three pertinent expressions of Q_s. For the non-Gaussian case, a cumulant matrix is chosen:

$$[C_s(M)]_{i\,j} = \sum_{kl} C_{i\,j\,k\,l}(\mathbf{s}) \cdot m_{k\,l} \tag{7.105}$$

(it is the same as in JADE, with a slight notation change)

This cumulant matrix can be computed as follows:

$$C_s(M) = E(\mathbf{s}^H M \, \mathbf{s}\,\mathbf{s}\,\mathbf{s}^H) \; - \; S_s \, trace(M \, S_s) \; - \; E(\mathbf{s}\,\mathbf{s}^T) \, M^T \, E(\mathbf{s}^* \, \mathbf{s}^H) \; - \; S_s M \, S_s \tag{7.106}$$

There is an equivalent definition for the mixed signals. And, if choose $M = I$, then:

$$C_x(I) \; = \; E(\mathbf{x}^H \, \mathbf{x}\,\mathbf{x}\,\mathbf{x}^H) \; - \; S_x \, trace(S_x) \; - \; E(\mathbf{x}\,\mathbf{x}^T) \; E(\mathbf{x}^* \, \mathbf{x}^H) \; - \; S_x \, S_x \tag{7.107}$$

This last expression is used as Q_x.

Therefore, the source separation routine in MATLAB can be summarized with the following fragment:

Fragment 7.24 Source separation

```
L=X*X';
Q=((ones(N,1)* sum(abs(X).^2)).*X)*X'- L*trace(L)/T-...
  (X*X.')*conj(X*X.')/T-L*L/T;
[W,D]= eig(L,Q);
S=W'*X;
```

In accord with the observations of other specialists, [250] comments that this method is very sensitive to estimation errors and the spread of kurtosis of the sources; therefore, it is recommended to use several matrices M.

More details on the algorithms already described can be found in [133, 142], together with some other algorithms. The article [141] presents recent advances of ICA. See [62] for an impressive review.

There are many extensions and variants of ICA and the BSS problem. An important review of nonlinear ICA is [158]. A much cited work on kernel ICA is [12], which is related to the fundamental [273].

7.4.8 Application Examples

As reviewed in [1, 236] there are many applications of ICA, with a lot of related papers [219]. One of the most important fields of application is biomedical signals [152, 157]. A specific problem that has received a lot of attention is face recognition [15, 331]. Other interesting applications are, structural dynamics [168], rotating machinery monitoring [111], radar [97], speech processing [191].

In general, the most representative ICA applications are: audio BSS [308], the analysis of images [22, 305], and image denoising [7].

7.4.8.1 Sounds

In this example two sounds are mixed and then separated using Infomax. Figure 7.33 shows the histograms of the two sound signals.

The Program 7.25 is in charge of mixing and then separating the two sounds. All figures of this example have been generated with this program. The code is based on programs published in the web page of James V. Stone.

Figure 7.34 shows the evolution of entropy, and the entropy gradient, along the algorithm iterations; they practically stabilize after 100 steps.

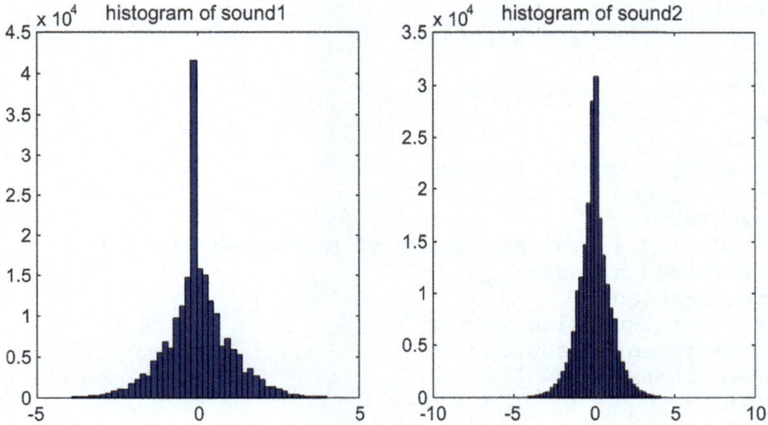

Fig. 7.33 Histograms of sound1 and sound2

Fig. 7.34 Evolution of entropy and entropy gradient along iterations

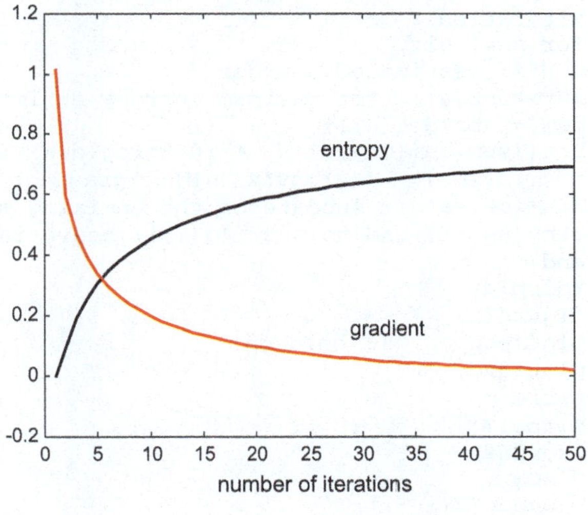

Program 7.25 Blind Source Separation using ICA, music

```
% Blind Source Separation using ICA
% Example of two music records
%read source signals
[s1,fs1]=wavread('wind1.wav'); %read wav file
[s2,fs2]=wavread('b1.wav'); %read wav file
s1=s1-mean(s1); %zero-mean
s2=s2-mean(s2); %" " "
vr1=var(s1); s1=s1/sqrt(vr1); %variance=1
vr2=var(s2); s2=s2/sqrt(vr2); %" " "
% Plot histogram of each source signal -
% this approximates pdf of each source.
figure(1);
subplot(1,2,1); hist(s1,50);
title('histogram of sound1');
subplot(1,2,2); hist(s2,50);
title('histogram of sound2');
drawnow;
s=[s1,s2]'; %combine sources
%mix of sources
N=length(s1);
M=[0.6 0.4; 0.4 0.6]; %example of mixing matrix
x=M*s; %mixed signals
%initialization
W=eye(2,2); %unmixing matrix
y=W*x; %estimated sources
%prepare iterations
nit=50;
delta=0.5;
etpy=zeros(1,nit); %for entropies
grd=zeros(1,nit); %for gradients
disp('working...'); %ask for patience
%iterations
for nn=1:nit,
y=W*x; %estimated sources
mhy=tanh(y); %for maximum entropy estimation
deW = abs(det(W));
h =((sum(sum(mhy)))/N) + (0.5*log(deW)); %entropy
g = inv(W') - ((2*mhy*x')/N); %gradient matrix
W=W+(delta*g); %update of the unmixing matrix
etpy(nn)=h; grd(nn)=norm(g(:)); %save intermediate values
end;
%display
figure(2)
plot(etpy,'k'); hold on;
plot(grd,'r')
title('entropy, and norm of gradient matrix');
xlabel('number of iterations');
%sounds
disp('one of the sound mixes');
soundsc(x(1,:));
disp('press bar when finished');
pause %hit a key!
```

```
disp('first extracted sound');
soundsc(y(1,:));
disp('press bar when finished');
pause %hit a key!
disp('second extracted sound');
soundsc(y(2,:));
%print correlations between sources and estimations
cr=corrcoef([y' s']);
cr(3:4,1:2)
```

The last lines of Program 7.25 are included for audio checking of the results, and for listing the correlations between original and estimated sources. The reader may wish to edit these lines (and the other for loading sound files) for experimentation purposes. It would be interesting to try different types of sounds, music, speech, natural, noises, etc.

7.4.8.2 Images

In this second example two images will be mixed and then separated, using the same algorithm of the previous example.

Figure 7.35 shows the histograms of the images: a lily and the Gioconda (not in color).

The mixing of images, the recovery of originals, and the generation of all figures of this example has been done with the Program 7.26.

Figure 7.36 shows the result of mixing the flower and the famous painting.

Now, the result of source separation is shown in Fig. 7.37.

As in the previous example, the evolution of entropy and the gradient of entropy is shown in Fig. 7.38.

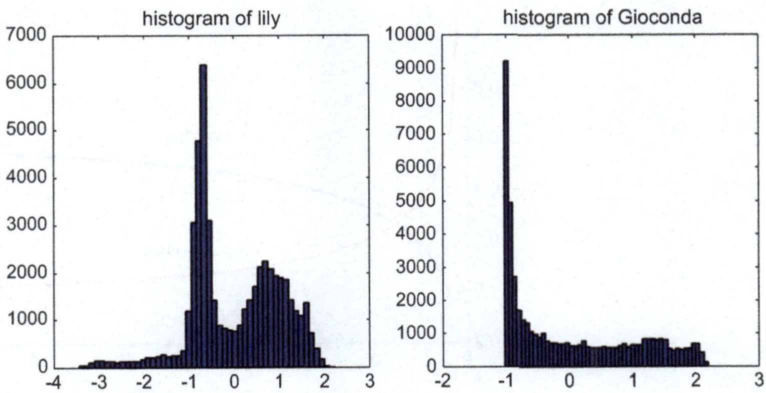

Fig. 7.35 Histogram of the photographs

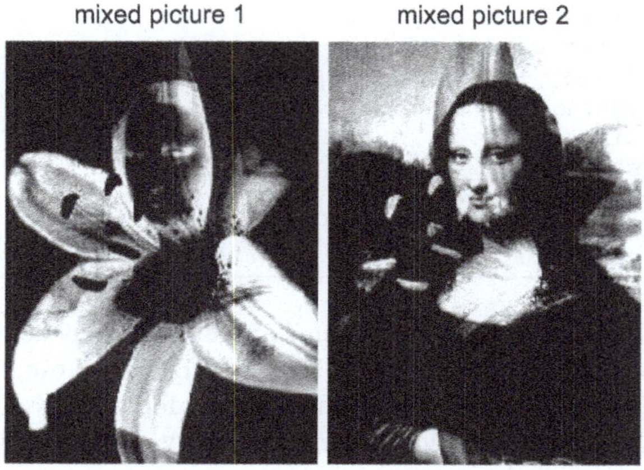

Fig. 7.36 Mixing of images

Fig. 7.37 Recovered pictures

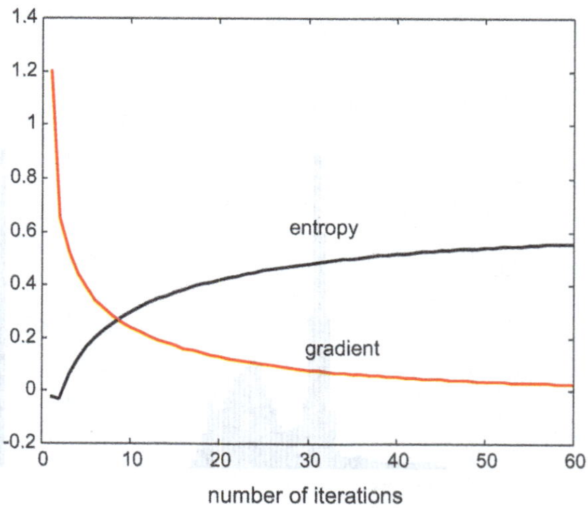

Fig. 7.38 Evolution of entropy and entropy gradient along iterations

Program 7.26 Blind Source Separation using ICA, images

```
% Blind Source Separation using ICA
% Example of two images
p1=imread('lily1.jpg'); %read the image file into a matrix
p2=imread('gioc1.jpg'); %read the image file into a matrix
N = 259*194; %picture size
%convert to vectors
s1=zeros(N,1); s2=zeros(N,1);
nn=1;
for ni=1:259,
for nj=1:194,
s1(nn)=p1(ni,nj);
s2(nn)=p2(ni,nj);
nn=nn+1;
end;
end;
%convert to +- 0.xxx float
ms1=mean(s1); ms2=mean(s2);
s1=s1-ms1; s1=s1/(max(abs(s1)));
s2=s2-ms2; s2=s2/(max(abs(s2)));
vr1=var(s1); s1=s1/sqrt(vr1); %variance=1
vr2=var(s2); s2=s2/sqrt(vr2); %" " "
% Plot histogram of each source signal -
% this approximates pdf of each source.
figure(1);
subplot(1,2,1); hist(s1,50);
title('histogram of lily');
subplot(1,2,2); hist(s2,50);
title('histogram of Gioconda');
s=[s1,s2]'; %combine sources
%mix of sources
M=[0.6 -0.4; -0.4 0.6]; %example of mixing matrix
x=M*s; %mixed signals
%mixed pictures
px1=zeros(259,194); nn=1;
for ni=1:259,
for nj=1:194,
px1(ni,nj)=x(1,nn);
nn=nn+1;
end;
end;
px2=zeros(259,194); nn=1;
for ni=1:259,
for nj=1:194,
px2(ni,nj)=x(2,nn);
nn=nn+1;
end;
end;
figure(2)
subplot(1,2,1); imshow(px1);
title('mixed picture 1');
%initialization
W=eye(2,2); %unmixing matrix
y=W*x; %estimated sources
```

```
%prepare iterations
nit=60;
delta=0.4;
etpy=zeros(1,nit); %for entropies
grd=zeros(1,nit); %for gradients
disp('working...'); %ask for patience
%iterations
for nn=1:nit,
y=W*x; %estimated sources
mhy=tanh(y); %for maximum entropy estimation
deW = abs(det(W));
h =((sum(sum(mhy)))/N) + (0.5*log(deW)); %entropy
g = inv(W') - ((2*mhy*x')/N); %gradient matrix
W=W+(delta*g); %update of the unmixing matrix
etpy(nn)=h; grd(nn)=norm(g(:)); %save intermediate values
end;
%display
figure(3)
plot(etpy,'k'); hold on;
plot(grd,'r')
title('entropy, and norm of gradient matrix');
%separated pictures
py1=zeros(259,194); nn=1;
for ni=1:259,
for nj=1:194,
py1(ni,nj)=y(1,nn);
nn=nn+1;
end;
end;
py2=zeros(259,194); nn=1;
for ni=1:259,
for nj=1:194,
py2(ni,nj)=y(2,nn);
nn=nn+1;
end;
end;
figure(4)
subplot(1,2,1)
imshow(py1);
title('recovered image 1');
subplot(1,2,2)
imshow(py2);
title('recovered image 2');
%print correlations between sources and estimations
cr=corrcoef([y' s']);
cr(3:4,1:2)
```

7.5 Clusters. Discrimination

You can easily classify people into two classes, those with age over 18, and those with age under or equal 18. This is a simple, one-dimensional classification example. There are also easy bi-dimensional classification cases in which two classes can be clearly distinguished, as in Fig. 7.39, where the circles correspond to the items being classified. A separating line and two labels, A and B, were added to the plot.

If one projects the circles on the separating line we added, the resulting points would be intermingled (or even overlapping), as depicted in Fig. 7.40.

A question of interest is whether there was a line such that projections of A and B circles on that line were separated. Figure 7.41 gives a positive answer.

Therefore, using projections onto a discriminating line, it is possible a one-dimensional classification, which means a dimension reduction from two to one for classification purposes.

Figure 7.41 suggests also that a 'distance' between A and B could be considered.

Notice that it was possible in Fig. 7.39 to draw many different separation lines. Likewise, it was also possible to draw many discriminating lines.

In the n-dimensional case the classification into two classes could be done using a separating hyperplane:

$$f(\mathbf{x}) = \mathbf{w}^T \mathbf{x} + b \tag{7.108}$$

Then, $\mathbf{x}$ belongs to the first class if $f(\mathbf{x}) > 0$ and $\mathbf{x}$ belongs to the second class if $f(\mathbf{x}) < 0$. The function $f(\mathbf{x})$ is called a *linear discriminant function*.

Of course, the reader may now suspect that classification matters are, in general, not as easy as in the examples already given. For the moment, let us emphasize that the examples were simple, low-dimensional, with just two groups of data, which admit

Fig. 7.39 Example of two groups of data

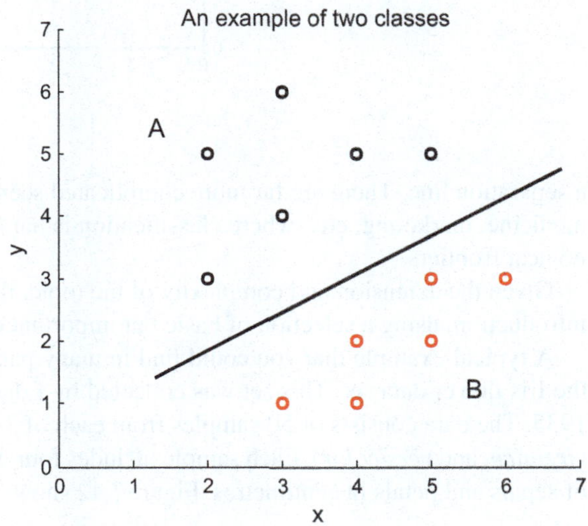

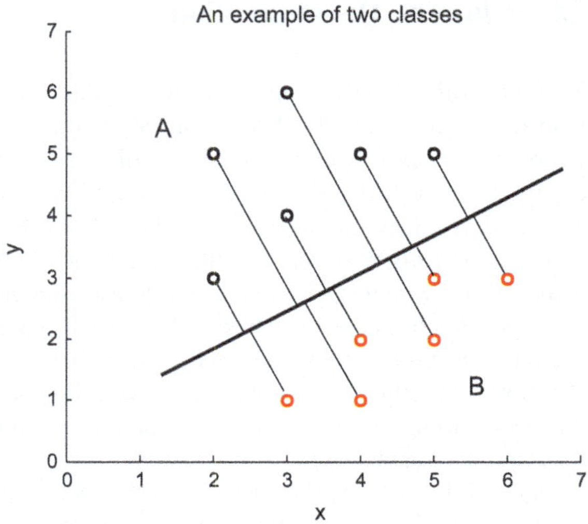

Fig. 7.40 Projection onto separating line

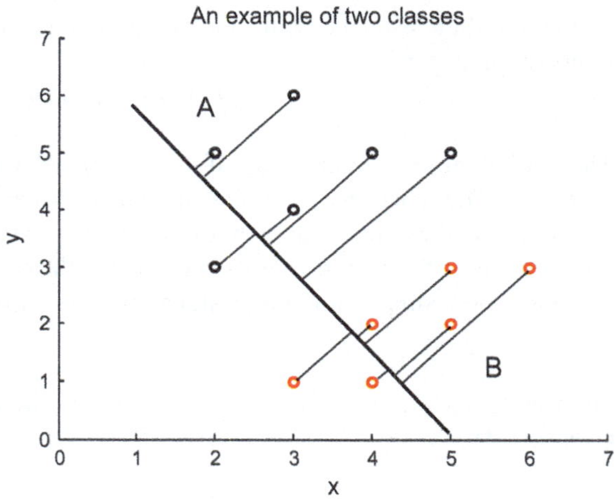

Fig. 7.41 Projection onto discriminating line

a separation line. There are far more complicated scenarios, for instance in biology, medicine, marketing, etc., where classification is hard to achieve, and perhaps with no neat frontiers.

Given the extension and complexity of the topic, this section can only be a short introduction, using a selection of basic but important concepts and tools.

A typical example that you could find in many papers related to classification is the Iris flower data set. This set was collected by Edgar Anderson and published in 1935. The data consists of 50 samples from each of 3 species of Iris flower (*cetosa, virginica*, and *versicolor*). Each sample includes four measures: the length and width of sepals and petals in centimetres. Figure 7.42 shows an Iris flower.

Fig. 7.42 Iris flower: petals and sepals

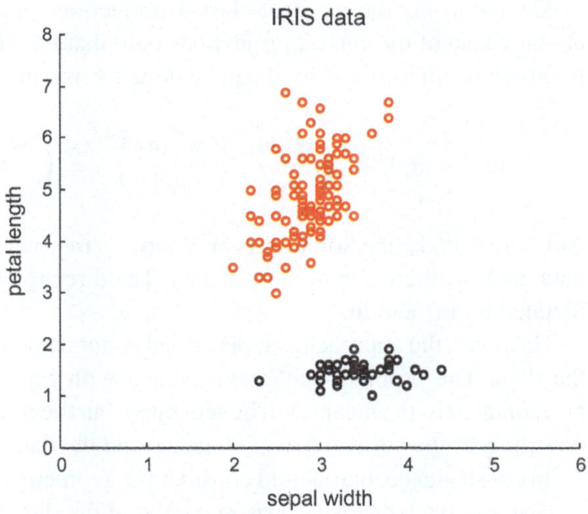

Fig. 7.43 IRIS data: two clusters

Figure 7.43 shows Iris data on a sepal width—petal length plane. The group at the bottom right corresponds to Iris cetosa, the other, larger group includes both Iris virginica and Iris versicolor. The figure has been generated with the Program 7.27, which uses the MATLAB *scatter( )* function.

Program 7.27 Scatterplot of IRIS data

```
% Scatterplot of IRIS data
% x=sepal width, y=petal length
D=dlmread('iris.data')'; %read IRIS data
```

```
%display
scatter(D(2,1:50),D(3,1:50),32,'k'); hold on;
scatter(D(2,51:150),D(3,51:150),32,'r');
axis([0 6 0 8]);
title('IRIS data');
xlabel('sepal width'); ylabel('petal length');
```

7.5.1 *Discrimination*

7.5.1.1 Linear Discriminant Analysis (LDA)

The linear discriminant analysis (LDA) was introduced by Sir Ronald Fisher in 1936, using as application example the Iris data set. The idea is to project the data onto a suitable discriminating line, which corresponds to a linear discriminant function.

Let us focus on binary classification, that is: classification into two different classes.

Denote as $\tilde{m}_1$ the mean of class-1 projections onto the discriminating line, and $\tilde{m}_2$ the mean of the class-2 projections onto that line. If one wishes to separate these points as much as possible, it can be done by maximization of:

$$(\tilde{m}_1 - \tilde{m}_2)^2 = \left(\frac{\mathbf{w}^T \mathbf{m}_1}{\|\mathbf{w}\|} - \frac{\mathbf{w}^T \mathbf{m}_2}{\|\mathbf{w}\|} \right)^2 = \left(\frac{\mathbf{w}^T}{\|\mathbf{w}\|} (\mathbf{m}_1 - \mathbf{m}_2) \right)^2 \tag{7.109}$$

Not surprisingly the solution is $\mathbf{w} = \mathbf{m}_1 - \mathbf{m}_2$, where $\mathbf{m}_1$ is the mean of class-1 data, and $\mathbf{m}_2$ the mean of class-2 data. The direction of the discriminating line was dictated by $\mathbf{m}_1$ and $\mathbf{m}_2$.

However, the approach just described is not a good idea if one wishes to separate the data. The problem can be visualized with Fig. 7.44. If one projects onto the horizontal axis, the means will be separated, but the projected data overlap. If, instead, one chooses the vertical axis, the means and the data will be separated.

In consequence, one should consider the geometry of variances of the data groups.

The approach of Fisher was maximizing the distance between $\tilde{m}_1$ and $\tilde{m}_2$ while minimizing the variance of the projected data points. Typically, specialized papers use here the concept of *data scatter*, which is:

$$S = \sum_{\mathbf{x} \in D} (\mathbf{x} - \mathbf{m})\,(\mathbf{x} - m)^T \tag{7.110}$$

Given class-1 and class-2 data groups, a *within-class scatter* can be considered:

$$S_w = S_1 + S_2 \tag{7.111}$$

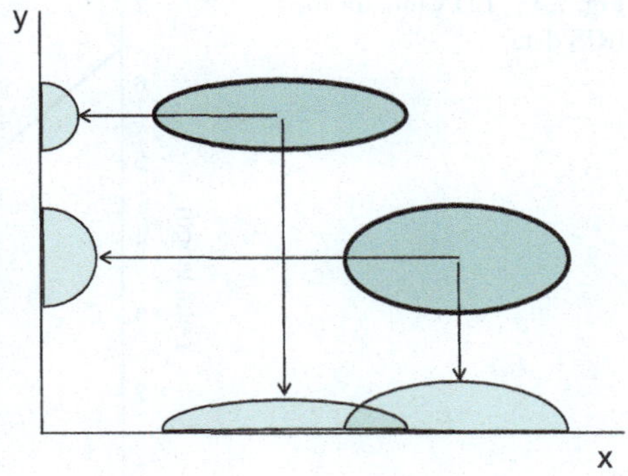

Fig. 7.44 Two projection scenarios

and a *between-class scatter*:

$$S_b = (\mathbf{m}_1 - \mathbf{m}_2)\,(\mathbf{m}_1 - \mathbf{m}_2)^T \tag{7.112}$$

Looking now at the projected data, the projected class centers would be:

$$\tilde{m}_i = \mathbf{w}^T \mathbf{m}_i \tag{7.113}$$

The projected within-class scatter:

$$\tilde{S}_w = \mathbf{w}^T S_w \mathbf{w} \tag{7.114}$$

The projected between-class scatter:

$$\tilde{S}_b = \mathbf{w}^T S_b \mathbf{w} \tag{7.115}$$

With these elements, Fisher forms the following linear discriminant:

$$J(\mathbf{w}) = \frac{\left|\tilde{S}_b\right|}{\left|\tilde{S}_w\right|} = \frac{\mathbf{w}^T S_b \mathbf{w}}{\mathbf{w}^T S_w \mathbf{w}} \tag{7.116}$$

which is a Rayleigh's quotient to be maximized.

The maximizing solution is obtained from the following eigensystem:

$$S_b\,\mathbf{w} = \lambda\, S_w \mathbf{w} \tag{7.117}$$

Then, the solution $\mathbf{w}$ (the direction of the discriminating line) is the leading eigenvector of $S_w^{-1} S_b$. In case of S_w being singular, a previous data reduction using PCA is recommended.

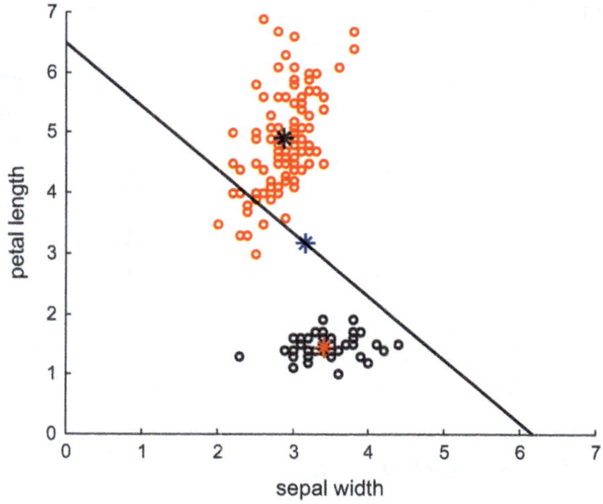

Fig. 7.45 LDA line for the IRIS data

Program 7.28 provides an example of LDA line calculation for the Iris data that were shown in Fig. 7.43. The program generates the Fig. 7.45, which shows the obtained discriminating line, which was positioned over the centroid of the total data. The centroids of the two data groups are also shown.

Program 7.28 LDA line

```
% LDA line
% x=sepal width, y=petal length
%read IRIS data:
D=dlmread('iris.data')';
D1=[D(2,1:50);D(3,1:50)]; %class 1, x-y data
D2=[D(2,51:150);D(3,51:150)]; %class 2, x-y data
%within-class mean
meanD1=mean(D1,2);
meanD2=mean(D2,2);
%data mean
meanD=(meanD1+meanD2)/2;
%within-class covariance matrix
mcovD1=cov(D1');
mcovD2=cov(D2');
%within-class scatter matrix
Sw=(mcovD1+mcovD2)/2;
%between-class scatter matrix
Sb1=(meanD1-meanD)*(meanD1-meanD)';
Sb2=(meanD2-meanD)*(meanD2-meanD)';
Sb=Sb1+Sb2;
%compute direction vector
A=inv(Sw)*Sb;
[V,E]=eig(A);
%in this case, the second eigenvector is the largest,
%so we take the second column of V
%LDA line: y=mx + b,
m=V(2,2)/V(1,2);
```

```
b=meanD(2)-(m*meanD(1)); %this line crosses the data centroid
%display
figure(1)
scatter(D1(1,:),D1(2,:),32,'k'); hold on;
scatter(D2(1,:),D2(2,:),32,'r');
axis([0 7 0 7]);
%centroid of set1:
plot(meanD1(1),meanD1(2),'r*','MarkerSize',12);
%centroid of set2:
plot(meanD2(1),meanD2(2),'k*','MarkerSize',12);
%data centroid:
plot(meanD(1),meanD(2),'b*','MarkerSize',12);
len=7; %line length, arbitrary value
plot([0 len],[b ((m*len)+b)],'k'); %LDA line
title('IRIS data: LDA line');
xlabel('sepal width'); ylabel('petal length');
```

In order to complete the view, Fig. 7.46 shows the projections of the two data groups onto the LDA line. The corresponding program has been included in the Appendix on long programs.

There is a lot of bibliography about LDA, like for instance the [173] classic book, or tutorials like [14]. A recommended reading for a practical use and judgement is [231]. The method can be extended for multi-class classification, [197]. Many applications have been reported, related to cancer diagnosis [56], face recognition [20, 93, 204], analysis of DNA [121], etc.

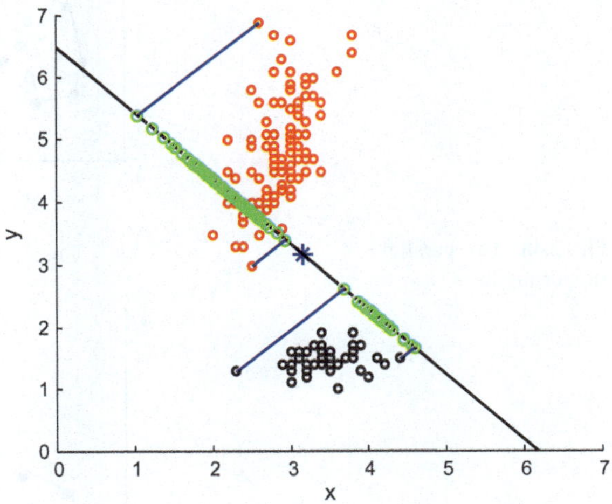

Fig. 7.46 Projections on LDA line

7.5.1.2 Support Vector Machine (SVM)

The support vector machine (SVM) was introduced by Cortes and Vapnik in 1995, [70], for binary classification. Figure 7.47 depicts the basic situation: there are two data groups, A and B, and one wants to establish a good separating line. Convex hulls of the data sets have been also highlighted by connecting points on the group boundaries with lines. Also, the two closest points, $a \in A$ and $b \in B$, have been emphasized by a double arrow. The convex hull of a set of points is the smallest convex set containing the points.

The two closest points, a and b, can be algebraically found by solving an optimization problem. A good separating line candidate would be the line bisecting the line between a and b.

Another point of view is related to the distance between two parallel supporting planes. A plane supports a class if all points of the class are on one side of the plane. Figure 7.48 depicts two parallel supporting lines (in 2D the supporting planes are lines) that have been separated as far as possible, so they intersect with the convex hulls.

Fig. 7.47 Two data groups and a separating line

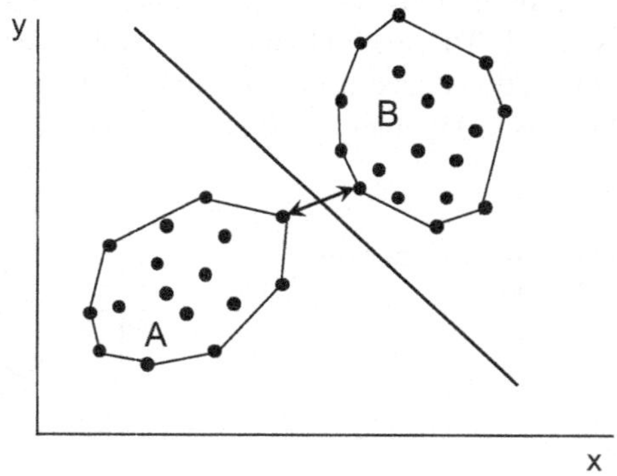

Fig. 7.48 Two parallel supporting lines

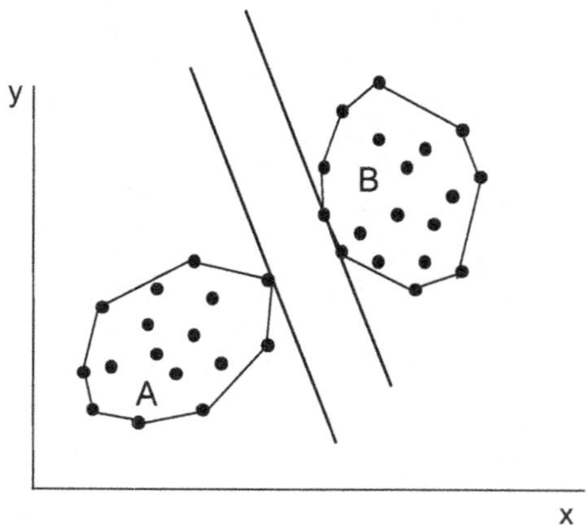

The distance between the two supporting lines is called *the margin*. It is convenient to find a direction such that the margin was maximized. Then, a good separating line would be the line in the middle between the supporting lines. As before, the margin maximization can be obtained by solving an optimization problem. It happens that this problem is the dual of finding the closest points, and so both optimization problems find the same separating line.

Notice that in Fig. 7.48 there are three points that intersect with the supporting lines; these points are called *support points* or also *support vectors*.

Let us come to the mathematical details. A first observation is that the equation of a supporting line, $\mathbf{w}^T\mathbf{x} + b = 0$, does not change if $\mathbf{w}$ and b are multiplied by the same constant. It is possible to scale the values of $\mathbf{w}$ and b such that.

$$\mathbf{w}^T\mathbf{x}_i + b \geq +1 \ for \ \mathbf{x}_i \in A \tag{7.118}$$

and

$$\mathbf{w}^T\mathbf{x}_i + b \geq -1 \ for \ \mathbf{x}_i \in B \tag{7.119}$$

The support vectors are the $\mathbf{x}_i$ satisfying the equality.

The distance or margin between the supporting lines corresponding to the two equations above, is:

$$\gamma = \frac{2}{\|\mathbf{w}\|_2} \tag{7.120}$$

Then, the margin maximization problem can be stated as follows:

minimize: $\mathbf{w}^T\mathbf{w}$

subject to: $y_i(\mathbf{w}^T\mathbf{x}_i + b) \geq 1$

where $y_i = +1 \ \ if \ \mathbf{x}_i \in A$ and $y_i = -1 \ \ if \ \mathbf{x}_i \in B$

This is a quadratic programming optimization problem that can be solved employing standard methods.

Program 7.29 provides an example of finding the widest corridor between the two Iris data classes that were also considered in Fig. 7.45. The program uses the *monqp( )* function of the Alaa Tharwat ToolBox (available from MATLAB Central), with the name changed to *qpg( )*.

Figure 7.49 shows the results obtained with the Program 7.29, emphasizing the separation line, and the supporting lines and points.

Program 7.29 SVM classification example

```
% SVM classification example
% IRIS data
% x=sepal width, y=petal length
%read IRIS data
D=dlmread('iris.data'); %columns
X=[D(:,2), D(:,3)]; %two columns (x,y)
Y=ones(150,1); Y(1:50)=-Y(1:50); %labels:1 or -1
%prepare quadratic optimization
lambda = 1e-7;
```

```
c = 1000;
ps=X*X'; %scalar product
A=Y;
b=0;
H =ps.*(Y*Y');
e = ones(size(Y));
%quadratic prog.:
[alpha , lambda , pos] = qpg(H,e,A,b,c,lambda,0,X,ps,[]);
%pos is position in X of the support vectors
%slope of separation line
mx=alpha'*(X(pos,1).*Y(pos));
my=alpha'*(X(pos,2).*Y(pos));
m=-mx/my; %
%compute line parameters
b1=X(pos(1),2)-(m*X(pos(1),1));
b2=X(pos(3),2)-(m*X(pos(3),1));
bm=(b1+b2)/2;
%display
figure(1)
%the data
scatter(X(1:50,1),X(1:50,2),32,'k'); hold on;
scatter(X(51:150,1),X(51:150,2),32,'r');
%the support vectors
scatter(X(pos,1),X(pos,2),64,'bd');
%lines for visualizing separation margin
plot([0 6],[b1 (m*6)+b1],'b--');
plot([0 6],[b2 (m*6)+b2],'b--');
plot([0 6],[bm (m*6)+bm],'r');
axis([0 6 0 8]);
title('Linear SVM on IRIS data');
xlabel('sepal width'); ylabel('petal length');
```

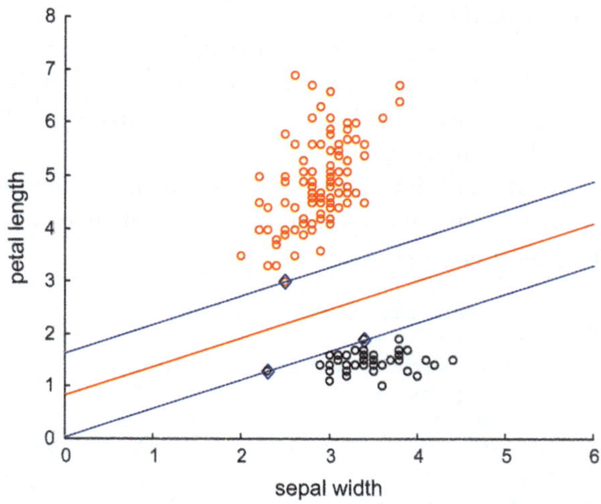

Fig. 7.49 Using SVM to find the separation line for the IRIS data

The bibliography on SVM is abundant, with books like [73, 306], tutorial papers [24, 44, 284], perspectives [245], and technical implementation reviews [38]. The article [147] presents a review of SVM applications in chemistry. Other important applications are: text classification [299], bioinformatics [312], pattern recognition [45], face recognition [131], image classification [57], medicine [91], etc.

7.5.2 *Clustering*

It is typical of descriptive sciences to use hierarchical decompositions. For instance, a first classification of animals is vertebrates or invertebrates, and then the classification continues with further divisions. This is a way of grouping data, and it is one example of clustering.

Another type of clustering activity is related to the discovery of groups or certain structures in the data. For instance, in relation with marketing it is important to identify groups of people that could be interested in a certain product. Other examples belong to medical diagnosis, genomics, environmental studies, etc. where it is important to establish relevant symptoms, features, characteristics that help to classify. In some sense, this could be similar to find where is Wally, or a hidden alphabet or message, only that many times you do not know a priori what structures are there and what labels should be used.

In this part of the section, two important, although simple, algorithms will be introduced: K-means and K-nearest neighbors.

The literature on data clustering, which is substantially linked to data mining, is quite large. Books like [2, 107, 164, 321], historical reviews [150], overviews and surveys [25, 151, 243], introductions [172], philosophy [309], etc.

Among most important applications, nowadays everything related to browsers and searching engines is of prime importance [74]. A most cited comparison of document clustering techniques is presented in [288]. Biomedical applications are also relevant [322]. Other applications will be mentioned in the next pages.

7.5.2.1 K-Means

People would notice that in Fig. 7.50 there are four groups of points. Humans are good for pattern recognition tasks; but computers need algorithms to try to perform these tasks. In the case of the K-means algorithm, the algorithm would start from a number k of centroids, at certain (usually at random) initial positions, and iteratively, step by step, it will move these centroids until satisfactory positions were found. Of course, in the example displayed in Fig. 7.49, it is best to try $k = 4$.

An important target for pattern recognition is to assign labels to data points. For instance, continuing with the example, the label attached to a point would say that this point belongs to group A, or B, etc. In the case of K-means, the distance to the

Fig. 7.50 Example of four data groups

centroids would say to which group the point belongs to. If the closest centroid is C, the point belongs to C, and so on with all points and centroids.

The user starts the algorithm giving a value to k, and, depending on the a priori knowledge, some convenient initial centroid positions or just positions at random. Then, the K-means algorithm repeats the following steps:

1. All points are labelled, according with distances to the centroids. (point 1 belongs to B, point 2 belongs to A, etc.)
2. The positions of the centroids are updated, using data from established groups. (centroid A is computed using the points belonging to A; etc.).

Usually, after not much iteration, the centroids do not move significantly and so the algorithm could be stopped.

Figure 7.51 shows an example with three groups of Gaussian random data. The case has been treated with the Program 7.30, which implements a K-means algorithm. The results of the algorithm are the three centroids that have been visualized in the figure, and the labelling of the data points in three colours indicating group A, B or C.

Program 7.30 Example of K-means

```
% Example of K-means
% Three clusters of Gaussian data
N=100;
%cluster 1
mux1=3; muy1=3; sigmax1=1; sigmay1=0.4;
x1=normrnd(mux1,sigmax1,1,N);
y1=normrnd(muy1,sigmay1,1,N);
%cluster 2
mux2=10; muy2=4; sigmax2=1; sigmay2=0.8;
x2=normrnd(mux2,sigmax2,1,N);
y2=normrnd(muy2,sigmay2,1,N);
%cluster 3
mux3=6; muy3=7; sigmax3=1; sigmay3=0.6;
x3=normrnd(mux3,sigmax3,1,N);
```

```
y3=normrnd(muy3,sigmay3,1,N);
% complete data set
D=zeros(2,3*N);
D(1,:)=[x1,x2,x3];
D(2,:)=[y1,y2,y3];
% centroids: initial guess (3 clusters)
mo=zeros(2,3);
mo(1,1)=D(1,1);mo(2,1)=D(2,1);
mo(1,2)=D(1,120); mo(2,2)=D(2,120);
mo(1,3)=D(1,215); mo(2,3)=D(2,215);
d=zeros(1,3);
%K-means
for it=1:100,
mn=zeros(2,3); %new centroid
nk=zeros(1,3);
for nn=1:(3*N),
k=1;
d(1)=(D(1,nn)-mo(1,1))^2+(D(2,nn)-mo(2,1))^2; %distance^2
d(2)=(D(1,nn)-mo(1,2))^2+(D(2,nn)-mo(2,2))^2; %distance^2
d(3)=(D(1,nn)-mo(1,3))^2+(D(2,nn)-mo(2,3))^2; %distance^2
if d(2)<d(1), k=2; end;
if (k==1 & d(3)<d(1)), k=3; end;
if (k==2 & d(3)<d(2)), k=3; end;
mn(:,k)=mn(:,k)+D(:,nn); nk(k)=nk(k)+1; %accumulate
end;
mn(1,:)=mn(1,:)./nk(1,:); %average
mn(2,:)=mn(2,:)./nk(1,:); % " "
mo=mn; %change of centroid
end;
%display
figure(1)
scatter(D(1,:),D(2,:),32,'ro'); hold on;
axis([0 14 0 9]);
title('Estimation of cluster centroids with K-means')
xlabel('x'); ylabel('y');
plot(mn(1,:),mn(2,:),'b*','MarkerSize',16);
plot(mn(1,:),mn(2,:),'ks','MarkerSize',16);
%Marker colors to discern clusters
for nn=1:(3*N),
k=1;
d(1)=(D(1,nn)-mo(1,1))^2+(D(2,nn)-mo(2,1))^2; %distance^2
d(2)=(D(1,nn)-mo(1,2))^2+(D(2,nn)-mo(2,2))^2; %distance^2
d(3)=(D(1,nn)-mo(1,3))^2+(D(2,nn)-mo(2,3))^2; %distance^2
if d(2)<d(1), k=2; end;
if (k==1 & d(3)<d(1)), k=3; end;
if (k==2 & d(3)<d(2)), k=3; end;
%
if k==1, plot(D(1,nn),D(2,nn),'bx'); end;
if k==2, plot(D(1,nn),D(2,nn),'gx'); end;
if k==3, plot(D(1,nn),D(2,nn),'mx'); end;
end;
```

Depending on the data, the K-means algorithm can suffer some difficulties. Of course, the value initially given to k is important, and perhaps it is not evident when one looks at the data points. Another issue is that the solution may depend on the

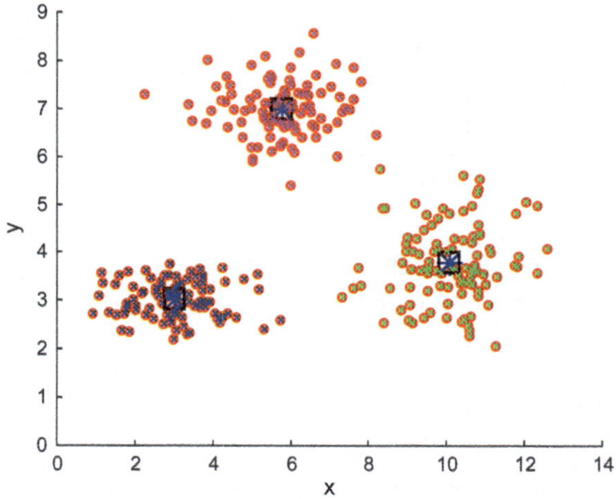

Fig. 7.51 Cluster centroids are obtained with K-means

initial values of the centroids. Actually, the K-means method tries to minimize a quadratic criterion, and it may happen that the algorithm becomes stuck in a local minimum. It also may happen that the algorithm results in empty clusters. The general recommendation is to try several initializations and several values of k.

The algorithm is also susceptible to outliers. Although the temptation would be to remove outliers beforehand, these outliers could be interesting cases deserving special study.

The criterion being minimized by K-means, is:

$$J = \sum_{i=1}^{k} \sum_{\mathbf{x} \in D_i}^{d(\mathbf{x}, c_i)^2} \tag{7.121}$$

where $d(\cdot)$ is distance, D_i the data clusters, and c_i are the centroids.

First ideas pertaining the algorithm come from 1957, Hugo Steinhaus. This same year, 1957, Stuart Lloyd proposed it as a method for pulse-code modulation, but it had to wait until 1982 to be published in a journal. In 1965, E.W. Forgy published the algorithm. The term "K-means" was coined by James MacQueen in 1967. Several authors remark that K-means can be regarded as a potential function reducing algorithm, taking J in the equation above as a potential.

The K-means algorithm is quite popular. Several improvements have been proposed, like the contributions of Hartigan and Wong in 1975, 1979. See [115] for a history of K-means type of algorithms, with appropriate references; an interesting comparison of algorithm versions, using Iris data, is [228]. A most cited K-means implementation is [162].

Variants and extensions of the algorithm are, *bisecting K-means* [288], *k-medoids* (see [249] for a fast version), a realization of k-medoids is *partitioning around medoids* (PAM) [164]. The *X-means* algorithm, [251], includes and efficient estimation of the number of clusters. *K-means++* tries to tackle large scale applications [13]. Other variants include fuzzy logic, different notions of distance, heuristic methods, etc.

Since the algorithm is easy and intuitive, it has found many applications. Important examples are document clustering [282], image segmentation [59] (this is relevant in medicine, [239]), weather forecasting [188], market segmentation (see [60] for a modern alternative), etc. The K-means algorithm can be easily parallelized, as in [185] for an environmental application.

Let us now refer briefly to '*vector quantization*' (VQ). This is a natural application for K-means. When you use a digital to analog converter (AD), say of 8 bits, you apply sampling along time, and quantization of each sample into 256 levels along signal amplitude. This encoding of amplitude into 8-bit words is an example of 'scalar quantization'. In certain cases, it is also possible to encode fragments of the signal (a number of samples) into a limited number of codes; this is '*vector quantization*'. In the fundamental article on vector quantization, [218], vector quantization is also called 'pattern-matching quantization'. For example, in the case of speech coding, the speech is divided into short segments—called 'frames'—with a duration of about 20 ms., and then codes are assigned to each frame. One of these codes could correspond to silence, another to an 'e' sound, etc. In this way, the speech is compressed and its transmission is more robust. Note that phonemes take around 500 ms each, so they are longer than frames.

For VQ the codes are taken from a set of codes called 'codebook'. In terms of clusters, the cluster indexes, 1, 2…N, corresponds to the codes 1, 2 ,…N. A first task for VQ encoding is to establish the codebook, which could be done using K-means.

In the case of images, VQ is typically applied dividing the image into blocks, and then encoding each block, [167, 214].

7.5.2.2 Classifying a New Datum using the K-Nearest Neighbour (K-nn) Rule

Suppose that a new Iris flower has grown in your garden. If you are curious about to which class it belongs to, you could plot its petal length and sepal width, as in Fig. 7.43, and try to guess an answer.

The idea of K-nn is to take a number k of nearest neighbors, and use majority voting. For instance, if you take 10 neighbors, 6 of them could be Iris cetosa, 3 of them Iris virginica, and 1 of them Iris versicolor. The K-nn rule would tell you that your new flower is Iris cetosa.

To illustrate the K-nn classification rule, an example has been devised using three synthetic data clouds, with labels 1, 2 and 3. Program 7.31 generates these clouds with Gaussian random data. Then, a new datum is added, and the K-nn rule is applied for $k = 12$. Figure 7.52 depicts the example. The program gives you the label that was assigned for the new datum, and the votes for each class.

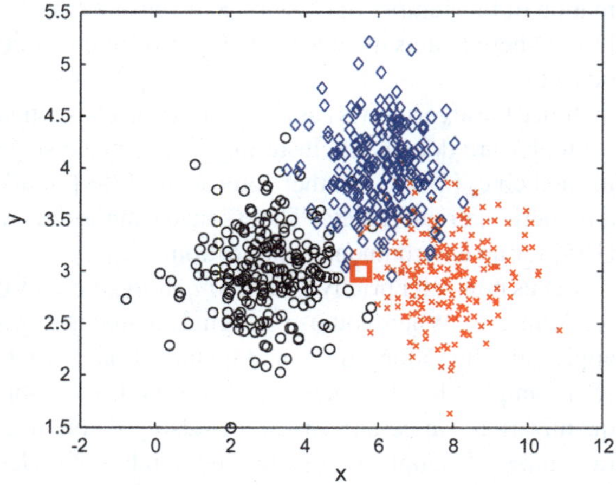

Fig. 7.52 Labeling of new datum is made with K-nn

Program 7.31 Example of K-nearest neighbor

```
% Example of K-nearest neighbor
% Three clusters of Gaussian data
K=12; %number of nearest neighbors to consider
N=200;
X=zeros(3*N,2); Y=zeros(3*N);
%cluster 1
mux1=3; muy1=3; sigmax1=1; sigmay1=0.4;
X(1:N,1)=normrnd(mux1,sigmax1,N,1);
X(1:N,2)=normrnd(muy1,sigmay1,N,1);
Y(1:N)=1;
%cluster 2
mux2=8; muy2=3; sigmax2=1; sigmay2=0.4;
X((N+1):(2*N),1)=normrnd(mux2,sigmax2,N,1);
X((N+1):(2*N),2)=normrnd(muy2,sigmay2,N,1);
Y((N+1):(2*N))=2;
%cluster 3
mux3=6; muy3=4; sigmax3=1; sigmay3=0.4;
X((2*N+1):(3*N),1)=normrnd(mux3,sigmax3,N,1);
X((2*N+1):(3*N),2)=normrnd(muy3,sigmay3,N,1);
Y((2*N+1):(3*N))=3;
%new datum to be classified
nx=5.5; ny=3;
%distances from new datum to all (vectorized code)
d=zeros(3*N,1);
nn=1:(3*N);
d(nn)=(X(nn,1)-nx).^2+(X(nn,2)-ny).^2; %distance^2
[od,idx]=sort(d); %sort in ascending order
%see the most voted class
kix=idx(1:K); %select K neighbors
cl=zeros(3,1);
```

```
aux1=0; aux2=0;
for nn=1:K,
aux1=kix(nn);
aux2=Y(aux1); %class value
cl(aux2)=cl(aux2)+1; %increase votes
end;
[oc,icx]=max(cl); %the most voted class
%display
figure(1)
%cluster 1, black:
plot(X(1:N,1),X(1:N,2),'ko'); hold on;
%cluster 2, red:
plot(X((N+1):(2*N),1),X((N+1):(2*N),2),'rx');
%cluster 3, blue:
plot(X((2*N+1):(3*N),1),X((2*N+1):(3*N),2),'bd');
% display classified new datum
switch icx
case 1,
plot(nx,ny,'ks','MarkerSize',12);
case 2
plot(nx,ny,'rs','MarkerSize',12);
case 3
plot(nx,ny,'bs','MarkerSize',12);
end;
title('K-NN example, with 3 data clusters')
xlabel('x'); ylabel('y');
%print
disp('assigned class')
icx
disp('the votes for each class')
cl
```

The K-nn rule is simple and powerful, it does not need modeling efforts. There is also no training involved, and therefore it is said that it is a 'lazy' algorithm. As a rule of thumb, $k < sqrt(N)$ where N is the number of samples. Depending on k the algorithm could become computationally expensive, so it is recommended to remove redundancies in the data (this is called 'condensing'). From time ago, the research is looking for techniques to speed up the nearest neighbour search, see for example [232, 270, 326]. The doctoral thesis [5] contains extensive information on nearest neighbour searching. A concise survey of K-nn techniques is [28].

Since there are applications with huge data amounts, part of the research has proposed approximated K-nn methods [9, 146].

Actually, the K-nearest neighbor can be considered as a general method [84], with several applications, not only for classification of a new datum. Anyway, there are many interesting examples of the primary application of K-nn, like for instance breast cancer diagnosis [223], speaker recognition [96], delineation of forest or non-forest land [124], machine learning and vision [276], electricity market price forecasting [210], etc.

7.5.3 Kernels

Even in the simple scenario of two classes, it is not always possible to find a separating line. Figure 7.53 shows an example where you cannot draw a separating line. It is said that data are not linearly separable. However, this same figure suggests that it would be convenient to have a separating curve. This can be done using kernels.

As a prelude for the study of classification based on kernels, is opportune to make an observation. Consider the example shown in Fig. 7.54. There are two data sets on a line, one is in the middle, and the other occupies places on the left and on the right. Data are not linearly separable.

However, if one applies a basic transformation, $x_i \rightarrow (x_i)^2$, then the two data sets can be linearly separated, as depicted in Fig. 7.55.

Notice that the transformation took us from 1D to 2D. The dimension of the problem increased.

Fig. 7.53 Example of two non-separable data classes

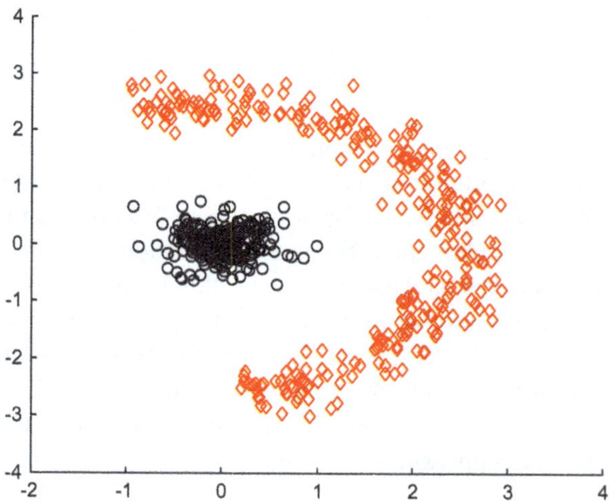

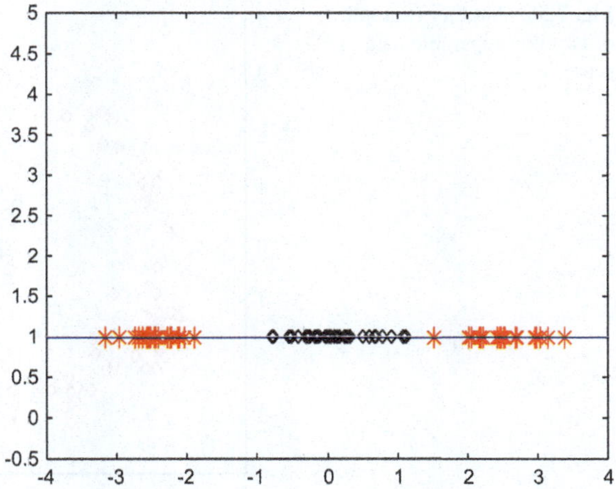

Fig. 7.54 Example of two non separable data sets

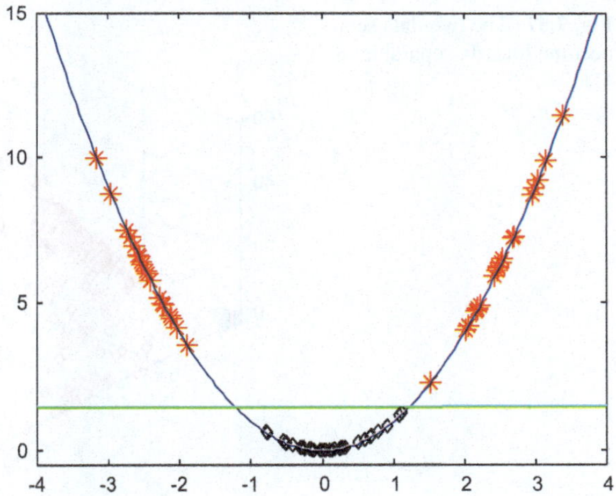

Fig. 7.55 The two data sets can now be separated with a line

Consider now a second example frequently used in the literature on kernels. The example is represented in Fig. 7.56. This time there are two data sets in 2D, which are clearly not linearly separable.

The literature recommends the following transformation, from 2D to 3D:

$$\begin{aligned} &\phi\,:\,\Re^2 \rightarrow \Re^3 \\ &(x_1, x_2) \;\rightarrow\; (z_1,\, z_2,\, z_3) \;=\; (x_1^2,\, x_2^2,\, \sqrt{2}\,x_1\,x_2) \end{aligned} \tag{7.122}$$

It is said that the data have been mapped to a feature space.

The results of the feature space mapping are shown in Fig. 7.57. It is now possible to draw a separating plane between the two classes.

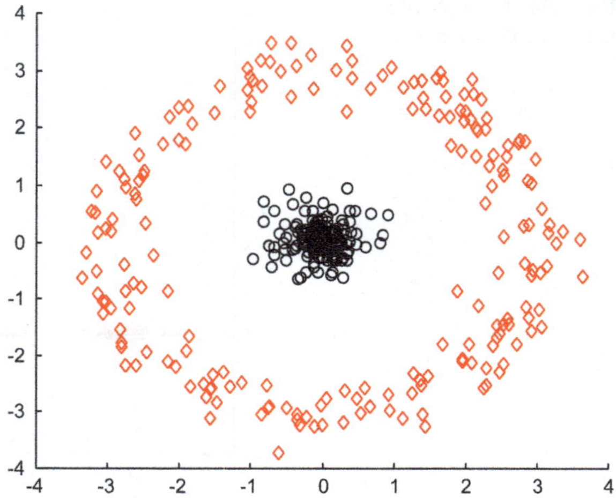

Fig. 7.56 Another example of two non separable data sets

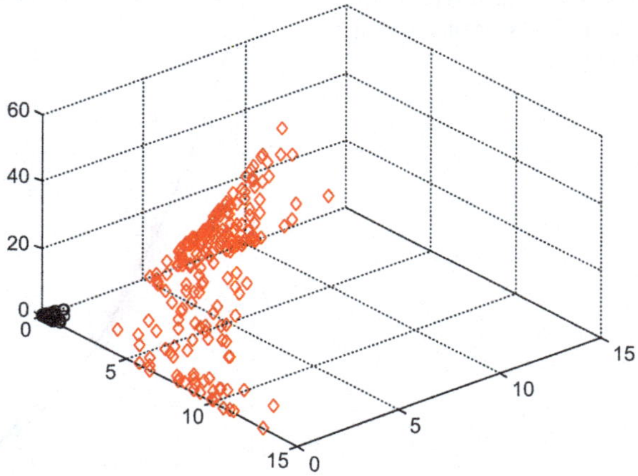

Fig. 7.57 The two data sets become linearly separable in 3D

Program 7.32 From 2D to 3D, data separability

```
% From 2D to 3D, data separability
% 2 data clusters (non linearly separable) on 2D
N=200;
X1=zeros(N,2); X2=zeros(N,2);
mu1=0; sigma1=0.4; mu2=0; sigma2=0.3;
for nn=1:N,
R1=normrnd(mu,sigma1);
phi1=rand(1)*2*pi;
X1(nn,1)=R1*cos(phi1); X1(nn,2)=R1*sin(phi1);
R2=3+normrnd(mu,sigma2);
phi2=rand(1)*2*pi;
X2(nn,1)=R2*cos(phi2); X2(nn,2)=R2*sin(phi2);
end;
```

```
figure(1)
scatter(X1(:,1),X1(:,2),'k'); hold on;
scatter(X2(:,1),X2(:,2),'rd');
title('two classes, non-separable data (2D)');
figure(2)
kq=sqrt(2);
x=X1(:,1).^2; y=X1(:,2).^2; z= kq*x.*y;
plot3(x,y,z,'ko'); hold on;
view(50,40);
x=X2(:,1).^2; y=X2(:,2).^2; z= kq*x.*y;
plot3(x,y,z,'rd'); hold on;
grid;
title('separable classes (3D)');
```

The separating plane can be expressed as:

$$\mathbf{w}^T \mathbf{z} + b = 0 \tag{7.123}$$

that corresponds to:

$$w_1 x_1^2 + w_2 x_2^2 \; + \; w_3 \sqrt{2}\, x_1 x_2 = 0 \tag{7.124}$$

which is the equation of an ellipse (it is not a surprise looking at Fig. 7.56).

7.5.3.1 Kernels, and the 'Kernel Trick'

Denote as $< \mathbf{x}, \mathbf{y} >$ the inner product of vectors **x** and **y**. For example, if $\mathbf{x} = (2, 5)$ and $\mathbf{y} = (1, 7)$, then $< \mathbf{x}, \mathbf{y} >= (2 * 1) + (5 * 7) = 37$. This inner product is also called 'dot product'.

Kernels are similarity functions that return inner products between feature mapping images:

$$K(\mathbf{x}, \mathbf{y}) \; = < \phi(\mathbf{x}), \phi(\mathbf{y}) > \tag{7.125}$$

It is possible to directly select a convenient $K(\mathbf{x}, \mathbf{y})$; choosing $K()$ is equivalent to choosing $\phi()$.

The kernel that corresponds to the example depicted in Figs. 7.56 and 7.57, can be easily computed:

$$\begin{aligned} K(\mathbf{x}, \mathbf{y}) \; &= < \phi(\mathbf{x}),\; \phi(\mathbf{y}) \; > = < (x_1^2, x_2^2,\; \sqrt{2}\, x_1 x_2),\; (y_1^2, y_2^2,\; \sqrt{2}\, y_1 y_2) = \\ &= \; x_1^2\, y_1^2 \; + \; x_2^2\, y_2^2 \; + \; 2\, x_1 y_1 x_2 y_2 \; = \; < \mathbf{x}, \mathbf{y} >^2 \end{aligned} \tag{7.126}$$

It has been found that important methods, like the SVM, can be expressed in function of inner products $< \mathbf{x}, \mathbf{y} >$, and that they can become more flexible if one directly replace those inner products by kernel functions (this is called the 'kernel trick'). In this case, there is no need of knowing $\phi()$, and no need to explicitly compute a feature space mapping. Examples of the kernel trick will be given later on, when introducing kernelized discrimination algorithms, etc.

If you wish to design a kernel function, the conditions that should be satisfied are given by the Mercer's theorem: $K(\mathbf{x}, \mathbf{y})$ must be symmetric and positive semidefinite, i.e.:

$$\int K(\mathbf{x}, \mathbf{y}) g(\mathbf{x}) g(\mathbf{y})\, d\mathbf{x} d\mathbf{y} \geq 0, \quad \forall g() \tag{7.127}$$

or, equivalently (kernel matrix):

$$\begin{pmatrix} K(x_1, x_1) & K(x_1, x_2) \ldots & K(x_1, x_n) \\ K(x_2, x_1) & K(x_2, x_2) \ \ldots & \\ \vdots & & \\ K(x_n, x_1) \ldots & & K(x_n, x_n) \end{pmatrix} \tag{7.128}$$

is positive semidefinite for any set $\{x_1, x_2, \ldots, x_n\}$.

Some typical kernels are the following:

- Linear kernel: $< \mathbf{x}, \mathbf{y} >$
- Polynomial kernel: $(\mathbf{x}^T \mathbf{y} + b)^p$
- Gaussian kernel: $\exp(-\frac{1}{2\sigma^2} \|\mathbf{x} - \mathbf{y}\|^2)$

Kernels can be combined to obtain new ones. For instance, the sum of kernels is a kernel, the product of kernels is a kernel, etc.

7.5.3.2 Kernel SVM

Returning to the quadratic optimization problem stated for SVM, as described in (7.5.1), it can be re-formulated with Lagrange multipliers $\alpha_i \geq 0$ as follows:

minimize, with respect to $\mathbf{w}$ and b:

$$L(\mathbf{w}, b, \alpha) = \|\mathbf{w}\|^2 - \sum_i \alpha_i (y_i (< \mathbf{w}, \mathbf{x}_i > + b) - 1) \tag{7.129}$$

Then, there are two equations:

$$\frac{\partial L}{\partial \mathbf{w}} = 0 \tag{7.130}$$

$$\frac{\partial L}{\partial b} = 0 \tag{7.131}$$

From the first equation, one obtains:

$$\mathbf{w} = \sum_i \alpha_i y_i \mathbf{x}_i \tag{7.132}$$

And, from the second equation:

$$\sum_i \alpha_i y_i = 0 \tag{7.133}$$

The dual problem should have the same solution. After some algebra, this problem can be formulated as follows:
maximize:

$$L(\boldsymbol{\alpha}) = \sum_i \alpha_i - \sum_i \sum_j \alpha_i \alpha_j y_i y_j < \mathbf{x}_i, \mathbf{x}_j > \tag{7.134}$$

subject to:

$$\alpha_i \geq 0 \; ; \quad \sum_i \alpha_i y_i = 0 \tag{7.135}$$

To complete the scenario, the separating line can be expressed in two ways:

$$f(\mathbf{x}) = < \mathbf{w}, \mathbf{x} > + b \tag{7.136}$$

$$f(\mathbf{x}) = \sum_i \alpha_i y_i < \mathbf{x}_i, \mathbf{x} > + b \tag{7.137}$$

The kernel SVM is obtained by direct application of the kernel trick. It suffices to use the following changes:

$$L(\boldsymbol{\alpha}) = \sum_i \alpha_i - \sum_i \sum_j \alpha_i \alpha_j y_i y_j \; K(\mathbf{x}_i, \mathbf{x}_j) \tag{7.138}$$

$$f(\mathbf{x}) = \sum_i \alpha_i y_i \; K(\mathbf{x}_i, \mathbf{x}) + b \tag{7.139}$$

By using kernels, SVM becomes a powerful tool for non linear discrimination.

As a first example, we prepared two non-linearly separable data clusters in 2D, and then devised a kernel SVM, with a Gaussian kernel, for non-linear discrimination. Based on the information provided by *insa-rouen* on the web, the Program 7.33 was developed. The satisfactory result is displayed in Fig. 7.58. Labeled contour lines have been included by using *clabel( )*. The program calls the function *qpg( )* for quadratic programming.

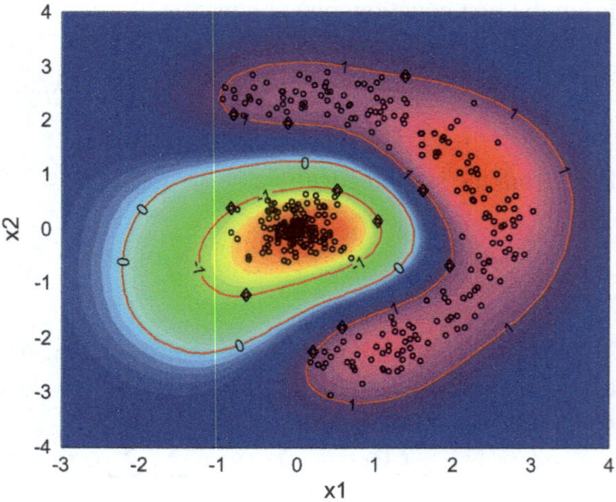

Fig. 7.58 Using kernel-SVM for a non separable case

Program 7.33 Kernel-SVM classification example

```
% kernel-SVM classification example
% 2 data clusters (non linearly separable) on 2D
%data
N=200;
X1=zeros(N,2); X2=zeros(N,2);
mu1=0; sigma1=0.35; mu2=0; sigma2=0.25;
for nn=1:N,
R1=normrnd(mu1,sigma1);
phi1=rand(1)*2*pi;
X1(nn,1)=R1*cos(phi1); X1(nn,2)=R1*sin(phi1);
R2=2.5+normrnd(mu2,sigma2);
phi2=-1.5+(1.1*rand(1)* pi);
X2(nn,1)=R2*cos(phi2); X2(nn,2)=R2*sin(phi2);
end;
X=[X1;X2];
Y=ones(2*N,1); Y(1:N,1)=-Y(1:N,1); %labels
%-------------------------------------------------------
%prepare quadratic optimization
lambda = 1e-7;
c = 1000;
ps=X*X'; %scalar product
normx=sum(X.^2,2);
[nps,mps]=size(ps);
aux=-2*ps+repmat(normx,1,mps)+repmat(normx',nps,1);
K=exp(-aux/2); %gaussian kernel
A=Y;
b=0;
H =K.*(Y*Y');
e = ones(size(Y));
%quadratic prog.:
[alpha , lambda , pos] = qpg(H,e,A,b,c,lambda,0,X,ps,[]);
%pos is position in X of the support vectors
```

```
xsup=X(pos,:);
ysup=Y(pos);
w=alpha.*ysup;
%-----------------------------------------------------
% Prepare visualization
[xg1,xg2]=meshgrid([-3:0.2:4],[-4:0.2:4]);
[nl,nc]=size(xg1); q=nl*nc;
xt1=reshape(xg1,1,q); xt2=reshape(xg2,1,q);
xt=[xt1;xt2]'; %set of test points
xtest=xt;
ps=xt*xsup'; %scalar product
normxt=sum(xt.^2,2);
normxsup=sum(xsup.^2,2);
[nps,mps]=size(ps);
yt=zeros(nps,1);
aux=-2*ps+repmat(normxt,1,mps)+repmat(normxsup',nps,1);
Kt=exp(-aux/2); %gaussian kernel
yt(:)=yt(:)+Kt*w(1:mps);
yt=yt+lambda;
yt2d=reshape(yt,nl,nc);
%-----------------------------------------------------
%Display
figure(1)
%background levels
colormap('hsv')
contourf(xg1,xg2,yt2d,50);shading flat; hold on
%separating line and margins
[L,A]=contour(xg1,xg2,yt2d,[-1 0 1],'r');
clabel(L,A); %plot of lines with labels
%data
scatter(X(:,1),X(:,2),16,'k')
%support vectors
scatter(xsup(:,1),xsup(:,2),40,'kd');
xlabel('x1');ylabel('x2');
title('Kernel-SVM, non-separable data');
axis([-3 4 -4 4]);
```

In order to compare with other methods already introduced in this chapter, it is interesting to apply kernel SVM to the IRIS data. A program has been written for this purpose, based on simple changes in the Program 7.33. This new program has been included in Appendix A.

Figure 7.59 shows the results of kernel SVM for the IRIS data.

For the reader interested on using kernel SVM methods, it would be recommended to read the guide provided by [23]. For a more extended material, including implementation details and a reference to the SVM MATLAB Toolbox, see [118]. It would also be convenient to consult [166], which focuses on the behaviours associated to Gaussain kernels.

The kernel SVM has been well received in scientific and technical ambients. It has fuelled new research activities, with quite a lot of applications. Many types of kernels have been proposed, and the field has extended towards machine learning and big data contexts, [6, 138]. The door for handling texts has been opened by considering

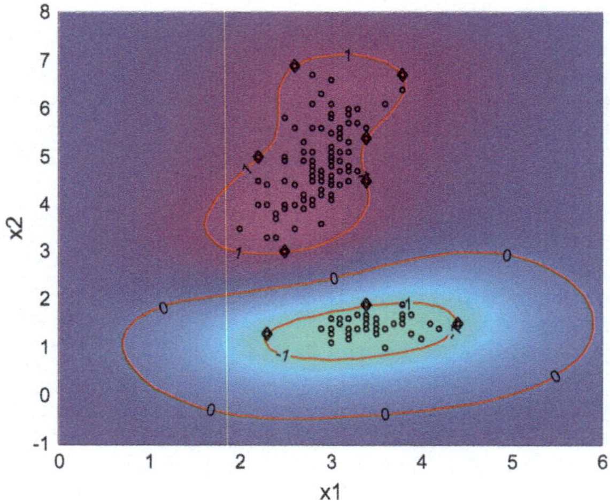

Fig. 7.59 Using kernel-SVM for IRIS data

string kernels [47, 209]. Texts can be regarded as a case of structured data, as well as genomic sequences, trees, etc. [110, 192, 301].

Let us mention some promising applications. In [18] a kernel SVM is proposed for vision-based pedestrian detection (which is good for autonomous cars). An application for diagnosis of breast cancer is described in [205]. Large scale genomic sequence classifiers are introduced in [285]. A method for multimedia semantic indexing is proposed in [11].

7.5.3.3 Kernel LDA

Recall that LDA obtains a discriminating line with an optimal direction $\mathbf{w}$. This direction maximizes the following objective function:

$$J(\mathbf{w}) = \frac{\mathbf{w}^T S_b \mathbf{w}}{\mathbf{w}^T S_w \mathbf{w}} \tag{7.140}$$

The maximization leads to the eigensystem:

$$S_b\,\mathbf{w} = \lambda\, S_w \mathbf{w} \tag{7.141}$$

And the solution $\mathbf{w}$ was the leading eigenvector of $S_w^{-1} S_b$.

In certain cases, it would be convenient to use, instead of a discriminating line, something else: a curve, a set of line segments, a set of curved segments, etc. A feature space mapping ϕ can be used for this purpose.

Let us define between-class and within-class scatters in the feature space F:

$$\widehat{S}_w = \sum_{i=1}^{2} \sum_{\mathbf{x} \in D} (\phi(\mathbf{x}) - \widehat{m}_i)\,(\phi(\mathbf{x}) - \widehat{m}_i)^T \tag{7.142}$$

$$\widehat{S}_b = (\widehat{m}_1 - \widehat{m}_2)\,(\widehat{m}_1 - \widehat{m}_2)^T \tag{7.143}$$

Now, the objective function is:

$$J(\mathbf{v}) = \frac{\mathbf{v}^T \widehat{S}_b \mathbf{v}}{\mathbf{v}^T \widehat{S}_w \mathbf{v}} \tag{7.144}$$

which is maximized by a leading eigenvector $\mathbf{v}$ that belongs to F. This vector can be expressed as the following linear combination

$$\mathbf{v} = \sum_i \alpha_i \phi(\mathbf{x}_i) \tag{7.145}$$

Then, let us do some algebraic work, following the derivation in [112]. Multiply by $\widehat{m}_i$:

$$\begin{aligned} \mathbf{v}^T \widehat{m}_i &= \left(\sum_{j=1}^{N} \alpha_j \phi(\mathbf{x}_j) \right)^T \left(\tfrac{1}{N_i} \sum_{k=1}^{N_i} \phi(\mathbf{x}_k^i) \right) = \\ &= \tfrac{1}{N_i} \sum_{j=1}^{N} \sum_{k=1}^{N_i} \alpha_j\, K(\mathbf{x}_j,\, \mathbf{x}_k^i) = \boldsymbol{\alpha}^T M_i \end{aligned} \tag{7.146}$$

where the kernel has been introduced, replacing dot products.

In the expression above, the following notation was defined:

$$(M_i)_j := \frac{1}{N_i} \sum_{k=1}^{N_u} K(\mathbf{x}_j, \mathbf{x}_k^i) \tag{7.147}$$

Then, it is possible to write:

$$\mathbf{v}^T \widehat{S}_b \mathbf{v} = \boldsymbol{\alpha}^T M\, \boldsymbol{\alpha} \tag{7.148}$$

with: $M = (M_1 - M_2)\,(M_1 - M_2)^T$

Also:

$$\mathbf{v}^T \widehat{S}_w \mathbf{v} = \boldsymbol{\alpha}^T N\, \boldsymbol{\alpha} \tag{7.149}$$

with:

$$N = \sum_{j=1}^{2} K_j (I - 1_{N_j})\, K_j^T \tag{7.150}$$

where 1_{N_j} is a $N_j \times N_j$ matrix with all entries being $1/N_j$; and K_j is a $N \times N_j$ matrix such that:

$$(K_j)_{n,m} = K(\mathbf{x}_n, \mathbf{x}_m^j) \tag{7.151}$$

Gathering together the expressions just obtained, one gets the following Rayleigh's quotient:

$$J(\alpha) = \frac{\alpha^T M \alpha}{\alpha^T N \alpha} \tag{7.152}$$

which is maximized by the leading eigenvector of $N^{-1} \cdot M$.

The projection of a new datum $\mathbf{y}$ is given by:

$$(\mathbf{v} \cdot \phi(\mathbf{y})) = \sum_{i=1}^{N} \alpha_i K(\mathbf{x}_i, \mathbf{y}) \tag{7.153}$$

In order to avoid numerical difficulties, it would be recommended to regularize the N matrix, replacing it with: $N + \mu I$.

Detailed background literature on kernel LDA is [17, 224]. There is a number of articles that propose improvements, like the regularized version introduced by [213], or the faster implementation of [46]. The optimization of the kernel is studied in [171, 327]. A program called *KDA.m* is available from MATLAB Central. The book [278] links to MATLAB code of kernel-based methods.

7.5.3.4 Kernel PCA

A common application of PCA is the reduction of data dimension, by projecting onto principal axes. Now, let us extend this approach taking advantage of kernels.

Recall that in PCA one writes:

$$S_X \mathbf{u} = \lambda \mathbf{u} \tag{7.154}$$

a series of eigenvectors and eigenvalues are then obtained, and one projects the data onto eigenvectors with largest eigenvalues.

In the case of using kernels, a feature space mapping is (implicitily) used. Once in the feature space F, one would compute the covariance matrix:

$$\widehat{S}_X = \frac{1}{N} \sum_{j=1}^{N} \phi(\bar{x}_j)\, \phi(\bar{x}_j)^T \tag{7.155}$$

And then, the principal components can be obtained from:

$$\widehat{S}_X \mathbf{v} = \lambda \mathbf{v} \tag{7.156}$$

Usually there is no need of knowing about ϕ when using a kernel. Therefore, let us obtain convenient expressions, much on track with the previous derivation corresponding to kernel-LDA.

From the eigensystem, one gets:

$$\mathbf{v} = \left(\frac{1}{\lambda N}\sum_{i=1}^{N}\phi(\overleftarrow{x}_i)\,\phi(\overleftarrow{x}_i)^T\right)\cdot\mathbf{v} = \sum_{i=1}^{N}\frac{(\phi(\overleftarrow{x}_i)^T\mathbf{v})}{\lambda N}\phi(\overleftarrow{x}_i) = \sum_{i=1}^{N}\alpha_i\cdot\phi(\overleftarrow{x}_i) \quad (7.157)$$

Also, multiplying both sides of $\phi(\mathbf{x}_k)$ by $\phi(\mathbf{x}_k)$:

$$\lambda(\phi(\mathbf{x}_k)\,\mathbf{v}) = \phi(\mathbf{x}_k)\,\widehat{S}_X\,\mathbf{v} \quad (7.158)$$

If one combines this with $\mathbf{v} = \sum_{i=1}^{N}\alpha_i\phi(\mathbf{x}_i)$:

$$\lambda\left\{\phi(\mathbf{x}_k)\sum_{i=1}^{N}\alpha_i\phi(\mathbf{x}_i))\right\} = \phi(\mathbf{x}_k)\left\{\frac{1}{N}\sum_{j=1}^{N}\phi(\overleftarrow{x}_j)\,\phi(\overleftarrow{x}_j)^T\right\}\sum_{i=1}^{N}\alpha_i\phi(\mathbf{x}_i) \quad (7.159)$$

Reordering terms:

$$\lambda\sum_{i=1}^{N}\alpha_i\,\phi(\mathbf{x}_k)\,\phi(\mathbf{x}_i) = \frac{1}{N}\sum_{i=1}^{N}\alpha_i\left\{\phi(\mathbf{x}_k)\cdot\sum_{j=1}^{N}\phi(\mathbf{x}_j)\right\}(\phi(\mathbf{x}_j)\,\phi(\mathbf{x}_i)) \quad (7.160)$$

This last equation can be expressed as:

$$(N\cdot\lambda)\;K\,\boldsymbol{\alpha} = K^2\boldsymbol{\alpha} \quad (7.161)$$

where K is the kernel matrix.

Obvoiusly the eigensystem to solve is, therefore:

$$(N\cdot\lambda)\,\boldsymbol{\alpha} = K\,\boldsymbol{\alpha} \quad (7.162)$$

In order to ensure that the eigenvectors $\mathbf{v}$ are orthonomal, the values of $\boldsymbol{\alpha}$ (which are the eigenvectors of K) are scaled such that:

$$\lambda_k(\boldsymbol{\alpha}^{(k)T}\cdot\boldsymbol{\alpha}^{(k)}) = 1 \quad (7.163)$$

When there is a new datum $\mathbf{y}$, its kth principal component can be obtained as follows:

$$\mathbf{v}^{(k)}.\,\phi(\mathbf{y}) = \left(\sum_{i=1}^{N}\alpha_i^{/k)}\phi(\mathbf{x}_i)\right)\phi(\mathbf{y}) = \sum_{i=1}^{N}\alpha_i^{/k)}K(\mathbf{x}_i,\mathbf{y}) \quad (7.164)$$

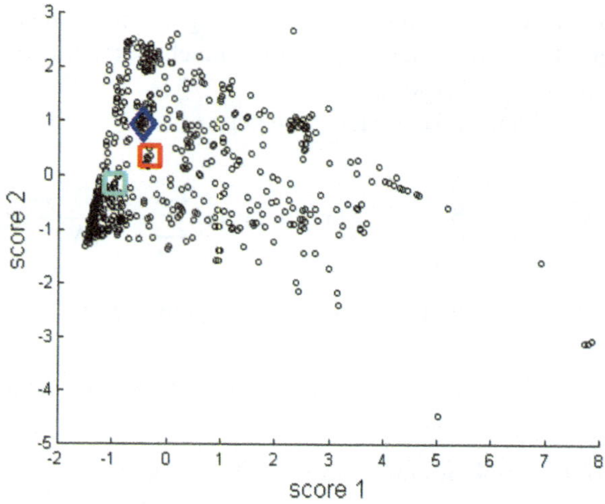

Fig. 7.60 PCA two first scores of food data

Practical applications would perhaps require data centering in F; see [112] for details.

Kernel PCA can be a convenient tool for data interpretation. One interesting example of this has been considered in [149], concerning a database of nutritional value of 961 food items, like cakes, yogurths, etc. The first 7 columns of the database contain data about fat, food energy, carbohydrates, protein, cholesterol, weight, and saturated fat. All quantities in grams, except for cholesterol (milligrams), and food energy (calories).

Some data processing is required before doing a principal component analysis. First, we divide each variable by its weight. Therefore, one obtains a database in the form of a 961×6 matrix. Then, we subtract from each column its mean. Finally, we divide each column by its standard deviation. We denote as DF the data matrix ready for analysis.

The first task has been to apply PCA to the matrix DF. Figure 7.60 shows a scatterplot of the two first scores. In order to speed up our example, only 700 of the 961 food items have been taken (the reader may wish to explore the complete data, with a little more patience). For illustrative purposes, three food items have been marked in the figure, corresponding to maximum calories, or protein, or saturated fat. Notice that these cases are difficult to discern, as they are closely surrounded by other data points.

A Gaussian kernel has been used for our Kernel PCA. Figure 7.61 shows a scatterplot with the results. Our target has been to obtain a more clear, distinctive location of the cases of interest.

The two Figs. 7.60 and 7.61, have been generated with the Program 7.34, which contains relevant implementation details, like, for instance, the centering of the kernel.

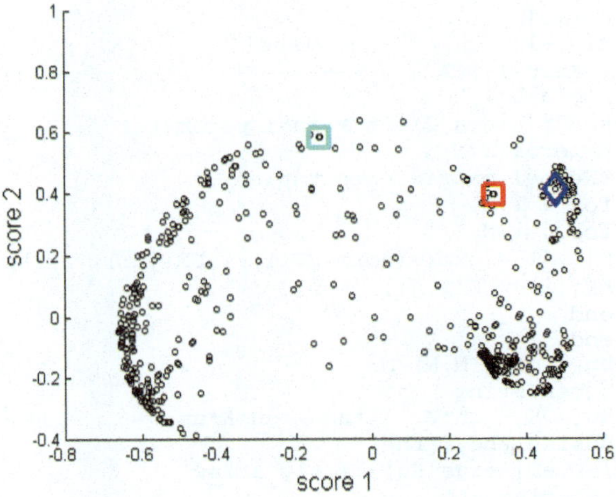

Fig. 7.61 Kernel PCA two first scores of food data

Program 7.34 Kernel PCA example

```
% Two overlapped data sets
% Kernel PCA example
% Nutritional value of food
N=700; %number of data (N<962)
% Prepare data ---------------------------------
Gfood=dlmread('food.txt',' ' );
Dfood=Gfood(:,1:7);%extract first 7 columns
OF=Dfood(1:N,:); %save original data for punctual analysis
DF=zeros(N,6); %space for data matrix
%divide by the 6th column (weight):
for n=1:5,
DF(:,n)=Dfood(1:N,n)./Dfood(1:N,6);
end;
DF(:,6)=Dfood(1:N,7)./Dfood(1:N,6); %" "
%subtract mean in each column
me=mean(DF,1);
aux1=DF-repmat(me,N,1);
%divide by standard deviation in each column
aux2=std(aux1,0,1);
DF=aux1./repmat(aux2,N,1);
[v,ix]= max(OF); %foods with peak values
% Principal components (no kernel) -------------------------
[U,S,V]=svd(DF); % V contains the 6 principal components
P=V'*DF'; % Scores
%scatter plot of 2 first scores
figure(1)
scatter(P(1,:),P(2,:),16,'k'); hold on;
% special food cases:
%Max. calories:
plot(P(1,ix(2)),P(2,ix(2)),'rs','MarkerSize',16);
%Max. protein:
plot(P(1,ix(4)),P(2,ix(4)),'bd','MarkerSize',16);
%Max. sat. fat:
plot(P(1,ix(6)),P(2,ix(6)),'cs','MarkerSize',16);
```

```
title('Scatter plot of 2 first PCA scores');
xlabel('score 1'); ylabel('score 2');
% Kernel PCA ------------------------------------------------
sigma=0.7;
kp=2*(sigma^2); % kernel parameter
K=zeros(N,N);
%Kernel matrix computation
for i=1:N,
for j=i:N,
K(i,j) = exp(-(norm(DF(i,:)-DF(j,:))^2)/kp);
K(j,i) = K(i,j);
end
end
un = ones(N,N)/N;
% centering
Ku = K - un*K - K*un + un*K*un;
%Using the Kernel
[eV,eD]=eigs(Ku); %only a few
reD=diag(eD);
nc=6; %choose a subset
neV=zeros(N,nc);
%normalize:
for i=1:nc,
neV(:,i)=eV(:,i)/sqrt(reD(i));
end;
sco=zeros(N,nc);
sco=Ku*neV; %KPCA scores
% scatter plot of 2 first scores
figure(2)
scatter(sco(:,1),sco(:,2),16,'k'); hold on;
% special food cases:
plot(sco(ix(2),1),sco(ix(2),2),'rs','MarkerSize',16); %Max.CAL
plot(sco(ix(4),1),sco(ix(4),2),'bd','MarkerSize',16); %Max.PRT
plot(sco(ix(6),1),sco(ix(6),2),'cs','MarkerSize',16); %Max.SATF
title('Scatter plot of 2 first KPCA scores');
xlabel('score 1'); ylabel('score 2');
```

The kernel PCA was proposed in 1997, [272]. A year after a more complete article followed, including a denoising application [225]. A robust version has been proposed by [240]. Fast, iterative implementations have been introduced in [119, 170] (both papers show interesting examples).

Representative application examples are provided by [313] for face recognition, [137] for novelty detection, [207] for gene expression data classification, and [268] for shape recognition.

Before closing this subsection, we should also have a look to the scientific literature that have a more general view: comparing methods, combining them, etc. For instance, let us mention once again [273], which is a most cited paper. Another frequently cited introduction is [234]. In a brief note, [324] establishes that the so-called kernel Fisher discriminant is equivalent to kernel PCA plus LDA. A tutorial on kernel methods is provided by [72, 104]. Linear and kernel methods for face recognition are compared by [122]. The article [175] studies several kernel-based methods with application to the classification of phonemes (speech processing).

7.5.4 *Other Approaches*

In addition to the methods already described, there are other approaches that would be opportune in certain types of applications. For instance, sequential methods take the data records one after one, so it may be suitable for on-line applications. Another type of methods is based on using ensembles of classifiers, which is more or less equivalent to consult a team of experts.

7.5.4.1 Sequential Methods

A simple clustering algorithm is the "*Basic Sequential Algorithmic Scheme*" (BSAS), [295]. Let us introduce it in a concise way.

Given a set of ordered data, one can define and use a distance $d(x, C)$ between a datum x and a cluster C, and then start the following sequential procedure:

- Take the first datum x_1 and consider it as belonging to the first cluster C_1
- Take the second datum x_2 and compute $d(x_2, C_1)$. Then, if this distance is less or equal than a threshold θ you establish that x_2 belongs to C_1. If not, then there is a new cluster C_2.
- When you take a third datum, it may belong to any of the already existing clusters, or to a new cluster. In the first case, you decide which cluster taking the minimum distance $\min_i d(x_3, C_i)$.
- The process continues on, taking $x_4, \ldots, x_n$ one after one.

The BSAS method belongs to the class of sequential clustering algorithms. It is fast algorithm; however, the result depends on the order by which data are sequentially presented.

The BSAS procedure can be also described with the following pseudo-code, [247]:

Pseudo-code 7.35 BSAS pseudo-code example

```
m=1
C_m ={x_1}
for i=2 to N
find C_k such that d(x_i ,C_k )= min [for{1<j<m}] d(x_i ,C_j )
if d(x_i ,C_k )> theta then
m=m+1
C_m ={x_i}
else
C_k =C_k union {x_i}
end
end
```

There are different versions of the method, depending on the choice of $d(x, C)$. This distance could be Euclidean, or just $d(\mathbf{x}, \mathbf{y}) = \sum_i |x_i - y_i|$, etc.

In case that the representative of C_l is a mean vector, this mean should be upgrade each time a new member of C_l is admitted.

A motivating example of application is [320] for background reconstruction during the identification of moving objects in a surveillance video. Likewise, [294] applies an improved BSAS for fast codebook generation for object class recognition in images.

7.5.4.2 Ensemble Methods

There are cases difficult to classify where one tries to get the opinion of several people. For instance, you want to install a certain background music in a waiting room and you wish to know whether that music is pleasant or not. Then, you ask to different persons according with some selection criteria. Many other cases like these are related to medical diagnosis, marketing, political decisions, etc.

Ensemble methods for classification imitate this way of judging. A set of classifiers are combined. Let us try to visualize the intuition behind this, [325]: suppose there is a set of data as represented in Fig. 7.62. A closed curve H_b has been used to represent a very good classifier for this case (the letter H is chosen for meaning '*hypothesis*').

Now, the situation is that the good hypothesis is unknown. Then, one asks for a number of classifiers, $H_1, H_2, \ldots, H_n$, and a possible result could be as represented in Fig. 7.63.

According with Fig. 7.63, one could tell of a certain consensus in the overlapping region. Each of the classifiers may be not good enough, but a combination of the classifiers would give a satisfactory classifier. This is the idea of ensemble methods. The combination can be obtained by averaging, majority counting, etc.

Clearly, combining several identical classifiers gives no gain. The ensemble approach would be useful only if there is disagreement between them.

The classifiers being used by ensemble methods are typically named as '*base classifiers*'. A simple example would be the following scenario [201]:

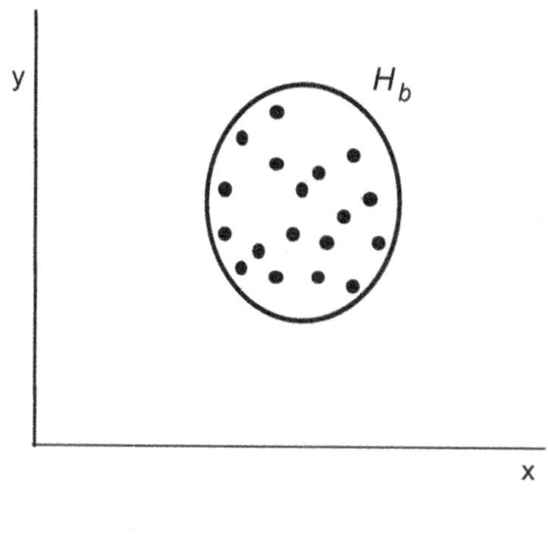

Fig. 7.62 A data set being classified with the hypothesis H_b

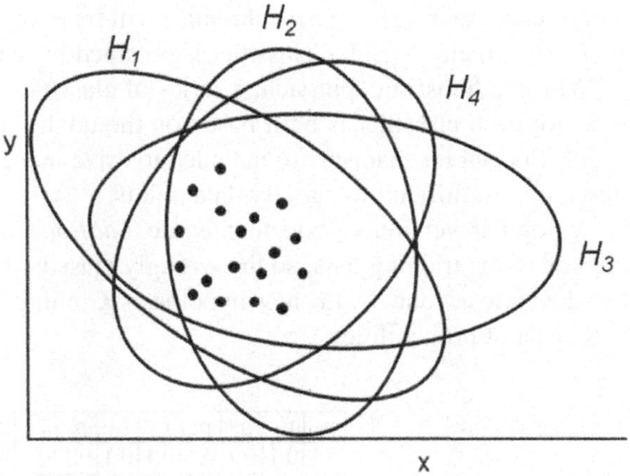

Fig. 7.63 A number of classifiers are applied to the case

Data index	0	1	2	3	4	5	6	7	8	9
x value	0	1	2	3	4	5	6	7	8	9
label	1	1	1	-1	-1	-1	1	1	1	-1

denote as $f(x < T)$ a function that uses the threshold T, and takes the value 1 if $x < T$ or -1 if $x > T$.

The best threshold is between 2 and 3, so one would use the following base classifier:

$$h_1(x) = f(x < 2.5) \tag{7.165}$$

With this classifier one has 3 mistakes.

We will come back to this example later on. In the terminology of ensemble methods, the classifier just introduced is considered as a '*weak classifier*', due to its limited capability for classification purposes.

Important types of ensemble methods are '*bagging*' classifiers, '*boosting*' classifiers, and '*random forests*'. These methods are extensively treated by [149], in its chapter on Committee Machines.

With the **bagging** approach, [41], several training data sets are created from the original data set, by bootstrapping (drawn with replacement). For instance, in the example just introduced, one could obtain the following training data sets:

Original data (x values)	0	1	2	3	4	5	6	7	8	9
Training set 1	2	6	2	7	1	3	6	7	9	9
Training set 2	0	1	2	9	5	5	6	1	9	1
Training set 3	3	4	1	9	8	6	7	4	2	4
etc										

For each training set a best classifier will be found (with a different threshold in each set). Finally, a global classifier is obtained by averaging.

With the **boosting** approach, a series of classifiers is produced. The training set used for each classifier is built based on the quality of the earlier classifiers in the series. Special mechanism are included to drive more and more classification effort into the most difficult to classify data points.

A popular version of boosting is the *AdaBoost* method, [99]. A weighting is applied to the training data, so the wrongly classified data points are more likely to be taken into account by the new hypothesis. Coming back to the example, one starts with a set of probabilities:

p_0	p_1	p_2	p_3	p_4	p_5	p_6	p_7	p_8	p_9
0.1	0.1	0.1	0.1	0.1	0.1	0.1	0.1	0.1	0.1

recall that with a threshold of 2.5 there are 3 mistakes. A weighted training error is computed as follows:

$$\varepsilon_1 = \sum_{wrong} p_i \; = \; 0.3 \tag{7.166}$$

Now, a "goodness" weight of $h_1(x)$ is computed:

$$\alpha_1 = \frac{1}{2} \ln \frac{1-\varepsilon_t}{\varepsilon_t} = 0.423649 \tag{7.167}$$

In order to update probabilities, a set of factors q_i is computed as follows:

$$q_i = \begin{cases} \exp(-\alpha_1) = 0.654654 \; for \; good \; clf. \\ \exp(\alpha_1) \; = 1.52753 \; for \; wrong \; clf. \end{cases} \tag{7.168}$$

The new probabilities are calculated with:

$$p_i = \frac{p_i \, q_i}{Z_t} \tag{7.169}$$

(Z_t is a normalization factor for $\sum p_i = 1$)

The new probabilities are 0.07143 for well classified data points, and 0.16667 for the 3 mistakes.

The next step is where emphasis is put on the wrongly classified data points. Another training data set is drawn with replacement from the original training data set. The *i-th* data point is selected with probability p_i.

Suppose for instance that with the new training set, the best threshold is 8.5, with $\varepsilon_2 = 0.214$ and $\alpha_2 = 0.6496$, and the table becomes:

Data index	0	1	2	3	4	5	6	7	8	9
Correct.	y	y	y	n	n	n	y	y	y	y
New p_i	0.045	0.045	0.045	0.167	0.167	0.167	0.106	0.106	0.106	0.045

Based on the new probabilities, another training set would be generated. For instance, suppose that now the best classifier is $f(x > 5.5)$, with $\varepsilon_3 = 0.1818$ and $\alpha_3 = 0.7520$; the table becomes:

Data index	0	1	2	3	4	5	6	7	8	9
Correct.	n	n	n	y	y	y	y	y	y	n
New p_i	0.125	0.125	0.125	0.102	0.102	0.102	0.064	0.064	0.064	0.125

The reader can confirm that the following classifier:

$$h(x) = 0.423649 f(x < 2.5) + 0.6496\, f(x < 8.5) + 0.752\, f(x > 5.5) \quad (7.170)$$

works properly, with no mistakes.

The **random forests**, [42], use a set of decision trees, and so the name. Before introducing random forests, it is helpful to consider a simple example of decision tree. Depending on how is the day, one decides to play tennis or not. The decision tree would be based on 3 variables: temperature, wind, rain. Figure 7.64 shows the tree.

Note that the most important question (rain?) has been put first.

There are two parameters, B and A, that should have certain values. If we move now to a *machine learning* context, we would consider a training approach. Suppose that along the year one has taken the decision of playing or not, in a number of days, say for instance 60 times. Then, you have a set of 60 records of the 3 variables and the outcomes of the decision. Let us denote as N the number of cases (records), and p the number of variables. Using the N cases, a training procedure could be applied

Fig. 7.64 Simple example of decision tree

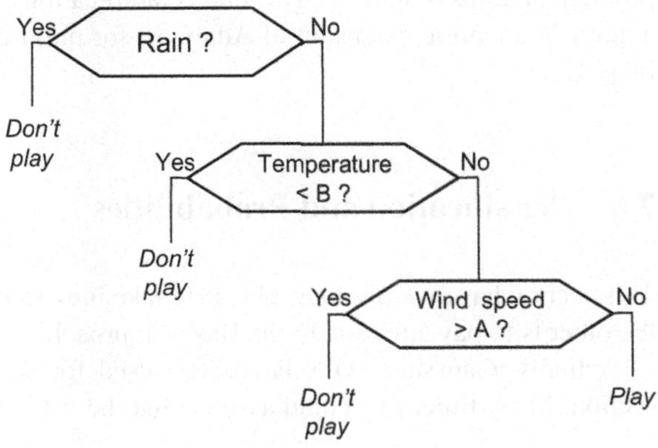

to deduce suitable values of A and B, so the decision tree will correspond as close as possible to your judgement.

Denote as S the set of N cases. The first node of the tree (rain?) would split S into two subsets, S_{1y} and S_{1n}. The second node would split S_{1n} into S_{2y} and S_{2n}; and so on with the next node.

Like the bagging approach, random forests use several training data sets that are built from the original data set by bootstrapping. Suppose that a number L of training sets are devised. For each training set a new decision tree is built, so one obtains L trees.

Each of the L trees is made as follows:

- for some number $m < p$, select at random m variables.
- choose for the first node of the tree, the variable that provides the best split (here one could establish suitable parameter values)
- at the next node select at random another m variables, and do the same
- an so on with the following nodes.

When a new case is considered, the random forest run all the trees, and then combines the results (for instance, in the case of categorical variables, by majority voting).

An interesting review of ensemble methods is provided by [105], which also deals with classification difficulties when there are large classes and, at the same time, small classes (for instance with a 1:100 imbalance ratio).

Regarding Adaboost, the method was introduced in 1996, [99], by Freund and Schapire. A more detailed article by same authors was published one year later, [100]. Both papers have been cited thousand of times. There has been a lot of research on what weak classifiers would be better. Some diversity in classifications is required for adequate work. One interesting point that has been addressed is whether support vector machines (SVM) could be used as weak classifiers, although it seems somewhat contradictory. A positive answer has been given, like for instance in [198] (see also [314]).

Originally, Adaboost focused on two-class classification problems. Some algorithms have been suggested for multi-class problems, mostly based on stating this problem in terms of multiple two-class classifications. Instead of this decomposition approach, a natural extension of Adaboost for multi-class cases has been proposed by [332].

7.6 Classification and Probabilities

This section has two missions. One is to take into account probability densities, and the other is to pay attention to the Bayes approach.

Actually, Gaussian PDFs have been used for some examples in the previous section. Many times one could assume that the data on hand have this type of PDF, but it should not be taken for granted in all cases. Anyway, if one knows that in a particular application, the PDF is of a certain type, one should have the opportunity to exploit this knowledge.

The Bayesian methods provide optimal references, which are at least interesting for comparison purposes in order to judge popular algorithms.

The section adds more items to the list of data processing tools, like for instance the EM algorithm. Nevertheless, it also considers again, under new perspectives, some of the methods already introduced, like the LDA.

One has to admit that the use of probabilities is natural for classification tasks. For example, it could be not completely clear what happens to your car when there is a strange noise. Similar things may happen with medical diagnosis, etc.

As you will see, there is a varied assortment of topics and problems to be introduced. A certain order, from simple to more complicated, has been tried, although this tree has many, intrincated, branches.

7.6.1 The Expectation-Maximization Algorithm (EM)

The expectation-maximization (EM) algorithm can be used for several purposes. Actually, some authors say that it is a meta-algorithm. One of the most successful applications is classification in the context of finite mixtures of Gaussians. This kind of application is, in turn, a good way for introducing the EM algorithm.

When we saw the K-means algorithm (in Sect. 7.5.2), each point was assigned to the closest centroid. Now, in a probabilistic context, we have to consider that each point has a certain probability of belonging to a particular class; in other words, one has to handle membership probabilities. Then, the PDF of each class becomes important.

Given a mixture of classes (each one with a certain PDF), if one knews which datum belongs to which class (labelled data), then it would be possible to study class by class, estimating the PDF parameters of each class. But, if data are not labelled, one has to assign labels to data and to estimate class PDF parameters simultaneously. This is the problem to be tackled by the EM algorithm.

Actually, the EM algorithm is simple. It repeats iteratively two steps:

- *Expectation*: estimate the cluster C_j of each data point,

$$p(C_j \,|\, \mathbf{x}_i) \tag{7.171}$$

- *Maximization*: estimate again the cluster PDF parameters (for instance mean and variance in the case of Gaussian clusters).

Evident similarities can be noticed with respect to K-means.

In Bayesian terms, the two steps can be expressed as follows:

- Expectation step:

$$p(C_j|\,\mathbf{x}_i) = \frac{p(\mathbf{x}_i|C_j)\,p(C_j)}{p(\mathbf{x}_i)} = \frac{p(\mathbf{x}_i|C_j)\,p(C_j)}{\sum_j p(\mathbf{x}_i|C_j)\,p(C_j)} \tag{7.172}$$

- Maximization step:

$$\boldsymbol{\mu}_j = \frac{\sum_i p(C_j|\mathbf{x}_i)\cdot \mathbf{x}_i}{\sum_i p(C_j|\mathbf{x}_i)} \tag{7.173}$$

$$\boldsymbol{\Sigma}_j = \frac{\sum_i p(C_j|\,\mathbf{x}_i)\cdot(\mathbf{x}_i - \boldsymbol{\mu}_j)\,(\mathbf{x}_i - \boldsymbol{\mu}_j)^T}{\sum_i p(C_j|\,\mathbf{x}_i)} \tag{7.174}$$

$$p(C_j) = \frac{\sum_i p(C_j|\,\mathbf{x}_i)}{N} \tag{7.175}$$

Given initial values of $\boldsymbol{\mu}_j$, $\boldsymbol{\Sigma}_j$ and $p(C_j)$, the algorithm computes the E step based on the current PDFs, and then in the M step it updates $\boldsymbol{\mu}_j$, $\boldsymbol{\Sigma}_j$ and $p(C_j)$.

A simple example of EM implementation is given in Program 7.36 for a case of three Gaussian clusters. The result of the EM algorithm is shown in Fig. 7.65, where probability densities have been highlighted with contours.

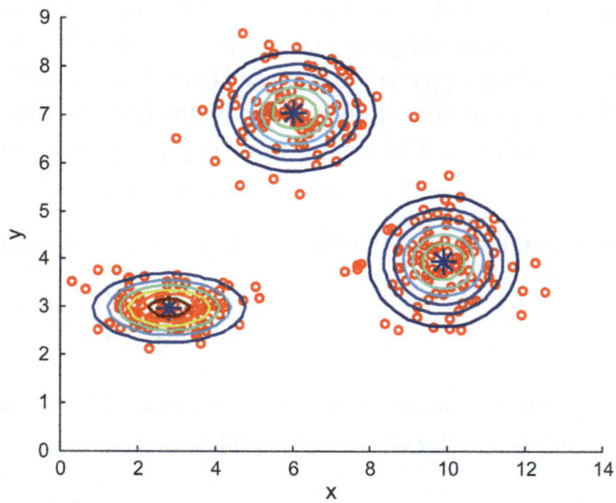

Fig. 7.65 Classification with EM

Program 7.36 Example of E-M algorithm

```
% Example of E-M algorithm
% Three clusters of Gaussian data
N=100;
%cluster 1
mux1=3; muy1=3; sigmax1=1; sigmay1=0.4;
x1=normrnd(mux1,sigmax1,1,N);
y1=normrnd(muy1,sigmay1,1,N);
%cluster 2
mux2=10; muy2=4; sigmax2=1; sigmay2=0.8;
x2=normrnd(mux2,sigmax2,1,N);
y2=normrnd(muy2,sigmay2,1,N);
%cluster 3
mux3=6; muy3=7; sigmax3=1; sigmay3=0.6;
x3=normrnd(mux3,sigmax3,1,N);
y3=normrnd(muy3,sigmay3,1,N);
% complete data set
D=zeros(2,3*N);
D(1,:)=[x1,x2,x3];
D(2,:)=[y1,y2,y3];
% centroids: initial guess (3 clusters)
mu=zeros(2,3);
mu(1,1)=D(1,1);mu(2,1)=D(2,1);
mu(1,2)=D(1,120); mu(2,2)=D(2,120);
mu(1,3)=D(1,215); mu(2,3)=D(2,215);
%variance: initial guess
si=2*ones(2,3);
pd=zeros(3,3*N); %for PDF value comparison
mL=zeros(3*N,1); %membership label
%iterations-------------------------------------------
for it=1:100,
mn=zeros(2,3); %new centroid
nk=zeros(1,3);
%step E
%values of PDFs at each datum (vectorized code)
pd(1,:)=(normpdf(D(1,:),mu(1,1),si(1,1))).*normpdf(D(2,:),...
mu(2,1),si(2,1));
pd(2,:)=(normpdf(D(1,:),mu(1,2),si(1,2))).*normpdf(D(2,:),...
mu(2,2),si(2,2));
pd(3,:)=(normpdf(D(1,:),mu(1,3),si(1,3))).*normpdf(D(2,:),...
mu(2,3),si(2,3));
for nn=1:(3*N),
k=1; v1=pd(1,nn); v2= pd(2,nn); v3= pd(3,nn);
if v1<v2, k=2; end;
if(k==1 & v1<v3), k=3; end;
if(k==2 & v2<v3), k=3; end;
mL(nn)=k; %assign membership label;
mn(:,k)=mn(:,k)+D(:,nn); nk(k)=nk(k)+1; %accumulate
end;
%step M
%new centroids
mn(1,:)=mn(1,:)./nk(1,:); %average
mn(2,:)=mn(2,:)./nk(1,:); % " "
%new variances
for nn=1:(3*N),
k=mL(nn); %read label
si(1,k)=si(1,k)+((D(1,nn)-mn(1,k))^2);
si(2,k)=si(2,k)+((D(2,nn)-mn(2,k))^2);
```

```
end;
for n=1:3,
si(1,n)=sqrt(si(1,n)/nk(n)); si(2,n)=sqrt(si(2,n)/nk(n));
end;
mu=mn; %change of centroid
end;
%-------------------------------------------------------------
%prepare contour display
p=0:0.2:100;
px1=normpdf(p,mu(1,1),si(1,1));py1=normpdf(p,mu(2,1),si(2,1));
pz1=px1'*py1; %matrix
px2=normpdf(p,mu(1,2),si(1,2));py2=normpdf(p,mu(2,2),si(2,2));
pz2=px2'*py2; %matrix
px3=normpdf(p,mu(1,3),si(1,3));py3=normpdf(p,mu(2,3),si(2,3));
pz3=px3'*py3; %matrix
%display
figure(1)
scatter(D(1,:),D(2,:),32,'ro'); hold on; %the data
axis([0 14 0 9]);
plot(mu(1,:),mu(2,:),'b*','MarkerSize',16); hold on; %centroids
contour(p,p,pz1',6); %gaussian PDF
contour(p,p,pz2',6); %" " "
contour(p,p,pz3',6); %" " "
title('Classification with EM')
xlabel('x'); ylabel('y');
%print mu and sigma for each cluster
mu
si
```

The algorithm was named as Expectation-Maximization (EM) by a famous article, [78], in 1977. In this article, the authors acknowledged that the algorithm had been already proposed, many times, by others. It is really a popular algorithm, as made clear by the thousands of citations of that article. Actually, as in a chain reaction, many papers that cite [78] have in turn thousands of citations. By the way, one of these articles is the tutorial [227].

Background literature on EM is provided by books like [123, 222]. From some time ago, [238], incremental, sparse, and other EM variants have been proposed. A fast implementation is introduced in [257]. The number of published applications is quite large; so we shall only mention a few remarkable examples. An interesting method for image segmentation and object recognition is introduced in [177]. Concerning image segmentation itself, a most cited paper is [52]. The application for laser radar is described in [300]. In the case of mobile robots, [206] uses the EM algorithm to learn 3D models of indoor environments. In the medical context, there is a number of publications on using the EM algorithm and Gaussian mixture models for diagnosis, like for instance [113] in relation with Alzheimer disease. In what concerns economics, clustering and segmentation are of evident interest; it would be recommended to look at [81] and references therein, which applies EM to market segmentation and customer heterogeneity.

7.6.2 Naïve Bayes Classifier

Suppose there is a new datum $\mathbf{y}$ and you want to label it (to classify it as belonging to a certain class). That is, you want to determine the most probable $p(C_j\,|\,\mathbf{y})$.

The idea of the naïve Bayes classifier is to estimate $p(\mathbf{y}|C_j)$ and $p(C_j)$, and use the Bayes rule. Typically, $\mathbf{y}$ is high-dimensional: $\mathbf{y} = [y_1, y_2, \ldots, y_n]$.

The naïve Bayes assumption is that the components of $\mathbf{y}$ are conditionally independent given C_j. Therefore:

$$p(\mathbf{y}|C_j) = \prod_i p(y_i|C_j) \tag{7.176}$$

And the classification is done with:

$$C_M = \arg\max_{C_j} \; p(C_j) \cdot \prod_i p(y_i|C_j) \tag{7.177}$$

A popular example in academic presentations is the following. There is one person that plays tennis when certain weather conditions are met. These preferences are described with the following table:

Day	Outlook	Temperature	Humidity	Wind	Play
1	Sunny	Hot	High	Weak	No
2	Sunny	Hot	High	Strong	No
3	Overcast	Hot	High	Weak	Yes
4	Rain	Mild	High	Weak	Yes
5	Rain	Cool	Normal	Weak	Yes
6	Rain	Cool	Normal	Strong	No
7	Overcast	Cool	Normal	Strong	Yes
8	Sunny	Mild	High	Weak	No
9	Sunny	Cool	Normal	Weak	Yes
10	Rain	Mild	Normal	Weak	Yes
11	Sunny	Mild	Normal	Strong	Yes
12	Overcast	Mild	High	Strong	Yes
13	Overcast	Hot	Normal	Weak	Yes
14	Rain	Mild	High	Strong	No

There are 9 yes, and 5 noes:

P(yes) = 9/14 = 0.64

P(no) = 5/14 = 0.36

Today is a sunny, cool, humid day with strong wind. Is it a day for tennis?

Take for instance what happens when the wind is strong:

P(strong|yes) = 3/9 = 0.33

P(strong|no) = 3/5 = 0.60

Completing the pertinent calculations, one has:

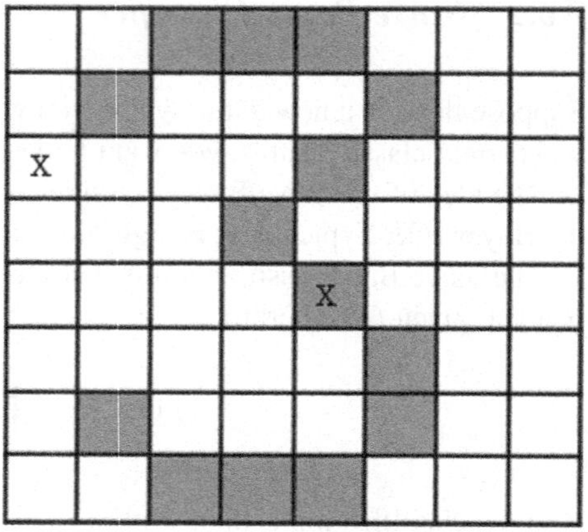

Fig. 7.66 Digit example

P(yes) P(sunny|yes) P(cool|yes) P(high|yes) P(strong|yes) = 0.0053
P(no) P(sunny|no) P(cool|no) P(high|no) P(strong|no) = 0.0206
Therefore, the answer is "no".

Other typical examples are related to medical diagnosis, recognition of digits, text classification (including the detection of e-mail spam), etc.

The case of digit recognizer could be treated as follows (Fig. 7.66).

We marked with an x the cells (3, 1) and (5, 5).

Probability that a digit (0..9) fills the cell (3, 1):

1	2	3	4	5	6	7	8	9	0
0.01	0.05	0.05	0.30	0.80	0.90	0.05	0.60	0.50	0.80

Probability that a digit (0..9) fills the cell (5, 5):

1	2	3	4	5	6	7	8	9	0
0.05	0.01	0.90	0.80	0.90	0.90	0.25	0.85	0.60	0.80

And so on, with all cells.

The naïve Bayes classifier is treated in many books, like for instance [3]. A most cited paper is [263] with an empirical study of how well this classifier works in practical applications. An interesting paper on text classification is [261]. An application in veterinary medicine is treated in the Thesis [135].

7.6.3 *Quadratic Discriminant Analysis (QDA)*

In practical terms, QDA is more flexible than LDA, and therefore it is applied to a wider range of problems. In order to present QDA, we shall focus again on a scenario with two classes.

As a previous and important step, let us introduce the Bayes's rule classifier. Denote:

$$p(\mathbf{x} \in C_i) = \pi_i \tag{7.178}$$

the prior probability that $\mathbf{x}$ belongs to class i. The posterior probability would be:

$$p(C_i|\mathbf{x}) = \frac{f_i(\vec{x})\pi_i}{f_1(\vec{x})\pi_1 + f_2(\vec{x})\pi_2} \tag{7.179}$$

According with the Bayes's rule classifier, $\mathbf{x}$ will be assigned to the class with the higher posterior probability. Therefore, $\mathbf{x}$ will be assigned to class 1 if:

$$\frac{f_1(\vec{x})}{f_2(\vec{x})} > \frac{\pi_2}{\pi_1} \tag{7.180}$$

an it will be assigned to class 2 otherwise.

Now, let us apply this classification rule for two classes with Gaussian PDFs. The ratio of the two densities is the following:

$$\frac{f_1(\mathbf{x})}{f_2(\mathbf{x})} = \frac{(2\pi)^{-n/2}|\Sigma_1|^{-1/2}\exp(-\frac{1}{2}(\mathbf{x}-\boldsymbol{\mu}_1)^T\,\Sigma_1^{-1}\,(\mathbf{x}-\boldsymbol{\mu}_1))}{(2\pi)^{-n/2}|\Sigma_2|^{-1/2}\exp(-\frac{1}{2}(\mathbf{x}-\boldsymbol{\mu}_2)^T\,\Sigma_2^{-1}\,(\mathbf{x}-\boldsymbol{\mu}_2))} \tag{7.181}$$

Let us consider, as a first case, that both Σ_1 and Σ_2 are equal. The normalization factors in numerator and denominator will cancel. Taking now logarithms:

$$\ln\frac{f_1(\mathbf{x})}{f_2(\mathbf{x})} = (\boldsymbol{\mu}_1-\boldsymbol{\mu}_2)^T\,\Sigma^{-1}\,\mathbf{x} - \frac{1}{2}(\boldsymbol{\mu}_1-\boldsymbol{\mu}_2)^T\,\Sigma^{-1}\,(\boldsymbol{\mu}_1+\boldsymbol{\mu}_2) \tag{7.182}$$

The last term on the equation can also be written as follows:

$$(\boldsymbol{\mu}_1-\boldsymbol{\mu}_2)^T\,\Sigma^{-1}\,(\boldsymbol{\mu}_1+\boldsymbol{\mu}_2) = \boldsymbol{\mu}_1^T\,\Sigma^{-1}\,\boldsymbol{\mu}_1 - \boldsymbol{\mu}_2^T\,\Sigma^{-1}\,\boldsymbol{\mu}_2 \tag{7.183}$$

Then, it is possible to write (log-likelihood ratio):

$$L(\mathbf{x}) = \ln\left(\frac{f_1(\mathbf{x})\pi_1}{f_2(\mathbf{x})\pi_2}\right) = \mathbf{w}^T\,\mathbf{x} + b \tag{7.184}$$

where:

$$\mathbf{w} = (\boldsymbol{\mu}_1-\boldsymbol{\mu}_2)^T\,\Sigma^{-1} \tag{7.185}$$

$$b = \frac{1}{2} [\mu_1^T \Sigma^{-1} \mu_1 - \mu_2^T \Sigma^{-1} \mu_2] + \ln(\frac{\pi_2}{\pi_1}) \tag{7.186}$$

According with the Bayes's rule, one assigns $\mathbf{x}$ to class 1 if $L(\mathbf{x}) > 0$, otherwise to class 2. This method is called '*Gaussian linear discriminant analysis*'.

The term:

$$\mathbf{w}^T \mathbf{x} = (\mu_1 - \mu_2)^T \Sigma^{-1} \mathbf{x} \tag{7.187}$$

is called the '*Fisher's linear discriminant function*'.

Now, let us consider, as a second case, that Σ_1 and Σ_2 are not equal. Then:

$$\begin{aligned} \ln \frac{f_1(\mathbf{x})}{f_2(\mathbf{x})} &= k_0 - \tfrac{1}{2} \left((\mathbf{x} - \mu_1)^T \Sigma_1^{-1} (\mathbf{x} + \mu_1) - (\mathbf{x} - \mu_2)^T \Sigma_2^{-1} (\mathbf{x} + \mu_2) \right) = \\ &= k_1 - \tfrac{1}{2} \mathbf{x}^T (\Sigma_1^{-1} - \Sigma_2^{-1}) \mathbf{x} + (\mu_1 \Sigma_1^{-1} - \mu_2 \Sigma_2^{-1}) \mathbf{x} \end{aligned} \tag{7.188}$$

where the constants k_0 and k_1 only depend on PDF parameters.

Then, it is possible to write the log-likelihood ratio as follows:

$$Q(\mathbf{x}) = \ln \left(\frac{f_1(\mathbf{x}) \pi_1}{f_2(\mathbf{x}) \pi_2} \right) = \mathbf{x}^T B \mathbf{x} + \mathbf{w}^T \mathbf{x} + b \tag{7.189}$$

where the symmetric matrix B is:

$$B = -\frac{1}{2} (\Sigma_1^{-1} - \Sigma_2^{-1}) \tag{7.190}$$

and the other variables are:

$$\mathbf{w} = (\Sigma_1^{-1} \mu_1 - \Sigma_2^{-1} \mu_2) \tag{7.191}$$

$$b = -\frac{1}{2} \left(\ln \frac{|\Sigma_1|}{|\Sigma_2|} + \mu_1^T \Sigma_1^{-1} \mu_1 - \mu_2^T \Sigma_2^{-1} \mu_2 \right) - \ln \frac{\pi_2}{\pi_1} \tag{7.192}$$

As before, one assigns $\mathbf{x}$ to class 1 if $Q(\mathbf{x}) > 0$, otherwise to class 2. This method is called '*Quadratic discriminant analysis (QDA)*'. The function $Q(\mathbf{x})$ is called a '*quadratic discriminant function*'.

Notice that, if both Σ_1 and Σ_2 were equal then the quadratic term $\mathbf{x}^T B \mathbf{x}$ would disappear, and the discrimination becomes linear.

The fact that QDA can be applied to cases where Σ_1 and Σ_2 are not equal is regarded by many papers as the peculiar characteristic of QDA, which makes QDA more generally applicable than LDA.

A rigorous view of QDA can be found in [149]. It is convenient to read the article [102], where a regularized version is proposed and where we can enjoy the recognized experience of Prof. Friedman as statistician. Recently, a generalization of QDA has been proposed by [37]. Also, a Bayesian version has been presented in [286]. As an intuitive help for analysis, graphical tools have been introduced in [248];

this paper includes an example concerning ionosphere radar returns. An interesting medical application is described in [76], about EEG synchrony analysis. A program called *QDA.m* is available from MATLAB Central. There is a Discriminant Analysis Toolbox, by Michael Kiefte, also available from File Exchange at MATLAB Central.

7.6.4 Logistic Discriminantion

The output of a binary classifier can be described with a step function, which changes from 0 to 1. For example, the value 1 would mean that a datum belongs to class 1; and the value 0 that the datum belongs to class 2. In certain contexts it is more convenient to use a continuous approximation of the step function, using sigmoid functions (they draw a 'S'). Figure 7.67 depicts the approximation concept. As it will be seen later on, this is typical in the context of artificial neural networks.

As before, let us focus on discrimination between two classes. The logistic approach assumes that the log-likelihood ratio can be linearly modelled, and so:

$$L(\mathbf{x}) = \ln\left(\frac{p(C_1|\mathbf{x})}{1 - p(C_1|\mathbf{x})}\right) = \mathbf{w}^T\mathbf{x} + b \tag{7.193}$$

(notice that $p(C_2|\mathbf{x}) = 1 - p(C_1|\mathbf{x})$)

Therefore:

$$p(C_1|\mathbf{x}) = \frac{e^{L(\mathbf{x})}}{1 + e^{L(\mathbf{x})}} = \frac{1}{1 + e^{-L(\mathbf{x})}} = \sigma(L(\mathbf{x})) \tag{7.194}$$

where $\sigma(\cdot)$ is a sigmoid function.

Denote as y the output of the discrimination. This variable has only two values, 1 if $\mathbf{x} \in C_1$, or 0 if $\mathbf{x} \in C_2$. As discrimination proceeds with a series of data, the variable y could be described with a Bernouilli distribution, which has the following log-likelihood:

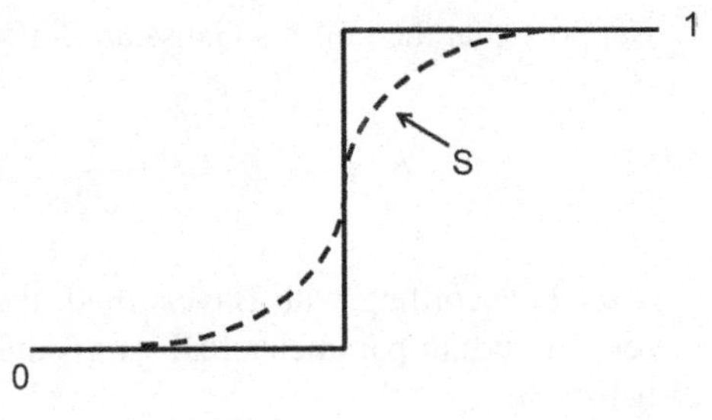

Fig. 7.67 Smooth approximation to step function

$$l(\mathbf{w}, b) = \sum_i y_i(\mathbf{w}^T \mathbf{x}_i + b) - \ln(1 + \exp(\mathbf{w}^T \mathbf{x}_i + b)) \tag{7.195}$$

For the optimization of the likelihood derivatives are taken, with respect to the model parameters, and equaled to zero. Usually the resulting equations are iteratively solved, using for instance Newton-Raphson.

Once the model is established, one assigns $\mathbf{x}$ to class 1 if $L(\mathbf{x}) > 0$, otherwise to class 2.

The book [149] give more mathematical details of logistic regression. An interesting comparison with LDA is presented in [258]. A robust version of logistic regression is used by [128] for the prediction of bankruptcy. A sparse logistic regression is proposed in [266] to identify multiregional brain activity based on fMRI data.

7.6.5 Bayesian Linear Regression. Prediction

Classical linear regression is a well-known procedure, which can be extended to data fitting with a proposed function. It can be regarded in certain applications as a parameter estimation method.

Like in other statistical methods, there is a Bayesian version. It provides a convenient basis to take into account uncertainties. This subsection is a concise introduction to Bayesian regression and some illustrative applications.

Since Gaussian PDFs will be extensively considered, a typical short-hand expression will be used to denote a Gaussian PDF: $N\,(\mu,\ var)$. Also, PDFs will be denoted as $P(\cdot)$. This subsection is based on [85, 90].

7.6.5.1 Bayesian Linear Regression

Given a data set $D = [x_i,\ y_i],\ i = 1,\ \ldots, n$, the standard Bayesian linear regression would be:

$$\mathbf{y} = \theta^T\,\mathbf{x} + \varepsilon$$

where $\boldsymbol{\theta}$ is a set of parameters, and ε is 'noise'.

Suppose that the noise is Gaussian, $N(0,\ \sigma^2)$. Then:

$$P(\mathbf{y}\,|\,\mathbf{x},\ \theta) = \frac{1}{\sigma\sqrt{2\pi}}\exp\left(\frac{-(\mathbf{y} - \theta^T\,\mathbf{x})^2}{2\,\sigma^2}\right)) \tag{7.196}$$

Of course, according with Bayes mind, there is a probability distribution of parameters. A certain parameter PDF is assumed, for instance $N(0,\ \gamma^2 I)$. It is a prior distribution.

Now, let us consider the data. A parameter posterior distribution is obtained with the Bayes's rule:

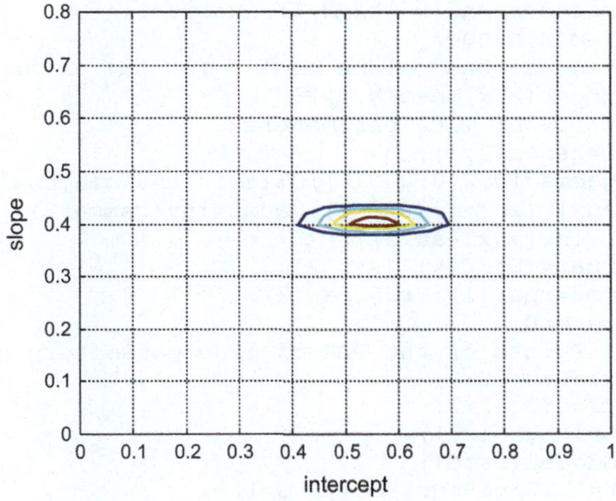

Fig. 7.68 PDF of estimated line parameters

$$P(\boldsymbol{\theta} \mid D) = \frac{P(\boldsymbol{\theta})\, P(D|\boldsymbol{\theta})}{P(D)} \tag{7.197}$$

Using Gaussian PDFs, it is possible to get explicit solutions for the equations above:

$$P(\boldsymbol{\theta} \mid D) \ \propto \ N(\frac{Q\, \mathbf{x}^T \mathbf{y}}{\sigma^2},\ Q) \tag{7.198}$$

with:

$$Q = \left(\frac{\mathbf{x}^T \mathbf{x}}{\sigma^2} + \frac{I}{\gamma^2} \right)^{-1} \tag{7.199}$$

Next two figures show an example of Bayesian linear regression. Figure 7.68 presents a contour plot of the PDF of the estimated line parameters: intercept and slope. Using the mean of these parameters a line can be plot, which corresponds to a Bayesian regression of the data.

The figures have been generated with the Program 7.37. It is important to notice the use of a variable, z, such that for any given x_i then $z_i = [x_i, 1]$. The 1 is added in order to consider, according with the linear regression model, the line intercept (Fig. 7.69).

Program 7.37 Bayesian regression example

```
% Bayesian regression example
% generate a data set
m=0.4; %slope
b=0.6; %intercept
N=25; %number of data points
x=10*rand(N,1);
std=0.2;
```

```
nse=normrnd(0,std,N,1); %noise
y=m*x+b+nse;
%add second column of 1's for the intercept:
z=cat(2,x,ones(N,1));
% PDF of line parameters
D=zeros(2,1);
gamma=[0.2 0;0 0.6]; %(edit the diagonal numbers)
aux1=(z'*z)/(std^2); aux2=inv(gamma^2);
D=inv(aux1+aux2);
rpar=(D*z'*y)/(std^2);
rmu=rpar(1); rb=rpar(2);
rstd=D;
% Points of the PDF of line parameters
x1=0:0.02:2;
x2=0:0.02:2;
L=length(x1);
dd=det(rstd);
K=1/(2*pi*sqrt(dd)); Q=1/2;
ypdf=zeros(L,L); %space for the PDF
for ni=1:L,
for nj=1:L,
aux=(((x1(ni)-rmu)^2)/rstd(1,1))+(((x2(nj)-rb)^2)/rstd(2,2));
ypdf(ni,nj)= K*exp(-Q*aux);
end;
end;
% display ------------------
figure(1)
contour(x1,x2,ypdf);
axis([0 1 0 0.8]);
grid;
title('PDF of line parameters');
xlabel('intercept'); ylabel('slope');
figure(2)
plot(x,y,'r*'); hold on;
bx0=0; by0=rb;
bxf=10; byf=rmu*bxf+rb;
plot([bx0 bxf],[by0 byf],'k');
title('Bayesian regression');
xlabel('x'); ylabel('y');
```

7.6.5.2 Prediction

As a motivating example, suppose that you have a set of data (the 'training data') $S = [f(x_1),\ f(x_2),\ \ldots,\ f(x_n)]$, and you want to predict the value of $f(x^*)$ (where x^* is a 'testing point'). The situation is depicted in Fig. 7.70.

With the Bayesian point of view, the prediction of $f(x^*)$ should be a probability distribution with a mean and a variance.

If there is a testing point x^*, the prediction about y^* would be the following PDF:

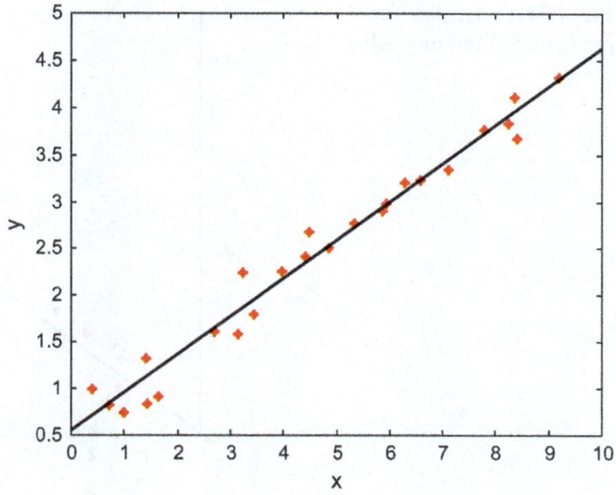

Fig. 7.69 Result of Bayesian regression

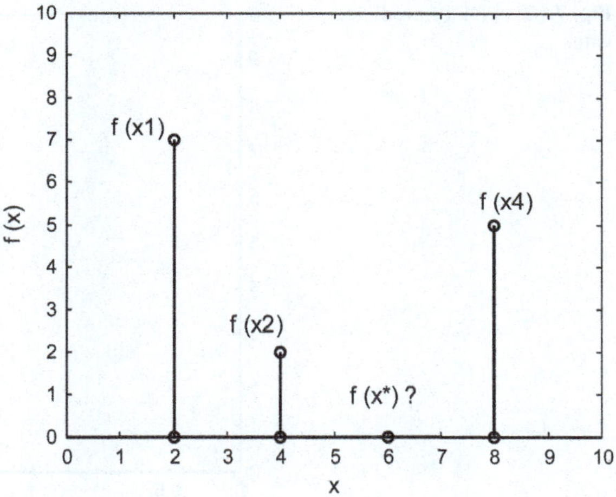

Fig. 7.70 A set of training data, and a testing point

$$P(y^* \mid x^*, D) = \int P(y^* \mid x^*, \boldsymbol{\theta})\, P(\boldsymbol{\theta} \mid D)\, d\boldsymbol{\theta} \tag{7.200}$$

Notice that the prediction is based on $\boldsymbol{\theta}$.

As before, if one assumes that the PDF is a Gaussian, then:

$$P(y^* \mid x^*, D) \propto N(\frac{x^* Q\, \mathbf{x\,y}}{\sigma^2},\ x^* Q\, x^* + \sigma^2) \tag{7.201}$$

Continuing with the example introduced before, now it is supposed that there is a missing datum at $x = 5$. Figure 7.71 shows the result of Bayesian prediction for this case. It has been represented with a diamond mark. Notice that it lies on the regression line. The next Fig. 7.72, shows the PDF corresponding to the prediction.

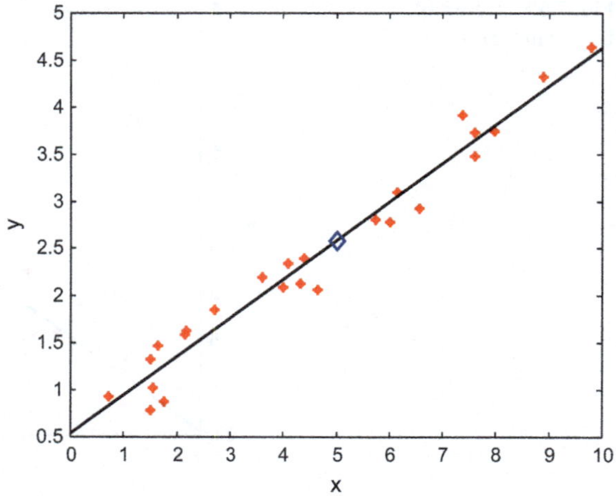

Fig. 7.71 Bayesian prediction of missing value

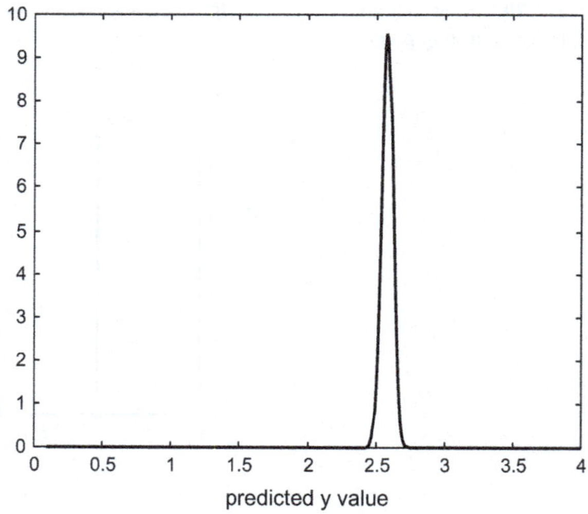

Fig. 7.72 PDF of predicted value

Both figures have been generated with the Program 7.38, which is based on code from JV Stone and code from Donders Institute (see the section on Resources).

Note that the chosen example can be also regarded as an interpolation case.

Program 7.38 Bayesian prediction/interpolation example

```
% Bayesian prediction/interpolation example
% generate a data set
m=0.4; %slope
b=0.6; %intercept
N=25; %number of data points
x=10*rand(N,1);
std=0.2;
```

```
nse=normrnd(0,std,N,1); %noise
y=m*x+b+nse;
%add second column of 1's for the intercept:
z=cat(2,x,ones(N,1));
% PDF of line parameters
D=zeros(2,1);
gamma=[0.2 0;0 0.6]; %(edit the diagonal numbers)
aux1=(z'*z)/(std^2); aux2=inv(gamma^2);
D=inv(aux1+aux2);
rpar=(D*z'*y)/(std^2);
rmu=rpar(1); rb=rpar(2);
rstd=D;
% new datum
nz=[5;1];
% PDF of predicted ny
ny=(nz'*D*z'*y)/(std^2);
rstd=(nz'*D*nz)+(std^2);
% Points of the PDF of predicted ny
x1=0:0.01:4;
L=length(x1);
K=1/(rstd*sqrt(2*pi)); Q=0.5;
ypdf=zeros(L,1); %space for the PDF
for ni=1:L,
aux=((x1(ni)-ny)^2)/(rstd^2);
ypdf(ni,1)= K*exp(-Q*aux);
end;
% display ------------------
figure(1)
plot(x,y,'r*'); hold on;
bx0=0; by0=rb;
bxf=10; byf=rmu*bxf+rb;
plot([bx0 bxf],[by0 byf],'k-');
plot(nz(1),ny,'bd','MarkerSize',10); %the predicted point
title('Bayesian prediction');
xlabel('x'); ylabel('y');
figure(2)
for ni=10:L-1,
plot([x1(ni) x1(ni+1)],[ypdf(ni) ypdf(ni+1)],'b'); hold on;
end;
title('PDF of the predicted point');
xlabel('predicted y value')
```

Inside the Bayesian regression methodology it is also posible to consider that each datum has its own confidence weight and that the noise has a specific PDF, etc. The model to be fitted could be non-linear as well [311]. The regression could be based on a set of basis functions [176]. An extensive treatment of Bayesian Methods can be found in the book [80].

A main application field for Bayesian methods is econometry [194]. There are also applications related to image processing, for counting people [55], denoising [66], background subtraction [298], etc.

7.6.6 Sets of Random Variables. Kriging

It is natural in certain cases to handle sets of random variables. For instance, if you study the weather in a region, it would be pertinent to take into account pluviometric data from several villages; each pluviometer will give you a sequence of data. Each sequence could be considered as a random variable. Then, the set of villages provide a set of random variables.

Note that if in a certain day you are in one of these villages, and you hear on the radio that it is starting to rain in nearby villages, you may suspect well that probably rain is coming to your place. Here, we are speaking about correlation, and that information from neighbours would be useful for prediction, especially if neighbours were near.

Another example could be an Electroencephalogram (EEG) being obtained with several electrodes. Then, a number of channels are recorded in parallel, giving a set of signals, which can be considered as random variables.

Let us invoke a more general and formal framework. A ***stochastic process*** is an indexed collection of random variables. An example could be a set of noisy signals along time; the index would be the time.

Figure 7.73 offers a visual example with several signals running in parallel. If you study any of the signals along time, in the horizontal direction, you could obtain a PDF of samples of this signal. If you study all the signals at a given time, in the vertical direction, as indicated by the vertical line that has been included in the figure, you could get another PDF of samples belonging to all signals.

Of course, it is also possible to represent the set of signals on same axes. The result would be like the example represented in Fig. 7.74.

Imagine what would happen if there were many curves, or an infinite number of them. Figure 7.75 shows a representation of this case, where the random curves

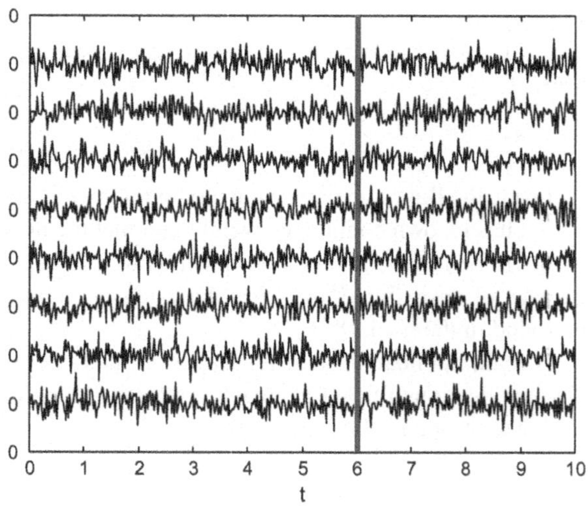

Fig. 7.73 An example of stochastic process

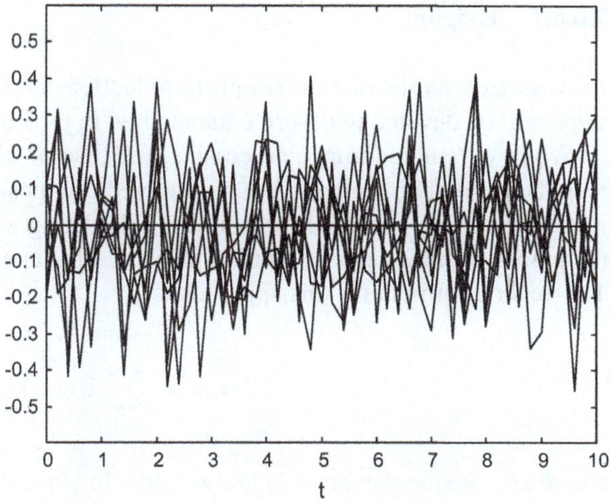

Fig. 7.74 An example of stochastic process

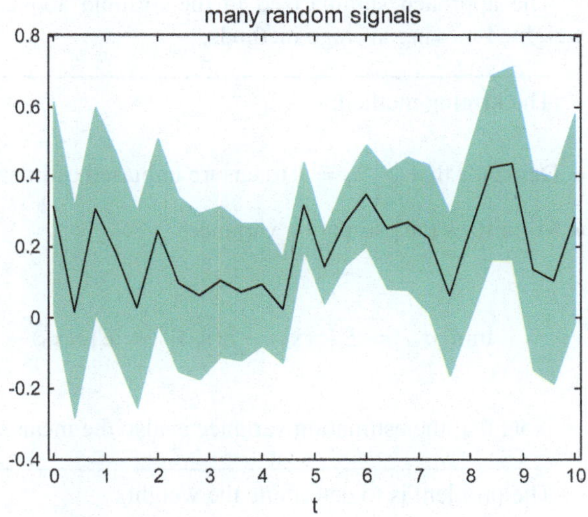

Fig. 7.75 The set has an infinite number of random signals

form a continuum. The curve in black corresponds to the mean, which can evolve along time. Notice that this figure could be interpreted as a weather prediction with confidence intervals.

Clearly, as the reader may easily guess, there are many possible types of stochastic processes. In fact, the books about stochastic processes cover many aspects, with chapters on Markov chains, martingales, Brownian motion, birth and death processes, etc. Some of the books that could be recommended for the interested reader are [88, 165, 190].

In view of matters that will be considered soon, it seems now opportune to focus on a popular interpolation method called 'kriging'.

7.6.6.1 Kriging

As a first step, an abstract and simple introduction of kriging will be made. Subsequent steps will be devoted to describe the method in typical application contexts.

When we studied image processing, the case of filtering by taking into account neighbour points (or pixels) was considered. Suppose there is a hole in the image, just a point. One could fill the point by calculating a distance weighted average of the points within a certain radius. This would be an example of local interpolation. The general formula for this approach is:

$$\hat{z}(\mathbf{x}_0) = \sum_{i=1}^{N} \lambda_i z(\mathbf{x}_i) \tag{7.202}$$

where $z(\cdot)$ are the data, and λ_i are weights. In general, the influence of close points would be higher than the influence of far away points.

The approach summarized in the formula above, corresponds to a family of weighted moving average methods.

The kriging method:

- Demands that $\sum_{i=1}^{N} \lambda_i = 1$ to ensure unbiased estimation
- Minimizes the estimation variance:

$$\min\ \sigma_E^2 = E\,[z(\mathbf{x}_0) - \hat{z}(\mathbf{x}_0)]^2 = E[z(\mathbf{x}_0) - \sum_{i=1}^{N} \lambda_i z(\mathbf{x}_i)]^2 \tag{7.203}$$

Note that the estimation variance is also the mean squared error (MSE).

The problem is to determine the weights.
Using a little notational freedom [217], one could write:

$$\begin{aligned} MSE = Var(z_0 - \sum \lambda_i z_i) = Var(z_0) - 2\sum \lambda_j\, Cov(z_0, z_j) + \\ + \sum\sum \lambda_j \lambda_i Cov(z_i, z_j) \end{aligned} \tag{7.204}$$

More briefly:

$$MSE = C(0,0) - 2\sum \lambda_j\, C(0,j) + \sum\sum \lambda_j \lambda_i C(i,j) \tag{7.205}$$

In order to minimize the MSE, derivatives are taken and made equal to zero:

$$\frac{\partial MSE}{\partial \lambda_k} = -2C(0,k) + 2\lambda_k C(k,k) + 2\sum_{j \neq k} \lambda_j C(k,j) = 0\ ; \quad k = 1, 2, \ldots, N \tag{7.206}$$

Then, one has to solve the following system of equations:

$$\sum_{j=1}^{N} \lambda_j C(k,j) = C(0,k)\ ;\ k = 1, 2, \ldots, N \tag{7.207}$$

This last equation has the form:

$$C\boldsymbol{\lambda} = \mathbf{b} \tag{7.208}$$

Therefore, it is easy to compute the weights. The optimal MSE will be:

$$MSE* = C(0,0) - \sum \lambda_j^* C(0,j) \tag{7.209}$$

(notice that in other parts of this chapter, the covariance matrix C has been denoted as S with a subindex).

For interpretation purposes it would be convenient to formulate the problem as follows:

$$\underbrace{\begin{pmatrix} C_{11} & \ldots C_{1N} & 1 \\ C_{21} & \ldots C_{2N} & 1 \\ \vdots & \vdots & \vdots \\ C_{N1} & \ldots C_{NN} & 1 \\ 1 & \ldots 1 & 0 \end{pmatrix}}_{F} \cdot \underbrace{\begin{pmatrix} \lambda_1 \\ \lambda_2 \\ \vdots \\ \lambda_N \\ \mu \end{pmatrix}}_{L} = \underbrace{\begin{pmatrix} C_{10} \\ C_{20} \\ \vdots \\ C_{N0} \\ 1 \end{pmatrix}}_{D} \tag{7.210}$$

That is:

$$F \cdot L = D \quad \Rightarrow \quad L = F^{-1} D \tag{7.211}$$

The covariances in the vector D would be high for points close to $\mathbf{x}_0$ or low for farther points. The matrix F provides information on distances between all sample pairs, and this includes information on clustering of available data (this degrades the quality of the estimation). The multiplication of D by the inverse of F has a declustering effect that improves the estimation [36].

In order to illustrate the use of kriging for interpolation, a simple example has been generated. The Program 7.39 generates a series of (x_i, y_i) data points; and then it applies kriging to find interpolated values on a regular grid of locations. Figure 7.76 shows the result in 3D. The program is based on academic software available on Internet (see section on Resources).

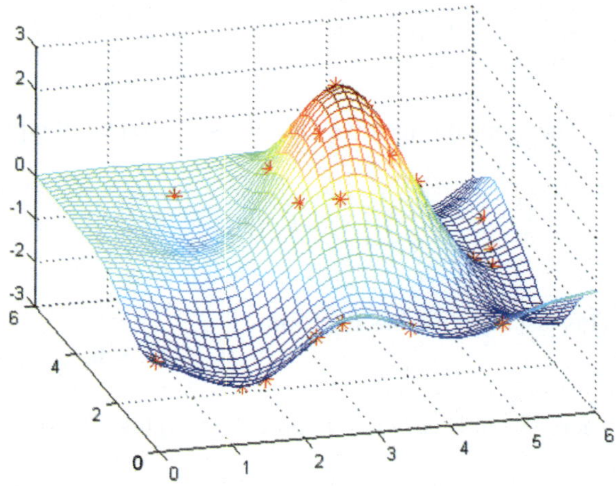

Fig. 7.76 Use of kriging for 2D function interpolation

Program 7.39 Kriging example

```
% Kriging example
% 2D spatial scenario
% Place observation points
N=25; %number of points
R=6; %distance parameter
x=R*rand(N,1); y=R*rand(N,1);
% Compute C matrix
sd=1; %standard deviation
ksd=(sd*sd)/2;
for i=1:N,
for j=i:N,
C(i,j)=exp(-ksd*((x(i)-x(j))^2+(y(i)-y(j))^2)); %upper triangle
C(j,i)=C(i,j); %symmetric lower triangle
end;
end;
ivC=inv(C); %the inverse of C
% Assign values to the observation points
% according with a 2D function
f=100; A=randn(1,f)*sqrt(2/f);
beta=2*pi*rand(f,1); k=sd*randn(f,2);
for nn=1:N,
z(nn,1)=A*cos(k*[x(nn);y(nn)]+ beta);
end;
% Set a grid of points where values will be predicted
gs=[40 40]; gr=[0 R 0 R]; %size and range of the grid
ivl=(gr(2)-gr(1))/(gs(1)-1); ivh=(gr(4)-gr(3))/(gs(2)-1);
[xp,yp]=ndgrid(gr(1):ivl:gr(2), gr(3):ivh:gr(4));
% kriging computations
for i=1:gs(1),
for j=1:gs(2),
%values of 2D function:
zf(i,j)=A*cos(k*[xp(i,j); yp(i,j)]+beta);
for nn=1:N,
Co(1,nn)=exp(-ksd*((x(nn)-xp(i,j))^2+(y(nn)-yp(i,j))^2));
```

```
end;
zp(i,j)=Co*ivC*z; % predicted values at (xp,yp)
end;
end;
% display
figure(1)
plot3(x,y,z,'r*','MarkerSize',10); hold on;
mesh(xp,yp,zp);
grid;
view(-20,30);
title('Kriging: observed values and interpolated surface')
```

7.6.6.2 Kriging and Spatial Statistics

Spatial statistics is an important area, with applications in management of natural resources, environmental science [195], weather prediction, sociopolitic analysis, transportation, etc. Like other topics that naturally emerge along this chapter, it is not possible to consider this matter in detail here, but it would be interesting to mention some aspects in relation with kriging.

What is the reason for the word 'kriging'? In 1963, G. Matheron, a French mathematician, published the method under the name *kriging* in honour of D.G. Krige, a South African mining engineer. The Master's Thesis of Mr. Krige gave the basis for the method, which was originally aimed to the determination of gold grades at a particular region of interest in South Africa.

In mathematical terms, we are dealing with '*random fields*', which is a kind of stochastic process where the index is the spatial location. In other words, for each fixed $\mathbf{x}_j$, $z(\mathbf{x}_j)$ is a random variable.

Denote as $|\mathbf{h}|$ the Euclidean distance between two points; then, the random field is *isotropic* if $C(\mathbf{h}) = C(|\mathbf{h}|)$, i.e. a function of only a distance.

Consider the following expression:

$$E[z(\mathbf{x}) - z(\mathbf{x}+\mathbf{h})]^2 = Var(z(\mathbf{x}) - z(\mathbf{x}+\mathbf{h})) = 2\gamma(\mathbf{h}) \tag{7.212}$$

The function $\gamma(\mathbf{h})$ is known as the semi-variance. It can be computed for different values of $|\mathbf{h}|$. The plot of $\gamma(\mathbf{h})$ versus $|\mathbf{h}|$ is called '*variogram*' (or semi-variogram).

The semi-variance and covariance are related:

$$\gamma(h) = C(0) - C(h) \tag{7.213}$$

Figure 7.77 compares the values of the variogram and the covariance as distance increases.

In using the variogram what is wanted is to identify structure, relationships. In case of any connections, instead of just random data one would recognize a certain pattern. In general, one would expect that nearby data pairs would be more related

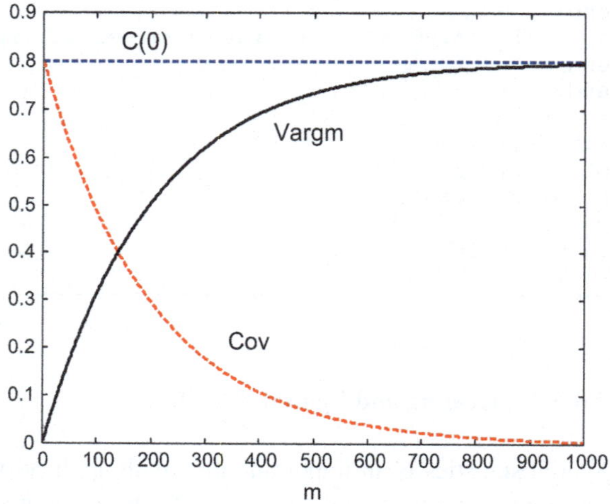

Fig. 7.77 Relationship of variogram and covariance

and have a smaller semi-variance than distant data pairs. Usually the variogram tends to a flat level as distance increases.

Usually one obtains from experiments a cloud of variogram data points; then, a model is chosen to fit the data. Typical models are the following:

- Spherical:

$$g(h) = \begin{cases} c\left[1.5\left(\frac{h}{a}\right) - 0.5\left(\frac{h}{a}\right)^3\right] & if\ h \le a \\ c\ ,\ otherwise \end{cases}$$

- Exponential:

$$g(h) = c\left(1 - \exp\left(\frac{-3h}{a}\right)\right)$$

- Gaussian:

$$g(h) = c\left(1 - \exp\left(\frac{-3h^2}{a^2}\right)\right)$$

- Power:

$$g(h) = c\,h^p \text{ with } 0 < p < 2$$

Figure 7.78 shows three characteristics commonly found in empirical variograms. In theory the value at the origin should be zero, but it is not: it takes instead a value called '*the nugget*'. This value represents a noticeable variation at distances smaller than the sample spacing. After a distance called '*range*', the variogram reaches '*the sill*': a flat level.

In spatial application scenarios, one of the first problems is to state equations from experimental data. Variograms are a preferred tool for that. The final kriging equation can be expressed in function of the variogram as follows:

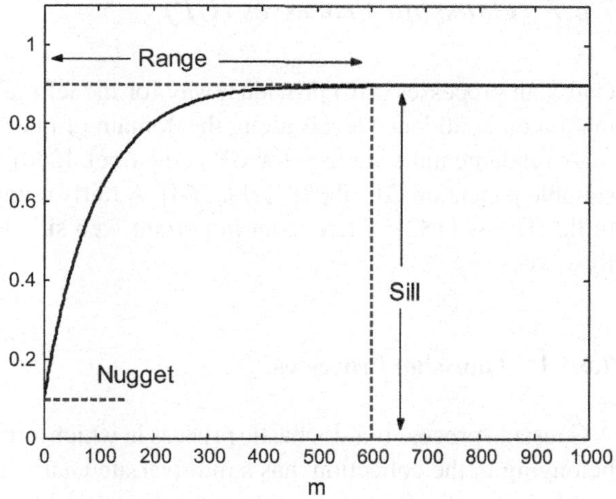

Fig. 7.78 Characteristics of a variogram

$$\underbrace{\begin{pmatrix} 0 & \gamma_{12} & \ldots \gamma_{1N} & 1 \\ \gamma_{21} & 0 & \ldots \gamma_{2N} & 1 \\ \vdots & \vdots & \vdots & \\ \gamma_{N1} & \gamma_{N2} & \ldots 0 & 1 \\ 1 & \ldots 1 & 0 & \end{pmatrix}}_{\Gamma} \cdot \underbrace{\begin{pmatrix} \lambda_1 \\ \lambda_2 \\ \vdots \\ \lambda_N \\ -\mu \end{pmatrix}}_{L} = \underbrace{\begin{pmatrix} \gamma_{10} \\ \gamma_{20} \\ \vdots \\ \gamma_{N0} \\ 1 \end{pmatrix}}_{D} \tag{7.214}$$

That is:

$$\Gamma \cdot L = D \quad \Rightarrow \quad L = \Gamma^{-1} D \tag{7.215}$$

A detailed and practical treatment of kriging can be found in [199]. There are several versions of kriging, namely simple, ordinary, and universal In addition, there are related methods, like co-kriging, blind kriging, stochastic kriging, etc. A convenient door to these versions and extensions is the manuals of a number of toolboxes, like the MATLAB Kriging Toolbox, DACE, ooDACE, mGstat, BMElib, EasyKrig, and others.

Among the proposed kriging versions, let us cite empirical Bayesian kriging [181], and factorial kriging [216].

A related topic is the use of surrogate models [109, 179]. An aerodynamics application example is described in [184]. The application of kriging for image impainting is described in [153].

There is abundant literature on geostatistics and variograms, like [35, 116, 283]. The section on Resources includes a link to a convenient academic site with MATLAB code.

7.6.7 *Gaussian Processes (GP)*

Gaussian processes (GP) provide a way for modelling that can easily include information on confidence levels along the domain of interest.

A fundamental reference for GP is the book [260]. In addition, there are several suitable papers on GP, like [85, 90, 264]. A fairly complete exposition can be found in the Thesis [182]. There is an important web site devoted to GP (see section on Resources).

7.6.7.1 Gaussian Processes

A Gaussian process is a stochastic process in which any finite set of random variables, belonging to the collection, has a multivariate Gaussian distribution.

More specifically, a collection of random variables $h(x)$ is said to be drawn from a GP with mean function $\mu(\cdot)$ and covariance function $k(\cdot,\ \cdot)$, if for any finite set of locations $\{x_1,\ x_2,\ \ldots,\ x_n\}$, the multivariate distribution of $h(x_1),\ h(x_2), \ldots,\ h(x_n)$ is proportional to $N(\boldsymbol{\mu},\ K)$, with:

$$\boldsymbol{\mu} = \begin{bmatrix} \mu(x_1) \\ \vdots \\ \mu(x_n) \end{bmatrix} \tag{7.216}$$

$$K = \begin{bmatrix} k(x_1, x_1) & k(x_1, x_2) & \ldots k(x_1,\ x_n) \\ \vdots & \vdots & \vdots \\ k(x_n, x_1) & k(x_n, x_2) & \ldots k(x_n,\ x_n) \end{bmatrix} \tag{7.217}$$

Gaussian processes are characterized by the mean and covariance functions, and so GPs are usually abbreviated with:

$$GP\ (\mu(\cdot),\ k(\cdot, \cdot)) \tag{7.218}$$

The covariance function should give a valid covariance matrix K for any set of locations. To be valid, the matrix K should be positive semidefinite. Several covariance functions (kernels) have been proposed; a popular one is the 'squared exponential (SE)':

$$k(x_1,\ x_2) = \sigma_k^2 \exp\left(\frac{-(x_1 - x_2)^2}{2\gamma^2}\right) \tag{7.219}$$

According with the remarks of [90], if x_1 and x_2 are close, then $k(x_1, x_2)$ approaches a maximum value, which means a large correlation between $h(x_1)$ and $h(x_2)$. On the contrary if the two locations are distant, then $k(x_1, x_2) \approx 0$, corresponding to

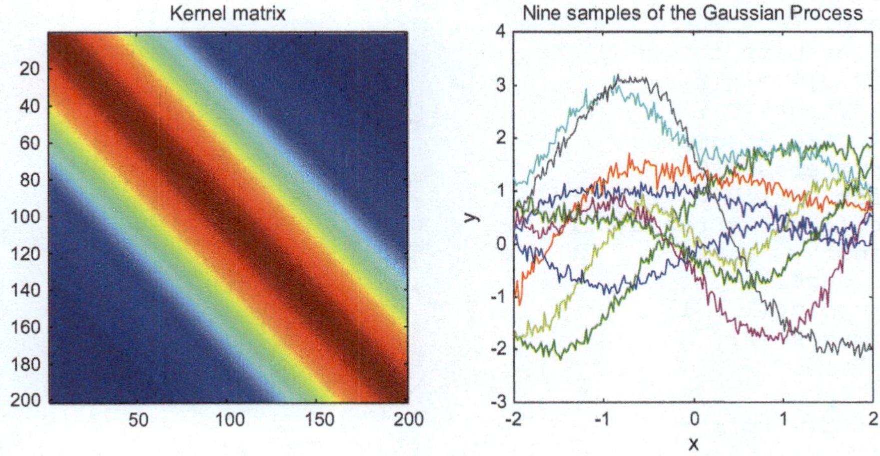

Fig. 7.79 A view of the Kernel and nine members of the Gaussian Process

negligible correlation between $h(x_1)$ and $h(x_2)$. The effect of separation on $k(x_1, x_2)$ depends on parameter γ.

In order to take into account measurement errors, it is useful to include the noise into the kernel:

$$k(x_1, x_2) = \sigma_k^2 \exp\left(\frac{-(x_1 - x_2)^2}{2\gamma^2}\right) + \sigma^2 \delta(x_1, x_2) \tag{7.220}$$

Figure 7.79 depicts a pseudocolor representation of a SE + noise kernel, and, in the right side, a plot of nine samples of a GP. The figure has been generated by the Program 7.40, which is based on code available from the web (provided by M.R. Martin; see the section of Resources).

Program 7.40 Gauss Process samples

```
% Gauss Process samples
% Build a kernel matrix
sigmaf=1.2;
gammaf=0.9; qf=2*(gammaf^2);
sigman=0.1;
x=-2:0.02:2; L=length(x);
K=zeros(L,L);
for i=1:L,
for j=i:L,
nse=(sigman^2)*(x(i)==x(j)); %noise term
K(i,j)=((sigmaf^2)*exp(-((x(i)-x(j))^2)/qf))+nse;
K(j,i)=K(i,j);
end;
```

```
end;
% prepare for sampling
[V,D]= eig(K);
A=V*sqrt(D);
% take 9 samples
rv=zeros(L,9);
for nn=1:9,
rv(:,nn)=A*randn(L,1);
end;
figure(1)
subplot(1,2,1)
imagesc(K);
title('Kernel matrix');
subplot(1,2,2)
plot(x,rv);
title('Nine samples of the Gaussian Process');
xlabel('x'); ylabel('y');
```

7.6.7.2 GP Regression. Prediction

Suppose you already obtained a series of observations of your problem, at locations $\{x_1, x_2, \ldots, x_n\}$. Now, it is desired to devise a prediction of values at another set of locations: $x^* = \{x_1^*, x_2^*, \ldots, x_k^*\}$. Using GP, this prediction can be easily computed as follows:

Use the kernel function and obtain:

$$K^* = [k(x^*, x_1), k(x^*, x_2), \ldots, k(x^*, x_n)] \tag{7.221}$$

$$K^{**} = k(x^*, x^*) \tag{7.222}$$

Then:

$$\begin{bmatrix} \mathbf{y} \\ y^* \end{bmatrix} \propto N\left(0, \begin{bmatrix} K & K^{*T} \\ K^* & K^{**} \end{bmatrix}\right) \tag{7.223}$$

Therefore, the prediction is:

$$P(y^* \,|\, \mathbf{y}) \propto N(K^* K^{-1} \mathbf{y},\ K^{**} - K^* K^{-1} K^{*T}) \tag{7.224}$$

A simple example has been built, in which a series of "observations" is generated on the basis of a sine function with added noise. In addition, a grid of points for $x^* = \{x_1^*, x_2^*, \ldots, x_k^*\}$ is generated. Then, GP regression is applied. The results are shown in Fig. 7.80. The variance is used to depict the confidence levels along the predictions. All this is done with the Program 7.41, which is based on the same source as before.

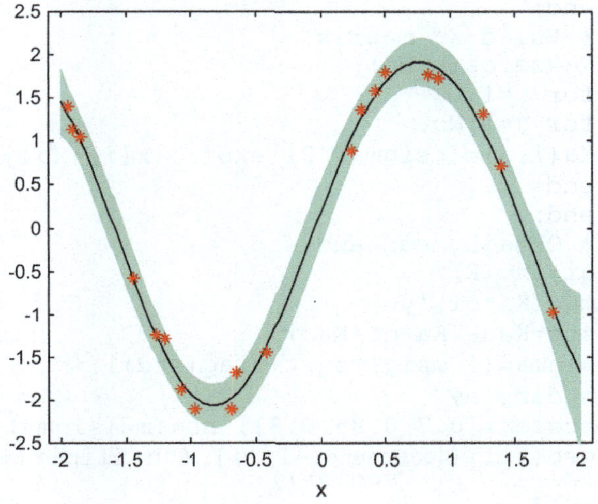

Fig. 7.80 Gauss Process regression result

Program 7.41 Gauss Process regression

```
% Gauss Process regression
% Generate observed data
No=20; %number of observations
xo=-2+(4*rand(No,1));
msig=0.2
%suppose there is some measurement:
yo=(2*sin(2*xo))+(msig*randn(No,1)); noise
% Build a kernel matrix wit the observations
sigmaf=1.2;
gammaf=0.9; qf=2*(gammaf^2);
sigman=0.1;
K=zeros(No,No);
for i=1:No,
for j=i:No,
nse=(sigman^2)*(xo(i)==xo(j)); %noise term
K(i,j)=((sigmaf^2)*exp(-((xo(i)-xo(j))^2)/qf))+nse;
K(j,i)=K(i,j);
end;
end;
% generate x coordinates of points to be predicted
x=-2:0.02:2; L=length(x);
% Build K** matrix for data to be predicted
Kaa=zeros(L,L);
for i=1:L,
for j=i:L, %(no noise term)
Kaa(i,j)=((sigmaf^2)*exp(-((x(i)-x(j))^2)/qf));
Kaa(j,i)=Kaa(i,j);
end;
```

```
end;
% Build K* matrix
Ka=zeros(L,No);
for i=1:L,
for j=1:No,
Ka(i,j)=((sigmaf^2)*exp(-((x(i)-xo(j))^2)/qf));
end;
end;
% Obtain mean and std
qi=inv(K);
mu=(Ka*qi)*yo;
std=Kaa-(Ka*qi*Ka');
sigma=(1/msig)*sqrt(diag(std));
% display
scolor=[0.7,0.85,0.8]; hh=(mu+sigma)'; ll=(mu-sigma)';
set(fill([x,x(end:-1:1)],[hh,fliplr(ll)],scolor,...
'EdgeColor',scolor));
hold on;
plot(x,mu,'k'); hold on;
plot(xo,yo,'r*','MarkerSize',8);
axis([-2.1 2.1 -2.5 2.5]);
title('Gaussian Process regression');
xlabel('x');
```

7.6.7.3 About Kernels

One of the chapters of [260] is entirely devoted to covariance functions, with many examples and detailed comments. It contains a table of kernel types, from which the following table is extracted.

Constant	σ_0^2
Linear	$\sum_j \sigma_j^2 x_j x'_j$
Polynomial	$(x \cdot x' + \sigma_0^2)^p$
Squared exponential	$\exp\left(-\frac{r^2}{2\gamma^2}\right)$
Exponential	$\exp\left(-\frac{r}{\gamma}\right)$
γ-exponential	$\exp\left(-\left(\frac{r}{\gamma}\right)^{\Upsilon}\right)$
Rational quadratic	$\left(1 + \frac{r^2}{2\alpha\gamma^2}\right)^{-\alpha}$

where $r = |x - x'|$.

Another useful kernel, frequently used in spatial statistics, is the Matérn covariance function:

$$\sigma^2 \frac{1}{\Gamma(\nu)2^{\nu-1}} \left(\frac{d}{\rho}\sqrt{2\nu}\right)^{\nu} K_{\nu}\left(\frac{d}{\rho}\sqrt{2\nu}\right) \tag{7.225}$$

where Γ is the gamma function, K_ν is the modified Bessel function of the second kind, ν and ρ are non-negative parameters, and d is the Euclidean distance between x and x'.

The web page of D. Duvenaud presents a series of typical GP kernel functions, with intuitive comments. Here is a brief list of these kernels:

- Squared exponential (SE):

$$k_{SE}(x_1, x_2) = \sigma_k^2 \exp\left(\frac{-(x_1 - x_2)^2}{2\gamma^2}\right) \tag{7.226}$$

- Rational quadratic:

$$k_{RQ}(x_1, x_2) = \sigma_k^2 \left(1 + \frac{(x_1 - x_2)^2}{2\alpha\gamma^2}\right)^{-\alpha} \tag{7.227}$$

- Periodic:

$$k_{Per}(x_1, x_2) = \sigma_k^2 \exp\left(-\frac{2\sin^2(\pi \cdot |x_1 - x_2|/p)}{\gamma^2}\right) \tag{7.228}$$

- Linear:

$$k_{Lin}(x_1, x_2) = \sigma_b^2 + \sigma_v^2 (x_1 - c)(x_2 - c) \tag{7.229}$$

In certain cases it would be convenient to combine several kernels. This is also discussed in the mentioned web page. Also, some hints were given in [90], as. for instance, to take into account a priori knowledge on data periodic fluctuations, so it is convenient to add a periodic kernel term to the complete kernel.

An interesting remark of [85] is that the positive semidefiniteness requirement for the K matrix is identical to Mercer's condition for kernels. For those that like deep mathematics, it would be recommended to read the paper [95] on positive definite kernels.

7.6.7.4 Other Aspects

In general, the covariance functions have parameters that should be specified. They are usually called '*hyperparameters*'. Commonly, these hyperparameters should be tuned to the problem at hand. There is a number of methods to do so, like simple systematic exploration, or based on gradients.

According with the regression equations, it is possible to obtain the probability of the data given the hyperparameters $p(\mathbf{y}|\overleftarrow{x}, \theta)$. The log likelihood is given by:

$$\log p(\mathbf{y}|\mathbf{x}, \theta) = -\frac{1}{2}\mathbf{y}^T K^{-1}\mathbf{y} - -\frac{1}{2}\log|K| - \frac{n}{2}\log 2\pi \tag{7.230}$$

As observed in [34], the first term could be interpreted as a data fit term, the second term as a complexity penalty, and the last term as a normalizing constant. In order to optimize the likelihood, the derivatives with respect to the hyperparameters are:

$$\frac{\partial}{\partial\theta_j}\log p(\mathbf{y}|\mathbf{x}, \theta) = \frac{1}{2}tr\left((\alpha\,\alpha^T - K^{-1})\frac{\partial K}{\partial\theta_j}\right) \tag{7.231}$$

where $\alpha = K^{-1}\mathbf{y}$.

See [254] for the application of genetic algorithms for hyperparameter tuning.

An intensive research is being devoted to the use of GP for classification. One of the chapters of [260] is dedicated to this aspect. The basic idea for classification is to use the GP regression to estimate the probability that a new datum belongs to a class with already known members.

Consider a basic classification scenario, with two classes. The label to be assigned would be $y \in \{0, 1\}$. We use a sigmoid function $s(\cdot)$ which converts $h(x)$ into:

$$p(y_i = 1|x_i) = s(h(x_i)) \tag{7.232}$$

The classification inference is done in two steps. The first is to compute:

$$p(h^*|x^*, x_{1:n}, y_{1:n}) = \int p(h^*|x^*, x_{1:n}, h_{1:n})\,p(h_{1:n}|x_{1:n}, y_{1:n})\,dh_{1:n} \tag{7.233}$$

And the second:

$$p(y^* = 1|x^*, x_{1:n}, y_{1:n}) = \int s(h^*)\,p(h^*|x^*, x_{1:n}, y_{1:n})\,dh^* \tag{7.234}$$

Due to non-Gaussianity, both integrals are analytically intractable. Several approximations have been proposed [186, 242].

7.6.7.5 Applications

Let us refer now to GP applications. A good account of significant examples is given in the already mentioned GP web page. In addition, [39] reports applications in the optimization field, and [174] focuses on modelling and control in process engineering. Machine learning is a popular target, considered in [260] and [92, 274]. Also, there are robotics applications, for model inversion or trajectories [126, 241].

Likewise, trajectories and motion edition are important for synthetic video actors [145]. Some image processing GP applications have been reported, like in [130] for super-resolution, or [163] for image categorization. An interesting approach for clustering using GP has been proposed in [169]. In the medical context, GP regression is applied for heart rate in [287], and some epidemiological studies have been described in several information sources. In this sense, it should be noted that recent scientific meetings have dealt with spatiotemporal modelling with GP. The adaptation of GP to large datasets is the subject of present-day research, like for example in [132] about GP for big data. In some cases, one could take advantage of certain structures associated to the data under study, and such is the scenario considered in [40] about structured prediction (with applications to text, DNA sequences, video processing, etc.).

7.7 Entropy, Divergence, and Related Aspects

When one tries to recognize a face in the photograph of a crowd, one is looking for similarity. When the problem is to classify data according with clusters, one is interested in divergence. And this type of criteria is pertinent in function approximation situations; for instance when one tries to approximate a given PDF with a Gaussian PDF. The need of similarity/divergence measures is there.

One of the ingredients of common divergence measures is entropy.

7.7.1 Entropy

In the late 1940s Shannon introduced a logarithmic measure of information, and a theory which included information entropy (the literature shows that it is related with Boltzmann entropy in statistical mechanics).

The Shannon's entropy can be regarded as a particular case of a more general concept: the Rényi entropy, which is defined as follows:

$$H_\alpha(X) = \frac{1}{1-\alpha} \ln\,(\sum_{i=1}^{n} p_i^\alpha) \qquad (7.235)$$

where $\alpha > 0,\ \alpha \neq 1$.

7.7.2 Divergence

Suppose a 2D data set distributed inside an ellipse. There is a new datum $\mathbf{x}$, and the question is if it belongs to the data set. In this case, it is opportune to use the *Mahalanobis distance*, which is defined as follows:

$$D_M(\mathbf{x}) = \sqrt{(\mathbf{x} - \boldsymbol{\mu})^T \, S^{-1} (\mathbf{x} - \boldsymbol{\mu})} \tag{7.236}$$

where $\boldsymbol{\mu}$ is the mean (the center of mass) of the data set and S its covariance matrix. The matrix is used to estimate the width of the ellipse in the direction of the new datum. Then, D_M is the Euclidean distance between the datum and the center of mass, divided by that width.

Two years before the Shannon's article, Jeffrey introduced *divergence* in terms of logarithms. Unlike the mathematical concept of metric distance, divergence could be non symmetric and not satisfying the triangle inequality.

Actually, several families of divergences have been proposed [83]. In particular, the (continuous) *f-divergence* of two PDFs has the following expression:

$$D_f(p \,||\, q) = \int p(x) \, f\left(\frac{q(x)}{p(x)}\right) dx \tag{7.237}$$

Here are some instances of f-divergences:

- Kullback-Leibler

$$D_{KL}(p \,||\, q) = \int p(x) \, (\ln \, p(x) \, - \, \ln \, q(x)) \, dx \tag{7.238}$$

- Squared Hellinger

$$H^2(p, \, q) = 2 \int (\sqrt{p(x)} \, - \, \sqrt{q(x)})^2 \, dx \tag{7.239}$$

- Jeffrey

$$D_J(p \,||\, q) = \int (p(x) - q(x) \, (\ln \, p(x) \, - \, \ln \, q(x)) \, dx \tag{7.240}$$

- Chernoff (α-divergence)

$$D^{\alpha}(p \,||\, q) = \frac{4}{1 - \alpha^2} \, (1 - \int p(x)^{(1-\alpha)\,/2} q(x)^{(1+\alpha)\,/2} \, dx) \tag{7.241}$$

The popular *Bhattacharyya divergence* is a particular case of Chernoff divergence (for $\alpha = 0.5$):

$$D_B(p\,||\,q) = -\ln \int \sqrt{p(x)\,q(x)}\; dx \tag{7.242}$$

An application of the Bhattacharyya divergence is for the estimation of overlapping between two data sets.

The KLD is a *relative entropy*, as it can be seen from its usual expression:

$$D_{KL}(p\,||\,q) = \int p(x)\,\ln\,\frac{p(x)}{q(x)}\; dx \tag{7.243}$$

The KLD is a nonnegative. It is also non symmetric:

$$D_{KL}(p\,||\,q) \neq D_{KL}(q\;||\,p) \tag{7.244}$$

A symmetrized distance has been introduced, with the following expression:

$$D_{SY}(p||q) = D_{KL}(p\,||\,q) + D_{KL}(q\;||\,p) \tag{7.245}$$

The *Jensen-Shannon divergence* is $0.5 * D_{SY}(p||q)$.

See [160, 202], which are most cited articles, for more details on divergence. A mathematical short compendium can be found in [16]. In the article [160] several interesting cases of noisy communication channels are discussed. A comprehensive survey of distance measures can be found in [267]

Let us consider an example that helps to see important aspects of the KLD. A certain mixture $p(x)$ of two Gaussians is given, with two separated peaks so $p(x)$ has two modes. One tries now to approximate this with one Gaussian $q(x)$, having the same area: $\int q(x)dx = \int p(x)dx$.

Figure 7.81 shows three approximation alternatives. What happens with the KLD between $p(x)$ and $q(x)$?

Two distinct behaviours can be observed. If one computes $D_{KL}(q||p)$, then low values are obtained for the approximations placed on mode 1, or in mode 2; however a large value is obtained for the moment-matched approximation

On the other hand, if one computes $D_{KL}(p||q)$, then a low value is obtained for the moment-matched approximation, while large values result for the approximations on moments.

It could be said then, that minimising $D_{KL}(q||p)$ produces mode-seeking; and that minimising $D_{KL}(p||q)$ produces moment matching.

This observation is of special interest for a topic to be studied below: variational Bayes.

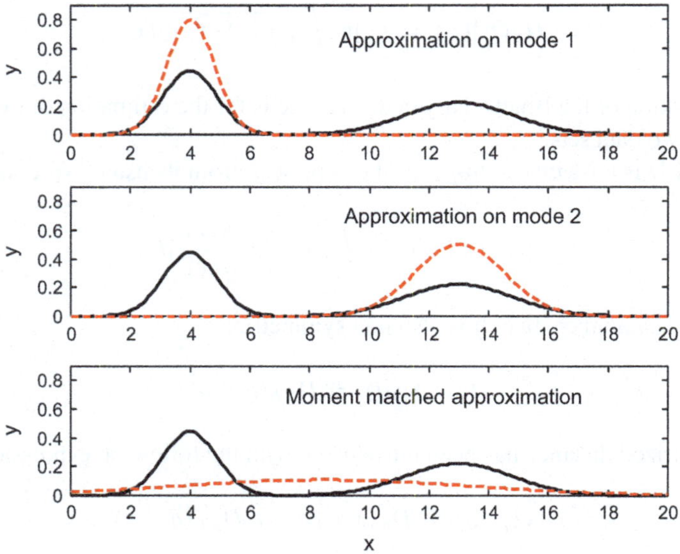

Fig. 7.81 Examples of approximations to a Gaussian mixture

7.7.3 Jensen's Inequality

The Jensen's inequality for convex functions is frequently invoked along theory developments.

A function $f(x)$ is *convex* over an interval (a, b), if for every $u,\ v \in (a, b)$ and $\lambda \in [0,\ 1]$:

$$f\left(\lambda u + (1-\lambda)\, v\right) \leq \lambda\, f(u) + (1-\lambda)\, f(v) \tag{7.246}$$

A convex function is below any chord. A concave function is above any chord.

The Jensen's inequality establishes that:

$$E\left(f(X)\right) \geq f(E(X)) \tag{7.247}$$

where X is a random variable, f is a convex function, $E()$ is the expected value.

Note that the entropy of a probability distribution $H(p(x))$ is a concave function.

Also, the mutual information MI(X,Y) is a concave function of $p(x)$ for fixed *p(x|y),* and is a convex function of *p(x|y)* for fixed $p(x)$.

Suppose you have a $p(x)$. If you choose a certain $q(\theta)$ (perhaps a more tractable function), the Jensen's inequality can be used as follows:

$$\begin{aligned} &\ln\ p(x) = \ln \int p(x,\theta)\, d\theta = \\ &= \ln \int q(\theta)\ \frac{p(x,\theta)}{q(\theta)}\, d\theta \geq \int q(\theta)\ \ln\ \frac{p(x,\theta)}{q(\theta)}\, d\theta \equiv -F \end{aligned} \tag{7.248}$$

where:

$$F = -\int q(\theta)\ \ln \frac{p(x,\theta)}{q(\theta)}\, d\theta \tag{7.249}$$

is the *variational negative free energy* (well known in statistical physics).

7.7.4 Variational Bayes Methodology

Given a probabilistic model of some data, the log of the evidence can be expressed as follows:

$$\ln\ p(x) = \int q(\theta)\ \ln p(x)\, d\theta = \int q(\theta)\ \ln \tfrac{p(x,\theta)}{p(\theta|x)}\, dz = \int q(\theta)\ \ln \tfrac{p(x,\theta) q(\theta)}{q(\theta) p(\theta|x)}\, d\theta\ =$$
$$= \int q(\theta)\ \ln \tfrac{p(x,\theta)}{q(\theta)}\, d\theta + \int q(\theta)\ \ln \tfrac{q(\theta)}{p(\theta|x)}\, d\theta$$

Therefore:

$$\ln\ p(x) = -F + D_{KL}(q(\theta)\, ||\, p(\theta|x)) \tag{7.250}$$

The aim of variational Bayes is to minimize the Kullback-Leibler divergence, in order to obtain a suitable approximation $q(\theta)$ for the posterior $p(\theta|x)$. The positive quantity $-F$ must be maximized [252].

Figure 7.82 depicts the situation.

The problem we are considering may correspond to estimation of model parameters, taking into account a set of observed data. The Bayesian estimation would choose the Maximum A Posteriori (MAP) value, while variational Bayes would opt for a minimum KLD.

Figure 7.83 compares an approximation to $p(\theta|x)$ based on MAP, so it focuses on a mode, and an approximation based on variational Bayes, which is moment-matching.

to $p(z|\, x)$

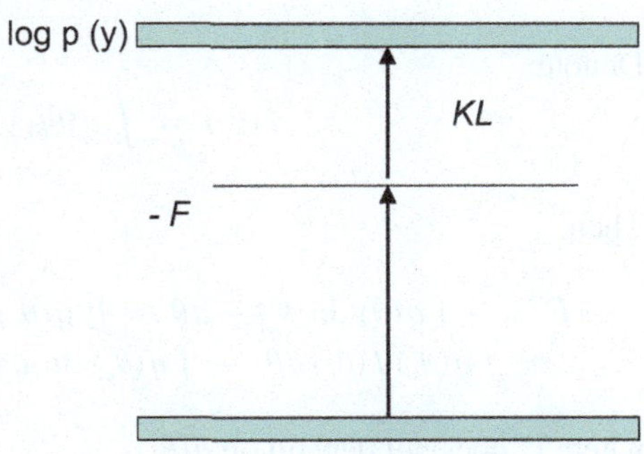

Fig. 7.82 Decomposition of model evidence

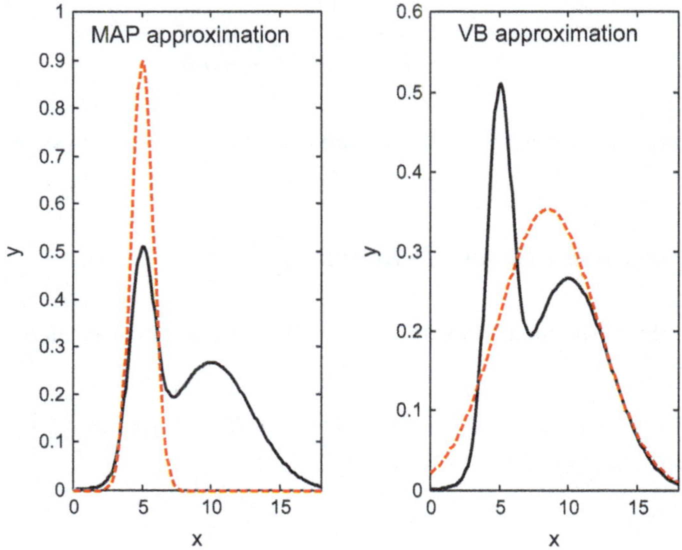

Fig. 7.83 MAP and Variational Bayes approximations

Let us focus on maximization of $-F$. A usual procedure to devise tractable integrals is to assume that $q(\theta)$ factorizes over groups of parameters. This corresponds to a *mean field* approach in physics. Hence:

$$q(\theta) = \prod_i q(\theta_i) \tag{7.251}$$

where θ_i is the *i-th* group of parameters. Some aspects of the next derivation are better expressed if we also use $\theta_{\triangleright i}$ to denote all parameters *not* in the group θ_i. Then, the factorization can be written as follows:

$$q(\theta) = q(\theta_i)\, q(\theta_{\triangleright i}) \tag{7.252}$$

Denote:

$$I(\theta_i) = \int q(\theta_{\triangleright i}) \ln\, p(x, \theta)\, d\theta_{\triangleright i} \tag{7.253}$$

Then:

$$\begin{aligned} -F &= -\int q(\theta) \ln \frac{p(x,\theta)}{q(\theta)}\, d\theta = \iint q(\theta_i) q(\theta_{\triangleright i}) \ln \frac{p(x,\theta)}{q(\theta_i) q(\theta_{\triangleright i})} d\theta_i\, d\theta_{\triangleright i} = \\ &= \int q(\theta_i)\, I(\theta_i)\, d\theta_i - \int q(\theta_i) \ln q(\theta_i)\, d\theta_i + C \end{aligned} \tag{7.254}$$

where C does not depend on $q(\theta_i)$.

With the simple tautology: $I(\theta_i) = \ln \exp I(\theta_i)$, it is possible to write:

$$-F = \int q(\theta_i) \ln \frac{\exp(I(\theta_i))}{q(\theta_i)} d\theta_i + C = D_{KL}[q(\theta_i) || \exp(I(\theta_i))] + C \quad (7.255)$$

This is optimized when:

$$q(\theta_i) = \frac{\exp |I(\theta_i)|}{k} \quad (7.256)$$

where k is a normalization factor for $q(\theta_i)$ to be a valid PDF.

On the basis of this result, a basic algorithm would be to cycle over parameters. revising each given the current estimate of the others. There are versions that iterate as the E-M algorithm [19, 26].

7.8 Neurons

Neurons receive signals from other neurons. To say so, the inputs to the neuron are the dendrites. In response to inputs, the neuron generates signals that are transmitted through the axon. The connections between axon and dendrites are the synapses (ions are exchanged trough a membrane). Figure 7.84 shows a sketch of one neuron and its connections.

In 1943, McCulloch and Pitts suggested a model of how the neuron works. According with this model, the neuron fires if the sum of the received signals exceeds a threshold and if no inhibitory input is received. Figure 7.85 shows a simple block

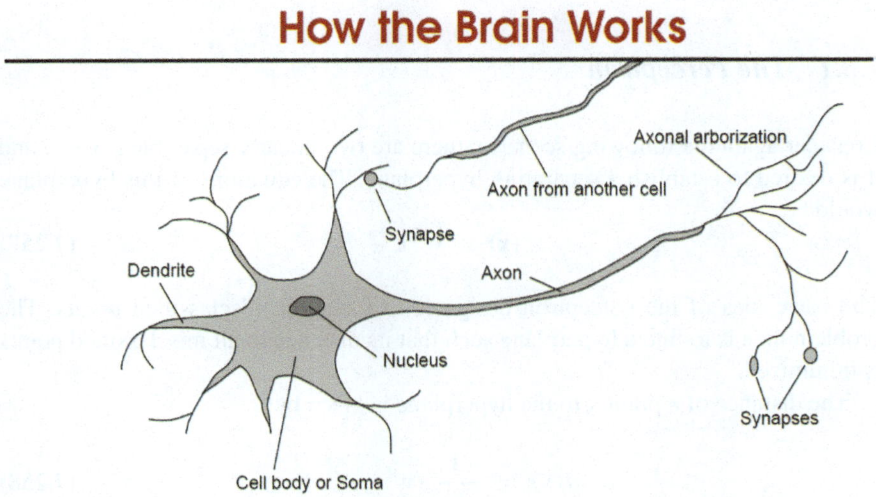

Fig. 7.84 A neuron: dendrites, axon, synapses

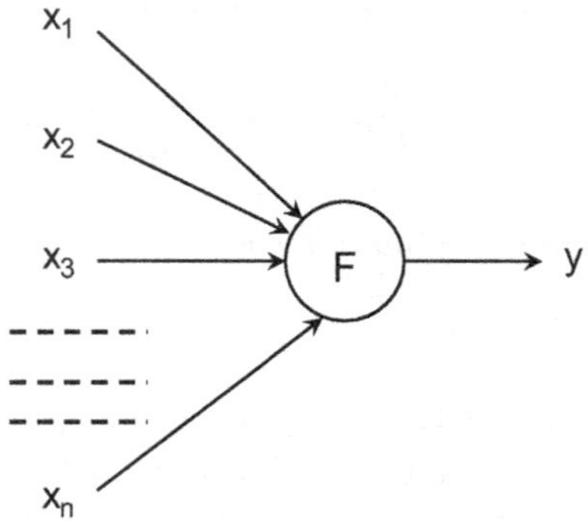

Fig. 7.85 McCulloch and Pitts neuron model

diagram with the model; the function F includes a summation of inputs and a processing of the result (for instance, a threshold) which is called the '*activation function*'. Several formulations of the activation function have been proposed.

The book of D. Hebb on '*The Organization of Behavior*', published in 1949, had a profound impact in the incipient world of machine learning. It postulates (Hebbian learning) that a synapse between two neurons becomes more and more effective as the activations of the second neuron by the first one—through this synapse—are repeated. Therefore, in general, artificial neural networks do use iterative learning procedures.

One of the main applications of neural networks is classification, so they ought to be included in this chapter. The main difficulty is, however, that nowadays the extension of this theme is impressively large. It is only possible to briefly include a selection of concepts and methods.

7.8.1 The Perceptron

Consider again the following scenario: there are two linearly separable classes, and it is desired to establish a separating hyperplane. The equations of this hyperplane would be:

$$f(\mathbf{x}) = \mathbf{w}^T \mathbf{x} + b \tag{7.257}$$

The basic idea of the perceptron design is to focus on misclassified points. The problem then is to find a hyperplane such that its distance to all misclassified points is minimized.

The distance of a point $\mathbf{x}$ to the hyperplane is given by:

$$d(\mathbf{x}) = \frac{1}{\| \mathbf{w} \|}(\mathbf{w}^T \mathbf{x} + b) \tag{7.258}$$

Let us employ a label y with two values, $+1$ and -1, to indicate if a point belongs to class 1 or to class 2. For points correctly classified, the product $y_i(\mathbf{w}^T\mathbf{x}_i + b)$ is positive. Joining the misclassified points, the distance to be minimized is:

$$D_M = -\sum_{i \in M} y_i(\mathbf{w}^T\mathbf{x}_i + b) \tag{7.259}$$

where M is the set of misclassified points.

An iterative method was proposed for reaching a solution (there are infinite solutions). The method is a kind of stochastic gradient. In fact, the gradients of D_M are easy to compute; the gradient with respect to $\mathbf{w}$ is $-\sum_{i \in M} y_i\mathbf{x}_i$, and the gradient with respect to b is $-\sum_{i \in M} y_i$.

The proposed algorithm is quite simple. Choose any of the misclassified points, for instance $\mathbf{x}_k$, then update the hyperplane parameters as follows:

$$\mathbf{w}^{(new)} = \mathbf{w}^{(old)} + y_k\mathbf{x}_k \tag{7.260}$$

$$b^{(new)} = b^{(old)} + y_k\,\mu \tag{7.261}$$

where μ is the learning rate parameter.

Once the hyperplane was updated, a new checking of misclassified points is done. If there is none, the algorithm successfully stops. If not, the updating procedure is repeated.

Figure 7.86 shows an example of perceptron application related again to the IRIS data. A discrimination line has been quickly found. Figure 7.87 shows the evolution of the perceptron iterative algorithm in terms of classification fails. Both figures have been obtained with the Program 7.42.

The reader is invited to run several times the program, changing the initial conditions each time. It would then become clear that the results depend on these initial conditions.

Program 7.42 Perceptron example

```
% Perceptron example
% Scatterplot of IRIS data
% x=sepal width, y=petal length
%read IRIS data
D=dlmread('iris.data')';
%Perceptron weights
w1=1; w2=0; %for data
w3=0.5; %for line threshold
%training constant
eta=0.01;
%number of epochs
N=10;
%failures counter
nfail=zeros(1,N);
```

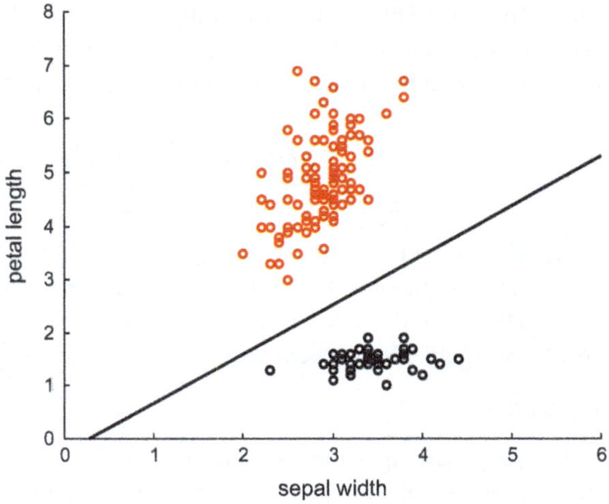

Fig. 7.86 Perceptron discrimination example

```
%training iterations
for it=1:N,
for nn=1:50, %data with label -1
x=D(2,nn); y=D(3,nn);
c=(w1*x)+(w2*y)+w3;
if(c>0), nfail(it)=nfail(it)+1; end; %failure count
delta=-1-c;
mf=eta*delta;
w1=w1+(mf*x); w2=w2+(mf*y); w3=w3+mf; %change of weights
end;
for nn=51:150, %data with label 1
x=D(2,nn); y=D(3,nn);
c=(w1*x)+(w2*y)+w3;
if(c<0), nfail(it)=nfail(it)+1; end; %failure count
delta=1-c;
mf=eta*delta;
w1=w1+(mf*x); w2=w2+(mf*y); w3=w3+mf; %change of weights
end;
end;
%result: line parameters
m=-w1/w2; b=-w3/w2;
%display
figure(1)
scatter(D(2,1:50),D(3,1:50),32,'k'); hold on;
scatter(D(2,51:150),D(3,51:150),32,'r');
axis([0 6 0 8]);
len=7; %arbitrary value
plot([0 len],[b (m*len)+b],'k');
title('Perceptron example (IRIS data)');
xlabel('sepal width'); ylabel('petal length');
figure(2)
plot(nfail,'k');
title('classification fails along training');
xlabel('epoch')
```

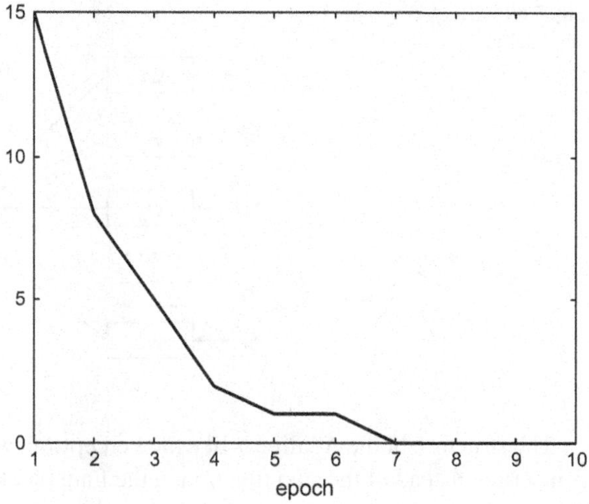

Fig. 7.87 Classification fails along training

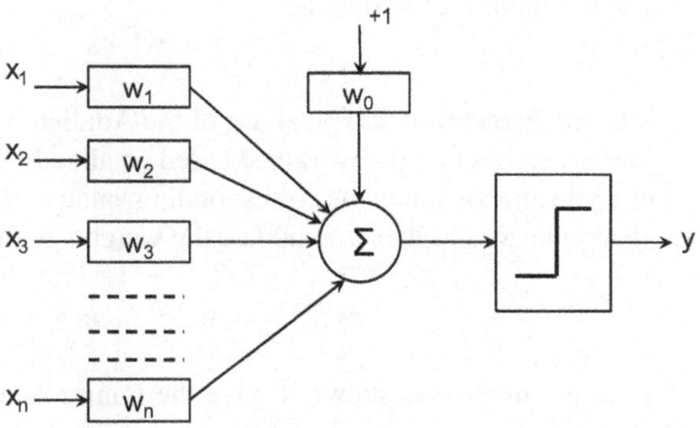

Fig. 7.88 Block diagram of the Perceptron

The Perceptron and its training procedure were introduced by F. Rosenblatt in 1958. It is a McCulloch-Pitts neuron with weights on the inputs, and with a step activation function to obtain y. Figure 7.88 presents a block diagram of the Perceptron, which follows the concept expressed in Fig. 7.85. An extra input is added with a weight $w_0 = b$. The output y ($+1$ or -1) is the result of the classification.

A large margin classification method using the Perceptron has been introduced in [101]. Also, kernelized versions of the Perceptron have been proposed [244].

7.8.2 *The Adaline*

In 1960, B. Widrow and M. Hoff introduced the '*Adaptive Linear Neuron (Adaline)*' and its training algorithm. This algorithm is a *Least Mean Square Algorithm (LMS)*, also called '*the delta rule*'.

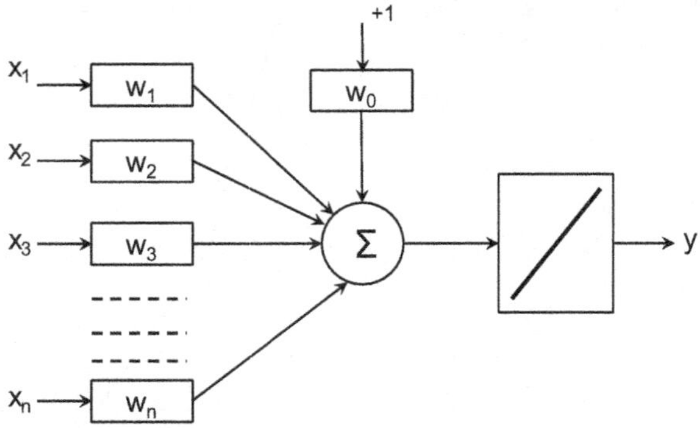

Fig. 7.89 Block diagram of the Adaline

The structure of the Adaline is like the Perceptron, with a change: a linear function is inserted instead of the step function in the final block. Figure 7.89 shows the block diagram.

The output of the Adaline is:

$$y = \mathbf{w}^T \cdot \mathbf{x} \tag{7.262}$$

Like the Perceptron, a typical use of the Adaline is binary classification.

The Adaline is iteratively trained based on already labeled data, so for a particular input $\mathbf{x}$ you already know the corresponding value of y, but the Adaline is giving you another value $\hat{y}$, and then you update the weights with:

$$\mathbf{w}^{(new)} = \mathbf{w}^{(old)} + \eta(y - \hat{y})\,\mathbf{x} \tag{7.263}$$

This simple method is shown to give the minimization of the mean square of the error:

$$mse = \sum_i e_i^2 = \sum_i (y - \hat{y})^2 \tag{7.264}$$

7.8.3 *Multilayer Neural Networks*

With proper training, Perceptrons could serve as logical gates of '*and*', '*or*', '*nand*', '*nor*', types. However, as it was shown in 1969, they cannot work as '*xor*' gate. This limitation, which corresponds to not linearly separable cases (where the Perceptron cannot discriminate), was regarded as a serious drawback of artificial neurons, and so the research abandoned artificial neurons for many years. Nevertheless, since the *xor* gate can be implemented with a structure of *and*, *nand* and *or* gates, two layers of Perceptrons can implement the *xor* gate. But, in general, it was unknow how to compute the weights in layered structures.

In 1986, Rumelhart, Hinton and Williams presented a method for training layered neural networks. It was called '*backward propagation of errors*'. Nowadays, the name

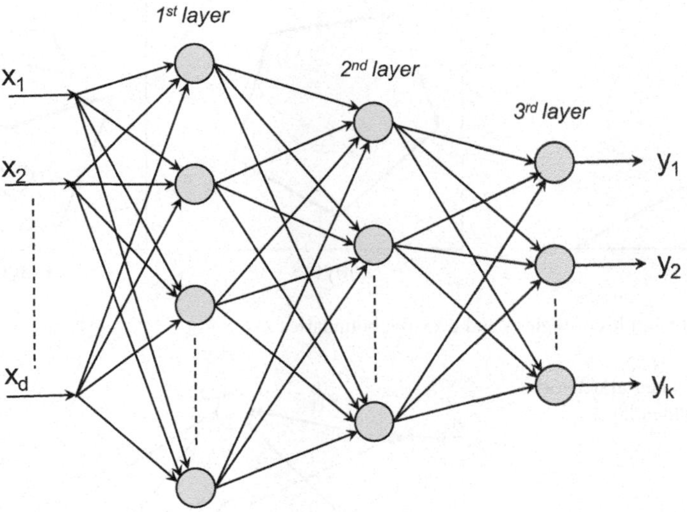

Fig. 7.90 Diagram of a layered neural network

used is backpropagation, or even shorter: backprop. This announcement injected new life in the topic, and since then the research has worked extensively and so, today, the published background is really impressive. It is important to say that, afterwards, it has been recognized that backprop was introduced by P. Werbos in his Ph.D. Thesis, Harvard, 1974.

It is time now to show a first, conceptual example of a neural network with three layers. Each circle represents a neuron. It is assumed that each neuron has inside a summation function followed by an activation function. It is also assumed that the arrows include the weighting factors (Fig. 7.90).

Neural networks can have one, two, etc., layers. More layers give more discrimination flexibility. In particular, if you use Perceptrons, one single layer would be able to place one discrimination hyperplane; two layers can place several hyperplanes, and so, for example, a convex region could be surrounded by a decision boundary; and three layers would be able to create arbitrary regions [106]. Figure 7.91 depicts examples of these cases.

In order to derive mathematical expressions, a simple network with one hidden layer would be considered. Figure 7.92 shows the network, which also serves to introduce some notation concerning weights and hidden variables.

Choose for example the output y_1, it can be obtained as follows:

$$y_1 = f(b_1^o + \sum_{j=1}^{m} w_{j1}^o \cdot v_j) \tag{7.265}$$

where $f(\cdot)$ is the activation function of output neurons.

Since v_j are the outputs of the hidden layer neurons:

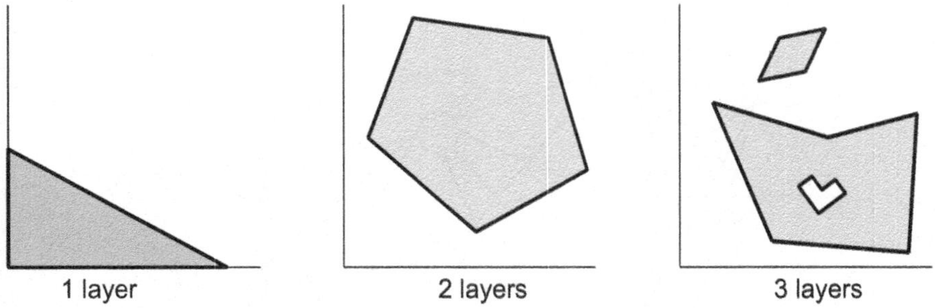

Fig. 7.91 Examples of regions and decision boundaries

Fig. 7.92 A neural network with one hidden layer

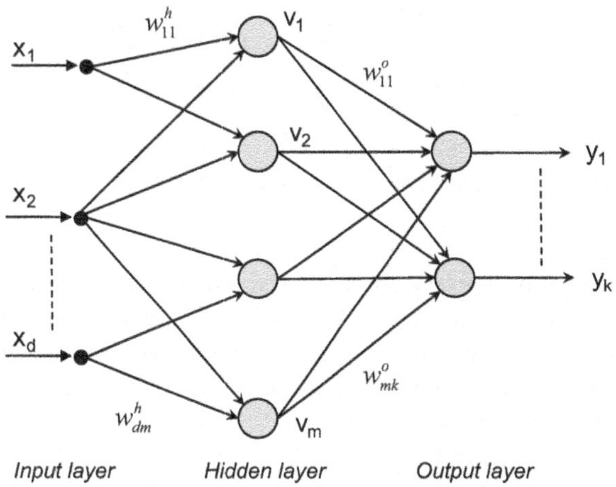

$$y_1 = f(b_1^o + \sum_{j=1}^{m} w_{j1}^o \cdot g(b_j^h + \sum_{n=1}^{d} w_{nj}^h \cdot x_n)) \tag{7.266}$$

where $g(\cdot)$ is the activation function of hidden layer neurons.

A typical option for the activation functions is to use sigmoidal functions. Two popular sigmoidal functions are the following:

(a)

$$f(x) = \frac{1}{1 + \exp(-x)} \tag{7.267}$$

its output range is [0, 1], and its derivative is $f'(x) = f(x)(1 - f(x))$.

(b)

$$f(x) = \tanh(x) \tag{7.268}$$

its output range is [−1, 1], and its derivative is $f'(x) = (1 - f(x)^2)$.

An important fact is that, according with Kolmogorov's universal approximation theorem, any continuous function defined on a compact subset of $\Re^d$ can be uniformly approximated by the expression above (see [149]).

When using a neural network, a first step is to use a data set for training, and subsequent steps would be to apply the trained network for new data to be classified.

Thus, during training, a learning set $X = \{\mathbf{x}_1, \mathbf{x}_2, \ldots \mathbf{x}_N\}$; where each vector $\mathbf{x}_i$ has d components, is used. One already knows what true responses correspond to these data. The set of true responses is $Y = \{\mathbf{y}_1, \mathbf{y}_2, \ldots \mathbf{y}_N\}$; where each vector $\mathbf{y}_i$ has k components.

The training process is iterative, along a number of steps. A basic alternative is to use the complete training data set for every step. Suppose we choose one of the steps; the error of the output node r when the input to the network is $\mathbf{x}_i$, would be:

$$e_{i,r} = y_{i,r} - \hat{y}_{i,r} \tag{7.269}$$

where $\hat{y}_{i,r}$ is the actual output of the neuron.

If one considers the responses of all output neurons to $\mathbf{x}_i$, an *error function* can be defined as follows:

$$E_i = \frac{1}{2} \sum_{r=1}^{k} e_{i,r}^2 \tag{7.270}$$

Using now the complete learning set X, an averaged sum of squares can be formed:

$$E = \frac{1}{N} \sum_{i=1}^{N} E_i \tag{7.271}$$

The training process is aimed to minimize this learning error E. This can be done by minimizing each E_i separately. The idea is to modify the weights, layer by layer, starting from the output layer toward the input layer. Gradient descent would be applied for adequate weight modifications.

Denote as $\mathbf{w}^o$ the set of weights between the hidden layer and the output layer. They are changed as follows:

$$\mathbf{w}^o_{new} = \mathbf{w}^o_{old} - \eta \frac{\partial E_i}{\partial \, \mathbf{w}^o_{old}} \tag{7.272}$$

Similarly, for the weights between the input node and the hidden node:

$$\mathbf{w}^h_{new} = \mathbf{w}^h_{old} - \eta \frac{\partial E_i}{\partial \, \mathbf{w}^h_{old}} \tag{7.273}$$

The updating of weights can be done in sequential way: using first $\mathbf{x}_1$, then $\mathbf{x}_2$, and so till the last datum $\mathbf{x}_N$. Alternatively, a batch mode could be devised for updating the weights only after all learning data have been taken into account. In both cases, the process is repeated along several epochs, each of them corresponding to the use of a complete learning data set. The computation of the jacobians $\partial E_i / \partial \mathbf{w}$ is usually done using the chain rule; for instance:

$$\frac{\partial E_i}{\partial w^o_{i\,j}} = \frac{\partial \frac{1}{2}(y_j - \hat{y}_j)^2}{\partial w^o_{i\,j}} = \frac{1}{2} \frac{\partial\, (y_j - f(b^o_j + \sum\limits_{n=1}^{m} w^o_{n\,j} \cdot v_n))^2}{\partial w^o_{i\,j}} = \\ = -e_j\, f'(b^o_j + \sum_{n=1}^{m} w^o_{n\,j} \cdot v_n)\, v_i \tag{7.274}$$

and:

$$\frac{\partial E_i}{\partial w^h_{i\,j}} = \sum_{n=1}^{k} \frac{\partial E_i}{\partial y_n} \cdot \frac{\partial y_n}{\partial v_j} \cdot \frac{\partial v_j}{\partial w^h_{i\,j}} = \\ = \sum_{n=1}^{k} -e_n \cdot \frac{\partial\, f(b^o_n + \sum\limits_{l=1}^{m} w^o_{l\,n} \cdot v_l)}{\partial v_j} \cdot \frac{\partial\, g(b^h_j + \sum\limits_{s=1}^{d} w^h_{s\,j} \cdot x_s)}{\partial w^h_{i\,j}} = \\ = \sum_{n=1}^{k} -e_n \cdot f'\,(b^o_n + \sum_{l=1}^{m} w^o_{l\,n} \cdot v_l)\, w^o_{jn} \cdot g'(\,b^h_j + \sum_{s=1}^{d} w^h_{s\,j} \cdot x_s)\; x_i \tag{7.275}$$

A simple trick frequently employed to simplify the expressions, is to consider the bias terms, b^o_i, b^h_j, as weights for inputs with value 1.

In order to illustrate the use of neural networks for classification, the case of three clusters of data has been chosen as a simple example. Figure 7.94 displays these data clusters. A neural network with a hidden layer of two neurons and an output layer of three neurons has been tried. The target is to classify data as belonging to class 1, class 2 or class 3. In case a datum belongs to class 1, the output neuron 1 (ON1) should give a value near 1, while the other two output neurons should give values near 0. Similarly, when a datum belongs to class 2 then ON2 $\approx$ 1, ON1 $\approx$ 0, ON3 $\approx$ 0. And, when a datum belongs to class 3, then ON3 $\approx$ 1, ON1 $\approx$ 0, ON2 $\approx$ 0.

Figure 7.93 depicts the neural network that has been used.

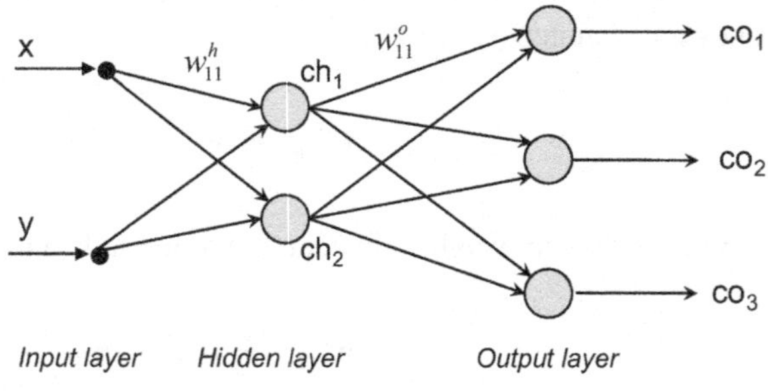

Fig. 7.93 Example of neural network

A program has been developed for this case, and has been included in the Appendix for long programs. Indeed, the code is not optimized and some patience was required to run it. It contains interesting implementation details. Figure 7.95 shows

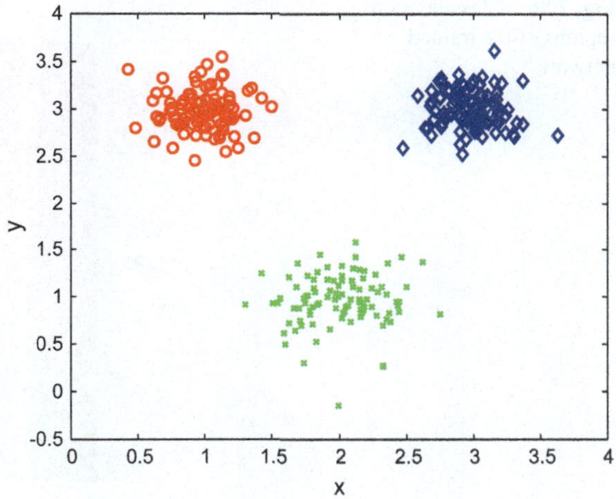

Fig. 7.94 Data input for training

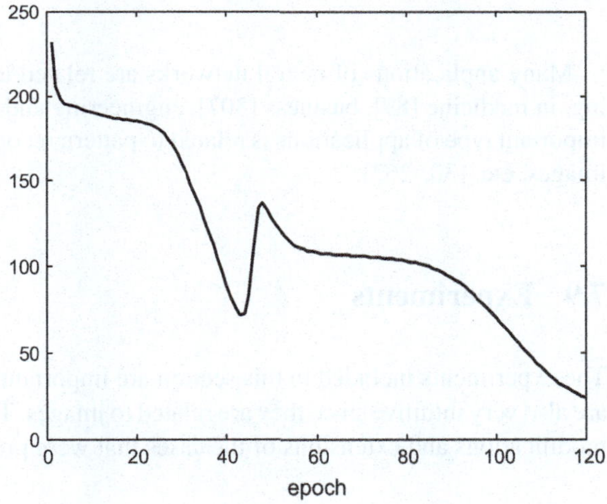

Fig. 7.95 Evolution of error during learning

how the total error evolved along a training experiment. During several training runs it happened, in a few cases, that the curve started to increase by the end of the process; it is one of the phenomena that neural network experts know it can occur.

In order to visualize the classification behaviour of the trained neural network, a regular set of testing data, covering a square 2D region, has been applied, and the results have been interpreted as RGB colors (ON1 gives R, ON2 gives G, ON3 gives B). Figure 7.96 displays the result.

If you search on the web for the term '*neural network*', you will get more than 13.000.000 entries, most of them documents in pdf. There are plenty of books, tutorials, reviews, etc. In addition, there are several scientific journals related to this topic. There is a MATLAB Neural Network Toolbox, and a number of neural network simulators (see the web page from grey.colorado.edu).

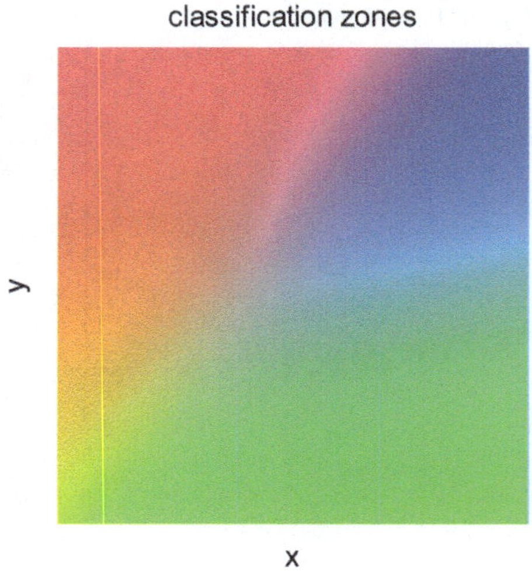

Fig. 7.96 Classification regions of the trained network

Many applications of neural networks are related to diagnosis and decision taking, in medicine [89], business [307], engineering and industry [269, 290]. Another important type of applications is related to pattern recognition: handwritting, speech, images, etc. [30, 262].

7.9 Experiments

The experiments included in this section are important application examples, which are also very intuitive since they are related to images. The reader is invited to explore modifications and extensions of the cases that were proposed next.

7.9.1 Face Detection

From time ago automatic face recognition has attracted a lot of attention, with many applications in mind [196, 329, 331].

In our experiments, it has been considered that one knows 6 people. Ten photographs of each person's face have been taken: 9 will be used for training, and 1 for testing. Therefore, there is an original database of 6×9 faces, and a test set of 6 faces. Once a certain training method was applied, one wants to choose any of the test faces and detect (correctly) that it corresponds to one of the 6 known people.

Moreover, one wants to choose any face of unknown people and detect (correctly) that it does not correspond to any of the 6 known people. For this purpose, a second test set of 6 unkown faces was considered.

In general, the type of experiments just described is not easy. Some of the typical difficulties in this context are differences of pose, illumination, expressions, face size, use of glasses, etc. Internet offers access to some face databases (see links to homepages in the section on Resources). In our case, a subset of the ORL database has been selected (AT&T Laboratories Cambridge).

The approaches selected for the experiments belong to important methods already studied in this chapter.

7.9.1.1 Eigenfaces Method

In a most cited article [303] the use of PCA for face recognition was proposed. It was argued that previous work fell into three categories, based on features, on geometry, or on using neural networks. These alternatives ignore the question of which features are important for classification, and which are not. Instead, the analysis via PCA reduces the dimensionality, leaving those features that are relevant for face recognition. According with [220], it seems that PCA outperforms LDA when the training set is small.

Two MATLAB programs have been developed to deal with a simplified example. The first program is devoted to PCA analysis, and the second for running some tests.

Figure 7.97 shows the original database of 6×9 faces. Each face has a size of 92×112 pixels.

Before the application of PCA there is a face preparation process. Each face is represented as a column vector. An average face is computed as follows:

$$MF = \frac{1}{N} \begin{bmatrix} a_1 & b_1 & \dots & r_1 \\ a_2 & b_2 & \dots & r_2 \\ \vdots & \vdots & & \vdots \\ a_L & b_L & \dots & r_L \end{bmatrix} \tag{7.276}$$

(each of the columns in the matrix is a face; there are N faces; the average face is also a column)

In our example, $N = 54$, $L = 10304$.

Figure 7.98 shows the average face.

The average face is now subtracted from each of the faces. Now, faces are prepared for PCA. A matrix P is formed with the prepared faces as columns.

The covariance matrix would be:

$$C = P\,P^T \tag{7.277}$$

The size of matrix C is 10304×10304. There are 10304 eigenvalues.

Fortunately, it can be shown, [303], that the first N eigenvalues can be obtained from:

$$Q = P^T\,P \tag{7.278}$$

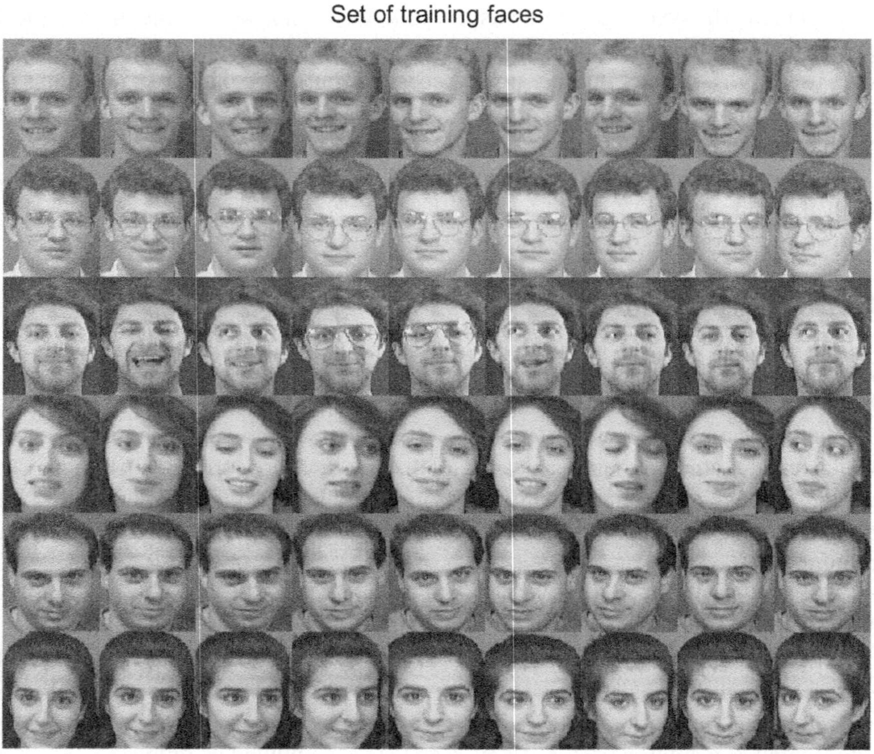

Fig. 7.97 A set of 6 × 9 training faces

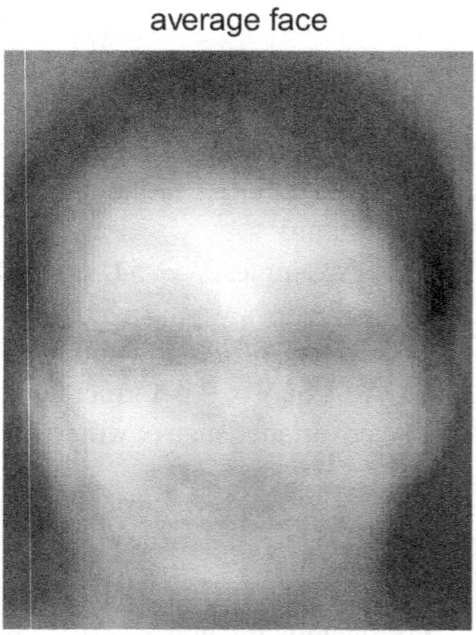

Fig. 7.98 Average face

The size of matrix Q is 54×54. There are 54 eigenvalues.

Evidently, the computational effort is much less now, based on Q, than based on C.

Figure 7.99 shows the eigenvalues of Q in our example. It can be noticed that eigenvalues tend to negligible values, so it suffices to consider 54 or less eigenvalues. A 54×54 matrix V is formed with the 54 eigenvectors of Q.

The first 54 eigenvectors of C can be obtained with:

$$U = P\,V \tag{7.279}$$

These eigenvectors are called '*Eigenfaces*'. Each Eigenface is a column vector with 10304 components. The columns of the matrix U are the 54 Eigenfaces.

Figure 7.100 shows the first 10 Eigenfaces in our example. Notice that Eigenfaces have the same size as the original faces.

All four figures already presented have been generated with the Program 7.43.

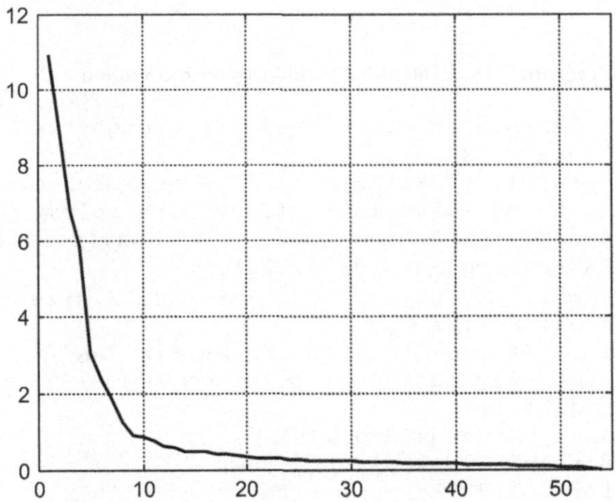

Fig. 7.99 The eigenvalues

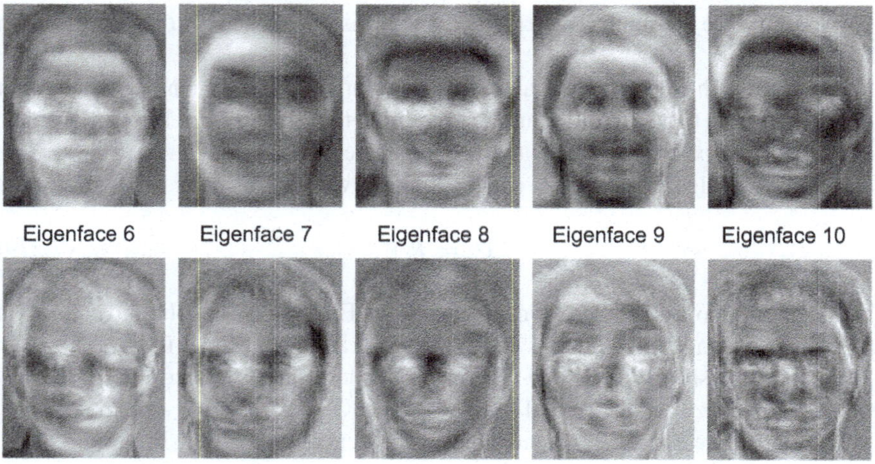

Fig. 7.100 The 10 first eigenfaces

Program 7.43 Example of Eigenfaces for recognition

```
% example of Eigenfaces for recognition
% analysis part
%read face database (120 faces, of 112x92=10304 length each)
AA=zeros(10304,120); %faces are columns of the matrix
fid=fopen('AAface.bin','r'); aux=fread(fid);
AA=reshape(aux,10304,120);
% will choose 6x9 faces of 92x112 size
W=92; H=112;
% faces are contained in matrix "AA"
% display mosaic of selected faces---------------------------
figure(1)
nn=1; M=zeros(H*6,W*9);
for L=5:10, %select 6 rows
for C=1:9, %choose 9 faces in each row
nn= 10*(L-1); nn=nn+C;
fa=reshape(AA(:,nn),H,W);
a=1+((L-5)*H); b=a+H-1; xl=a:b;
c=1+((C-1)*W); d=c+W-1; xc=c:d;
M(xl,xc)=fa;
end;
end;
imshow(M,[]);
title('Set of training faces');
% Eigenfaces algorithm---------------------------------
% form matrix with image columns
a=6*9; Or=zeros(H*W,a); %54 columns
nn=41;ns=1; %select 60 faces
for nc=1:a,
aux=AA(:,nn);
Or(:,nc)=aux;
```

```
if ns==9 %skip next face
ns=1; nn=nn+2;
else
ns=ns+1; nn=nn+1;
end;
end;
% get average face
MF=mean(Or,2);
% subtract average face from each face
% and normalize
PF=zeros(H*W,a);
for nc=1:a,
aux=Or(:,nc)-MF;
PF(:,nc)=aux/norm(aux);
end;
% display average face
figure(2)
aux=reshape(MF,H,W);
imshow(aux,[]);
title('average face');
% get eigenvectors
[V D]=eig(PF'*PF);
% sort in descending order
aux=diag(D);
[aux2, sx]=sort(-1.*aux);
eigL=aux(sx); eigV=V(:,sx);
% display Eigenvalues------------------------------
figure(3)
plot(eigL,'k');
title('Eigenvalues')
axis([0 55 0 12]); grid;
% project to obtain eigenfaces
aux=PF*eigV;
eigF=aux/norm(aux);
% display 10 first Eigenfaces
figure(4)
for nf=1:10,
ef=reshape(eigF(:,nf),H,W);
subplot(2,5,nf)
imshow(ef,[]);
tt=['Eigenface ', num2str(nf)]; title(tt);
end;
```

The set of Eigenfaces provides a basis for a '*face space*'. Now, the prepared faces can be projected onto this face space as follows:

$$\Omega = U^T P \tag{7.280}$$

The 54×54 matrix Ω means a significant dimensionality reduction.

One of the measures that can be of interest is the '*threshold*', which is:

$$\theta = \frac{1}{2} \max \left\| \Omega_i - \Omega_j \right\|, \; for \; i, j = 1, 2, \ldots, N \tag{7.281}$$

where each Ω_i is a projected face.

A reconstruction of the prepared faces can be done with:

$$S = U\,\Omega \tag{7.282}$$

Another measure that can be of interest is the distance between the prepared face and its reconstruction:

$$\zeta_i = \|P_i - S_i\| \,, \quad for\ i = 1, 2, \cdots, N \tag{7.283}$$

Next figures have been generated with a program that has been included in the Appendix on long programs. The program calls two times a function, also included in the Appendix.

Figure 7.101 depicts the distances corresponding to the 54 faces in our example.

Part of the preparation of faces is devoted to normalization, so a normalization constant has been applied to each face. The program keeps a record of the 54 normalization constants, so it has been possible to properly add the averaged face to each reconstructed face. Figure 7.102 shows the reconstruction results, which invite to extend the analysis effort: more faces, more people.

Now, one of the experiments is to take one more photograph from the 6 known persons, and see what happens with their projection onto the face space, and how is the reconstruction.

A suitable measure in this case, when a certain Ω_i is tested, would be:

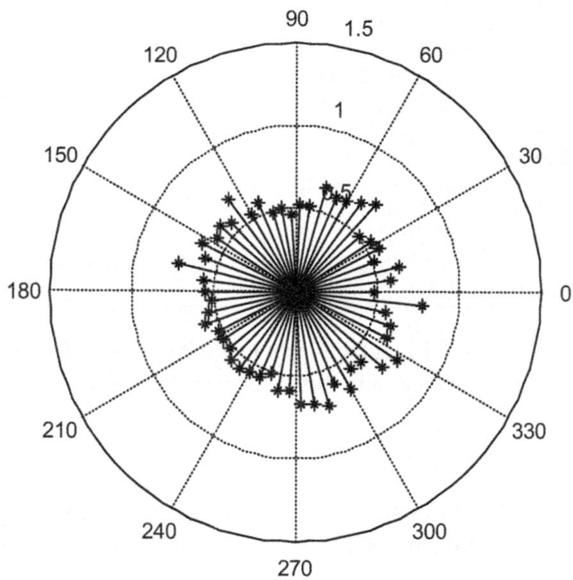

Fig. 7.101 Distance between prepared and reconstructed faces

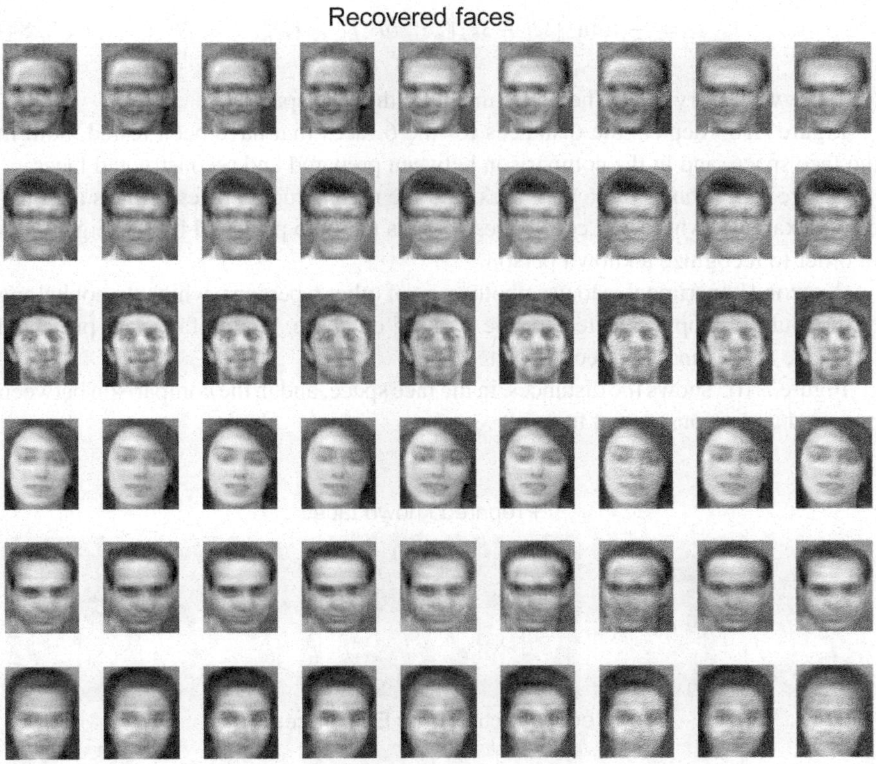

Fig. 7.102 Recovered faces (adding the average face)

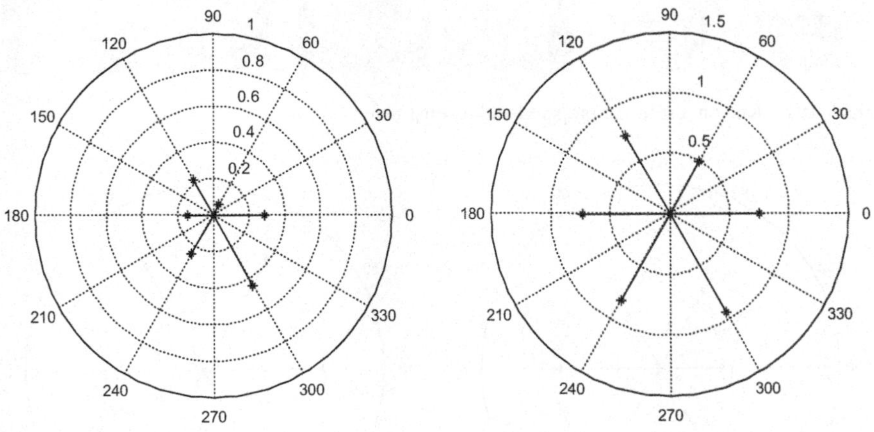

Fig. 7.103 Distances for known faces (*left* face space; *right* prepared faces)

$$\varepsilon_i = \frac{1}{2} \min \left\| \Omega_i - \Omega_j \right\|, \quad for \ j = 1, 2, \ldots, N \tag{7.284}$$

(in other words: try to get the better match in the face space).

Figure 7.103 depicts the distances for the 6 faces that have been tested, both in the face space, and in the comparison between prepared and reconstructed faces.

Figure 7.104 shows the 6 test faces, and the reconstructed faces. This is an interesting example, where one could guess how is the help provided by the Eigenfaces in order to recognize a known person.

A second experiment is to use photograps of other 6 persons, which do not belong to the club of people selected for the training database. Again, faces are projected onto face space, and then reconstructed.

Figure 7.105 shows the distances, in the face space, and in the comparison between prepared and reconstructed faces.

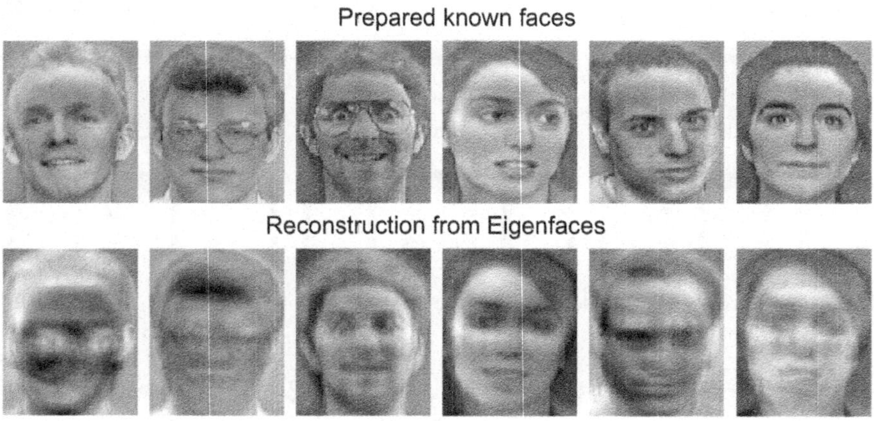

Fig. 7.104 Known test faces: prepared and reconstructed

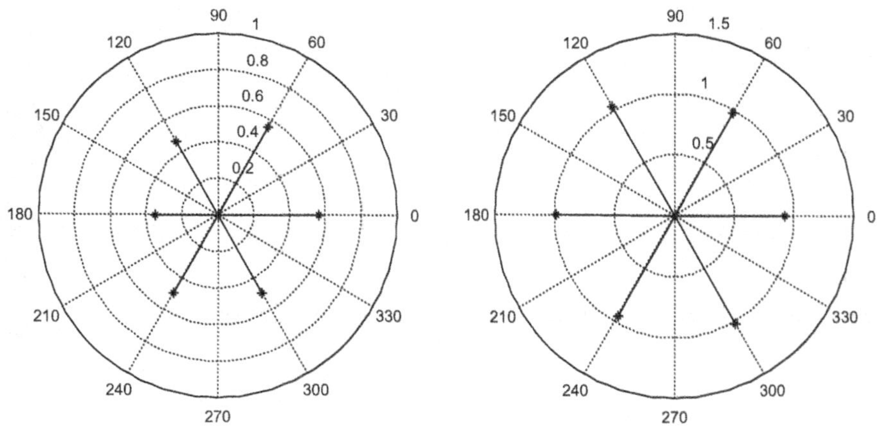

Fig. 7.105 Distances for unknown faces (*left* face space; *right* prepared faces

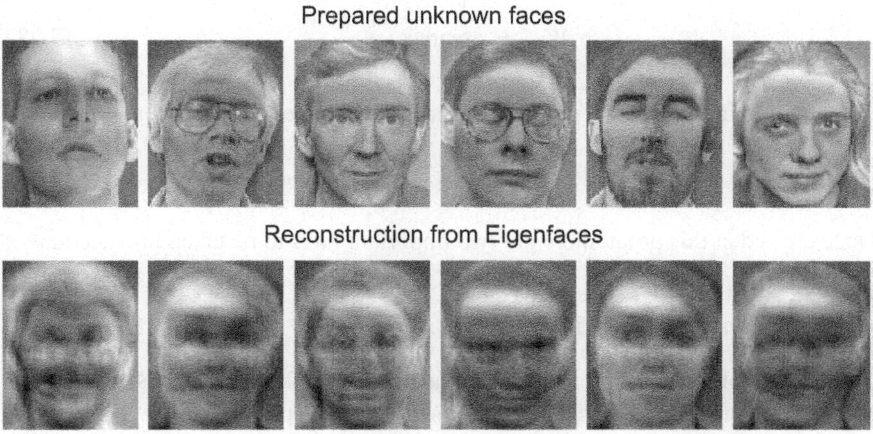

Fig. 7.106 Unknown test faces: prepared and reconstructed

Finally, Fig. 7.106 shows the 6 test faces, and their reconstruction based on the Eigenfaces obtained before with the training database. The results are very illustrative of how the test faces are clearly out of the space spanned by the established Eigenfaces.

7.9.1.2 Fisherfaces Method

When the target is to discriminate between classes, a suitable choice would be to use LDA. In our example, with $N = 54$ faces, which belong to $C = 6$ classes (6 persons), each class including $R = 9$ pictures, the LDA method computes:

- The between-class scatter matrix:

$$S_b = \sum_{i=1}^{C} R(\boldsymbol{\mu}_i - \boldsymbol{\mu})(\boldsymbol{\mu}_i - \boldsymbol{\mu})^T \tag{7.285}$$

where $\boldsymbol{\mu}_i$ is the mean image of class Cl_i, and $\boldsymbol{\mu}$ is the mean image of the total training set.

- The within-class scatter matrix:

$$S_w = \sum_{i=1}^{C} \sum_{\mathbf{x}_k \in Cl_i} (\mathbf{x}_k - \boldsymbol{\mu}_i)(\mathbf{x}_k - \boldsymbol{\mu}_i)^T \tag{7.286}$$

where $\mathbf{x}_k$ are the pictures in column format.

Now, the Fisher linear discriminant should be maximized:

$$W_{opt} = \arg\max_{W} \frac{\mathbf{w}^T S_b \mathbf{w}}{\mathbf{w}^T S_w \mathbf{w}} \tag{7.287}$$

The maximizing solution is obtained from the generalized eigensystem:

$$S_b\, \mathbf{w} = \lambda\, S_w \mathbf{w} \tag{7.288}$$

Once arrived at this point, there are two important facts to be taken into account:

- There are at most $C - 1$ eigenvectors
- S_w is always singular, its rank at most $(N - C)$.

(in our example, at most 5 eigenvectors, and the S_w rank is at most 48)

Then, it makes sense to reduce the dimension of the problem to 48. The proposal of [20] was to use PCA for this reduction. A matrix W_p is formed with the first 48 eigenfaces.

Based on W_p, a PCA face space, with 54 faces, is obtained by projection. Each PCA face is a column vector with 48 components.

Then, LDA is applied. Both S_b and S_w are computed on the PCA face space. As a result of the generalized eigensystem, 48 eigenvalues are obtained, but only the 5 largest values are of interest.

A matrix W_l is formed with the 5 eigenvectors corresponding to the 5largest eigenvalues. Now, the following matrix is obtained:

$$W_F = W_p\, W_l \tag{7.289}$$

The five columns of W_F are the five '*Fisherfaces*'. Based on this matrix, a Fisher face space can be obtained by projection. This will be a set of 54 vectors, each with 5 components. Taking the two most significant components of each vector, it is possible to depict a scatterplot that is most interesting, since LDA aimed to get data clusters (corresponding to classes) maximally separated.

In general, the Fisherfaces method is not suitable for face reconstruction after the analysis process. It is more oriented to discrimination.

A program was developed as an example of Fisherfaces analysis (Program 7.44). Four figures were generated. The fist one, Fig. 7.107, shows the 5 Fisherfaces corresponding to the same example studied in the previous subsection. Each of these images contains traits that are relevant for face discrimination.

As before, a set of 6 known faces has been used for testing. Figure 7.108 shows the distances in the Fisher face space.

Similarly, a set of 6 unknow faces has also been used for testing. Figure 7.109 shows the distances in the Fisher face space.

Finally, Fig. 7.110 shows the data clusters obtained with the two most significant components of each vector in the Fisher face space. The clusters corresponds to the 6 persons. From inspection of this figure, one could imagine why some faces in our training set were easy to discern, while others might be difficult for neat identification.

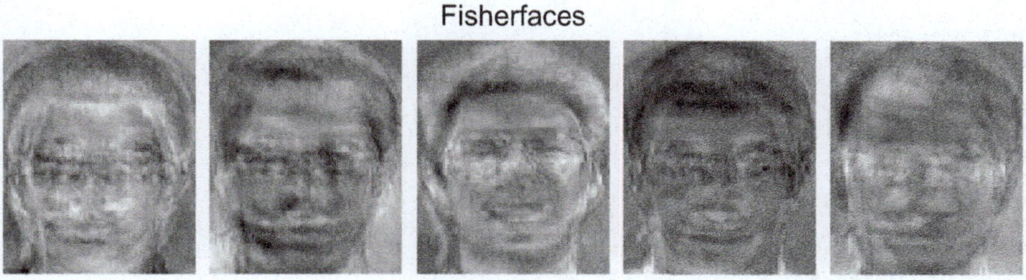

Fig. 7.107 Fisherfaces

Fig. 7.108 Distances for known faces

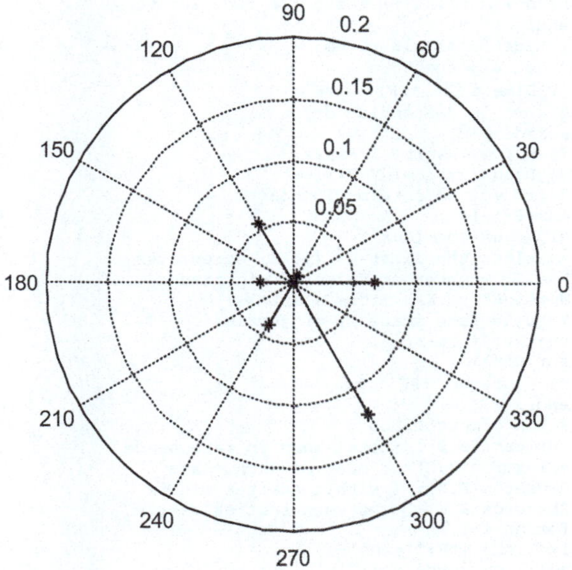

Like in other programs, the coding was not optimized and, in certain parts, it certainly can be done in a better way, but we preferred clarity.

Program 7.44 example of Fisherfaces for recognition

```
% example of Fisherfaces for recognition
%read face database (120 faces, of 112x92=10304 length each)
AA=zeros(10304,120); %faces are columns of the matrix
fid=fopen('AAface.bin','r'); aux=fread(fid);
AA=reshape(aux,10304,120);
% will choose 6x9 faces of 92x112 size
W=92; H=112;
C=6; %number of classes (6 people)
R=9; %number of faces per class
% faces are contained in matrix "AA"
% Preparing faces ------------------------------
% form matrix with image columns
a=C*R; Or=zeros(H*W,a); %54 columns
nn=41;ns=1; %select 60 faces
for nc=1:a,
aux=AA(:,nn);
```

```
Or(:,nc)=aux;
if ns==9 %skip next face
ns=1; nn=nn+2;
else
ns=ns+1; nn=nn+1;
end;
end;
% get average face
MF=mean(Or,2);
% subtract average face from each face
% and normalize faces
PF=zeros(H*W,a); nZ=zeros(1,a);
for nc=1:a,
aux=Or(:,nc)-MF;
PF(:,nc)=aux/norm(aux);
nZ(nc)=norm(aux); %keep for face recovery
end;
% Eigenfaces algorithm ----------------------------
% get eigenvectors
[V D]=eig(PF'*PF);
% sort in descending order
aux=diag(D);
[aux2, sx]=sort(-1.*aux);
eigL=aux(sx); eigV=V(:,sx);
% project to obtain eigenfaces
aux=PF*eigV;
eigF=aux/norm(aux);
% Select the first 48 (54-6) eigenfaces
K=a-C; %number of selected eigenfaces
Wp=eigF(:,1:K); %form a matrix
% build face space by projection
fs=zeros(K,a);
for nf=1:a,
fs(:,nf)=Wp'*PF(:,nf);
end;
% LDA computations
mu=mean(fs,2); %total mean in face space
mcl=zeros(K,C); % mean of each class
Sw=zeros(K,K); % within scatter matrix
Sb=zeros(K,K); % between scatter matrix
for nn=1:C,
ix=((nn-1)*R+1):(nn*R);
mcl(:,nn)=mean(fs(:,ix),2);
aux=zeros(K,K);
for j=ix,
aux=aux+((fs(:,j)-mcl(:,nn))*(fs(:,j)-mcl(:,nn))');
end;
Sw=Sw+aux;
Sb=Sb+((mcl(:,nn)-mu)*(mcl(:,nn)-mu)');
end;
% solve general eigenvalue problem
[Wx D]=eig(Sb,Sw);
% sort in descending order
aux=diag(real(D));
[aux2, sx]=sort(-1.*aux);
Waux=real(Wx(:,sx));
Wl=Waux(:,1:(C-1)); % select C-1 largest eigenvectors
WF=Wp*Wl; %Fisherfaces
% show Fisherfaces
figure(1)
for nn=1:C-1,
subplot(1,5,nn)
FF=reshape(WF(:,nn),H,W);
imshow(FF,[]);
end;
title('Fisherfaces');
% Testing --------------------------------
% build (Fisher) face space by projection
```

```
Ffs=zeros(C-1,a);
for nf=1:a,
Ffs(:,nf)=WF'*PF(:,nf);
end;
% reconstruction
RF=zeros(H*W,a);
for nn=1:a,
aux=WF*Ffs(:,nn); %reconstruction
rf=aux/norm(aux);
RF(:,nn)=rf; %save reconstructed faces
end;
% plot face space clusters, taking the two larger components
figure(2)
for nf=1:a,
aux1=abs(Ffs(:,nf));
[aux2,ix]=sort(aux1);
aux=Ffs(ix,nf);
if nf<10, plot(aux(5),aux(4),'ko'); hold on; end;
if (nf>9 & nf<19),plot(aux(5),aux(4),'bo'); end;
if (nf>18 & nf<28),plot(aux(5),aux(4),'ro'); end;
if (nf>27 & nf<37),plot(aux(5),aux(4),'go'); end;
if (nf>36 & nf<46),plot(aux(5),aux(4),'mo'); end;
if (nf>45 & nf<55),plot(aux(5),aux(4),'co'); end;
end;
axis([-0.2 0.2 -0.15 0.15]);
title('clusters in Fisher face space');
xlabel('component 1');
ylabel('component 2');
%------------------------------------------------------
% test the 10th face for each row
% (these 6 faces have not been used during training)
% extract faces
to=zeros(H*W,6);
to(:,1)=AA(:,50); to(:,2)=AA(:,60); to(:,3)=AA(:,70);
to(:,4)=AA(:,80);to(:,5)=AA(:,90); to(:,6)=AA(:,100);
numfig=3;
ANfish(H,W,C,to,MF,WF,a,Ffs,numfig);
%------------------------------------------------------------
% test 6 new faces, corresponding to new people
% not yet considered
% extract faces
to=zeros(H*W,6);
to(:,1)=AA(:,9); to(:,2)=AA(:,19); to(:,3)=AA(:,29);
to(:,4)=AA(:,39);to(:,5)=AA(:,109); to(:,6)=AA(:,119);
numfig=4;
ANfish(H,W,C,to,MF,WF,a,Ffs,numfig);
```

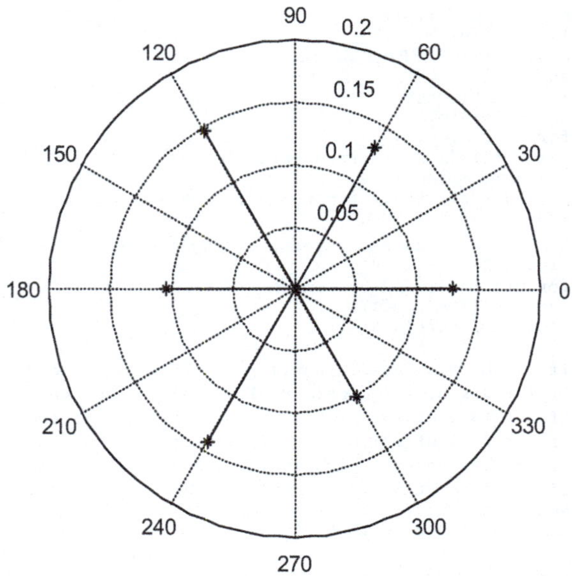

Fig. 7.109 Distances for unknown faces

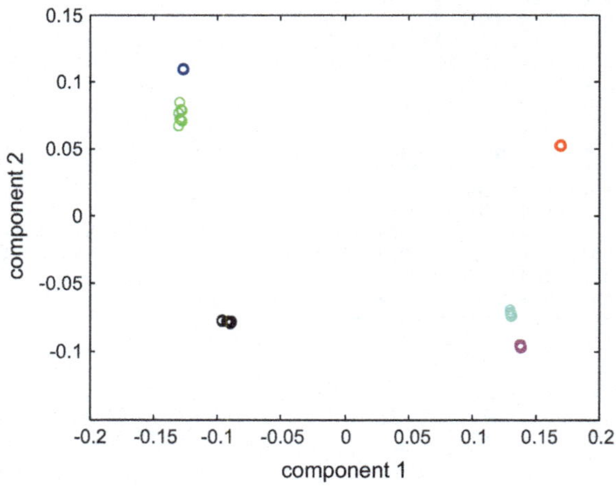

Fig. 7.110 Clusters in (Fisher) face space

Function 7.45 FunctionANeigf()

```
% test 6 faces after Fisherface analysis
function ANfish(H,W,C,to,MF,WF,a,Ffs,numfig);
tp=zeros(H*W,6); tfs=zeros(C-1,6);
% prepare faces and project onto face space
for nt=1:6,
aux=to(:,nt)-MF; %subtract average face and normalize
tp(:,nt)=aux/norm(aux);
```

```
tfs(:,nt)=WF'*tp(:,nt); %project onto face space
end;
% compute minimum distance respect established faces
% in face space
Tmd=zeros(6,1); Tix=zeros(6,1);
for nt=1:6,
ips=100; mdn=1;
for ni=1:a,
aux=norm(tfs(:,nt)-Ffs(:,ni));
if aux<ips, ips=aux; mdn=ni; end;
end;
Tmd(nt)=ips; Tix(nt)=mdn; %results for 1 of 6 faces
end;
% display minimum distances in face space
figure(numfig)
polar(0,0.2,'y'); hold on;
for nt=1:6,
ag=((nt-1)*2*pi)/6; %angle
polar([ag ag],[0 Tmd(nt)],'k*-');
end;
title('distances in face space: 6 faces');
```

Several papers have compared Eigenfaces and Fisherfaces methods [187, 275]. The Fisherfaces algorithm is described in [212], and implementation examples can be found in [310] and in the MATLAB Central file exchange site (contribution of A. Omidvarnia). An interesting procedure has been proposed in [277] for face recognition from a single image per person.

7.9.2 Color Reduction Using K-Means

As you recall from a previous chapter, RGB color images assign three 8-bit unsigned integers to each pixel. In the next example a color reduction scheme would be tried. The idea is to identify a palette of 16 colors, and re-paint the target image using only these 16 colors. Based on this, data can be enormously compressed, since only 4 bits would be assigned to each pixel (a pointer to a particular color of the palette), instead of 24 bits.

The palette will be identified using K-means. What we called centroids, would be now the colors of the palette. The K-means algorithm iterates two steps: to assign to each pixel its nearest centroid, and then to move the centroids using distance averages (each centroid according with its assigned pixels).

Figure 7.111 presents the picture that has been chosen for our exercise.

The K-means algorithm starts with an initial set of centroids. Figure 7.112 depicts the palette we set for this initial phase. A simple inspection of the picture says that white, blue, black, and other colors would be good candidates for the palette.

Program 7.46 applies K-means for this example, and generates the four figures of this subsection. It calls two times a simple function for palette display.

After some processing time, the program obtains the palette shown in Fig. 7.113.

Based on this palette, a re-painted version of the picture has been generated. The result is shown in Fig. 7.114.

Program 7.46 Color reduction using K-means

```
% Color reduction using K-means
F=imread('houses1','jpg');
[hF,wF,cF]=size(F);
L=zeros(hF,wF); %reserve for labels assigned to pixels
% show original painting
figure(1)
imshow(F,[]);
title('original picture');
% initial centroids
```

Fig. 7.111 Original picture

Fig. 7.112 Initial palette

Fig. 7.113 Palette found by K-means

Fig. 7.114 Re-colored picture

```
C=zeros(3,16);
C(:,1)=[0.9;0.9;0.9]; C(:,2)=[0.1;0.1;0.1];
C(:,3)=[0.8;0.9;0.3];
C(:,4)=[0.6;0.7;0.6]; C(:,5)=[0.8;0.8;0.8];
C(:,6)=[0.9;0.7;0.7];
C(:,7)=[0.4;0.4;0.5]; C(:,8)=[0.5;0.3;0.2];
C(:,9)=[0.5;0.5;0.2];
C(:,10)=[0.7;0.5;0.5]; C(:,11)=[0.5;0.7;0.5];
C(:,12)=[0.5;0.6;0.6];
C(:,13)=[0.6;0.6;0.6]; C(:,14)=[0.8;0.8;0.95];
C(:,15)=[0.5;0.6;0.3];
C(:,16)=[0.3;0.3;0.3];
SD=zeros(3,16); % for adding distances
ND=zeros(1,16); % counter of set members for each centroid
% show initial palette------------------------------------
[PL]=Palette(C);
figure(2);imshow(PL);
title('initial palette');
disp('wait for next 10 iterations');
%========================================================
```

```
% K-means algorithm
for it=1:10,
%--------------------------------------------------------
% assign centroids to image pixels
D=zeros(1,16);
for nl=1:hF,
for nc=1:wF,
aux1=im2double(F(nl,nc,:)); aux2=squeeze(aux1);
for mm=1:16,
D(mm)=norm(aux2-C(:,mm)); %see distance to centroid mm
end;
[dist,imin]=min(D); % select centroid with shortest distance
L(nl,nc)=imin; % assign label to pixel
SD(:,imin)=SD(:,imin)+aux2; % add values of pixel
ND(imin)=ND(imin)+1; % increase members counter
end;
end;
disp('***');
%--------------------------------------------------------
% move centroids
for nn=1:16,
aux=ND(nn);
if aux>0,
avg=SD(:,nn)/aux;
C(:,nn)=avg;
end;
end;
%--------------------------------------------------------
end;
disp('the end')
% form new palette------------------------------------------
[PL]=Palette(C);
% display re-colored image
RF=im2double(F); %init
for nl=1:hF,
for nc=1:wF,
mm=L(nl,nc); % retrieve centroid number
aux=C(:,mm);
RF(nl,nc,:)=aux;
end;
end;
%--------------------------------------------------------
% show the results of K-means
figure(3)
% show the palette found by the algorithm
imshow(PL);
title('Palette found by K-means');
figure(4)
% show re-colored image
imshow(RF);
title('re-colored picture');
```

Function 7.47 Function Palette()

```
% Show the palette of colors
function [pal]=Palette(C)
```

```
%small square
sqR=zeros(32,32); sqG=zeros(32,32);sqB=zeros(32,32);
%large square
WW=4*32;
pltR=zeros(WW,WW); pltG=zeros(WW,WW);pltB=zeros(WW,WW);
pal=zeros(WW,WW,3);
nc=1; nl=1; % line and column counters
% square filling
for nsq=1:16,
for i=1:32,
for j=1:32,
sqR(i,j)=C(1,nsq); sqG(i,j)=C(2,nsq); sqB(i,j)=C(3,nsq);
end;
end;
% pointers:
bl=1+((nl-1)*32); el=bl+31; il=bl:el;
bc=1+((nc-1)*32); ec=bc+31; ic=bc:ec;
% putting small square in large square
pltR(il,ic)=sqR; pltG(il,ic)=sqG; pltB(il,ic)=sqB;
% counters update
if nc==4,
nc=1; nl=nl+1;
else
nc=nc+1;
end;
end;
pal(:,:,1)=pltR; pal(:,:,2)=pltG; pal(:,:,3)=pltB;
```

Since the results of K-means depend on initial conditions, probably better results could be attained (see examples from Mathworks). Anyway, it was interesting to propose such an experiment. Similar activities could be also addressed to color quantization [54], image segmentation [215], etc.

7.10 Some Pointers to Related Topics

Probably the reader would like to know more on the topics related to this chapter. This last, short section, is devoted to bring some pointers that may help to start personal explorations.

In certain applications is becoming popular to use the Karhunen-Loeve transform for signal decorrelation and image processing [193]. As explained in a number of publications, like for instance [53], this transform is closely related to PCA. Other aspects of interest of the mentioned article, [53], is that it includes a review and an application of component analysis for electrocardiogram signal processing.

Factor analysis, [114, 233], is a widely used statistical technique, with a long tradition in economical, social, medical, etc. research fields. It is based on a statistical model that predicts observed variables from theoretical latent factors. The relations between factor analysis and PCA are discussed in [71] and many other publications.

Although methodologies related to sparse representations will be treated in the last chapter of this book, it seems opportune to mention now some of these methods. In the case of conventional PCA, the principal components are frequently linear combinations of all original variables; the idea of '*Sparse PCA*' is to find linear combinations that use much less variables, [334]. Several approaches for obtaining sparse PCA combinations have been introduced, like in [75, 279]; see [291] for a survey of methods. A structured extension of sparse PCA is proposed in [154], including experiments for face recognition and for the study of the dynamics of a protein complex. A most cited article on face recognition via sparse representation is [319].

The '*Sparse ICA*' method has been introduced in [333] for blind source separation. An interesting application example is [43] for blind separation of transmitted and reflected images.

In addition, a '*Sparse Discriminant Analysis*' has also been proposed, [64]. An interesting application for breast cancer biomarker identification and classification is described in [280]. Likewise, a "*Sparse SVM*" was introduced in [29]; an application example is provided by [302] concerning hyperspectral sensing and aerial image classification.

More sparse versions of the methods described in this chapter have been proposed. However, let us stop here on this aspect. Instead, we will turn now to more general contexts, namely pattern recognition, machine learning and artificial intelligence.

Pattern recognition is constantly present in our daily life: when we maintain a conversation, when we read something, when we recognize a traffic sign, etc. It is an important objective for new technologies to incorporate recognition abilities.

Most books on pattern recognition, like for instance [31, 82, 87, 278, 295], include the topics contemplated in the present chapter. The same should be said about the MATLAB Statistical Pattern Recognition Toolbox, which contains many illustrative examples.

Quite a lot of pattern recognition methods and applications have been published. Let us cite just some reviews and illustrative examples related to the topics covered in this chapter. A survey of pattern recognition research between 1968 and 1974 is provided by [161]. A classical book on pattern recognition with neural networks is [30]. The review presented in [8] on speech recognition by machine includes a fairly detailed history of the research on this aspect, and lists applications of vector quantization, SVM, neural networks, etc. Some experimental results of SVM application for speech recognition are given in [108]. A comprehensive survey of handwriting recognition is [256]. In [208], a specific neural network is applied for handwriting recognition. The doctoral Thesis [98] deals with traffic and road sign recognition, with emphasis on SVM. Another approach for sign recognition by intelligent vehicles is introduced in [77], using neural networks. The use of K-means for learning feature representations is suggested in [65]. Indeed, distance measures are important for pattern recognition, [58].

Along our life we learned to eat, to walk, to speak, …; then we learned geography, history, mathematics, etc.; then we learned (or perhaps not) to ride a bike, to drive a car,…, even to teach. Learning has many facets. It exerts our main faculties: reasoning, memory, senses, pattern recognition, etc. When trying to develop machine learning, one has to use adaptivity, plasticity, discrimination, perhaps functional hierarchies, etc. It is not a surprise that neural networks and adaptive schemes constitute main pillars of machine learning theory. In more or less extent, the contents of the present chapter is also considered in machine intelligence books, [127, 235]. With respect to published applications, there are many related to medical diagnosis, like the Thesis [120] on magnetic resonance imaging analysis, based on ICA, Gaussian processes, and SVM. A tutorial overview of machine learning classifiers and functional magnetic resonance is provided by [253]. In other non-medical contexts, two examples of '*deep learning*' studies are [61, 292], where deep means to use several layers, either with neural networks or with SVMs; likewise, the paper [330] introduces a deep neural SVM for speech recognition. A tutorial survey of deep learning is given by [79].

Learning and adaptation may focus on obtaining suitable parameters of conventional algorithms. For instance, to learn the k in *K-means*, [125], or for confidence weighted linear classification [86].

Thanks to a series of fundamental contributions, main constituents of the artificial intelligence domain, nowadays, are expert systems, fuzzy logic, neural networks, and biologically-inspired heuristic methods. Many combinations of these constituents and data analysis/classification methods have been investigated. For instance, the genetic K-means algorithm proposed by [180], the fuzzy SVMs introduced in [200], or the boosted mixture of experts suggested by [10].

Turning now the attention towards the Bayesian context, the research has proposed several new versions of data analysis/classification methods. In [63] a Bayesian ICA is introduced for blind source separation; in [183], some Bayesian variations of K-means are suggested. A most cited article is [297] on sparse Bayesian learning and the relevance SVM. A Bayesian technique for face recognition is proposed by [226].

Let us look, briefly, at another important domain. Brain research has a number of objectives, including medical aspects, or brain-computer interfacing, etc. A vast number of publications refer to brain data analysis/classification, where the methods explained in this chapter have a significant role. As already mentioned in previous sections, there are image segmentation (IMS) applications based on clustering methods, like for example [318] for dynamic PET images, or [230] for Alzheimer detection. See [229] for a review of machine learning for brain IMS. Other most cited reviews are [255] on medical IMS, and [27, 316] on magnetic resonance IMS. The possibility of decoding mental states is suggested in [129]. There is a number of frequently cited articles on the use of electroencephalographic signals for brain-computer interfacing (BCI), like for instance [271, 317]. This interfacing can be employed for the control of robots or prosthetic devices, [94, 136, 293]. A review of classification algorithms for EEG-based BCI is provided by [211]. In many cases, component analysis is applied for event-detection. There is a lot of interest on BCI, indicated by the proliferation of MATLAB Toolboxes and companies offering BCI products.

7.11 Resources

7.11.1 MATLAB

7.11.1.1 Toolboxes

- ICALAB Toolboxes:
  http://www.bsp.brain.riken.jp/ICALAB/
- LYCIA Toolbox:
  http://mlsp.umbc.edu/lycia/lycia.html
- MISEP Linear and Nonlinear ICA Toolbox:
  http://www.lx.it.pt/~lbalmeida/ica/mitoolbox.html
- Toolbox for separation of convolutive mixtures:
  http://www-public.int-evry.fr/~castella/toolbox/
- Toolbox for dimensionality reduction:
  http://homepage.tudelft.nl/19j49/Matlab_Toolbox_for_Dimensionality_Reduction.html
- Statistical pattern recognition toolbox for MATLAB:
  http://cmp.felk.cvut.cz/cmp/software/stprtool/
- Alaa Tharwat Toolbox:
  http://www.mathworks.com/matlabcentral/fileexchange/23315-alaa-tharwat-toolbox/
- MATLAB SVM Toolbox:
  https://www.uea.ac.uk/computing/matlab-svm-toolbox
- LS-SVMlab:
  http://www.esat.kuleuven.be/sista/lssvmlab/
- Classification Toolbox:
  http://matlab-classification-toolbox.soft112.com/
- Clustering Toolbox:
  http://www.mathworks.com/matlabcentral/fileexchange/7486-clustering-toolbox
- Fuzzy Clustering and Data Analysis Toolbox:
  http://www.abonyilab.com/software-and-data/fclusttoolbox
- SOM Toolbox:
  http://www.cis.hut.fi/somtoolbox/package/docs2/somtoolbox.html
- Machine Learning Toolbox:
  http://mirlab.org/jang/matlab/toolbox/machineLearning/
- Matlab MLTOOLS Toolbox (machine learning):
  http://ml.sheffield.ac.uk/people/N.Lawrence/mltools/
- DACE (Kriging Toolbox):
  http://www.imm.dtu.dk/~hbni/dace/dace.pdf
- ooDACE:
  http://www.sumo.intec.ugent.be/ooDACE
- Matlab Kriging Toolbox:
  http://globec.whoi.edu/software/kriging/V3/english.html

- STK (Kriging Toolbox):
  http://sourceforge.net/projects/kriging/
- BMElib (Kriging, etc.):
  http://www.unc.edu/depts/case/BMELIB/
- EasyKriging:
  ftp://ftp.cmarz.org/pub/software/kriging/easy_krig/V1.0/README/easy_krig.html
- SaGA (MIT: Kriging, etc.):
  http://puddle.mit.edu/~glenn/kirill/saga.html

7.11.1.2 Matlab Code

Dimensionality reduction:

- Deng's Cai:
  http://www.cad.zju.edu.cn/home/dengcai/Data/
  http://www.DimensionReduction.html

About ICA:

- FastICA:
  http://research.ics.aalto.fi/ica/fastica/code/dlcode.shtml
- P. Comon source codes:
  http://www.gipsa-lab.grenoble-inp.fr/~pierre.comon/matlab.html
- ICA and natural images:
  http://research.ics.aalto.fi/ica/imageica/
- An implementation of JADE:
  http://perso.telecom-paristech.fr/~cardoso/guidesepsou.html
- Sam Roweis page, ICA, PCA, etc.:
  http://www.cs.nyu.edu/~roweis/code.html
- ICA-CNL
  http://cnl.salk.edu/~tewon/ica_cnl.html
- EFICA
  http://itakura.ite.tul.cz/zbynek/efica.htm
- RADICAL ICA
  http://people.cs.umass.edu/~elm/ICA/
- C-FICA (convolutive ICA)
  http://www.ast.obs-mip.fr/article595.html

About discrimination and clusters:

- LIBSVM:
  http://www.csie.ntu.edu.tw/~cjlin/libsvm/
- SVM and Kernel Methods:
  http://asi.insa-rouen.fr/enseignants/~arakoto/toolbox/

- Kernel LDA:
  https://research.cs.washington.edu/istc/lfb/software/FS-KFDA.htm
- Handwriting recognition with kernel LDA:
  http://www.codeproject.com/Articles/74348/Handwriting-Recognition-using-Kernel-Discriminant

Images and signals:

- Natural image collection for ICA experiments:
  http://www.cs.helsinki.fi/u/phoyer/NCimages.html
- Some trials of blind source separation:
  http://www.ism.ac.jp/~shiro/research/blindsep.html

Classification and probabilities:

- Bayesian regression:
  http://www.jim-stone.staff.shef.ac.uk/BookBayes2012/BayesRuleMatlab Code.html
  and:
  www.cs.ru.nl/~ali/index_files/bayesian_regression_example.m
- Variograms:
  http://www.colorado.edu/geography/class_homepages/geog_4023_s07/labs/
- Kriging:
  http://www-math.bgsu.edu/~zirbel/rfim/
- Gaussian Processes:
  http://mrmartin.net/?p=223

7.11.2 Internet

7.11.2.1 Web Sites

The web site of V. Kepuska, Florida Inst. Tech., contains extensive material related to signal and speech processing:

http://www.my.fit.edu/~vkepuska/ece5525/

See the Paolo D'Incau's blog for watermarking code and details.

It is also interesting to visit the web site of the PHYDYAS Project on cognitive radio.

- Hyvärinen, A.:
  http://www.cs.helsinki.fi/u/ahyvarin/
- James V. Stone:
  http://www.jim-stone.staff.shef.ac.uk/
- Jung, T-P:
  http://sccn.ucsd.edu/~jung/Site/Home.html

- Petr Tichavski:
  http://si.utia.cas.cz/downloadPT.htm
- Tutorial on ICA:
  http://cis.legacy.ics.tkk.fi/aapo/papers/IJCNN99_tutorialweb/
- ICA central:
  http://perso.telecom-paristech.fr/~cardoso/icacentral/
- An ICA page:
  http://cnl.salk.edu/~tony/ica.html
- LVA Central:
  http://lvacentral.inria.fr/tiki-index.php
- SISEC 2013 (signal separation campaigns):
  http://sisec.wiki.irisa.fr/tiki-index.php
- Spatial Statistics Links:
  http://www.spatial-statistics.com/spatial_links_index.htm
- Gaussian Processes:
  http://www.gaussianprocess.org
- D. Duvenaud:
  http://mlg.eng.cam.ac.uk/duvenaud/cookbook/index.html
- Neural Network Simulators:
  https://grey.colorado.edu/emergent/index.php/Comparison_of_Neural_Network_Simulators
- The ORL database of faces, AT&T Laboratories Cambridge:
  http://www.cl.cam.ac.uk/research/dtg/attarchive/facedatabase.html
- Face recognition homepage:
  http://www.face-rec.org/
- Face detection homepage:
  http://www.facedetection.com/
- Face databases, MIT:
  http://web.mit.edu/emeyers/www/face_databases.html

7.11.2.2 Links

There is a web page with links to SVM tutorials:
www.svms.org/tutorials

- For a list of links to PCA, etc., MATLAB codes see:
  www.cad.zju.edu.cn/.../DimensionReduction.html
- ICA link collection:
  http://research.ics.aalto.fi/ica/book/links.html
- BSP-GRID, Downloads (links):
  http://bsp.teithe.gr/members/downloads.html
- For a list of links to SVM code, see:
  http://www.svms.org/software.html
  and:
  http://www.support-vector.net/software.html
  and:
  http://www.support-vector-machines.org/SVM_soft.html
- Signals and data for ICA (links):
  http://perso.telecom-paristech.fr/~cardoso/icacentral/base_multi.html
- Data for ICA (links):
  http://www.tech.plym.ac.uk/spmc/links/ica/ica_data.html
- Paris page (links):
  http://web.media.mit.edu/~paris/ica.html

References

1. D.P. Acharya, G. Panda, A review of independent component analysis techniques. IETE Tech. Rev. **25**(6), 320–332 (2008)
2. C.C. Aggarwal, C.K. Reddy, *Data Clustering: Algorithms and Applications* (Chapman and Hall, 2013)
3. C.C. Aggarwal, (ed.), *Data Classification: Algorithms and Applications* (CRC Press, 2015)
4. S.I. Amari, Natural gradient works efficiently in learning. Neural Comput. **10**(2), 251–276 (1998)
5. A. Andoni, Nearest neighbor search: the old, the new, and the impossible. Ph.D. thesis, MIT, 2009
6. S. Andrews, I. Tsochantaridis, T. Hofmann, Support vector machines for multiple-instance learning, in Advances in Neural Information Processing Systems, pp. 561–568 (2002)
7. P. Anjali, S. Ajay, S.D. Sapre, A review on natural image denoising using independent component analysis (ica) technique. Adv. Comput. Res. **2**(1), 06–14 (2010)
8. M.A. Anusuya, S.K. Katti, *Speech Recognition by Machine, a Review* (Department of Computer Science and Engineering Sri Jayachamarajendra College of Engineering Mysore, India, 2010). arXiv preprint arXiv:1001.2267
9. S. Arya, D.M. Mount, N.S. Netanyahu, R. Silverman, A.Y. Wu, An optimal algorithm for approximate nearest neighbor searching fixed dimensions. J. ACM **45**(6), 891–923 (1998)
10. R. Avnimelech, N. Intrator, Boosted mixture of experts: an ensemble learning scheme. Neural Comput. **11**(2), 483–497 (1999)

11. S. Ayache, G. Quénot, J. Gensel, Classifier fusion for SVM-based multimedia semantic indexing, in *Advances in Information Retrieval*, pp. 494–504 (Springer, 2007)
12. F.R. Bach, M.I. Jordan, Kernel independent component analysis. J. Mach. Learn. Res. **3**, 1–48 (2002)
13. B. Bahmani, B. Moseley, A. Vattani, R. Kumar, S. Vassilvitskii, Scalable k-means++. Proc. VLDB Endowment **5**(7), 622–633 (2012)
14. S. Balakrishnama, A. Ganapathiraju, *Linear Discriminant Analysis-A Brief Tutorial* (Institute for Signal and information Processing, Dept. Electrical and Computer Engineering, Mississippi State University, 1998). https://www.researchgate.net/publication/240093048_Linear_Discriminant_Analysis-A_Brief_Tutorial
15. M.S. Bartlett, J.R. Movellan, T.J. Sejnowski, Face recognition by independent component analysis. IEEE T. Neural Netw. **13**(6), 1450–1464 (2002)
16. M. Basseville, *Divergence Measures for Statistical Data Processing* (HAL, INRIA, France, 2010). http://hal.inria.fr/docs/00/54/23/37/PDF/PI-1961.pdf
17. G. Baudat, F. Anouar, Generalized discriminant analysis using a Kernel approach. Neural Comput. **12**(10), 2385–2404 (2000)
18. S. Bauer, S. Köhler, K. Doll, U. Brunsmann, FPGA-GPU architecture for kernel SVM pedestrian detection, in *Proceedings IEEE Computer Vision and Pattern Recognition Workshops (CVPRW)*, pp. 61–68 (2010)
19. M.J. Beal, Variational algorithms for approximate Bayesian inference. Ph.D. thesis, University of London, 2003
20. P.N. Belhumeur, J.P. Hespanha, D. Kriegman, Eigenfaces vs. fisherfaces: recognition using class specific linear projection. IEEE Trans. Pattern Anal. Mach. Intell. **19**(7), 711–720 (1997)
21. A.J. Bell, T.J. Sejnowski, An information-maximization approach to blind separation and blind deconvolution. Neural Comput. **7**, 1129–1159 (1995)
22. A.J. Bell, T.J. Sejnowski, The "independent components" of natural scenes are edge filters. Vision Res. **37**(23), 3327–3338 (1997)
23. A. Ben-Hur, J. Weston, A user's guide to support vector machines, in *Data Mining Techniques for the Life Sciences*, pp. 223–239 (Humana Press, 2010)
24. K.P. Bennett, C. Campbell, Support vector machines: hype or hallelujah? ACM SIGKDD Explor. Newslett. **2**(2), 1–13 (2000)
25. P. Berkhin, A survey of clustering data mining techniques, in *Grouping Multidimensional Data*, pp. 25–71 (Springer, 2006)
26. J.M. Bernardo, M.J. Bayarri, J.O. Berger, A.P. Dawid, D. Heckerman, A.F.M. Smith, M. West, The variational Bayesian EM algorithm for incomplete data: with application to scoring graphical model structures. Bayesian Stat. **7**, 453–464 (2003)
27. J.C. Bezdek, L.O. Hall, L. Clarke, Review of MR image segmentation techniques using pattern recognition. Med. Phys. **20**(4), 1033–1048 (1992)
28. N. Bhatia, *Survey of Nearest Neighbor Techniques* (Department of Computer Science DAV College Jalandhar, India, 2010). arXiv preprint arXiv:1007.0085
29. J. Bi, K. Bennett, M. Embrechts, C. Breneman, M. Song, Dimensionality reduction via sparse support vector machines. J. Mach. Learn. Res. **3**, 1229–1243 (2003)
30. C.M. Bishop, *Neural Networks for Pattern Recognition* (Clarendon Press, Oxford, 1995)
31. C.M. Bishop, *Pattern Recognition and Machine Learning* (Springer Verlag, 2010)
32. R. Blahut, *Principles and Practices of Information Theory* (Addison-Wesley, 1987)
33. T. Blaschke, L. Wiskott, Cubica: independent component analysis by simultaneous third- and fourth-order cumulant diagonalization. IEEE T. Sign Process. **52**(5), 1250–1256 (2004)
34. M. Blum, M. Riedmiller, Optimization of Gaussian process hyperparameters using Rprop, in *Proceedings European Symposium on Artificial Neural Networks, Computational Intelligence and Machine Learning, 2013*, pp. 1–6
35. G. Bohling, Introduction to geostatistics and variogram analysis. Kansas Geol. Survey 1–20 (2005)
36. G. Bohling, Kriging. *C&PE 940*, 2005. http://people.ku.edu/~gbohling/cpe940/Kriging.pdf

37. S. Bose, A. Pal, R. SahaRay, J. Nayak, Generalized quadratic discriminant analysis. Pattern Recogn. **48**(8), 2676–2684 (2015)
38. L. Bottou, C.J. Lin, Support vector machine solvers, eds. by L. Bottou, O. Chapelle, D. DeCoste, J. Weston. *Large Scale Kernel Machines*, pp. 1–17 (MIT Press, 2007)
39. P. Boyle, Gaussian processes for regression and optimisation. Ph.D. thesis, Victoria University of Wellington, 2007
40. S. Bratieres, N. Quadrianto, Z. Ghahramani, *Bayesian Structured Prediction Using Gaussian Processes* (Department of Engineering, University of Cambridge, 2013). arXiv preprint arXiv:1307.3846
41. L. Breiman, Bagging predictors. Mach. Learn. **24**(2), 123–140 (1996)
42. L. Breiman, Random forests. Mach. Learn. **45**(1), 5–32 (2001)
43. A.M. Bronstein, M.M. Bronstein, M. Zibulevsky, Y.Y. Zeevi, Sparse ICA for blind separation of transmitted and reflected images. Int. J. Imag. Syst. Technol. **15**(1), 84–91 (2005)
44. C.J. Burges, A tutorial on support vector machines for pattern recognition. Data Min. Knowl. Disc. **2**(2), 121–167 (1998)
45. H. Byun, S.W. Lee, Applications of support vector machines for pattern recognition: a survey, in *Pattern Recognition with Support Vector Machines*, pp. 213–236 (Springer, 2002)
46. D. Cai, X. He, J. Han, Speed up kernel discriminant analysis. The VLDB J. **20**(1), 21–33 (2011)
47. N. Cancedda, E. Gaussier, C. Goutte, J.M. Renders, Word sequence kernels. J. Mach. Learn. Res. **3**, 1059–1082 (2003)
48. J.F. Cardoso, Infomax and maximum likelihood for blind source separation. IEEE Sign. Process. Lett. **4**, 109–111 (1997)
49. J.F. Cardoso, High-order contrasts for independent component analysis. Neural Comput. **11**, 157–192 (1999)
50. J.F. Cardoso, B. Laheld, Equivariant adaptive source separation. IEEE T. Sign. Process. **45**(2), 434–444 (1996)
51. J.F. Cardoso, A. Souloumiac, Blind beamforming for non Gaussian signals. IEE Proc.-F **140**, 362–370 (1993)
52. C. Carson, S. Belongie, H. Greenspan, J. Malik, Blobworld: image segmentation using expectation-maximization and its application to image querying. IEEE T. Pattern Anal. Mach. Intell. **24**(8), 1026–1038 (2002)
53. F. Castells, P. Laguna, L. Sörnmo, A. Bollmann, J.M. Roig, Principal component analysis in ECG signal processing. EURASIP J. Appl. Sign. Process. **2007**(1), 1–21 (2007)
54. M.E. Celebi, Improving the performance of k-means for color quantization. Image Vision Comput. **29**(4), 260–271 (2011)
55. A.B. Chan, N. Vasconcelos, Counting people with low-level features and Bayesian regression. IEEE T. Image Process. **21**(4), 2160–2177 (2012)
56. H.P. Chan, D. Wei, M.A. Helvie, B. Sahiner, D.D. Adler, M.M. Goodsitt, N. Petrick, Computer-aided classification of mammographic masses and normal tissue: linear discriminant analysis in texture feature space. Phys. Med. Biol. **40**(5), 857–876 (1995)
57. O. Chapelle, P. Haffner, V.N. Vapnik, Support vector machines for histogram-based image classification. IEEE T. Neural Netw. **10**(5), 1055–1064 (1999)
58. C.H. Chen, On information an distance measures, error bounds, and feature selection. Inf. Sci. **10**(2), 159–173 (1976)
59. C.W. Chen, J. Luo, K.J. Parker, Image segmentation via adaptive k-mean clustering and knowledge-based morphological operations with biomedical applications. IEEE Trans. Image Process. **7**(12), 1673–1683 (1998)
60. C.Y. Chiu, Y.F. Chen, I. Kuo, H.C. Ku, An intelligent market segmentation system using k-means and particle swarm optimization. Expert Syst. Appl. **36**(3), 4558–4565 (2009)
61. Y. Cho, L.K. Saul, Kernel methods for deep learning, in *NIPS Proceedings: Advances in Neural Information Processing Systems*, pp. 342–350 (2009)
62. S. Choi, A. Cichocki, H.M. Park, S.Y. Lee, Blind source separation and independent component analysis: a review. Neural Inf. Process.-Lett. Rev. **6**(1), 1–57 (2005)

63. A. Choudrey, S.J. Roberts, Flexible Bayesian independent component analysis for blind source separation, in *Proceedings International Conference on Independent Component Analysis and Signal Separation, (ICA2001)*, pp. 90–95 (2001)
64. L. Clemmensen, T. Hastie, D. Witten, B. Ersbøll, Sparse discriminant analysis. Technometrics **53**(4), 1–25 (2011)
65. A. Coates, A.Y. Ng, Learning feature representations with k-means, in *Neural Networks: Tricks of the Trade*, pp. 561–580 (Springer Berlin Heidelberg, 2012)
66. S. Cohen, R. Ben-Ari. Image de-noising by Bayesian regression, in *Proceedings Image Analysis and Processing, ICIAP 2011*, pp. 19–28 (Springer Berlin Heidelberg, 2011)
67. P. Comon, Independent component analysis, a new concept? Sign. Process. **36**(3), 287–314 (1994)
68. D. Cook, A. Buja, J. Cabrera, C. Hurley, Grand tour and projection pursuit. J. Comput. Graph. Stat. **4**(3), 155–172 (1995)
69. G. Coombe, An introduction to principal component analysis and online singular value decomposition. Ph.D. thesis, Dept. of Computer Science, University of North Carolina, 1993
70. C. Cortes, V. Vapnik, Support-vector network. Mach. Learn. **20**, 1–25 (1995)
71. A.B. Costello, J. Osborno, Best practices in exploratory factor analysis: four recommendations for getting the most from your analysis. Pract. Assess. Res. Eval. **10**(7) (2005). http://pareonline.net/getvn.asp?v=10&n=7
72. N. Cristianini, *Kernel Methods for General Pattern Analysis* (Lecture Presentation, University of California at Davis, 2004). http://www.kernel-methods.net/tutorials/KMtalk.pdf
73. N. Cristianini, J. Shawe-Taylor, *An Introduction to Support Vector Machines* (Cambridge University Press, 2000)
74. D.R. Cutting, D.R. Karger, J.O. Pedersen, J.W. Tukey, Scatter/gather: a cluster-based approach to browsing large document collections, in *Proceedings of the 15th Annual International ACM SIGIR Conference on Research and Development in Information Retrieval, ACM*, pp. 318–329 (1992)
75. A. d'Aspremont, L. El Ghaoui, M.I. Jordan, G.R. Lanckriet, A direct formulation for sparse PCA using semidefinite programming. SIAM Rev. **49**(3), 434–448 (2007)
76. J. Dauwels, K. Srinivasan, M. Ramasubba Reddy, T. Musha, F.B. Vialatte, C. Latchoumane, A. Cichocki, Slowing and loss of complexity in Alzheimer's EEG: Two sides of the same coin? Int. J. Alzheimer's Disease 1–9 (2011)
77. A. De la Escalera, J.M. Armingol, M. Mata, Traffic sign recognition and analysis for intelligent vehicles. Image Vis. Comput. **21**(3), 247–258 (2003)
78. A.P. Dempster, N.M. Laird, Maximum likelihood from incomplete data via the EM algorithm. J. Roy. Stat. Soc. Ser. R **39**(1), 1–38 (1977)
79. L. Deng, A tutorial survey of architectures, algorithms, and applications for deep learning. APSIPA Trans. Sign. Inf. Process. **3**(e2), 1–29 (2014)
80. D.G.T. Denison, C.C. Holmes, B.K. Mallick, A.F.M. Smith, *Bayesian Methods for Nonlinear Classification and Regression* (Wiley, 2002)
81. W. DeSarbo, A. Ansari, P. Chintagunta, C. Himmelberg, K. Jedidi, R. Johnson, M. Wedel, Representing heterogeneity in consumer response models 1996 choice conference participants. Mark. Lett. **8**(3), 335–348 (1997)
82. L. Devroye, L. Györfi, G. Lugosi, *A Probabilistic Theory of Pattern Recognition*, vol. 31 (Springer Science & Business Media, 2013)
83. M.M. Deza, E. Deza, *Encyclopedia of Distances* (Springer Verlag, 2013)
84. P.M. Dixon, Nearest neighbor methods, in *Encyclopedia of Environmetrics* (Wiley Online Library, 2002)
85. C.B. Do, *Gaussian Processes* (Stanford University, 2007). http://www.see.stanford.edu/materials/aimlcs229/cs229-gp.pdf
86. M. Dredze, K. Crammer, F. Pereira, Confidence-weighted linear classification, in *Proceedings of the 25th ACM International Conference Machine Learning*, pp. 264–271 (2008)
87. R.O. Duda, P.E. Hart, D.G. Stork, *Pattern Classification* (Wiley, 2012)
88. R. Durrett, *Essentials of Stochastic Processes* (Springer, 2012)

89. R. Dybowski, V. Gant, *Clinical Applications of Artificial Neural Networks* (Cambridge University Press, 2007)
90. M. Ebden. *Gaussian Processes for Regression: A Quick Introduction* (Robotics Research Group, University of Oxford, 2008). www.robots.ox.ac.uk/~mebden/reports/GPtutorial.pdf
91. I. El-Naqa, Y. Yang, M.N. Wernick, N.P. Galatsanos, R.M. Nishikawa, A support vector machine approach for detection of microcalcifications. IEEE T. Med. Imag. **21**(12), 1552–1563 (2002)
92. Y. Engel, S. Mannor, R. Meir, Reinforcement learning with Gaussian processes, in *Proceedings of the ACM 22nd International Conference on Machine Learning*, pp. 201–208 (2005)
93. K. Etemad, R. Chellappa, Discriminant analysis for recognition of human face images. JOSA A **14**(8), 1724–1733 (1997)
94. L.A. Farwell, E. Donchin, Talking off the top of your head: toward a mental prosthesis utilizing event-related brain potentials. Electroencephalogr. Clin. Neurophysiol. **70**(6), 510–523 (1988)
95. G.E. Fasshauer, Positive definite kernels: past, present and future. Dolomite Res. Notes Approximation **4**, 21–63 (2011)
96. L. Feng, Speaker recognition. Ph.D. thesis, Technical University of Denmark, DTU, DK-2800 Kgs. Lyngby, Denmark, 2004
97. S. Fiori, Overview of independent component analysis technique with an application to synthetic aperture radar (SAR) imagery processing. Neural Netw. **16**(3–4), 453–467 (2003)
98. H. Fleyeh, Traffic and road sign recognition. Ph.D. thesis, Napier University, 2008
99. Y. Freund, R.E. Schapire, Experiments with a new boosting algorithm, in *Proceedings 13th International Conference Machine Learning*, vol. 96, pp. 148–156 (1996)
100. Y. Freund, R.E. Schapire, A decision-theoretic generalization of on-line learning and an application to boosting. J. Comput. Syst. Sci. **55**(1), 119–139 (1997)
101. Y. Freund, R.E. Schapire, Large margin classification using the perceptron algorithm. Mach. Learn. **37**(3), 277–296 (1999)
102. J.H. Friedman, Regularized discriminant analysis. J. Am. Stat. Assoc. **84**(405), 165–175 (1989)
103. J.H. Friedman, J.W. Tukey, A projection pursuit algorithm for exploratory data analysis. IEEE T. Comput. **23**(9), 881–890 (1974)
104. F. Fukumizu, *Methods with Kernels* (Lecture Presentation, The Institute of Statistical Mathematics, Tokyo, 2008). http://www.ism.ac.jp/~fukumizu/H20_kernel/Kernel_3_methods.pdf
105. M. Galar, A. Fernandez, E. Barrenechea, H. Bustince, F. Herrera, A review on ensembles for the class imbalance problem: Bagging-, boosting-, and hybrid-based approaches. IEEE T. Syst. Man Cybern. Part C: Appl. Rev. **42**(4), 463–484 (2012)
106. M. Gales, *Multi-Layer Perceptrons* (University of Cambridge, 2011). Handout 6, Module 4F10, Engineering Part II B. http://www.mi.eng.cam.ac.uk/~mjfg/local/4F10/lect6.pdf
107. G. Gan, C. Ma, J. Wu, *Data Clustering: Theory, Algorithms, and Applications* (SIAM, 2007)
108. A. Ganapathiraju, J.E. Hamaker, J. Picone, Applications of support vector machines to speech recognition. IEEE T. Sign. Process. **52**(8), 2348–2355 (2004)
109. S.E. Gano, H. Kim, D.E. Brown, Comparison of three surrogate modeling techniques: Datascape, kriging, and second order regression, in *Proceedings 11th AIAA/ISSMO Multidisciplinary Analysis and Optimization Conference*, pp. 1–18 (2006). AIAA-2006–7048 Portsmouth, Virginia
110. T. Gärtner, A survey of kernels for structured data. ACM SIGKDD Explor. Newslett. **5**(1), 49–58 (2003)
111. G. Gelle, M. Colas, C. Serviere, Blind source separation: a tool for rotating machine monitoring by vibrations analysis? J. Sound Vibr. **248**(5), 865–885 (2001)
112. R. Gonzalez Osuna, *Pattern Recognition, Lecture Notes, Course 666* (Texas A&M University, 2014). http://psi.cse.tamu.edu/teaching/lecture_notes/
113. J.M. Górriz, F. Segovia, J. Ramírez, A. Lassl, D. Salas-Gonzalez, GMM based SPECT image classification for the diagnosis of Alzheimer's disease. Appl. Soft Comput. **11**(2), 2313–2325 (2011)
114. R.L. Gorsuch, *Factor Analysis* (Lawrence Erlbaum Associates, 1983)

115. R.M. Gray, D.L. Neuhoff, Quantization. IEEE Trans. Inf. Theory **44**, 2325–2384 (1998)
116. E. Gringarten, C.V. Deutsch, Teacher's aide variogram interpretation and modeling. Math. Geol. **33**(4), 507–534 (2001)
117. Gaithersburg Statistics Group, *NIST/SEMATECH Engineering Statistics Handbook* (NIST Information Technology Lab., 2010)
118. S.R. Gunn, Support vector machines for classification and regression. Technical Report 14, ISIS, 1998
119. S. Günter, N.N. Schraudolph, S.V.N. Vishwanathan, Fast iterative kernel principal component analysis. J. Mach. Learn. Res. **8**, 1893–1918 (2007)
120. C. Guo, Machine learning methods for magnetic resonance imaging analysis. Ph.D. thesis, The University of Michigan, 2012
121. Y. Guo, T. Hastie, R. Tibshirani, Regularized linear discriminant analysis and its application in microarrays. Biostatistics **8**(1), 86–100 (2007)
122. H. Gupta, A.K. Agrawal, T. Pruthi, C. Shekhar, R. Chellappa, An experimental evaluation of linear and kernel-based methods for face recognition, in *Proceedings IEEE 6th Workshop Applications of Computer Vision,(WACV)*, pp. 13–18 (2002)
123. M.R. Gupta, Y. Chen, *Theory and Use of the EM Algorithm* (Now Publishers Inc, 2011)
124. R. Haapanen, A.R. Ek, M.E. Bauer, A.O. Finley, Delineation of forest/nonforest land use classes using nearest neighbor methods. Remote Sens. Environ. **89**(3), 265–271 (2004)
125. G. Hamerly, C. Elkan, Learning the k in k-means. NIPS Proc Adv. Neural Inf. Process. Syst. **16**, 281–288 (2004)
126. C. Hartmann, J. Boedecker, O. Obst, S. Ikemoto, M. Asada, Real-time inverse dynamics learning for musculoskeletal robots based on echo state Gaussian process regression. Robot.: Sci. Syst. (2012)
127. T. Hastie, R. Tibshirani, J. Friedman, *The Elements of Statistical Learning* (Springer, 2013)
128. R.P. Hauser, D. Booth, Predicting bankruptcy with robust logistic regression. J. Data Sci. **9**(4), 565–584 (2011)
129. J.D. Haynes, G. Rees, Decoding mental states from brain activity in humans. Nat. Rev. Neurosci. **7**(7), 523–534 (2006)
130. H. He, W.C. Siu, Single image super-resolution using Gaussian process regression, in *Proceedings IEEE Conference on Computer Vision and Pattern Recognition, (CVPR)*, pp. 449–456 (2011)
131. B. Heisele, P. Ho, T. Poggio, Face recognition with support vector machines: global versus component-based approach. Proc. IEEE Intl. Conf. Comput. Vision **2**, 688–694 (2001)
132. J. Hensman, N. Fusi, N.D. Lawrence, *Gaussian Processes for Big Data* (Dept. Computer Science The University of Sheffield, 2013). arXiv preprint arXiv:1309.6835
133. J. Herault, J. Jutten, Space or time adaptive signal processing by neural network models, ed. by J.S. Denker. *Neural Networks for Computing: AIP Conference Proceedings 151* (American Institute of Physics, 1986)
134. G.G. Herrero, E. Huuppone, *Blind Source Separation Techniques for Processing Electroencephalographic Recordings* (Tampere University of Technology, 2004). http://www.kasku.org/projects/bss/review/review.pdf
135. Z.S.J. Hoare, Feature selection and classification of non-traditional data. Examples from veterinary medicine. Ph.D. thesis, University of Wales, Bangor, 2006
136. L.R. Hochberg, M.D. Serruya, G.M. Friehs, J.A. Mukand, M. Saleh, A.H. Caplan, J.P. Donoghue, Neuronal ensemble control of prosthetic devices by a human with tetraplegia. Nature **442**(7099), 164–171 (2006)
137. H. Hoffmann, Kernel pca for novelty detection. Pattern Recogn. **40**(3), 863–874 (2007)
138. T. Hofmann, B. Schölkopf, A.J. Smola, Kernel methods in machine learning. Ann. Stat. 1171–1220 (2008)
139. P.J. Huber, Projection pursuit. Ann. Stat. **13**, 435–475 (1974)
140. A. Hyvärinen, Survey of independent component analysis. Neural Comput. Surv. **2**, 94–128 (1999)

141. A. Hyvärinen, Independent component analysis: recent advances. Philos. Trans. Roy. Soc. 1–19 (2013). Open Access
142. A. Hyvärinen, J. Karhunen, E. Oja, *Independent Component Analysis* (Wiley-Interscience, 2001)
143. A. Hyvärinen, E. Oja, A fast fixed-point algorithm for independent component analysis. Neural Comput. **9**(7), 1483–1492 (1997)
144. A. Hyvärinen, E. Oja, Independent component analysis: algorithms and applications. Neural Netw. **13**(4–5), 411–430 (2000)
145. L. Ikemoto, O. Arikan, D. Forsyth, Generalizing motion edits with Gaussian processes. ACM Trans. Graph. (TOG) **28**(1), 1–12 (2009)
146. P. Indyk, R. Motwani, Approximate nearest neighbors: towards removing the curse of dimensionality, in *Proceedings of the 30th Annual ACM Symposium on Theory of Computing*, pp. 604–613 (ACM, 1998)
147. O. Ivanciuc, Applications of support vector machines in chemistry. Rev. Comput. Chem. **23**(291) (2007)
148. A.J. Izenman, *What is independent component analysis?* (Temple University, 2003). http://astro.temple.edu/~alan/files/ICA.PDF
149. A.J. Izenman, *Modern Multivariable Statistical Techniques* (Springer, 2008)
150. A.K. Jain, Data clustering: 50 years beyond k-means. Pattern Recogn. **31**, 651–666 (2010)
151. A.K. Jain, M.N. Murty, P.J. Flynn, Data clustering: a review. ACM Comput. Surveys (CSUR) **31**(3), 264–323 (1999)
152. C.J. James, C.W. Hesse, Independent component analysis for biomedical signals. Physiol. Measur. **26**(1), 15–39 (2005)
153. F.A. Jassim, *Image Inpainting by Kriging Interpolation Technique* (Faculty of Administrative Sciences, Management Information Systems Department, Irbid National University, Jordan, 2013). arXiv preprint arXiv:1306.0139
154. R. Jenatton, G. Obozinski, F. Bach, *Structured Sparse Principal Component Analysis* (INRIA, France, 2009). arXiv preprint arXiv:0909.1440
155. A. Jin, B. Yin, G. Morren, H. Duric, R.M. Aarts, Performance evaluation of a tri-axial accelerometry-based respiration monitoring for ambient assisted living, in *Proceedings IEEE 31st Annual International Conference EMBS*, pp. 5677–5680 (2009)
156. I. Jolliffe, Principal component analysis, ed. by Everitt. *Encyclopedia of Statistics in Behavioral Science* (Wiley, 2005)
157. T.P. Jung, S. Makeig, T.W. Lee, M.J. McKeown, G. Brown, A.J. Bell, T.J. Sejnowski, Independent component analysis of biomedical signals, in *Proceedings International Workshop on Independent Component Analysis and Signal Separation*, pp. 633–644 (2000)
158. C. Jutten, J. Karhunen, Advances in nonlinear blind source separation, in *Proceedings 4th International Symposium Independent Component Analysis and Blind Signal Separation, ICA*, pp. 245–256 (2003)
159. C. Jutten, A. Taleb, Source separation: From dusk till dawn, in *Proceedings 2nd International Workshop on Independent Component Analysis and Blind Source Separation, (ICA2000)*, pp. 15–26 (Helsinki, 2000)
160. T. Kailath, The divergence and Bhattacharyya distance measures in signal selection. IEEE T. Commun. Technol. **15**(1), 52–60 (1967)
161. L. Kanal, Patterns in pattern recognition: 1968–1974. IEEE T. Inf. Theory **20**(6), 697–722 (1974)
162. T. Kanungo, D.M. Mount, N.S. Netanyahu, C.D. Piatko, R. Silverman, A.Y. Wu, An efficient k-means clustering algorithm: analysis and implementation. IEEE T. Patt. Anal. Mach. Intell. **24**(7), 881–892 (2002)
163. A. Kapoor, K. Grauman, R. Urtasun, T. Darrell, Active learning with Gaussian processes for object categorization, in *Proceedings IEEE 11th International Conference on Computer Vision, ICCV* 2007
164. L. Kaufman, P. Rousseau, *Finding Groups in Data* (Wiley, 1990)
165. S. Kay, *Intuitive Probability and Random Processes Using MATLAB* (Springer, 2006)

166. S.S. Keerthi, C.J. Lin, Asymptotic behaviors of support vector machines with Gaussian kernel. Neural Comput. **15**(7), 1667–1689 (2003)
167. H.B. Kekre, T.K. Sarode, *New Clustering Algorithm for Vector Quantization Using Rotation of Error Vector* (Computer Engineering Mukesh Patel School of Technology Management and Engineering, NMIMS University, Vileparle(w), India, 2010). arXiv preprint arXiv:1004.1686
168. G. Kerschen, F. Poncelet, J.C. Golinval, Physical interpretation of independent component analysis in structural dynamics. Mech. Syst. Sign. Process. **21**, 1561–1575 (2007)
169. H.C. Kim, J. Lee, Clustering based on Gaussian processes. Neural Comput. **19**(11), 3088–3107 (2007)
170. K.I. Kim, M.O. Franz, B. Schölkopf, Iterative kernel principal component analysis for image modeling. IEEE T. Pattern Anal. Mach. Intell. **27**(9), 1351–1366 (2005)
171. S.J. Kim, A. Magnani, S. Boyd, Optimal kernel selection in kernel Fisher discriminant analysis, in *Proceedings ACM 23rd International Conference Machine Learning*, pp. 465–472 (2006)
172. R.S. King, *Cluster Analysis and Data Mining* (Trasatlantic Publishers, 2014)
173. W.R. Klecka, *Discriminant Analysis* (Sage Publications, 1980)
174. J. Kocijan, A. Grancharova, Application of Gaussian processes to the modelling and control in process engineering, in *Innovations in Intelligent Machines-5*, pp. 155–190 (Springer, 2014)
175. A. Kocsor, L. Tóth, Kernel-based feature extraction with a speech technology application. IEEE T. Sign. Process. **52**(8), 2250–2263 (2004)
176. R. Kohn, M. Smith, D. Chan, Nonparametric regression using linear combinations of basis functions. Stat. Comput. **11**(4), 313–322 (2001)
177. I. Kokkinos, P. Maragos, Synergy between object recognition and image segmentation using the expectation-maximization algorithm. IEEE T. Pattern Anal. Mach. Intell. **31**(8), 1486–1501 (2009)
178. Z. Koldovsky, Fast and accurate methods for independent component analysis. Ph.D. thesis, Czech Technical University in Prague, 2005
179. S. Koziel, D.E. Ciaurri, L. Leifsson, Surrogate-based methods, in *Computational Optimization, Methods and Algorithms*, pp. 33–59 (Springer, 2011)
180. K. Krishna, M.N. Murty, Genetic k-means algorithm. IEEE T. Syst. Man Cybern. Part B Cybern. **29**(3), 433–439 (1999)
181. K. Krivoruchko, *Empirical Bayesian Kriging* (Software Development Team, Esri, 2012). http://www.esri.com/news/arcuser/1012/empirical-byesian-kriging.html
182. M. Kuβ, Gaussian process models. Ph.D. thesis, Technischen Darmstadt, 2006
183. B. Kulis, M.I. Jordan, *Revisiting K-means: New Algorithms Via Bayesian Nonparametrics* (Department of CSE, Ohio State University, Columbus, 2011). arXiv preprint arXiv:1111.0352
184. T. Kumano, S. Jeong, S. Obayashi, Y. Ito, K. Hatanaka, H. Morino, Multidisciplinary design optimization of wing shape for a small jet aircraft using kriging model. AIAA Paper **932**, 9–12 (2006)
185. J. Kumar, R.T. Mills, F.M. Hoffman, W.W. Hargrove, Parallel k-means clustering for quantitative ecoregion delineation using large data sets. Proc. Comput. Sci. **4**, 1602–1611 (2011)
186. M. Kuss, C.E. Rasmussen, Assesing approximate inference for binary Gaussian process classification. J. Mach. Learn. Res. **6**, 1679–1704 (2005)
187. N. Kwak, Feature extraction for classification problems and its application to face recognition. Pattern Recogn. **41**(5), 1701–1717 (2008)
188. V. Lakshmanan, R. Rabin, V. DeBrunner, Multiscale storm identification and forecast. Atmos. Res. **67**, 367–380 (2003)
189. L.D. Lathauwer, B.D. Moor, J. Vandewalle, An introduction to independent component analysis. J. Chemom. **14**, 123–149 (2000)
190. G.F. Lawler, *Introduction to Stochastic Processes* (Chapman and Hall, 2006)
191. J.H. Lee, H.Y. Jung, T.W. Lee, S.Y. Lee, Speech feature extraction using independent component analysis, in *Proceedings IEEE International Conference Acoustics, Speech, and Signal Processing, ICASSP'00*, vol. 3, pp. 1631–1634 (2000)

192. C.S. Leslie, E. Eskin, A. Cohen, J. Weston, W.S. Noble, Mismatch string kernels for discriminative protein classification. Bioinformatics **20**(4), 467–476 (2004)
193. A. Levey, M. Lindenbaum, Sequential Karhunen-Loeve basis extraction and its application to images. IEEE T. Image Process. **9**(8), 1371–1374 (2000)
194. E. Ley, M.F. Steel, On the effect of prior assumptions in Bayesian model averaging with applications to growth regression. J. Appl. Econ. **24**(4), 651–674 (2009)
195. J. Li, A.D. Heap, A review of spatial interpolation methods for environmental scientists. Geosci. Australia Record **23** (2008)
196. S.Z. Li, A.K. Jain, *Handbook of Face Recognition* (Springer, 2005)
197. T. Li, S. Zhu, M. Ogihara, Using discriminant analysis for multi-class classification: an experimental investigation. Knowl. Inf. Syst. **10**(4), 453–472 (2006)
198. X. Li, L. Wang, E. Sung, Adaboost with SVM-based component classifiers. Eng. Appl. Artif. Intell. **21**(5), 785–795 (2008)
199. A. Lichtenstern, *Kriging Methods in Spatial Statistics* (Bachelor's Thesis, Technische Universität München, 2013). http://ww.mediatum.ub.tum.de/doc/1173364/1173364.pdf
200. C.F. Lin, S.D. Wang, Fuzzy support vector machines. IEEE T. Neural Netw. **13**(2), 464–471 (2002)
201. H.T. Lin, *Adaptive Boosting – AdaBoosting* (Lecture Notes, Machine Learning, National Taiwan University, 2008). https://www.csie.ntu.edu.tw/~b92109/course/
202. J. Lin, Divergence measures based on the Shannon entropy. IEEE T. Inf. Theory **37**(1), 145–151 (1991)
203. R. Linsker, Local synaptic learning rules suffice to maximize mutual information in a linear network. Neural Comput. **4**, 691–702 (1992)
204. C. Liu, H. Wechsler, Gabor feature based classification using the enhanced Fisher linear discriminant model for face recognition. IEEE Trans. Image Process. **11**(4), 467–476 (2002)
205. H.X. Liu, R.S. Zhang, F. Luan, X.J. Yao, M.C. Liu, Z.D. Hu, B.T. Fan, Diagnosing breast cancer based on support vector machines. J. Chem. Inf. Comput. Sci. **43**(3), 900–907 (2003)
206. Y. Liu, R. Emery, D. Chakrabarti, W. Burgard, S. Thrun, Using EM to learn 3D models of indoor environments with mobile robots, in *Proceedingd 18th International Conference Machine Learning*, vol. 1, pp. 329–336 (2001)
207. Z. Liu, D. Chen, H. Bensmail, Gene expression data classification with Kernel principal component analysis. BioMed. Res. Int. **2005**(2), 155–159 (2005)
208. M. Liwicki, A. Graves, H. Bunke, J. Schmidhuber, A novel approach to on-line handwriting recognition based on bidirectional long short-term memory networks, in *Proceedings 9th International Conference on Document Analysis and Recognition*, vol. 1, pp. 367–371 (2007)
209. H. Lodhi, C. Saunders, J. Shawe-Taylor, N. Cristianini, C. Watkins, Text classification using string kernels. J. Mach. Learn. Res. **2**, 419–444 (2002)
210. A.T. Lora, J.M.R. Santos, A.G. Expósito, J.L.M. Ramos, J.C.R. Santos, Electricity market price forecasting based on weighted nearest neighbors techniques. IEEE T. Power Syst. **22**(3), 1294–1301 (2007)
211. F. Lotte, M. Congedo, A. Lécuyer, F. Lamarche, A review of classification algorithms for EEG-based brain-computer interfaces. J. Neural Eng. **4**, 1–25 (2007)
212. J. Lu, K.N. Plataniotis, A.N. Venetsanopoulos, Face recognition using LDA-based algorithms. IEEE T. Neural Netw. **14**(1), 195–200 (2003)
213. J. Lu, K.N. Plataniotis, A.N. Venetsanopoulos, J. Wang, An efficient kernel discriminant analysis method. Pattern Recogn. **38**(10), 1788–1790 (2005)
214. T.C. Lu, C.Y. Chang, A survey of VQ codebook generation. J. Inf. Hiding Multimed. Sign. Process. **1**(3), 190–203 (2010)
215. M. Luo, Y.F. Ma, H.J. Zhang, A spatial constrained k-means approach to image segmentation, in *Proceedings IEEE 4th Conference Information, Communications and Signal Processing*, vol. 2, pp. 738–742 (2003)
216. Y.Z. Ma, J.J. Royer, H. Wang, Y. Wang, T. Zhang, Factorial kriging for multi-scale modeling. J. Southern African Instit. Mining Metall. **114**, 651–657 (2014)

217. O. Makhnin, *Introduction to Kriging* (Lecture 10, Math 586, New Mexico Institute of Mining and Technology, Department of Mathematics, 2013). http://infohost.nmt.edu/~olegm/586/HYD10.pdf
218. J. Makhoul, S. Roucos, H. Gish, Vector quantization in speech coding. Proc. IEEE **73**(11), 1551–1588 (1985)
219. A. Mansour, M. Kawamoto, ICA papers classified according to their applications and performances. IEICE T. Fundam. E86-A(3), 620–633 (2003)
220. A.M. Martínez, A.C. Kak, PCA versus LDA. IEEE T. Pattern Anal. Mach. Intell. **23**(2), 228–233 (2001)
221. B. Matei, *A Review of Independent Component Analysis Techniques* (Rutgers University School of Engineering, 2000). http://coewww.rutgers.edu/riul/research/tutorials/tutorialica.pdf
222. G. McLachlan, T. Krishnan, *The EM Algorithm and Extensions* (Wiley, 2007)
223. S.A. Medjahed, T.A. Saadi, A. Benyettou, Breast cancer diagnosis by using k-nearest neighbor with different distances and classification rules. Intl. J. Comput. Appl. **62**(1) (2013)
224. S. Mika, J. Ratsch, G. Weston, B. Scholkopf, Fisher discriminant analysis with kernels, in *Proceedings IEEE Workshop Neural Networks for Signal Processing IX*, vol. 1, pp. 41–48 (1999)
225. S. Mika, B. Schölkopf, A.J. Smola, K.R. Müller, M. Scholz, G. Rätsch, Kernel PCA and de-noising in feature spaces. NIPS **4**(5), 1–7 (1998)
226. B. Moghaddam, T. Jebara, A. Pentland, Bayesian face recognition. Pattern Recogn. **33**(11), 1771–1782 (2000)
227. T.K. Moon, The expectation-maximization algorithm. IEEE Sign. Process. Mag. **13**(6), 47–60 (1996)
228. L. Morissette, S. Chartier, The k-means clustering technique: general considerations and implementation in mathematica. Tutorials Quant. Meth. Psychol. **91**(1), 15–24 (2013)
229. J. Morra, Z. Tu, A. Toga, P. Thompson, Machine learning for brain image segmentation, eds. by Gonzalez and Romero. *Biomedical Image Analysis and Machine Learning Technologies: Applications and Techniques* (Medical Information Science Reference, 2009)
230. J.H. Morra, Z. Tu, L.G. Apostolova, A.E. Green, A.W. Toga, P.M. Thompson, Comparison of AdaBoost and support vector machines for detecting Alzheimer's disease through automated hippocampal segmentation. IEEE T. Med. Imag. **29**(1), 30–43 (2010)
231. D.G. Morrison, On the interpretation of discriminant analysis. J. Market. Res. 156–163 (1969)
232. M. Muja, D.G. Lowe, Fast approximate nearest neighbors with automatic algorithm configuration. Proc. VISAPP **1**, 331–340 (2009)
233. S.A. Mulaik, *Foundations of Factor Analysis* (CRC Press, 2009)
234. K. Muller, S. Mika, G. Ratsch, K. Tsuda, B. Scholkopf, An introduction to kernel-based learning algorithms. IEEE T. Neural Netw. **12**(2), 181–201 (2001)
235. K. Murphy, *Machine Learning: A Probabilistic Perspective* (MIT Press, 2012)
236. G.R. Naik, D.K. Kumar, An overview of independent component analysis and its applications. Informatica **35**, 63–81 (2011)
237. G.P. Nason, Design and choice of projection indices. Ph.D. thesis, University of Bath, UK, 1992
238. R.M. Neal, G.E. Hinton, A view of the EM algorithm that justifies incremental, sparse, and other variants, in *Learning in Graphical Models*, pp. 355–368 (Springer, 1998)
239. H.P. Ng, S.H. Ong, K.W.C. Foong, P.S. Goh, W.L. Nowinski, Medical image segmentation using k-means clustering and improved watershed algorithm, in *Proceedings IEEE Southwest Symposium o Image Analysis and Interpretation*, pp. 61–65 (2006)
240. M.H. Nguyen, F. Torre, Robust kernel principal component analysis, in *Proceedings Advances in Neural Information Processing Systems*, pp. 1185–1192 (2009)
241. D. Nguyen-Tuong, M. Seeger, J. Peters, Model learning with local Gaussian process regression. Adv. Robot. **23**(15), 2015–2034 (2009)
242. H. Nickisch, C.E. Rasmussen, Approximations for binary Gaussian process classification. J. Mach. Learn. Res. **9**, 2035–2078 (2008)

243. M.G. Omran, A.P. Engelbrecht, A. Salman, An overview of clustering methods. Intell. Data Anal. **11**(5), 583–605 (2007)
244. F. Orabona, J. Keshet, B. Caputo, The projectron: a bounded kernel-based perceptron, in *Proceedings of the ACM 25th International Conference Machine Learning*, pp. 720–727 (2008)
245. E. Osuna, R. Freund, F. Girosi, Support vector machines: training and applications. Technical report, MIT, 1997. AI Memo 1602
246. A. Oursland, J. De Paula, N. Mahmood, Case Studies of Independent Component Analysis (2013). Numerical Analysis of Linear Algebra, CS383C. http://www.oursland.net/tutorials/ica/ica-report.pdf
247. B. Pardo, *Machine Learning, Topic 6: Clustering* (Lecture Presentation, Northwestern University, 2009). http://www.cs.northwestern.edu/~pardo/courses/eecs349/lectures/NUEECS349Falltopic6-clustering.pdf
248. I. Pardoe, X. Yin, R.D. Cook, Graphical tools for quadratic discriminant analysis. Technometrics **49**(2) (2007)
249. H.S. Park, C.H. Jun, A simple and fast algorithm for k-medoids clustering. Expert Syst. Appl. **36**(2), 3336–3341 (2009)
250. L. Parra, P. Sajda, Blind source separation via generalized eigenvalue decomposition. J. Mach. Learn. Res. **4**, 1261–1269 (2003)
251. D. Pelleg, A.W. Moore, X-means: extending k-means with efficient estimation of the number of clusters, in *Proceedings ICML*, pp. 727–734 (2000)
252. W. Penny, S. Kiebel, K. Friston, Variational bayes, eds. by K. Friston, J. Ashburner, S. Kiebel, T. Nichols, W. Penny. *Statistical Parametric Mapping: The Analysis of Functional Brain Images* (Elsevier, 2006)
253. F. Pereira, T. Mitchell, M. Botvinick, Machine learning classifiers and FMRI: a tutorial overview. Neuroimage **45**(1), S199–S209 (2009)
254. D. Petelin, B. Filipi, J. Kocijan, Optimization of Gaussian process models with evolutionary algorithms, in *Adaptive and Natural Computing Algorithms*, pp. 420–429 (Springer, 2011)
255. D.L. Pham, C. Xu, J.L. Prince, Current methods in medical image segmentation 1. Ann. Rev. Biomed. Eng. **2**(1), 315–337 (2000)
256. R. Plamondon, S.N. Srihari, Online and off-line handwriting recognition: a comprehensive survey. IEEE T. Pattern Anal. Mach. Intell. **22**(1), 63–84 (2000)
257. J.H. Plasse, The EM algorithm in multivariate Gaussian mixture models using Anderson acceleration. Master's thesis, Worcester Polytechnic Institute, 2013
258. M. Pohar, M. Blas, S. Turk, Comparison of logistic regression and linear discriminant analysis. Metodoloki Zvezki **1**(1), 143–161 (2004)
259. C. Posse, Tools for two-dimensional exploratory projection pursuit. J. Comput. Graph. Stat. **4**(2), 83–100 (1995)
260. C.E. Rasmussen, C.K.I. Williams, *Gaussian Processes for Machine Learning* (MIT Press, 2006)
261. J.D. Rennie, L. Shih, J. Teevan, D.R. Karger, Tackling the poor assumptions of naive Bayes text classifiers. ICML **3**, 616–623 (2003)
262. B.D. Ripley, *Pattern Recognition and Neural Networks* (Cambridge University Press, 2008)
263. I. Rish, An empirical study of the naive Bayes classifier, in *Proceedings IJCAI 2001 Workshop on Empirical Methods in Artificial Intelligence*, vol. 3, pp. 41–46 (2001)
264. S. Roberts, M. Osborne, M. Ebden, S. Reece, N. Gibson, S. Aigrain, Gaussian processes for time-series modeling. Philos. Trans. Royal Soc. A: Math. Phys. Eng. Sci. **371**(1984), 20110550 (2013)
265. D.N. Rutledge, D.J-R Bouveresse, Independent component analysis with the JADE algorithm. Trends Anal. Chem. **50**, 22–32 (2013)
266. S. Ryali, K. Supekar, D.A. Abrams, V. Menon, Sparse logistic regression for whole-brain classification of FMRI data. NeuroImage **51**(2), 752–764 (2010)
267. S. Cha, Comprehensive survey on distance/similarity measures between probability density functions. Int. J. Math. Models Meth. Appl. Sci. **1**(4), 300–307 (2007)

268. H. Sahbi, Kernel PCA for similarity invariant shape recognition. Neurocomputing **70**(16), 3034–3045 (2007)
269. S. Samarasinghe, *Neural Networks for Applied Sciences and Engineering* (Auerbach Publications, 2006)
270. J. Sankaranarayanan, H. Samet, A. Varshney, A fast all nearest neighbor algorithm for applications involving large point-clouds. Comput. Graph. **31**(2), 157–174 (2007)
271. G. Schalk, D.J. McFarland, T. Hinterberger, N. Birbaumer, J.R. Wolpaw, BCI2000: a general-purpose brain-computer interface (BCI) system. IEEE T. Biomed. Eng. **51**(6), 1034–1043 (2004)
272. B. Schölkopf, A. Smola, K.R. Müller, Kernel principal component analysis, in *Artificial Neural Networks—ICANN'97*, pp. 583–588 (Springer, 1997)
273. B. Schölkopf, A. Smola, K.R. Müller, Nonlinear component analysis as a kernel eigenvalue problem. Neural Comput. **10**(5), 1299–1319 (1998)
274. M. Seeger, Gaussian processes for machine learning. Int. J. Neural Syst. **14**(2), 69–106 (2004)
275. N. Seo, *Eigenfaces and Fisherfaces* (University of Maryland, ENEE633 Pattern Recognition, 2007). http://note.sonots.com/SciSoftware/FaceRecognition.html
276. G. Shakhnarovich, P. Indyk, T. Darrell, *Nearest Neighbor Methods in Learning and Vision: Theory and Practice* (MIT Press, 2006)
277. S. Shan, B. Cao, W. Gao, D. Zhao, Extended Fisherface for face recognition from a single example image per person, in *Proceedings IEEE International Symposium Circuits and Systems, ISCAS 2002*, vol. 2 (2002). II-81
278. J. Shawe-Taylor, N. Cristianini, *Kernel Methods for Pattern Analysis* (Cambridge University Press, 2004)
279. H. Shen, J.Z. Huang, Sparse principal component analysis via regularized low rank matrix approximation. J. Multivar. Anal. **99**(6), 1015–1034 (2008)
280. Y. Shi, D. Dai, C. Liu, H. Yan, Sparse discriminant analysis for breast cancer biomarker identification and classification. Progr. Nat. Sci. **19**(11), 1635–1641 (2009)
281. J. Shlens, A tutorial on principal component analysis. J. Comput. Graph. Stat. **4**(2), 83–100 (2003)
282. V.K. Singh, N. Tiwari, S. Garg, Document clustering using k-means, heuristic k-means and fuzzy c-means, in *Proceedings International Conference Computational Intelligence and Communication, Networks, 2011*, pp. 297–301
283. T.E. Smith, *Notebook on Spatial Data Analysis* (SEAS, Penn Engineering, 2014). http://www.seas.upenn.edu/~ese502/#notebook
284. A.J. Smola, B. Schölkopf, A tutorial on support vector regression. Stat. Comput. **14**(3), 199–222 (2004)
285. S. Sonnenburg, G. Rätsch, B. Schölkopf, Large scale genomic sequence SVM classifiers, in *Proceedings 22nd ACM International Conference Machine Learning*, pp. 848–855 (2005)
286. S. Srivastava, M.R. Gupta, B.A. Frigyik, Bayesian quadratic discriminant analysis. J. Mach. Learn. Res. **8**(6), 1277–1305 (2007)
287. O. Stegle, S.V. Fallert, D.J. MacKay, S. Brage, Gaussian process robust regression for noisy heart rate data. IEEE T. Biomed. Eng. **55**(9), 2143–2151 (2008)
288. M. Steinbach, G. Karypis, V. Kumar, A comparison of document clustering techniques, in *Proceedings KDD Workshop on Text Mining*, vol. 400, pp. 525–526 (2000)
289. J.V. Stone, *Independent Component Analysis* (MIT Press, 2004)
290. K. Suzuki, *Artificial Neural Networks – Industrial and Control Engineering Applications* (InTech, 2011)
291. R. Tandon, *A Survey of Sparse PCA* (The University of Texas at Austin, 2012). http://www.cs.utexas.edu/~rashish/sparse_pca.pdf
292. Y. Tang, *Deep Learning Using Linear Support Vector Machines* (Department of Computer Science, University of Toronto, 2013). arXiv preprint arXiv:1306.0239
293. D.M. Taylor, S.I.H. Tillery, A.B. Schwartz, Direct cortical control of 3D neuroprosthetic devices. Science **296**(5574), 1829–1832 (2002)

294. A. Teynor, H. Burkhardt, Fast codebook generation by sequential data analysis for object classification, in *Advances in Visual Computing*, pp. 610–620 (Springer, 2007)
295. S. Theodoridis, A. Pikrakis, K. Koutroumbas, D. Cavouras, *Introduction to Pattern Recognition* (Academic Press, 2010)
296. P. Tichavsky, Z. Koldovsky, E. Oja, Performance analysis of the FastICA algorithm and Cramer-Rao bounds for linear independent component analysis. IEEE T. Sign. Process. **54**(4), 1189–1197 (2006)
297. M.E. Tipping, Sparse Bayesian learning and the relevance vector machine. J. Mach. Learn. Res. **1**, 211–244 (2001)
298. F. Tombari, L. Di Stefano, A. Lanza, S. Mattoccia, Non-linear parametric Bayesian regression for robust background subtraction, in *Proceedings IEEE Workshop onMotion and Video Computing, WMVC'09*, pp. 1–7 (2009)
299. S. Tong, D. Koller, Support vector machine active learning with applications to text classification. J. Mach. Learn. Res. **2**, 45–66 (2002)
300. A. Tsai, J. Zhang, A.S. Willsky, Expectation-maximization algorithms for image processing using multiscale models and mean-field theory, with applications to laser radar range profiling and segmentation. Opt. Eng. **40**(7), 1287–1301 (2001)
301. I. Tsochantaridis, T. Joachims, T. Hofmann, Y. Altun, Large margin methods for structured and interdependent output variables. J. Mach. Learn. Res. 1453–1484 (2005)
302. D. Tuia, M. Volpi, M. Dalla Mura, A. Rakotomamonjy, R. Flamary, Automatic feature learning for spatio-spectral image classification with sparse SVM. IEEE T. Geosci. Remote Sens. **52**(10), 6062–6074 (2014)
303. M. Turk, A. Pentland, Eigenfaces for recognition. J. Cogn. Neurosci. **3**(1), 71–86 (1991)
304. I.Y. Turner, E.M. Huff, Principal components analysis of triaxial vibration data from helicopter transmissions, in *Proceedings 56th Meeting of the Society for Machinery Failure Prevention Technology*, 2002
305. J.H. van Hateren, A. van der Schaaf, Independent component filters of natural images compared with simple cells in primary visual cortex. Proc. Roy. Soc. London. Series B: Biol. Sci. **265**(1394), 359–366 (1998)
306. V. Vapnik, *The Nature of Statistical Learning Theory* (Springer, 2000)
307. A. Vellido, P.J. Lisboa, J. Vaughan, Neural networks in business: a survey of applications (1992–1998). Expert Syst. Appl. **17**(1), 51–70 (1999)
308. E. Vincent, R. Gribonval, C. Févotte, Performance measurement in blind audio source separation. IEEE T. Audio, Speech, Lang. Process. **14**(4), 1462–1469 (2006)
309. U. Von Luxburg, R.C. Williamson, I. Guyon, *Clustering: Science or art?* ICML Unsupervised and Transfer, Learning, pp. 65–80 (2012)
310. P. Wagner, *Face Recognition with GNU Octave/MATLAB* (Cracow University of Technology, Poland, 2012). http://mars.iti.pk.edu.pl/~chmaj/APSC/facerec_octave.pdf
311. J. Wakefield, Non-linear regression modelling and inference. Meth. Models Stat. 119–153 (2004)
312. J.Y. Wang, Application of support vector machines in bioinformatics. Ph.D. thesis, National Taiwan University, 2002
313. Q. Wang, *Kernel Principal Component Analysis and Its Applications in Face Recognition and Active Shape Models* (Rensselaer Polytechnic Institute, 2012). arXiv preprint arXiv:1207.3538
314. R. Wang, Adaboost for feature selection, classification and its relation with SVM, a review. Phys. Proc. **25**, 800–807 (2012)
315. K. Wayne, *Tutorial 2: Numerical Linear Algebra* (Computer Science Dept., Princeton University, 2007. SEAS Short Course Programming in MATLAB). https://www.cs.princeton.edu/~wayne/teaching/linear-algebra.pdf
316. W.M. Wells III, W.E.L. Grimson, R. Kikinis, F.A. Jolesz, Adaptive segmentation of MRI data. IEEE T. Med. Imag. **15**(4), 429–442 (1996)
317. J.R. Wolpaw, N. Birbaumer, D.J. McFarland, G. Pfurtscheller, T.M. Vaughan, Brain-computer interfaces for communication and control. Clin. Neurophysiol. **113**(6), 767–791 (2002)

318. K.P. Wong, D. Feng, S.R. Meikle, M.J. Fulham, Segmentation of dynamic PET images using cluster analysis. IEEE T. Nucl. Sci. **49**(1), 200–207 (2002)
319. J. Wright, A.Y. Yang, A. Ganesh, S.S. Sastry, Y. Ma, Robust face recognition via sparse representation. IEEE T. Pattern Anal. Mach. Intell. **31**(2), 210–227 (2009)
320. M. Xiao, An improved background reconstruction algorithm based on basic sequential clustering. Inf. Technol. J. **7**(3), 522–527 (2008)
321. R. Xu, D. Wunsch, *Clustering*, vol. 10 (Wiley, 2008)
322. R. Xu, D.C. Wunsch, Clustering algorithms in biomedical research: a review. IEEE Rev. Biomed. Eng. **3**, 120–154 (2010)
323. I. Yamaguchi, T. Kuzuyoshi, An algebraic solution to independent component analysis. Opt. Commun. **178**, 59–64 (2000)
324. J. Yang, Z. Jin, J.Y. Yang, D. Zhang, A.F. Frangi, Essence of kernel Fisher discriminant: KPCA plus LDA. Pattern Recogn. **37**(10), 2097–2100 (2004)
325. P. Yang, Y. Hwa, Yang, B.B Zhou, A.Y. Zomaya, A review of ensemble methods in bioinformatics. Curr. Bioinf. **5**(4), 296–308 (2010)
326. P.N. Yianilos, Data structures and algorithms for nearest neighbor search in general metric spaces, in *Proceedings of the fourth annual ACM-SIAM Symposium on Discrete algorithms*, pp. 311–321 (Society for Industrial and Applied Mathematics, 1993)
327. D. You, O.C. Hamsici, A.M. Martinez, Kernel optimization in discriminant analysis. IEEE T. Pattern Anal. Mach. Intell. **33**(3), 631–638 (2011)
328. V. Zarzoso, P. Comon, M. Kallel, How fast is FastICA? in *Proc. EUSIPCO-2006* (2006)
329. C. Zhang, Z. Zhang, A survey of recent advances in face detection. Technical report, Microsoft Research, 2010
330. S.X. Zhang, C. Liu, K. Yao, Y. Gong, *Deep Neural Support Vector Machines for Speech Recognition* (Microsoft Research, 2015). http://research.microsoft.com/pubs/244711/0004275.pdf
331. W. Zhao, R. Chellappa, P.J. Phillips, A. Rosenfeld, Face recognition: a literature survey. ACM Comput. Surv. (CSUR) **35**(4), 399–458 (2003)
332. J. Zhu, H. Zou, S. Rosset, T. Hastie, Multi-class AdaBoost. Stat. Interf. **2**(3), 349–360 (2009)
333. M. Zibulevsky, B. Pearlmutter, Blind source separation by sparse decomposition in a signal dictionary. Neural Comput. **13**(4), 863–882 (2001)
334. H. Zou, T. Hastie, R. Tibshirani, Sparse principal component analysis. J. Comput. Graph. Stat. **15**(2), 265–286 (2006)

Appendix
Long Programs

A.1 Introduction

Some of the programs made for the chapters of this book are long. Aiming at simplifying the use of the book, it has been preferred to put together these programs in the present Appendix.

Samll versions of the figures generated by the programs have been included, to help identifying each program

A.2 Chapter 1: Filter Banks

A.2.1 Modulated Filter Bank (1.2.2)

The next figures correspond to a modulated filter bank depicted in Fig. 1.8. An example of filter bank with only three channels has been considered (Figs. A.1, A.2, A.3, A.4, A.5, A.6, A.7, A.8 and A.9).

Program A.1 Modulated filters: analysis and synthesis

```
% Modulated analysis and synthesis
fs=20; %sampling frequency
tiv=1/fs; %sampling period
Ns=120; %number of samples
t=0:tiv:((Ns-1)*tiv); %time vector,
%signal components
y0=cos(2*2*pi*t);
y1=cos(6*2*pi*t);
y2=cos(10*2*pi*t);
%added signal
y=(1*y0)+(4*y1)+(3*y2);
%prototype filters (Kaiser, FIR)
fc=3/(fs/2); %cut-off frequency at 3 Hz
L=50;beta=5;
hw=kaiser(L+1,beta); %Kaiser window
Hnum=fir1(L,fc,hw); %FIR coeffs
Hden=[1]; %denominator
```

© Springer Science+Business Media Singapore 2017
J.M. Giron-Sierra, *Digital Signal Processing with Matlab Examples, Volume 2*,
Signals and Communication Technology, DOI 10.1007/978-981-10-2537-2

```
Fnum=Hnum;
Fden=Hden;
%modulations for 3 filters
m0=y.*exp(-j*(2*pi*0*t));
m1=y.*exp(-j*(2*pi*4*t));
m2=y.*exp(-j*(2*pi*8*t));
%low-pass filtering and decimation
a0=filter(Hnum,Hden,m0); b0=a0(1:3:Ns);
a1=filter(Hnum,Hden,m1); b1=a1(1:3:Ns);
a2=filter(Hnum,Hden,m2); b2=a2(1:3:Ns);
%upsampling
c0=zeros(1,Ns);c1=zeros(1,Ns);c2=zeros(1,Ns);
c0(1:3:Ns)=b0(1:end);
c1(1:3:Ns)=b1(1:end);
c2(1:3:Ns)=b2(1:end);
%second low-pass filtering
M=3;
d0=M*filter(Fnum,Fden,c0);
d1=M*filter(Fnum,Fden,c1);
d2=M*filter(Fnum,Fden,c2);
tp=t-((50)*tiv); %delay compensation
%demodulations for 3 filters
dm0=d0.*exp(j*(2*pi*0*tp));
dm1=d1.*exp(j*(2*pi*4*tp));
dm2=d2.*exp(j*(2*pi*8*tp));
%output
x=dm0+dm1+dm2;
%display
figure(1)
plot(t,real(y),'k');
title('the input signal');
axis([0 6 -9 9]);
figure(2)
ff=logspace(0,1);
G=freqz(Hnum,Hden,ff,fs);
semilogx(ff,abs(G),'k'); grid;
title('Frequency response of the prototype filter');
xlabel(`Hz');
figure(3)
subplot(1,3,1)
plot(t,real(m0));
title('ch.1, modulated input');
axis([0 6 -9 9]);
subplot(1,3,2)
plot(t,real(m1));
title('ch.2, modulated input');
axis([0 6 -9 9]);
subplot(1,3,3)
plot(t,real(m2));
title('ch.3, modulated input');
axis([0 6 -9 9]);
figure(4)
subplot(1,3,1)
```

```
plot(t,real(a0));
title('out.1, analysis filter');
axis([0 6 -6 6]);
subplot(1,3,2)
plot(t,real(a1));
title('out.2, analysis filter ');
axis([0 6 -6 6]);
subplot(1,3,3)
plot(t,real(a2));
title('out.3, analysis filter');
axis([0 6 -6 6]);
figure(5)
subplot(1,3,1)
plot(t(1:3:Ns),real(b0));
title('decimated 1');
axis([0 6 -6 6]);
subplot(1,3,2)
plot(t(1:3:Ns),real(b1));
title('decimated 2');
axis([0 6 -6 6]);
subplot(1,3,3)
plot(t(1:3:Ns),real(b2));
title('decimated 3');
axis([0 6 -6 6]);
figure(6)
subplot(1,3,1)
plot(t,real(c0));
title('upsampled 1');
axis([0 6 -6 6]);
subplot(1,3,2)
plot(t,real(c1));
title('upsampled 2');
axis([0 6 -6 6]);
subplot(1,3,3)
plot(t,real(c2));
title('upsampled 3');
axis([0 6 -6 6]);
figure(7)
subplot(1,3,1)
plot(t,real(d0));
title('out.1, synthesis filter');
axis([0 6 -6 6]);
subplot(1,3,2)
plot(t,real(d1));
title('out.2, synthesis filter');
axis([0 6 -6 6]);
subplot(1,3,3)
plot(t,real(d2));
title('out.3, synthesis filter');
axis([0 6 -6 6]);
figure(8)
subplot(1,3,1)
plot(t,real(dm0));
```

```
title('out.1, demodulation');
axis([0 6 -6 6]);
subplot(1,3,2)
plot(t,real(dm1));
title('out.2, demodulation');
axis([0 6 -6 6]);
subplot(1,3,3)
plot(t,real(dm2));
title('out.3, demodulation');
axis([0 6 -6 6]);
figure(9)
plot(t,real(x),'k');
title('the output signal');
axis([0 6 -9 9]);
```

Fig. A.1 The input signal (Fig. 1.9)

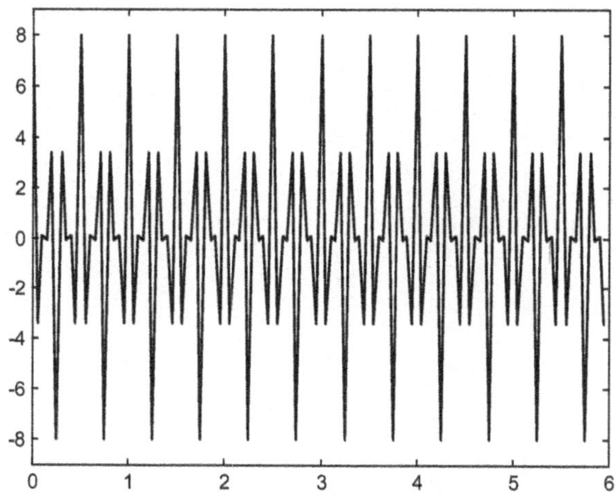

Fig. A.2 The frequency response of the prototype filter

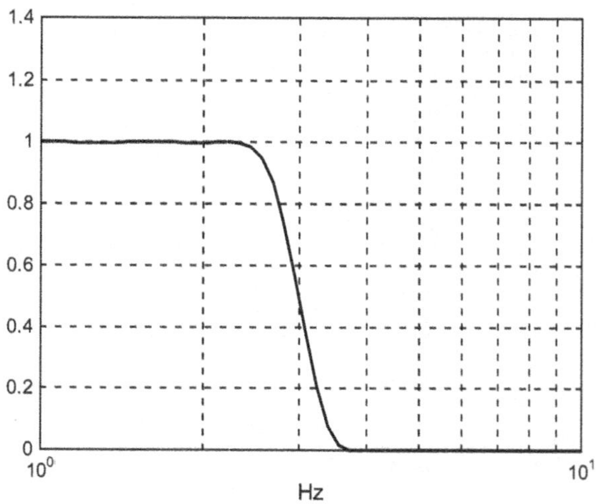

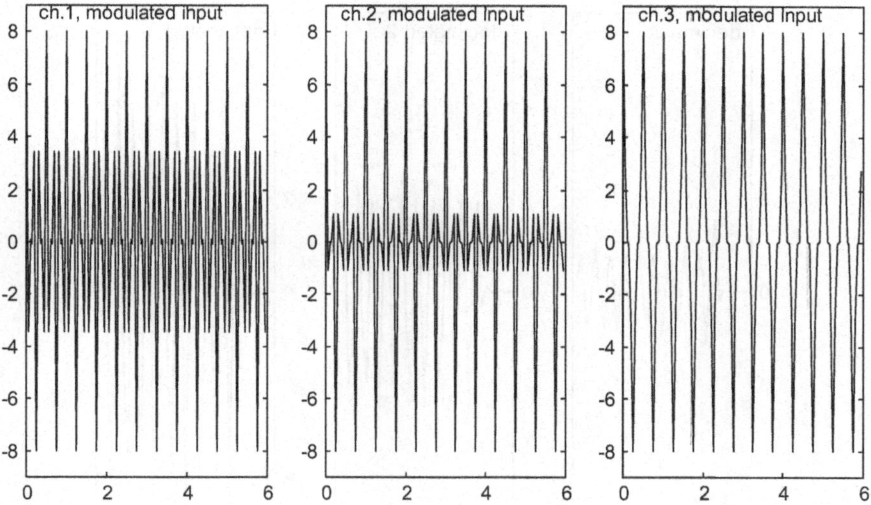

Fig. A.3 The outputs of modulations

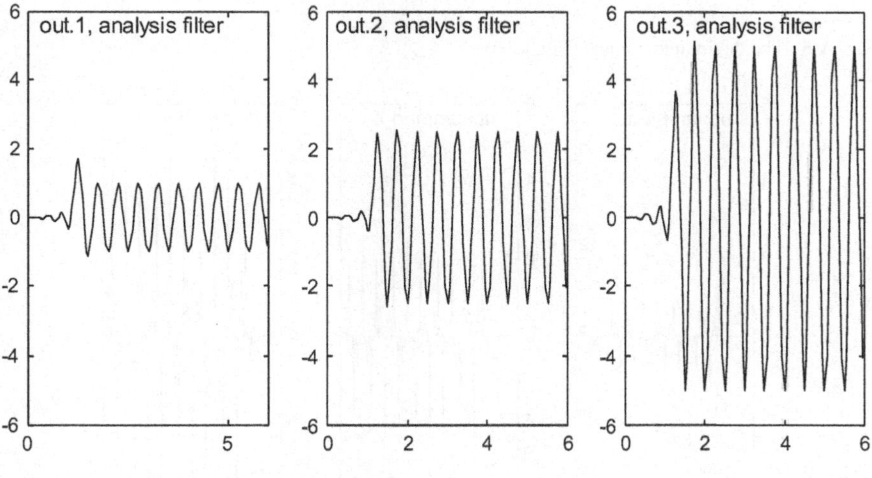

Fig. A.4 The outputs of the analysis filters

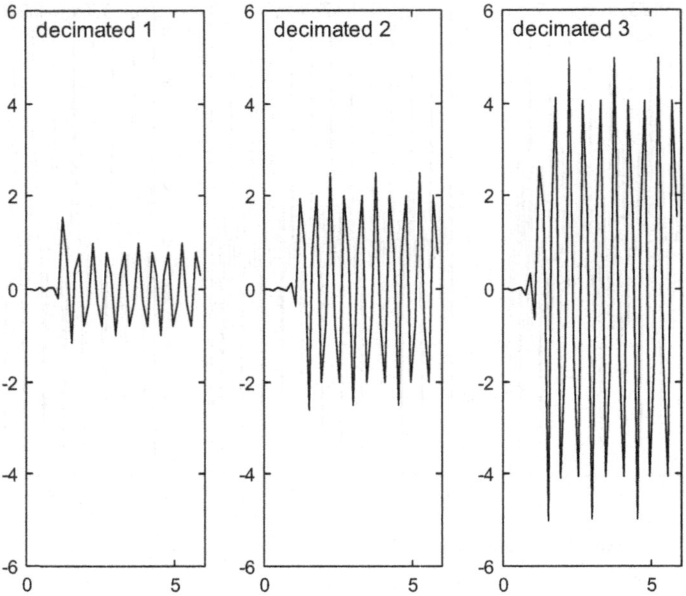

Fig. A.5 The decimated signals (Fig. 1.10)

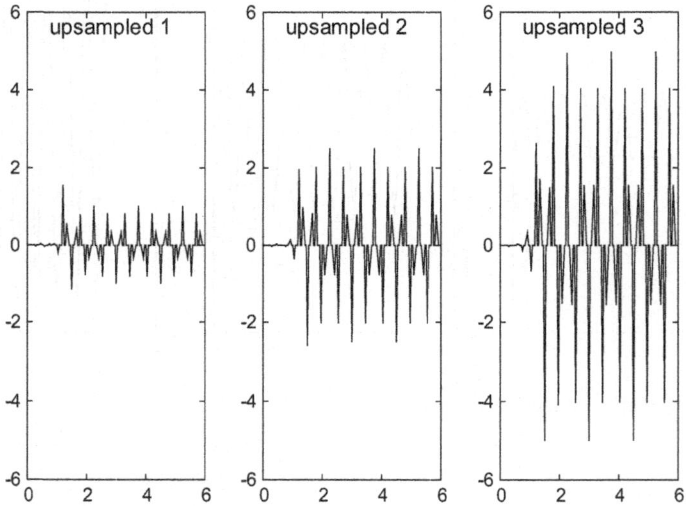

Fig. A.6 The upsampled signals (Fig. 1.11)

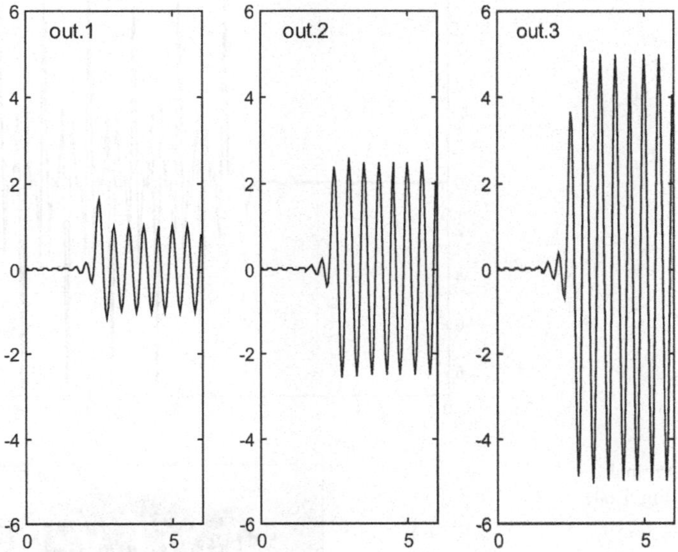

Fig. A.7 The outputs of the synthesis filters (Fig. 1.12)

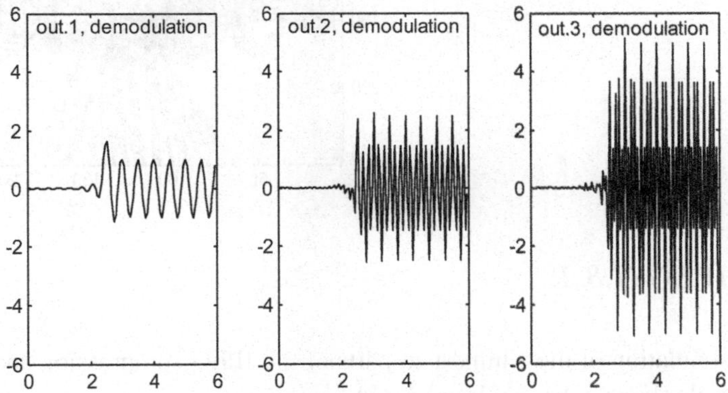

Fig. A.8 The demodulated signals

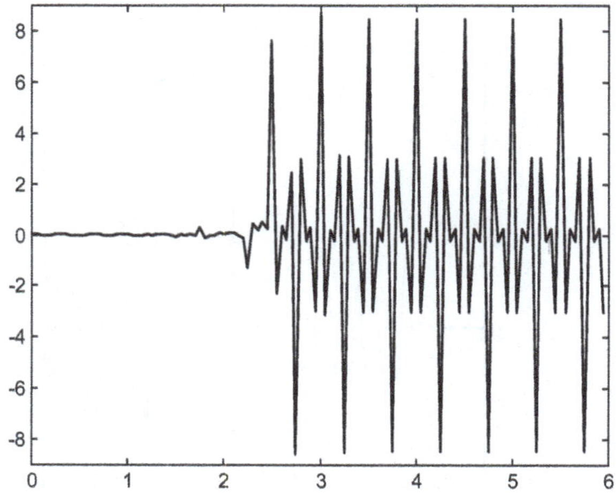

Fig. A.9 The output signal (Fig. 1.13)

Fig. A.10 Original photograph (Fig. 1.64)

A.2.2 JPEG (1.8.2)

This is a simulation of most important parts of the JPEG compression and decompression cycle (Figs. A.10, A.11 and A.12).

Fig. A.11 Recovered photograph (Fig. 1.65)

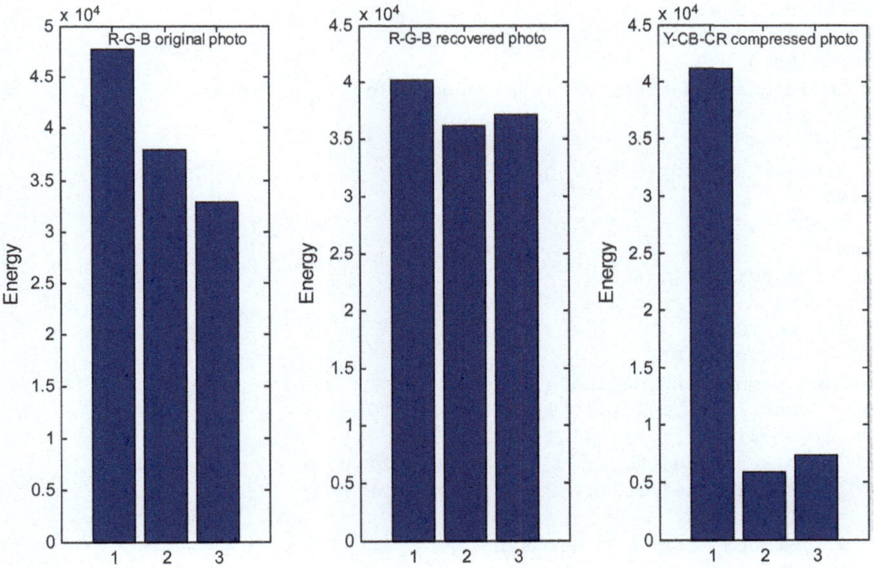

Fig. A.12 "Energies" of representations (Fig. 1.66)

Program A.2 JPEG simulation

```
% JPEG simulation
%
P=imread('flowers.jpg'); %read image
P=P(1:256,1:256,:); %256x256 size
N=256;
F=im2double(P);
M=dctmtx(8);
% Quantization tables
qmax = 255;
% quality factors for Y components
```

```
QY = ...
[16 11 10 16 124 140 151 161;
12 12 14 19 126 158 160 155;
14 13 16 24 140 157 169 156;
14 17 22 29 151 187 180 162;
18 22 37 56 168 109 103 177;
24 35 55 64 181 104 113 192;
49 64 78 87 103 121 120 101;
72 92 95 98 112 100 103 199];
% quality factors for chroma components
QC = ...
[17 18 24 47 99 99 99 99;
18 21 26 66 99 99 99 99;
24 26 56 99 99 99 99 99;
47 66 99 99 99 99 99 99;
99 99 99 99 99 99 99 99;
99 99 99 99 99 99 99 99;
99 99 99 99 99 99 99 99;
99 99 99 99 99 99 99 99];
% Scale quantization matrices based on quality factor
qf = 20;
if qf < 50
  q_scale = floor(5000 / qf);
else
  q_scale = 200 - 2 * qf;
end
QY = round(QY * q_scale / 100);
QC = round(QC * q_scale / 100);
% RGB to YCbCr
YC = rgb2ycbcr(F);
% Down-sample and decimate chroma
cb = conv2(YC(:,:,2), [1 1; 1 1]) ./ 4.0;
cr = conv2(YC(:,:,3), [1 1; 1 1]) ./ 4.0;
CB = cb(2:2:size(cb, 1), 2:2:size(cb, 2));
CR = cr(2:2:size(cr, 1), 2:2:size(cr, 2));
Y = YC(:,:,1);
L=N/2; %both CB and CR are LxL
% 2D DCT, and scaling
DY = blkproc(Y,[8 8],'P1*x*P2',M,M' ) .* qmax;
DCB = blkproc(CB,[8 8], 'P1*x*P2',M,M') .* qmax;
DCR = blkproc(CR,[8 8], 'P1*x*P2',M,M') .* qmax;
% Quantize DCT coefficients
qDY = blkproc(DY,[8 8],'round(round(x)./P1)',QY);
qDCB = blkproc(DCB,[8 8],'round(round(x)./P1)',QC);
qDCR = blkproc(DCR,[8 8],'round(round(x)./P1)',QC);
%end of compression
```

```
%=============================================================
%start decompression
% Dequantize DCT coefficients
dDY = blkproc(qDY,[8 8],'x.*P1',QY);
dDCB = blkproc(qDCB,[8 8],'x.*P1',QC);
dDCR = blkproc(qDCR,[8 8],'x.*P1',QC);
% Inverse DCT
iY = blkproc(dDY./qmax, [8 8],'P1*x*P2',M',M);
iCB = blkproc(dDCB./qmax, [8 8],'P1*x*P2',M',M);
iCR = blkproc(dDCR./qmax, [8 8],'P1*x*P2',M',M);
% Up-sample chroma
uf1=[1 3 3 1] / 4; %1D filter
uf2=uf1'*uf1; %2D filter
% array padding (replicate borders)
aCB=zeros(L+2,L+2); aCB(2:L+1,2:L+1)=iCB;
aCB(2:L+1,1)=iCB(:,1); aCB(2:L+1,L+2)=iCB(:,L);
aCB(1,:)=aCB(2,:); aCB(L+2,:)=aCB(L+1,:);
aCR=zeros(L+2,L+2); aCR(2:L+1,2:L+1)=iCR;
aCR(2:L+1,1)=iCR(:,1); aCR(2:L+1,L+2)=iCR(:,L);
aCR(1,:)=aCR(2,:); aCR(L+2,:)=aCR(L+1,:);
% upsampling
M=2*(L+2);
uCB=zeros(M,M);
uCB(1:2:M-1,1:2:M-1)=aCB;
rCB=conv2(uf2,uCB);
uCR=zeros(M,M);
uCR(1:2:M-1,1:2:M-1)=aCR;
rCR=conv2(uf2,uCR);
rCB = rCB(4:size(rCB,1)-4, 4:size(rCB,2)-4);
rCR = rCR(4:size(rCR,1)-4, 4:size(rCR,2)-4);
% Concatenate and convert to RGB
JF = ycbcr2rgb(cat(3, iY, rCB, rCR));
% ensure range (0..1)
m1=min(min(JF(:,:,1))); M1=max(max(JF(:,:,1)));
JF(:,:,1)=(JF(:,:,1)-m1)/(M1-m1);
m2=min(min(JF(:,:,2))); M2=max(max(JF(:,:,2)));
JF(:,:,2)=(JF(:,:,2)-m2)/(M2-m2);
m3=min(min(JF(:,:,3))); M3=max(max(JF(:,:,3)));
JF(:,:,3)=(JF(:,:,3)-m3)/(M3-m3);
% "Energies"
EO=round(sum(sum(abs(F)))); %original
ECY=sum(sum(abs(qDY))); %compressed Y
ECB=sum(sum(abs(qDCB))); %compressed CB
ECR=sum(sum(abs(qDCR))); %compressed CR
EJ=round(sum(sum(abs(JF)))); %recovered
%display
figure(1)
imagesc(F);
title('original photo');
```

```
figure(2)
imagesc(JF);
title('recovered photo');
figure(3)
subplot(1,3,1)
bar(squeeze(EO)); title('R-G-B original photo')
ylabel('Energy');
subplot(1,3,2)
bar(squeeze(EJ)); title('R-G-B recovered photo')
ylabel('Energy');
subplot(1,3,3)
bar([ECY ECB ECR]); title('Y-CB-CR compressed photo')
ylabel('Energy');
```

Depending on the version of MATLAB you are using, you may take advantage of *blockproc( )* instead of *blkproc( )*.

A.3 Chapter 2: Wavelets

A.3.1 The Multiresolution Analysis Equation (2.3)

Computation of the Daubechies 4 scaling function based on dyadic approach (Fig. A.13).

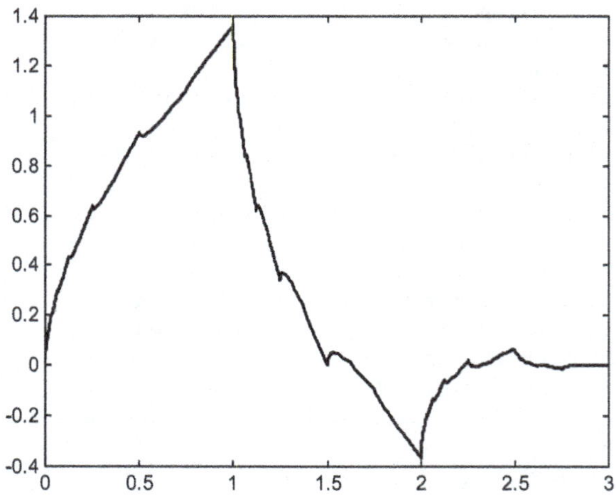

Fig. A.13 Daubechies 4 scaling function (Fig. 2.25)

Program A.3 Compute Daubechies 4 scaling function with dyadic approach

```
% Compute a scaling function with dyadic approach
% Daubechies 4,coeffs.
hden=4*sqrt(2); %coeff. denominator
hsq=sqrt(3); %factor
%Daubechies 4:
h=[(1+hsq)/hden, (3+hsq)/hden, (3-hsq)/hden, (1-hsq)/hden];
hN=(h*2)/sum(h); %normalization
K=length(hN);
hrev=hN(K:-1:1); %reverse hN
%M0 matrix
MA=[hrev,zeros(1,(2*K)-2)]; MB=MA;
for nn=1:K-1,
  MA=[0,0,MA(1:(3*K)-4)]; MB=[MB; MA];
end
M0=MB(:,K:(2*K)-1);
%Solving the first system of equations, for fi(0)..fi(3)
MF=M0-eye(K);
MG=[MF(1:K-1,:);ones(1,K)];
nfi=MG\[zeros(K-1,1);1];
%getting middle & quarter values
fi=nfi(2:length(nfi)-1); fi=conv(hN,fi);
aux=fi(1:2:length(fi)); %downsampling by 2
y=conv(hN,aux); %quarter values
%merge y and fi
aux=[y;fi,0]; aux=aux(:)'; fi=aux(1:length(aux)-1);
%iteration
hup=hN; Nx=9;
for nn=1:Nx,
  %upsample by 2
  L=length(hup); aux=[hup;zeros(1,L)]; aux=aux(:)';
  hup=aux(1:2*L-1);
  y=conv(hup,y); %intermediate terms
  %merge y and fi
  aux=[y;fi,0]; aux=aux(:)'; fi=aux(1:length(aux)-1);
end
%result
x=(1:length(fi))*(K-1)/length(fi);
plot(x,fi,'k')
title('Daubechies 4 scaling function');
```

A.3.2 Orthogonal Wavelets (2.4.3)

Figure A.14 shows the Daubechies scaling function (left) and the wavelet (right) for $N = 4, 8, 12,..26$.

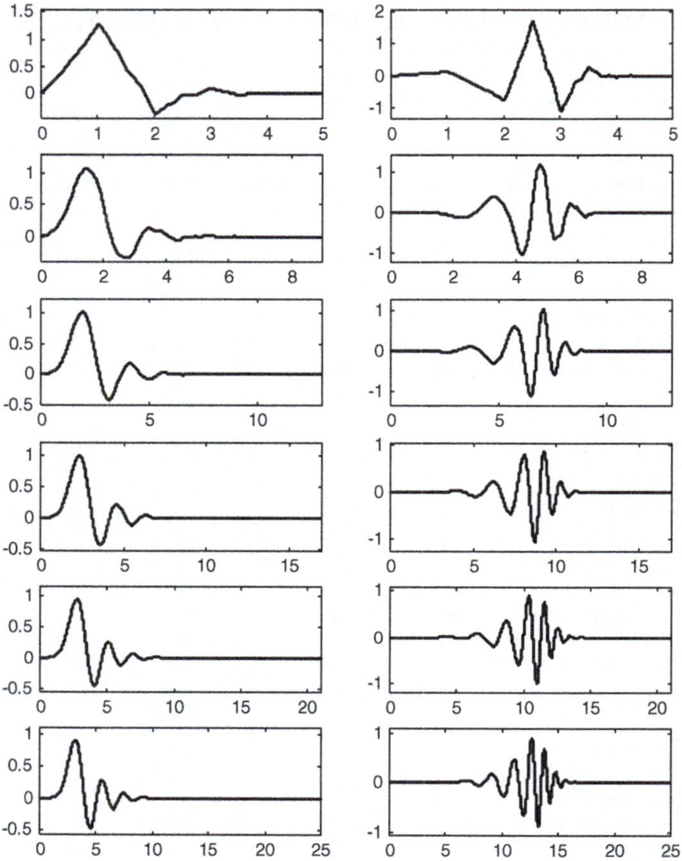

Fig. A.14 Daubechies scaling function (*left*) and wavelet (*right*) for N = 4, 8, 12, ..26 (Fig. 2.41)

Program A.4 Compute Daubechies phi(t), psi(t)

```
% Compute Daubechies phi(t), psi(t)
% for L=6,10,14..26
for MM=2:2:12,
  K=MM+1;
  a=1; p=1; q=1;
  h=[1 1];
  M=2*K; %length of the filter
  % the h0(n) coefficients
  for nn=1:K-1,
    h=conv(h,[1,1]);
    a=-a*0.25*(nn+K-1)/nn;
    p=conv(p,[1,-2,1]);
    q=[0 q 0] + a*p;
  end;
  q=sort(roots(q));
  aux=real(poly(q(1:K-1)));
```

```
  h=conv(h,aux);
  h0=(h*sqrt(2))/sum(h); %normalization
  %the scaling function phi(t), using cascade algorithm
  Ns=128; %number of fi samples
  hN=sqrt(2)*h0;
  phi=[ones(1,3*M*Ns),0]/(3*M); %initial iteration
  %upsample hN, inserting Ns-1 zeros between samples
  hup=[hN;zeros(Ns-1,M)];
  hup=hup(1:(Ns*M));
  %iteration
  for nn=0:12,
    aux=conv(hup,phi);
    phi=aux(1:2:length(aux)); %downsampling by 2
  end
  %the h1(n) coefficients
  h1=fliplr(h0); h1(1:2:end)=-h1(1:2:end);
  H1=fft(h1,512); %HP filter frequency response
  %the wavelet psi(t), using definition
  %upsample by K
  hN=-sqrt(2)*h1;
  h1up=[hN;zeros(Ns-1,M)];
  h1up=h1up(1:Ns*M-1);
  %downsample by 2
  aux=conv(h1up,phi);
  psi=aux(1:2:length(aux));
  %display
  subplot(6,2,(MM)-1)
  phi=phi(1:(M-1)*Ns); %the supported part
  t=(1:length(phi))/Ns;
  plot(t,phi,'k'); %plots the scaling function
  axis([0 max(t) 1.2*min(phi) 1.2*max(phi)]);
  subplot(6,2,(MM))
  psi=psi(1:(M-1)*Ns); %the supported part
  plot(t,psi,'k'); %plots the wavelet
  axis([0 max(t) 1.2*min(psi) 1.2*max(psi)]);
end;
```

A.3.3 Coiflets (2.4.3)

Figure A.15 shows the Coiflet1 scaling function and wavelet.

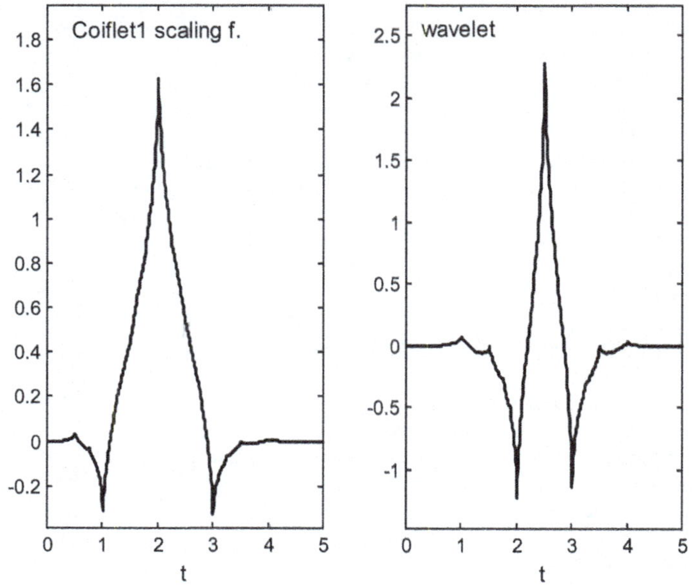

Fig. A.15 Coiflet1 scaling function (*left*) and wavelet (*right*) (Fig. 2.43)

Program A.5 Coiflet1 phi(t), psi(t)

```
Coiflet1 phi(t), psi(t)
%coefficients
h0=[-0.051429728471,0.238929728471,0.602859456942...
,0.272140543058...
,-0.051429972847,-0.011070271529]*sqrt(2);
Ns=128; %number of function samples
%scaling function
%using cascade algorithm
M=length(h0);
hN=sqrt(2)*h0;
phi=[ones(1,3*M*Ns),0]/(3*M); %initial iteration
%upsample hN, inserting Ns-1 zeros between samples
hup=[hN;zeros(Ns-1,M)];
hup=hup(1:(Ns*M));
%iteration
for nn=0:12,
  aux=conv(hup,phi);
  phi=aux(1:2:length(aux)); %downsampling by 2
end
%wavelet
%the h1(n) coefficients
h1=fliplr(h0); h1(1:2:end)=-h1(1:2:end);
%the wavelet psi(t), using definition
%upsample
hN=sqrt(2)*h1;
h1up=[hN;zeros(Ns-1,M)];
```

```
h1up=h1up(1:Ns*M-1);
%downsample by 2
aux=conv(h1up,phi);
psi=aux(1:2:length(aux));
%display
subplot(1,2,1)
phi=phi(1:(M-1)*Ns); %the supported part
t=(1:length(phi))/Ns;
plot(t,phi,'k'); %plots the scaling function
axis([0 max(t) 1.2*min(phi) 1.2*max(phi)]);
title('Coiflet1 scaling f.'); xlabel('t');
subplot(1,2,2)
psi=psi(1:(M-1)*Ns); %the supported part
t=(1:length(psi))/Ns;
plot(t,psi,'k'); %plots the wavelet
axis([0 max(t) 1.2*min(psi) 1.2*max(psi)]);
title('wavelet'); xlabel('t');
```

A.3.4 Biorthogonal Wavelets (2.5.1)

Figure A.16 shows the scaling functions and the wavelets of cdf 9/7.

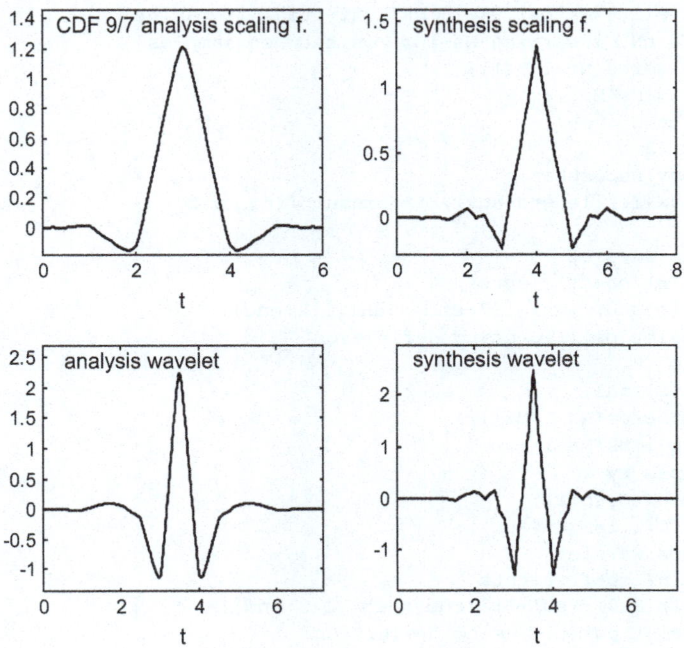

Fig. A.16 CDF 9/7 scaling functions and wavelets (Fig. 2.49)

Program A.6 CDF 9/7 phi(t), psi(t)

```
% CDF 9/7 phi(t), psi(t)
%coefficients
ah0=[-0.045635881557,-0.028771763114,0.295635881557...
,0.557543526229...
,0.295635881557,-0.028771763114,-0.045635881557];
sh0=[0.026748757411,-0.016864118443,-0.078223266529...
,0.266864118443,0.602949018236,0.266864118443...
,-0.078223266529,-0.016864118443,0.026748757411];
Ns=128; %number of function samples
%analysis scaling function
%using cascade algorithm
Ma=length(ah0);
ah0=2*ah0/sum(ah0);
aphi=[ones(1,3*Ma*Ns),0]/(3*Ma); %initial iteration
%upsample hN, inserting Ns-1 zeros between samples
hup=[ah0;zeros(Ns-1,Ma)];
hup=hup(1:(Ns*Ma));
%iteration
for nn=0:12,
  aux=conv(hup,aphi);
  aphi=aux(1:2:length(aux)); %downsampling by 2
end
%synthesis scaling function
%using cascade algorithm
Ms=length(sh0);
sh0=2*sh0/sum(sh0);
sphi=[ones(1,3*Ms*Ns),0]/(3*Ms); %initial iteration
%upsample hN, inserting Ns-1 zeros between samples
hup=[sh0;zeros(Ns-1,Ms)];
hup=hup(1:(Ns*Ms));
%iteration
for nn=0:12,
  aux=conv(hup,sphi);
  sphi=aux(1:2:length(aux)); %downsampling by 2
end
%analysis wavelet
%the ah1(n) coefficients
ah1=fliplr(sh0); ah1(1:2:end)=-ah1(1:2:end);
%the wavelet psi(t), using definition
%upsample
hN=-sqrt(2)*ah1;
h1up=[hN;zeros(Ns-1,Ms)];
h1up=h1up(1:Ns*Ms-1);
%downsample by 2
aux=conv(h1up,aphi);
apsi=aux(1:2:length(aux));
%synthesis wavelet
%the sh1(n) coefficients
sh1=fliplr(ah0); sh1(1:2:end)=-sh1(1:2:end);
%the wavelet psi(t), using definition
%upsample
hN=sqrt(2)*sh1;
```

```
h1up=[hN;zeros(Ns-1,Ma)];
h1up=h1up(1:Ns*Ma-1);
%downsample by 2
aux=conv(h1up,sphi);
spsi=aux(1:2:length(aux));
%display
subplot(2,2,1)
aphi=aphi(1:(Ma-1)*Ns); %the supported part
t=(1:length(aphi))/Ns;
plot(t,aphi,'k'); %plots the scaling function
axis([0 max(t) 1.2*min(aphi) 1.2*max(aphi)]);
title('CDF 9/7 analysis scaling f.'); xlabel('t');
subplot(2,2,3)
su=round(0.9*length(apsi));
t=(1:su)/Ns;
plot(t,apsi(1:su),'k'); %plots the wavelet
axis([0 max(t) 1.2*min(apsi) 1.2*max(apsi)]);
title('analysis wavelet'); xlabel('t');
subplot(2,2,2)
sphi=sphi(1:(Ms-1)*Ns); %the supported part
t=(1:length(sphi))/Ns;
plot(t,sphi,'k'); %plots the scaling function
axis([0 max(t) 1.2*min(sphi) 1.2*max(sphi)]);
title('synthesis scaling f.'); xlabel('t');
subplot(2,2,4)
su=round(0.9*length(spsi));
t=(1:su)/Ns;
plot(t,spsi(1:su),'k'); %plots the wavelet
axis([0 max(t) 1.2*min(spsi) 1.2*max(spsi)]);
title('synthesis wavelet'); xlabel('t');
```

A.4 Chapter 3: Images and 2D Signals

A.4.1 Design of 2D Filters (3.9.2)

Figure A.17 shows the frequency response of a 2D fan filter.

An example of image was chosen (a Big-Ben) and then it was filtered with the fan filter. Figure A.18 shows the result.

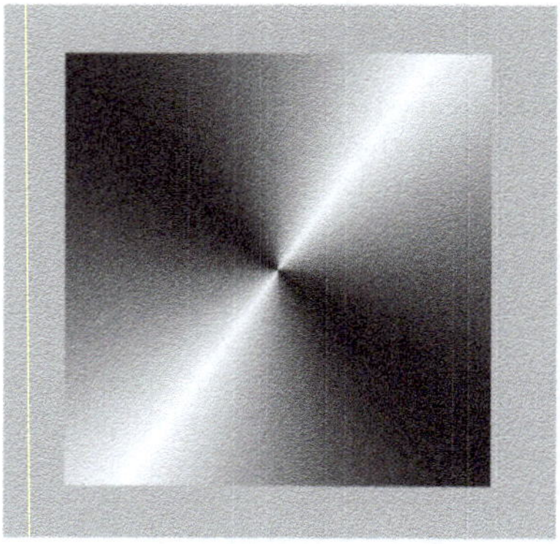

Fig. A.17 The general fan filter frequency response (Fig. 3.105)

Fig. A.18 The filtered picture (Fig. 3.106)

Program A.7 Example of general fan filter

```
% Example of general fan filter
% filtering on 2-D Fourier domain
%
%read the 256x256 image file into a matrix:
phot=imread('Ben1.tif');
N=256;
L=N/2;
%filter specification
alpha=1; beta=0.1; %prototype
theta=pi/3;
a=1/tan(theta/2); %aperture
phi=0.3*pi; %direction
cp=cos(phi); sp=sin(phi);
aux1=[-1 -3 -1; 0 0 0; 1 3 1];
aux2=[-1 0 1; -3 0 3; -1 0 1];
P=(cp*aux1)-(sp*aux2);
Q=(sp*aux1)+(cp*aux2);
```

```
B=alpha*Q;
aux3=(a*P)+(beta*Q);
A=B+(j*aux3);
H=zeros(N,N);
%frequency response of the fan filter
% (first quarter)
AX=zeros(1,L);
for ny=1:L,
  nl=L+1-ny;
  w2=(ny*pi)/N; z2=exp(j*w2);
  for nx=1:L,
    nc=L+nx;
    w1=(nx*pi)/N; z1=exp(j*w1);
    VZ1=[1 z1 z1^2]; VZ2=[1 z2 z2^2];
    hnum=VZ1*B*VZ2'; hden=VZ1*A*VZ2';
    if abs(hden)<1.0e-3, hnum=100; hden=1; end;
    H(nl,nc)=hnum/hden;
  end;
end;
BX=zeros(1,L);
% (second quarter, counterclockwise)
for ny=1:L,
  nl=L+1-ny;
  w2=(ny*pi)/N; z2=exp(j*w2);
  for nx=1:L,
    nc=L+1-nx;
    w1=-(nx*pi)/N; z1=exp(j*w1);
    VZ1=[1 z1 z1^2]; VZ2=[1 z2 z2^2];
    hnum=VZ1*B*VZ2'; hden=VZ1*A*VZ2';
    if abs(hden)<1.0e-3, hnum=100; hden=1; end;
    H(nl,nc)=hnum/hden;
  end;
end;
% (third quarter)
for ny=1:L,
  nl=L+ny;
  w2=-(ny*pi)/N; z2=exp(j*w2);
  for nx=1:L,
    nc=L+1-nx;
    w1=-(nx*pi)/N; z1=exp(j*w1);
    VZ1=[1 z1 z1^2]; VZ2=[1 z2 z2^2];
    hnum=VZ1*B*VZ2'; hden=VZ1*A*VZ2';
    if abs(hden)<1.0e-3, hnum=100; hden=1; end;
    H(nl,nc)=hnum/hden;
  end;
end;
```

```
% (fourth quarter)
for ny=1:L,
  nl=L+ny;
  w2=-(ny*pi)/N; z2=exp(j*w2);
  for nx=1:L,
    nc=L+nx;
    w1=(nx*pi)/N; z1=exp(j*w1);
    VZ1=[1 z1 z1^2]; VZ2=[1 z2 z2^2];
    hnum=VZ1*B*VZ2'; hden=VZ1*A*VZ2';
    if abs(hden)<1.0e-3, hnum=100; hden=1; end;
    H(nl,nc)=hnum/hden;
  end;
end;
%H=abs(H);
%filtering (Fourier domain)
Fph=fftshift(fft2(phot));
Fphfil=Fph.*H;
Iph=ifft2(Fphfil);
uIp=uint8(abs(Iph));
%display
figure(1)
uH=uint8(256-abs(256*H)); %invert color
imshow(uH);
title('Support of the general fan filter');
figure(2)
imshow(uIp);
title('Filtered image');
```

A.5 Chapter 4: Wavelet Variants for 2D Analysis

A.5.1 Laplacian Pyramid (4.2)

Figure A.19 depicts the sizes of images obtained with a Laplacian pyramid.

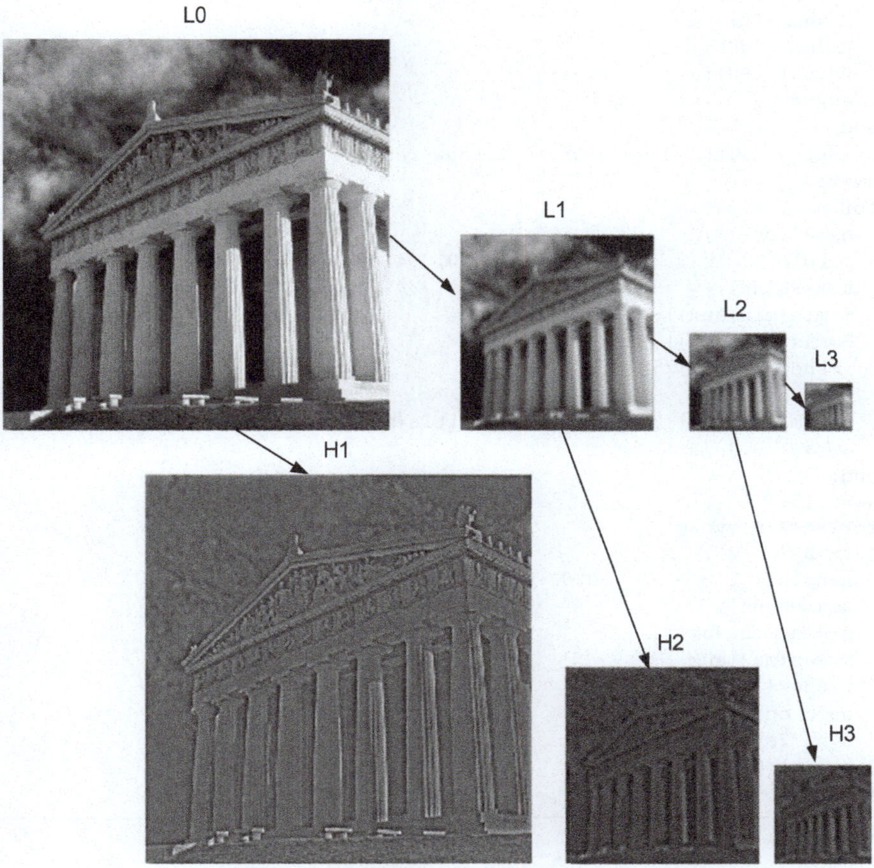

Fig. A.19 Laplacian pyramid: sizes of images shrink along analysis (Fig. 4.2)

Program A.8 Laplacian pyramid

```
%Laplacian pyramid
%get an image
ufg=imread('Parth1.tif');
fg=double(ufg); %convert to float
fg=fg-mean(mean(fg)); %zero mean
[Nr,Nc]=size(fg);
nlevels=4; %number of pyramid levels
%preparing indexed matrix sets
PH=cell(nlevels,1); %for high pass
PL=cell(nlevels,1); %for low pass
PL(1)={fg}; PH(1)={ones(Nr,Nc)};
aux=fg;
for nn=2:nlevels,
  fil=fspecial('gaussian',[16,16],4);
  FL=filter2(fil,aux);
  dFL=FL(1:2:end,1:2:end); %subsampling
```

```
  FH=aux-FL;
  PL(nn)={dFL};
  PH(nn)={FH};
  aux=dFL;
end;
%display (with conversion to imshow range)
aw=0;
for nn=1:nlevels
  ax=2^(nn-1);
  subplot('Position',[0.01+aw,0.50,0.44/ax,0.44/ax])
  aux=PL{nn};
  m=min(min(aux));
  M=max(max(aux));
  imshow((aux-m)/(M-m));
  hold on;
  s=num2str(nn-1); msg=['L',s]; title(msg);
  aw=aw+(0.49/ax);
end;
aw=0.15;
for nn=2:nlevels
  ax=2^(nn-2);
  subplot('Position',[0.01+aw,0.01,0.44/ax,0.44/ax])
  aux=PH{nn};
  m=min(min(aux));
  M=max(max(aux));
  imshow((aux-m)/(M-m));
  hold on;
  s=num2str(nn-1); msg=['H',s]; title(msg);
  aw=aw+(0.45/ax);
end;
```

A.5.2 Application of Wavelets to Images (4.4.1)

Figure A.20 shows the test image recovered from wavelet transform. It is generated with the Program A.9, which uses filters.

Fig. A.20 Image recovered from wavelet transform using filters

Program A.9 Haar wavelet transform of a basic image

```
% Haar wavelet transform of a basic image
% recovery of image
% square and rhombus
% Using filters
%
%Haar filter
c=1/sqrt(2);
h0=[c c]; %low-pass filter
h1=[-c c]; %high-pass filter
%
%----------------------------------------
% First the transformed images
%
%Original image
fg=ones(256,256); %white plane
%the rhombus
for n=1:32,
  fgx=64+(-n:n); fgy=96+n;
  fg(fgx,fgy)=0; %one triangle
  fg(fgx,256-fgy)=0; %the other triangle
end
%the square
fg(128:196,96:160)=0;
%
%Step1
%1 scale wavelet transform of rows
lfg1=conv2(fg,h0); %low-pass filtering of rows
lfg1=lfg1(:,2:2:end); %downsampling to get L
hfg1=conv2(fg,h1); %high-pass filtering of rows
hfg1=hfg1(:,2:2:end); %downsampling to get H
%Step 2
%1 scale wavelet transform of columns of previous step
llfg1=conv2(lfg1,h0'); %low-pass filtering of L columns
llfg1=llfg1(2:2:end,:); %downsampling
hlfg1=conv2(lfg1,h1'); %high-pass filtering of L columns
hlfg1=hlfg1(2:2:end,:); %downsampling
lhfg1=conv2(hfg1,h0'); %low-pass filtering of H columns
lhfg1=lhfg1(2:2:end,:); %downsampling
hhfg1=conv2(hfg1,h1'); %high-pass filtering of H columns
hhfg1=hhfg1(2:2:end,:); %downsampling
figure(1)
imshow(fg);
%----------------------------------------
% Second the recovery of original image
% from LL,LH, HL and HH
%
rfg=zeros(256,256); %space for image
auxL=zeros(258,128);
auxH=zeros(258,128);
auxO=zeros(256,258);
%recovery of L
auxL(1:2:258,:)=conv2(llfg1,h0')+conv2(hlfg1,h1');
```

```
auxL(2:2:258,:)=conv2(llfg1,h0')-conv2(hlfg1,h1');
lfg1=auxL(3:258,:);
%recovery of H
auxH(1:2:258,:)=conv2(lhfg1,h0')+conv2(hhfg1,h1');
auxH(2:2:258,:)=conv2(lhfg1,h0')-conv2(hhfg1,h1');
hfg1=auxH(3:258,:);
%recovery of original
auxO(:,1:2:258)=conv2(lfg1,h0)+conv2(hfg1,h1);
auxO(:,2:2:258)=conv2(lfg1,h0)-conv2(hfg1,h1);
rfg=auxO(:,1:256)/4;
figure(2)
imshow(rfg);
title('recovered image');
h=gca;ht=get(h,'Title'); set(ht,'FontSize',12);
```

A.5.3 Curvelets (Second Generation) (4.5.4)

Figure A.21 shows a 3D view of the basic curvelet obtained by the product of V and W. A top view is shown in Fig. A.22.

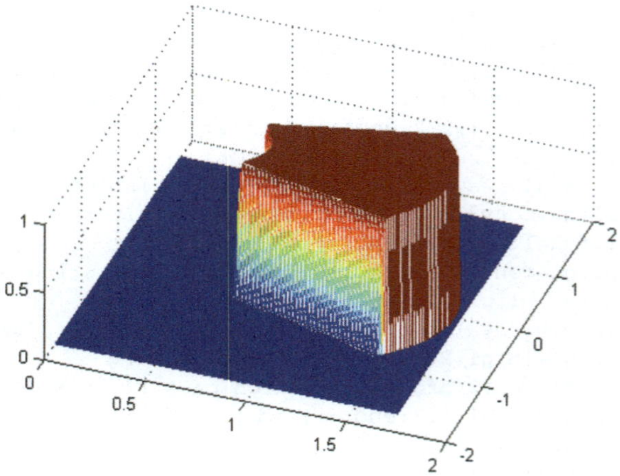

Fig. A.21 3D view of the basic curvelet example (Fig. 4.46)

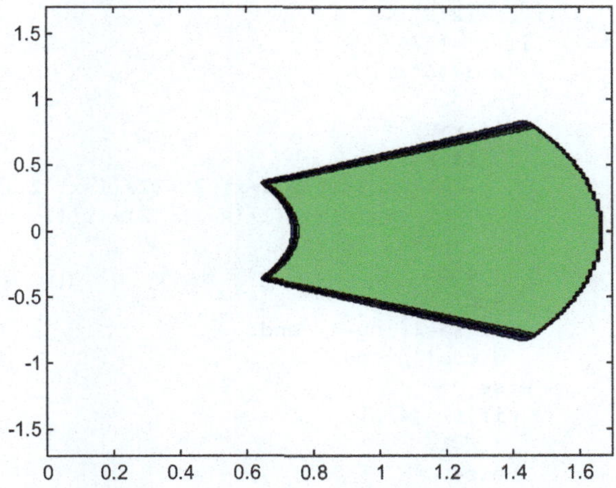

Fig. A.22 Top view of the curvelet (Fig. 4.47)

Program A.10 Using Meyer windows to build a curvelet

```
% Using Meyer windows to build a curvelet
L=170; %grid side length
X=zeros(L+1,1); Y=zeros(L+1,1);
CV=zeros(L+1,L+1); %space for curvelet
for nl=0:L,
  Y(nl+1)=nl/100; yaux=nl/100;
  for nc=0:L,
    X(nc+1)=nc/100; xaux=nc/100;
    r=sqrt(xaux^2+yaux^2);
    if xaux>0,
      theta=atan(yaux/xaux);
    else
      theta=pi/2;
    end;
    V=0;
    if theta<=(1/3),
      V=1;
    else
      if theta<=(2/3),
        x=(3*theta)-1;
        nu=0;
        if x>0,
          if x<1,
            s1=exp(-((1/(1+x)^2)+(1/(1-x)^2)));
            s2=exp(-((1/(1+(x-1))^2)+(1/(1-(x-1))^2)));
            nu=s2/(s2+s1);
          end;
        end;
        if x>=1, nu=1; end;
        V=cos((pi*nu)/2);
      end;
    end;
    W=0;
```

```
    if r>=(2/3),
      if r<=(5/6),
        x=5-(6*r);
        nu=0;
        if x>0,
          if x<1,
            s1=exp(-((1/(1+x)^2)+(1/(1-x)^2)));
            s2=exp(-((1/(1+(x-1))^2)+(1/(1-(x-1))^2)));
            nu=s2/(s2+s1);
          end;
        end;
        if x>=1, nu=1; end;
        W=cos((pi*nu)/2);
      else
        if r<=(4/3),
          W=1;
        else
          if r<=(5/3),
            x=3-(4*r);
            nu=0;
            if x>0,
              if x<1,
                s1=exp(-((1/(1+x)^2)+(1/(1-x)^2)));
                s2=exp(-((1/(1+(x-1))^2)+(1/(1-(x-1))^2)));
                nu=s2/(s2+s1);
              end;
            end;
            if x>=1, nu=1; end;
            W=cos((pi*nu)/2);
          end;
        end;
      end;
    end;
    CV(nl+1,nc+1)=W*V;
  end;
end;
Y=[flipud(Y);-Y]; %use symmetry around horizontal
Z=zeros(L+1,L+1);
CV=[flipud(CV);CV];
figure(1)
colormap('jet');
mesh(X,Y,CV);
view(20, 60);
title('Basic curvelet in the frequency plane: 3D view');
figure(2)
colormap('winter');
contourf(X,Y,CV,[0.1 0.3 0.5 0.7 0.9]);
view(2);
title('Basic curvelet in the frequency plane: support');
```

A.5.4 Complex Wavelets (4.6)

Figure A.23 shows an example of two wavelets that can be used to approximate a Hilbert pair. Most of the the spectrum of the complex wavelet formed with this pair lies on the right-hand side, as shown in Fig. A.24.

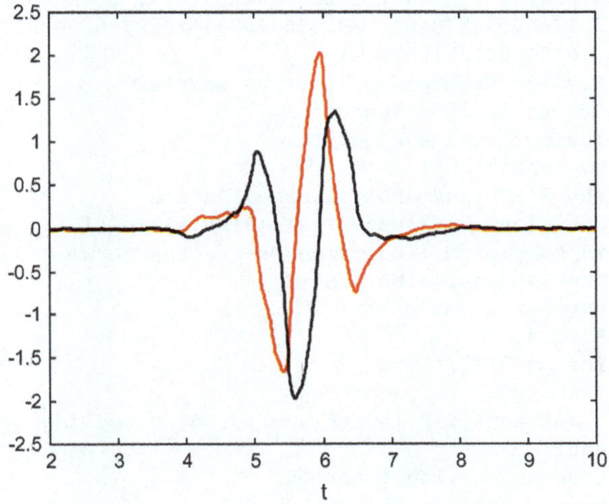

Fig. A.23 Example of wavelets for Hilbert pair approximation (Fig. 4.81)

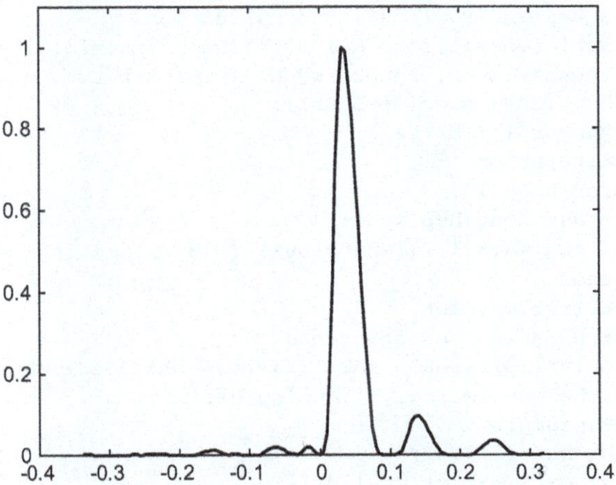

Fig. A.24 Spectrum of the complex wavelet (Fig. 4.82)

Program A.11 Dual-tree complex wavelet example

```
% Dual-tree complex wavelet example
%coefficients (scaling functions)
ah0=[0.00419528584157, -0.03976408134143, -0.08807084231507,...
0.28789890325798, 0.80289644768232, 0.50734324828341,...
-0.04438514804476, -0.05179712664076, 0.03247103802248,...
0.00342583762736];
sh0=[0.00051763584333, -0.00016716564000, -0.09187942035452,...
0.02408482114448, 0.61837942541527, 0.75480699639212,...
0.17400853530401, -0.09044673462008, 0.00608060497845,...
0.01882886391002, 0];
Ns=128; %number of function samples
%first scaling function
%using cascade algorithm
Ma=length(ah0);
ah0=2*ah0/sum(ah0); %normalization
aphi=[ones(1,3*Ma*Ns),0]/(3*Ma); %initial iteration
%upsample hN, inserting Ns-1 zeros between samples
hup=[ah0;zeros(Ns-1,Ma)];
hup=hup(1:(Ns*Ma));
%iteration
for nn=0:12,
  aux=conv(hup,aphi);
  aphi=aux(1:2:length(aux)); %downsampling by 2
end
%second scaling function
%using cascade algorithm
Ms=length(sh0);
sh0=2*sh0/sum(sh0); %normalization
sphi=[ones(1,3*Ms*Ns),0]/(3*Ms); %initial iteration
%upsample hN, inserting Ns-1 zeros between samples
hup=[sh0;zeros(Ns-1,Ms)];
hup=hup(1:(Ns*Ms));
%iteration
for nn=0:12,
  aux=conv(hup,sphi);
  sphi=aux(1:2:length(aux)); %downsampling by 2
end
%first wavelet
%the ah1(n) coefficients
ah1=fliplr(sh0); ah1(1:2:end)=-ah1(1:2:end);
%the wavelet psi(t), using definition
%upsample
hN=sqrt(2)*ah1;
h1up=[hN;zeros(Ns-1,Ms)];
h1up=h1up(1:Ns*Ms-1);
%downsample by 2
aux=conv(h1up,aphi);
ax=aux(1:2:length(aux));
apsi=fliplr(ax);
%second wavelet
%the sh1(n) coefficients
sh1=fliplr(ah0);sh1(1:2:end)=-sh1(1:2:end);
```

```
%the wavelet psi(t), using definition
%upsample
hN=-sqrt(2)*sh1;
h1up=[hN;zeros(Ns-1,Ma)];
h1up=h1up(1:Ns*Ma-1);
%downsample by 2
aux=conv(h1up,sphi);
ax=aux(1:2:length(aux));
spsi=fliplr(ax);
%spectrum of the complex wavelet
FA=fftshift(fft(apsi));
FS=fftshift(fft(spsi));
S=abs(FA+(i*FS));
S=S/max(S); %normalize
%display
figure(1)
su=1280;
ti=256;
t=(ti:su)/Ns;
plot(t,apsi(ti:su),'r'); hold on; %plots the wavelet
plot(t,spsi(ti:su),'k'); %plots the other wavelet
axis([ti/Ns max(t) -2.5 2.5]);
title('the wavelets'); xlabel('t');
figure(2)
w=-pi:(2*pi/(length(S)-1)):pi;
rg=600:740; %horizontal zoom
plot(w(rg),S(rg),'k');
axis([-0.4 0.4 0 1.1]);
title('spectrum');
```

A.5.5 Six Complex Wavelets (4.6)

Figure A.25 shows six complex wavelets: the first row is the real part, the second row is the imaginary part, and the third row is the magnitude (which has no oscillatory behavior). The program uses a series of functions that are are listed after the program.

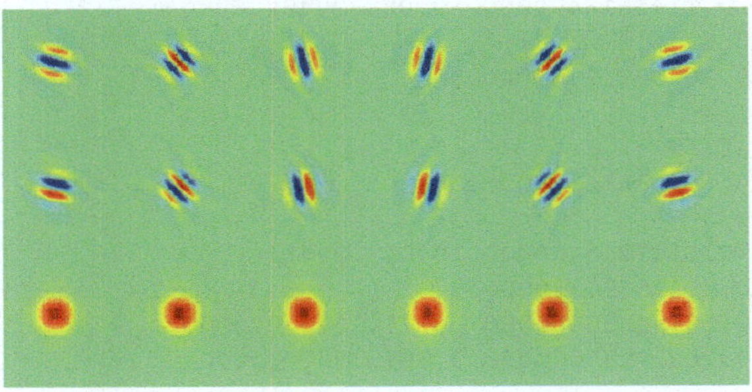

Fig. A.25 The six complex wavelets (Fig. 4.91)

Program A.12 Complex 2D dual-tree wavelets display

```
% Complex 2D dual-tree wavelets display
NS=4; %number of stages
L=3*(2^(NS+1));
M=L/(2^NS);
fg=zeros(2*L,6*L); %blank image
%First filter coeffs.----------------------------
af1{1} = [
0 0
-0.08838834764832 -0.01122679215254
0.08838834764832 0.01122679215254
0.69587998903400 0.08838834764832
0.69587998903400 0.08838834764832
0.08838834764832 -0.69587998903400
-0.08838834764832 0.69587998903400
0.01122679215254 -0.08838834764832
0.01122679215254 -0.08838834764832
0 0
];
sf1{1} = af1{1}(end:-1:1, :);
af1{2} = [
0.01122679215254 0
0.01122679215254 0
-0.08838834764832 -0.08838834764832
0.08838834764832 -0.08838834764832
0.69587998903400 0.69587998903400
0.69587998903400 -0.69587998903400
0.08838834764832 0.08838834764832
-0.08838834764832 0.08838834764832
0 0.01122679215254
0 -0.01122679215254
];
sf1{2} = af1{2}(end:-1:1, :);
%The other filter coeffs.----------------------------
aff{1} = [
0.03516384000000 0
0 0
-0.08832942000000 -0.11430184000000
0.23389032000000 0
0.76027237000000 0.58751830000000
0.58751830000000 -0.76027237000000
0 0.23389032000000
-0.11430184000000 0.08832942000000
0 0
0 -0.03516384000000
];
aff{2} = [
0 -0.03516384000000
0 0
-0.11430184000000 0.08832942000000
0 0.23389032000000
0.58751830000000 -0.76027237000000
0.76027237000000 0.58751830000000
```

```
0.23389032000000 0
-0.08832942000000 -0.11430184000000
0 0
0.03516384000000 0
];
sff{1} = aff{1}(end:-1:1, :);
sff{2} = aff{2}(end:-1:1, :);
%wavelet coeffs as cells:-------------------------------
% w{s}{p}{d1}{d2}
% s=1:NS, scale; p=1(real part),2(imaginary part)
% orientations d1=1..2, d2=1..3
% w(NS+1}{m}{n} - low-pass coeffs
%Complex 2D Dual-tree transform --------------
fg=fg/2; %normalize
for m=1:2,
  for n=1:2,
    [LO w{1}{m}{n}]=D_afilt(fg,af1{m},af1{n});
    for nj=2:NS,
      [LO w{nj}{m}{n}]=D_afilt(LO,aff{m},aff{n});
    end;
    w{NS+1}{m}{n}=LO; %low-pass coeffs.
  end;
end;
q=sqrt(2);
for nj=1:NS,
  for m=1:3,
    auxa=w{nj}{1}{1}{m};
    auxb=w{nj}{2}{2}{m};
    w{nj}{1}{1}{m}=(auxa+auxb)/q;
    w{nj}{2}{2}{m}=(auxa-auxb)/q;
    auxa=w{nj}{1}{2}{m};
    auxb=w{nj}{2}{1}{m};
    w{nj}{1}{2}{m}=(auxa+auxb)/q;
    w{nj}{2}{1}{m}=(auxa-auxb)/q;
  end;
end;
%Prepare for inversion experiment---------------
w{NS}{1}{2}{2}(M/2,M/2)=1;
w{NS}{1}{1}{3}(M/2,M/2+M)=1;
w{NS}{1}{2}{1}(M/2,M/2+2*M)=1;
w{NS}{1}{1}{1}(M/2,M/2+3*M)=1;
w{NS}{1}{2}{3}(M/2,M/2+4*M)=1;
w{NS}{1}{1}{2}(M/2,M/2+5*M)=1;
w{NS}{2}{2}{2}(M/2+M,M/2)=1;
w{NS}{2}{1}{3}(M/2+M,M/2+M)=1;
w{NS}{2}{2}{1}(M/2+M,M/2+2*M)=1;
w{NS}{2}{1}{1}(M/2+M,M/2+3*M)=1;
w{NS}{2}{2}{3}(M/2+M,M/2+4*M)=1;
w{NS}{2}{1}{2}(M/2+M,M/2+5*M)=1;
%Inverse complex 2D Dual-tree transform --------------
for nj=1:NS,
  for m=1:3,
    auxa=w{nj}{1}{1}{m};
```

```
    auxb=w{nj}{2}{2}{m};
    w{nj}{1}{1}{m}=(auxa+auxb)/q;
    w{nj}{2}{2}{m}=(auxa-auxb)/q;
    auxa=w{nj}{1}{2}{m};
    auxb=w{nj}{2}{1}{m};
    w{nj}{1}{2}{m}=(auxa+auxb)/q;
    w{nj}{2}{1}{m}=(auxa-auxb)/q;
  end;
end;
y=zeros(size(w{1}{1}{1}{1})*2);
for m=1:2,
  for n=1:2,
    LO=w{NS+1}{m}{n};
    for nj=NS:-1:2,
      LO=D_sfilt(LO,w{nj}{m}{n},sff{m},sff{n});
    end;
    LO=D_sfilt(LO,w{1}{m}{n},sf1{m},sf1{n});
    y=y+LO;
  end;
end;
y=y/2; %normalize
%display-----------------------------------------------
figure(1)
aux=(y(1:L,:).^2+y(L+[1:L],:).^2);
y=[y; sqrt(aux)];
imagesc(y);
title('the Complex 2D wavelets');
axis image;
axis off;
```

Here are the functions used by the Program A.12.

Program A.13 D_afilt

```
function [lo,hi]=D_afilt(x,af1,aff)
% 2D analysis filtering
%columns filtering
[l,h]=D_1af(x,af1,1);
%rows filtering
[lo, hi{1}]=D_1af(l,aff,2);
[hi{2}, hi{3}]=D_1af(h,aff,2);
```

Program A.14 D_1af

```
function [lo,hi]=D_1af(x,af,d)
% 1D analysis filtering
lpf=af(:,1); %low-pass filter
hpf=af(:,2); %high-pass filter
if d==2, x=x'; end;
N=size(x,1);
L=size(af,1)/2;
%2D circular shift
```

```
[nn,mm]=size(x); m=-L; n=0:nn-1;
n=mod(n-m,nn);
x=x(n+1,:);
lo=upfirdn(x,lpf,1,2);
lo(1:L,:)=lo(1:L,:)+lo([1:L]+N/2,:);
lo=lo(1:N/2,:);
hi=upfirdn(x,hpf,1,2);
hi(1:L,:)=hi(1:L,:)+hi([1:L]+N/2,:);
hi=hi(1:N/2,:);
if d==2, lo=lo'; hi=hi'; end;
```

Program A.15 D_sfilt

```
function y=D_sfilt(lo,hi,sf1,sf2)
% 2D synthesis filtering
%rows filtering
lo=D_1sf(lo,hi{1},sf2,2);
hi=D_1sf(hi{2},hi{3},sf2,2);
%columns filtering
y=D_1sf(lo,hi,sf1,1);
```

Program A.16 D_1sf

```
function y=D_1sf(lo,hi,sf,d)
% 1D synthesis filtering
lpf=sf(:,1); %low-pass filter
hpf=sf(:,2); %high-pass filter
if d==2, lo=lo'; hi=hi'; end;
N=2*size(lo,1);
L=length(sf);
y=upfirdn(lo,lpf,2,1)+upfirdn(hi,hpf,2,1);
y(1:L-2,:)=y(1:L-2,:)+y(N+[1:L-2],:);
y=y(1:N,:);
%2D circular shift
[nn,mm]=size(y); m=1-L/2; n=0:nn-1;
n=mod(n-m,nn);
y=y(n+1,:);
if d==2, y=y'; end;
```

A.6 Chapter 5: Adaptive Filters and Observers

A.6.1 Observer in the Presence of Noise (5.8.2)

Figure A.26 shows the evolution of the system and the observer states in the presence of noise.

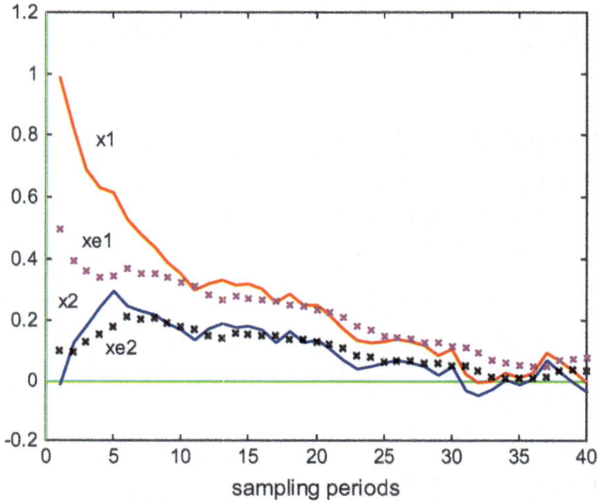

Fig. A.26 System and observer state evolution in presence of noise (Fig. 5.66)

Figure A.27 shows the evolution of the observation error.

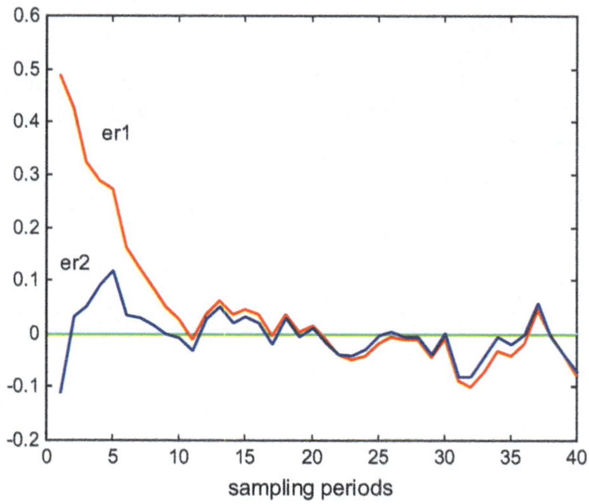

Fig. A.27 Observation error evolution (Fig. 5.67)

Program A.17 Observer example, in noisy conditions

```
%Observer example, in noisy conditions
%state space system model (2 tank system):
A1=1; A2=1; R1=0.5; R2=0.4;
A=[-1/(R1*A1) 1/(R1*A1); 1/(R1*A2) -(1/A2)*((1/R1)+(1/R2))];
B=[1/A1; 0]; C=[0 1]; D=0;
Ts=0.1; %sampling period
csys=ss(A,B,C,D); %setting the continuous time model
dsys=c2d(csys,Ts,'zoh'); %getting the discrete-time model
[a,b,c,d]=ssdata(dsys); %retrieves discrete-time model matrices
% noise sizes:
fiw=0.03; %state noise standard deviation
fiv=0.02; %output noise standard deviation
% noise influence on next state:
Jw=[1;1];
% system simulation preparation
Nf=40; %simulation horizon
x1=zeros(1,Nf); % for x1(n) record
x2=zeros(1,Nf); % for x2(n) record
y=zeros(1,Nf); % for y(n) record
x=[1;0]; % state vector with initial tank levels
u=0.1; %constant input
sn=fiw*randn(1,Nf+1); %state noise along simulation
on=fiv*randn(1,Nf+1); %output noise along simulation
x=x+(Jw*sn(1)); %initial system state, with noise
% observer simulation preparation
K=[0.3; 0.3]; %observer constants
er=zeros(2,Nf); % for error record
xe1=zeros(1,Nf); % for xe1(n) record
xe2=zeros(1,Nf); % for xe2(n) record
xe=[0.5; 0.1]; % observer state vector with initial values
%behaviour of the system and the observer after initial state
% with constant input u
for nn=1:Nf,
  x1(nn)=x(1); x2(nn)=x(2); %recording the system state
  y(nn)=(c*x)+on(nn); %recording the system output
  xe1(nn)=xe(1); xe2(nn)=xe(2); %recording the observer state
  er(:,nn)=x-xe; %recording the error
  xn=(a*x)+(b*u)+(Jw*sn(nn)); %next system state
  x=xn; %system state actualization
  xen=(a*xe)+(b*u)+(K*(y(nn)-(c*xe))); %next observer state
  xe=xen; %observer state actualization
end;
% display of states evolution
figure(1)
plot([0 Nf],[0 0],'g'); hold on; %horizontal axis
plot([0 0],[-0.2 1.2],'k'); %vertical axis
plot(x1,'r'); %plots x1
plot(x2,'b'); %plots x2
plot(xe1,'mx'); %plots xe1
plot(xe2,'kx'); %plots xe2
xlabel('sampling periods');
title('system and observer states');
```

```
% display of error evolution
figure(2)
plot([0 Nf],[0 0],'g'); hold on; %horizontal axis
plot([0 0],[-0.2 0.6],'k'); %vertical axis
plot(er(1,:),'r'); %plots x1 error
plot(er(2,:),'b'); %plots x2 error
xlabel('sampling periods');
title('observation error');
```

A.7 Chapter 6: Experimental Modelling

A.7.1 Obtaining a Transfer Function Model from Impulse Response (6.4.2)

Figure A.28 compares the impulse response of $G2(z)$ and the impulse response of the estimated transfer function $G2(z)$ and Figure A.29 compares the frequency responses of original and estimated $G2(z)$.

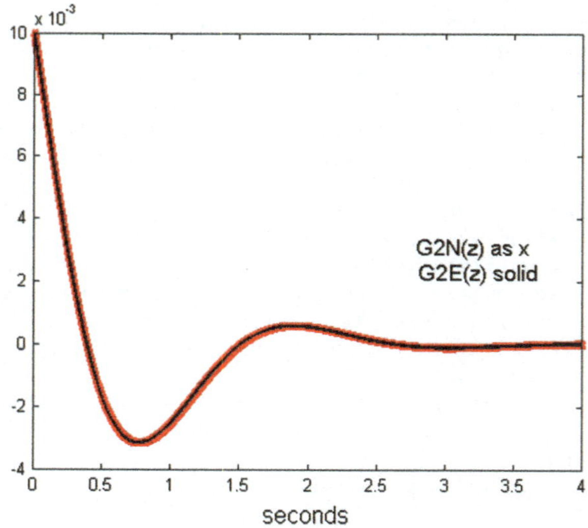

Fig. A.28 Comparison of impulse responses of original and estimated G2(z) (Fig. 6.17)

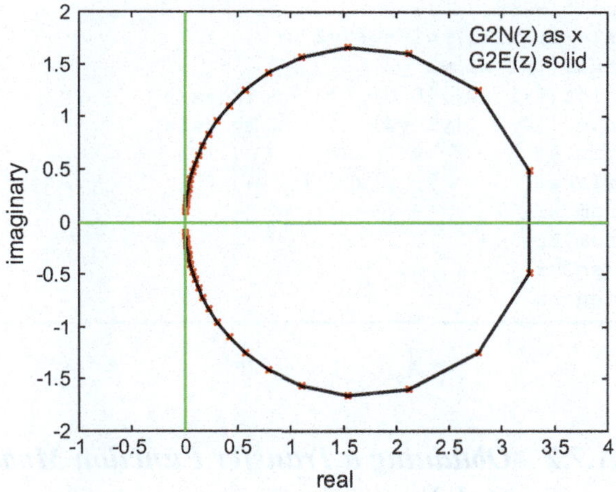

Fig. A.29 Comparison of frequency responses of original and estimated G2(z) (Fig. 6.18)

Program A.18 Obtain discrete transfer function (DTrF) from impulse response

```
%Obtain discrete transfer function (DTrF) from impulse response
%DTrF case 2
%no noise
% continuous time transfer function:
num2Ns=[10 0]; den2Ns=[1 3 10]; %G2(s)=10s/(s^2+3s+10);
fs=1000; %sampling frequency in Hz
Ts=1/fs; %sampling period in seconds
%discrete transfer function (from the continuous case)
[num2Nz,den2Nz]= impinvar(num2Ns,den2Ns,fs); %G2(z)
%impulse response of G2(z)
h2Nz=impz(num2Nz,den2Nz,128,fs);
%using stmcb to obtain the DTrF
na=2; %denominator degree
nb=1; %numerator degree
[num2Ez,den2Ez]=stmcb(h2Nz,nb,na); %DTrF computation
%comparing impulse responses
figure(1)
t=0:Ts:(4-Ts); %sampling times data set (4 second)
Ns=length(t);
h2Nz=impz(num2Nz,den2Nz,Ns); %impulse response of G2(z)
h2Ez=impz(num2Ez,den2Ez,Ns); %impulse response of ^G2(z)
plot(t,h2Nz,'xr',t,h2Ez,'k'); %plots both impulse responses
title('impulse response, G2N(z)as x & G2E(z) solid');
xlabel('seconds');
%comparing frequency responses
figure(2)
wr=logspace(-1,2); %frequency values for response (rad/s)
%G2(z) frequency response:
H2Nz=freqz(num2Nz,den2Nz,wr/(2*pi),fs);
plot(H2Nz,'xr'); hold on;
%^G2(z) frequency response:
H2Ez=freqz(num2Ez,den2Ez,wr/(2*pi),fs);
```

```
plot(H2Ez,'k');
x1=-1; x2=4; y1=-2; y2=2;
axis([x1 x2 y1 y2]);
plot([x1 x2],[0 0],':g'); %x axis
plot([0 0],[y1 y2],':g'); %y axis
title('complex frequency response, G2N(z)as x & G2E(z) solid');
xlabel('real'); ylabel('imaginary');
num2Nz
num2Ez
den2Nz
den2Ez
```

A.7.2 *Obtaining a Transfer Function Model from Sine Sweep (6.4.3)*

Figure A.30 compares on the complex plane the frequency response of the estimated transfer function $G2(s)$, and the experimental frequency response data.

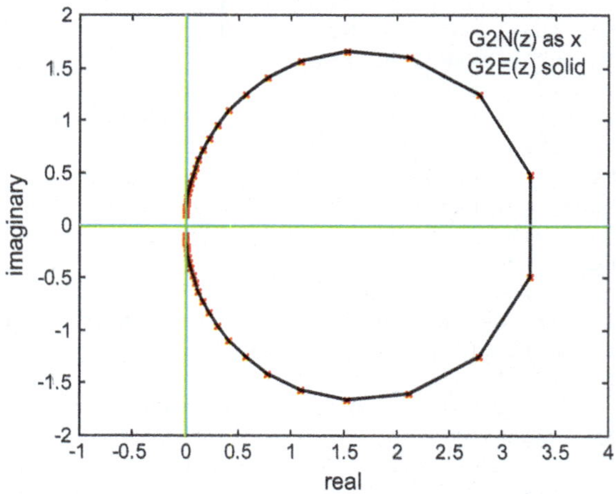

Fig. A.30 Comparison of frequency responses of original and estimated G2(z) (Fig. 6.21)

Program A.19 Obtain transfer function (TrF) from continuous frequency response

```
%Obtain transfer function (TrF)
% from continuous frequency response
%TrF case 2
%no noise
% continuous time transfer function:
num2Ns=[10 0]; den2Ns=[1 3 10]; %G2(s)=10s/(s^2+3s+10);
wr=logspace(-1,2); %frequency values for response (rad/s)
%continuous frequency response of G2(s)
H2Ns=freqs(num2Ns,den2Ns,wr); %G2(s) frequency response
%using invfreqs to obtain the TrF
na=2; %denominator degree
nb=1; %numerator degree
[num2Es,den2Es]=invfreqs(H2Ns,wr,nb,na); %TrF computation
H2Es=freqs(num2Es,den2Es,wr); %^G2(s) frequency response
%comparing frequency responses
plot(H2Ns,'xr'); hold on;
plot(H2Es,'k');
x1=-1; x2=4; y1=-2; y2=2;
axis([x1 x2 y1 y2]);
plot([x1 x2],[0 0],':g'); %x axis
plot([0 0],[y1 y2],':g'); %y axis
title('complex frequency response, G2N(z)as x & G2E(z) solid');
xlabel('real'); ylabel('imaginary');
num2Ns
num2Es
den2Ns
den2Es
```

A.7.3 Obtaining a Transfer Function Model from Noise Response (6.4.4)

Figure A.31 shows the $G2(z)$system input and output.

Figure A.32 compares the frequency response of the original $G2(z)$and the result of estimating this response from noisy input and output data.

The estimated frequency response data are used to estimate the transfer function of the plant. Figure A.33 compares the frequency response of the original $G2(z)$and the estimated $G2(z)$.

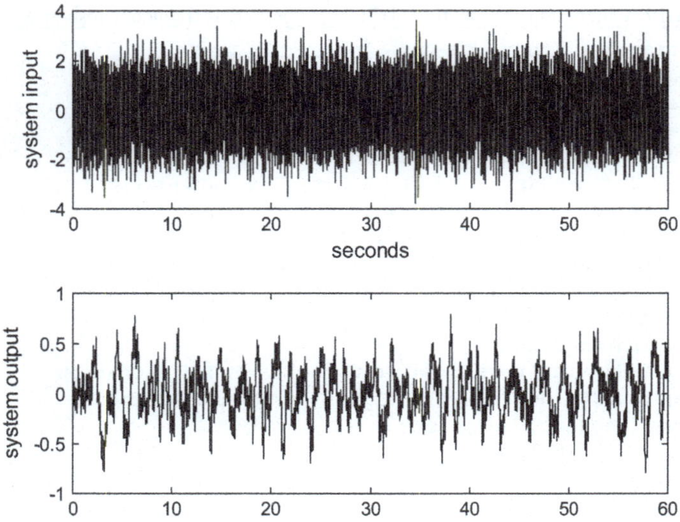

Fig. A.31 Noise input and response of G2(z) (Fig. 6.26)

Fig. A.32 Comparison of original G2(z) frequency response, and estimated frequency response (Fig. 6.27)

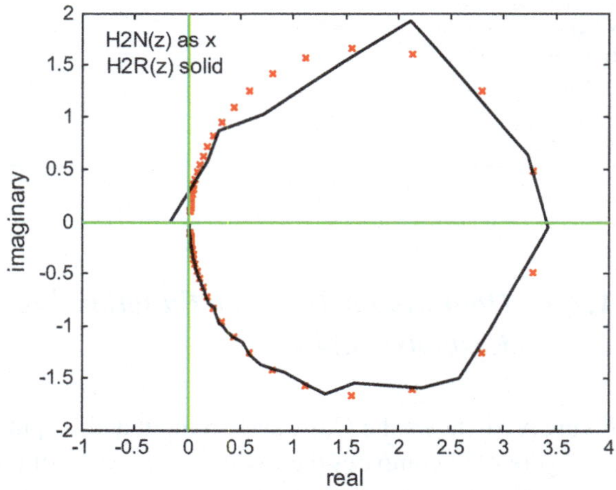

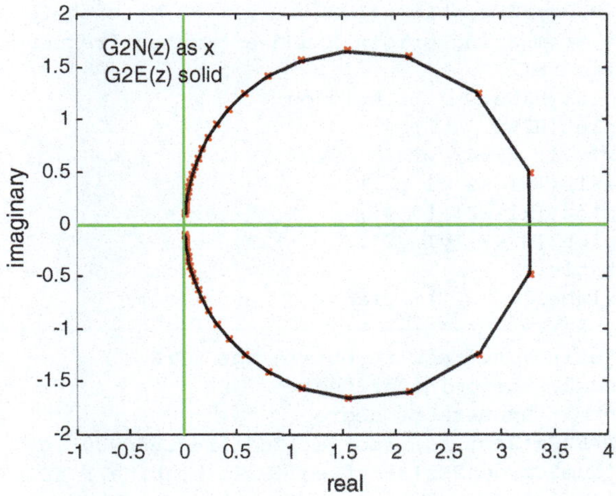

Fig. A.33 Comparison of frequency responses of original and estimated G2(z) (Fig. 6.28)

Program A.20 Obtain discrete transfer function (DTrF) from discrete frequency response to noise

```
%Obtain discrete transfer function (DTrF)
% from discrete frequency response to noise
%DTrF case 2
% continuous time transfer function:
num2Ns=[10 0]; den2Ns=[1 3 10]; %G2(s)=10s/(s^2+3s+10);
fs=300; %sampling frequency in Hz
%discrete transfer function (from the continuous case)
[num2Nz,den2Nz]= impinvar(num2Ns,den2Ns,fs); %G2(z)
%discrete frequency response of G2(z)
wr=logspace(-1,2); %frequency values for response (rad/s)
%G2(z) frequency response:
H2Nz=freqz(num2Nz,den2Nz,wr/(2*pi),fs);
%response of G2 to noise
tiv=1/fs; %time interval between samples;
t=0:tiv:(60-tiv); %time intervals set (60 seconds)
N=length(t); %number of data points
u=randn(N,1); %random input signal data set
y=filter(num2Nz,den2Nz,u); %G2(z) response to noise
%display of input and output signals
figure(1)
subplot(2,1,1)
plot(t,u,'k'); %input u plot
xlabel('seconds'); ylabel('system input');
title('input and output signals: case 2')
subplot(2,1,2)
plot(t,y,'k'); %output y plot
ylabel('system output');
%-------------------------------------------------
%frequency response estimate,using tfe
nfft=4096; %length of FFT
window=hanning(nfft); %window function
%frequency response estimate:
```

```
[H2Rz,F2Rz]=tfe(u(1000:N),y(1000:N),nfft,fs,window);
% %comparing original and estimated frequency responses
figure(2)
plot(H2Nz,'xr'); hold on;
plot(H2Rz,'k');
x1=-1; x2=4; y1=-2; y2=2;
axis([x1 x2 y1 y2]);
plot([x1 x2],[0 0],':g'); %x axis
plot([0 0],[y1 y2],':g'); %y axis
title('complex frequency response, H2N(z)as x & H2R(z) solid');
xlabel('real'); ylabel('imaginary');
%-----------------------------------------------
%using invfreqz to obtain the DTrF
na=2; %denominator degree
nb=1; %numerator degree
W=F2Rz*2*pi/fs; %normalized frequency 0..pi
[num2Ez,den2Ez]=invfreqz(H2Rz,W,nb,na); %DTrF computation
%^G2(z) frequency response:
H2Ez=freqz(num2Ez,den2Ez,wr/(2*pi),fs);
%comparing G2(z) and ^G2(z) frequency responses
figure(3)
plot(H2Nz,'xr'); hold on;
plot(H2Ez,'k');
axis([x1 x2 y1 y2]);
plot([x1 x2],[0 0],':g'); %x axis
plot([0 0],[y1 y2],':g'); %y axis
title('complex frequency response, G2N(z)as x & G2E(z) solid');
xlabel('real'); ylabel('imaginary');
num2Nz
num2Ez
den2Nz
den2Ez
```

A.7.4 Transfer Functions with Delay. Responses of Case 2d (6.5.3)

As it can be observed in Fig. A.34, the impulse response of case 2d is just a time-shifted version of the impulse response of case 2 (see Fig. 6.12).

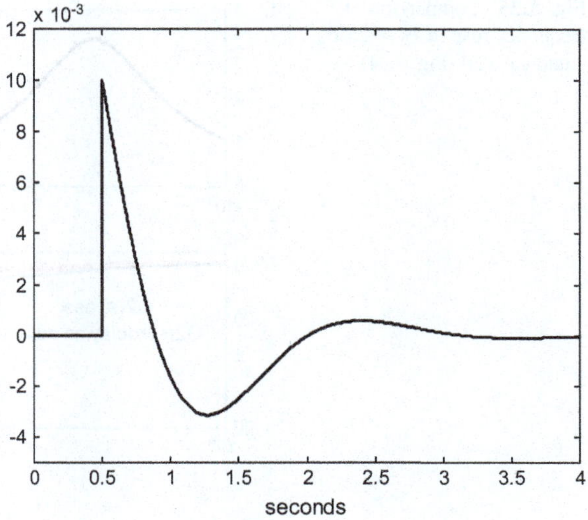

Fig. A.34 Impulse response of case 2d (Fig. 6.33)

Program A.21 Transfer function + delay, for study: impulse response

```
% Transfer function + delay, for study: impulse response
%case 2d
Td=0.5; %pure delay in seconds
% continuous time transfer function:
num2Ns=[10 0]; den2Ns=[1 3 10]; %G2(s)=10s/(s^2+3s+10)
fs=1000; %sampling frequency in Hz
Ts=1/fs; %sampling period in seconds
Nd=ceil(Td/Ts); %number of samples corresponding to the delay
%discrete transfer function (from the continuous cases)
[num2Nz,den2Nz]= impinvar(num2Ns,den2Ns,fs); %G2(z)
num2Nz=[zeros(1,Nd) num2Nz]; %adding delay to G2(z)
%impulse response
t=0:Ts:(4-Ts); %sampling times data set (4 seconds)
Ns=length(t); %number of samples
[h2Nz,t1]=impz(num2Nz,den2Nz,Ns); %G2(z) impulse response
plot(t,h2Nz,'b'); hold on;
title('impulse response of G2(z)');
xlabel('seconds');
axis([0 4 -0.005 0.012]);
```

The frequency response of case 2d is compared in Fig. A.35 with the frequency response of case 2.

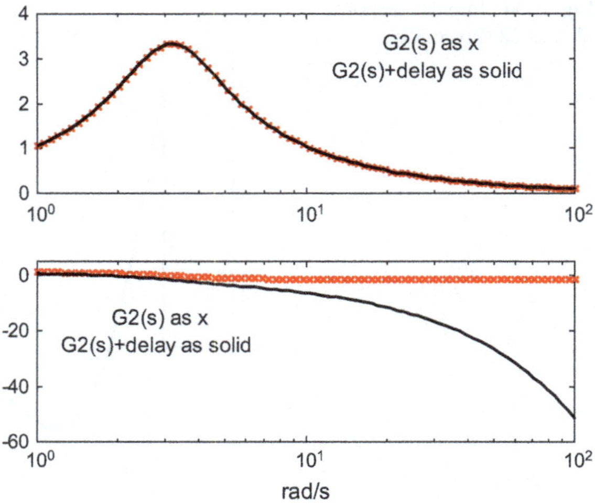

Fig. A.35 Comparison of frequency responses of case 2 and case 2d (Fig. 6.34)

Program A.22 Transfer function + delay, for study: frequency response (Bode d.)

```
%Transfer function + delay, for study:
% frequency response (Bode diagram)
%case 2d
Td=0.5; %pure delay in seconds
% continuous time transfer function:
num2Ns=[10 0]; den2Ns=[1 3 10]; %G2(s)=10S/(s^2+3S+10)
%frequency response
wr=logspace(0,2,80); %frequency values for response (rad/s)
H2Ns=freqs(num2Ns,den2Ns,wr); %G2(s) frequency response
H2Nsd=H2Ns.*exp(-j*Td*wr); %adding delay to G2(s)
subplot(2,1,1)
semilogx(wr,abs(H2Ns),'xr'); hold on;
semilogx(wr,abs(H2Nsd),'k'); hold on;
title('G2(s) as x & G2(s)+delay as solid');
axis([1 100 0 4]);
subplot(2,1,2)
semilogx(wr,angle(H2Ns),'xr'); hold on;
semilogx(wr,unwrap(angle(H2Nsd)),'k'); hold on;
xlabel('rad/s'); axis([1 100 -60 5]);
```

The frequency response of case 2d is also plotted on the complex plane, Fig. A.36

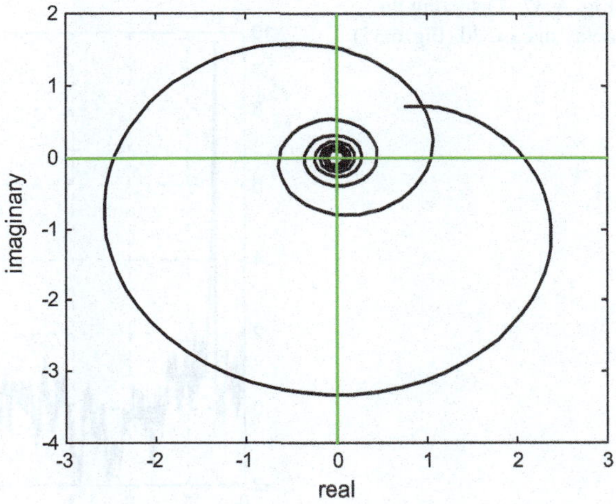

Fig. A.36 Frequency response of case 2d in the complex plane (Fig. 6.35)

Program A.23 Transfer function + delay, for study: frequency response (complex plane)

```
%Transfer function + delay, for study:
% frequency response (complex plane)
%case 2d
Td=0.5; %pure delay in seconds
% continuous time transfer function:
num2Ns=[10 0]; den2Ns=[1 3 10]; %G2(s)=10s/(s^2+3s+10)
%frequency response
wr=logspace(0,2,200); %frequency values for response (rad/s)
H2Ns=freqs(num2Ns,den2Ns,wr); %G2(s) frequency response
H2Nsd=H2Ns.*exp(-j*Td*wr); %adding delay to G2(s)
%display frequency response
plot(H2Nsd,'k'); hold on;
x1=-3; x2=3; y1=-4; y2=2;
axis([x1 x2 y1 y2]);
plot([x1 x2],[0 0],':g'); %x axis
plot([0 0],[y1 y2],':g'); %y axis
title('G2(s)+delay complex frequency response');
xlabel('real'); ylabel('imaginary');
```

A.7.5 Detecting the Delay (6.5.4)

There is a significant peak of the cross-correlation at 0.5 s (Fig. A.37).

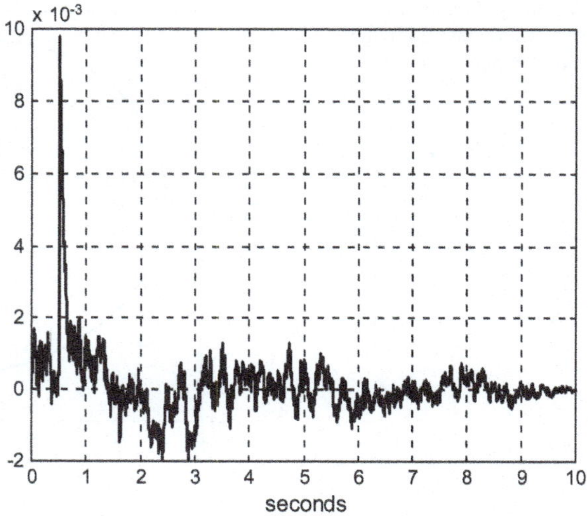

Fig. A.37 Detecting the delay in case 2d (Fig. 6.37)

Program A.24 Transfer function + delay, for study: noise response

```
% Transfer function + delay, for study: noise response
%case 2d
Td=0.5; %pure delay in seconds
% continuous time transfer function:
num2Ns=10; den2Ns=[1 10]; %G2(s)=10/(s+10)
fs=1000; %sampling frequency in Hz
Ts=1/fs; %sampling period in seconds
Nd=ceil(Td/Ts); %number of samples corresponding to the delay
%discrete transfer function (from the continuous cases)
[num2Nz,den2Nz]= impinvar(num2Ns,den2Ns,fs); %G2(z)
num2Nz=[zeros(1,Nd) num2Nz]; %adding delay to G2(z)
%response of G2 to noise
tiv=1/fs; %time interval between samples;
t=0:tiv:(10-tiv); %time intervals set (10 seconds)
N=length(t); %number of data points
u=randn(N,1); %random input signal data set
y=filter(num2Nz,den2Nz,u); %G2(z) response to noise
ac=xcorr(u,y); %cross-correlation of u and y
ac=ac/N; %normalization
Nac=length(ac); mNac=ceil(0.5*Nac); %to plot half ac
%display cross-correlation
plot(t,ac(mNac:Nac),'k');
grid;
title('G2(z)+delay noise response cross-correlation');
xlabel('seconds');
```

A.7.6 Getting Strange Models (6.5.5)

Figure A.38 compares the frequency response of case 2d and the frequency response of the obtained model.

Figure A.39 shows the pole-zero map of the obtained model. Delays may cause a lot of pole-zero pairs.

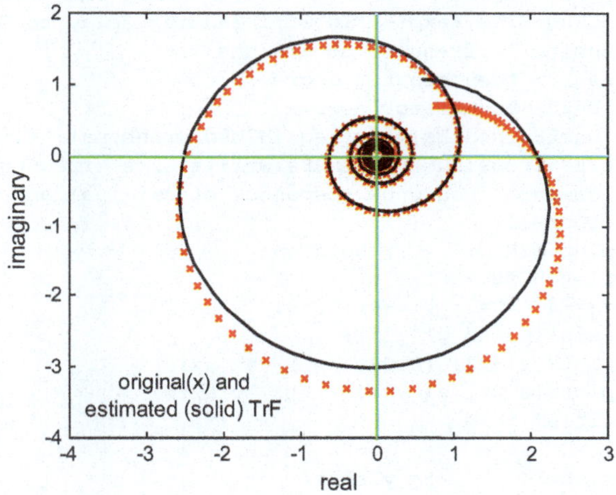

Fig. A.38 Comparison of frequency responses of original and estimated case 2d (Fig. 6.40)

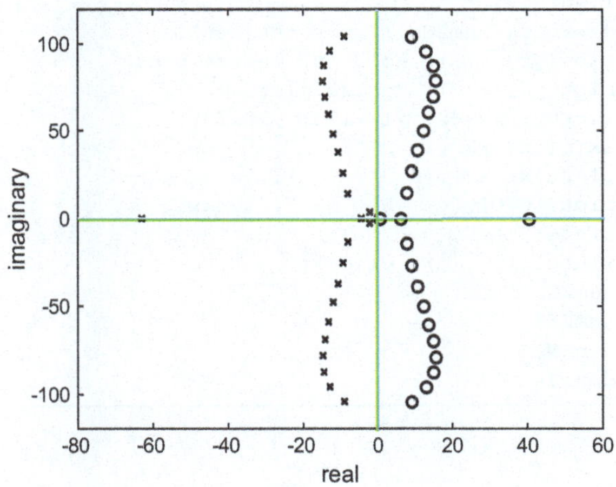

Fig. A.39 Pole-zero map of estimated TrF for case 2d (Fig. 6.41)

Program A.25 Strange model from frequency response of transfer function with delay

```
%Strange model from frequency response of transfer
% function with delay
%case 2d
Td=0.5; %pure delay in seconds
% continuous time transfer function:
num2Ns=[10 0]; den2Ns=[1 3 10]; %G2(s)=10s/(s^2+3s+10)
%frequency response
wr=logspace(0,2,200); %frequency values for response (rad/s)
H2Ns=freqs(num2Ns,den2Ns,wr); %G2(s) frequency response
H2Nsd=H2Ns.*exp(-j*Td*wr); %adding delay to G2(s)
%using invfreqs to obtain the TrF
na=29; %denominator degree
nb=27; %numerator degree
[num2Es,den2Es]=invfreqs(H2Nsd,wr,nb,na); %TrF computation
H2Es=freqs(num2Es,den2Es,wr); %^G2(s) frequency response
%compare frequency responses of original and estimated TrF
figure(1)
plot(H2Nsd,'xr'); hold on;
plot(H2Es,'k');
x1=-4; x2=3; y1=-4; y2=2;
axis([x1 x2 y1 y2]);
plot([x1 x2],[0 0],':g'); %x axis
plot([0 0],[y1 y2],':g'); %y axis
title('frequency responses of original(x)
  and estimated (solid) TrF');
xlabel('real'); ylabel('imaginary');
%pole-zero map of ^G2(s)
figure(2)
gz=roots(num2Es); gp=roots(den2Es);
plot(gz,'ok'); hold on; %zeros plot
plot(gp,'xk'); %poles plot
x1=-80; x2=60; y1=-120; y2=120;
axis([x1 x2 y1 y2]);
plot([x1 x2],[0 0],':g'); %x axis
plot([0 0],[y1 y2],':g'); %y axis
title('zeros and poles of estimated TrF');
xlabel('real'); ylabel('imaginary');
num2Ns
num2Es
den2Ns
den2Es
```

A.7.7 *The Vector Fitting (VF) Approach (6.6.3)*

Figure A.40 shows the good result of model fitting for the chosen example

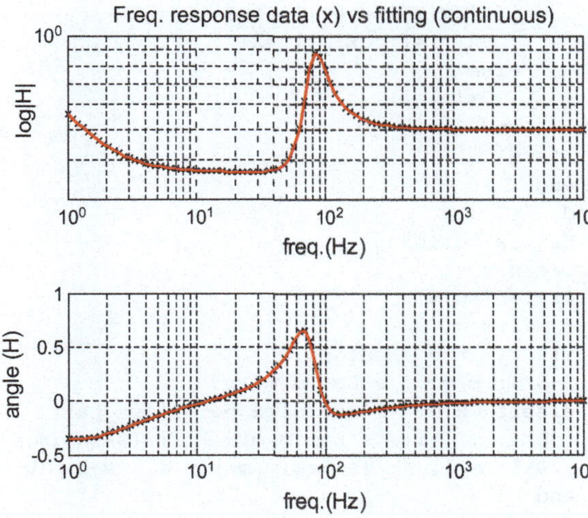

Fig. A.40 Frequency domain modelling using vector fitting (Fig. 6.43)

Program A.26 Example of Vector Fitting

```
% Example of Vector Fitting
clear all
Ns=101; %number of frequency samples
s=2*pi*j*logspace(0,4,Ns); %set of frequency values
H=zeros(1,Ns); %space for frequency responses
offs=2;
% Fequency response of a plant (edit)
for k=1:Ns,
  sk=s(k);
  H1=2/(sk+5);
  H2=(30+j*40)/(sk-(-100+j*500));
  H3=(30-j*40)/(sk-(-100-j*500));
  H(1,k)=H1+H2+H3+0.5;
end;
%xxxxxxxxxxxxxxxxxxxxxxxxxxxxxxxxxxxxxxxxxxxxxxxxxxxxxxxxxxxxx
% Start the algorithm
% Initial poles
N=3; %number of poles
pa=-2*pi*logspace(0,4,N); %initial poles
Lam=diag(pa);
%========================================================
% Identification of poles
% pole labelling
% Cx=0, real pole;
% Cx=1, real part of complex pole;
% Cx=2, imag part of complex pole
Cx=zeros(1,N);
for m=1:N,
  if imag(Lam(m,m))~=0,
    if m==1, Cx(m)=1; end;
    if m~=1,
```

```
      if (Cx(m-1)==0 || Cx(m-1)==2)
        Cx(m)=1; Cx(m+1)=2;
      else
        Cx(m)=2;
      end
    end
  end
end
% Matrix building-------------------------------
% factors
Dk=zeros(Ns,N);
for m=1:N
  if Cx(m)==0 %real pole
    Dk(:,m)=1./(s-Lam(m,m));
  elseif Cx(m)==1 %complex pole, 1st part
    Dk(:,m) =1./(s-Lam(m,m)) + 1./(s-Lam(m,m)');
    Dk(:,m+1)=j./(s-Lam(m,m)) - j./(s-Lam(m,m)');
  end
end
Dk(:,N+1)=1;
Dk(:,N+2)=s;
% scaling
aux=norm(H)^2;
scl=sqrt(aux)/Ns;
% combined matrix
M=N+1; No=N+offs;
AA=zeros(M,M); bb=zeros(M,1); Escl=zeros(1,M);
A=zeros(Ns,No+M);
%left part
for m=1:No,
  A(1:Ns,m)=Dk(1:Ns,m);
end
%right part
for m=1:M,
  A(1:Ns,No+m)=-Dk(1:Ns,m).*H.';
end
A=[real(A);imag(A)];
%Integral criterion
for m=1:N+1
  A(2*Ns+1,No+m)=real(scl*sum(Dk(:,m)));
end
[Q,R]=qr(A,0); %QR decomposition
ix1=No+1;
ix2=No+M;
R22=R(ix1:ix2,ix1:ix2);
AA(1:M,:)=R22;
bb(1:M,1)=Q(end,No+1:end)'*Ns*scl;
for col=1:M
  Escl(col)=1/norm(AA(:,col));
  AA(:,col)=Escl(col).*AA(:,col);
end
x=AA\bb;
x=x.*Escl.';
```

```
C=x(1:end-1);
D=x(end);
% get back a complex C
for m=1:N
  if Cx(m)==1
    r1=C(m); r2=C(m+1);
    C(m)=r1+j*r2; C(m+1)=r1-j*r2;
  end
end
% Zeros of sigma---------------------------
% modify Lam and C if complex poles
B=ones(N,1);
m=0;
for n=1:N
  m=m+1;
  if m<N
    if imag(Lam(m,m) ~=0 %complex number
      Lam(m+1,m)=-imag(Lam(m,m)); Lam(m,m+1)=imag(Lam(m,m));
      Lam(m,m)=real(Lam(m,m));Lam(m+1,m+1)=Lam(m,m);
      B(m,1)=2; B(m+1,1)=0;
      aux=C(m); C(m)=real(aux); C(m+1)=imag(aux);
      m=m+1;
    end
  end
end
Zer=Lam-B*C.'/D;
rt=eig(Zer).';
% take care of unstable roots
unst=real(rt)>0;
rt(unst)=rt(unst)-2*real(rt(unst)); %force stable roots
rt=sort(rt);
N=length(rt);
% sorting
for n=1:N
  for m=n+1:N
    if imag(rt(m))==0 && imag(rt(n))~=0
      aux=rt(n); rt(n)=rt(m); rt(m)=aux;
    end
  end
end
Nr=0; %number of real roots
for m=1:N
  if imag(rt(m))==0, Nr=m; end
end
if Nr<N, rt(Nr+1:N)=sort(rt(Nr+1:N)); end
rt=rt-2*j*imag(rt);
SERA=rt.';
%==========================================================
% Identification of residues
% Use of the zeros (rt) as new poles ---------------------
Lam=rt;
% pole labelling
% Cx=0, real pole;
```

```
% Cx=1, real part of complex pole;
% Cx=2, imag part of complex pole
Cx=zeros(1,N);
for m=1:N,
  if imag(Lam(m))~=0
    if m==1, Cx(m)=1; end;
    if m~=1,
      if (Cx(m-1)==0 || Cx(m-1)==2)
        Cx(m)=1;Cx(m+1)=2;
      else
        Cx(m)=2;
      end
    end
  end
end
% Matrix building--------------------------------
A=zeros(2*Ns,N+2); BB=zeros(2*Ns,1);
wg=ones(1,Ns);
% factors
Dk=zeros(Ns,N);
for m=1:N
  if Cx(m)==0 %real pole
    Dk(:,m)=1./(s-Lam(m));
  elseif Cx(m)==1 %complex pole, 1st part
    Dk(:,m) =1./(s-Lam(m)) + 1./(s-Lam(m)');
    Dk(:,m+1)=j./(s-Lam(m)) - j./(s-Lam(m)');
  end
end
% combined matrix
A(1:Ns,1:N)=Dk; A(1:Ns,N+1)=wg; A(1:Ns,N+2)=wg.*s;
BB(1:Ns)=H;
A(Ns+1:2*Ns,:)=imag(A(1:Ns,:));
A(1:Ns,:)=real(A(1:Ns,:));
BB(Ns+1:2*Ns)=imag(BB(1:Ns));
BB(1:Ns)=real(BB(1:Ns));
% re-scale
aux=length(A(1,:));
Escl=zeros(1,aux);
for col=1:aux
  Escl(col)=norm(A(:,col),2);
  A(:,col)=A(:,col)./Escl(col);
end
X=A\BB;
X=X./Escl.';
X=X.';
C=X(1:N);
SERE=X(N+2);
SERD=X(N+1);
% get back a complex C
for m=1:N
  if Cx(m)==1
    r1=C(m); r2=C(m+1);
    C(m)=r1+j*r2; C(m+1)=r1-j*r2;
```

```
  end
end
%================================================
% Final fitting
Hf=zeros(Ns); %reserve space for fitted data
Dk=zeros(Ns,N);
for m=1:N
  Dk(:,m)=1./(s-Lam(m));
end
Hf=(Dk*C.').';
Hf=Hf+SERD+SERE*s;
%================================================
% Display
freq=s./(2*pi*j);
figure(3)
subplot(2,1,1)
loglog(freq,abs(H),'kx'); hold on;
loglog(freq,abs(Hf),'r');
axis([1e0 1e4 0.3e0 1e0]);
grid
title('Freq. response data (x) vs fitting (continuous)')
xlabel('freq.(Hz)')
ylabel('log|H|');
subplot(2,1,2)
semilogx(freq,angle(H),'kx'); hold on;
semilogx(freq,angle(Hf),'r');
grid
xlabel('freq.(Hz)')
ylabel('angle(H)');
```

A.8 Chapter 7: Data Analysis and Classification

A.8.1 Determination of Non-Gaussianity (7.4.4)

A projection pursuit study has been done with respect to the mixed speeches using negentropy instead of kurtosis. Figure A.41 shows the result

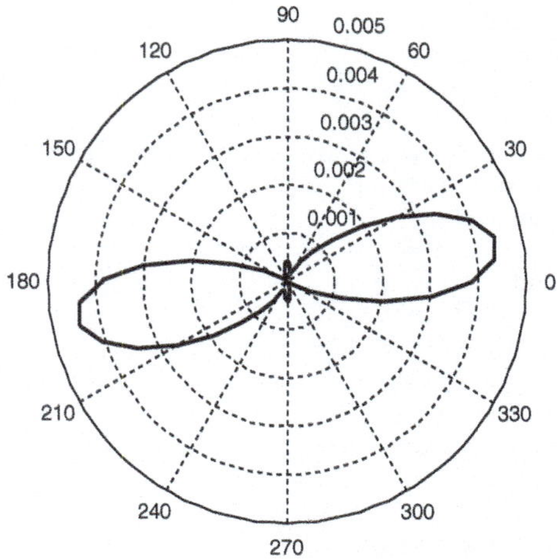

Fig. A.41 Negentropy pursuit for the mix of two speeches (Fig. 7.25)

Program A.27 Negentropy projection pursuit

```
% Negentropy projection pursuit
% example of two mixed speeches
%read two sound files
[a,fs1]=wavread('spch1.wav'); %read wav file
[b,fs1]=wavread('spch2.wav'); % " " "
s1=(a-mean(a))'; %zero-mean
s2=(b-mean(b))'; % " " "
vr1=var(s1); s1=s1/sqrt(vr1); %variance=1
vr2=var(s2); s2=s2/sqrt(vr2); %" " "
s=[s1;s2]; %combine sources
%mix of sources
N=length(s1);
M=[0.7 0.3; 0.3 0.7]; %example of mixing matrix
x=M*s; %mixed signals
N=length(a);
A=60; %number of circle partitions
kur=zeros(1,A+1);
ag=zeros(1,A+1);
i=1:N; %vectorized iterations
for nn=0:A,
  alpha=(2*pi*nn)/A; %angle of projection axis in radians
  p=zeros(1,N);
  %projection of scatterplot on the inclined axis
  p(i)=(x(1,i)*cos(alpha)) + (x(2,i)*sin(alpha));
  moment4=mean(p.^4); %fourth moment
  moment2=mean(p.^2); %second moment
  moment3=mean(p.^3);
```

```
  kst=moment4-(3*(moment2.^2)); %kurtosis
  ng=((moment3^2)/12)-((kst^2)/48); %negentropy
  negt(nn+1)=ng; %save result
  ag(nn+1)=alpha;
end;
%display pursuit
figure(1)
polar(ag,negt,'k');
title('negentropy as projection axis rotates');
```

A.8.2 Clusters. Discrimination (7.5.1)

Projections of the two data groups onto the LDA line (Fig. A.42).

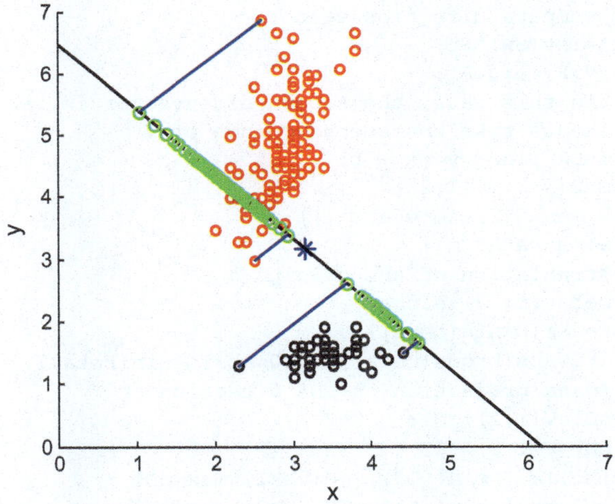

Fig. A.42 Projections on LDA line (Fig. 7.46)

Program A.28 Projection on the LDA line

```
% Projection on the LDA line
% x=sepal width, y=petal length
%read IRIS data
D=dlmread('iris.data')';
D1=[D(2,1:50);D(3,1:50)]; %class 1, x-y data
D2=[D(2,51:150);D(3,51:150)]; %class 2, x-y data
%within-class mean
meanD1=mean(D1,2);
meanD2=mean(D2,2);
%data mean
meanD=(meanD1+meanD2)/2;
%within-class covariance matrix
mcovD1=cov(D1');
mcovD2=cov(D2');
%within-class scatter matrix
Sw=(mcovD1+mcovD2)/2;
%between-class scatter matrix
Sb1=(meanD1-meanD)*(meanD1-meanD)';
Sb2=(meanD2-meanD)*(meanD2-meanD)';
Sb=Sb1+Sb2;
%compute direction vector
A=inv(Sw)*Sb;
[V,E]=eig(A);
%in this case, the second eigenvector is the largest,
%so we take the second column of V
%LDA line: y=mx + b,
m=V(2,2)/V(1,2);
b=meanD(2)-(m*meanD(1));
alpha=atan(m);
%traslation of the origin
beta=(pi/2)-alpha;
L=abs(b*sin(beta));
xt0=abs(L*cos(beta)); yt0=abs(L*sin(beta));
%data projection on LDA direction
pD1=V(:,2)'*D1;
pD2=V(:,2)'*D2;
[m1,mk1]=min(pD1); [M1,Mk1]=max(pD1);
[m2,mk2]=min(pD2); [M2,Mk2]=max(pD2);
dist=m1-M2; %free distance between projections
%x,y coordinates
xD1=xt0+(pD1*cos(alpha)); yD1=yt0+(pD1*sin(alpha));
xD2=xt0+(pD2*cos(alpha)); yD2=yt0+(pD2*sin(alpha));
%display
```

```
figure(1)
scatter(D1(1,:),D1(2,:),32,'k'); hold on;
scatter(D2(1,:),D2(2,:),32,'r');
axis([0 7 0 7]);
plot(meanD(1),meanD(2),'b*','MarkerSize',12); %data centroid
len=7; %line length, arbitrary value
plot([0 len],[b ((m*len)+b)],'k'); %LDA line
plot(xD1,yD1,'go'); %projections of D1
plot(xD2,yD2,'go'); %projections of D2
%visualize some projections
plot([D1(1,mk1),xD1(mk1)],[D1(2,mk1),yD1(mk1)],'b');
plot([D1(1,Mk1),xD1(Mk1)],[D1(2,Mk1),yD1(Mk1)],'b');
plot([D2(1,mk2),xD2(mk2)],[D2(2,mk2),yD2(mk2)],'b');
plot([D2(1,Mk2),xD2(Mk2)],[D2(2,Mk2),yD2(Mk2)],'b');
title('Projections on LDA line');
xlabel('x'); ylabel('y');
dist
```

Figure A.43 shows the results of kernel SVM for the IRIS data.

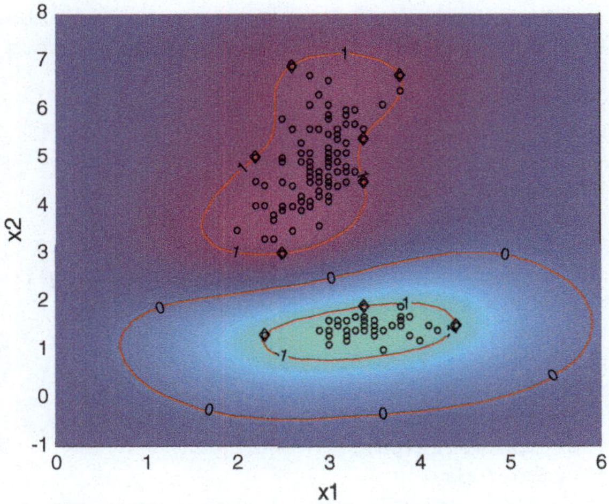

Fig. A.43 Using Kernel-SVM for IRIS data (Fig. 7.59)

Program A.29 Kernel-SVM classification example

```
% Kernel-SVM classification example
% IRIS data
% x=sepal width, y=petal length
%read IRIS data
D=dlmread('iris.data'); %columns
X=[D(:,2), D(:,3)]; %two columns (x,y)
Y=ones(150,1); Y(1:50)=-Y(1:50); %labels:1 or -1
%--------------------------------------------------
%prepare quadratic optimization
lambda = 1e-7;
c = 1000;
ps=X*X'; %scalar product
normx=sum(X.^2,2);
[nps,mps]=size(ps);
aux=-2*ps+repmat(normx,1,mps)+repmat(normx',nps,1);
K=exp(-aux/2); %gaussian kernel
A=Y;
b=0;
H =K.*(Y*Y');
e = ones(size(Y));
%quadratic prog.
[alpha , lambda , pos] = qpg(H,e,A,b,c,lambda,0,X,ps,[]);
%pos is position in X of the support vectors
xsup=X(pos,:);
ysup=Y(pos);
w=alpha.*ysup;
%-------------------------------------------------
% Prepare visualization
[xg1,xg2]=meshgrid([0:0.2:6],[-1:0.2:8]);
[nl,nc]=size(xg1); q=nl*nc;
xt1=reshape(xg1,1,q); xt2=reshape(xg2,1,q);
xt=[xt1;xt2]'; %set of test points
xtest=xt;
ps=xt*xsup'; %scalar product
normxt=sum(xt.^2,2);
normxsup=sum(xsup.^2,2);
[nps,mps]=size(ps);
yt=zeros(nps,1);
aux=-2*ps+repmat(normxt,1,mps)+repmat(normxsup',nps,1);
Kt=exp(-aux/2); %gaussian kernel
yt(:)=yt(:)+Kt*w(1:mps);
yt=yt+lambda;
yt2d=reshape(yt,nl,nc);
%-------------------------------------------------
%Display
```

```
figure(1)
%background levels
colormap('cool')
contourf(xg1,xg2,yt2d,50);shading flat; hold on
%separating line and margins
[L,A]=contour(xg1,xg2,yt2d,[-1 0 1],'r');
clabel(L,A); %plot of lines with labels
%data
scatter(X(:,1),X(:,2),16,'k')
%support vectors
scatter(xsup(:,1),xsup(:,2),40,'kd');
xlabel('x1');ylabel('x2');
title('Kernel-SVM, IRIS data');
axis([0 6 -1 8]);
```

A.8.3 Multilayer Neural Networks (7.8.3)

In order to illustrate the use of neural networks for classification, the case of three clusters of data has been chosen as a simple example. Figure A.44 displays these data clusters. Figure A.45 shows the evolution of error during learning and Figure A.46 depicts the training result.

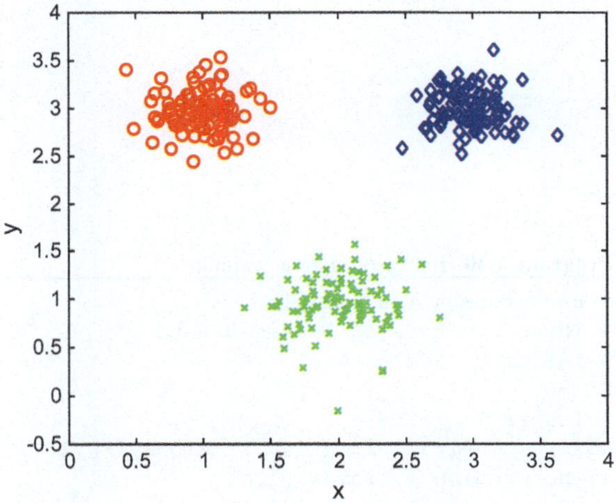

Fig. A.44 Data input for training (Fig. 7.94)

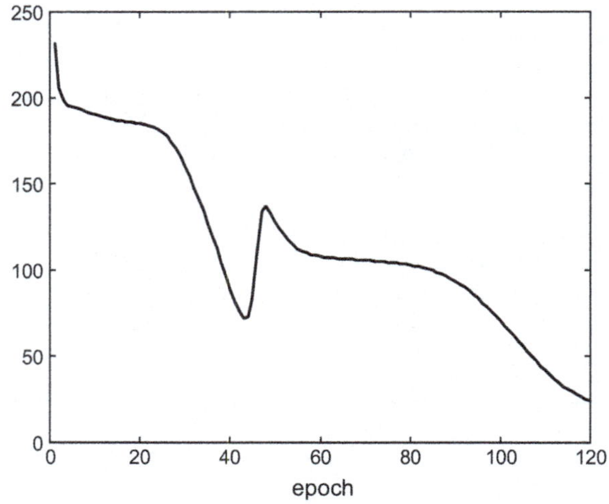

Fig. A.45 Evolution of error during learning (Fig. 7.95)

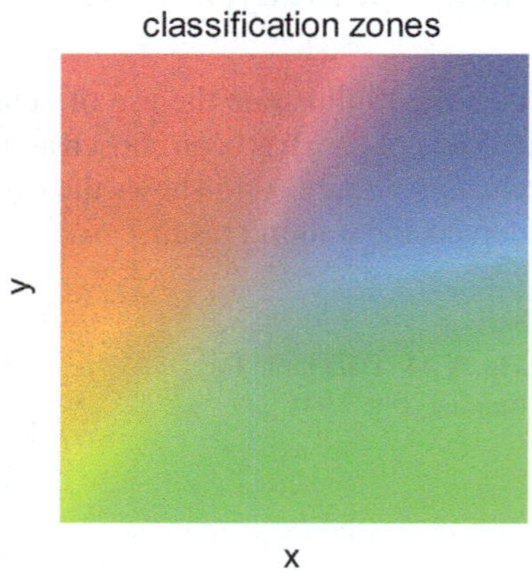

Fig. A.46 Classification regions of the trained network (Fig. 7.96)

Program A.30 Backpropagation example

```
% Backpropagation example
% Three clusters of Gaussian data
% the data (3 clusters)---------------------------
N=100;
%cluster 1
mux1=1; muy1=3; sigmax1=0.2; sigmay1=0.2;
x1=normrnd(mux1,sigmax1,1,N);
y1=normrnd(muy1,sigmay1,1,N);
%cluster 2
mux2=2; muy2=1; sigmax2=0.3; sigmay2=0.3;
x2=normrnd(mux2,sigmax2,1,N);
y2=normrnd(muy2,sigmay2,1,N);
%cluster 3
mux3=3; muy3=3; sigmax3=0.2; sigmay3=0.2;
x3=normrnd(mux3,sigmax3,1,N);
```

```
y3=normrnd(muy3,sigmay3,1,N);
% complete data set
D=zeros(2,3*N);
D(1,:)=[x1,x2,x3];
D(2,:)=[y1,y2,y3];
% neural network init------------------------------
% (w: weights; c: outputs; delta: error)
% 2 input neurons
ci=zeros(2,1);
% 2 hidden neurons
wh=0.05*ones(2,2); bh=0.05*ones(2,1);
ch=zeros(2,1); deltah=zeros(2,1);
% 3 output neurons
wo=0.05*ones(2,3); bo=0.05*ones(3,1);
co=zeros(3,1); deltao=zeros(3,1);
%training constant
eta=0.001;
Nte=120; %number of training epochs
Er=zeros(Nte,1); %error record
% neural network training------------------------------
%training iterations
Ef=1000; %for info after search
for it=1:Nte,
  for in=1:300,
  %----------------------------------------
    %neuron inputs
    nn=in;
    x=D(1,nn); y=D(2,nn);
    %neuron outputs
    ci(1)=x; ci(2)=y; ch=tanh(bh+(wh'*ci));
    co=tanh(bo+(wo'*ch));
    %errors (output):
    deltao=(1-(co.^2));
    if in>200,
      %data with label co1=0, co2=0, co3=1
      deltao(1)=deltao(1)*(-co(1));
      deltao(2)=deltao(2)*(-co(2));
      deltao(3)=deltao(3)*(1-co(3));
      %training error
      Er(it)=Er(it)+((co(1))^2)+((co(2))^2)+((co(3)-1)^2);
    elseif in>100,
      %data with label co1=0, co2=1, co3=0
      deltao(1)=deltao(1)*(-co(1));
      deltao(2)=deltao(2)*(1-co(2));
      deltao(3)=deltao(3)*(-co(3));
      %training error
      Er(it)=Er(it)+((co(1))^2)+((co(2)-1)^2)+((co(3))^2);
    else
      %data with label co1=1, co2=0, co3=0
```

```
      deltao(1)=deltao(1)*(1-co(1));
      deltao(2)=deltao(2)*(-co(2));
      deltao(3)=deltao(3)*(-co(3));
      %training error
      Er(it)=Er(it)+((co(1)-1)^2)+((co(2))^2)+((co(3))^2);
    end;
    %change of weights wo
    aux=eta*deltao*ch'; wo=wo+aux'; bo=bo+(eta*deltao);
    %errors (hidden)
    aux=deltao'*wo'; deltah=(1-(ch.^2)); deltah=deltah.*aux';
    %change of weights wh
    aux=eta*deltah*ci'; wh=wh+aux; bh=bh+(eta*deltah);
  end;
  Ef=Er(it);
end;
Ef
% display-------------------------------------
figure(1)
plot(Er,'k');
title('Training error evolution');
xlabel('epoch')
% 3 data clusters
figure(2)
plot(x1,y1,'ro'); hold on;
plot(x2,y2,'gx');
plot(x3,y3,'bd');
title('Neural network example (3 clusters)');
xlabel('x'); ylabel('y');
% test of the trained network---------------------------
% test points:
figure(3)
zon=zeros(80,80,3); %space for RGB image
x=0; y=0;
%make zones using test points
for nl=1:80,
  y=0.05*nl;
  for nc=1:80,
    x=0.05*nc;
    %neuron outputs
    ci(1)=x; ci(2)=y; ch=tanh(bh+(wh'*ci));
    co=tanh(bo+(wo'*ch));
    aux=co'; colr=0.5+(aux/2); %force into 0..1 range
    zon(nl,nc,:)=colr;
  end;
end;
imshow(zon);
axis xy;
title('classification zones')
xlabel('x'); ylabel('y');
```

A.8.4 Face Detection (7.9.1)

Figure A.47 depicts the distances corresponding to the 54 faces in our example.

Figure A.48 shows the reconstruction results.

Figure A.49 depicts the distances for the 6 faces that have been tested.

Figure A.50 shows the 6 test faces, and the reconstructed faces

Figure A.51 shows the distances, in the face space, and in the comparison between prepared and reconstructed faces.

Figure A.52 shows the 6 test faces, and their reconstruction.

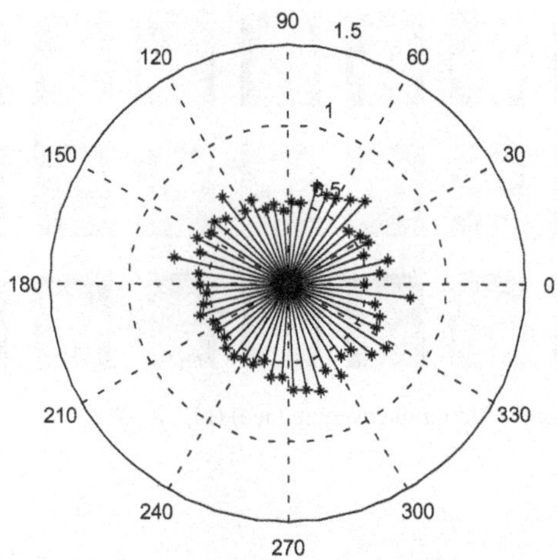

Fig. A.47 Distance between prepared and reconstructed faces (Fig. 7.101)

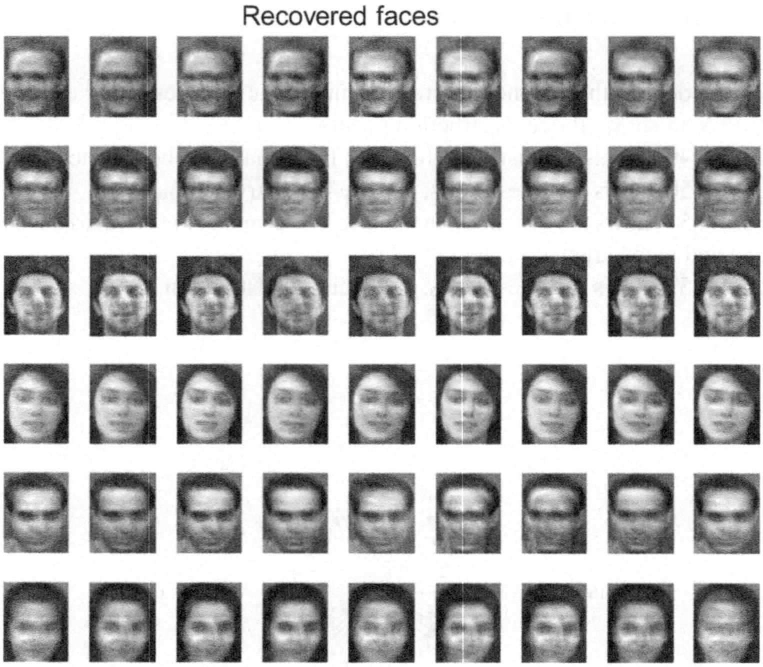

Fig. A.48 Recovered faces (adding the average face) (Fig. 7.102)

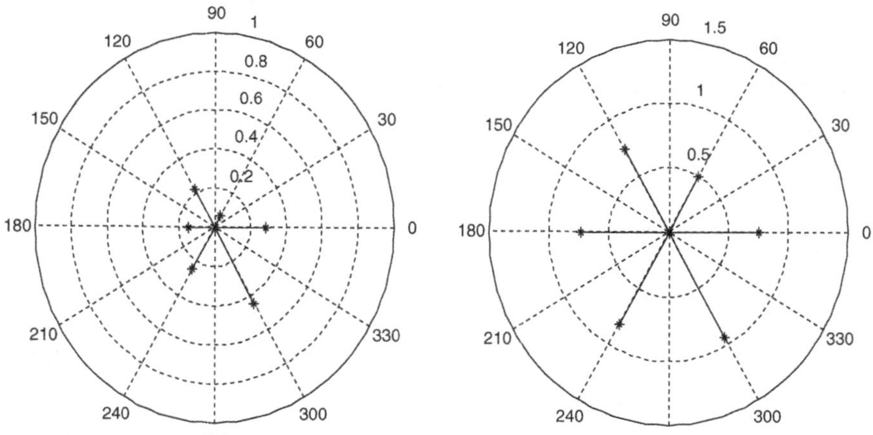

Fig. A.49 Distances for known faces (*left* face space; *right* prepared faces) (Fig. 7.103)

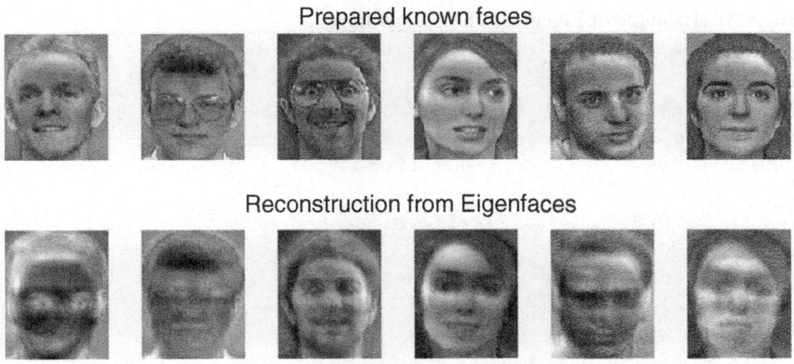

Fig. A.50 Known test faces: prepared and reconstructed (Fig. 7.104)

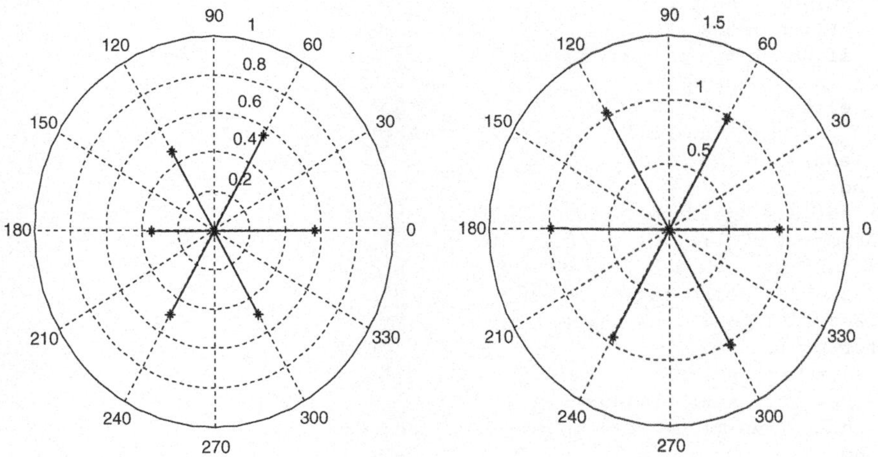

Fig. A.51 Distances for unknown faces (*left* face space; *right* prepared faces) (Fig. 7.105)

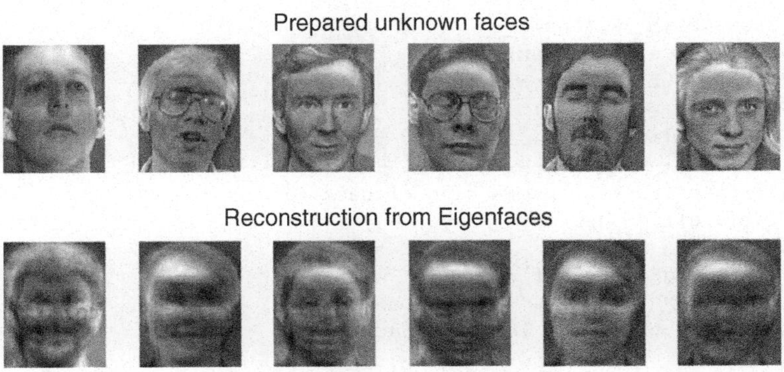

Fig. A.52 Unknown test faces: prepared and reconstructed (Fig. 7.106)

Program A.31 Example of Eigenfaces for recognition

```
% Example of Eigenfaces for recognition
% test part
%read face database (120 faces, of 112x92=10304 length each)
AA=zeros(10304,120); %faces are columns of the matrix
fid=fopen('AAface.bin','r'); aux=fread(fid);
AA=reshape(aux,10304,120);
% will choose 6x9 faces of 92x112 size
W=92; H=112;
% faces are contained in matrix "AA"
% Eigenfaces algorithm------------------------------
% form matrix with image columns
a=6*9; Or=zeros(H*W,a); %54 columns
nn=41;ns=1; %select 60 faces
for nc=1:a,
  aux=AA(:,nn);
  Or(:,nc)=aux;
  if ns==9 %skip next face
    ns=1; nn=nn+2;
  else
    ns=ns+1; nn=nn+1;
  end;
end;
% get average face
MF=mean(Or,2);
% subtract average face from each face
% and normalize faces
PF=zeros(H*W,a); nZ=zeros(1,a);
for nc=1:a,
  aux=Or(:,nc)-MF;
  PF(:,nc)=aux/norm(aux);
  nZ(nc)=norm(aux); %keep for face recovery
end;
% get eigenvectors
[V D]=eig(PF'*PF);
% sort in descending order
aux=diag(D);
[aux2, sx]=sort(-1.*aux);
eigL=aux(sx); eigV=V(:,sx);
% project to obtain eigenfaces
aux=PF*eigV;
eigF=aux/norm(aux);
%-----------------------------------------------
% build face space by projection of each face
fs=zeros(a,a);
for nf=1:a,
  fs(:,nf)=eigF'*PF(:,nf);
end;
% compute the threshold in face space
th=0;
for ni=1:a,
  for nj=ni:a,
    aux=norm(fs(:,ni)-fs(:,nj));
```

```
    if aux>th,th=aux; end;
  end;
end;
TH=0.5*th; %threshold
% reconstruction from eigenfaces -------------------------
RF=zeros(H*W,a);
for nn=1:a,
  aux=eigF*fs(:,nn); %reconstruction
  rf=aux/norm(aux);
  RF(:,nn)=rf; %save reconstructed faces
end;
% the worst reconstruction
aux=0; won=1;
for nn=1:a,
  R=norm(PF(:,nn)-RF(:,nn)); %distance for one face
  if R>aux, aux=R; won=nn; end;
end;
wR=aux; %worst distance
% display distances between prepared and reconstructed faces
figure(1)
polar(0,1.2,'y'); hold on;
for nn=1:a,
  ag=((nn-1)*2*pi)/a; %angle
  R=norm(PF(:,nn)-RF(:,nn)); %distance for one face
  polar([ag ag],[0 R],'k*-');
end;
title('distances between prepared and reconstructed faces');
% display complete reconstruction (adding average face) -------
figure(2)
cRF=zeros(H*W,a); nc=1;
for nn=1:a,
  aux=nZ(nn); %recover normalization factor
  cRF(:,nn)=aux*RF(:,nn)+MF; %add average face
  subplot(6,9,nn)
  aux=reshape(cRF(:,nn),H,W);
  imshow(aux,[])
end;
%-------------------------------------------------------
% test the 10th face for each row
% (these 6 faces have not been used during training)
%extract faces
to=zeros(H*W,6);
to(:,1)=AA(:,50); to(:,2)=AA(:,60); to(:,3)=AA(:,70);
to(:,4)=AA(:,80);to(:,5)=AA(:,90); to(:,6)=AA(:,100);
ANeigf(H,W,to,MF,eigF,a,fs,3);
%-------------------------------------------------------
% test 6 new faces, corresponding to new people
% not yet considered
% (this part of the program is a duplicate of previous part)
%extract faces
to=zeros(H*W,6);
to(:,1)=AA(:,9); to(:,2)=AA(:,19); to(:,3)=AA(:,29);
to(:,4)=AA(:,39);to(:,5)=AA(:,109); to(:,6)=AA(:,119);
```

```
ANeigf(H,W,to,MF,eigF,a,fs,6);
%--------------------------------
% print worst recosntruction and Threshold
wR
TH
```

Here is the function used in the previous program.

Function A.32 ANeigf

```
% test 6 faces after eigenface analysis
function ANeigf(H,W,to,MF,eigF,a,fs,numfig);
tp=zeros(H*W,6); tfs=zeros(a,6);
% prepare faces and project onto face space
for nt=1:6,
  aux=to(:,nt)-MF; %subtract average face and normalize
  tp(:,nt)=aux/norm(aux);
  tfs(:,nt)=eigF'*tp(:,nt); %project onto face space
end;
% compute minimum distance respect established faces
% in face space
Tmd=zeros(6,1); Tix=zeros(6,1);
for nt=1:6,
  ips=100; mdn=1;
  for ni=1:a,
    aux=norm(tfs(:,nt)-fs(:,ni));
    if aux<ips, ips=aux; mdn=ni; end;
  end;
  Tmd(nt)=ips; Tix(nt)=mdn; %results for 1 of 6 faces
end;
% display minimum distances in face space
figure(numfig)
polar(0,1,'y'); hold on;
for nt=1:6,
  ag=((nt-1)*2*pi)/6; %angle
  polar([ag ag],[0 Tmd(nt)],'k*-');
end;
title('distances in face space: 6 faces');
% display distances between prepared and reconstructed faces
trec=zeros(H*W,6);
figure(numfig+1)
polar(0,1.2,'y'); hold on;
for nt=1:6,
  aux=eigF*tfs(:,nt); %reconstruction
  trec(:,nt)=aux/norm(aux);
  ag=((nt-1)*2*pi)/6; %angle
  R=norm(tp(:,nt)-trec(:,nt)); %distance for one face
  polar([ag ag],[0 R],'k*-');
end;
title('distances between prepared and reconstructed 6 faces');
% display of 6 test faces and their reconstruction
figure(numfig+2)
for nt=1:6,
```

```
  subplot(2,6,nt)
  aux=reshape(tp(:,nt),H,W);
  imshow(aux,[]);
  subplot(2,6,nt+6)
  aux=reshape(trec(:,nt),H,W);
  imshow(aux,[]);
end;
```

Index

© Springer Science+Business Media Singapore 2017

J.M. Giron-Sierra, *Digital Signal Processing with Matlab Examples, Volume 2*,
Signals and Communication Technology, DOI 10.1007/978-981-10-2537-2